NIELS BOHR
COLLECTED WORKS
VOLUME 1

HARALD AND NIELS BOHR AS STUDENTS

ERRATA

	instead of	read
p. XII line 13 from bottom	(1912)	(1911)
p. [7] line 2	consist	consists
p. [14] line 20	Wi	Vi
p. [14] line 24	Forsøgsanording	Forsøgsanordning
p. [15] line 2 from bottom	mag	mag.
p. [97] line 15	phenomocena nnected	phenomena connected
p. [145] lines 2 and 15	Jäger	Jaeger
p. [338] line 8 from bottom	eq (5)	eq. (33)
p. [339] lines 3 and 8 of footnote	Jaeger	Jäger
p. [348] line 10	Jaeger	Jäger
p. [348] second equation	$\grave{\mu}$	μ
p. [348] lines 1 and 2 of footnote	Jaeger	Jäger
p. [357] line 7	depe ds	depends
p. [424] line 1 of footnote	jet	jeg
p. [483] line 1	bu	but
p. [483] line 5	afte	after
p. [483] line 6	reaso	reason
p. [483] line 7	Kelvin'	Kelvin's
p. [509] line 10	square is	squares are

ERRATA

	instead of	read
p. [513] replace text of footnote 18		Allusion to a legendary episode of Danish history, related by Saxo Grammaticus, *Gesta Danorum*, lib. IV, cap. IV, and known to every Danish schoolboy. Skræp is the name of the sword with which Uffe, son of the blind king Wermund, fought, single-handed against the son of the Saxon king and a Saxon warrior, and slew them both in succession. On hearing the second blow, Wermund exclaimed: "I recognize Skræp again!"
p. [528] line 6	frade	fra de
p. [542] line 2	kunde	havde
p. [603] right-hand column, replace lines 5 and 6:		Jaeger W. 145, 222, 225, 336, 339 Jäger G. 225, 236, 339, 348

NIELS BOHR

COLLECTED WORKS

GENERAL EDITOR

L. ROSENFELD

PROFESSOR AT
THE NORDIC INSTITUTE FOR THEORETICAL ATOMIC PHYSICS
COPENHAGEN

VOLUME 1
EARLY WORK (1905-1911)

EDITED BY

J. RUD NIELSEN

GEORGE LYNN CROSS RESEARCH PROFESSOR EMERITUS
UNIVERSITY OF OKLAHOMA

NORTH-HOLLAND
AMSTERDAM · OXFORD · NEW YORK · TOKYO

PUBLISHERS:

NORTH-HOLLAND PHYSICS PUBLISHING

A DIVISION OF

ELSEVIER SCIENCE PUBLISHERS B.V.
P.O. BOX 103, 1000 AC AMSTERDAM, THE NETHERLANDS

SOLE DISTRIBUTORS FOR THE U.S.A. AND CANADA:

ELSEVIER SCIENCE PUBLISHING COMPANY, INC.
52 VANDERBILT AVENUE, NEW YORK, N.Y. 10017, USA

Library of Congress Catalog Card Number: 70-126498
ISBN North-Holland, Collected Works: 0 7204 1800 3
ISBN North-Holland, Volume 1: 0 7204 1801 1

Transferred to digital printing 2006
Printed and bound by CPI Antony Rowe Ltd, Eastbourne

First edition 1972
Reprinted 1986

GENERAL INTRODUCTION

by

L. ROSENFELD

For a deeper study of Niels Bohr's contributions to our knowledge of the atomic constituents of matter, and to our view of science and other human concerns, abundant material is available. The primary source is of course the succession of his published papers; Bohr was uncommonly painstaking in the writing of these papers, and each of them accordingly contains – often in very concentrated form – the most carefully weighed expression of his thought at the time of publication. However, the constant solicitation of new investigations prevented him more than once from giving the finishing touch even to papers having reached an advanced stage of completion, and his reluctance to publish anything which fell short of his exacting standards led him in these cases to renounce publication altogether. Clearly, the extant manuscripts of such unpublished papers have the same importance to the students of his work as the published ones, the more so as these manuscripts were, at the time of their preparation, usually communicated to his near collaborators and thus exerted, albeit indirectly, an influence on the development of the subjects with which they were concerned.

Besides the published and unpublished papers in which are condensed the considered conclusions at which Bohr arrived in the course of his work, there are two further types of documents which may throw light on the progress, as well as the hesitations, of his thought: the preliminary drafts and accessory documents pertaining to the papers, and the correspondence. As regards the former, it turns out that, with few exceptions, the extant material is not as informative as one might wish. In the first place, this preliminary material was not systematically preserved, and what we have is a very casual mass of mostly incomplete stray notes to which Bohr himself attached no importance. Moreover, although Bohr made great efforts to reach the most precise expression for the ideas he wanted to convey, he was not pedantic with regard to the particular wording he finally adopted: for instance, he allowed—and even introduced himself—slight stylistic variations in the versions of his papers in different languages; by the same token, one should beware of suspecting hidden meanings in mere differences in wording between a surviving draft and the final version of some sentence. With the letters, the situation is quite different: Bohr bestowed as much care on the writing of a letter as on that of a paper, and right from the beginning of his career he kept both the letters he received and drafts of those he sent. This voluminous correspondence is a major source of information not only about Bohr's own ideas concerning a great variety of problems, but also about the whole scientific activity of his time, in which he occupied such a central position.

All this unpublished material, deposited in the Niels Bohr Archive in Copenhagen, is a substantial part of the still more extensive collection of letters and documents which was assembled, under the able direction of Professor Thomas S. Kuhn, by the Sources for History of Quantum Physics Project sponsored by the American Physical Society and the American Philosophical Society in the course of a vast enquiry carried out in the years 1962–1964*. This collection is being continually enlarged by new accessions. All the documents are accessible to *bona fide* scholars for consultation, either in the original or on microfilm, in two large American libraries as well as in the Niels Bohr Archive. Even so, there is no doubt that there is a need for an edition of Bohr's collected works, including not only a complete reissue of his published papers, but also a comprehensive, but critical, selection of additional documents. Such a publication would make at any rate the essential sources more widely available, and would facilitate the use of the complete collection of documents by providing the user with a general orientation and a basic store of auxiliary information. This is the background for the present undertaking and the aim which has guided its organization.

The first consideration goes to the published papers and the manuscripts of unpublished papers in a state of near completion, which (as indicated above) should be handled on the same basis. They will be distributed in a number of volumes, in which they will be grouped according to subjects. It turns out that, with few exceptions, this classification coincides with the chronological sequence, and it has obvious advantages over the latter. Each paper (or group of closely related papers) will be accompanied by those documents—drafts, notes, letters and other items—which illustrate the circumstances in which the work thus reported was undertaken, the repercussions of its publication, or any other information pertaining to it. An editorial introduction will have the function of coordinating

* The project is described, and its results listed, in the following report, which in particular includes a catalogue of the Niels Bohr Archive: THOMAS S. KUHN, JOHN L. HEILBRON, PAUL L. FORMAN and LINI ALLEN, *Sources for History of Quantum Physics*, An Inventory and Report (Memoirs of the American Philosophical Society, vol. 68, Philadelphia 1967).

all this material and supplying any additional pieces of information which may be deemed helpful for the assessment of the role and significance of the work. The decision to make a selection from the extant material was not taken lightheartedly: with regard to the correspondence, it was thought more convenient to insert the relevant pieces in their logical place, rather than to refer the reader to separate volumes in which they would appear in chronological order (this does not preclude such a separate issue of the correspondence at a later date); as to the other documents, a selection appeared justified by the state in which they are found, as described above. The inevitable arbitrariness all selection implies is mitigated by the consideration that the whole material is in any case available for exhaustive examination. For the introductions, the general policy is to restrict them to an informative and explanatory function, and to exclude comments of more interpretative character (except such as might remove obscurities). However, no editor can be expected to abide by such a policy too rigidly, nor will the reader reproach him for an expression of personal opinion if he clearly draws the line between fact and conjecture.

In the fulfilment of the programme just outlined, the general editor is seconded by a number of eminent collaborators, to whom the preparation of the different volumes is entrusted. The editor of each volume is primarily responsible for the arrangement and selection of the material as well as for writing the introductions. However, decisions about all these points are never taken without careful deliberation with the general editor, who accordingly accepts joint responsibility for them. In order to ensure uniformity of presentation, at a high, but not unreasonable, standard of scholarship, the following rules have been agreed upon among the editors. The leading language of the edition will be English, in the sense that all important documents and accompanying editorial texts will appear in this language, the only exception being secondary documents (mainly letters) written in German or French, since one may well assume that those interested in such documents will be sufficiently versed in these languages. This ruling entails, in particular, that all Danish and Swedish texts are given in English translation; in most cases (but not always) the original Scandinavian text is also reproduced. Many of Bohr's papers were published in more than one language; whenever there exists an English version, it is the one selected for reproduction. While references to other versions are of course given, differences from the English text are only reported if they happen to be significant, which (as pointed out above) is not usually the case. If no English text of a paper exists, the paper is reproduced in its original language and accompanied by an English translation. In the editing of the unpublished texts, scrupulous care has been bestowed on their accurate reproduction, but in many cases the editor has used his discretion—and common sense—in tacitly correcting misspellings and trivial grammatical mistakes. When this could be done conveniently, the simple *addition* of missing letters or words is explicitly indicated by putting the addition between angular brackets ⟨...⟩. Alterations requiring justification are pointed out with the necessary explanations; all editorial comments in the text and in footnotes are placed between square brackets [...]. In the introductions, biographical information concerning the persons mentioned in the papers or the correspondence is only given when it is not readily available in standard sources; in particular, notices are systematically given of all Scandinavian persons, since their biographies are mostly written in one of the Scandinavian languages. In order to help the reader to situate more easily each of the papers in the development of Bohr's investigations, and relate it to the circumstances of his life, a biographical sketch of Niels Bohr has been inserted at the beginning of this first volume. This is based on a (somewhat more condensed) article written for the *Dictionary of Scientific Biography*, edited by Professor Charles C. Gillispie under the auspices of the American Council of Learned Societies and published by Charles Scribner's Sons in New York. Thanks are due to the editor and the publishers of this scholarly work for their willingness to grant the requested authorization to make extensive use of the Niels Bohr article in the present biographical sketch.

The foundation of the Niels Bohr Archive, its participation in the Sources for History of Quantum Physics Project and the undertaking of this edition of Niels Bohr's collected works would not have been possible without the generous, continued support of the Carlsberg Foundation in Copenhagen. Permission to reproduce articles and documents has been graciously granted by all institutions and persons concerned. The general editor and his collaborators wish to record their appreciation of the unfailing assistance of the staff members of the Niels Bohr Archive and of the Niels Bohr Institute who are responsible for the safe keeping and classification of the various documents. Mrs. Betty Schultz, who had been Niels Bohr's secretary from the very creation of the Copenhagen Institute for Theoretical Physics, was active as the Archive's secretary until her recent retirement after more than forty years' service. She has been succeeded by Mrs. Sophie Hellmann, Niels Bohr's private secretary for many years, who is also in charge of the Institute's collection of Bohr's published papers. Mr. Sven Holm adds to his heavy administrative duties at the Niels Bohr Institute the care of all technical and personnel questions at the Archive. The former scientific collaborator at the Archive, Erik Rüdinger, initiated the ordering of the correspondence and other documents deposited in the Archive and set his successors in this task the highest standards of efficiency and accuracy; moreover, he started a detailed chronological analysis of the content of the Archive, which will remain for the period it covers (until the end of 1929) an invaluable help to all its users, and in particular to the editors of the earlier volumes of the Collected Works. The North-Holland Publishing Company, and especially its director M. D. Frank, who is taking a personal interest in this edition, is sparing no effort to find satisfactory solutions to the technical problems it raises and to make the typographical appearance of the volumes worthy of their contents.

It is hoped that this edition will come to include all Niels Bohr's writings: in the first place his great creative work in atomic and nuclear physics and his no less fundamental contributions to epistemology, which he so anxiously wished to be considered in the same scientific spirit in which they were conceived; but also his occasional writings on public affairs, which illustrate the width of his interests and the generosity and optimism with which he approached all human problems. Bringing together Niels Bohr's writings should not merely provide historians of science with a serviceable tool; it should above all give all those who value the spirit of science easy access to a life-work entirely devoted, with uncommon power and earnestness of purpose, to the rational analysis of the laws of nature and of the singular character of their meaning for us.

FOREWORD TO VOLUME 1

This volume covers the early period of Niels Bohr's scientific career, essentially until his arrival at Manchester, in Rutherford's laboratory. It consists of three parts. The first contains Bohr's earliest work – an experimental determination of the surface tension of water accompanied by a thorough theoretical discussion of the method; the second is devoted to his important work on the electron theory of metals – it makes for the first time his doctoral dissertation accessible in an English translation. The third part offers a substantial portion of the correspondence exchanged during these early years between Niels Bohr and his brother Harald, together with some letters from Niels to his mother: besides yielding precious information about the progress of his studies, these intimate letters, in which he candidly – and not without humour – describes his moods and impressions, give a vivid picture of the young man's sensitive personality. In order to illustrate as fully as possible the first stages of Bohr's intellectual development, we have decided to present a very wide choice of the available material; the complete inventory at the end of the volume will inform the reader of what has been left out and enable him to form an idea of the scope of our selection. Special thanks are due, for their various co-operation to the present volume, to Professor J. Brookes Spencer, Dr. Joan Bromberg and Mrs. Lise Madsen, as well as to the typing service of the University of Oklahoma. The publishers have spared no effort to solve our typographical problems; Mr. A. C. Pouwels has supervised the set-up and the printing with uncommon care and resourcefulness.

L.R.

CONTENTS

PART III: SELECTED FAMILY CORRESPONDENCE 1909–1916

INVENTORY OF MANUSCRIPTS IN THE NIELS BOHR ARCHIVE

INDEX

ABBREVIATED TITLES OF PERIODICALS

Ann. Chim. et Phys.	Annales de chimie et de physique (Paris)
Ann. d. Phys.	Annalen der Physik (Leipzig)
Ann. d. Phys. u. Chem.	Annalen der Physik und Chemie (Leipzig)
Anz. d. Akad. d. Wiss., Krakau, math.-nat. Kl.	Anzeiger der Akademie der Wissenschaften in Krakau. Mathematisch-naturwissenschaftliche Klasse
Arch. d. sciences phys. et nat. Archives d. Sc. Phys. et Nat.	Archives des Sciences physiques et naturelles (Genève)
Arch. Néerl.	Archives néerlandaises des Sciences exactes et naturelles (Haarlem)
Ber. d. D. phys. Ges.	Berichte der deutschen physikalischen Gesellschaft (Braunschweig)
Berl. Ber.	Sitzungsberichte der Königlichen Akademie der Wissenschaften zu Berlin
Camb. Trans.	Transactions of the Cambridge Philosophical Society (Cambridge)
Dan. Vid. Selsk., mat.-fys. Medd.	Matematisk-fysiske Meddelelser udgivet af Det Kongelige Danske Videnskabernes Selskab (København)
Dan. Vid. Selsk. Oversigt	Det Kongelige Danske Videnskabernes Selskab. Oversigt over Selskabets Virksomhed (København)
Dan. Vid. Selsk. Skrifter, naturvid.-mat. Afd.	Det Kongelige Danske Videnskabernes Selskab. Skrifter. Naturvidenskabelig og mathematisk Afdeling (København)

Jahrb. d. Rad. u. El. *Jahrb. d. Rad. u. Elek.*	Jahrbuch der Radioaktivität und Elektronik (Leipzig)
J. d. Physique	Journal de physique (Paris)
Mat. es Természettud.	Matematikai és természettudományi értesitö (Budapest)
Nach. d. Kgl. Ges. d. Wiss. *Göttingen, math.-phys. Kl.*	Nachrichten der Königlichen Gesellschaft der Wissenschaften zu Göttingen. Mathematisch-physikalische Klasse
Phil. Mag.	Philosophical Magazine (London)
Phil. Trans. *Phil. Trans. Roy. Soc.*	Philosophical Transactions of the Royal Society (London)
Phys. Rev.	The Physical Review (New York)
Phys. Zeitschr.	Physikalische Zeitschrift (Leipzig)
Pogg. Ann.	Annalen der Physik (Leipzig)
Prace mat.-fiz. Warszawa	Prace matematyczno-fizyczne (Warszawa)
Proc. Acad. Amsterdam	Proceedings, Koninklijke Nederlandse Akademie van Wetenschappen (Amsterdam)
Proc. Cambridge Phil. Soc.	Proceedings of the Cambridge Philosophical Society (Cambridge)
Proc. Phys. Soc. (London)	Proceedings of the Physical Society (London)
Proc. Roy. Soc.	Proceedings of the Royal Society (London)
Proc. Tokyo Math.-Phys. Soc.	Proceedings of the Physico-Mathematical Society of Japan (Tokyo)
Roy. Soc. Proc.	Proceedings of the Royal Society (London)
Sitzungsber. d. Berliner Akad. d. Wiss.	Sitzungsberichte der Königlichen Akademie der Wissenschaften zu Berlin
Sitzungsber. d. Heidelberger Akad. d. Wiss., Math.-nat. Kl.	Sitzungsberichte der Heidelberger Akademie der Wissenschaften (Heidelberg). Mathematisch-naturwissenschaftliche Klasse
Sitzungsber. d. Wiener Akad. d. Wiss., math.-nat. Kl.	Sitzungsberichte der Akademie der Wissenschaften in Wien. Mathematisch-naturwissenschaftliche Klasse
Trans. Cambridge Phil. Soc.	Transactions of the Cambridge Philosophical Society (Cambridge)

Trans. Roy. Soc., Edinburgh	Transactions of the Royal Society of Edinburgh (Edinburgh)
Verh. d. Deutsch. Phys. Ges. *Verh. d. D. Phys. Ges.*	Verhandlungen der Deutschen Physikalischen Gesellschaft (Braunschweig)
Wied. Ann.	Annalen der Physik (Leipzig)
Wiss. Abh. d. phys.-tech. *Reichsanstalt, Berlin*	Wissenschaftliche Abhandlungen der Physikalisch-technischen Reichsanstalt (Berlin)
Zeitschr. f. Elektrochem.	Zeitschrift für Elektrochemie (Halle)
Zeitschr. f. physik. Chem.	Zeitschrift für physikalische Chemie (Leipzig)
Z. Phys. *Z. Physik*	Zeitschrift für Physik (Braunschweig)

NIELS BOHR

BIOGRAPHICAL SKETCH *

by

L. ROSENFELD

In the history of physical science, the twentieth century stands out as a period of tremendous progress in the exploration of nature—from the widest expanses of the cosmos to the innermost recesses of the constitution of matter; and in the broad prospect of this history, Einstein and Bohr emerge as the two giants whose thinking has given the whole development its orientation and significance. Their role may be compared to those of Galilei and Newton, who from the first beginnings assigned to science its goal and method, and of their followers of the nineteenth century who carried on the tradition by opening new domains of knowledge and initiating new ways of thinking: Laplace and Faraday in the first half of the century, Maxwell and Boltzmann in the latter. Common to all these pioneers is a combination of achievement in actual discovery of natural laws and philosophical reflection on the nature of scientific thinking and the foundations of scientific truth— a combination essential in the sense that epistemological considerations played a decisive part in the success of their investigations, and that, conversely, the results of the latter led them to deeper understanding of the theory of knowledge. Niels Bohr in particular was very conscious of this twofold aspect of his scientific activity, deep-rooted as it was in the environment in which he grew up and received his education.

* This biography is mainly based on personal experience and conversations with Niels Bohr and his closest collaborators, as well as on the correspondence and documents in the Niels Bohr Archive. Detailed biographical material is published in the collective book *Niels Bohr, his life and work as seen by his friends and colleagues*, edited by S. Rozental (North-Holland Publishing Co., Amsterdam 1967). See also the report of the Niels Bohr Memorial Session held in Washington, D.C., on 22 April 1963, published in Physics Today **16**, no. 10 (1963) p. 21–62, and an earlier essay of a more personal character: L. Rosenfeld, *Niels Bohr: an Essay* (North-Holland Publishing Co., Amsterdam 1945; revised edition 1961). There is much autobiographical material in Niels Bohr's Rutherford Memorial Lecture, *Reminiscences of the founder of nuclear science and of some developments based on his work* (Proc. Phys. Soc. (London) **78** (1961) 1083).

1. FAMILY AND EDUCATION (1885–1910)

Niels Henrik David Bohr was born in Copenhagen on the 7th October 1885. His whole life was centered upon his native city: there he spent his childhood and youth, founded his family and pursued his whole scientific career; although he travelled widely, Copenhagen was always the seat of his work and of his home, where he died on the 18th November 1962. The family in which he was the second of three children belonged to the well-to-do intellectual circles of the city of Copenhagen; his father, Christian Bohr, was a talented physiologist, professor in the University *; his mother, Ellen Adler, came from a wealthy Jewish family, prominent in such varied activities as banking, politics, classical philology and pedagogy of a distinctly progressive character. It is clear that in such a favourable environment, the children's native gifts were allowed the fullest development and their formal education at school could be supplemented at every stage by example and stimulation in the home. Niels was not as brilliant a pupil as his younger brother Harald **, for whom he felt throughout his life unbounded affection and trust; but they both showed all-round interests, also for sportive pursuits.

At the University, where his personality reached its full unfolding, Niels stood out

* Christian Bohr was a pupil of Carl Ludwig, in whose laboratory at Leipzig he worked in 1881 and again in 1883, after he had taken his doctor's degree in Copenhagen in 1880. His main work was devoted to the elucidation of the physical and chemical processes governing pulmonary respiration. He founded a brilliant school, in which his methods of accurate investigation of the physico-chemical aspects of physiological functions were extended in various directions, and which is still flourishing today. Christian Bohr was born on 14th February 1855; he was appointed to the chair of physiology in the University of Copenhagen in 1886; at the height of his research and teaching activity he died on 3rd February 1911. See V. Henriques, *Chr. Bohrs videnskabelige Gerning* (Dan. Vid. Selsk. Oversigt (1911) 395).

** Harald Bohr was born on 22nd April 1887; he died on 22nd January 1951. He concluded his mathematical studies at the University of Copenhagen in 1910 with a remarkable doctor's thesis on Dirichlet series. This at once established his reputation, especially with the Göttingen and the Cambridge mathematicians, who held him in high esteem; his early acquaintance with Edmund Landau, Courant, Hardy and Littlewood developed into a lasting friendship, soon shared by his elder brother. Harald Bohr's outstanding contribution to mathematics was his conception of the almost periodic functions, which he defined and studied in three great papers published in the Acta Mathematica in 1924–26, and which formed the brilliant conclusion of his prolonged investigations of the Dirichlet series and the functions representable by such series. He had uncommon didactic talent; his teaching at the Polytechnical School and the University of Copenhagen worthily upheld the traditions of the Danish school of mathematics and by the deep influence he exerted on his students was instrumental in keeping the standing of both pure and applied mathematics at a high level throughout the country. He had wide interests, a lively and cheerful temperament, and was witty and outspoken in his comments of current affairs – not least those in which his brother was involved and about which he sought his advice. Obituary by N. E. Nørlund, Dan. Vid. Selsk. Oversigt (1950–51) 62.

as an investigatoɪ of quite uncommon power *. His first research was completed in 1906 when he was still a student, and won him a gold medal from the highest scientific institution of the country, the Royal Danish Academy of Sciences and Letters **. It was a precision measurement of the surface tension of water by a method consisting in the observation of a regularly vibrating jet. It is a fully mature piece of work, remarkable by the minute care and the thoroughness with which both the experimental and theoretical parts of the problem were handled.

His doctoral dissertation ***, which followed in 1911, was a purely theoretical work, again exhibiting a sovereign mastery of the vast subject he had chosen, the electron theory of metals. This theory, which pictures the metallic state as a gas of electrons moving more or less freely in the potential created by the positively charged atoms disposed in a regular lattice, accounts qualitatively for the most varied properties of metals, but it ran into many difficulties as soon as a quantitative treatment was attempted on the basis of the then accepted principles of classical electrodynamics. In order to throw light on the nature of these difficulties, Bohr developed general methods allowing him to derive the main features of the phenomena from the fundamental assumptions in a very direct way. He could thus clearly exhibit the deep-lying nature of the failures of the theory, which he in fact referred to an insufficiency of the classical principles themselves. Thus, he showed that the magnetic properties of the metals could in no way be derived from a consistent application of these principles. The very rigour of his analysis gave him, at this early stage, the firm conviction of the necessity of a radical departure from classical electrodynamics for the description of atomic phenomena.

The study of physics, even carried to such unusual depth, did not absorb all the

---

* Niels was introduced to both experimental and theoretical physics by Professor C. Christiansen, who was one of his father's close friends. At that time Christiansen was actively investigating electrocapillary phenomena and following with interest the new developments of the theory of electrons. He was an all-round physicist of the old school. Born on 9th October 1843, he spent a happy childhood in a remote village as a shepherd, but his father, noticing his studious dispositions, wisely directed him to higher education, and after another happy time in which he just followed his inclination, he completed the study of physics at the University of Copenhagen in 1866. In the following 20 years, he devoted himself to research, in spite of a heavy teaching burden, and undaunted by the rudimentary conditions under which he had to carry it out. From this time dates the discovery of anomalous dispersion (1870), by which he is best remembered, although he also made a number of very original contributions to the investigation of radiative emission, gaseous diffusion and electricity. In 1886 he was appointed to the chair of physics in the University of Copenhagen, which he held until his retirement in 1912. He died on 14th December 1917. Obituary by K. Prytz (with a contribution by M. Knudsen), Dan. Vid. Selsk. Oversigt (1917–18) 31.
** See Part I of the present volume.
*** *Studier over Metallernes Elektrontheori* (Thaning & Appel, København 1911). See Part II of the present volume.

activity of the adolescent, whose intellectual curiosity knew no bounds. With his characteristic earnestness and thoroughness, he took up the hints which circumstances offered as starting points for highly original philosophical reflections. His father's scientific work, dealing with the quantitative analysis of physical processes underlying physiological functions, required on the one hand closest attention to the elaboration of refined techniques of physical measurements, and raised on the other hand deep philosophical issues about the relationship between physical and biological phenomena. At the time of Niels' adolescence, the philosophical trend in scientific circles was a reaction against the mechanistic materialism of the preceding generation; however, in the liberal atmosphere of the group of Christian Bohr's friends, to which belonged in particular the physicist C. Christiansen and the philosopher Harald Høffding *, this reaction took a moderate and thoughtful form: the master in the investigation of the physical basis of the physiological processes insisted on the practical necessity of also considering these processes from the teleological point of view of their function in the organism in order to arrive at a complete description **. The two brothers, Niels and Harald, were admitted as silent listeners to philosophical conversations of their father and his friends, and this first confrontation with the epistemological problem of biology, in which apparently conflicting views were found equally indispensable for a full understanding of the phenomena, made a lasting impression upon Niels' mind.

He also soon came to share the negative attitude of the progressive bourgeoisie to which his family belonged towards the church and religious beliefs in general;

---

* Harald Høffding won universal recognition by his uncommon mastery of all aspects of philosophy, from ethics and psychology to the theory of knowledge. His attitude was liberal and broadminded, averse to every form of dogmatism, and he recognized the paramount importance of the development of science for the philosophical outlook. He was born on 11th March 1843; he studied theology at a time when, on the parochial Danish scene, a struggle was raging about the respective merits of "faith and science". Høffding solved the crisis for himself by forsaking both the official church and the personal christianity preached by Kierkegaard, and he started in quite a systematic way a life-long enquiry into the sources of human knowledge and ethical behaviour: from 1875 to 1887 he was occupied with psychology and ethics, from 1887 to 1895 with the history of philosophy and from 1895 to his death, which occurred on 2nd July 1931, with the philosophy of religion (which he conceived in a very wide sense) and epistemology. After taking his doctor's degree in 1870 and working for some years as a teacher, he became lecturer in the University of Copenhagen in 1880 and professor in 1883. Among his abundant production (his last paper dates from 1930), there is a series of textbooks, widely appreciated abroad in various translations. Obituaries by V. Grønbech and by Niels Bohr, Dan. Vid. Selsk. Oversigt (1931–32) 57, 131. Article by S. V. Rasmussen in *Dansk Biografisk Leksikon* (Schultz, Copenhagen 1937).
** A concise statement of these views was published by Chr. Bohr in an article *Om den patologiske Lungeudvikling*, Universitetets Festskrift (1910). Niels Bohr often referred to it in conversation, and included a translation of it in his paper *Physical science and the problem of life*, published in *Atomic physics and human knowledge* (John Wiley and Sons, New York 1958).

but it is characteristic of his candour and independence of judgement that he only arrived at this conclusion after he had convinced himself that the church upheld doctrines that were logically untenable and shunned the pressing task, which at the time preoccupied all liberal minds, of alleviating a still widespread pauperism. He never found any occasion in later life to depart from the position of the free-thinker, which he maintained with tolerance and humanity. His approach to these questions, even at such an early stage, was marked by the same logical rigour and breadth of vision as his scientific thinking.

It was in the course of his meditations on the human condition that, considering the role of language as a means of communication, he first came across a situation of great generality whose recognition was at the source of his later decisive contribution to the epistemology of physics. He was struck by the fact that the same word usually serves to denote a state of our consciousness and the concomitant behaviour of our body. In trying to describe this fundamental ambiguity of every word referring to our mental activity, he had recourse to an analogy drawn from the mathematical theory of multiform functions: each such word, he said, belongs to different "planes of objectivity", and we must be careful not to allow it to glide from one plane of objectivity to another. However, it is an inherent property of language that there is one word only for the different aspects of a given psychical activity; there is no point in trying to remove such ambiguities, we must rather recognize their existence and live with them *.

2. STUDIES IN ENGLAND (1911–1912)

After the termination of his studies in Copenhagen, Bohr went to Cambridge, in the hope of pursuing his work on the electron theory under the guidance of J. J. Thomson. However, when he arrived there at the beginning of October 1911, he soon found that Thomson had lost interest in the subject, and could not be persuaded to examine the thesis, of which he had been at great pains to make an English translation. His attempts at getting this translation published were equally unsuccessful. This grievous disappointment did not prevent the studious youth from making the most of his stay in Cambridge, but as soon as he conveniently could, he moved to Manchester, where Rutherford had established a flourishing laboratory.

There, in the short span of four months (from March to July 1912), working with incredible intensity and concentration, he laid the foundations of his greatest achievement in physics, the theory of atomic constitution. It would be difficult to imagine two temperaments more different than those of Bohr and Rutherford; but

* See L. Rosenfeld, *Niels Bohr's contribution to epistemology*, Physics Today **16**, no. 10 (1963) 47.

this first contact, besides a new epoch in science, initiated a lifelong friendship of rare quality between the two men, made of filial affection on Bohr's part and of frank and warm cordiality, tinged with respect, on the part of the jovial New Zealander. With his shrewd judgement of people, Rutherford had indeed soon sensed the genius in the shy, unassuming "boy" (as he called him), and unwittingly, by the sheer display of his immense strength, imaginative insight and directness of approach, he set him into the right mood.

Some time before Bohr's arrival, towards the end of 1910, Rutherford had proposed a "nuclear" model of the atom in order to account for the large angle scattering of α-rays observed in his laboratory. Since the discovery of the electron, as a carrier of an elementary unit of negative electric charge, the atom was thought of as a system of a certain number of such electrons, kept together by an equivalent positive charge, somehow attached to the massive substance of the atom (the electron itself being nearly two thousand times lighter than the lightest atom). If this positive substance were spread over the whole atom, the α-rays, or positively charged helium atoms, impinging upon it would mostly undergo small deviations from their course; the frequent occurrence of large angle deviations suggested direct hits with a strongly concentrated positive substance. A quantitative check of the idea fully confirmed this inference and revealed indeed that the massive, positively charged, nucleus of the atom had linear dimensions a hundred thousand times smaller than those of the whole atomic structure.

Bohr eagerly took up the new model and soon recognized its far-reaching implications. In particular, he pointed out that the nuclear model of the atom implied a sharp separation between the chemical properties, ascribed to the peripheral electrons, and the radioactive properties, which affected the nucleus itself. This immediately suggested a close relation between the atomic number, indicating the position of an element in Mendeleev's periodic table, and the number of its electrons, or its nuclear charge, which should thus be more fundamental than its atomic weight. Indeed, the periodic table showed one or two irregularities in the sequence of atomic weights, and it became increasingly difficult to accomodate in it the newly discovered radioactive products; Bohr showed how all these anomalies could be eliminated if one admitted the occurrence of atomic nuclei of the same charge and different masses, so that there could be more than one species of atoms occupying the same place in the periodic table—hence the name "isotope" coined somewhat later for such chemically indistinguishable atomic species of different weights.

According to the nuclear model, radioactive transformations had to be conceived as actual transmutations of the atomic nucleus. Thus, Bohr argued, by the emission of an α-ray, the nucleus lost two units of charge and became an isotope of the element two places back in the periodic table. In the β-decay, on the other hand, the emission of a negative electron resulted in the gain of one unit of charge, and the product nucleus occupied the next higher place in the periodic table. Simple as it may seem, the inference leading to these "displacement laws" of radioactive elements was far from obvious at that time. The only one in the laboratory who followed Bohr's thoughts with deep interest and genuine understanding, and was able to help him in the discussion of the empirical information, was a young Hungarian chemist, G. Hevesy—himself on the verge of the discovery of the use of radioactive isotopes as tracers, which consecrated his fame. Indeed, Rutherford, insensible to the logical cogency of Bohr's argument, dissuaded him from publishing such hazardous deductions from his own atomic model, to which he was not prepared to ascribe the fundamental significance that Bohr gave it; and when, a few months later, the displacement laws could be discerned by mere inspection of the accumulated experimental evidence, Fajans (one of those who then enunciated them) so little understood their meaning that he actually presented them as evidence against the Rutherford atomic model!

Bohr's clear-sighted survey of the implications of Rutherford's atomic model did not stop at the recognition of the existence of a relation between the atomic number (which summarizes the whole physico-chemical behaviour of the element) and the number of electrons in the atom. He resolutely attacked the much harder problem of determining the exact nature of this relation, which amounts to a dynamical analysis of the atomic structure represented by the nuclear model. Following J. J. Thomson's example, Bohr assumed that the electrons would be symmetrically distributed around the nucleus in concentric circular rings. He had then to face the problem, not present in Thomson's model, of how to account for the stability of such ring configurations, which could not be maintained by the electrostatic forces alone.

Bohr had acquired the conviction, from his study of the behaviour of electrons in metals, that the validity of classical electrodynamics would be subject to a fundamental limitation in the atomic domain, and he had no doubt that this limitation would somehow be governed by Planck's quantum of action; one knew already how to "quantize" the motion of a harmonic oscillator, i.e. to select from the infinity of possible motions a discrete series characterized by energy values increasing by finite steps of magnitude $h\nu$, where h is Planck's universal constant and ν the frequency of the oscillator. One could try to apply a similar quantization to the motions of the electrons in an atom, whose "frequencies" might be identified with the resonance frequencies observed in the scattering of light by the atom. Thus an allowed state of motion characterized by a frequency ω_n would have a binding energy of the form $W_n = Knh\omega_n$, where n is an integer numbering the state and K some numerical factor possibly depending on the type of motion. Such a formula could be combined with the relation given by the classical theory between the binding energy and the amplitude of the motion, so as to obtain a relation between the latter quantity, whose order of magnitude is known from various evidence about the atomic dimensions, and the corresponding resonance frequency, obtained from optical measurements. It was easy to ascertain that the numerical value of Planck's constant, entering such a relation, did lead to the expected orders of magnitude, but this rough check, however encouraging, was clearly insufficient to establish the precise form of the quantum condition.

At this juncture, Bohr obtained a much deeper insight into the problem by a brilliant piece of work, which—working, as he himself says, "day and night"—he carried to completion with astonishing speed. The problem was one of immediate interest for Rutherford's laboratory: in their passage through a material medium, α-particles continually lose energy by ionizing the atoms they encounter, at a rate depending on their velocity; this energy loss limits the depth to which the particles can penetrate into the medium, and the relation between this depth, or "range", and the velocity offers a way of determining the latter. What Bohr did was to analyse the ionizing process on the basis of the Rutherford model of the atom and thus express the rate of energy loss in terms of the velocity by a much more accurate formula than had so far been achieved—a formula, in fact, to which modern quantum mechanics only adds unessential refinements *. Bohr's interest in atomic collision problems never faltered. In the early thirties, when the modern theory of these processes was being elaborated, especially by H. Bethe, F. Bloch and E. J. Williams, he took an active part in the work, a good deal of which took place in Copenhagen; and as late as 1948 he wrote a masterly synthetic account of the whole subject, in which one still finds, in modernized form, the arguments of his early analysis **.

What the success of this analysis showed him, however, was that the classical theory, while completely failing to account for the stability of the periodic motions of the atomic electrons, could deal with undiminished power with the aperiodic motions of charged particles traversing a region in which there is an electric field. This means that, however radical the break with classical ideas implied by the existence of the quantum of action, one must expect a gradual merging of the quantum theory into the classical one for motions of lower and lower frequencies. Moreover, one may expect that the effect of a very slow and gradual modification of the forces acting upon or within an atomic system will be correctly estimated by the classical theory. These were shrewd remarks, of which Bohr made skilful use and which he eventually developed into powerful heuristic principles.

---

* Bohr's paper was published in the Philosophical Magazine **25** (1913) 10.
** N. Bohr, *The penetration of atomic particles through matter*, Dan. Vid. Selsk., mat.-fys. Medd. **18**, no. 8 (1948).

An immediate application of the second remark helped him to discuss simple models of atomic and molecular structures, which reproduced at least in order of magnitude a number of features derived from various experiments and thus further illustrated the fruitfulness of the Rutherford atomic model. Indeed, this model was the first to permit a clearcut distinction to be made between atom and molecule—a molecule being defined as a system with more than one nucleus—and thereby to open the way to an understanding of the nature of chemical binding. The models considered by Bohr were all characterized by the arrangement of the electrons in one or more ring configurations, either disposed around the nucleus as common centre in an atom, or in a symmetrical way with respect to the nuclei in molecules. While the absolute dimensions of these configurations depended on quantum conditions which he could only roughly guess, their stability, according to the argument mentioned above, could be examined by classical methods: thus, he could explain why hydrogen could form a diatomic molecule, while helium could not. Although these considerations were crude—and are completely superseded by the modern conceptions—they were remarkably successful; in fact, they do embody an important feature of the chemical bond that is part of the modern theory, namely the fact that this bond is due to the formation of a configuration of electrons shared by the combining atoms: the hydrogen molecule, for instance, was well represented by a ring of two electrons perpendicular to the line joining the two nuclei.

With regard to the determination of the states of motion allowed by the quantum condition mentioned above, Bohr found that the Rutherford model leads to remarkably simple results, at any rate for the type of configuration he considered. In general, the classical theory of the motion furnishes an additional relation between the binding energy and the frequency, which allows one to eliminate the frequency from the quantum condition and thus obtain for the binding energy W_n an expression depending only on the integer n, with a coefficient that, besides Planck's constant, contains the parameters characterizing the system and the type of motion. Thus, to take the simplest example of the hydrogen atom, consisting of a singly charged nucleus and an electron of mass m and charge e, the classical theory shows that there is proportionality between W_n^3 and ω_n^2; this leads, for the allowed states of binding, to the very simple law $W_n = A/n^2$, and the precise value of the coefficient A is $\pi^2 e^4 m/2K^2 h^2$; only the numerical factor K remains in doubt.

3. THE QUANTUM POSTULATES (1913)

When he left Manchester in July 1912, Bohr was teeming with ideas and projects for further exploration of this world of atoms that was displaying such wide prospects; but he had another ground to be in high spirits. Since 1911, shortly before his departure for England, he was engaged to a girl of great charm and sensibility, Margrethe Nørlund. The marriage that took place in Copenhagen on 1 August 1912 sealed the happiest and most harmonious union. Margrethe's role was not an easy one; she fulfilled it to perfection in all simplicity. Bohr was of a sensitive nature; he needed the stimulus of human sympathy and understanding, not only in these most creative years, when he was struggling single-handed with his arduous problems, but even later, after he had won universal fame. Margrethe was always at his side, helping him, with infinite devotion and patience, in the practical matters connected with his work and partaking in all his hopes and disappointments. When children came—they had six sons, two of whom died prematurely—Bohr never shunned his duties of *pater familias*; he took very seriously his position as head of a family whose fine traditions he was anxious to uphold. In this also he had his wife's full support; she adapted herself without apparent effort to the difficult task of hostess. Evenings at the Bohrs' were distinguished by the warm cordiality of the reception and the exhilarating tone of a conversation combining intellectual distinction with the kind of unmalicious wit and innocent fun that scientists enjoy.

In the autumn of 1912, Bohr took up modest assistant duties at the University of Copenhagen; he fulfilled them most conscientiously, and even used the privilege attached to the doctor title of giving a free course of lectures. Concurrently with these absorbing chores, he started to write up the account of his Manchester ideas. Then, at the beginning of 1913, the orientation of his thoughts took a sudden turn towards the problem of atomic radiation, which rapidly led him to the decisive step in the incorporation of the quantum of action into the theory of atomic constitution. The rest of the academic year was spent in hectic work reconstructing the whole theory upon the new foundation and expounding it in a large treatise, which was immediately published, in three parts, in the *Philosophical Magazine* *.

It had been known since Kirchhoff's pioneering work that the spectral composition of the light emitted by atoms is characteristic for the chemical species; a whole science of "spectroscopy" had developed on this principle and accumulated an enormous material of extreme accuracy. Obviously, the tables of wave-lengths of the characteristic spectral lines must contain very precise information on the structure of the emitting atoms; but since atomic spectra consist of apparently capricious sequences of thousands of lines, it looked pretty hopeless to try and decipher such complicated codes. It therefore came as a surprise to Bohr to learn from a

---

* *On the constitution of atoms and molecules.* Papers of 1913 reprinted from the Philosophical Magazine with an introduction by L. Rosenfeld (Munksgaard, Copenhagen 1963).

casual conversation with his colleague H. M. Hansen * that spectroscopists had managed to discover regularities behind the chaos; in particular, Rydberg **, of the nearby University of Lund, had found a formula of very simple and remarkable structure expressing the frequencies of several "series" of spectral lines which recurred, with different values of the parameters, in the spectra of different atoms. The striking feature of Rydberg's formula was that the frequencies were represented by differences of two "terms", each of which depended in a simple way on a number which could take a sequence of integral values; a series corresponded to the sequence obtained by keeping one of the terms fixed and varying the other.

Thus, the frequencies v_{nm} of the lines of the hydrogen spectrum could be represented in the simplest possible form in terms of two integers n, m as

$$v_{nm} = R\left(\frac{1}{n^2} - \frac{1}{m^2}\right)$$

with a single parameter R of accurately known numerical value. As soon as Bohr saw this formula ***, he immediately recognized that it gave him the missing clue

* Hans Marius Hansen was assistant at the physics laboratory of the Polytechnical School since 1908; he was one year younger than Bohr (being born on 7th September 1886). In 1911 he had worked under W. Voigt in Göttingen on the inverse Zeeman effect—the subject of his doctor's thesis of 1913; he was accordingly well versed in spectroscopy when he met Niels Bohr on the latter's return from Manchester. Hansen became lecturer in the University of Copenhagen in 1918 and professor in 1923. He displayed a great interest and talent in organisational and administrative questions and played an important role in the development of Danish institutions whose work depended on the application of physical processes. He was elected rector of the University in 1948 and remained in this responsible function until his death on 13th June 1956. Obituaries by J. K. Bøggild and J. A. Christiansen, Universitetets Festskrift (1956) 187, 191.

** Rydberg's great spectroscopic work, published in 1890, was part of the programme of research he pursued with remarkable singleness of purpose during his whole life: to derive from the experimental data the relationships between the various properties of the elements and the atomic number indicating their position in the periodic system. It is clear that such a programme could only lead to very limited success, and was not likely to further his academic career. Born on 8th November 1854, Johannes Robert Rydberg was appointed lecturer in mathematics in the University of Lund in 1880, after having obtained the doctorate the preceding year. When the physics laboratory was founded in 1882, he was attached to it, but did not obtain the professorship until 1901. He died on the 28th December 1919, but after 1914 his failing health had prevented him from following the new developments of atomic theory, which gave such prominence to his pioneering investigations. He was a gentle and modest man and bore the disappointments of his life with touching resignation. See articles by G. Borelius, Fysisk Tidsskrift 21 (1923) 65; B. Edlén, Kosmos 32 (1954) 9; N. Bohr and W. Pauli in Proceedings of the Rydberg Centennial Conference on Atomic Spectroscopy (Lunds Universitets Årsskrift 50, no. 21 (1955) 15, 22); Sister St. John Nepomucene, Chymia 6 (1960) 127.

*** There is evidence, published in the last article quoted in the preceding footnote, that by 1899 Professor Christiansen knew and appreciated Rydberg's work; and one may surmise that Bohr must have been aware of it at an early period of his studies. Its significance, however, could only become apparent from the properties of Rutherford's atomic model.

to the correct way in which the quantum of action had to be introduced into the description of atomic systems. The formal similarity between the "terms" of the Rydberg formula R/n^2 and the expression for the energies $W_n = A/n^2$ of the possible stationary states of the atom suggested to him, in the spirit of Planck's conception of the quanta of radiation, that the emission by the atom of light of frequency v_{nm} occurred in the form of single quanta of energy hv_{nm}; Rydberg's formula then indicated that in this process the atom passed from an initial stationary state W_n to another stationary state W_m. An immediate check on this interpretation offered itself: according to it, the value of Rydberg's constant should be given by $Rh = A$, i.e. by $R = \pi^2 e^4 m / 2K^2 h^3$. Inserting in this expression the known values of e, m and h, and taking for K the value $\frac{1}{2}$ (which would give the correct potential energy W_n for a harmonic oscillator of frequency ω_n), Bohr obtained a value of R as near the experimental one as the errors in the determinations of the other constants allowed.

However convincing such a stringent quantitative test could appear, there was in this new conception of the radiation process a feature that must be felt as very strange, not to say scandalous: the frequencies v_{nm} of the emitted light did not coincide with any of the allowed frequencies of revolution ω_n of the electron or their harmonics—a feature of the classical theory of radiation so immediate and elementary that it seemed unthinkable to abandon it. That Bohr was not deterred by this consideration is due essentially to the dialectical turn of mind he had acquired in his youthful philosophical reflections. The conflict between the classical picture of the atomic phenomena and their quantal features was so acute that no hopes (such as Planck was still expressing) could be entertained to solve it by reducing the latter to the former; one had rather to accept the co-existence of these two aspects of our experience, and the real problem was to integrate them into a rational synthesis. The example of the Rutherford model of the hydrogen atom (for which there was no problem of dynamical instability) was particularly striking in this respect: the classical mechanism of continuous radiation due to the accelerated motion of the electron around the nucleus would rapidly lead to a collapse of the atom, and there was therefore in this case an imperious necessity, independent of any arbitrary assumption, to secure the stability of the system by some element foreign to classical electromagnetism. As Bohr later said, the clue offered by Rydberg's formula was so transparent as to lead uniquely to the quantal description of the radiation process he proposed; this gave him the conviction that it was right, in spite of the radical break with classical ideas it implied.

In order to clinch the argument, however, Bohr went a very important step further. He knew that the quantal behaviour of a system, whatever it was, had to satisfy the requirement of going over to the corresponding classical one in the limiting case of motions involving large numbers of quanta of action. Applying this test to his interpretation of Rydberg's formula, he found that the condition could only be fulfilled by ascribing to the numerical coefficient K just the value $\frac{1}{2}$, for which the right value of Rydberg's constant was obtained. Indeed, for large values of the number n, the frequencies $v_{n,n+p}$ are then seen to tend to the values of the frequency of revolution $\omega_n = 2R/n^3$ and its successive harmonics $p\omega_n$. Thus, as Bohr expressed it, "the most beautiful analogy" was established—in the sense just indicated—between classical electrodynamics and the quantum theory of radiation.

In his great papers of 1913, Bohr presented his theory as founded upon two "postulates", whose formulation he refined in later papers. The first postulate enunciates the existence of stationary states of an atomic system, the behaviour of which may be described in terms of classical mechanics; the second postulate states that the transition of the system from one stationary state to another is a non-classical process, accompanied by the emission or absorption of one quantum of homogeneous radiation, whose frequency is connected with its energy by Planck's equation. As to the principle by which the possible stationary states are selected, Bohr was still very far from a general formulation; indeed, he was keenly aware of the necessity of extending the investigation to other configurations than the simple ones to which he had restricted himself. The search for sufficiently general quantum conditions defining the stationary states of atomic systems was going to be a major problem in the following period of development of the theory.

A remark in Bohr's first paper gave rise to a controversy which soon ended in triumph for the new theory and in no small degree contributed to its swift success. On the strength of his interpretation of Rydberg's formula, Bohr had pointed out that a certain series of spectral lines attributed to hydrogen ought actually be ascribed to helium: it had been fitted to the formula for hydrogen with half-integral values of the numbers n, m; in Bohr's view, which required integral values for these numbers, this could only mean that the Rydberg constant for this series was four times that for hydrogen, corresponding to a doubly charged nucleus. An experienced spectroscopist like Fowler received the suggestion with humanly understandable scepticism, but control experiments, which were at once performed, with impressive efficiency, in Rutherford's laboratory, confirmed Bohr's prediction. Fowler's last ditch resistance, in the form of the pointed objection that Rydberg's coefficient for the contested series was not exactly $4R$ (R being the hydrogen value), was brilliantly countered by Bohr: he showed that the slight difference was just to be expected as an effect of the motion of the nucleus, which he had neglected in first approximation. There is no doubt that this dramatic incident had a decisive influence, especially on Anglo-Saxon minds like Rutherford's and Fowler's, in convincing them that there was something after all in this young foreigner's cloudy theorizing.

This was also Jeans' attitude, when, in the thorough report of Bohr's work he gave at the British Association meeting at Birmingham in September 1913, he pointed out that the only justification of his postulates "is the very weighty one of success". At Göttingen, that high place of mathematics and physics, where the sense for propriety was strong, the prevailing impression was one of scandal, or at least bewilderment, before the undeserved success of such high-handed disregard of the canons of formal logic; but the significance of Bohr's ideas did not escape those who had themselves most searchingly pondered over the problems of quantum theory, Einstein and Sommerfeld.

4. ATOMS AND RADIATION (1914–1925)

No one realized more keenly than Bohr himself the provisional character of his first conclusions, and above all the need for a deeper analysis of the logical relationship between the classical and quantal aspects of the atomic phenomena, embodied in the two postulates. At the same time, he was faced with an overwhelming programme of generalization of the theory and unfolding of all its consequences. He was more and more dissatisfied with his job at the University of Copenhagen. In 1913, he had been promoted to the rank of lecturer (docent), but this left him little time for research and (since he had mainly to deal with medical students) little hope of forming pupils able to assist him in his work. The academic authorities were slow in realizing that an exceptional situation had arisen, and when Rutherford offered him a lectureship in Manchester, Bohr was glad to avail himself of such an opportunity to pursue his work in the most favourable conditions. He remained two years in Manchester, from 1914 to 1916. In the meantime, the Danish authorities had moved so far as to offer Bohr a professorship, which he accepted; and three years later, thanks to the active intervention of a group of friends, who donated the ground, they were at last persuaded to build Bohr a laboratory: this was the famous Institute for Theoretical Physics, of which he assumed the directorship, and which remained the centre of his life-long activity in pursuing and inspiring research. The foundation of the institute came in fact just in time to retain Bohr in his native country; for Rutherford, who had just been called to the directorship of the Cavendish laboratory, had already invited Bohr to join him in Cambridge.

The new institute was meant to be primarily a physical laboratory; the phrase "theoretical physics" in its name conveys the idea that we would now express as "fundamental physics". Bohr did not draw any sharp distinction between theoretical and experimental research; on the contrary, he visualized the two aspects of research conducted in such a way as to give each other support and inspiration, and he wanted the outfit of the laboratory to be such as to make it possible to test new theoretical developments or conjectures by appropriate experiments. He fully managed to put this conception into effect; the experimental investigations carried out at the institute have not been numerous, but always of high quality, some of them indeed of pioneering importance, and all bearing the mark of direct relevancy to the theoretical questions under debate. In order to keep up with the changing outlook of the theory, it was imperative, in this conception, to expand and even to renew the experimental equipment in order to adapt it to entirely new lines of research; this Bohr did with remarkable foresight as well as persuasive tenacity in securing the necessary funds. It was a part of his activity to which he devoted much care and attached much importance, and the tradition he thus founded continues to bear fruit today.

Bohr's atomic theory inaugurated two of the most adventurous decades in the history of science, a period in which the efforts of the elite among the younger generation of physicists were concentrated on the numerous problems raised by the theory and on experimental investigations which further stimulated the theoretical developments or provided the required tests of theoretical predictions. Three experimental advances which furthered the progress of the theory were made as early as 1913 and 1914. The domain of X-ray spectroscopy was opened up by Moseley's brilliant work in Manchester and its significance for atomic theory pointed out by W. Kossel on the basis of Bohr's ideas. The experiments of Franck and Hertz on the excitation of radiation from atoms by collisions with electrons, and those of Stark on the modification of the atomic spectra by strong electric fields, offered a new approach to the study of the dynamical behaviour of atomic systems; their interpretation was soon outlined by Bohr himself. Optical spectroscopy, whose importance had been suddenly enhanced, was actively developed, especially by the school established at Tübingen under Paschen's leadership; with his collaborators Back, Landé and others, Paschen analysed in great detail the fine structure of the line spectra and the further splitting of the lines under the action of magnetic fields of increasing strength, and formulated the regularities obeyed by the frequencies and intensities of the lines in terms of sets of "quantum numbers" attached to the spectroscopic terms and taking integral or half-integral values.

On the theoretical side also, the scene was rapidly changing. The isolation in which Bohr had hitherto found himself gave place to a lively intercourse with a growing number of fellow-workers, all striving towards the common goal, freely exchanging ideas, discussing results and conjectures, sharing the thrill of success and the expectation of further progress. By tacit consent, Bohr was the leader towards whom all turned for guidance and inspiration; he won everyone's affection by his modesty and candour. There were other great schools of theoretical physics, the foremost being those newly established by Sommerfeld in Munich and by

Born in Göttingen; they pursued their own lines of research in complete harmony with the Copenhagen group. The first to join Bohr in Copenhagen was a young Dutchman, H. A. Kramers, who arrived on his own initiative in 1916 and was for the next ten years Bohr's unflagging helper and talented collaborator. During this period, many others came to Bohr's institute for shorter or longer stays; among them were Bohr's faithful friend Hevesy, as well as younger men, O. Klein, W. Pauli and W. Heisenberg, who soon were tied to him by the same bonds of friendship and affection.

The first of the main problems requiring consideration was the generalization of the quantum conditions defining the stationary states. Bohr did not at first attempt to make use of the general methods of classical mechanics; this was not his way of tackling problems. He preferred to handle concrete cases, and to develop ingenious arguments, which, although lacking generality, had the advantage of clearly bringing out the physical features of essential importance. In the present instance, he again started from the remark that slow deformations of a system would not change its quantal state, and developed it into a "principle of mechanical transformability", which proved quite efficient within a limited scope. The idea was to transform one type of motion continuously into another by slow variation of some parameter; if the determination of the stationary states had been accomplished for one of the two motions, one could in this way derive it for the other. To this end, one could take advantage of the existence of dynamical quantities, the "adiabatic invariants", which (as the name indicates) have the property of remaining unchanged under slow mechanical transformations. As early as 1911, Ehrenfest had emphasized the important role played by adiabatic invariants in the quantum theory of radiation in thermodynamic equilibrium; but neither he nor Bohr at first succeeded in extending this conception to modes of motion more complicated than simple periodic ones. Decisive progress in this problem was made by Sommerfeld, who at the end of 1915 succeeded in formulating a full set of quantum conditions for the general Keplerian motion, including even the relativistic precession of the elliptic trajectory. Sommerfeld's work not only supplied an explanation (a partial one, as it turned out) of the fine doublet structure of the lines of the hydrogen spectrum, but showed the way to the desired generalization of the rules of quantization to more complex atomic systems, whose motions were not simply periodic.

Bohr eagerly followed this new line of attack; he now made full use of the powerful methods of Hamiltonian dynamics, especially in the form adapted to the wide class of motions known as "multiply periodic", to which the motions of the electrons in atoms belonged. It was a fortunate circumstance that Kramers, a master in these questions, formed in Ehrenfest's school, was at hand to help him; even so, it took years of strenuous effort to bring the work to completion. In their general form, the quantum conditions stated that a certain set of adiabatic invariants should be integral multiples of Planck's constant; but in the process of establishing this result, a formidable hurdle was the occurrence of "degeneracies" of the motions into simple periodic ones, leading to ambiguities in the formulation of the corresponding quantum conditions. This difficulty was eventually overcome by another ingenious application of the principle of mechanical transformability.

The theory of multiply periodic systems offered the possibility of a more rational treatment of the question which Bohr had tackled in his very first reflections on the nuclear atomic model: the gradual building up of atoms of increasing complexity and the origin of the periodicities in the atomic structures revealed by Mendeleev's table. The starting point was the consideration of the individual stationary orbits of each single electron in the electrostatic field of the nucleus, "screened" by the average field of the other electrons; the residual interaction of the electrons could then be treated by the perturbation methods originally developed for the use of astronomers. By the quantum conditions these individual stationary states received a characterization which was directly comparable to that of the spectroscopic terms by quantum numbers in the case of spectra originating from quantum transitions of a single electron, usually the most weakly bound one. The confrontation of the theory with the relevant spectroscopic evidence from this point of view led only to partial success: the main features of the empirical term sequences were well reproduced by the theory, and the spectroscopic quantum numbers on which these features depended accordingly acquired a simple mechanical interpretation (except for the occurrence of half-integral values, which appeared as an arbitrary modification of the quantum conditions); but the finer structure of the term sequences presented a complexity for which the atomic model offered no mechanical counterpart.

In spite of this imperfection, the model could be expected to give reliable guidance at least in the investigation of the broader outlines of atomic structures. The primitive ring configurations of Bohr's previous attempt were now replaced by groupings of individual electron orbits in "shells" specified by definite sets of quantum numbers, according to rules which were inferred, with uncommon sagacity, from the spectroscopic data. This conception of the shell structure of atomic systems did not only account for the main classification of the stationary states; its scope could be extended to include also the interpretation of the empirical rules established by the spectroscopists for the intensities of the quantal transitions between these states. This was a much more difficult problem than that of the formulation of quantum conditions for the stationary states; the complete breakdown of classical electrodynamics, reflected in Bohr's quantum postulates, seemed at first sight to remove the very foundation on which a comprehensive theory of atomic radiation could rest. It was just in taking up this challenge that Bohr was led to one of his deepest and most powerful conceptions, the idea of a general "correspondence" between the classical and the quantal description of the atomic phenomena.

Bohr seized upon the only link that subsisted between the emission of light in a quantal transition and the classical process of radiation, namely the requirement that the classical description should be valid in the limiting case of transitions between states with very large quantum numbers. If the atom was treated as a multiply periodic system, its states of motion could be represented as superpositions of harmonic oscillations of specified frequencies and their integral multiples, each occurring with a definite amplitude; it was indeed possible to verify that the frequencies of quantal transitions between states of very large quantum numbers tended to become equal to those multiples of the classical frequencies given by the differences between these quantum numbers; in this limit, then, the classical amplitudes could be directly used to calculate the intensities of the quantal transitions. Now, Bohr boldly postulated that such a correspondence should persist, at least approximately, even for transitions between states of small quantum numbers; in other words, the amplitudes of the harmonics of the classical motion should in all cases give an estimate of the corresponding quantal amplitudes. The power of this "correspondence argument" was immediately illustrated by the application Kramers made of it, in a brilliant paper, to the splitting of the hydrogen lines in an electric field *. Not only did the correspondence argument, by want of a more precise formulation, play an indispensable part in the interpretation of the spectroscopic data; it eventually gave the decisive clue to the mathematical structure of a consistent quantum mechanics.

By 1918 Bohr had visualized at least in outline the whole theory of atomic phenomena whose main points have been reviewed in the preceding sections. He of course realized that he was still very far from a logically consistent framework wide enough to incorporate both the quantum postulates and those aspects of classical mechanics and electrodynamics which seemed to retain some validity. Nevertheless, he at once started writing up a synthetic exposition of his arguments and of all the evidence upon which they could have any bearing; this was his usual way of working: in trying out how well he could "summarize what we know" he found occasion to check the soundness of his ideas and to improve their formulation. In the present case, however, he could hardly keep pace with the growth of the subject; the paper he had in mind at the beginning developed into a large treatise in four parts, the publication of which dragged over four years without reaching completion; the three first parts appeared between 1918 and 1922, while the fourth

---

* H. A. Kramers, *Intensities of spectral lines*, Dan. Vid. Selsk. Skrifter, naturvid.-mat. Afd. **3**, no. 3 (1919).

was unfortunately never published*. Thus it came that the full impact of Bohr's views remained confined to the small, but brilliant circle of his disciples, who indeed managed better than the master to make them more widely known by the prompter publication of their own results.

Bohr's theory of the periodic system of the elements, essentially based on the analysis of the evidence of spectra, renovated the science of chemistry, by putting at the chemists' disposal rational spectroscopic methods much more refined than the traditional ones. This was dramatically illustrated, in 1922, by the identification, in Bohr's institute, of the element of atomic number 72. This discovery was made by Coster and Hevesy under the direct guidance of Bohr's theoretical predictions of the properties of this element; they gave it the name "hafnium", from the latinized name of Copenhagen. The conclusive results were obtained just in time to be announced by Bohr in the lecture he delivered in Stockholm when he received the Nobel prize for that year **.

There was no question for Bohr of resting on these well-deserved laurels. He was not misled by the apparent triumph of the quantum theory of atomic systems into believing that the model used to describe these systems—simple point charges interacting by electrostatic forces according to the laws of classical mechanics—bore any close resemblance to reality. In fact, the fine structure of the spectroscopic classification manifested an essential insufficiency of this model, whose nature was not yet elucidated; but above all, the peculiar character of the correspondence between the quantal radiation processes and their classical counterpart strongly suggested that the classical model was no more than an auxiliary framework in the application of quantum conditions and correspondence considerations. After Kramers had succeeded in extending the scope of the correspondence argument to the theory of optical dispersion, thus rounding off a treatment of the interaction of atomic systems with radiation that accounted for all emission, absorption and scattering processes, Bohr ventured to propose a systematic formulation of the whole theory, in which what he called the "virtual" character of the classical model was emphasized. This formulation, in which he was aided by Kramers, was stimulated by very original ideas put forward by a young American visitor, Slater, and the new theory was published, in 1924, under threefold authorship ***. The most striking feature of this remarkable paper was the renouncement of the classical form of causality in favour of a purely statistical description. Even the distribution of energy and momentum between the radiation field and the "virtual

* N. Bohr, *On the quantum theory of line spectra*, Dan. Vid. Selsk. Skrifter, naturvid.-mat. Afd. **4**, no. 1 (1918–22).
** N. Bohr, *The structure of the atoms*, Nobel Lecture (1922).
*** N. Bohr, H. A. Kramers and J. C. Slater, *The quantum theory of radiation*, Phil. Mag. **47** (1924) 785.

oscillators" constituting the atomic systems was assumed to be statistical, the conservation laws being fulfilled only on the average. This was going too far: hardly was the paper in print, than direct experiments established the strict conservation of energy and momentum in an individual process of interaction between atom and radiation. Nevertheless, this short-lived attempt exerted a profound influence on the course of events; what persisted after its failure was the conviction that the classical mode of description of the atomic processes had to be entirely relinquished.

This conviction was strengthened by the outcome of the other line of investigation most actively pursued in Copenhagen in those years: the search for the missing dynamical element of the atomic model. Pauli approached this arduous problem by trying to unravel the spectroscopic rules governing the fine structure of the terms and the splitting of the spectral lines in an external magnetic field—the "anomalous Zeeman effect". He at length recognized that the whole material could be reduced to great simplicity by attributing to the individual stationary states of each electron an additional quantum number, susceptible of two values only, and combining with the other quantum numbers according to definite rules. This conclusion at once threw light on the systematics of the shell structure of the elements, which Bohr had left incomplete, but which had lately been quite essentially improved by Stoner. In fact, Pauli was now able, in 1925, to formulate the simple underlying principle of this systematics: each individual stationary state—including the specification of the new quantum number—cannot be occupied by more than one electron *. This "exclusion principle" has since received considerable extension, and has in fact turned out to be one of the most fundamental laws of nature.

In the same year, decisive progress was made in the interpretation of the new quantum number by two of Ehrenfest's young pupils, Goudsmit and Uhlenbeck: they pointed out that it could be ascribed to a proper rotation, or "spin", of the electron, and that an intrinsic magnetic moment, related to the spin, could then account for the anomalous Zeeman effect. However, the quantization of the spin was at variance with that expressed by the quantum conditions; this circumstance, as well as the exclusion principle, which was obviously quite unaccountable in classical terms, showed in the most striking fashion that not only the radiation field but also the atomic constituents were out of reach of the conceptions of classical physics.

---

* W. Pauli, *Über den Zusammenhang des Abschlusses der Elektronengruppen im Atom mit der Komplexstruktur der Spektren*, Z. Physik **31** (1925) 765.

5. QUANTUM MECHANICS AND COMPLEMENTARITY (1925–1935)

The crisis to which the attempt to treat the atom as a dynamical system of the classical type had led was not of long duration. By the summer of 1925 Heisenberg had found the clue to the elaboration of a consistent mathematical scheme embodying the quantum postulates. This momentous progress was the direct outcome of the line of investigation of the optical dispersion theory initiated by Kramers. Heisenberg had taken an active part in this work and had been much impressed by the stand taken by Bohr, Kramers and Slater. If classical conceptions no longer could be relied upon to supply at least a framework for the quantum theory, he concluded, what we have to look for is an abstract formal scheme expressing only relations between directly observable quantities, like the energies of the stationary states and the amplitudes whose absolute squares should express the probabilities of quantal transitions between these states. The correspondence between classical and quantal amplitudes established in the theory of dispersion, when envisaged from this point of view, took the shape of a set of algebraic rules that these quantal amplitudes had to obey. This defined a mathematical algorism adapted to the rational formulation of laws of motion and quantum conditions, as well as to the precise calculation of radiation amplitudes.

Heisenberg's programme was eagerly taken up in Göttingen, where Born had immediately recognized that the non-commutative algebra involved in Heisenberg's relations was the matrix calculus; at the same time a young Cambridge physicist, Dirac, was developing even more abstract and elegant methods *. While in the high places of mathematics the formal scheme of the new "quantum mechanics" was thus being rapidly built up, a more critical attitude prevailed in Copenhagen. Pauli pointed out that by limiting the observable quantities to stationary states and radiation amplitudes Heisenberg was unduly restricting the scope of the theory, since it was an essential part of the correspondence argument that the new theory should contain as a limiting case, for large quantum numbers, the more detailed description of the motion in classical terms.

The fulfilment of this essential requirement necessitated a considerable extension of the mathematical framework of the theory, allowing it to accommodate both discontinuous and continuous aspects of the atomic phenomena. The decisive contribution was unexpectedly made by the "outsiders" de Broglie and Schrödinger, who were just at that time exploring the conjecture that the constituents of matter might be governed, like radiation, by a law of propagation of continuous wave fields. Although the idea in this one-sided form was at once seen to be untenable, it nevertheless provided the missing element; as Born especially empha-

---

* The history of this first phase of the development of quantum mechanics is told by B. L. van der Waerden, *Sources of quantum mechanics* (North-Holland Publishing Co., Amsterdam 1967).

sized, the wave fields associated with the particles give the probability distributions of the variables specifying the state of motion of these particles. Thus, the required formal completion of quantum mechanics could be carried out already at the beginning of 1927, when Dirac laid the crown to the edifice by indicating the most general representation of the operators belonging to the physical quantities, and the way to pass at will from any representation to any other according to definite prescriptions which guaranteed the fulfilment of all correspondence requirements. However, such classical features of the motion of particles as a sequence of positions forming a uniquely determined trajectory appeared only as a limiting case of a more general mode of description of essentially statistical character.

The quantum conditions were found to impose a peculiar restriction on the statistical distributions of the values of physical quantities. If, as a consequence of these conditions, the operators representing two such quantities do not commute, the average spreads in the assignment of the values they may take under given circumstances are reciprocal to each other; their product exceeds a limit depending on the degree of non-commutation and proportional to Planck's constant. Thus, if in definite experimental circumstances the position of an electron, relative to some fixed frame of reference, is confined within narrow limits, its momentum will have a correspondingly wide range of possible values, each with its definite probability of occurrence, depending on the experimental conditions.

Heisenberg, who in 1927 discovered these remarkable "indeterminacy relations", realized their epistemological significance. In fact, the novelty of quantum mechanics in this respect is that it allows for the possibility of using all classical concepts, even though their precise determinations may be mutually exclusive,—as is the case for the concept of a particle localized at a point in space and time, and that of a wave field of precisely given momentum and energy, whose space-time extension is infinite. Indeterminacy relations between such concepts then indicate to what extent they may nevertheless be used concurrently in statistical statements. Heisenberg saw that the origin of these reciprocal limitations must lie in quantal features of the processes in which the quantities in question are observable, and he attempted to analyse from this point of view such idealized processes of observation [*].

This was the occasion for Bohr to re-enter the scene. His role so far had been to inspire and orient the creative effort of the younger men, especially Heisenberg and Pauli, and he could legitimately consider the new theory as the attainment of the goal towards which he had so long been striving. On the one hand, the radical break with classical physical theories, which he had felt from the very beginning

---

[*] W. Heisenberg, *Über den anschaulichen Inhalt der quantentheoretischen Kinematik und Mechanik*, Z. Physik **43** (1927) 172.

to be inescapable, was now formally accomplished by the substitution of abstract relations between operators for the simple numerical relations of classical physics. On the other hand, just this abstract character of the new formalism made it at last possible to fulfil the requirement he had always emphasized: not to sacrifice any aspect of the phenomena, but to retain every element of the classical description within the limits suggested by experience. The peculiar form of limitation of the validity of classical concepts expressed by the indeterminacy relations demanded, however, a more thorough analysis than that Heisenberg had initiated. For this challenging task Bohr was, of course, not unprepared. The occurrence of conflicting, yet equally indispensable, representations of the phenomena strikingly evoked the ambiguities of our account of mental processes over which he had pondered in his student days. Now, however, similar dilemmas confronted him in an incomparably simpler form: for the description of atomic phenomena operated with only a few physical idealizations. Bohr hoped that the study of such a transparent case would lead him to a formulation of the epistemological situation sufficiently general to find application to the deeper lying problems of life and mind, and he devoted to it all his energy. Although he very soon was able to elucidate the essential features, he spent most of the following decade patiently refining the formulation of the fundamental ideas and exploring all their implications.

In any investigation of the scope of physical concepts, the method to follow is prescribed by the nature of the problem: one has to go back to the definition of the concepts by means of apparatus—real or idealized—suited to the measurement of the physical quantities they represent; the analysis of such measuring operations should then reveal any limitation in the use of these concepts resulting from the laws of physics. This had been the method followed by Einstein in establishing the relativity of simultaneity; the same method was followed by Heisenberg and Bohr to elucidate the indeterminacy relations. It emerged from Bohr's analysis that the decisive element brought in by the quantum of action is what Bohr called the "individual" character of quantal processes: any such process—for instance the emission of radiation by an atom—occurs as a whole; it is only well-defined when it is completed, and it cannot be subdivided into a sequence of gradual changes of the system, like the processes dealt with in classical physics, which involve immense numbers of quanta.

In particular, the measurement of a physical quantity pertaining to an atomic system can only be regarded as completed when its result has been recorded as some permanent mark left upon a registering device. Now, such a recording cannot be performed without some irreversible loss of control of the quantal interaction between the atomic system and the apparatus. Thus, if we record the position of an electron by a spot on a rigidly fixed photographic plate, we lose the possibility of ascertaining the exchange of momentum between the electron and the plate.

Conversely, an apparatus suited to the determination of the momentum of the electron must include a mobile part, completely disconnected from the rigid frame of spatial reference, and whose position when it exchanges momentum with the electron therefore necessarily escapes our control. Here is the root of the mutual exclusion of the application of such concepts as position and momentum in the extreme case of their ideally precise determination. More generally, by relaxing the accuracy requirements, it is possible to check that the reciprocal exclusion is limited to the extent indicated by the indeterminacy relations, thus allowing for the concurrent use of the two concepts in a description which is then necessarily of a statistical character.

It thus appears that, in order to reach full clarity in such a novel situation, the very notion of physical phenomenon is first of all in need of a more careful definition, embodying the feature of individuality or wholeness typical of quantal processes. This is achieved by inserting in the definition the explicit specification of all the relevant experimental arrangement, including the recording devices. Between phenomena occurring under such strictly specified conditions of observation, there may then arise the type of mutual exclusion for which an indeterminacy relation is the formal expression. It is this relationship of mutual exclusion between two phenomena that Bohr designated as "complementarity"; by this denomination he wanted to stress that two complementary phenomena belong to aspects of our experience which, although mutually exclusive, are nevertheless both indispensable for a full account of experience.

The introduction of the notion of complementarity finally solved the problem of the consistent incorporation of the quantum of action into the conceptual framework of physics,—the problem with which Bohr had struggled so long. Complementarity was not any arbitrary creation of Bohr's mind, but the precise expression, won after patient efforts demanding a tremendous mental concentration, for a state of affairs entirely grounded in nature's laws and that, according to Bohr's familiar exhortation, we had to learn from her. It consecrated the recognition of a statistical form of causality as the only possible link between phenomena presenting quantal individuality, but made it plain that the statistical mode of description of quantum mechanics was perfectly adapted to these phenomena and gave an exhaustive account of all their observable aspects.

From the epistemological point of view, the discovery of the new type of logical relationship that complementarity represents is a major advance, which radically changes our whole view of the role and meaning of science. In contrast to the 19th century ideal of a description of the phenomena from which every reference to their observation would be eliminated, we have now the much wider and truer prospect of an account of the phenomena in which due regard is paid to the conditions under which they can actually be observed,—thereby securing the full objectivity of the description, since the latter is based on purely physical operations intelligible and verifiable by all observers. The role of the classical concepts in this description is obviously essential, since those concepts are the only ones adapted to our possibilities of observation and unambiguous communication. In order to establish a link between these concepts and the behaviour of atomic systems, we have to use measuring instruments composed – like ourselves – of large numbers of atoms, and this unavoidably leads to complementary relations and a statistical type of causality.

These are the main lines of the new structure of scientific thought that gradually unfolded itself as Bohr, with uncompromising consistency, pursued his epistemological analysis to its last consequences. That a few of the greatest representatives of the type of physical thinking with which he was so decisively breaking refused to follow him is humanly understandable; that Einstein should be among them was always a matter of surprise and regret to Bohr. On the other hand, the progress of his work owed much to Einstein's opposition; indeed, its successive stages are marked by the refutation of Einstein's subtle objections. Bohr has himself retraced the dramatic course of this long controversy in an article of 1949, which gives the most complete and systematic exposition of his argumentation he ever came to writing, and which will remain one of the great classics in the history of science *.

The role of complementarity in the edifice of quantum mechanics is above all to provide a logical frame sufficiently wide to ensure the consistent application of classical concepts whose unrestricted use would lead to contradictions. Obviously, such a function is of universal scope, and indeed an occasion soon presented itself to put its usefulness to the test. In the early thirties, the extension of the mathematical methods of quantum mechanics to electrodynamics was beset with considerable formal difficulties, which raised doubts regarding the possibility of upholding the concept of the electromagnetic field in quantum theory. This was clearly a point of crucial importance, since it bore upon the fundamental issue of a possible limit to the validity of the correspondence argument, hitherto unchallenged. One had to enquire, according to Bohr's point of view, whether every component of the electromagnetic field could be measured, in principle, with unlimited accuracy, and whether the measurements of more than one component were only subject to the reciprocal limitations resulting from their complementary relationships. Bohr took up this investigation, which occupied him, with Rosenfeld's collaboration, during most of the years 1931 to 1933. He succeeded in devising idealized measuring procedures, satisfying all requirements of relativity, and by means of which all

---

* N. Bohr, *Discussion with Einstein on epistemological problems in atomic physics*, in *Albert Einstein: philosopher-scientist* (The Library of Living Philosophers Inc., Evanston 1949).

consequences of the quantization of the electromagnetic field could be confirmed *. In view of the significance of the issue at stake, this work had a wider repercussion than its immediate effect in establishing the consistency of quantum electrodynamics: it showed how essential a part Bohr's epistemological standpoint played in our conception of the quantal phenomena.

6. NUCLEAR PHYSICS (1936–1943)

By the middle thirties, the main interest had shifted, in Copenhagen as elsewhere, to the rapidly expanding field of nuclear physics. On the theoretical side, the results of the experiments on the reactions induced by the impact of slow neutrons on nuclei, carried out by Fermi and his school at Rome, created quite a critical situation. In discussing the processes involving the impact of charged particles, α-particles or protons, onto a nucleus, it had been found sufficient to represent the effect of the proper nuclear forces acting between the nucleus and the impinging particle schematically by an attractive potential well extending over the volume of the nucleus; to this was added the repulsive electrostatic potential, forming a "Coulomb barrier" around the nucleus. It was therefore natural to analyse the neutron reactions with the help of the same potential, without the Coulomb barrier; and it was a surprise that this model did not even qualitatively account for the observed effects. In particular, it was impossible to understand on this basis the very large probabilities with which the capture of the neutron by the nucleus occurred for a sequence of "resonance" energies.

Faced with this puzzling problem, Bohr proceeded in his characteristic way to look for cases of capture processes occurring under a simpler form than in the range of low energies, in which they appeared to be tied to resonance conditions. As it happened, he had only to return to Chadwick's earliest experiments, performed with neutrons of higher energy; he noticed that the different reactions induced by these neutrons all occurred at any energy with about the same probability, whose order of magnitude indicated that almost every neutron hitting the nucleus was captured by it. This was indeed a strikingly simple result, which suggested to him a reaction mechanism radically different from the distortion of neutron waves by a potential well; indeed, in contrast to the quantal character of the latter model, the analogy Bohr proposed was completely classical. He visualized the nucleus as an assembly of nucleons held together by short-range forces, and thus effectively behaving as the assembly of the molecules forming a droplet of liquid. The energy of a particle impinging upon such a system of similar particles moving about and

---

* N. Bohr and L. Rosenfeld, *Zur Frage der Messbarkeit der elektromagnetischen Feldgrössen*, Dan. Vid. Selsk., mat.-fys. Medd. **12**, no. 8 (1933).

continually colliding with each other will be rapidly distributed among all of them, with the result that none has enough energy to leave the system: the impinging neutron is captured, and a "compound nucleus" is thus formed in a state of high excitation. This state will persist for a time which is long on the nuclear scale, i.e., which corresponds to many crossings of the nuclear volume by any single nucleon. It will decay as soon as some random fluctuation in the energy distribution will have concentrated a sufficient amount of energy on some nucleon, or group of nucleons, to allow it to escape—a process comparable to evaporation from the heated droplet. It was further easy to understand that the density of possible states of the compound nucleus would rapidly increase with the energy of excitation: this explained the absence of resonance effects at high energies as well as their presence in the low energy range.

In working out the details of this "droplet model" of nuclear reactions *, Bohr was helped by one of the younger research students of the institute, F. Kalckar, whose promising career was unfortunately cut short by an untimely death **. Refined in various ways since it was proposed in 1936, this theory still holds good as the adequate mode of description of one of the most important types of nuclear processes. It is of course an idealized model, and its basic assumptions are not always sufficiently well fulfilled to ensure its validity. Thus, another type of reaction has been found to occur, in which the interaction of the impinging particle with a single mode of motion of the target nucleus directly leads to a transfer of energy large enough to complete the process, without formation of a compound nucleus; and these "direct interaction" processes are successfully treated with the help of the old method of the potential well, in which provision is made for the possibility of capture by adding to the potential a small imaginary part—a formal device similar to that by which the absorption of light is taken into account in classical optics. Compound nucleus and "optical" potential have now shed all apparent opposition and are blended into a comprehensive theory.

The most considerable application of Bohr's theory was the interpretation of the phenomenon of nuclear fission. This is a type of reaction that may be initiated by

---

* N. Bohr and F. Kalckar, *On the transmutation of atomic nuclei by impact of material particles,* Dan. Vid. Selsk., mat.-fys. Medd. **14**, no. 10 (1937).

** Fritz Kalckar, born on the 13th February 1910, started research in theoretical physics in 1933, after completing his study at the University. He had published two papers, one of them in collaboration with E. Teller, when he collaborated with Bohr. In 1937, he accompanied Bohr to the United States and prolonged his stay there after the latter's departure. He spent a few months at Berkeley and Pasadena, where he worked together with Oppenheimer and Serber on the theory of the nuclear photoeffect and on the interpretation of some peculiarities observed in proton reactions with light nuclei. Shortly after his return, he died suddenly on the 6th January 1938. Obituary by C. Møller, Fysisk Tidsskrift **36** (1938) 1.

impact of a neutron on a very heavy nucleus: the compound nucleus formed by the capture of the neutron has so little stability that it can split into two fragments of about the same mass and charge. It was Hahn and Strassmann's chemical identification of such fragments as decay products of uranium under neutron bombardment that suggested to Frisch and Lise Meitner the fission mechanism as the only conceivable interpretation. The first experiments actually showing the emission of the fragments were performed in Copenhagen by Frisch in January 1939. By then, Bohr had left for Princeton, where he had been invited to spend a few months. It was just on his departure that he had heard of Frisch's idea and project of experiment; during the voyage and shortly after his arrival, in the same month of January 1939, he outlined the whole theory of the process. In the following months, this theory was refined and elaborated in great detail thanks to Wheeler's collaboration *.

A point that at first looked surprising was that such a splitting of the nucleus into two parts, obviously initiated by a relative oscillation of these parts with increasing amplitude, could occur with a probability comparable to that of more familiar processes, such as the emission of a γ-ray, due to a stable motion affecting only a very few nucleons. However, as Bohr pointed out, this is a direct consequence of the statistical law of energy distribution among the various modes of motion of the compound nucleus. It seemed harder to explain the differences in the efficacy of slow and fast neutrons in inducing fission in different nuclei, but Bohr solved this problem also as soon as he was confronted with the experimental data. By one of his most brilliant feats of rigorous induction from experiment, he unravelled the complex case of uranium, concluding that only the rare isotope of mass number 235 was fissile by slow neutrons, while the abundant isotope of mass 238 was not; and he showed by a very simple argument that this difference of behaviour was only due to the fact that the numbers of neutrons in the two isotopes were respectively odd and even.

The discovery that the highly unstable fission fragments emitted neutrons immediately raised the question of the possibility of a chain reaction leading to the liberation of huge amounts of energy of nuclear origin. The answer to this question was not long in doubt, and coming as it did at a critical moment in the social and political evolution of the world, the unfolding of its consequences was precipitated with unprecedented violence. If this was a fateful development in the history of mankind, it also deeply affected Bohr's individual fate. The work with fission, continued after his return to Copenhagen during the first three years of the war, was the last piece of research he carried to completion in the quiet and serene atmosphere he had himself contributed so much to create. Only much later, during

* N. Bohr and J. A. Wheeler, *The mechanism of nuclear fission*, Phys. Rev. **56** (1939) 426.

the two last summers of his life, did he for a while manage to concentrate again on a deep-lying phenomenon very near to those with which he had started his scientific career: the superconductivity of metals, in which the quantum of action manifests itself, so to speak, by macroscopic effects; he tried, without success, to put the somewhat abstract theory of these effects on a more physical basis. In 1943, however, he was dragged into the turmoil of the war, and when he later came back to Copenhagen, he had to cope with profoundly changed conditions of scientific work, that banished from his institute the intimacy of bygone years.

7. PUBLIC AFFAIRS AND EPISTEMOLOGY (1943–1962)

Bohr did not fare too well among blundering statesmen and politically immature physicists. To the former, his candour and directness looked strange and suspicious, and his clearsightedness was beyond their grasp. The latter, while showing more understanding for his ideas, were no match for politicians and military men. It was the physicists, desperately striving, under great moral and intellectual stress, towards the dark goal of the nuclear weapon, who felt the need of calling Bohr to their support. Transported, in 1943, from Copenhagen to England by way of Sweden, not without danger to his life, Bohr was suddenly faced, to his surprise and dismay, with the advanced stage of progress of a project he had deemed beyond the realm of technical accomplishment. Although he did take part, both in England and in the United States, in discussions on the physical problems related to the development of nuclear weapons, his main concern was to make the statesmen as well as the physicists aware of the political and human implications of the new source of power.

It is a striking example of his optimism that, besides the obvious dangers, he also stressed the potential advantages of the situation: the existence of a weapon equally threatening to all nations, he argued, offered a unique opportunity for reaching a universal agreement never to use it, which could become the foundation of an era of durable peace. The condition for setting up such an agreement, he added with his wonted logic, was universal knowledge of the issue at stake, and therefore full publicity in this as well as other vital questions affecting the relations between nations. More concretely, he urged the Western leaders to initiate contacts with the Russians with the view of creating the climate of mutual trust necessary to the establishment of a lasting peace between the West and the East. Although these thoughtful considerations were appreciated by some of the men in key positions, his attempts to put them before Roosevelt and Churchill ended in failure. The fulfilment of his darkest predictions in the following years did not prevent him from persevering, and in 1950, he decided to publish an *Open Letter to the United Nations*, in which he repeated his plea for an "open world" as a precondition for peace. The timing for such an appeal to a raving world to behave rationally was unfortunately the worst possible; it is now as relevant as ever, and may still perhaps some day find a response.

Apart from this unhappy excursion into the realm of world politics, Bohr devoted much time and energy to the more immediate tasks he was called upon to fulfil. In Denmark, the expansion of his institute occupied him to the last, and he also took a leading part in the foundation and organization, in 1955, of a Danish establishment for the constructive applications of nuclear energy. When the European Organization for Nuclear Research was founded in 1952, its theoretical division was installed in Bohr's institute, until it could usefully move nearer to the experimental divisions at Geneva in 1957; it was then replaced in Copenhagen by a similar institution of more restricted scope, the Nordic Institute for Theoretical Atomic Physics, created with Bohr's active participation by the five Nordic governments to accommodate young theoretical physicists from their respective countries. In these years of unprecedented expansion of scientific research all over the world, Bohr's advice and support was sought on many more occasions, and never in vain. He was now more than ever a public figure, used and misused for all kinds of ceremonies at home and abroad, and honours were poured over him from every quarter.

Unaffected by this lionizing, Bohr had always tried to make the best of it. An invitation to give a lecture at some official function was the occasion for him to orient his thought towards the particular aspect of science which would be familiar to his audience, and to reflect on the possible bearing on it of the new epistemological conceptions he had developed in quantum theory. Thus, in the thirties, he had discussed *Light and Life* before a congress of phototherapists, and spoken of the complementary features of human cultures in an assembly of anthropologists. In the post-war period, he went on in this vein and gave expression to the deep thoughts about the human condition which for him were inseparable from a proper understanding of the aim and meaning of science.

His writings on such topics were collected in three booklets *, successively published in 1934, 1958 and posthumously in 1963. These books have been translated in several languages, and one must hope that in spite of the difficulty of a style which makes no concession to the reader they may exert the same influence on the philosophical attitude of coming generations as on the mind of those who have been privileged to hear Bohr discuss his ideas in conversation. In fact, the form of publication of Bohr's essays is not felicitous: they unavoidably involve some amount of repetition, especially in elementary expositions of the physical back-

---

* *Atomic theory and the description of nature* (Cambridge University Press 1934; reprint 1961); *Atomic physics and human knowledge* (John Wiley and Sons, Inc., New York 1958); *Essays 1958–1962 on atomic physics and human knowledge* (John Wiley and Sons, Inc., New York 1963).

ground, and the main points are then often allusively suggested to the reader rather than plainly stated; long and involved sentences try to embrace all the shades of an uncommonly subtle dialectical form of thinking. Such obstacles ought not deter those who are genuinely concerned with the problems, but just the unprepared audiences to whom the message was addressed have too often failed to appreciate its true character. Bohr put an enormous amount of work into the composition of his essays, and they certainly contain the most carefully weighed expression of his philosophy.

What Bohr's essays most strikingly illustrate is the continuity of his thought. He was of course striving at all times to find more precise formulations and to disclose new aspects of the complementary relationships he was exploring, but the basic conception remained the same in all essentials from his youth to his last days. Critics endeavouring to trace foreign influences on his thinking are quite misguided: he was no doubt interested when analogies were pointed out to him between his own conceptions and those of others, but such comparisons never led to any modification of his argumentation,—the reason being that his argumentation, in contrast to the other, was so solidly founded in the analysis of the clear and precise situation offered by the development of quantum theory that there was no need for any firmer foundation. Indeed, Bohr repeatedly stressed the fortunate circumstance that it was just the simplicity of the physical issue that made it possible for him to arrive at an adequate formulation of the relations of complementarity he perceived in all aspects of human knowledge.

The domain in which complementary situations manifest themselves most immediately is the realm of psychical phenomena—that had been the starting point for Bohr's early observations. He was now able to express in terms of complementarity the peculiar relation between the description of our emotions as revealed by our behaviour and our consciousness of them; in such considerations he liked to imagine (on slender evidence, it must be said) that sayings of old philosophers and prophets were groping expressions for complementary aspects of human existence. In the development of human societies, he emphasized the dominating role of tradition over the complementary aspect of hereditary transmission in determining the essential elements of what we call "culture"; this he did in opposition to the racial theories then propagated in Germany.

Nearer to physics, he pointed out that the two modes of description of biological phenomena which were usually put in absolute opposition to each other—the physical and chemical analysis on the one hand, the functional analysis on the other—ought actually to be regarded as complementary. Altogether, he saw in complementarity a rational means of avoiding the exclusion of any line of thought which had in any way proved fruitful, and of always keeping an open mind to new possibilities of development. Thus, in his last years, he followed with the deepest satisfaction the spectacular advances of molecular biology. In the last essay he wrote, under the title *Light and Life Revisited*, he made it quite clear that in upholding the use of functional concepts in biology, he had not in mind any insuperable limitation of the scope of the physical description; on the contrary, he saw in the recent progress the unlimited prospect of a full account of biological processes in physical terms, without prejudice to an equally full account of their functional aspect.

The origin of Bohr's epistemological ideas in a purely scientific situation confers on them the character of scientific soundness and certainty. Bohr was always careful to stress both the necessity in epistemological investigations of divesting oneself of all preconceived opinions and of seeking guidance exclusively in the data of experience, and the equally stringent necessity of recognizing in every case the limitations inherent in the concepts used in the account of the phenomena. In order to understand the unique significance of his contribution to epistemology, it is necessary to realize that complementarity is a logical relationship, referring to our way of describing and communicating our experience about a universe in which we occupy the singular position of being at the same time, and inseparably, spectators and actors. Far from excluding from our reach any aspect of the universe, complementarity gives us, so far as we can judge, a logical frame wide enough for a comprehensive, rational and objective account of all aspects of the phenomena. By the rigour of his rational thinking, the universality of his outlook and his deep humanity, Bohr will for ever rank, in the judgement of history, among the few fortunate men to whom it was given to help the human mind to a decisive step towards a fuller harmony with nature.

PART I

SURFACE TENSION
OF WATER

INTRODUCTION

by

J. RUD NIELSEN

1. EARLIEST WORK

Niels Bohr entered the University of Copenhagen in the autumn of 1903 and immediately began the study of physics, with mathematics, astronomy and chemistry as secondary subjects. Like all students at the University of Copenhagen, he attended lectures on philosophy during the first year. His interest in this subject is evidenced by the fact that his teacher, the well-known philosopher Harald Høffding, when preparing a new edition of a textbook on formal logic in 1906, accepted his criticism of a certain point (concerning the principle of excluded middle) and asked him to go over the manuscript of the entire book.

With few lectures to attend, he began already as a young student to carry out original research in physics. The field of his earliest work, surface tension and surface waves on liquids, was determined more by the interest of his teachers than by his own choice. In the Niels Bohr Archive are found three sheets marked "Kurver vedrørende Overfladespænding" (Curves pertaining to surface tension). It is indicated that they were worked out at the suggestion of Professor C. Christiansen, with the view of possible use for an experimental determination of surface tension. The sheets are not dated, but the envelope is marked "1905–1906?". It is believed that they represent Bohr's earliest work on surface tension. This work was never published, nor was it used in connection with experimental determinations of surface tension.

2. DANISH ACADEMY PRIZE ESSAY

In those days the Royal Danish Academy of Sciences and Letters (Det Kongelige Danske Videnskabernes Selskab) awarded each year gold or silver medals for monographs on topics specified by the Academy two years previously. Among the prize problems announced in February 1905 was one for physics which read as follows:

Physics prize problem
Danish text p. [13]

"In the Proceedings of the Royal Society XXIX, 1879, Lord Rayleigh has developed the theory of the vibrations which a liquid jet carries out about the cylindric shape when it is somehow made to take another cross section. From the theory, as well as from the experiments which Lord Rayleigh has performed on these vibrations, it appears that they could serve to determine the surface tension of a liquid. The Academy, therefore, offers its gold medal for a more detailed investigation of the vibrations of liquid jets with special reference to the application mentioned. It is desired that the investigation be extended to a fairly large number of liquids. The results are to be compared with those previously found in other ways."

Solutions were to be submitted anonymously by October 30, 1906. Although most winners of these prizes had been mature scholars, the 19-year-old Niels Bohr decided to try his hand on the problem. He carried out the experimental work in the physiological laboratory of the University of Copenhagen, of which his father, Professor Christian Bohr, was the director. His father had to put pressure on him to terminate the experiments and cease making new and time-consuming corrections to the theory. The paper was written at his grandmother's estate, Nærumgaard, a few miles North of Copenhagen. The fair copy was handwritten by Niels Bohr's two years younger brother, Harald. Consisting of 114 pages and 19 figures, it was submitted on the very day of the deadline. We only reproduce two pages of it as a sample.

On November 2, 1906, Bohr submitted an eleven-page addendum with the following note:

Note accompanying
Addendum
Danish text p. [13]

"It is respectfully requested that the enclosed addendum, which, owing to an accident in the copying, was not submitted with the main essay, may be appended to the paper with the motto $\beta\gamma\delta$ which was submitted as physics prize essay."

That there was difficulty in meeting the deadline is indicated by the fact that this addendum was handwritten in part by Harald, in part by Niels himself, and in part by their mother.

3. DISCUSSION ON A QUESTION OF MECHANICS

In January and February 1907, while awaiting the decision of the Royal Academy, Bohr had a discussion and exchanged a series of letters with Fr. Johannsen*, the director of the Copenhagen Telephone Company. In 1897 Johannsen had published a paper in which he claimed that the stresses in mechanical systems with supernumerary constraints, and the equilibrium of such systems, can be uniquely

* Johannsen's letters and copies of Bohr's letters are found in the Niels Bohr Archive.

[4]

determined by requiring that all acting forces balance and that the internal work of elastic deformations be a minimum. Bohr challenged this contention and pointed out that this procedure does not in all cases give the correct solution, but that the applicability of the principle of minimum work of deformation must be investigated for each type of mechanical system in question. He illustrated his view by simple examples, but failed to convince the 30-years older Johannsen.

4. JUDGEMENT OF PRIZE ESSAYS

The physics prize essays were judged by Professors C. Christiansen and K. Prytz*. Their report to the Royal Academy reads:

In answer to the Physics Prize Problem set by the Society for 1905, in which a detailed investigation of the vibrations of liquid jets was requested, two essays have been submitted. Judgement of Prize essays
Danish text pp. [13]–[15]

One of these, having the motto "The preparations are the worst", sticks closely to the problem set and must be said to give a thorough solution of it in all respects. After a brief account of Lord Rayleigh's theory of the vibrations of liquid jets, the methods applied by the author to measure the cross-sectional area of the jet, and its velocity and wavelength, are described. By a kind of swing, he cuts off segments of a horizontal jet, and, from the total weight and length of these, he determines the cross-sectional area. The velocity is obtained by measuring the amount of liquid flowing out in a certain time.

In order that waves may form on the jet, it must flow out of a hole that is not circular. However, when this is the case there will generally occur at the same time waves of different wavelengths, and this makes reliable measurements difficult. Hence, the author has prepared orifices of different shapes, corresponding to the simplest vibrations which the jet is capable of executing, and in this way he has succeeded in obtaining so simple experimental conditions that he could get single forms of vibration. In order that viscosity may not influence the result, and to insure a sufficiently good agreement with the theory, the cross section of the jet must only deviate very little from being circular; that the author accomplishes this, he verifies by showing that the surface tension finally becomes independent of the deviations from the circular form.

However, under these circumstances the determination of the wavelength becomes difficult. The author has therefore devised an optical method, the theory of which is the same as that of the rainbow. He illuminates the jet with a linear

* Peter Kristian Prytz (1851–1929) was docent in physics from 1886 to 1894 and professor from 1894 to 1921 at the Polytekniske Læreanstalt (Polytechnic School, founded by Oersted in 1829, now the Technical University of Denmark). He developed precision methods of measurement in many fields of physics.

light source which is parallel to the jet; when the jet is cylindrical the light twice refracted and once reflected will then give a straight luminous line corresponding to the rainbow; however, when there are waves on the jet, this line will be serpentine. A series of beautifully made photographs illustrate this method.

The author has first of all applied the method to water, but thereafter also to toluene, aniline, aqueous solutions of ammonia and copper sulfate, and to a series of alcohol and water mixtures.

The whole work gives evidence of great ingenuity and experimental energy and skill. It follows from the investigation that Lord Rayleigh's method, so applied, is greatly superior to those previously used.

Hence, we do not hesitate to recommend that the Society reward this highly satisfactory solution of the prize problem with its gold medal*.

The author of the other essay, who designates himself by the mark $\beta\gamma\delta$**, has only managed to investigate the surface tension of water, on account of the experimental arrangement used. On the other hand, he has carried out a very extensive investigation of the conditions in the water jet. To produce a sufficiently long, regular, undivided and untwisted jet, the author let the water flow out through a long narrow glass tube whose orifice was made elliptic to bring about the vibratory motion. The jet was examined at distances of about 25 cm from the orifice to permit the viscosity to smooth out irregularities in the motion. By selecting from among a large number of prepared tubes, a few were found which gave the jet a symmetrical shape with respect to two mutually perpendicular planes through the axis. This symmetry was tested by an optical method which was used also to measure the wavelength. The amplitude of the vibration was found by measurement on an enlarged photograph of the jet.

To determine the constant of capillarity, the author measures the amount of water flowing out in a given time, the velocity of the jet, and the wavelength. The velocity was found by a clever method which consists in cutting through the jet at a given place at two instants separated by a short time interval and measuring the length of the segment cut out and the corresponding time (ca. 1/50 s). The length of the segment cut out was found by photographing the cut jet by instantaneous illumination. The method gave very good results which could be checked by varying the time interval.

* [The author of this prize essay was P. O. Pedersen, an electrical engineer eleven years older than Bohr. In 1909 he became docent, and in 1912 professor, of electrical engineering at the Polytechnical School. He served as president of the Polytechnical School from 1922 until his death in 1941. The best known of his many scientific publications dealt with the propagation of radio waves. His results on surface tension were published in Phil. Trans. Roy. Soc. **A207** (1907) 341.]

** [This is Niels Bohr.]

[6]

To measure the wavelength, the aforementioned optical method was applied. It consist in reflecting a light source at the surface of the jet and finding the positions on the jet where the tangent planes are parallel to the axis.

The performance of a single determination by the author's method requires continued work through many hours. Hence, the jet must be maintained for a long time and under very constant conditions. This long time limits the applicability of the method in the case of liquids which change under exposure to air, and requires a comparatively large amount of liquid.

When the problem was formulated, it was thought that the theory given by Lord Rayleigh should form the basis for the investigation. However, this theory gives only the first approximation. The author of this essay has remedied this by extending the theory so as to take into account viscosity and amplitudes that are not infinitely small. It is obvious that these investigations are of great interest for judging the value of the method and finding under what conditions it may be expected to yield the best results.

This work does not solve the problem as completely as the former, in that it has studied only a single liquid, water. On the other hand, its author has earned so great merit by carrying the solution further at other points that we feel we must recommend that also this essay be awarded the Society's gold medal.

25 January 1907   C. Christiansen   K. Prytz

A month later Niels Bohr was notified of the award by the following letter:

Upon recommendation of the Section for Natural Sciences and Mathematics, the Royal Danish Academy of Sciences and Letters at its meeting on February 22 has awarded you its gold medal for your solution of the Physics Prize Problem set in 1905.

Notification of award
Danish text p. [15]

In sending you the medal, the Academy extends to you its congratulations.

Copenhagen, February 23, 1907

Julius Thomsen
President
H. G. Zeuthen
Secretary

Among the many other letters of congratulation which Bohr received was one from Harald Høffding, a close friend of his family:

Letter from H. Høffding
to Niels Bohr (25 Jan 07)
Danish text p. [16]

Friday evening January 25, 1907

Dear Niels Bohr

It was a great pleasure for me to learn this evening at the Academy that your paper was rewarded with a prize. I congratulate you on this fine distinction that you have won at such an early age, and I take this opportunity to thank you for your valuable collaboration.

Sincerely yours,
Harald Høffding

5. FIRST ROYAL SOCIETY PAPER

After receiving the gold medal, Bohr carried out additional measurements of the surface tension of water. At the same time he was occupied with the considerable task of preparing his work for publication. In the latter part of 1908 he submitted a paper entitled "Determination of the Surface-Tension of Water by the Method of Jet Vibration" to the Royal Society of London. This paper is not a simple translation of the prize essay but deviates from the latter at a number of points.

Thus, on p. 8 of the prize essay an exact equation is obtained which is so cumbersome that it must be solved approximately by a process of iteration. In the paper, on the other hand, approximations are introduced at an earlier stage, and a simpler equation, eq. (35) on p. 288, is derived.

On p. 16 of the prize essay, where the correction to the wavelength due to finite wave amplitudes is calculated to the second approximation, an error was found by Bohr after the essay was returned by the Royal Academy. This error, which also affected the third approximation, has been corrected in the paper. The description of the experiments was also revised considerably. The data presented in the paper were obtained in measurements carried out in February 1908. The paper refers to a few more experimental results of other investigators than did the prize essay. In particular, the results of P. O. Pedersen are discussed.

The paper was received by the Royal Society of London on January 2, 1909, and was read at a meeting on January 21. A week later the secretary of the Society wrote to Bohr:

Letter from J. Larmor
to Niels Bohr (28 Jan 09)

The Royal Society
Burlington House, London W.

Dear Professor Bohr,

Your paper on capillarity has been read formally at the Society. It is suggested by Prof. Lamb that you have perhaps not fully considered the point that up to the second order the corrections for finite amplitude and for viscosity are not

[8]

additive. Thus, he says, if $T = f(\mu, \alpha)$ then

$$T = T_0 + \frac{\partial f}{\partial \mu}\mu + \left[\frac{\partial f}{\partial \alpha}\alpha + \tfrac{1}{2}\frac{\partial^2 f}{\partial \mu^2}\mu^2 + \frac{\partial^2 f}{\partial \mu \partial \alpha}\mu\alpha\right] + \tfrac{1}{2}\frac{\partial^2 f}{\partial \alpha^2}\alpha^2$$

and that your procedure appears to neglect the bracket terms – that in fact your results agree without keeping them in.

It is suggested that you may wish to consider this point before the paper goes to press – by way of a small addition or otherwise.

Very faithfully yours,

J. Larmor
Sec. R.S.

Bohr replied on February 6 (draft in his own handwriting; errors of spelling, punctuation, or grammar, may have been corrected in the letter actually sent):

Copenhagen, Bredgade 62. 6–2–09

Letter from Niels Bohr
to J. Larmor (6 Feb 09)

Dear Professor Larmor!

I thank you very much for your kind letter containing the interesting suggestion of Professor Lamb, that it may not be justified to consider the corrections in question as additive.

It seems however to me, that the corrections mentioned must be additive, as the wavelength can be shown to be an even function of the amplitude also in the case, where the fluid is viscous. This can be shown as follows.

I find in my paper, that for a finite coefficient of viscosity the surface of the jet will, considering the wave-amplitude as infinitely small, be of the form

$$r = a(1 + \alpha \cos(n\vartheta)\mathrm{e}^{-\varepsilon z}\sin(kz)).$$

Now proceeding by successive approximations, and introducing the solution found in the terms of higher order partly in the equations of motion and partly in the surface conditions, etc. . . . we see, that we will get a solution of the form

$$\begin{aligned}
r = a(1 &+ \alpha \cos(n\vartheta)f_1(z) + \alpha^2(\cos^2(n\vartheta)f_{20}(z) \\
&+ \cos(n\vartheta)\sin(n\vartheta)f_{21}(z) + \sin^2(n\vartheta)f_{22}(z)) + \cdots \\
&- \alpha^k(\cos^k(n\vartheta)f_{k0}(z) + \cdots + \cos^{k-s}(n\vartheta)\sin^s(n\vartheta)f_{ks}(z) + \cdots) + \cdots)
\end{aligned}$$

in which the functions f* will be even functions of α, this follows directly from the progress of the calculations (in analogous manner as in the problem without viscosity) and is further indicated by the fact, that $\alpha^k \cos^{k-s}(n\vartheta)\sin^s(n\vartheta)$ will

---

* [Rather, the terms in the expansion of r.]

[9]

not be varied if we give α an opposite sign and in the same time change the system of coordinates in the way that z is unchanged and ϑ is replaced by $(\vartheta + \pi/n)$.

The addition to my paper, kindly proposed in your letter, could therefore be as follows:

"The corrections are* to be considered as additive, as it can be shown that the wave-length, also in the case of viscosity, will be an even function of the wave-amplitude."

This addition could be placed in the paper immediately before the collecting formula given just at the end of the calculation of the corrections**.

Thanking you for your kind trouble, I allow me to ask you to bring my compliments and thanks to professor Lamb.

> Believe me dear Sir
> Very faithfully yours,
> N. Bohr

Having received no reply to this letter, Bohr wrote to Larmor again two months later (draft in his own handwriting):

Letter from Niels Bohr
to J. Larmor (4 Apr 09)

Bredgade 62, Copenhagen 4–4–09

Dear Professor Larmor!

I hope you have in due time received my registered letter of 6–2–09 in answer to your letter of 28–1–09. I should be most thankfull if you with a few words kindly would let me know, at what time I may exspect to receive the proof-sheets of my paper.

Thanking you beforehand for your trouble

> I remain very faithfully yours
> Niels Bohr

P.S. Allow me to remark, that I am not a professor but am studying natural philosophy at the university of Copenhagen.

Bohr must have received the proofs shortly afterwards, for on June 9, 1909, he sent a reprint of the paper to his brother Harald, then in Göttingen. The paper appeared in the Philosophical Transactions of the Royal Society. Of all the papers published by Bohr, this is unique, not only in being his earliest publication, but also in being the only paper in which he reports experimental work carried out by himself.

---

* [The draft has "is".]
** [The addition appears in the published paper as a footnote on p. 297.]

6. ADDENDUM TO THE PRIZE ESSAY

The addendum submitted three days after the main part of the prize essay was an investigation of the influence of the finite amplitude upon waves on the surface of deep water, progressing without change in shape under the simultaneous influence of gravity and surface tension. Since it was never published*, an English translation of it is given; a few obvious writing slips in the formulas have been corrected.

7. SECOND ROYAL SOCIETY PAPER

In 1910, while Bohr was working on his doctor's dissertation, a paper was published by P. Lenard**, in which he claimed that a recently formed water surface has a high surface tension which rapidly decreases. Lenard stated that this was in agreement with Bohr's results. This led Bohr to re-examine the matter and, in particular, to test by new calculations the claim made by Lenard that a variation of the velocity over the different concentric parts of a jet will prolong the periods of vibration and increase the wavelengths of surface waves on the jet. He made a direct calculation of the wavelength when the velocity in the jet varies with the distance from the axis and concluded that his experiments do not support the claims made by Lenard. In August, 1910, he submitted to the Royal Society a paper entitled "On the Determination of the Tension in a recently formed Water-Surface", which was published in its Proceedings.

This was Bohr's last work on surface tension. The results of this paper, and the merit of the experimental method described, have apparently been appreciated by most later workers in the field. Thus, N. K. Adam writes in his book *The physics and chemistry of surfaces* (Clarendon Press, Oxford 1930, p. 324): "The various dynamic methods give the surface tension of more or less recently formed surfaces, and may yield results different from the static methods, if adsorption occurs, and is incomplete at the moment when the tension is actually measured. One factor in dynamic measurements which cannot be satisfactorily measured at present, is the time which has elapsed between the formation of the surface from the homogeneous interior liquid, and the actual measurement of the surface tension. If this could be varied, and measured with an accuracy of say 10^{-4} second, a valuable new weapon would be available for investigating the progress of adsorption. Bohr's work on oscillating jets is probably the best of any dynamic

---

* However, one of the "Theses" at the end of Bohr's doctor's dissertation is a brief statement of the conclusion that there is a range of wavelengths corresponding to which purely periodic waves cannot exist on the surface of a liquid subject to the simultaneous influence of gravity and surface tension.

** Sitzungsber. d. Heidelberger Akad. d. Wiss., Math.-nat. Kl., Jahrg. 1910, Abh. 18.

methods." What he learned by working in this field may have been a help to him when, more than a quarter of a century later, he showed that some of the properties of atomic nuclei can be understood by comparing them to liquid drops.

DANISH TEXTS OF QUOTED DOCUMENTS

Physics prize problem

I Proceedings of the Royal Society XXIX, 1879 har Lord Rayleigh udviklet
Theorien for de Svingninger, som en Vædskestraale udfører omkring Cylinder-
formen, naar man paa en eller anden Maade bringer den til at antage et andet
Tværsnit. Saavel af Theorien, som af de Forsøg, Lord Rayleigh har anstillet over
disse Svingninger, synes det at fremgaa, at man ved Hjælp af dem kan bestemme en
Vædskes Overfladespænding. Videnskabernes Selskab udsætter derfor sin Guldmedaille for en nærmere Undersøgelse af Vædskestraalernes Svingninger, hvorved
der særlig tages Sigte paa den nævnte Anvendelse af dem. Undersøgelsen ønskes
udstrakt til en større Gruppe af Vædsker; Resultaterne maa sammenlignes med
dem, som tidligere ere fundne ad andre Veie.

Note accompanying Addendum

Man tillader sig at andrage om at medfølgende Bilag, der paa Grund af Uheld
ved Afskrivningen ikke er fulgt med Hovedafhandlingen, maa blive vedlagt den
for den fysiske Prisopgave indleverede Afhandling med Motto $\beta\gamma\delta$.

Judgement of prize essays

Som Besvarelse af den af Selskabet udsatte fysiske Prisopgave for 1905, hvori
forlangtes en nærmere Undersøgelse af Vædskestraalers Svingninger (Oversigt
1905 S. $\langle 18 \rangle$–$\langle 19 \rangle$), er indkommen to Afhandlinger.

Den ene af disse, som har til Motto: "Forberedelserne er de værste", holder sig
nøje til den stillede Opgave og maa i alle Henseender siges at give en udtømmende
Besvarelse af den. Efter en kort Gengivelse af Lord Rayleighs Teori for Vædskestraalers Svingninger beskrives de af Forfatteren anvendte Metoder til Maaling
af Straalens Tværsnitsareal, Hastighed og Bølgelængde. Den førstnævnte Størrelse
finder han ved en Metode, der næppe har været anvendt tidligere, idet han ved en
Art Gynge afhugger Stykker af en vandret Straale; af disses samlede Vægt og
Længde findes Tværsnitsarealet. Idet man søger den i en vis Tid udstrømmende
Vædskemængde, bliver Hastigheden bestemt.

For at faa dannet Bølger paa Straalen maa den strømme ud ad et Hul, som ikke
er rundt. Men i dette Tilfælde vil der sædvanlig samtidig findes Bølger med
forskellig Bølgelængde, hvilket vanskeliggør en paalidelig Maaling. Forfatteren
har derfor tilvejebragt Udløbsaabninger af forskellig Form svarende til de sim-

pleste Svingninger, som Straalen kan udføre, og han har derved opnaaet saa simple Forhold at arbejde med, at han kunde faa enkelte Svingningsformer. Forat den indre Gnidning ikke skal faa Indflydelse paa Resultatet, og for at give fornøden Overensstemmelse med Teorien, maa Straalens Tværsnit kun afvige meget lidt fra Cirkelen; at Forfatteren opnaar dette, godtgør han ved at vise, at Overfladespændingen til sidst bliver uafhængig af Afvigelserne fra Cirkelformen.

Under disse Forhold bliver Bestemmelsen af Bølgelængden imidlertid vanskelig. Forfatteren har da funden paa at anvende en optisk Metode, hvis Teori er den samme som Regnbuens, han belyser Straalen med en linieformig Lysgiver, som er parallel med Straalen; det to Gange brudte og en Gang tilbagekastede Lys vil da, naar Straalen er cylindrisk, give en ret Lyslinie svarende til Regnbuen, men naar der er Bølger paa Straalen, vil denne Linie blive bugtet. En Række særdeles smukt udførte Fotografier illustrerer denne Metode.

Forfatteren har først og fremmest anvendt Metoden paa Vand men dernæst ogsaa paa Toluol, Anilin, vandige Opløsninger af Ammoniak og Kobbersulfat samt paa en Række af Blandinger af Alkohol og Vand.

Det hele Arbejde vidner om stor Opfindsomhed og eksperimental Energi og Dygtighed; det fremgaar af Undersøgelsen, at Lord Rayleighs Metode anvendt paa denne Maade langt overgaar alle andre hidtil anvendte.

Wi tage derfor ikke i Betænkning at indstille, at Selskabet belønner denne særdeles tilfredsstillende Løsning af Prisopgaven med sin Guldmedaille.

Den anden Afhandling, hvis Forfatter betegner sig med Mærket $\beta\gamma\delta$, har kun naaet at undersøge Vandets Overfladespænding, hvad der staar i Forbindelse med hans Forsøgsanording. Paa den anden Side har han udført en meget omfattende Undersøgelse af Forholdene i Vandstraalen. Forat skaffe en tilstrækkelig lang, regelmæssig, udelt og usnoet Straale tilveje lod Forfatteren Vandet løbe ud gennem et snævert, langt Glasrør, hvis Munding gjordes elliptisk af Hensyn til at frembringe Bølgebevægelsen. Straalen blev undersøgt i Afstande omkring 25 cm. fra Rørmundingen forat give den indre Gnidning Lejlighed til at udjævne Uregelmæssigheder i Bevægelsen. Ved at udsøge mellem en stor Mængde tildannede Rør blev der funden nogle faa, som gav Straalen en symmetrisk Form i Forhold til to paa hinanden vinkelrette Planer gennem Aksen, hvilken Symmetri blev prøvet ved en optisk Metode, der ogsaa blev anvendt til Maaling af Bølgelængden. Svingningsamplitudens Størrelse blev funden ved Udmaaling paa et forstørret fotografisk Billede af Straalen.

Til Bestemmelse af Haarrørskonstanten maaler Forf. den i given Tid udløbende Vandmængde, Hastigheden i Straalen og Bølgelængden. Hastigheden blev funden direkte ved en snild Metode, som bestaar i paa et givet Sted af Straalen at hugge denne over til to kort efter hinanden følgende Tidspunkter og maale Længden af det saaledes udhuggede Stykke og den tilhørende Tid (ca. 1/50 Sek.). Stykkets Længde fandtes ved Fotografering af den overhuggede Straale ved momentan Belysning. Metoden gav særdeles gode Resultater, kontrollerede ved at variere Tiden.

Til Maaling af Bølgelængden blev den før nævnte optiske Metode anvendt. Den bestaar i ved Spejling af en Lyskilde i Straalens Overflade at finde Steder af Straalen, hvor dennes Tangentplaner er parallele med dens Akse.

Udførelsen af en enkelt Bestemmelse efter Forfatterens Metode kræver fortsat Arbejde gennem mange Timer. Af den Grund maa Straalen holdes vedlige i lang Tid og under meget konstante Forhold; den lange Tid medfører en Begrænsning af Metodens Anvendelighed overfor Vædsker, som forandres i Luften, og kræver en forholdsvis stor Vædskemængde.

Da Opgaven blev stillet, var der nærmest tænkt paa, at den af Lord Rayleigh givne Teori skulde lægges til Grund for Undersøgelsen. Denne Teori giver imidlertid kun den første Tilnærmelse; herpaa har Forfatteren af denne Afhandling bødet ved at udvide Teorien saaledes, at der ogsaa tages Hensyn til indre Gnidning og til Amplituder, der ikke er uendelig smaa. Det er en Selvfølge, at disse Undersøgelser frembyde stor Interesse, naar det gælder om at bedømme Metodens Værdi og at finde, under hvilke Betingelser den kan ventes at give de bedste Resultater.

Skønt dette Arbejde forsaavidt ikke løser Opgaven saa fuldstændigt som det første, som der kun er arbejdet med en enkelt Vædske, Vandet, saa har dets Forfatter til Gengæld indlagt sig saa megen Fortjeneste ved at føre Løsningen videre paa andre Punkter, at vi mene at burde foreslaa, at ogsaa denne Afhandling belønnes med Selskabets Guldmedaille.

D. 25. Januar 1907.   C. Christiansen.   K. Prytz.

Notification of award

Det Kongelige Danske
Videnskabernes Selskab
     København

Det Kongelige Danske Videnskabernes Selskab har i sit Møde den 22de Februar, efter Indstilling af den naturvidenskabelig-mathematiske Klasse, tilkendt Dem sin Guldmedaille for Deres Besvarelse af den i 1905 stillede fysiske Prisopgave.
    Idet Medaillen hermed tilsendes, bringer Selskabet Dem sin Lykønskning.
    København, d. 23. Februar 1907

                                        Julius Thomsen
Til   *Præsident*
    Herr stud. mag Niels Hendrik David Bohr,   H. G. Zeuthen
                        København   *Sekretær*

Letter from H. Høffding to Niels Bohr (25 Jan 07)

Fredag Aften d. 25. Januar 1907

Kære Student Niels Bohr!

Det var mig en stor Glæde at høre i Aften i Videnskabernes Selskab, at Deres Afhandling er blevet prisbelønnet. Jeg lykønsker Dem til det smukke Resultat, De har vundet i Deres unge Alder, og benytter Lejligheden til at takke Dem for Deres gode Samarbejde.

Deres hengivne
Harald Høffding

[16]

I. CURVES PERTAINING TO SURFACE TENSION

(1905–06?)

[See Introduction, sect. 1. One curve, calculated and drawn from a classical formula*, represents the cross section of the surface of a liquid adhering to a long rectangular horizontal plate brought to touch a liquid which wets it and then raised, while remaining horizontal, until the liquid is about to break away. The other curve, for which Bohr has derived the formula, represents the cross section of the caustic surface formed when a parallel beam of light, incident horizontally and at right angles to the edge of the plate, is reflected from the surface of the adhering liquid.

A point on the first curve is denoted by x, y, a point on the second curve by ξ, η. The constant a, which in Bohr's formulas is put equal to unity, is related to the surface tension T in the following way:

$$a = \sqrt{T/g\rho},$$

where g is the acceleration due to gravity and ρ is the density of the liquid.]

---

* [The problem is treated by E. Mathieu, *Théorie de la capillarité* (Gauthier-Villars, Paris 1883), ch. V. See fig. 28 on p. 126; the first formula written by Bohr is derived on p. 128 (for a large disk), and the equation of the curve used by Bohr is easily derived from it and from the relation $dy/dx = -\tan \alpha$. Mathieu's book may be the source Bohr used, since it is one of the two detailed treatises quoted in the textbook by C. Christiansen and J. J. C. Müller, *Elemente der theoretischen Physik* (J. A. Barth, Leipzig 1903), p. 180, a copy of which was in Bohr's personal library. A copy of Mathieu's book belongs to the earliest nucleus of the Copenhagen Institute's library.]

$$z = 2a\sin\frac{\alpha}{2}$$

$$\partial\beta = dz = a\cos\frac{\alpha}{2}\,d\alpha$$

$$AB = \frac{\partial\beta}{\sin\alpha} = \frac{a}{2}\frac{d\alpha}{\sin\frac{\alpha}{2}}$$

$$AC = \sin\alpha\frac{AB}{2d\alpha} = \frac{a}{2}\cos\frac{\alpha}{2}$$

$$\xi = X - \frac{a}{2}\cos\frac{\alpha}{2}\cos 2\alpha$$

$$\Upsilon = z + \frac{a}{2}\cos\frac{\alpha}{2}\sin 2\alpha$$

$$\xi = X - \frac{a}{2}\cos\frac{\alpha}{2}(1-8\sin^2\frac{\alpha}{2}\cos^2\frac{\alpha}{2}) = X - \frac{a}{2}\sqrt{1-\sin^2\frac{\alpha}{2}}(1-8\sin^2\frac{\alpha}{2}+8\sin^4\frac{\alpha}{2}) = X - \frac{1}{4}\sqrt{4a^2-z^2}(1-2\frac{z^2}{a^2}+\frac{z^4}{2a^4}) = X + X'$$

$$\Upsilon = z + \frac{a}{2}\cos\frac{\alpha}{2}\cdot 4\cos^2\frac{\alpha}{2}\sin\frac{\alpha}{2}\cos\alpha = z + 2a\sin\frac{\alpha}{2}(1-\sin^2\frac{\alpha}{2})(1-2\sin^2\frac{\alpha}{2}) = z + z(1-\frac{3z^2}{2a^2}+\frac{z^4}{2a^4}) = z + z'$$

$$X = -\sqrt{4a^2-z^2} + a\cdot\int\frac{\sqrt{4a^2-z^2}+2a}{z}$$

$$a = 1 \qquad X = -\sqrt{4-z^2}+\int\frac{\sqrt{4-z^2}+2}{z} \qquad X_1 = -\frac{1}{4}\sqrt{4-z^2}(-2z^2+\frac{z^4}{2}) \qquad z' = z(1-\frac{3z^2}{2}+\frac{z^4}{2})$$

$$\Upsilon = 2a\sin\frac{\alpha}{2} + \frac{a}{2}\cos\frac{\alpha}{2}\sin 2\alpha = 2a(\sin\frac{\alpha}{2}+\sin\frac{\alpha}{2}(1-\sin^2\frac{\alpha}{2})(1-2\sin^2\frac{\alpha}{2})) = 2a(2\sin\frac{\alpha}{2}-3\sin^3\frac{\alpha}{2}+2\sin^5\frac{\alpha}{2})$$

$$\frac{d\Upsilon}{d\alpha} = a\cos\frac{\alpha}{2}(2-9\sin^2\frac{\alpha}{2}+10\sin^4\frac{\alpha}{2}) = 10a\cos\frac{\alpha}{2}(\sin^2\frac{\alpha}{2}-\frac{1}{2})(\sin^2\frac{\alpha}{2}-\frac{2}{5})$$

$$\frac{d\Upsilon}{d\alpha}=0 \qquad ^{1)}\sin\frac{\alpha}{2}=1\,.\ \Upsilon=2a\,.\quad ^{2)}\sin\frac{\alpha}{2}=\sqrt{\tfrac{1}{2}}\,.\ \Upsilon=a\sqrt{2}\,.\quad ^{3)}\sin\frac{\alpha}{2}=\sqrt{\tfrac{2}{5}}\,.\quad \Upsilon=2a\sqrt{\tfrac{2}{5}}\,.$$

[18]

y	X	X'	y'	ξ	η	
2	0	0	0	0	2	
1,97	-0,1742	-0,0665	-0,0544	-0,2407	1,9156	
1,9	-0,3014	-0,0462	-0,1492	-0,3476	1,7508	
1,8	-0,4046	0,0504	-0,2138	-0,3542	1,5862	
1,7	-0,4679	0,1591	-0,2100	-0,3088	1,4900	
1,6	-0,5068	0,2530	-0,1613	-0,2538	1,4387	
1,5	-0,5275	0,3204	-0,0820	-0,2071	1,4180	
$\sqrt{2}$ 1,4142	-0,5328	0,3536	0	-0,1792	1,4142	Minimum
$2\sqrt{\tfrac{1}{3}}$ 1,2649	-0,5175	0,3563	0,1518	-0,1612	1,4167	Maximum
1,1	-0,4653	0,2873	0,3030	-0,1780	1,4030	
1,0	-0,4150	0,2165	0,3750	-0,1985	1,3750	
0,9	-0,3494	0,1304	0,4271	-0,2190	1,3271	
0,8	-0,2662	0,0345	0,4570	-0,2317	1,2570	
0,7	-0,1626	-0,0656	0,4638	-0,2282	1,1638	
0,6	-0,0343	-0,1645	0,4477	-0,1988	1,0477	
0,5	0,1270	-0,2572	0,4101	-0,1302	0,9101	
0,4	0,3329	-0,3394	0,3533	-0,0065	0,7533	
0,3	0,6072	-0,4074	0,2801	0,1998	0,5801	
0,2	1,0032	-0,4581	0,1940	0,5451	0,3940	
0,1	1,6908	-0,4894	0,0993	1,2014	0,1993	
0	∞	-0,5	0	∞	0.	

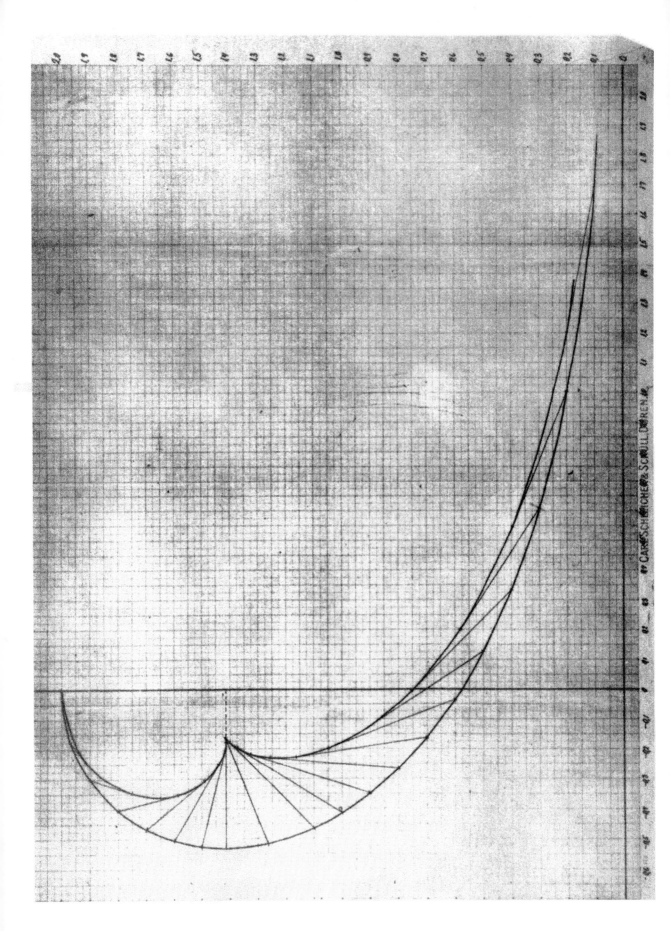

II. PRIZE ESSAY IN PHYSICS FOR 1905

[Facsimile Reproduction of pp. 7 and 8 of Niels Bohr's Prize Essay (1906), entitled "Besvarelse af det Kongelige Danske Videnskabernes Selskabs Prisopgave i Fysik for 1905".*]

* [See Introduction, sect. 2.]

Antages Overfladens Ligning at være $r-a-\zeta=f(\vartheta,t)=$

$D \cdot e^{int+ib\epsilon}$, bliver den almindelige Overfladebetingelse

$$\frac{D(r-a-\zeta)}{Dt}=\left(u\frac{\delta}{\delta x}+v\frac{\delta}{\delta y}+w\frac{\delta}{\delta z}\right)(r-a-\zeta)=\left(\alpha\frac{\delta}{\delta r}+\frac{\beta}{r}\frac{\delta}{\delta \vartheta}+(c+\omega)\frac{\delta}{\delta z}\right)(r-a-\zeta)=0$$

hvilket med samme Grad af Tilnærmelse som tidligere anvendt, giver $\alpha-c\,i\,b\,\zeta=0$.

Med samme Tilnærmelsesgrad haves endvidere

$$\frac{1}{R_1}+\frac{1}{R_2}=\frac{1}{a}+\frac{\zeta}{a^2}(n^2-1+b^2a^2)=\frac{1}{a}-\frac{i\,b\,\zeta}{a^2\,c\cdot b}(n^2-1+b^2a^2)$$, hvor R_1 og R_2 er de principale Krumningsradier.

De Betingelser, der skal være opfyldte for $r=a$ er følgende $P_z=0$ (1) $P_\vartheta=0$ (2) $Konst+P_r=\div T\left(\frac{1}{R_1}+\frac{1}{R_2}\right)$ (3). Altsaa

$$0=\frac{d\omega}{dr}+\frac{d\alpha}{dz}$$ (1) $$0=\frac{1}{r}\frac{d\alpha}{d\vartheta}+\frac{d\beta}{dr}-\frac{\beta}{r}$$ (2)   og $$0=-T\frac{i\alpha(n^2-1+b^2a^2)}{a^2\,c\cdot b}-p+2\mu\frac{d\alpha}{dr}$$ (3)

(1) $$0=-2A\frac{ib}{c\zeta}\,f_n'(iba)+ib\,B\left(\frac{1}{a}+\frac{b}{c}\right)f_n'(ida)+ib\,b\,\frac{1}{a}\,f_n(ida)$$

(2) $$0=-\frac{2A}{c\zeta}\left(\frac{in}{a}\,f_n'(iba)-\frac{n}{b}\cdot\frac{1}{a}\,f_n(iba)\right)+2B\frac{b}{a}\left(\frac{in}{a}\,f_n'(ida)-\frac{n}{a}\cdot\frac{1}{a}\,f_n(ida)\right)+$$
$$b\left(\frac{in}{a}\,f_n(ida)+\frac{d}{na}\,f_n'(ida)-\frac{id^2}{n}\,f_n''(ida)\right)$$

Sammentrækkes, faas:

(2) $$0=-\frac{2A}{c\zeta}\left(a\,f_n'(iba)+\frac{1}{b}\,f_n(iba)\right)+2B\frac{b}{a}\left(a\,f_n'(ida)+\frac{1}{a}\,f_n(ida)\right)+b\left(\left(n+\frac{d^2a}{n}\right)f_n(ida)-2\frac{ida}{n}\,f_n'(ida)\right)$$

(1) $$0=-\frac{2A}{c\zeta}\,a\,f_n'(iba)+B\,a\cdot\left(\frac{1}{a}+\frac{b}{c}\right)f_n'(ida)+b\,f_n(ida)$$

(2)+(1) $$0=-\frac{2A}{c\zeta}\frac{1}{b}\,f_n(iba)+B\left(a\left(\frac{1}{a}-\frac{b}{c}\right)f_n'(ida)+\frac{2ib}{a}\,f_n(ida)\right)+b\left(\left(1+\frac{d^2a}{n}\right)f_n(ida)-2\frac{ida}{n}\,f_n'(ida)\right)$$

Løses Ligningerne faar man:

$$B=\frac{1}{\Theta}\cdot\frac{2A}{c\zeta}\left(a\left(1+\frac{d^2a}{n}\right)f_n'(iba)f_n(ida)-2\frac{ida^2}{n}\,f_n'(iba)f_n'(ida)-\frac{1}{b}\,f_n(iba)f_n(ida)\right)$$

$$C=\frac{1}{\Theta}\cdot\frac{2A}{c\zeta}\left(\frac{a}{b}\left(\frac{1}{a}+\frac{b}{c}\right)f_n(iba)f_n'(ida)-a^2\left(\frac{1}{a}-\frac{b}{c}\right)f_n'(iba)f_n'(ida)-\frac{2ia}{a}\,f_n'(iba)f_n(ida)\right)$$

idet $$\Theta=f_n(ida)f_n'(ida)\left(2a\frac{1}{a}+\frac{d^2a^2}{n}\left(\frac{1}{a}+\frac{b}{c}\right)\right)-\frac{2ib}{a}\,f_n(ida)^2-\frac{2ida^2}{n}\left(\frac{1}{a}+\frac{b}{c}\right)\left(f_n'(ida)\right)^2$$

* [The denominator in the last term should be n^2 rather than n.]

$$(3)\quad 0 = -\int \frac{i\,(u^2-1+b^2a^2)}{c\,\xi\,a^2}\left(-\frac{A}{c\,\xi}\,\mathcal{P}_u'(iba) + \frac{B}{2}\,\mathcal{P}_u'(ida) + \frac{C}{a}\,\mathcal{P}_u(ida)\right) - A\,\mathcal{P}_u(iba) +$$

$$2\mu\left(-\frac{A}{c\,\xi}\,ib\,\mathcal{P}_u'(iba) + B\frac{\xi}{2}\,id\,\mathcal{P}_u'(ida) + C\left(\frac{id}{a}\,\mathcal{P}_u'(ida) - \frac{1}{a^2}\,\mathcal{P}_u(ida)\right)\right)$$

$$0 = \int \frac{iba\,\mathcal{P}_u'(iba)}{c\,\xi\,a^2\,\mathcal{P}_u(iba)}\,(u^2-1+b^2a^2)\left(-\frac{A}{c\,\xi}\,\mathcal{P}_u'(iba) + \frac{B}{2}\,\mathcal{P}_u'(ida) + \frac{C}{a}\,\mathcal{P}_u(ida)\right) : \left(c\frac{A}{\xi}\,\mathcal{P}_u'(iba)\right) - A\,b^2 -$$

$$\frac{2\mu}{a}\frac{b^2}{\mathcal{P}_u(iba)}\left(-\frac{A}{c\,\xi}\left(\mathcal{P}_u(iba) + iba\left(1+\frac{u^2}{b^2a^2}\right)\mathcal{P}_u(iba)\right) + B\frac{\xi}{2}\left(\mathcal{P}_u'(ida) + ida\left(1+\frac{u^2}{b^2a^2}\right)\mathcal{P}_u(ida)\right) + C\left(\frac{id}{a}\,\mathcal{P}_u'(ida) - id\,\mathcal{P}_u'(ida)\right)\right)$$

$$0 = \left(\int \frac{iba\,\mathcal{P}_u(iba)}{c\,\xi\,a^2\,\mathcal{P}_u(iba)}\,(u^2-1+b^2a^2) - b^2\right)\left(-\frac{A}{c\,\xi}\,\mathcal{P}_u'(iba) + B\frac{\xi}{2}\,\mathcal{P}_u'(ida) + \frac{C}{a}\,\mathcal{P}_u(ida)\right) +$$

$$\frac{2\mu}{c\,\xi}\frac{b^2}{a\,\mathcal{P}_u(iba)}\left(\mathcal{P}_u'(iab)\left(-\frac{A}{c\,\xi}\left(\mathcal{P}_u(iba) + iba\left(1+\frac{u^2}{b^2a^2}\right)\mathcal{P}_u(iba)\right) + B\frac{\xi}{2}\left(\mathcal{P}_u'(ida) + ida\left(1+\frac{u^2}{b^2a^2}\right)\mathcal{P}_u(ida)\right) + C\left(\frac{id}{a}\,\mathcal{P}_u(ida) - id\,\mathcal{P}_u'(ida)\right)\right)\right)$$

$$\left(a^2-b^2+b\frac{\partial}{\partial b}\right)$$

$$- \frac{id}{c\,\xi}(a^2-b^2)\,\mathcal{P}_u(iab)\left(\frac{C}{2}B\,\mathcal{P}_u'(ida) + \frac{C}{a}\,\mathcal{P}_u(ida)\right)\Big\}$$

Indsætter man nu for A og B g C de Udtryk, der blev fundet paa forrige Side, faar man efter Sammentrækning

$$b^2 + ib\,\frac{2\mu}{c\,\xi}\frac{iab\,\mathcal{P}_u'(iab)}{a^2\,\mathcal{P}_u(iab)}\cdot\frac{R}{S} - \int \frac{iba\,\mathcal{P}_u'(iba)}{c^2\,\xi\,a^3\,\mathcal{P}_u(iba)}\,(u^2-1+b^2a^2) = 0 \qquad (I), \quad hvor$$

$$R = \mathcal{P}_u'(iba)\,\mathcal{P}_u'(ida)\,\mathcal{P}_u(ida)\frac{d^2\xi^2}{u^2}\left(5\frac{c}{\xi} - \frac{\xi}{c}\right) + \mathcal{P}_u'(iba)(\mathcal{P}_u(ida))^2\left(2iba^2\left(2+\frac{d^2\xi^2}{u^2}+\frac{u^2}{b^2a^2}\right) - \frac{2ib}{c^2}\right) +$$

$$\mathcal{P}_u'(iba)(\mathcal{P}_u'(ida))^2\frac{id a^2}{u^2}\left(\frac{c}{\xi}-\frac{\xi}{c}\right)(u^2-1) + \mathcal{P}_u(iba)\,\mathcal{P}_u'(ida)\,\mathcal{P}_u(ida)\left(2\frac{id}{c}-2a^2id\left(1+\frac{u^2}{b^2a^2}\right)\left(1+\frac{d^2\xi^2}{u^2}\right)\right) -$$

$$\frac{4a^2d^2}{u^2}(\mathcal{P}_u(iba)(\mathcal{P}_u'(ida))^2) + \frac{a^4}{u^2}(d^2-b^2)(\mathcal{P}_u(iba))^2\,\mathcal{P}_u'(ida)\,\mathcal{P}_u(ida)(\mathcal{P}_u'(iba))$$

$$S = \mathcal{P}_u'(iba)\,\mathcal{P}_u'(ida)\,\mathcal{P}_u(ida)\frac{d^2\xi^2}{u^2}\left(\frac{c}{\xi}-\frac{\xi}{c}\right) - \frac{2ib}{a}\,\mathcal{P}_u'(iba)(\mathcal{P}_u(ida))^2 - \frac{2ida^2}{u^2}\left(d^2-\frac{c^2}{\xi^2}\right)\mathcal{P}_u'(iba)(\mathcal{P}_u'(ida))^2 +$$

$$\frac{2id}{c}\,\mathcal{P}_u(iba)\,\mathcal{P}_u(ida)\,\mathcal{P}_u'(ida)$$

Til Beregning af $\mathcal{P}_u(x)$ benytte for smaa numeriske Værdier af x Rækkeudviklingen

$$\mathcal{P}_u(x) = \frac{x^u}{2^u\,[u]} - \frac{x^{u+2}}{1\cdot 2^{u+2}\,[u+1]} + \frac{x^{u+4}}{1\cdot 2\cdot 2^{u+4}\,[u+2]} - \cdots$$

Rækken er konvergent for alle Værdier af x, men kun praktisk brugbar for smaa numeriske Værdier

[23]

* [The correction, changing *d* to *b* in the second term in the bracket, is in error.]

III. FIRST ROYAL SOCIETY PAPER

DETERMINATION OF THE SURFACE-TENSION
OF WATER BY
THE METHOD OF JET VIBRATION

(Phil. Trans. Roy. Soc. **209** (1909) 281 *)

* [See Introduction, sect. 5.]

PHILOSOPHICAL TRANSACTIONS
OF THE
ROYAL SOCIETY OF LONDON

SERIES A, VOL. 209, pp. 281–317.

DETERMINATION OF THE SURFACE-TENSION
OF WATER BY
THE METHOD OF JET VIBRATION

BY

N. BOHR,
COPENHAGEN.

XII. *Determination of the Surface-Tension of Water by the Method of Jet Vibration.**

By N. Bohr, *Copenhagen.*

Communicated by *Sir* William Ramsay, *K.C.B., F.R.S.*

Received January 12,—Read January 21, 1909.

Introduction.

It has been shown that one of the most important and difficult questions in regard to the determination of the surface-tension of water is to produce a sufficiently pure surface, and in later investigations great importance has therefore been attached to this point.

In 1879 Lord Rayleigh,[†] however, indicated a method which solves the above-mentioned difficulty in a far more perfect manner than any other method used hitherto ; this method makes it possible to determine the surface-tension of an almost perfectly fresh and constantly newly formed surface.

In the paper cited above, Lord Rayleigh has developed the theory of the vibrations of a jet of liquid under the influence of surface-tension, and as appears from this theory, it is possible to determine the surface-tension of a liquid when the velocity and cross-section of a jet of liquid, and the length of the waves formed on the jet, are known.

Lord Rayleigh has attached a series of experiments to the theoretical development. By these experiments, however, it was more especially intended to give illustrations of the theory rather than to give an exact determination of the surface-tension.

If, however, this is the problem, it is necessary to consider more closely some questions which are not discussed in Lord Rayleigh's investigation, for it is necessary

---

* Based on a response to Det Kongl. Danske Videnskabernes Selskabs (The Royal Danish Scientific Society's) Problem in Physics for 1905 ; delivered October 30, 1906 ; awarded the Society's Gold Medal. (The investigation has since been completed with a number of experiments.)

† Lord Rayleigh, ' Roy. Soc. Proc.,' vol. XXIX., p. 71, 1879.

to be sure, first, that the theoretical treatment is sufficiently developed, and secondly, that the phenomenon satisfies, to a sufficient degree, the assumptions on which the theoretical treatment rests.

The main purpose of the present investigation is to try to show how this can be done.

In spite of the great advantages of the above-mentioned method for the determination of surface-tension, it has, however, not been very much used. Except by Lord RAYLEIGH,[*] the method has till recently been used only by F. PICCARD[†] and G. MEYER[‡] for relative measurements. During the completion of this investigation a treatise on this subject has been published by P. O. PEDERSEN.[§]

The Theory of the Vibrations of a Jet.

The theory of the vibrations of a jet of liquid about its cylindrical form of equilibrium has been developed by Lord RAYLEIGH for the case in which the amplitudes of the vibrations are infinitely small and the liquid has no viscosity.

The equations found by Lord RAYLEIGH can, when the amplitudes have small values and the viscosity coefficient is small, be considered as a good approximation; but if the equations are to be used for exact determination of the surface-tension, it is of importance to know how great the approximation is under the given circumstances. In the first part of this investigation we will therefore attempt to supplement the theory with corrections both for the influence of the finite amplitudes and for the viscosity.

Calculation of the Effect of the Viscosity.

Under the influence of the viscosity the jet will execute damped vibrations. If the problem is to find the law according to which the amplitudes decrease, this can, when the viscosity-coefficient is small, be done with approximation by a simple consideration of the energy dissipated. Some authors[||] are of opinion that the correction on the wave-length (time of vibration) due to the viscosity for a problem of this kind can be found directly from the logarithmic decrement of the wave-amplitudes δ by means of the formula $T_1 = T (1 + \delta^2/4\pi^2)^{1/2}$, where T_1 is the time of vibration with damping, T is the time of vibration without it. This application of the formula given does not, however, seem to me to be correct. For the formula is established for a problem by which the only difference between the equation of motion

---

* RAYLEIGH, 'Roy. Soc. Proc.,' vol. XLVII., p. 281, 1890.

† PICCARD, 'Archives d. Sc. Phys. et Nat.' (3), XXIV., p. 561, 1890 (Genève).

‡ MEYER, 'WIED. Ann.,' LXVI., p. 523, 1898.

§ PEDERSEN, 'Phil. Trans. Roy. Soc.,' A, 207, p. 341, 1907.

|| P. O. PEDERSEN (loc. cit., p. 346) ; and PH. LENARD ('WIED. Ann.,' XXX., p. 239, 1887) in his paper about the analogous problem, the vibrations of a drop.

in normal co-ordinates for the conservative system $\left(a \dfrac{\partial q^2}{\partial t^2} + cq = 0. \right.$ One degree of freedom. Small oscillations. Free motion$\Big)$ and the equation of motion for the dissipative system consists in the addition of a frictional term $\left(a \dfrac{\partial q^2}{\partial t^2} + b \dfrac{\partial q}{\partial t} + cq = 0 \right).$

In the present problem—as in all fluid-problems in which a velocity-potential exists for the conservative system, but not for the dissipative one—the coefficient of inertia a will not be the same for the two systems, since a in the dissipative system is dependent on the coefficient of viscosity.

As will be seen from what follows, the correction on the wave-length will not be proportional to δ^2 but to $\delta^{3/2}$.

In order to find the variation of the wave-length due to the viscosity, the problem must be treated in greater detail. Such an investigation is given by Lord RAYLEIGH[*] for the vibrations of a cylinder of viscous fluid under capillary force, in the case where the original symmetry about the axis of the cylinder is maintained. In the development to be found in the paper cited above, the assumption mentioned (the symmetry) is, however, from the outset used in such a manner that the calculation cannot be extended to treat the more general vibrations which will be mentioned here. The result of our development does not include the problem investigated by Lord RAYLEIGH, as, in order to simplify the calculation, special precautions relative to the limiting case ($n = 0$) are not taken.

The general equations of motion of an incompressible viscous fluid, unaffected by extraneous forces, are

$$\mu \nabla u - \rho \frac{Du}{Dt} = \frac{\partial p}{\partial x}, \quad \mu \nabla v - \rho \frac{Dv}{Dt} = \frac{\partial p}{\partial y}, \quad \mu \nabla w - \rho \frac{Dw}{Dt} = \frac{\partial p}{\partial z} \quad . \quad . \quad . \quad (1)$$

and

$$\frac{\partial u}{\partial x} + \frac{\partial v}{\partial y} + \frac{\partial w}{\partial z} = 0, \quad . \quad . \quad . \quad . \quad . \quad . \quad (2)$$

in which u, v, w are the components of the velocity, p the pressure, ρ the density, μ the coefficient of viscosity, and

$$\nabla = \frac{\partial^2}{\partial x^2} + \frac{\partial^2}{\partial y^2} + \frac{\partial^2}{\partial z^2}, \quad \frac{D}{Dt} = \frac{\partial}{\partial t} + u \frac{\partial}{\partial x} + v \frac{\partial}{\partial y} + w \frac{\partial}{\partial z}.$$

In the problem in question the motion will be steady. Putting $w = c + \omega$, and supposing that u, v, and ω have the form $f(x, y) e^{ibz}$, and that u, v, and ω are so small that products of them and quantities of the same order of magnitude can be neglected in the calculations, we get from the equations (1)

$$\left(\nabla - ib \frac{c\rho}{\mu} \right) u = \frac{1}{\mu} \frac{\partial p}{\partial x}, \quad \left(\nabla - ib \frac{c\rho}{\mu} \right) v = \frac{1}{\mu} \frac{\partial p}{\partial y}, \quad \left(\nabla - ib \frac{c\rho}{\mu} \right) \omega = \frac{1}{\mu} \frac{\partial p}{\partial z}; \quad . \quad (3)$$

* Lord RAYLEIGH, 'Phil. Mag.,' XXXIV., p. 145, 1892.

2 o 2

[31]

from (3) and (2) it follows that

$$\nabla p = 0. \qquad \qquad \qquad \qquad (4)$$

Putting

$$u = \frac{i}{cb\rho}\frac{\partial p}{\partial x} + u_1, \quad v = \frac{i}{cb\rho}\frac{\partial p}{\partial y} + v_1, \quad \omega = \frac{i}{cb\rho}\frac{\partial p}{\partial z} + \omega_1 \qquad (5)$$

we get

$$\left(\nabla - ib\frac{c\rho}{\mu}\right)u_1 = 0, \quad \left(\nabla - ib\frac{c\rho}{\mu}\right)v_1 = 0, \quad \left(\nabla - ib\frac{c\rho}{\mu}\right)\omega_1 = 0, \qquad (6)$$

and

$$\frac{\partial u_1}{\partial x} + \frac{\partial v_1}{\partial y} + \frac{\partial \omega_1}{\partial z} = 0. \qquad \qquad (7)$$

Now introducing polar co-ordinates r and ϑ ($x = r\cos\vartheta$, $y = r\sin\vartheta$), and the radial and tangential velocity α and β, we get, by help of the following relations,

$$u = \alpha\cos\vartheta - \beta\sin\vartheta, \quad u_1 = \alpha_1\cos\vartheta - \beta_1\sin\vartheta, \quad \frac{\partial}{\partial x} = \cos\vartheta\frac{\partial}{\partial r} - \sin\vartheta\frac{1}{r}\frac{\partial}{\partial\vartheta},$$

$$v = \alpha\sin\vartheta + \beta\cos\vartheta, \quad v_1 = \alpha_1\sin\vartheta + \beta_1\cos\vartheta, \quad \frac{\partial}{\partial y} = \sin\vartheta\frac{\partial}{\partial r} + \cos\vartheta\frac{1}{r}\frac{\partial}{\partial\vartheta}, \qquad (8)$$

from (5)

$$\alpha = \frac{i}{cb\rho}\frac{\partial p}{\partial r} + \alpha_1, \quad \beta = \frac{i}{cb\rho}\frac{1}{r}\frac{\partial p}{\partial\vartheta} + \beta_1, \qquad \qquad (9)$$

and from (6) and (7), considering $\nabla = \dfrac{\partial^2}{\partial r^2} + \dfrac{1}{r}\dfrac{\partial}{\partial r} + \dfrac{1}{r^2}\dfrac{\partial^2}{\partial\vartheta^2} + \dfrac{\partial^2}{\partial z^2}$,

$$\left(\nabla - ib\frac{c\rho}{\mu}\right)\alpha_1 - \frac{\alpha_1}{r^2} - \frac{2}{r^2}\frac{\partial\beta_1}{\partial\vartheta} = 0, \quad \left(\nabla - ib\frac{c\rho}{\mu}\right)\beta_1 - \frac{\beta_1}{r^2} + \frac{2}{r^2}\frac{\partial\alpha_1}{\partial\vartheta} = 0, \qquad (10)$$

and

$$\frac{\partial\alpha_1}{\partial r} + \frac{\alpha_1}{r} + \frac{1}{r}\frac{\partial\beta_1}{\partial\vartheta} + \frac{\partial\omega_1}{\partial z} = 0. \qquad \qquad (11)$$

Now supposing that p, α, β, ω, and consequently α_1, β_1, ω_1, have the form $f(r)e^{in\vartheta + ibz}$, we get from (4)

$$\nabla p = \frac{\partial^2 p}{\partial r^2} + \frac{1}{r}\frac{\partial p}{\partial r} - p\left(\frac{n^2}{r^2} + b^2\right) = 0,$$

of which the solution, subject to the condition to be imposed when $r = 0$, is

$$p = A J_n(ibr) e^{in\vartheta + ibz}, \qquad \qquad (12)$$

in which J_n is the symbol of the BESSEL's function of nth order.

From (6) we get

$$\left(\nabla - ib\frac{c\rho}{\mu}\right)\omega_1 = \frac{\partial^2\omega_1}{\partial r^2} + \frac{1}{r}\frac{\partial\omega_1}{\partial r} - \omega_1\left(\frac{n^2}{r^2} + d^2\right) = 0, \qquad \left(d^2 = b^2 + ib\frac{c\rho}{\mu}\right), \qquad (13)$$

which gives

$$\omega_1 = B J_n(idr) e^{in\vartheta + ibz}. \qquad \qquad (14)$$

Eliminating β_1 from (10) and (11), we get

$$r\left(\nabla - ib\frac{c\rho}{\mu}\right)\alpha_1 + 2\frac{\partial\alpha_1}{\partial r} + \frac{\alpha_1}{r} = -2\frac{\partial\omega_1}{\partial z},$$

which gives

$$\left(\nabla - ib\frac{c\rho}{\mu}\right)(r\alpha_1) = -2ibJ_n(idr)\,e^{in\vartheta + ibz}. \quad . \quad . \quad . \quad . \quad . \quad (15) \qquad *$$

We have, however,

$$\left(\nabla - ib\frac{c\rho}{\mu}\right)\left(r\frac{\partial}{\partial r}\right) = \left(r\frac{\partial}{\partial r}\right)\left(\nabla - ib\frac{c\rho}{\mu}\right) + 2\frac{\partial^2}{\partial r^2} + \frac{2}{r}\frac{\partial}{\partial r} + \frac{2}{r^2}\frac{\partial^2}{\partial\vartheta^2}$$

$$= \left(r\frac{\partial}{\partial r} + 2\right)\left(\nabla - ib\frac{c\rho}{\mu}\right) - 2\left(\frac{\partial^2}{\partial z^2} - ib\frac{c\rho}{\mu}\right),$$

whence we get

$$\left(\nabla - ib\frac{c\rho}{\mu}\right)\left[r\frac{\partial}{\partial r}J_n(idr)e^{in\vartheta + ibz}\right] = 2\left(b^2 + ib\frac{c\rho}{\mu}\right)J_n(idr)\,e^{in\vartheta + ibz} = 2d^2J_n(idr)\,e^{in\vartheta + ibz};\quad (16)$$

from (15) and (16) we get

$$\alpha_1 = \left[\frac{b}{d}BJ'_n(idr) + C\frac{1}{r}J_n(idr)\right]e^{in\vartheta + ibz}, \quad . \quad . \quad . \quad . \quad . \quad (17)$$

and from (11) we get

$$-\frac{1}{r}\frac{\partial\beta_1}{\partial\vartheta} = \frac{\partial\alpha_1}{\partial r} + \frac{\alpha_1}{r} + \frac{\partial\omega_1}{\partial z}$$

$$= \left\{B\left[ibJ''_n(idr) + \frac{b}{d}\frac{1}{r}J'_n(idr) + ibJ_n(idr)\right] + Cid\frac{1}{r}J'_n(idr)\right\}e^{in\vartheta + ibz}. \quad . \quad (18)$$

With the help of the relation

$$J''_n(x) + \frac{1}{x}J'_n(x) + \left(1 - \frac{n^2}{x^2}\right)J_n(x) = 0 \quad . \quad . \quad . \quad . \quad . \quad (19)$$

(18) gives

$$\beta_1 = \left[B\frac{nb}{d^2}\frac{1}{r}J_n(idr) - C\frac{d}{n}J'_n(idr)\right]e^{in\vartheta + ibz}. \quad . \quad . \quad . \quad . \quad (20)$$

Introducing in the equations (9) and (5) the values of p, α_1, β_1, and ω_1, found by (12), (14), (17), and (20), we get

$$\alpha = \left[-A\frac{1}{c\rho}J'_n(ibr) + B\frac{b}{d}J'_n(idr) + C\frac{1}{r}J_n(idr)\right]e^{in\vartheta + ibz},$$

$$\beta = \left[-A\frac{n}{bc\rho}\frac{1}{r}J_n(ibr) + B\frac{bn}{d^2}\frac{1}{r}J_n(idr) - C\frac{d}{n}J'_n(idr)\right]e^{in\vartheta + ibz}, \quad . \quad . \quad (21)$$

$$w = c + \omega = c + \left[-A\frac{1}{e\rho}J_n(ibr) + BJ_n(idr)\right]e^{in\vartheta + ibz}.$$

* [A factor B is missing on the right side of eq. (15).]

[33]

Let us suppose that the equation of the surface is

$$r - a = \zeta = D e^{in\vartheta + ibz}.$$

The general surface-condition gives

$$\frac{D}{Dt}(r - a - \zeta) = \left(\alpha \frac{\partial}{\partial r} + \frac{\beta}{r} \frac{\partial}{\partial \vartheta} + w \frac{\partial}{\partial z} \right)(r - a - \zeta) = 0,$$

whence we get, neglecting quantities of the same order of magnitude as above,

$$\alpha - c \frac{\partial \zeta}{\partial z} = 0, \quad \zeta = -\frac{i}{cb}\alpha. \quad . \quad . \quad . \quad . \quad . \quad . \quad (22)$$

In the same manner we get further, if the principal radii of curvature are R_1 and R_2,

$$\frac{1}{R_1} + \frac{1}{R_2} = \frac{1}{a} - \frac{\zeta}{a^2} - \frac{1}{a^2} \frac{\partial^2 \zeta}{\partial \vartheta^2} - \frac{\partial^2 \zeta}{\partial z^2} = \frac{1}{a} - \alpha \frac{i(n^2 - 1 + b^2 a^2)}{a^2 cb}. \quad . \quad . \quad (23)$$

Let Pr, $P\vartheta$, Pz be respectively the radial, tangential, and axial component of the traction, per unit area, exerted by the viscous fluid across a surface-element perpendicular to radius-vector. Taking the radius-vector concerned as X-axis, and using the notation generally employed, we have

$$Pr = p_{x,x} = -p + 2\mu \frac{\partial u}{\partial x}, \quad P\vartheta = p_{x,y} = \mu \left(\frac{\partial v}{\partial x} + \frac{\partial u}{\partial y} \right), \quad Pz = p_{x,z} = \mu \left(\frac{\partial w}{\partial x} + \frac{\partial u}{\partial z} \right).$$

Using the relations (8), and after the differentiation setting $\vartheta = 0$, we get

$$Pr = -p + 2\mu \frac{\partial \alpha}{\partial r}, \quad P\vartheta = \mu \left(\frac{\partial \beta}{\partial r} + \frac{1}{r} \frac{\partial \alpha}{\partial \vartheta} - \frac{\beta}{r} \right), \quad Pz = \mu \left(\frac{\partial \alpha}{\partial z} + \frac{\partial w}{\partial r} \right). \quad . \quad (24)$$

Calling the surface-tension T and assuming that there is no " superficial viscosity," the dynamical surface-conditions will be, using the same rate of approximation as before,

$$T \left(\frac{1}{R_1} + \frac{1}{R_2} \right) + Pr = \text{const.}, \quad P\vartheta = 0, \quad Pz = 0; \quad . \quad . \quad . \quad . \quad (25)$$

from (25) we get, using (23) and (24),

$$\left[-T\alpha \frac{i(n^2 - 1 + a^2 b^2)}{a^2 cb} - p + 2\mu \frac{\partial \alpha}{\partial r} \right]_{r=a} = 0, \quad . \quad . \quad . \quad . \quad . \quad (26)$$

$$\left(\frac{1}{r} \frac{\partial \alpha}{\partial \vartheta} + \frac{\partial \beta}{\partial r} - \frac{\beta}{r} \right)_{r=a} = 0, \quad \left(\frac{\partial \alpha}{\partial z} + \frac{\partial w}{\partial r} \right)_{r=a} = 0. \quad . \quad . \quad . \quad . \quad (27)$$

Introducing in these conditions the values of p, α, β, w found by (12) and (21), we get, after the elimination of B/A and C/A, an equation for the determination of b.

[34]

As these calculations will be rather long and the result unmanageable, we will not perform the elimination exactly, but only with an approximation which takes regard of the application of the results.

In the experiments the numerical value of iab will be small—the wave-length large in comparison to the diameter of the jet—and the numerical value of iad great—the coefficient of viscosity small—(in all the experiments $|iab| < 0.24$ and $|iad| > 20$).

For every value of x we have

$$J_n(x) = \frac{x^n}{2^n [n]} - \frac{x^{n+2}}{2^{n+2} [n+1]} + \frac{x^{n+4}}{1.2.2^{n+4} [n+2]} - \dots \qquad . \quad . \quad . \quad (28)$$

The series converges rapidly for small numerical values of x, but very slowly for great values. From (28) it follows that

$$J'_n(x) = \frac{n}{x} J_n(x) \left[1 - \frac{x^2}{2n(n+1)} - \frac{x^4}{2^3 n (n+1)^2 (n+2)} \dots \right],$$

and further, by (19), that

$$J''_n(x) = \frac{n(n-1)}{x^2} J_n(x) \left[1 - \frac{x^2 (2n+1)}{2(n-1)n(n+1)} + \frac{x^4}{2^3 (n-1) n (n+1)^2 (n+2)} \dots \right].$$

Referring to the above, by the calculation of the frictional terms in the equation for the determination of b, we will therefore put

$$J'_n(iab) = -\frac{in}{ab} J_n(iab) \left[1 + \frac{a^2 b^2}{2n(n+1)} \right]$$

$$\text{and} \quad J''_n(iab) = -\frac{n(n-1)}{a^2 b^2} J_n(iab) \left[1 + \frac{a^2 b^2 (2n+1)}{2(n-1)n(n+1)} \right]. \quad . \quad . \quad (29)$$

For calculating $J_n(x)$ for great values of x the asymptotical expression

$$J_n(x) \frown (2\pi x)^{-\frac{1}{2}} \left\{ [P_n(x) + iQ_n(x)] e^{i\left(x - \frac{2n+1}{4}\pi\right)} + [P_n(x) - iQ_n(x)] e^{-i\left(x - \frac{2n+1}{4}\pi\right)} \right\} \quad . \quad (30)$$

is used, in which

$$P_n(x) = 1 - \frac{(4n^2 - 1^2)(4n^2 - 3^2)}{1.2 (8x)^2} + \frac{(4n^2 - 1^2)(4n^2 - 3^2)(4n^2 - 5^2)(4n^2 - 7^2)}{1.2.3.4 (8x)^4} - \dots,$$

and

$$Q_n(x) = \frac{(4n^2 - 1^2)}{8x} - \frac{(4n^2 - 1^2)(4n^2 - 3^2)(4n^2 - 5^2)}{1.2.3 (8x)^3} + \dots.$$

A few terms of the formula (30), which is only correct when the real component of x is positive, will for a great numerical value of x give an excellent approximation for $J_n(x)$. By our application of the formula (30), x can be given the form $a - ib$,

[35]

where both a and b are great positive quantities. Thereby the term with e^{ix} will be quite predominant; we therefore get, neglecting the term with e^{-ix},

$$J'_n(x) = iJ_n(x)\left[1 + \frac{i}{2x} - \frac{4n^2-1}{8x^2} + \frac{i(4n^2-1)}{8x^3} \cdots\right],$$

and further, by (19),

$$J''_n(x) = -J_n(x)\left[1 + \frac{i}{x} - \frac{2n^2+1}{2x^2} - \frac{i(4n^2-1)}{8x^3} \cdots\right].$$

In the following calculations we will therefore put

$$J'_n(iad) = iJ_n(iad)\left(1 + \frac{1}{2ad} + \frac{4n^2-1}{8a^2d^2}\right) \text{ and } J''_n(iad) = \div J_n(iad)\left(1 + \frac{1}{ad} + \frac{2n^2+1}{2a^2d^2}\right). \quad (31)$$

From (27) we get now, using (29) and (31),

$$A\frac{1}{c\rho}J_n(iab)\frac{2n(n-1)}{a^2b}\left[1 + \frac{a^2b^2}{2(n-1)(n+1)}\right]$$
$$+ BJ_n(iad)\frac{2nb}{ad}\left(1 + \frac{3}{2ad} + \frac{4n^2-1}{8a^2d^2}\right) - CJ_n(iad)\frac{id^2}{n}\left(1 + \frac{2}{ad} + \frac{2n^2+1}{a^2d^2}\right) = 0, \quad (32)$$

and

$$A\frac{1}{c\rho}J_n(iab)\frac{2n}{a}\left[1 + \frac{a^2b^2}{2n(n+1)}\right] + BJ_n(iad)d\left(1 + \frac{1}{2ad} + \frac{4n^2-1}{8a^2d^2}\right) - CJ_n(iad)\frac{ib}{a} = 0; \quad (33)$$

from (32) and (33) we get

$$BJ_n(iad) = \div A\frac{1}{c\rho}J_n(iab)\frac{2n}{ad}\left[1 + \frac{a^2b^2}{2n(n+1)}\right]\left(1 - \frac{1}{2ad} - \frac{12n^2-8n-3}{8a^2d^2}\right)$$

and

$$CJ_n(iad) = \div A\frac{1}{c\rho}J_n(iab)\frac{i2n^2(n-1)}{a^2d^2b}\left[1 + \frac{a^2b^2}{2(n^2-1)}\right]\left(1 - \frac{2}{ad} - \frac{2n^2-3}{a^2d^2}\right) \quad (34)$$

From (26) we get now, using (12), (21), (29), (31), (34) and (13),

$$b^2 - ib\frac{\mu}{\rho}\frac{4n(n-1)}{a^2c}\left[1 + \frac{a^2b^2}{n(n-1)}\right]\left[1 + \frac{n-1}{ad} + \frac{(n-1)(2n-3)}{2a^2d^2}\right]$$
$$- T\frac{ibaJ'_n(iab)}{\rho c^2 a^3 J_n(iab)}(n^2 - 1 + a^2b^2) = 0. \quad (35)$$

Putting $\mu = 0$ in (35), we get the solution of Lord RAYLEIGH,[*]

$$b_0^2 = T\frac{iab_0 J'_n(iab_0)}{\rho c^2 a^3 J_n(iab_0)}(n^2 - 1 + a^2b_0^2)$$
$$= \frac{T(n^3-n)}{\rho c^2 a^3}\left[1 + \frac{(3n-1)a^2b_0^2}{2n(n^2-1)} + \frac{3(n+3)a^4b_0^4}{8n(n-1)(n+1)^2(n+2)} + \cdots\right]. \quad (36)$$

In the following we will denote the positive root of this equation by k_0.

* Lord RAYLEIGH, 'Roy. Soc. Proc.,' vol. XXIX., p. 94, 1879.

[36]

* [The symbol $\div$ should be changed to $-$.]
** [To verify eq. (35), one must calculate the terms to an accuracy of not less than one part in $a^2d^2 \approx 400$ and apply the formula $d^2 = b^2 + ibc\rho/\mu \approx ibc\rho/\mu$.]

From (35) and (36) we get, with the same approximation as used hitherto,

$$b^2 - ib\frac{\mu}{\rho}\frac{4n(n-1)}{a^2c}\left[1+\frac{(5n+1)a^2k_0{}^2}{2n(n^2-1)}\right]\left[1+\frac{n-1}{ad}+\frac{(n-1)(2n-3)}{2a^2d^2}\right] - k_0{}^2 = 0. \quad . \quad (37)$$

With the same approximation it will be permissible, by use of (13), to put in (37)

$$iad = ia\left(ik_0\frac{c\rho}{\mu}\right)^{1/2} = (1-i)\left(\frac{a^2k_0c\rho}{2\mu}\right)^{1/2},$$

choosing the sign for iad so that the real component is positive [see (30)].

The equation (37) now becomes

$$b^2 - ib\frac{\mu}{\rho}\frac{4n(n-1)}{a^2c}\left[1+\frac{(5n+1)a^2k_0{}^2}{2n(n^2-1)}\right]$$
$$\left[1-(1-i)\frac{n-1}{2}\left(\frac{2\mu}{\rho ca^2k_0}\right)^{1/2} - i\frac{(n-1)(2n-3)}{4}\left(\frac{2\mu}{\rho ca^2k_0}\right)\right] - k_0{}^2 = 0. \quad . \quad (38)$$

Now solving (38) with regard to b, and setting $b = k+i\epsilon$, we get

$$k = k_0\left[1-\frac{n(n-1)^2}{2}\left(\frac{2\mu}{\rho ca^2k_0}\right)^{3/2} - \frac{3n(n-1)^2}{4}\left(\frac{2\mu}{\rho ca^2k_0}\right)^2\right] \quad . \quad . \quad . \quad (39)$$

and

$$\epsilon = \frac{\mu}{\rho}\frac{2n(n-1)}{a^2c}\left[1+\frac{(5n+1)a^2k_0{}^2}{2n(n^2-1)}\right]\left[1-\frac{n-1}{2}\left(\frac{2\mu}{\rho ca^2k_0}\right)^{1/2}\right]. \quad . \quad . \quad (40)$$

As all the equations used are linear, it will be seen that the physical meaning of the above calculation is the proof of the existence of a real fluid-motion corresponding to a surface of the form

$$r = a+be^{-\epsilon z}\cos kz \cos n\vartheta,$$

where k and ϵ are expressed by the equations (39) and (40).

The correction, which on account of the effect of the viscosity is to be introduced in the expression for the surface-tension, can be found by (36) and (39); we get

$$T = k^2\frac{\rho c^2 a^3 J_n(iak)}{iakJ'_n(iak)(n^2-1+a^2k^2)}\left[1+n(n-1)^2\left(\frac{2\mu}{\rho ca^2k}\right)^{3/2}+\frac{3n(n-1)^2}{2}\left(\frac{2\mu}{\rho ca^2k}\right)^2\right]. \quad . \quad (41)$$

Calculation of the Influence of the Finite Wave-amplitudes.

We will now calculate the correction of the wave-length due to the magnitude of the wave-amplitudes. The method of approximation which will be used is on the principle indicated by G. G. STOKES.*

The following calculation will only treat of the vibrations in two dimensions of a fluid-cylinder without viscosity. The problem in three dimensions could be treated

* G. G. STOKES, 'Camb. Trans.,' VIII., p. 441, 1847,

[37]

in a corresponding manner; but the calculations would in this case be very extensive, and, with regard to the present investigation, it would not be of any practical importance. Using jets the diameter of which is small in proportion to the wavelength, the motion will differ so little from the motion in two dimensions that the small correction of the wave-length due to the finite values of the amplitudes can be considered the same in both cases.

In the present problem the existence of a velocity-potential ϕ can be supposed. Using polar co-ordinates and calling the radial and tangential velocity respectively α and β, we get

$$\alpha = -\frac{\partial \phi}{\partial r}, \quad \beta = -\frac{1}{r}\frac{\partial \phi}{\partial \vartheta}.$$

Considering the fluid as incompressible, we get

$$0 = \frac{\partial \alpha}{\partial r} + \frac{\alpha}{r} + \frac{1}{r}\frac{\partial \beta}{\partial \vartheta} = -\left(\frac{\partial^2 \phi}{\partial r^2} + \frac{1}{r}\frac{\partial \phi}{\partial r} + \frac{1}{r^2}\frac{\partial^2 \phi}{\partial \vartheta^2}\right). \quad \cdots \quad (1)$$

The solution of (1), subject to the condition that the velocity shall be finite for $r = 0$, can be written (n being a positive integer)

$$\phi = \Sigma\Sigma A_{n,q}\, r^n \cos\left(n\vartheta + \tau_n\right)\sin\left(qt + \epsilon_q\right). \quad \cdots \quad \cdots \quad (2)$$

The equation of the surface can be written

$$r = a + \zeta, \quad \zeta = \psi\left(\vartheta,\, t\right).$$

The surface-conditions are, using the same notations as above and calling the radius of curvature of the surface R,

$$\frac{D}{Dt}(a + \zeta - r) = \left(\frac{\partial}{\partial t} + \alpha\frac{\partial}{\partial r} + \frac{\beta}{r}\frac{\partial}{\partial \vartheta}\right)(a + \zeta - r) \quad \text{and} \quad p - \frac{T}{R} = 0. \quad \cdots \quad (3)$$

From (3) we get

$$\left(\frac{\partial \zeta}{\partial t} - \frac{1}{r^2}\frac{\partial \phi}{\partial \vartheta}\frac{\partial \zeta}{\partial \vartheta} + \frac{\partial \phi}{\partial r}\right)_{r = a + \zeta} = 0 \quad \cdots \quad \cdots \quad \cdots \quad (4)$$

and

$$\rho\left\{\frac{\partial \phi}{\partial t} - \frac{1}{2}\left[\left(\frac{\partial \phi}{\partial r}\right)^2 + \frac{1}{r^2}\left(\frac{\partial \phi}{\partial \vartheta}\right)^2\right]\right\}_{r = a + \zeta}$$

$$-T\left[(a + \zeta)^2 + 2\left(\frac{\partial \zeta}{\partial \vartheta}\right)^2 - (a + \zeta)\frac{\partial^2 \zeta}{\partial \vartheta^2}\right]\left[(a + \zeta)^2 + \left(\frac{\partial \zeta}{\partial \vartheta}\right)^2\right]^{-3/2} + F(t) = 0. \quad \cdots \quad (5)$$

Considering only small vibrations of the surface about the position of equilibrium $r = a$, ζ is a small quantity which we will consider as being of the first order. From (2), (4), and (5) it can be seen that ϕ must also be of the first order, when $F(t)$ is defined in such a manner that ϕ does not contain terms independent of r or ϑ.

From the equations (4) and (5) we get, by help of TAYLOR's theorem,

$$\frac{\partial \zeta}{\partial t} + \left[\left(1 + \zeta \frac{\partial}{\partial r} + \frac{\zeta^2}{2} \frac{\partial^2}{\partial r^2} \cdots\right)\left(\frac{\partial \phi}{\partial r} - \frac{1}{r^2} \frac{\partial \phi}{\partial \vartheta} \frac{\partial \zeta}{\partial \vartheta}\right)\right]_{r=a} = 0, \quad \ldots \ldots \quad (6)$$

and

$$\rho \left\{\left(1 + \zeta \frac{\partial}{\partial r} + \frac{\zeta^2}{2} \frac{\partial^2}{\partial r^2} \cdots\right)\left[\frac{\partial \phi}{\partial t} - \frac{1}{2}\left(\frac{\partial \phi}{\partial r}\right)^2 - \frac{1}{2r^2}\left(\frac{\partial \phi}{\partial \vartheta}\right)^2\right]\right\}_{r=a}$$

$$- T \left[(a+\zeta)^2 + 2\left(\frac{\partial \zeta}{\partial \vartheta}\right)^2 - (a+\zeta)\frac{\partial^2 \zeta}{\partial \vartheta^2}\right]\left[(a+\zeta)^2 + \left(\frac{\partial \zeta}{\partial \vartheta}\right)^2\right]^{-3/2} + F(t) = 0. \quad \ldots \quad (7)$$

From (2), (6), and (7) ζ can be found except for a constant, which can be determined by the condition

$$\int_0^{2\pi} \int_0^{a+\zeta} r\, dr\, d\vartheta = \int_0^{2\pi} \tfrac{1}{2}(a+\zeta)^2\, d\vartheta = \pi a^2. \quad \ldots \ldots \ldots \quad (8)$$

First Approximation.

(Solution of the problem neglecting all terms of higher order than the first.)

From (6) and (7) we get

$$\frac{\partial \zeta}{\partial t} + \left[\frac{\partial \phi}{\partial r}\right]_{r=a} = 0, \quad \ldots \ldots \ldots \ldots \quad (9)$$

and

$$\rho \left[\frac{\partial \phi}{\partial t}\right]_{r=a} - T \left(\frac{1}{a} - \frac{\zeta}{a^2} - \frac{1}{a^2}\frac{\partial^2 \zeta}{\partial \vartheta^2}\right) + F(t) = 0. \quad \ldots \ldots \quad (10)$$

Eliminating ζ from (9) and (10), we get

$$\left[\rho \frac{\partial^2 \phi}{\partial t^2} - \frac{T}{a^2}\left(\frac{\partial \phi}{\partial r} + \frac{\partial^3 \phi}{\partial r\, \partial \vartheta^2}\right)\right]_{r=a} + F'(t) = 0. \quad \ldots \ldots \quad (11)$$

Putting $F'(t) = 0$, (11) will be satisfied by

$$\phi = A r^n \cos n\vartheta \sin qt \qquad \text{when} \qquad q^2 = \frac{T}{\rho a^3}(n^3 - n). \quad \ldots \ldots \quad (12)$$

Introducing this in (9), we get

$$\frac{\partial \zeta}{\partial t} = -n a^{n-1} A \cos n\vartheta \sin qt, \qquad \zeta = \frac{n}{q} a^{n-1} A \cos n\vartheta \cos qt + f(\vartheta). \quad \ldots \quad (13)$$

From (10) we get, using (12) and (13),

$$f(\vartheta) + f''(\vartheta) = \text{const.},$$

which is satisfied by

$$f(\vartheta) = C.$$

From (8) we get in this case

$$C = 0.$$

2 P 2

[39]

To a first approximation we get as the general form of the vibrations

$$r = a + \Sigma b_n \cos (n\vartheta + \tau_n) \cos (q_n t + \epsilon_n), \qquad \text{where} \qquad q_n^2 = \frac{T}{\rho a^3} (n^3 - n).$$

As to the higher approximations, it is not possible to find the form for the general vibrations in a corresponding manner, the single types of vibration only being independent of each other to a first approximation.

We shall now determine the next approximations of the pure periodical type of vibration, of which the first approximation is defined by

$$\phi = A r^n \cos n\vartheta \sin qt, \qquad \zeta = \frac{n}{q} a^{n-1} A \cos n\vartheta \cos qt, \qquad q^2 = \frac{T}{\rho a^3} (n^3 - n). \quad (14)$$

Second Approximation.

From (6) and (7) we get

$$\frac{\partial \zeta}{\partial t} + \left[\frac{\partial \phi}{\partial r} + \zeta \frac{\partial^2 \phi}{\partial r^2} - \frac{1}{r^2} \frac{\partial \phi}{\partial \vartheta} \frac{\partial \zeta}{\partial \vartheta} \right]_{r=a} = 0 \quad . \quad . \quad . \quad . \quad . \quad . \quad (15)$$

and

$$\rho \left[\frac{\partial \phi}{\partial t} + \zeta \frac{\partial^2 \phi}{\partial r \partial t} - \frac{1}{2} \left(\frac{\partial \phi}{\partial r} \right)^2 - \frac{1}{2r^2} \left(\frac{\partial \phi}{\partial \vartheta} \right)^2 \right]_{r=a}$$

$$- T \left[\frac{1}{a} - \frac{\zeta}{a^2} - \frac{1}{a^2} \frac{\partial^2 \zeta}{\partial \vartheta^2} + \frac{\zeta^2}{a^3} + \frac{1}{2a^3} \left(\frac{\partial \zeta}{\partial \vartheta} \right)^2 + \frac{2\zeta}{a^3} \frac{\partial^2 \zeta}{\partial \vartheta^2} \right] + F(t) = 0. \quad . \quad . \quad (16)$$

Introducing the values of ϕ, ζ, q, defined by (14), in the terms of second order, we get

$$\frac{\partial \zeta}{\partial t} + \left[\frac{\partial \phi}{\partial r} \right]_{r=a} = - \frac{n^2 (2n-1)}{4q} a^{2n-3} A^2 \cos 2n\vartheta \sin 2qt + \frac{n^2}{4q} a^{2n-3} A^2 \sin 2qt, \quad . \quad (17)$$

and

$$\rho \left[\frac{\partial \phi}{\partial t} \right]_{r=a} - T \left(\frac{1}{a} - \frac{\zeta}{a^2} - \frac{1}{a^2} \frac{\partial^2 \zeta}{\partial \vartheta^2} \right) + F(t)$$

$$= - \rho \frac{n (2n+1) (n^2 + 2n - 2)}{8 (n^2 - 1)} a^{2n-2} A^2 (\cos 2n\vartheta \cos 2qt + \cos 2n\vartheta)$$

$$- \rho \frac{n (4n^3 + 3n^2 - 4n - 2)}{8 (n^2 - 1)} a^{2n-2} A^2 \cos 2qt - \rho \frac{n (3n^2 - 2)}{8 (n^2 - 1)} a^{2n-2} A^2. \quad . \quad . \quad (18)$$

Eliminating ζ from (17) and (18), we get

$$\left[\rho \frac{\partial^2 \phi}{\partial t^2} - \frac{T}{a^2} \left(\frac{\partial \phi}{\partial r} + \frac{\partial^3 \phi}{\partial r \partial \vartheta^2} \right) \right]_{r=a} + F'(t)$$

$$= - \rho \frac{3qn (n-1) (2n+1)}{4 (n+1)} a^{2n-2} A^2 \cos 2n\vartheta \sin 2qt + \frac{\rho}{4} qn (4n+3) a^{2n-2} A^2 \sin 2qt. \quad (19)$$

Putting

$$F'(t) = \frac{\rho}{4} qn (4n+3) a^{2n-2} A^2 \sin 2qt,$$

(19) will be satisfied by

$$\phi = A r^n \cos n\vartheta \sin qt - \frac{3n (n-1)^2 (2n+1)}{8 (2n^2+1) qa^2} A^2 r^{2n} \cos 2n\vartheta \sin 2qt, \quad . \quad . \quad (20)$$

when, as by first approximation,

$$q^2 = \frac{T}{\rho a^3} (n^3 - n). \qquad . \qquad . \qquad . \qquad . \qquad . \qquad . \qquad (21)$$

Introducing this in (17), we get

$$\frac{\partial \zeta}{\partial t} = -na^{n-1} A \cos n\vartheta \sin qt + A^2 \frac{n^2}{4q} \frac{2n^3 - 7n^2 - 2n + 4}{2n^2+1} a^{2n-3} \cos 2n\vartheta \sin 2qt$$
$$+ A^2 \frac{n^2}{4q} a^{2n-3} \sin 2qt ; \quad . \quad . \quad (22)$$

from (22) we get

$$\zeta = \frac{n}{q} A a^{n-1} \cos n\vartheta \cos qt - A^2 \frac{n^2}{8q^2} \frac{2n^3 - 7n^2 - 2n + 4}{2n^2+1} a^{2n-3} \cos 2n\vartheta \cos 2qt$$
$$- A^2 \frac{n^2}{8q^2} a^{2n-3} \cos 2qt + f(\vartheta). \quad . \quad (23)$$

Introducing in (18) the values found for ϕ, q, ζ, and $F'(t)$, we get

$$f(\vartheta) + f''(\vartheta) = -A^2 \frac{n^2}{8q^2} (2n+1) (n^2 + 2n - 2) a^{2n-3} \cos 2n\vartheta + \text{const.},$$

which is satisfied by

$$f(\vartheta) = A^2 \frac{n^2}{8q^2} \frac{n^2 + 2n - 2}{2n-1} a^{2n-3} \cos 2n\vartheta + C. \qquad . \qquad . \qquad . \qquad (24)$$

By (8) we get in this case

$$C = -A^2 \frac{n^2}{8q^2} a^{2n-3}. \qquad . \qquad . \qquad . \qquad . \qquad . \qquad (25)$$

From (23), (24), and (25) we get

$$\zeta = \frac{n}{q} A a^{n-1} \cos n\vartheta \cos qt - A^2 \frac{n^2}{8q^2} \frac{2n^3 - 7n^2 - 2n + 4}{2n^2+1} a^{2n-3} \cos 2n\vartheta \cos 2qt$$
$$+ A^2 \frac{n^2}{8q^2} \frac{n^2 + 2n - 2}{2n-1} a^{2n-3} \cos 2n\vartheta - A^2 \frac{n^2}{8q^2} a^{2n-3} \cos 2qt - A^2 \frac{n^2}{8q^2} a^{2n-3}. \quad . \quad (26)$$

Third Approximation.

From (6) and (7) we get

$$\frac{\partial \zeta}{\partial t} + \left[\frac{\partial \phi}{\partial r} - \frac{1}{r^2} \frac{\partial \phi}{\partial \vartheta} \frac{\partial \zeta}{\partial \vartheta} + \zeta \frac{\partial^2 \phi}{\partial r^2} + \frac{\zeta^2}{2} \frac{\partial^3 \phi}{\partial r^3} + \frac{2\zeta}{r^3} \frac{\partial \phi}{\partial \vartheta} \frac{\partial \zeta}{\partial \vartheta} - \frac{\zeta}{r^2} \frac{\partial^2 \phi}{\partial r \partial \vartheta} \frac{\partial \zeta}{\partial \vartheta} \right]_{r=a} = 0, \quad . \quad (27)$$

[41]

and

$$\rho\left[\frac{\partial\phi}{\partial t}+\zeta\frac{\partial^2\phi}{\partial r\partial t}+\frac{\zeta^2}{2}\frac{\partial^3\phi}{\partial r^2\partial t}-\frac{1}{2}\left(\frac{\partial\phi}{\partial r}\right)^2-\frac{1}{2r^2}\left(\frac{\partial\phi}{\partial\vartheta}\right)^2-\zeta\frac{\partial^2\phi}{\partial r^2}\frac{\partial\phi}{\partial r}-\frac{\zeta}{r^2}\frac{\partial^2\phi}{\partial r\partial\vartheta}\frac{\partial\phi}{\partial\vartheta}+\frac{\zeta}{r^3}\left(\frac{\partial\phi}{\partial\vartheta}\right)^2\right]_{r=a}$$

$$-T\left[\frac{1}{a}-\frac{\zeta}{a^2}-\frac{1}{a^2}\frac{\partial^2\zeta}{\partial\vartheta^2}+\frac{\zeta^2}{a^3}+\frac{1}{2a^3}\left(\frac{\partial\zeta}{\partial\vartheta}\right)^2+\frac{2\zeta}{a^3}\frac{\partial^2\zeta}{\partial\vartheta^2}\right.$$

$$\left.-\frac{\zeta^3}{a^4}-\frac{3\zeta}{2a^4}\left(\frac{\partial\zeta}{\partial\vartheta}\right)^2-\frac{3\zeta^2}{a^4}\frac{\partial^2\zeta}{\partial\vartheta^2}+\frac{3}{2a^4}\frac{\partial^2\zeta}{\partial\vartheta^2}\left(\frac{\partial\zeta}{\partial\vartheta}\right)^2\right]+F(t)=0.\quad\dots\quad(28)$$

Introducing the values of ϕ, ζ, q, defined by (20), (21), and (26), we get (in order not to make the calculations more extensive than necessary, we will only calculate the terms which have references to the determination of q)

$$\frac{\partial\zeta}{\partial t}+\left[\frac{\partial\phi}{\partial r}\right]_{r=a}=\frac{n^3(n^2-1)(28n^3-42n^2+35n-6)}{32q^2(2n^2+1)(2n-1)}A^3a^{3n-5}\cos n\vartheta\sin qt$$

$$+P_1\cos 2n\vartheta\sin 2qt+P_2\sin 2qt+P_3\cos 3n\vartheta\sin 3qt$$

$$+P_4\cos 3n\vartheta\sin qt+P_5\cos n\vartheta\sin 3qt\quad\dots\quad\dots\quad(29)$$

and

$$\rho\left[\frac{\partial\phi}{\partial t}\right]_{r=a}-T\left(\frac{1}{a}-\frac{\zeta}{a^2}-\frac{1}{a^2}\frac{\partial^2\zeta}{\partial\vartheta^2}\right)+F(t)$$

$$=-\rho\frac{n^2(n^2-1)(40n^3-24n^2+65n-30)}{32q(2n^2+1)(2n-1)}A^3a^{3n-4}\cos n\vartheta\cos qt$$

$$+Q_1\cos 2n\vartheta\cos 2qt+Q_2\cos 2n\vartheta+Q_3\cos 2qt+Q_4$$

$$+Q_5\cos 3n\vartheta\cos 3qt+Q_6\cos 3n\vartheta\cos qt+Q_7\cos n\vartheta\cos 3qt.\quad\dots\quad(30)$$

Eliminating ζ from (29) and (30), we get

$$\left[\rho\frac{\partial^2\phi}{\partial t^2}-\frac{T}{a^2}\left(\frac{\partial\phi}{\partial r}+\frac{\partial^3\phi}{\partial r\partial\vartheta^2}\right)\right]_{r=a}+F'(t)$$

$$=\rho\frac{n^2(n^2-1)(34n^3-33n^2+50n-18)}{16(2n^2+1)(2n-1)}A^3a^{3n-4}\cos n\vartheta\sin qt$$

$$+S_1\cos 2n\vartheta\sin 2qt+S_2\sin 2qt+S_3\cos 3n\vartheta\sin 3qt$$

$$+S_4\cos 3n\vartheta\sin qt+S_5\cos n\vartheta\sin 3qt.\quad\dots\quad\dots\quad\dots\quad(31)$$

Setting $F'(t)=S_2\sin 2qt$, (31) will be satisfied by

$$Q=Ar^n\cos n\vartheta\sin qt+A_1r^{2n}\cos 2n\vartheta\sin 2qt+A_2r^{3n}\cos 3n\vartheta\sin 3qt$$

$$+A_3r^{3n}\cos 3n\vartheta\sin qt+A_4r^n\cos n\vartheta\sin 3qt,\quad\dots\quad(32)$$

when

$$q^2=\frac{T}{a^3\rho}(n^3-n)\left[1-A^2a^{2n-4}\frac{n^2(n^2-1)(34n^3-33n^2+50n-18)}{16q^2(2n^2+1)(2n-1)}\right].\quad\dots\quad(33)$$

[42]

Proceeding in the same manner as in the second approximation, we get

$$\zeta = A\frac{n}{q}\,a^{n-1}\left[1 - A^2\frac{n^2}{q^2}\,a^{2n-4}\frac{(n^2-1)\,(28n^3-42n^2+35n-6)}{32\,(2n^2+1)\,(2n-1)}\right]\cos n\vartheta\,\cos qt$$

$$+ B_1\cos 2n\vartheta\,\cos 2qt + B_2\cos 2n\vartheta + B_3\cos 2qt + B_4$$

$$+ B_5\cos 3n\vartheta\,\cos 3qt + B_6\cos 3n\vartheta\,\cos qt + B_7\cos n\vartheta\,\cos 3qt, \qquad . \quad (34)$$

where the coefficients B_1, B_2, B_3, B_4 are the same as in the second approximation, and the coefficients B_5, B_6, B_7 are of the same order as A^3.

As the result of the calculation we note [putting the coefficient of $\cos n\vartheta \cos qt$ in the equation (34) equal to b], that the surface of a fluid-cylinder, executing pure periodical vibrations in two dimensions, can be expressed by

$$r = a + b\cos n\vartheta\,\cos qt + \frac{b^2}{a}\left[-\frac{2n^3-7n^2-2n+4}{8\,(2n^2+1)}\cos 2n\vartheta\,\cos 2qt\right.$$

$$\left.+\frac{n^2+2n-2}{8\,(2n-1)}\cos 2n\vartheta - \frac{1}{8}\cos 2qt - \frac{1}{8}\right] + \frac{b^3}{a^2}(\ldots) + \ldots, \qquad . \quad . \quad . \quad (35)$$

where

$$q^2 = \frac{T}{\rho a^3}\,(n^3-n)\left[1 - \frac{b^2}{a^2}\frac{(n^2-1)\,(34n^3-33n^2+50n-18)}{16\,(2n^2+1)\,(2n-1)} + \frac{b^4}{a^4}(\ldots) + \ldots\right].$$

In the experiments, the jets produced (stationary waves) will execute vibrations in three dimensions—the cross-section will not be the same at different points of the jet. If, however, the velocity of the jet c is so great that the wave-length λ is great in comparison with the diameter of the jet, the motion in the single cross-sections will differ very little from the motion in two dimensions, and the equation (35) can, therefore, also in this case give information about the form of the surface of the jet.

The complete solution in three dimensions can be expressed by

$$r = a + b\cos n\vartheta\,\cos kz + N_1\frac{b^2}{a}\left[1 + \alpha_{1,1}\left(\frac{a}{\lambda}\right)^2 + \alpha_{1,2}\left(\frac{a}{\lambda}\right)^4 + \ldots\right]\cos 2n\vartheta\,\cos 2kz$$

$$+ N_2\frac{b^2}{a}\left[1 + \alpha_{2,1}\left(\frac{a}{\lambda}\right)^2 + \ldots\right]\cos 2n\vartheta + \ldots$$

and

$$k^2 = \frac{1}{c^2}\frac{T}{\rho a^3}(n^3-n)\left[1 + \beta_1\left(\frac{a}{\lambda}\right)^2 + \beta_2\left(\frac{a}{\lambda}\right)^4 + \ldots\right]\left\{1 + M_1\frac{b^2}{a^2}\left[1 + \gamma_1\left(\frac{a}{\lambda}\right)^2 + \ldots\right] + M_2\frac{b^4}{a^4}(1 + \ldots) + \ldots\right\},$$

where the constant N_1, N_2,... and M_1, M_2,... will be equal to the corresponding constants in the equations (35), putting in these equations $t = \dfrac{z}{c}$ and $q = 2\pi\dfrac{\lambda}{c} = kc$.

Neglecting the corrections in the terms of higher order in b/a, we get, using the formula of Lord RAYLEIGH for the wave-length of infinitely small vibrations in three

[43]

dimensions [see (36), p. 288], and for sake of simplicity putting $n = 2$, which corresponds to the experiments executed,

$$r = a + b \cos 2\vartheta \cos kz + \frac{b^2}{6a} \cos 4\vartheta \cos 2kz + \frac{b^2}{4a} \cos 4\vartheta - \frac{b^2}{8a} \cos 2kz - \frac{b^2}{8a} \dots . \quad (36)$$

and

$$k^2 = \frac{Tiak\,J'_2(iak)}{\rho c^2 a^3 J_2(iak)} \,(3 + a^2 k^2)\left(1 - \frac{b^2}{a^2}\frac{37}{24}\right). \quad . \quad . \quad . \quad . \quad (37)$$

The equation (37) gives the correction on the wave-length sought.

The equation (36) permits some further applications.

Putting $z = 0$, we get

$$r = a - \frac{b^2}{4a} + b \cos 2\vartheta + \frac{5}{12}\frac{b^2}{a} \cos 4\vartheta\dots . \quad . \quad . \quad . \quad . \quad (38)$$

(38) is the equation for the form of the orifice (supposing that the velocity, at every point of the cross-section of the jet at the orifice, has the same magnitude and direction), when the jet is to execute pure periodical vibrations. We see from this that the opinion of P. O. PEDERSEN,* according to which a jet issuing from an orifice of the form $(r = a + \beta \cos 2\vartheta)$ must be expected to execute much purer vibrations than a jet from an elliptical orifice $\left(r = a + \beta \cos 2\vartheta + \frac{3}{4}\frac{\beta^2}{a} \cos 4\vartheta\dots\right)$, is not correct.

Putting $\vartheta = 0$, we get

$$r = a + \frac{b^2}{8a} + b \cos kz + \frac{1}{24}\frac{b^2}{a} \cos 2kz\dots . \quad . \quad . \quad . \quad . \quad (39)$$

(39) is the equation for the wave-profile, formed by intersecting the surface of the jet by one of the two perpendicular planes of symmetry. Maximum- and minimum-values of r are obtained by putting in (39) respectively $z = 2n\frac{\pi}{k}$ and $z = (2n + 1)\frac{\pi}{k}$.

We thus get

$$\tfrac{1}{2}(r_{\text{max.}} + r_{\text{min.}}) = a\left(1 + \frac{1}{6}\frac{b^2}{a^2}\right) \quad \text{and} \quad \tfrac{1}{2}(r_{\text{max.}} - r_{\text{min.}}) = b. \quad . \quad . \quad . \quad (40)$$

These formulas will be used in the measuring of the jets.

Calculation of the Effect of the surrounding Air.

We have hitherto neglected the density of the air.† A sufficient approximation of the small correction on the wave-length, due to the inertia of the air, is however very simply obtained by the following calculation regarding infinitely small vibrations in two dimensions of a cylindrical surface, separating two fluids of different density.

* P. O. PEDERSEN, *loc. cit.*, p. 365.

† Lord RAYLEIGH ('Phil. Mag.,' XXXIV., p. 177, 1892) has investigated the corresponding problem in the case where the symmetry about the axis of the fluid-cylinder is maintained during the vibrations,

Considering the fluids as inviscid, we can suppose the existence of a velocity-potential ϕ.   Putting

$$\phi = f(r)\,e^{in\vartheta+iqt},$$

we get

$$\frac{\partial^2 f}{\partial r^2} + \frac{1}{r}\frac{\partial f}{\partial r} - \frac{n^2}{r^2}f = 0,$$

which gives

$$f(r) = \mathrm{A}r^n + \mathrm{B}r^{-n}.$$

As the velocity, as well inside as outside the cylinder-surface, must not be infinitely great, the potential inside the surface must be

$$\phi_1 = \mathrm{A}r^n e^{in\vartheta+iqt},$$

and outside

$$\phi_2 = \mathrm{B}r^{-n} e^{in\vartheta+iqt}.$$

Let the surface be expressed by

$$r - a = \zeta = \mathrm{C}e^{in\vartheta+iqt}.$$

For $r = a$ the following conditions must be satisfied :—

$$\frac{\partial \zeta}{\partial t} = -\frac{\partial \phi_1}{\partial r} = -\frac{\partial \phi_2}{\partial r} \quad . \quad . \quad . \quad . \quad (1) \qquad \text{and} \qquad p_1 - p_2 = \mathrm{T}\frac{1}{\mathrm{R}} \quad . \quad . \quad . \quad . \quad (2)$$

From (1) we get

$$\mathrm{B} = -\mathrm{A}a^{2n} \quad \text{and} \quad \mathrm{C} = i\mathrm{A}\frac{n}{q}a^{n-1},$$

from (2) we get

$$\rho_1 \frac{\partial \phi_1}{\partial t} - \rho_2 \frac{\partial \phi_2}{\partial t} + \mathrm{F}(t) = \mathrm{T}\left(\frac{1}{a} - \frac{\zeta}{a^2} - \frac{1}{a^2}\frac{\partial^2 \zeta}{\partial \vartheta^2}\right) \quad . \quad . \quad . \quad . \quad (3)$$

Introducing in (3) the values found for ϕ_1, ϕ_2, and ζ, we get

$$q^2 = \frac{\mathrm{T}}{\rho_1 + \rho_2}\frac{n^3 - n}{a^3}.$$

In the above we have investigated the influence on the phenomenon in question of the viscosity of the liquid, the magnitude of the wave-amplitudes, and the inertia of the air.* Collecting the results found, we get the following formula for determination of the surface-tension, setting $n = 2$, as will be the case in the experiments :—

$$\mathrm{T} = \frac{(\rho_1+\rho_2)\,k^2 a^3 c\;\mathrm{J}_2(iak)}{(3+a^2k^2)\,iak\,\mathrm{J}'_2(iak)}\left[1 + 2\left(\frac{2\mu}{\rho ca^2 k}\right)^{3/2} + 3\left(\frac{2\mu}{\rho ca^2 k}\right)^2\right]\left(1 + \frac{37}{24}\frac{b^2}{a^2}\right).$$

* The corrections are to be considered as additive, since it can be shown that the wave-length also in the case of viscosity will be an even function of b/a.

Before proceeding to the experimental part of the investigation, we will yet consider a question which may be of interest for the discussion that follows.

In the jets produced in the experiments the velocity must be supposed to be greater in the middle of the jet than closer to the surface.

We can, however, in the following manner, get an idea of the rate at which the velocity-differences will be extinguished by the viscosity of the liquid. For this purpose we will consider a circular fluid-cylinder, in which each part moves with a velocity parallel to the axis of the cylinder, and in which the velocities of the different parts are only functions of the distance from the axis and the time.

Using the axis of the cylinder as Z-axis, we have with the same notation as above,

$$\alpha = 0, \quad \beta = 0, \quad \text{and} \quad w = f(r, t).$$

From the two first equations it follows that $\dfrac{\partial p}{\partial r} = 0$, and, as further, $p = \text{const.}$ for $r = a$, we get $p = \text{const.}$

Supposing $w = \phi(r) e^{-\epsilon t}$, the equation of motion

$$\mu \nabla w - \rho \frac{Dw}{Dt} = \frac{\partial p}{\partial z} \quad \text{gives} \quad \frac{\partial^2 \phi}{\partial r^2} + \frac{1}{r} \frac{\partial \phi}{\partial r} + \frac{\rho \epsilon}{\mu} \phi = 0,$$

of which the solution, subject to the condition to be imposed, when $r = 0$, is

$$\phi = c J_0(kr), \quad \text{in which} \quad k^2 = \frac{\rho \epsilon}{\mu}.$$

The dynamical surface-condition gives $\left(\dfrac{\partial w}{\partial r}\right)_{r=a} = 0$; therefrom it follows that k must be a root of the equation

$$J_0'(ka) = 0, \quad (k_0 = 0, \; k_1 a = \pi 1\cdot 2197, \; k_2 a = \pi 2\cdot 2330, \; k_3 a = \pi 3\cdot 2383, \ldots).$$

The general expression for w is consequently

$$w = \Sigma c_n J_0(k_n r) e^{-\frac{\mu}{\rho} k_n^2 t}.$$

We see that the term of the expression for w containing $J_0(k_1 r)$ decreases much more slowly than the terms with higher index.

For the jets in question the term mentioned will furthermore be predominant already at the orifice, as $\partial w/\partial r$ must be supposed to have the same sign between 0 and a.

The velocity in a jet-piece must therefore be expressed with a high degree of approximation by

$$w = c_0 + c_1 J_0(k_1 r) e^{-\epsilon t}, \quad \text{where} \quad \epsilon = \frac{\mu}{\rho} \left(\frac{\pi 1\cdot 2197}{a}\right)^2,$$

t being the time which the jet-piece has taken to move from the orifice.

[46]

Production of the Jet.

The most important question in the experiments is to produce a jet which, while satisfying the suppositions made in the theoretical development, executes vibrations of a single type.

This demand, however, cannot be expected to be satisfied by a portion of a jet of liquid which lies at a short distance from the orifice. Apart from possible variations of the value of the surface-tension on account of the very rapid extension of the surface, it will be very difficult to obtain pure harmonic vibrations of the jet at this place, for this requires not only a definite form of the cross-section of the jet at the orifice, but also a definite velocity at every point of this section. While it might be possible to satisfy the first condition by suitable choice of the orifice (see p. 296, (38)), it would, no doubt, be very difficult to satisfy the last; among other reasons the velocity of the fluid will, for various causes, be greater in the middle of the jet than closer to the surface. It is, therefore, of great importance to produce a jet which is so stable that the vibrations can be examined at a considerable distance from the orifice where the viscosity of the liquid has had time to act.

A jet issuing from a hole in a thin plate is, however, not very stable, and therefore the jet rather rapidly falls into drops. If, however, drawn-out glass-tubes are used as orifice, very long and stable jets can be formed when the tube has a suitable shape.

In the experiments, jets were exclusively employed the qualities of which repeated themselves twice over the circumference.

The orifices of the glass-tubes employed were given an elliptic section by specially heating the tubes, before drawing them out, on two opposite sides. Twisting of the glass-tubes would produce a rotation of the jet about its axis, and the planes of vibration would not preserve the same direction at different distances from the orifice; to avoid such results it was necessary during the heating and drawing-out to have both ends of the tube fastened on slides which could be displaced along a metal-prism.

When the glass-tubes were drawn out and cut off, they were examined under a microscope, and only those whose orifice had a uniform elliptic section were used. After this the jets, which were formed by the tubes, were examined. The purpose of this examination, which will be mentioned later (p. 307), was to find out if the jet was symmetrical about two perpendicular planes passing through its axis.

It has been mentioned above that, because of the effect of viscosity, a portion of the jet will execute vibrations, which are in better conformity with those wanted the more removed it is from the orifice. It might here be of interest to illustrate this by an example.

As such can be employed an experiment carried out with tube I (see the table on p. 310).

2 Q 2

The jet in question, which had a mean radius $a = 0.0675$ cm. and a velocity $c = 425$ cm./sec., was so stable that the wave-length could be measured with great exactness up to a distance of about 35 cm. from the orifice. We will now examine such a jet at a distance of 30 cm. from the orifice.

The viscosity will firstly have the effect that an original difference in velocity at different points of the cross-section of the jet is rapidly extinguished.

The calculation on p. 298 shows that the differences mentioned must decrease approximately as $e^{-\epsilon t}$, where $\epsilon = \mu/\rho \, (\pi 1.2197/a)^2$. Let, now, $a = 0.0675$ and $\mu/\rho = 0.0125$ (temperature $11.8°$ C.), we get $\epsilon = 40.3$. Let, further, $t = \frac{30}{425}$, and we get $e^{-\epsilon t} = e^{-2.844} = 0.0582$. We see from this that the differences in velocity at the place in question must be about 17 times smaller than close by the orifice.

The viscosity will furthermore have the effect that also the waves on the surface of the jet tend to be of single types. We found above that the general form of the surface of the jet, considering the vibrations as infinitely small, can be expressed by

$$r = a + \Sigma b_n \cos \left(n\vartheta + \tau_n \right) \cos \left(k_n z + \gamma_n \right) e^{-\epsilon_n z},$$

where with approximation we have

$$\epsilon_n = \frac{\mu}{\rho} \frac{2n \, (n-1)}{ca^2}.$$

Let now $a = 0.0675$, $\mu/\rho = 0.0125$, $c = 425$ and $z = 30$, we get $e^{-\epsilon_2 z} = 0.461$, $e^{-\epsilon_3 z} = 0.098$, $e^{-\epsilon_4 z} = 0.0096$, $e^{-\epsilon_5 z} = 0.00043$, $e^{-\epsilon_6 z} = 0.000009$, &c.

If now the form of the surface of the jet is close to the orifice,

$$r = a + b_2 \cos 2\vartheta \cos k_2 z + b_3 \cos 3\vartheta \cos k_3 z + b_4 \cos 4\vartheta \cos k_4 z + \ldots,$$

the form of the surface will therefore be at a distance of 30 cm. from the orifice, approximately,

$$r = a + \tfrac{1}{2} (b_2 \cos 2\vartheta \cos k_2 z + \tfrac{1}{5} b_3 \cos 3\vartheta \cos k_3 z$$
$$+ \tfrac{1}{50} b_4 \cos 4\vartheta \cos k_4 z + \tfrac{1}{1000} b_5 \cos 5\vartheta \cos k_5 z + \ldots).$$

For the jet used, the term with $\cos 2\vartheta \cos k_2 z$ was already at the orifice quite predominant, and especially the quantities b_3, b_5, ..., were very small in proportion to b_2, as the jet at the examination mentioned was found to be very nearly symmetrical about two perpendicular planes through its axis.

We thus see that the jet in the experiment mentioned at a distance of 30 cm. from the orifice must have executed exceedingly pure vibrations.

In the experiments ordinary tap-water was used.

For the sake of the investigation it was important to get a jet which could run without variation (same velocity and temperature) as long as wanted. In order to give the water a suitable constant temperature, it was led from the tap through a long leaden spiral tube, placed in a water-bath, and a regulator connected with the gasburner heating the bath. In this way the temperature of the water could be kept constant at 0·01° C. as long as wanted.

The arrangement for keeping the pressure constant is shown in fig. 1. The water coming from the heating apparatus was led into a glass-bottle A, in which a constant water-level was maintained with help of an overflow B. From A the water was led down to the pressure-reservoir, consisting of two 5-litre glass-bottles C and D.

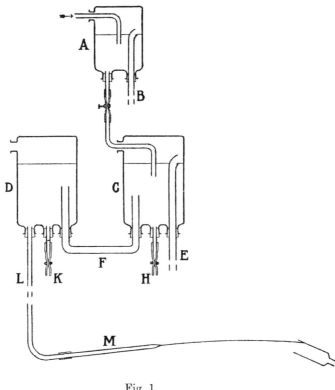

Fig. 1.

Inside C was placed an overflow E. C and D were connected by a wide bent-glass-tube F. H and K were two outlets through which the bottles could be emptied. From D the water was led down through a long glass-tube L to the jet-tube M. The whole arrangement was situated in a cellar, and the pressure-reservoirs as well as the jet-tube were supported by stone foundations. At the beginning of each experiment all the reservoirs and tubes were carefully cleaned and rinsed, whereupon the waterflow was adjusted so that a constant not particularly rapid flow ran through both the overflows.

With the arrangement mentioned, the water-surface in the bottle D was very steady and quite independent of the variations of the pressure in the supply pipe.

The temperature of the water was very near 12° C. in all experiments.

In order to calculate the surface-tension of a liquid, the following quantities had to be known :—(1) the density, ρ ; (2) the discharge per second, V ; (3) the velocity of the jet, c; (4) the mean radius of the jet, a (which four quantities are connected by the relation $V = \rho c \pi a^2$); (5) the wave-length, and, finally for the correction, (6) the amplitudes of the waves.

The density ρ of the tap-water used was at 12° found to be so near 1 ($\rho =$ about 1·0001) that by putting $\rho = 1$ only errors far below the exactness of the experiment were made.

The measuring of the discharge presented no difficulty, and could be executed to 0·02 per cent. of its value.

*Determination of the Velocity of the Jet.**

When the jet is formed by a glass-tube, the velocity cannot be exactly calculated by the height of pressure on account of the friction in the tubes. In the present

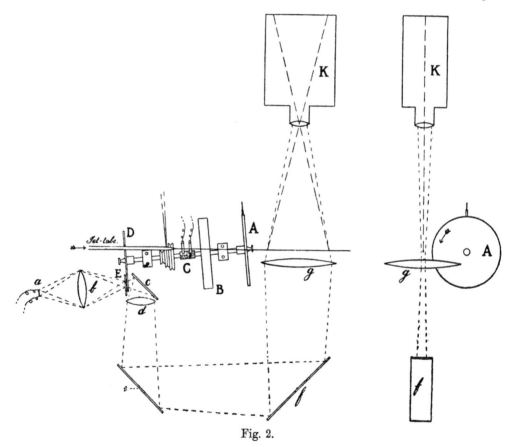

Fig. 2.

investigation a direct method was therefore used to measure the velocity of the jet, the main features of which were as follows: In a fixed point the jet was cut through, at constant time-intervals, by help of a sharp and thin knife, and at the

---

* A critical account of methods used in former investigations is to be found in the paper of P. O. Pedersen (*loc. cit.*, p. 352).

same time photographed instantaneously.   Let the distance between two cuts, measured by help of the photographic plate, be a, and the time-interval be t, we have $c = a/t$, c being the velocity of the jet.

Fig. 2 shows the arrangement seen from above and from the side.

The rotation-apparatus ABCD executes the cutting of the jet and the opening and closing of the light.   A is a metal disk, to the edge of which the knives were fastened in radial direction.   The knives, made from ground needles, measured about 0·4 mm. in width and were about 0·03 mm. thick.   The axis of the rotation-apparatus was not parallel to the jet, but formed a small angle with it, so that the knife cutting through the jet had the same velocity parallel to the axis of the jet as the water-particles.

D is a metal disk which has a radial slit close to the edge, which once at every revolution passes a corresponding slit in the screen E.   The apparatus was driven by an electric motor, the speed of which could be regulated by means of an adjustable resistance, and, in order to make the velocity steady, the axis of the rotation-apparatus was provided with a small fly-wheel B.   Further, to count the revolutions, the axis of the apparatus carried a contact C, which, completing the circuit of an electric current once at every revolution, marked a kymograph by help of an electro-magnet.   The kymograph was also marked every second by another electromagnet.

abcdefg provided for the illumination of the jet.   By help of a powerful lens-system b an image of the horizontal linear filament of a Nernst lamp a was formed on the slit of the screen E.   The mirrors c, e, f, and the lenses d and g, thereupon formed a magnified image of the slit on the jet, and from the lens g all the light was finally directed into the camera K.   In the figure the dotted lines show the limitation of the beam of light.

Every photograph was taken during about 12 seconds, which corresponded with about 600 revolutions of the apparatus with the following exposures of the plate· Some photographs are shown in the accompanying figure.   (The direction of the jet is

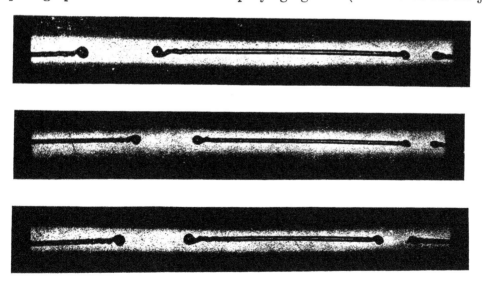

from right to left.)   We see how the ends of the jet set free by the knives very rapidly contract themselves into the corresponding jet-pieces, assuming a drop-like appearance.

The plates were measured by examining them pressed together with a glass-rule under a microscope, and reading on the scale the positions of the lines, perpendicular to the jet, touching the drop-like free ends of the jet-pieces.   Thereupon the mean of the results found for the two ends of each cut was calculated, and the difference between the means from two succeeding cuts, divided by the magnification of the plate, was equated to the distance which the jet had moved while the rotation-apparatus had made a revolution.   The conditions of the correctness of this were partly that the cuts moved independently of each other, partly that the ends of the jet contracted themselves equally into the respective jet-pieces during the time which a cut took to move from the one place where it was photographed to the other.   That these conditions were satisfied appears, first, from the fact that the part of the jet-pieces placed midway between the cuts was completely undisturbed by the cutting of the jet (see the photographs), and, secondly, from the symmetrical forms of the ends of the jet facing each other.

The magnification of the plate was found by taking a photograph of a glass-rule placed directly under the jet.

The interval of time between the cuts was determined as a mean from the number of revolutions per second during the time of exposure; the photographic plate also giving a sort of mean of the single exposures, very great accuracy might be obtained in this way.

At each determination of the velocity of the jet photographs were taken for the sake of the control with different times of revolution of the rotation-apparatus.

The following table shows the result of an experiment by which four photographs were taken :—

Magnification of the photographs. f.	Distance between two cuts. a.	Rotations per second. n.	Velocity of the jet. $v = \dfrac{na}{f}$.
	cm.		cm./sec.
0·8624	8·37	40·19	390·0
0·8624	6·735	49·92	389·8
0·8624	6·845	49·15	390·1
0·8624	6·54	51·41	389·9

The values found show a very good agreement, the largest mutual deviation being less than 0·1 per cent.

Determination of the Wave-Length.

In the experiments, jets were used with so small wave-amplitudes that the wave-length could not be measured directly with sufficient accuracy either on the jet itself or on a photograph of the same.

The method used to determine the wave-length consisted in finding out the summits of the jet (the points where the tangent-planes were parallel to the axis of the jet), using the jet as an optical image-forming system.

Fig. 3 represents a horizontal jet-piece S (placed so that the two perpendicular planes of symmetry of the jet are respectively horizontal and vertical), a telescope T,

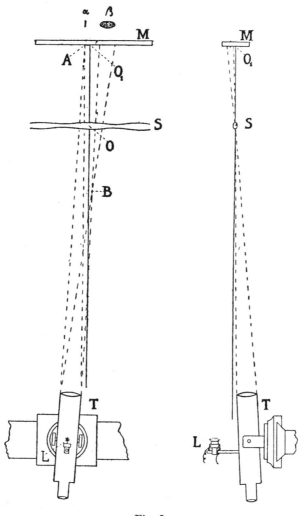

Fig. 3.

and a Nernst lamp L (the filament being vertical) fastened to the telescope in a position vertically over its axis, seen from above and from the side. OO_1 is a horizontal line, perpendicular to the jet, through a summit.

Seen from above, the jet acts as a lens with large focus-width, the front-surface of which will form a virtual image at A, and whose back-surface will form a real image, modified through the refraction during the double passage through the front-surface at B. Seen from the side, all the light reflected can, on account of the small diameter of the jet, be considered as intersecting the jet-axis. If, now, the telescope is focussed for the distance TA, a small vertical bright line α, and a less bright, but sharply limited ellipse β, with horizontal great axis will therefore be seen in a dark field.

When the telescope is displaced parallel to the jet, the distance between the bright line and the ellipse will vary, and the telescope being brought into a position quite opposite a summit, the bright line will fall together with the minor axis of the ellipse.

Parallel to the jet was placed a fine glass-rule M, divided into $\frac{1}{10}$ mm., which divisions could be seen sharply in the field of the telescope, together with the bright lines mentioned.

The measurements were executed in such a manner that the telescope was partly displaced parallel to the jet and partly turned around a vertical axis, until the vertical spider-line fell together with the bright line at the same time as this halved the ellipse.

Fig. 4 shows the appearance of the telescope-field. Every time when such an adjustment was obtained, the position of the spider-line was read on the glass-scale.

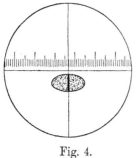

Fig. 4.

The adjustment and reading could be done with an accuracy of 0·01 mm.

In the above we have supposed that the jet-axis was horizontal. If, on the contrary, the jet formed an angle with the horizontal plane—and this must be the case at certain places of the jet-piece examined on account of the curvature of the jet—the bright line and the minor axis of the ellipse will form the same angle with the vertical spider-line. If, however, care was taken in the arrangement that the centre of the vertical line from the middle of the filament to the telescope-axis was at the same horizontal height as the jet, the vertical plane through the telescope-axis will here, too, as a closer examination shows, be perpendicular to the vertical plane containing the jet and go through a summit when the vertical spider-line goes through the middle of the bright line at the same time as this falls together with the minor axis of the ellipse.

The circumstance that the wave-amplitudes on account of the viscosity of the liquid are decreasing in direction from the orifice has the effect that the distance OA between the focus-lines and the jet is not the same everywhere. While this fact is not of great importance when measuring the wave-lengths on a short jet-piece, it will, when measuring on very long jet-pieces (as in the table, p. 310) have the effect that the focussing of the telescope cannot be kept constant during the measuring, and the readings of the single summits could therefore, in this case, not be executed with quite as great accuracy as mentioned above.

The differences between the readings indicate the distances between the projections of the summits on a horizontal plane. Dividing the mentioned differences by cos α, α being the slope of the jet at the place in question, we therefore get the distances between the summits. These distances can be directly put equal to the wave-lengths sought, for, considering the jet-axis as a straight line, the wave-profile can, apart from possible irregularities, be expressed with great approximation by (see p. 296)

$$r = a + b e^{-\epsilon z} \cos kz + \frac{1}{24} \frac{b^2}{a} e^{-2\epsilon z} \cos 2kz + \frac{1}{8} \frac{b^2}{a} e^{-2\epsilon z}.$$

Finding the summits z_n of this curve by putting $\partial r / \partial z = 0$, we get approximately

$$z_n - n \frac{\pi}{k} = -\frac{\epsilon}{k^2} - \frac{1}{6} \frac{b}{a} \frac{\epsilon}{k^2} e^{-\epsilon z_n}.$$

The first term on the right side is a constant, and with the values of ϵ, k, and b/a, which correspond to the experiments carried out, the second term is quite negligible as compared with the accuracy of the experiments.

P. O. PEDERSEN has also measured the wave-length on a jet with very small wave-amplitudes. The author mentions* that it has not been possible for him to produce jets with such regular vibrations that he could use a method for the determination of the wave-lengths which he describes and which in the main features is of the same nature as that described above. He therefore employed another method, which is, in the main, as follows: Illuminating the jet with a parallel beam of light, the rays twice refracted and once reflected form a wave-like image on a photographic plate, and as the amplitude of the image is much larger than that of the jet, the wave-length could be measured directly on the image. By this method the wave-length is determined as mean wave-length on a longer jet-piece. As however will appear from experiments, which will be described later, it is of the greatest importance to be able to determine the single wave-length with great enough accuracy for the variations of the wave-lengths to be examined.

The image-formation of the jet was also used in the examination of the jets mentioned on p. 299. If the tube was turned around its axis at the same time as the images were observed in the telescope, the appearances of these changed as the curvature of the profile of the jet seen from above gradually varied, the points A and B being displaced. The variations in the appearance of the images were most rapid in the moments in which the curvature of the profile was near to O. Every time the curvature became O, the points A and B fell together and a regular elliptic luminous spot without structure was seen in the telescope. To the tube was fastened a disk with a graduation, and every time the luminous spot mentioned appeared in the telescope during the revolution of the tube the graduation was read off. If the jet were symmetrical with respect to two perpendicular planes, the places read off must lie symmetrical on the circumference of the circle and also be the same at different distances from the orifice. This examination was very sensitive and it showed, too, that not all of the tubes examined satisfied the conditions to a sufficient

*

* PEDERSEN, *loc. cit.*, p. 368.

2 R 2

* [The letter O should be changed to 0 (zero).]

degree of accuracy. However, four satisfactory tubes were found. That the jets produced by these four tubes executed exceedingly pure vibrations appears also very distinctly from the measurement of the wave-length, which will be mentioned later.

In the experiments the jet had to be placed so that the two symmetrical planes were respectively horizontal and vertical. This was attained by turning the tube into a position midway between two of the above-mentioned readings.

The Photographing of the Jet.

To determine the magnitude of the amplitudes of the waves, magnified photographs of the jet were taken.

Using nearly monochromatic light, and a special limitation of the illuminating beams, the profile of the jet was brought to appear with very great sharpness on the photographic plates.

By help of an object-micrometer the diameter of the jet was measured at different places of the plate. The single diameters could in this way, on account of the sharpness of the outline, be measured with a relative exactness of about 0·03 per cent. of their value.

From the largest and smallest diameter of the jet ($2r_{max.}$ and $2r_{min.}$) the amplitude $\dfrac{b}{a} = \dfrac{r_{max.} - r_{min.}}{r_{max.} + r_{min.}}$ and the mean radius $a = \dfrac{1}{2}\left(r_{max.} + r_{min.}\right)\left[1 - \dfrac{1}{6}\left(\dfrac{b}{a}\right)^2\right]$ [see p. 296 (40)] were determined. (On account of the decreasing of the amplitudes for $r_{max.}$ and $r_{min.}$ respectively the mean of two succeeding $r_{max.}$ and the $r_{min.}$ lying between were used.)

The value of the mean radius obtained in this way showed a very great conformity with the value which could be calculated by help of the discharge and the velocity of the jet, measured in the manner described above.

In order to show this, two experiments (executed with the tubes I and IV) will be mentioned, by which the mean radius of the jet was determined in both ways :—

Tube I.

$r_{max.}$ 0·06929 cm., 0·06918 cm.   Discharge V = 6·274 cm.^{3}/sec.

$r_{min.}$ 0·06549 cm.   Velocity   c = 440·8 cm./sec.

$\dfrac{b}{a} = 0·0278, \quad a = 0·06736$ cm. $a = \sqrt{\dfrac{V}{\pi c}} = 0·06731$ cm.

Tube IV.

$r_{max.}$ 0·08263 cm., 0·08255 cm.   Discharge V = 7·862 cm.^{3}/sec.

$r_{min.}$ 0·07777 cm.   Velocity   c = 390·0 cm./sec.

$\dfrac{b}{a} = 0·0301, \quad a = 0·08017$ cm. $a = \sqrt{\dfrac{V}{\pi c}} = 0·08011$ cm.

We see that the mean radii a, determined in the two ways, are very nearly the same (mutual deviation less than 0·1 per cent.).   This conformity having been stated, the velocity-determination was omitted at the later experiments and a was determined by help of the photographs only, whereby the experiments became very much simplified.

The Results of the Experiments.

In the above we have described the methods used in the different measurements; it is further mentioned how it was possible, by the arrangement described on p. 301, to keep the pressure-height and temperature of the water constant during the comparatively long space of time taken to determine the discharges, the velocity, the mean radius, and the wave-length.

Before giving the results of the experiments we must, however, call attention to some special circumstances occurring in the determination of the wave-lengths sought, due to the fact that the wave-lengths found were not equal at different distances from the orifice.   In order to show plainly what is meant by this, we shall commence with mentioning four experiments (one executed with each of the four tubes) carried out at a pressure-height of about 100 cm., in which the single wave-lengths were determined immediately outside the orifice and as far out on the jet as its stability permitted.

The results can be seen in the table overleaf.

As it will be seen, the differences between the readings are not constant, but increase until they reach a maximum, whereupon they slowly decrease again.   The same can be seen from the table on p. 311, where the numbers in the column designated by "mean values" are calculated from the table overleaf by a simple adjustment.

The variation of the differences read off is, however, the result of many causes, among which are some the influence of which can be directly calculated.   The first cause is the curvature of the jet, the effect of which is partly that the differences found are not equal to the real wave-lengths (see p. 306), partly that velocity and cross-section are not the same at different places of the jet-piece examined.   The second cause is the decreasing of the wave-amplitudes, the influence of which appears from the equation (37) on p. 296.   The column of the table on p. 311 designated by "corrected values" therefore contains wave-lengths, at different distances from the orifice, belonging to a horizontal jet which has the same velocity and cross-section as the jet examined on the horizontal place and which executes vibrations with infinitely small wave-amplitudes.

We see that the numbers in the last-mentioned column increase until they reach a maximum, whereupon they keep very nearly constant.   This seems to show that all the causes of the variation of the wave-length, the influence of which is not corrected for, must originate in irregularities of the phenomenon which arise in the

	I.		II.		III.		IV.	
Tube								
Temperature. . .	11·82° C.		11·73° C.		11·76° C.		11·80° C.	
Distance from the orifice of the horizontal part of the jet	26·3 cm.		29·4 cm.		28·9 cm.		34·6 cm.	
Discharge . . .	6·100 cm.³/sec.		7·678 cm.³/sec.		7·720 cm.³/sec.		8·649 cm.³/sec.	
Mean radius of the jet on the horizontal place	0·06755 cm.		0·07554 cm.		0·07595 cm.		0·08010 cm.	
Orifice . . .	cm. 0·0		cm. 0·0		cm. 0·0		cm. 0·0	
Summit I. . .	0·99		2·39		2·375		2·555	
		2·03		2·545		2·525		2·76
„ II. . .	3·02		4·935		4·90		5·315	
		2·125		2·56		2·56		2·765
„ III. . .	5·145		7·495		7·46		8·08	
		2·155		2·575		2·58		2·795
„ IV. . .	7·30		*10·07		*10·04		10·875	
		2·17		2·605		2·605		2·83
„ V. . .	*9·47		12·675		12·645		13·705	
		2·18		2·635		2·62		2·85
„ VI. . .	11·65		15·31		15·265		*16·555	
		2·195		2·65		2·63		2·87
„ VII. . .	13·845		17·96		17·895		19·425	
		2·215		2·65		2·64		2·885
„ VIII. . .	16·06		20·61		20·535		22·31	
		2·215		2·655		2·645		2·895
„ IX. . .	18·275		23·265		23·18		25·205	
		2·22		2·66		2·65		2·90
„ X. . .	20·495		25·925		25·83		28·105	
		2·22		2·66		2·655		2·90
„ XI. . .	22·715		†28·585		†28·485		31·005	
		2·225		2·655		2·655		2·90
„ XII. . .	24·94		31·24		31·14		†33·905	
		2·225		2·66		2·655		2·90
„ XIII. . .	†27·165		33·90		33·795		36·805	
		2·225		2·655		2·655		2·90
„ XIV. . .	29·39		36·555		36·45		39·705	
		2·22		2·65		2·65		2·895
„ XV. . .	31·61		39·205		39·10		42·60	
		2·225		2·65		2·65		2·895
„ XVI. . .	33·835		41·855		41·75		45·495	
The amplitude $\frac{b}{a}$ on the summits designated by * and † *	0·0417		0·0699		0·0640		0·0382	
†	0·0258		0·0472		0·0432		0·0276	

(Readings on a horizontal rule carried out with an exactness of 0·005 cm.)

formation of the jet and which are rapidly extinguished.   We see, however, that the influence of these irregularities on the wave-length is not insignificant, even at a considerable distance from the orifice.   Thus, in the experiments mentioned, the wave-lengths are at a distance of 10 cm. from the orifice in the mean 2 per cent., and

Distance from the orifice in centimetres.	I.		II.		III.		IV.	
	Mean values.	Corrected values.	Mean values.	Corrected values.	Mean values.	Corrected values.	Mean values.	Corrected values.
5	2·14	2·148	2·55	2·553	2·54	2·545	2·76	2·782
10	2·185	2·189	2·59	2·591	2·595	2·597	2·815	2·829
15	2·210	2·211	2·640	2·638	2·625	2·624	2·850	9·859
20	2·221	2·221	2·654	2·650	2·642	2·639	2·880	2·884
25	2·225	2·224	2·658	2·653	2·652	2·648	2·896	2·897
30	2·224	2·224	2·658	2·654	2·656	2·652	2·900	2·899
35	—	—	2·656	2·653	2·654	2·652	2·901	2·899
40	—	—	2·651	2·652	2·650	2·652	2·898	2·898

at a distance of 20 cm., 0·3 per cent., smaller than is the wave-length at a distance of 30 cm. from the orifice ; if, therefore, the wave-length at 10 cm. or at 20 cm. distance from the orifice had been used for the calculation of the surface-tension, a value respectively 4 per cent. and 0·6 per cent. too great would have been obtained.

The four experiments mentioned furthermore illustrate the influence of the viscosity on the phenomenon, the magnitude of the wave-amplitudes at two places of the jet with considerable mutual distance being measured (see the table on p. 310).

Putting $\dfrac{b}{a} = A e^{-\epsilon z}$, we get for the four jets respectively

$$\epsilon = 0\cdot0271, \qquad \epsilon = 0\cdot0212, \qquad \epsilon = 0\cdot0213, \qquad \epsilon = 0\cdot0187.$$

In the above we have found (p. 289 (40))

$$\epsilon = \frac{\mu}{\rho}\frac{4}{ca^2}\left(1 + \frac{11}{12}a^2k^2\right)\left[1 - \frac{1}{2}\left(\frac{2\mu}{\rho ca^2 k}\right)^{1/2}\right].$$

From this formula we get the following results for μ,

$$\mu = 0\cdot0131, \qquad \mu = 0\cdot0129, \qquad \mu = 0\cdot0130, \qquad \mu = 0\cdot0129.$$

We see that these values do not differ much from the most generally adopted value for μ, namely, $\mu = 0\cdot0125$ (temperature, $11\cdot8^\circ$ C.).   That they, however, are all greater suggests, perhaps, a very small superficial viscosity.

The correction of the formula to calculate the surface-tension, due to the effect of the viscosity on the wave-length, is, according to the equation (41), on p. 289, determined by the coefficient

$$1 + 2\left(\frac{2\mu}{\rho ca^2 k}\right)^{3/2} + 3\left(\frac{2\mu}{\rho ca^2 k}\right)^2.$$

[59]

By introducing the values for μ, ρ, c, a, and k, which correspond to the experiments carried out, this correction becomes very small, about 0·1 per cent.*

As to the calculated correction for the influence of the finite wave-amplitudes, it may be mentioned that the values of the surface-tension in the table of the experiments on p. 313, which has been calculated according to the formula (37) on p. 296, does not show any systematic deviation due to the wave-amplitude.

As, however, the correction for the value of the wave-amplitude in the experiments carried out is rather small (from 0·10 per cent. to 0·33 per cent.), the agreement mentioned is not adapted to give an experimental verification of the formula theoretically developed. It may here be remarked that P. O. PEDERSEN (*loc. cit.*, p. 371) has experimentally investigated the influence of the value of the wave-amplitudes upon the calculated values of the surface-tension and has found results, by using greater wave-amplitudes, which can be shown to be in very good agreement with the formula in question.

In the other experiments the wave-length was measured only on a shorter jet-piece, which, however, was so far from the orifice that the value of the wave-lengths had become constant.

As an example of such a measurement, an experiment with tube I may be mentioned, which was carried out with a pressure-height of about 70 cm.

In the table below are quoted two sets of readings, obtained in succession, with their differences.

Readings.	Difference.	Readings.	Difference.
cm.	cm.	cm.	cm.
1·819		1·818	
	1·796		1·797
3·615		3·615	
	1·798		1·800
5·413		5·415	
	1·799		1·800
7·212		7·215	
	1·801		1·799
9·013		9·014	
	1·799		1·799
10·812		10·813	
	1·796		1·797
12·608		12·610	

The horizontal place of the jet was at a distance of 21·5 cm. from the orifice, which corresponded to a reading on the glass-rule of 7·5 cm. Corrections of the readings

---

* The smallness of the correction is due to the small coefficient of viscosity ($\mu = 0·0125$) and the great surface-tension (T = 74) of water. The correction mentioned can, however, become quite considerable for liquids in which these quantities have other values; if, for example, aniline ($\mu = 0·062$, T = 44) was used, the correction would have been more than 1 per cent. by corresponding experiments.

have not been introduced, as they were in this case very small; thus the correction for the decreasing of the wave-amplitudes would become completely imperceptible on account of the small value of the amplitudes. Besides, the very small decrease of the utmost differences corresponds to the one to be expected on account of the curvature of the jet.

We see that the wave-lengths are very constant, and that the jet at the place in question must have executed exceedingly pure vibrations, because only a small deviation from pure harmonic vibrations must involve considerable irregularities of the differences between the readings.

The other experiments carried out show results very similar to those described here. It may be remarked that, *à priori*, we could expect exceedingly pure vibrations of the jets from all the four tubes. The circumstance that the regular variation of the differences between the readings commenced directly past the orifice (see the table on p. 310) shows that already at this point we had to do with what was very nearly a single vibration only, and, as mentioned on p. 300, the vibrations of the jet must be much purer at a considerable distance from the orifice than close to it.

The table below contains the result of all the experiments carried out. The surface-tension is calculated according to the following equation (see p. 297):—

$$T_{12} = \frac{(\rho_1+\rho)\, k^2 c^2 a^3 J_2(iak)}{(3+a^2k^2)\, iak J'_2(iak)} \left[1 + 2\left(\frac{2\mu}{\rho c a^2 k}\right)^{3/2}\right]\left(1 + \frac{37}{24}\frac{b^2}{a^2}\right)[1 \div 0{\cdot}002\,(12-t)]. \qquad *$$

Tube.	Temperature.	Discharge.	Mean radius.	Wave-length.	Amplitude.	T_{12}.
	° C.	cm.³/sec.	cm.	cm.		dyne/cm.
I.	11·8	6·100	0·06755	2·225	0·026	73·24
I.	11·4	5·608	0·06758	2·039	—	73·41
I.	11·3	4·965	0·06767	1·800	—	73·34
II.	11·7	7·678	0·07554	2·658	0·046	73·01
II.	11·2	7·076	0·07567	2·443	—	72·98
II.	11·4	6·272	0·07587	2·154	—	73·26
III.	11·8	7·720	0·07595	2·656	0·042	73·45
III.	11·4	6·290	0·07604	2·157	—	73·28
IV.	11·8	8·649	0·08010	2·901	0·027	73·21
IV.	11·9	7·984	0·08014	2·677	—	73·09
Mean value of the experiments with Tube I ..						73·33
,, ,, ,, ,, II ..						73·08
,, ,, ,, ,, III ..						73·37
,, ,, ,, ,, IV ..						73·15
Mean value of all experiments ..						73·23

We see that the mutual agreement between the single determinations is very good (the greatest deviation from the mean value being about 0·35 per cent.).

It may be remarked that in the values found for T_{12} no indication of a distinct

* [In the formula for T_{12}, $(\rho_1+\rho)$ should be changed to $(\rho_1+\rho_2)$ and the sign $\div$ should be changed to $-$.]

influence originating either from the variation of the diameter of the jet, or of the discharge, or of the amplitudes of the waves can be found.

In all the experiments mentioned, tap-water was used. An investigation was, however, carried out to see if a different result would be obtained by using distilled water instead of tap-water. For this purpose two large reservoirs were filled respectively with distilled and tap-water. After the contents of the reservoirs had assumed the same temperature, measurements of wave-lengths in exactly the same conditions were undertaken on a jet of each of the two sorts of water by connecting first the one reservoir and then the other with the glass-bottle A, fig. 1, by a siphon. The experiment, which was repeated several times, showed that no sensible difference was to be found between the two jets.

This result was also to be expected from previous investigations on the surface-tension of water.

Now proceeding to compare the value found here with values found by previous determinations, we shall not try to give a complete account of the very extensive literature on this subject. The table opposite contains only the results of a few of the investigations of later years, which are generally considered the most important for the estimate of the value of the surface-tension.

The table shows rather considerable deviations between the values found by the different investigators. As an explanation of these deviations, the question of the purity of the surface has been among the most prominent, relying on the fact that the tension of a water-surface may decrease very considerably when the surface becomes contaminated with even an extremely small amount of foreign substances. This circumstance, however, does not seem sufficient to explain the deviations among the values found by authors who have used the same method for purifying the surface (e.g., GRUNMACH and KALÄHNE; FORCK and ZLOBICKI).

The fact that a number of authors (e.g., VOLKMANN, DORSEY, FORCK) who have worked with different methods have found such exceedingly good conformity among the results of their single experiments after all seems to show that the surface-tension of a carefully purified surface is a very constant quantity. This assumption is further confirmed by the circumstance that several authors (KALÄHNE, DORSEY, &c.) have not found any sensible diminishing of the surface-tension during the time of the experiment.

The results of the investigations by Miss A. POCKELS,[*] Lord RAYLEIGH,[†] and F. NANSEN[‡] on the influence of contaminations upon the tension of a water-surface seem also highly to point in this direction.

In consequence of the above-mentioned, it therefore seems that a great deal of the

---

[*] A. POCKELS, 'Nature,' XLIII., p. 437; XLVI., p. 418; XLVIII., p. 152. 'Ann. d. Phys.,' VIII., p. 854.

[†] RAYLEIGH, 'Phil. Mag.' XLVIII., p. 321, 1899.

[‡] F. NANSEN, 'Norweg. North Polar Exped. Scient. Results,' 10, 1900.

Authority.	Publication.	Method.	$T_{12}.$*
WEINSTEIN	'Metr. Beitr. d. K. Norm. Aich-Komm.,' VI, 1889	Capillary tubes	73·53
GOLDSTEIN	'Ztschr. Phys. Chem.,' V., p. 233, 1890	"	73·82
RAMSAY and SHIELDS	'Ztschr. Phys. Chem.,' XII, p. 433, 1893	"	71·67
QUINCKE	'WIED. Ann.,' LII., p. 1, 1894	"	73·3–77·8
VOLKMANN	'WIED. Ann.,' LVI., p. 457, 1895	"	73·72
DOMKE	'Wiss. Abh. d. K. Norm. Aich-Komm., III., 1902	"	73·92
GRABOWSKY	'Diss., Königsberg, 1904	"	73·71
SENTIS	'Thèse,' Grenoble, 1897	Capillary tubes (virtual)	74·24
HALL	'Phil. Mag.' (5), XXXVI., p. 385, 1893	Weighing of the tension	73·90
FORCK	'Ann. d. Phys.,' XVII., p. 744, 1905	Pressure in air-bubbles	77·25
ZLOBICKI	'Rozpr. Akad. Kraków,' S. 3, T. VI., A, p. 181, 1906	"	73·70
RAYLEIGH	'Phil. Mag.,' XXX., p. 386, 1890	Capillary ripples (advancing waves)	74·88
DORSEY	'Phil. Mag.,' XLIV., p. 369, 1897	"	74·08
WATSON	'Phys. Rev.,' XII., p. 257, 1901	"	75·15
KOLOWRAT-TSCHERWINSKI	'I. d. Russ. Phys.,' XXXVI., p. 265, 1904	"	73·22
KALÄHNE	'Ann. d. Phys.,' VII., p. 440, 1902	(standing waves)	74·67
GRUNMACH	'Wiss. Abh. d. K. Norm. Aich-Komm., III., 1902	"	76·35
BRÜMMER	'Diss.,' Rostock, 1903	"	75·39
LOEWENFELD	'Diss.,' Rostock, 1904	"	75·78
PEDERSEN	'Trans. Roy. Soc.,' A 207, p. 341, 1907	Jet-vibration	74·76

* From the papers in which the surface-tension is not given at 12° C, T_{12} is calculated by means of the formula $T_t = T_0 (1 - 0·0020t)$ (the temperature-coefficient being known with sufficient accuracy for this purpose).

2 s 2

deviations in question must be explained not by real differences of the surface-tension, but by the methods used in measuring this tension.

We now proceed to consider more closely some of the investigations mentioned, and compare the results with that found in the present paper.

We will commence with P. O. PEDERSEN's investigations, as his determination of the surface-tension of water is executed by the same method (jet-vibration) as that used by the author. PEDERSEN finds, as the table shows, a value which is considerably greater (about 2 per cent.) than the value here found. As, however, PEDERSEN has not examined the variations of the wave-length, but only determined the wave-length as mean wave-length on a jet-piece at a comparatively short distance from the orifice, the cause of the difference between the value found by PEDERSEN and the author may be that PEDERSEN probably has used too small a value for the wave-length (see p. 311).

Among the other methods to determine the surface-tension, the capillary-tube method and the method of capillary ripples are those mostly used and generally considered the most important.

Among the investigations carried out by the former methods, VOLKMANN's must be especially mentioned on account of the excellent agreement between the single experiments which he has obtained, taking great care in the measurement of the dimensions of the tubes and in their purification. This agreement, being independent of the dimensions of the tubes and of the nature of the glass, seems to have taken away the foundations of the criticism of the results which the capillary-tube method can give. VOLKMANN finds, as is seen, a value which lies rather near the author's, the difference being, however, about 0·7 per cent.

As is to be seen from the table, a great number of investigations have recently been executed by the method of capillary ripples. We see that the values found by this method are generally higher than the value found in this paper. The mutual conformity between the results of the different investigations is however not very great. In the author's opinion this disagreement depends on the fact that in many cases the conditions of the experimental investigations do not sufficiently correspond to the assumptions on which the theoretical development rests; in what follows, an attempt has been made to show what is meant by this.

The experiments executed by the method mentioned can be divided into two groups, according to advancing rectilinear waves, produced by help of the vibrations of a glass plate fastened to one prong of a tuning-fork, or standing waves formed by interference between two systems of advancing circular waves, generated by two pins fastened to both prongs of a tuning-fork, being used.

Among the authors who have used the former method, only DORSEY and KOLOWRAT-TSCHERWINSKI seem to have examined the magnitude of the wave-length at different distances from the generator. Both of the investigators found considerable irregularities near the generator, the wave-length here being dependent on the distance from the plate and first becoming constant at a greater distance from this. The authors mentioned, being aware of this fact, used for calculating the surface-tension only the length of waves which were at a certain distance from the glass-plate (DORSEY 4 cm., and KOLOWRAT-TSCHERWINSKI 8 cm.). As the wave-length near to the glass-plate was larger than further out, this may explain the fact that DORSEY and especially KOLOWRAT-TSCHERWINSKI have found lower values than other investigators who have used the same method, but as it seems have not taken precautions in this direction.

The other method, using the standing waves, suffers, as also KOLOWRAT-TSCHERWINSKI remarks, from certain defects, because the measuring of the wave-length taking place on the straight line which connects the above-mentioned pins, those waves only can be examined which are at so short a distance from the pins that there is no security of the phenomenon being sufficiently regular. On account of this the results found by this method do not seem to be very reliable, especially the very high values of the surface-tension, and the great deviations between the result of the single experiments found by GRUNMACH, BRÜMMER, and LOEWENFELD may probably be explained by the very small distance (1·8 cm.) between the pins used by these investigators. KALÄHNE, who employs the same method, but has a distance of 7 cm. between the pins, also finds a value considerably lower and with a much better mutual conformity than the above-named investigators.

In consequence of these considerations it does not seem necessary for the author to conclude that the method of capillary ripples in reality gives a value essentially higher than the one found by the method described in this paper.

Conclusions.

In the present determination of the surface-tension of water the method of jet-vibration proposed by Lord RAYLEIGH is used; this method has the fundamental advantage that a perfectly fresh new-formed surface can be examined.

In the first part of this investigation it is shown how Lord RAYLEIGH's theory of infinitely small vibrations of a jet of non-viscid liquid can be supplemented by corrections for the influence of the finite amplitudes as well as for the viscosity.

In the experimental part of this investigation an attempt has been made to show how in a simple manner it seems to be possible to secure that the jet-piece used for the measurement satisfies the assumptions on which the theoretical development rests.

As the final result of his experiments the author finds the surface-tension of water at 12° C. to be 73·23 dyne/cm.

IV. ADDENDUM TO PRIZE ESSAY

(TRANSLATION *)

ADDENDUM TO THE PHYSICS ESSAY MARKED BY THE MOTTO $\beta\gamma\delta$

Submitted Nov. 2, 1906.

The theory of waves on deep water progressing without change in shape under the simultaneous influence of gravity and surface tension has been given by Lord Kelvin.

The formulas developed in this theory have been used to determine the surface tension of liquids. One would think that corrections should be made in these formulas for the viscosity of the liquid and for the finite amplitude of the waves. An investigation of the influence of viscosity on the waves is found in Lamb's *Hydrodynamics* (p. 546). It is seen that, for a liquid like water, the corrections have no appreciable influence upon the results.

It appears, however, that corrections for a finite wave amplitude have nowhere been worked out. In the present addendum, an attempt has, therefore, been made to investigate this problem.

Investigations of waves on deep water progressing without change of shape and with finite amplitude (i.e., calculations to more than the first approximation) have been carried out by G. G. Stokes (Trans. Cambridge Phil. Soc. (1847) 441) and by Lord Rayleigh (Phil. Mag. **1** (1876) 257). However, in neither of these papers has the surface tension been taken into account. Stokes applies a very beautiful and straightforward method of approximation which can be extended to treat non-stationary motions and motions in three dimensions. (It has been applied to non-stationary motions in the main part of the present Prize Essay on the vibrations of a liquid jet.) Lord Rayleigh's method shows how, by making an assumption on the form of the potential, one can obtain the higher approximations with much shorter calculations. However, this method can only be applied to steady motions in two dimensions.

In the present investigation, two methods are applied which correspond to those mentioned. Stokes' method is used, since its straightforward procedure facilitates the investigation of certain exceptional cases met with; Lord Rayleigh's method is used also, because it enables one to ascertain, by brief computations, that the calculations have been correctly carried out.

Considering a stationary motion and proceeding in the same manner as Stokes, one has**

---

* [See Introduction, sect. 2 and 6.]

** [Most of the symbols used are not defined in the Addendum. They are in part the same as those

$$u = c - \frac{\partial \varphi}{\partial x}, \qquad v = -\frac{\partial \varphi}{\partial y},$$

$$\frac{\partial u}{\partial x} + \frac{\partial v}{\partial y} = -\left(\frac{\partial^2 \varphi}{\partial x^2} + \frac{\partial^2 \varphi}{\partial y^2}\right) = 0,$$

$$\varphi = \sum f(y) \sin kx, \qquad \frac{\partial^2 f}{\partial y^2} = k^2 f, \qquad f(y) = Ae^{-ky} + Be^{ky}.$$

Since $\partial \varphi / \partial y = 0$ for $y = \infty$, $B = 0$ and $\varphi = \sum_k A_k e^{-ky} \sin kx$.

The equation for the surface can be written $y = \zeta = f(x)$. The conditions which must be satisfied for $y = \zeta$ are

$$\frac{D(y-\zeta)}{Dt} = 0 \tag{1}$$

and

$$p = \frac{T}{R} + \text{const.}, \tag{2}$$

where

$$\frac{D}{Dt} = \frac{\partial}{\partial t} + u\frac{\partial}{\partial x} + v\frac{\partial}{\partial y} = \left(c - \frac{\partial \varphi}{\partial x}\right)\frac{\partial}{\partial x} - \frac{\partial \varphi}{\partial y}\frac{\partial}{\partial y}.$$

From (1) one obtains

$$0 = -\frac{\partial \varphi}{\partial y} - \left(c - \frac{\partial \varphi}{\partial x}\right)\frac{\partial \zeta}{\partial x}, \qquad \frac{\partial \zeta}{\partial x} = \left[-\frac{\partial \varphi}{\partial y}\left(c - \frac{\partial \varphi}{\partial x}\right)^{-1}\right]_{(y=\zeta)} = F(x),$$

$$\frac{\partial^2 \zeta}{\partial x^2} = \frac{\partial F(x)}{\partial x} = \left[\left(\frac{\partial}{\partial x} + \frac{\partial \zeta}{\partial x}\frac{\partial}{\partial y}\right)\left(-\frac{\partial \varphi}{\partial y}\left(c - \frac{\partial \varphi}{\partial x}\right)^{-1}\right)\right]_{(y=\zeta)},$$

$$\frac{\partial^2 \zeta}{\partial x^2} = -\frac{\partial^2 \varphi}{\partial x \partial y}\left(c - \frac{\partial \varphi}{\partial x}\right)^{-1} - \frac{\partial \varphi}{\partial y}\left(\frac{\partial^2 \varphi}{\partial x^2} - \frac{\partial^2 \varphi}{\partial y^2}\right)\left(c - \frac{\partial \varphi}{\partial x}\right)^{-2}$$

$$+ \left(\frac{\partial \varphi}{\partial y}\right)^2 \frac{\partial^2 \varphi}{\partial x \partial y}\left(c - \frac{\partial \varphi}{\partial x}\right)^{-3}.$$

used in the references and in the main part of the Prize Essay. Their definitions are as follows: c is the velocity of wave propagation; the water is regarded as moving with a velocity c in the opposite direction to the progressing waves which then appear stationary in a resting coordinate system x, y; x is measured horizontally in the direction of the moving water, and y is measured downwards from the surface;

u and v are the x- and y-components of the velocity, φ is the velocity potential and ψ the stream function;

ζ is the vertical displacement at the surface;

g is the acceleration due to gravity, T is the surface tension and ρ the density of the water; $T_1 = T/\rho$;

R is the finite radius of curvature of the surface and p the pressure. There is an obvious and unimportant inconsistency between the definitions of φ in the first and second parts of the Addendum.]

From (2) we get

$$\frac{p}{\rho} - \frac{T/\rho}{R} = \frac{p}{\rho} - \frac{T_1}{R} = gy - \tfrac{1}{2}(u^2+v^2) - T_1 \frac{\partial^2 \zeta}{\partial x^2}\left(1+\left(\frac{\partial \zeta}{\partial x}\right)^2\right)^{-\frac{3}{2}}_{(y=\zeta)} = \text{const.},$$

$$gy + c\frac{\partial \varphi}{\partial x} - \frac{1}{2}\left(\left(\frac{\partial \varphi}{\partial x}\right)^2 + \left(\frac{\partial \varphi}{\partial y}\right)^2\right)$$

$$-T_1\left[-\frac{\partial^2 \varphi}{\partial x\, \partial y}\left(c-\frac{\partial \varphi}{\partial x}\right)^{-1} - \frac{\partial \varphi}{\partial y}\left(\frac{\partial^2 \varphi}{\partial x^2} - \frac{\partial^2 \varphi}{\partial y^2}\right)\left(c-\frac{\partial \varphi}{\partial x}\right)^{-2}\right.$$

$$\left. + \left(\frac{\partial \varphi}{\partial y}\right)^2 \frac{\partial^2 \varphi}{\partial x\, \partial y}\left(c-\frac{\partial \varphi}{\partial x}\right)^{-3}\right] \times \left[\left(\frac{\partial \varphi}{\partial y}\left(c-\frac{\partial \varphi}{\partial x}\right)^{-1}\right)^2 + 1\right]^{-\frac{3}{2}} + K_{(y=\zeta)} = 0.$$

As an equation between x and y which is satisfied for $y = \zeta$, this equation may be regarded as the equation for the surface. Hence D/Dt of the left member of this equation must vanish. This gives

$$-g\frac{\partial \varphi}{\partial y} + c^2 \frac{\partial^2 \varphi}{\partial x^2} - 2c\frac{\partial^2 \varphi}{\partial x^2}\frac{\partial \varphi}{\partial x} - 2c\frac{\partial^2 \varphi}{\partial x\, \partial y}\frac{\partial \varphi}{\partial y} + \frac{\partial^2 \varphi}{\partial x^2}\left(\frac{\partial \varphi}{\partial x}\right)^2 + 2\frac{\partial^2 \varphi}{\partial x\, \partial y}\frac{\partial \varphi}{\partial x}\frac{\partial \varphi}{\partial y}$$

$$+ \frac{\partial^2 \varphi}{\partial y^2}\left(\frac{\partial \varphi}{\partial y}\right)^2 - T_1\left(\left(c-\frac{\partial \varphi}{\partial x}\right)\frac{\partial}{\partial x} - \frac{\partial \varphi}{\partial y}\frac{\partial}{\partial y}\right)\left\{\left(-\frac{\partial^2 \varphi}{\partial x\, \partial y}\left(c-\frac{\partial \varphi}{\partial x}\right)^{-1}\right.\right.$$

$$\left. - \frac{\partial \varphi}{\partial y}\left(\frac{\partial^2 \varphi}{\partial x^2} - \frac{\partial^2 \varphi}{\partial y^2}\right)\left(c-\frac{\partial \varphi}{\partial x}\right)^{-2} + \left(\frac{\partial \varphi}{\partial y}\right)^2 \frac{\partial^2 \varphi}{\partial x\, \partial y}\left(c-\frac{\partial \varphi}{\partial x}\right)^{-3}\right)$$

$$\left. \times \left(1+\left(\frac{\partial \varphi}{\partial y}\left(c-\frac{\partial \varphi}{\partial x}\right)^{-1}\right)^2\right)^{-\frac{3}{2}}\right\}_{(y=\zeta)} = 0. \tag{3}$$

Eq. (3) is exact. Dropping from now on in all calculations terms of order higher than the third, carrying out all calculations, and contracting the result with the aid of $\nabla^2 \varphi = 0$, one obtains the following expression for the coefficient of T_1 in (3):

$$\frac{\partial^3 \varphi}{\partial x^2\, \partial y} + \frac{3}{c}\frac{\partial^2 \varphi}{\partial x\, \partial y}\frac{\partial^2 \varphi}{\partial x^2} + \frac{3}{c}\frac{\partial^3 \varphi}{\partial x^3}\frac{\partial \varphi}{\partial y} + \frac{3}{c^2}\frac{\partial^2 \varphi}{\partial x\, \partial y}\frac{\partial^2 \varphi}{\partial x^2}\frac{\partial \varphi}{\partial x} + \frac{3}{c^2}\frac{\partial^3 \varphi}{\partial x^3}\frac{\partial \varphi}{\partial x}\frac{\partial \varphi}{\partial y}$$

$$- \frac{6}{c^2}\left(\frac{\partial^2 \varphi}{\partial x\, \partial y}\right)^2\frac{\partial \varphi}{\partial y} + \frac{6}{c^2}\left(\frac{\partial^2 \varphi}{\partial x^2}\right)^2\frac{\partial \varphi}{\partial y} - \frac{9}{2c^2}\frac{\partial^3 \varphi}{\partial x^2\, \partial y}\left(\frac{\partial \varphi}{\partial y}\right)^2.$$

When the terms in (3) are arranged according to order of magnitude, this equation becomes

$$-g\,\frac{\partial\varphi}{\partial y}+c^2\,\frac{\partial^2\varphi}{\partial x^2}+T_1\,\frac{\partial^3\varphi}{\partial x^2\partial y}-2c\left(\frac{\partial^2\varphi}{\partial x^2}\,\frac{\partial\varphi}{\partial x}+\frac{\partial^2\varphi}{\partial x\partial y}\,\frac{\partial\varphi}{\partial y}\right)$$

$$+\frac{3T_1}{c}\left(\frac{\partial^2\varphi}{\partial x\partial y}\,\frac{\partial^2\varphi}{\partial x^2}+\frac{\partial^3\varphi}{\partial x^3}\,\frac{\partial\varphi}{\partial y}\right)+\frac{\partial^2\varphi}{\partial x^2}\left(\frac{\partial\varphi}{\partial x}\right)^2+2\frac{\partial^2\varphi}{\partial x\partial y}\,\frac{\partial\varphi}{\partial x}\,\frac{\partial\varphi}{\partial y}+\frac{\partial^2\varphi}{\partial y^2}\left(\frac{\partial\varphi}{\partial y}\right)^2$$

$$+\frac{3T_1}{c^2}\left(\frac{\partial^2\varphi}{\partial x\partial y}\,\frac{\partial^2\varphi}{\partial x^2}\,\frac{\partial\varphi}{\partial x}+\frac{\partial^3\varphi}{\partial x^3}\,\frac{\partial\varphi}{\partial x}\,\frac{\partial\varphi}{\partial y}-2\left(\frac{\partial^2\varphi}{\partial x\partial y}\right)^2\frac{\partial\varphi}{\partial y}+2\left(\frac{\partial^2\varphi}{\partial x^2}\right)^2\frac{\partial\varphi}{\partial y}\right.$$

$$\left.-\frac{3}{2}\,\frac{\partial^3\varphi}{\partial x^2\partial y}\left(\frac{\partial\varphi}{\partial y}\right)^2\right)_{(y=\zeta)}=0. \tag{4}$$

Moreover, one has

$$\frac{\partial\zeta}{\partial x}=-\frac{\partial\varphi}{\partial y}\left(c-\frac{\partial\varphi}{\partial x}\right)^{-1}=-\frac{1}{c}\left(1+\frac{1}{c}\,\frac{\partial\varphi}{\partial x}+\frac{1}{c^2}\left(\frac{\partial\varphi}{\partial x}\right)^2\right)\frac{\partial\varphi}{\partial y}_{(y=\zeta)}. \tag{5}$$

Eqs. (4) and (5) hold for $y=\zeta$. However, with the aid of Taylor's series one can, without reducing the degree of approximation, write down the following two equations holding for $y=0$:

$$\left(1+\zeta\,\frac{\partial}{\partial y}+\tfrac{1}{2}\zeta^2\,\frac{\partial^2}{\partial y^2}\right)\left(-g\,\frac{\partial\varphi}{\partial y}+c^2\,\frac{\partial^2\varphi}{\partial x^2}+T_1\,\frac{\partial^3\varphi}{\partial x^2\partial y}\right)$$

$$+\left(1+\zeta\,\frac{\partial}{\partial y}\right)\left(-2c\left(\frac{\partial^2\varphi}{\partial x^2}\,\frac{\partial\varphi}{\partial x}+\frac{\partial^2\varphi}{\partial x\partial y}\,\frac{\partial\varphi}{\partial y}\right)+\frac{3T_1}{c}\left(\frac{\partial^2\varphi}{\partial x\partial y}\,\frac{\partial^2\varphi}{\partial x^2}+\frac{\partial^3\varphi}{\partial x^3}\,\frac{\partial\varphi}{\partial y}\right)\right)$$

$$+\frac{\partial^2\varphi}{\partial x^2}\left(\frac{\partial\varphi}{\partial x}\right)^2+2\frac{\partial^2\varphi}{\partial x\partial y}\,\frac{\partial\varphi}{\partial x}\,\frac{\partial\varphi}{\partial y}+\frac{\partial^2\varphi}{\partial y^2}\left(\frac{\partial\varphi}{\partial y}\right)^2+\frac{3T_1}{c^2}\left\{\frac{\partial^2\varphi}{\partial x\partial y}\,\frac{\partial^2\varphi}{\partial x^2}\,\frac{\partial\varphi}{\partial x}\right. \tag{6}$$

$$\left.+\frac{\partial^3\varphi}{\partial x^3}\,\frac{\partial\varphi}{\partial x}\,\frac{\partial\varphi}{\partial y}-2\left(\frac{\partial^2\varphi}{\partial x\partial y}\right)^2\frac{\partial\varphi}{\partial y}+2\left(\frac{\partial^2\varphi}{\partial x^2}\right)^2\frac{\partial\varphi}{\partial y}-\frac{3}{2}\,\frac{\partial^3\varphi}{\partial x^2\partial y}\left(\frac{\partial\varphi}{\partial y}\right)^2\right\}_{(y=0)}=0,$$

$$\frac{\partial\zeta}{\partial x}=-\frac{1}{c}\,\frac{\partial\varphi}{\partial y}-\frac{1}{c}\,\frac{\partial^2\varphi}{\partial y^2}\,\zeta-\frac{1}{c^2}\,\frac{\partial\varphi}{\partial x}\,\frac{\partial\varphi}{\partial y}-\frac{1}{2c}\,\frac{\partial^3\varphi}{\partial y^3}\,\zeta^2$$

$$-\frac{1}{c^2}\left(\frac{\partial^2\varphi}{\partial y^2}\,\frac{\partial\varphi}{\partial x}+\frac{\partial^2\varphi}{\partial x\partial y}\,\frac{\partial\varphi}{\partial y}\right)\zeta-\frac{1}{c^3}\,\frac{\partial\varphi}{\partial y}\left(\frac{\partial\varphi}{\partial x}\right)^2, \qquad (y=0). \tag{7}$$

Eqs. (6) and (7) are used to determine φ and ζ.

First approximation:

$$-g\frac{\partial\varphi}{\partial y}+c^2\frac{\partial^2\varphi}{\partial x^2}+T_1\frac{\partial^3\varphi}{\partial x^2\partial y}=0,\qquad(y=0).$$

Putting $\varphi = Ae^{-ky}\sin kx$, one gets

$$A(gk-k^2c^2+k^3T_1)\sin kx = 0,\qquad c^2 = \frac{g}{k}+T_1k,$$

$$\frac{\partial\zeta}{\partial x} = -\frac{1}{c}\left(\frac{\partial\varphi}{\partial y}\right)_{y=0} = \frac{A}{c}k\sin kx,\qquad \zeta = -\frac{A}{c}\cos kx.$$

Second approximation:

$$-g\frac{\partial\varphi}{\partial y}+c^2\frac{\partial^2\varphi}{\partial x^2}+T_1\frac{\partial^3\varphi}{\partial x^2\partial y}\Big._{(y=0)} = -\zeta\frac{\partial}{\partial y}\left(-g\frac{\partial\varphi}{\partial y}+c^2\frac{\partial^2\varphi}{\partial x^2}+T_1\frac{\partial^3\varphi}{\partial x^2\partial y}\right)$$

$$+2c\left(\frac{\partial^2\varphi}{\partial x^2}\frac{\partial\varphi}{\partial x}+\frac{\partial^2\varphi}{\partial x\partial y}\frac{\partial\varphi}{\partial y}\right)-\frac{3T_1}{c}\left(\frac{\partial^2\varphi}{\partial x\partial y}\frac{\partial^2\varphi}{\partial x^2}+\frac{\partial^3\varphi}{\partial x^3}\frac{\partial\varphi}{\partial y}\right)_{(y=0)} =$$

$$-\frac{A^2k}{c}(gk-k^2c^2+k^3T_1)\cos kx\sin kx+2cA^2(-k^3+k^3)\cos kx\sin kx$$

$$-\frac{3T_1}{c}A^2(k^4+k^4)\sin kx\cos kx = -\frac{3T_1}{c}k^4A^2\sin 2kx;$$

$$\varphi = Ae^{-ky}\sin kx+Be^{-2ky}\sin 2kx,$$

$$B(2kg-4c^2k^2+8k^3T_1)\sin 2kx = -\frac{3T_1}{c}k^4A^2\sin 2kx,$$

$$B = \frac{3}{2}\frac{A^2T_1k^3}{c(g-2T_1k^2)},$$

since

$$c^2 = \frac{g}{k}+T_1k;$$

$$\frac{\partial\zeta}{\partial x} = -\frac{1}{c}\frac{\partial\varphi}{\partial y}-\frac{\zeta}{c}\frac{\partial^2\varphi}{\partial y^2}-\frac{1}{c^2}\frac{\partial\varphi}{\partial x}\frac{\partial\varphi}{\partial y}\Big._{(y=0)} = \frac{A}{c}k\sin kx$$

$$+\frac{3A^2T_1k^4}{c^2(g-2T_1k^2)}\sin 2kx+\frac{A^2}{c^2}k^2\sin kx\cos kx+\frac{A^2k^2}{c^2}\sin kx\cos kx$$

$$= \frac{A}{c}k\sin kx+\frac{A^2k^3}{g-2T_1k^2}\sin 2kx;$$

[71]

$$\zeta = -\frac{A}{c}\cos kx - \frac{A^2}{2}\frac{k^2}{g-2T_1 k^2}\cos 2kx,$$

or

$$\zeta = \alpha \cos kx - \frac{\alpha^2}{2} k \frac{g+T_1 k^2}{g-2T_1 k^2}\cos 2kx, \qquad \left[\alpha = -\frac{A}{c}\right].$$

Hence, the second approximation gives

$$\varphi = Ae^{-ky}\sin kx + \frac{3}{2}\frac{A^2 T_1 k^3}{c(g-2T_1 k^2)}e^{-2ky}\sin 2kx,$$

and, since $c^2 = (g/k)+T_1 k$,

$$\zeta = -\frac{A}{c}\cos kx - \frac{A^2}{2}\frac{k^2}{g-2T_1 k^2}\cos 2kx.$$

Third approximation:

$$\left(-g\frac{\partial\varphi}{\partial y}+c^2\frac{\partial^2\varphi}{\partial x^2}+T_1\frac{\partial^3\varphi}{\partial x^2\partial y}\right)_{(y=0)}$$

$$= -\left(\zeta\frac{\partial}{\partial y}+\frac{\zeta^2}{2}\frac{\partial^2}{\partial y^2}\right)\left(-g\left(\frac{\partial\varphi}{\partial y}+c^2\frac{\partial^2\varphi}{\partial x^2}+T_1\frac{\partial^3\varphi}{\partial x^2\partial y}\right)\right.$$

$$+ \left(1+\zeta\frac{\partial}{\partial y}\right)\left\{2c\left(\frac{\partial^2\varphi}{\partial x^2}\frac{\partial\varphi}{\partial x}+\frac{\partial^2\varphi}{\partial x\partial y}\frac{\partial\varphi}{\partial y}\right)-\frac{3T_1}{c}\left(\frac{\partial^2\varphi}{\partial x\partial y}\frac{\partial^2\varphi}{\partial x^2}+\frac{\partial^3\varphi}{\partial x^3}\frac{\partial\varphi}{\partial y}\right)\right\}$$

$$-\frac{\partial^2\varphi}{\partial x^2}\left(\frac{\partial\varphi}{\partial x}\right)^2-2\frac{\partial^2\varphi}{\partial x\partial y}\frac{\partial\varphi}{\partial x}\frac{\partial\varphi}{\partial y}-\frac{\partial^2\varphi}{\partial y^2}\left(\frac{\partial\varphi}{\partial y}\right)^2$$

$$-\frac{3T_1}{c}\left(\frac{\partial^2\varphi}{\partial x\partial y}\frac{\partial^2\varphi}{\partial x^2}\frac{\partial\varphi}{\partial x}+\frac{\partial^3\varphi}{\partial x^3}\frac{\partial\varphi}{\partial x}\frac{\partial\varphi}{\partial y}-2\left(\frac{\partial^2\varphi}{\partial x\partial y}\right)^2\frac{\partial\varphi}{\partial y}\right.$$

$$+ 2\left(\frac{\partial^2\varphi}{\partial x^2}\right)^2\frac{\partial\varphi}{\partial y}-\frac{3}{2}\frac{\partial^3\varphi}{\partial x^2\partial y}\left(\frac{\partial\varphi}{\partial y}\right)^2\bigg)_{(y=0)}$$

$$= -\left(\zeta(-k)+\frac{\zeta^2}{2}(-k)^2\right)(gk-c^2k^2+T_1 k^3)A\sin kx$$

$$- \left(-\frac{A}{c}\cos kx\right)\frac{6T_1}{c}k^5 A^2\sin 2kx-\frac{3T_1}{c}k^4 A^2\sin 2kx$$

$$+ \left(-\frac{A}{c}\cos kx\right)\frac{6T_1}{c}k^5 A^2\sin 2kx-4ck^3\tfrac{3}{2}A^3\frac{T_1 k^3}{c(g-2T_1 k^2)}\sin kx$$

$$- \frac{3T_1}{c}3k^4\tfrac{3}{2}A^2\frac{T_1 k^3}{c(g-2T_1 k^2)}(3\sin 3kx-\sin kx)-A^3 k^4\sin kx$$

$$-\frac{3T_1}{c^2}\frac{A^3k^5}{8}(15\sin 3kx-13\sin kx)$$

$$=-k^3A^3\frac{8g^2+gT_1k^2+2T_1^2k^4}{8c^2(g-2T_1k^2)}\sin kx-\frac{3T_1}{c}k^4A^2\sin 2kx$$

$$-\frac{3T_1}{c^2}k^5A^3\frac{15g+78k^2T_1}{8(g-2T_1k^2)}\sin 3kx;$$

$$\varphi = Ae^{-ky}\sin kx+Be^{-2ky}\sin 2kx+Ce^{-3ky}\sin 3kx,$$

$$A(gk-c^2k^2+T_1k^3)\sin kx = -k^3A^3\frac{8g^2+gT_1k^2+2T_1^2k^4}{8c^2(g-2T_1k^2)}\sin kx.$$

Since A is approximately $-\alpha c$, we obtain

$$c^2 = \frac{g}{k}+T_1k+k\alpha^2\frac{8g^2+gT_1k^2+2T_1^2k^4}{8(g-2T_1k^2)}.$$

Since the coefficient to $\sin 2kx$ is the same as in the second approximation, we obtain again

$$B = \frac{3}{2}\frac{A^2T_1k^3}{c(g-2T_1k^2)}.$$

Also,

$$C(3gk-9c^2k^2+27T_1k^3)\sin 3kx = -\frac{3T_1}{c^2}k^5A^3\frac{15g+78T_1k^2}{8(g-2T_1k^2)}\sin 3kx,$$

$$C = \frac{3A^3T_1k^4(5g+26T_1k^2)}{16c^2(g-2T_1k^2)(g-3T_1k^2)};$$

$$\frac{\partial\zeta}{\partial x} = -\frac{1}{c}\frac{\partial\varphi}{\partial y}-\frac{1}{c}\frac{\partial^2\varphi}{\partial y^2}\zeta-\frac{1}{c^2}\frac{\partial\varphi}{\partial x}\frac{\partial\varphi}{\partial y}-\frac{1}{2c}\frac{\partial^3\varphi}{\partial y^3}\zeta^2$$

$$-\frac{1}{c^2}\left(\frac{\partial^2\varphi}{\partial y^2}\frac{\partial\varphi}{\partial x}+\frac{\partial^2\varphi}{\partial x\partial y}\frac{\partial\varphi}{\partial y}\right)\zeta-\frac{1}{c^3}\frac{\partial\varphi}{\partial y}\left(\frac{\partial\varphi}{\partial x}\right)^2_{(y=0)}$$

$$=\frac{kA}{c}\sin kx+\frac{3A^2T_1k^4}{c^2(g-2T_1k^2)}\sin 2kx+\frac{9A^3T_1k^5(5g+26T_1k^2)}{16c^3(g-2T_1k^2)(g-3T_1k^2)}\sin 3kx$$

$$+\frac{k^3A^3(5g+8T_1k^2)}{8c^3(g-2T_1k^2)}\sin kx+\frac{A^2k^2}{c^2}\sin 2kx+\frac{9A^3k^3(g+4T_1k^2)}{8c^3(g-2T_1k^2)}\sin 3kx;$$

$$\zeta = -\frac{A}{c}\left(1+\frac{A^2k^2(5g+8T_1k^2)}{8c^2(g-2T_1k^2)}\right)\cos kx-\frac{A^2k(g+T_1k^2)}{2c^2(g-2T_1k^2)}\cos 2kx$$

$$-\frac{3A^3k^2(2g^2+7gT_1k^2+2T_1^2k^4)}{16c^3(g-2T_1k^2)(g-3T_1k^2)}\cos 3kx.$$

Assuming that the velocity potential is of the form $\varphi = -cx + Ae^{-ky} \sin kx$, and neglecting surface tension, Lord Rayleigh has found the first three approximations for the waves treated here. However, if one takes surface tension into account, the problem cannot be solved with such a potential to more than the first approximation. On the other hand, if we put

$$\varphi = -cx + Ae^{-ky} \sin kx + Be^{-2ky} \sin 2kx + Ce^{-3ky} \sin 3kx,$$

where B and C are of the second and third order with respect to A, respectively, the problem can be solved to the first three approximations. Proceeding otherwise in exactly the same manner as Lord Rayleigh, we get

$$\varphi = -cx + Ae^{-ky} \sin kx + Be^{-2ky} \sin 2kx + Ce^{-3ky} \sin 3kx,$$

$$\psi = -cy - Ae^{-ky} \cos kx - Be^{-2ky} \cos 2kx - Ce^{-3ky} \cos 3kx.$$

For $\psi = 0$,

$$y = -\frac{A}{c} e^{-ky} \cos kx - \frac{B}{c} e^{-2ky} \cos 2kx - \frac{C}{c} e^{-3ky} \cos 3kx.$$

To the first approximation, $y_1 = -(A/C) \cos kx$.

To the second approximation, we get

$$y_2 = -\frac{A}{c}(1 - ky_1) \cos kx - \frac{B}{c} \cos 2kx,$$

or

$$y_2 = -\frac{A^2 k}{2c^2} - \frac{A}{c} \cos kx - \left(\frac{B}{c} + \frac{A^2 k}{2c^2}\right) \cos 2kx.$$

Putting $\psi = -A^2 k / 2c$, we find

$$y = \frac{A^2 k}{2c^2} - \frac{A}{c} e^{-ky} \cos kx - \frac{B}{c} e^{-2ky} \cos 2kx - \frac{C}{c} e^{-3ky} \cos 3kx.$$

The first approximation is the same as before. To the second approximation, one obtains

$$y_2 = -\frac{A}{c} \cos kx - \left(\frac{B}{c} + \frac{A^2 k}{2c^2}\right) \cos 2kx.$$

The third approximation gives

$$y = \frac{A^2 k}{2c^2} - \frac{A}{c}(1 - ky_2 + \tfrac{1}{2}k^2 y_1^2) \cos kx - \frac{B}{c}(1 - 2ky_1) \cos 2kx - \frac{C}{c} \cos 3kx = \zeta,$$

$$\zeta = -\left(\frac{A}{c} + \frac{5A^3 k^2}{8c^3} + \frac{3}{2} \frac{ABk}{c^2}\right) \cos kx - \left(\frac{B}{c} + \frac{A^2 k}{2c^2}\right) \cos 2kx$$

$$-\left(\frac{3A^3 k^2}{8c^3} + \frac{3ABk}{2c^2} + \frac{C}{c}\right) \cos 3kx,$$

$$\frac{1}{R} = \frac{\partial^2 \zeta}{\partial x^2} \left(1 + \left(\frac{\partial \zeta}{\partial x}\right)^2\right)^{-\frac{3}{2}} = k \left(\frac{Ak}{c} + \frac{A^3 k^3}{4c^3} + \frac{3ABk^2}{2c^2}\right) \cos kx$$

$$+ k \left(\frac{2A^2 k^2}{c^2} + \frac{4Bk}{c}\right) \cos 2kx + k \left(\frac{15A^3 k^3}{4c^3} + \frac{27ABk^2}{2c^2} + \frac{9Ck}{c}\right) \cos 3kx.$$

To the third approximation, one has

$$q^2 = \left(\frac{\partial \varphi}{\partial x}\right)^2 + \left(\frac{\partial \varphi}{\partial y}\right)^2 = c^2 - 2Acke^{-ky} \cos kx - 4Bcke^{-2ky} \cos 2kx$$

$$- 6Ccke^{-3ky} \cos 3kx + A^2 k^2 e^{-2ky} \cos 2kx + 4k^2 ABe^{-3ky} \cos 3kx;$$

for $y = \zeta$ one gets, when the calculations are carried out,

$$q^2 = c^2 - c^2 \left(\frac{2Ak}{c} - \frac{3A^3 k^3}{4c^3} + k^2 \frac{AB}{c^2}\right) \cos kx$$

$$- c^2 \left(\frac{A^2 k^2}{c^2} + \frac{4Bk}{c}\right) \cos 2kx - c^2 \left(\frac{3A^3 k^3}{4c^3} + \frac{5ABk^2}{c^2} + \frac{6Ck}{c}\right) \cos 3kx.$$

The condition is now $gy - T_1/R - \frac{1}{2}q^2 = $ constant.
This gives

$$\frac{g}{k} \left(\frac{Ak}{c} + \frac{5A^3 k^3}{8c^3} + \frac{3ABk^2}{2c^2}\right) + T_1 k \left(\frac{Ak}{c} + \frac{A^3 k^3}{4c^3} + \frac{3ABk^2}{2c^2}\right)$$

$$- c^2 \left(\frac{Ak}{c} - \frac{3A^3 k^3}{8c^3} + \frac{ABk^2}{c^2}\right) = 0, \quad \text{(I)}$$

$$\frac{g}{k} \left(\frac{Bk}{c} + \frac{A^2 k^2}{2c^2}\right) + T_1 k \left(\frac{4Bk}{c} + \frac{2A^2 k^2}{c^2}\right) - c^2 \left(\frac{2Bk}{c} + \frac{A^2 k^2}{2c^2}\right) = 0, \quad \text{(II)}$$

$$\frac{g}{k} \left(\frac{3A^3 k^3}{8c^3} + \frac{3ABk^2}{2c^2}\right) + T_1 k \left(\frac{15A^3 k^3}{4c^3} + \frac{27ABk^2}{2c^2} + \frac{9Ck}{c}\right)$$

$$- c^2 \left(\frac{3A^3 k^3}{8c^3} + \frac{5ABk^2}{2c^2} + \frac{3Ck}{c}\right) = 0. \quad \text{(III)}$$

From (I) one gets immediately $c^2 = g/k + T_1 k$ to within quantities of order A^2. When this is substituted in (II), one finds

$$B = \frac{3A^2 T_1 k^3}{2c(g - 2T_1 k^2)},$$

and substitution of this expression in (I) gives

$$c^2 = \frac{g}{k} + T_1 k + \frac{A^2 k(8g^2 + g T_1 k^2 + 2T_1^2 k^4)}{8c^2(g - 2T_1 k^2)}.$$

When these expressions for B and c are substituted in (III), one gets

$$C = \frac{3A^3 T_1 k^4(5g + 26T_1 k^2)}{16c^2(g - 2T_1 k^2)(g - 3T_1 k^2)}.$$

Substitution of the calculated constants in the expressions for φ and ζ gives the desired results:

$$\varphi = -cx + Ae^{-ky}\sin kx + \frac{3A^2 T_1 k^3}{2c(g - 2T_1 k^2)} e^{-2ky}\sin 2kx$$

$$+ \frac{3A^3 T_1 k^4(5g + 26T_1 k^2)}{16c^2(g - 2T_1 k^2)(g - 3T_1 k^2)} e^{-3ky}\sin 3kx,$$

$$\zeta = -\frac{A}{c}\left(1 + \frac{A^2 k^2(5g + 8T_1 k^2)}{8c^2(g - 2T_1 k^2)}\right)\cos kx - \frac{A^2 k(g + T_1 k^2)}{2c^2(g - 2T_1 k^2)}\cos 2kx$$

$$- \frac{3A^3 k^2(2g^2 + 7g T_1 k^2 + 2T_1^2 k^4)}{16c^3(g - 2T_1 k^2)(g - 3T_1 k^2)}\cos 3kx.$$

The latter equation can be written

$$\zeta = \alpha\left(1 + \frac{\alpha^2 k^2(5g + 8T_1 k^2)}{8(g - T_1 k^2)}\right)\cos kx - \frac{\alpha^2 k(g + T_1 k^2)}{2(g - 2T_1 k^2)}\cos 2kx$$

$$+ \frac{3\alpha^3 k^2(2g^2 + 7g T_1 k^2 + 2T_1^2 k^4)}{16(g - 2T_1 k^2)(g - 3T_1 k^2)}\cos 3kx,$$

where, as above, $\alpha = -A/c$, and where c is given by

$$c^2 = \frac{g}{k} + T_1 k + k\alpha^2 \frac{8g^2 + g T_1 k^2 + 2T_1^2 k^4}{8(g - 2T_1 k^2)}.$$

It is seen that the same results have been found in both ways. When T is put equal to zero, the expressions for φ, ζ, and c^2, become the same as those found by Stokes and Lord Rayleigh.

To see what effect the finite value of the amplitude would have upon a determination of the surface tension, we may put $g = 0$ (in fact, such small waves are always used that the effect of surface tension greatly predominates). We then obtain

$$c^2 = T_1 k - \frac{k\alpha^2}{8} T_1 k^2 = kT_1\left(1 - \frac{\alpha^2 k^2}{8}\right) = k\left(1 - \frac{4\pi^2}{8}\left(\frac{\alpha}{\lambda}\right)^2\right)T_1,$$

Thus, it is seen that, if the finite values of the amplitudes are neglected,

$$T_1(1 - \tfrac{1}{2}\pi^2(\alpha/\lambda)^2)$$

is found instead of T_1, i.e., one finds a value that is too low. How large the corrections are that should have been made in the experiments actually performed by Lord Rayleigh and others to determine the surface tension from such waves cannot be stated, however, since the amplitudes were not reported. (The formulas do not hold for standing waves, since two wave systems are not independent in the higher approximations.)

The expressions $g - 2T_1 k^2$ and $g - 3T_1 k^2$ are seen to occur in some denominators in the formulas developed here. These quantities vanish for certain definite wavelengths. Of course, when this is the case, the formulas cannot be applied, and it is necessary to investigate more closely what this means.

Returning to the equation

$$B(2kg - 4c^2 k^2 + 8k^3 T_1) \sin 2kx = -\frac{3T_1}{c} k^4 A^2 \sin 2kx,$$

we see that when $g - 2T_1 k^2 = 0$, and $c^2 = (g/k) + T_1 k$, the surface conditions cannot be satisfied in the second approximation. Moreover, it is seen that $g - 2T_1 k^2 = 0$ implies that waves with a wavelength equal to half that of the principal waves will have the same velocity of propagation as these. That the inverse holds can be inferred directly as follows.

When a system of waves satisfies the conditions for propagation without change of shape to the nth approximation (i.e., to the nth order of magnitude, the amplitudes being reckoned to be of the first order), but not to the $(n+1)$th approximation, it can be regarded as consisting of parts that propagate with different velocities. However, this cannot be remedied by adding to the wave system another system of the $(n+1)$th order which (to the first approximation) has the same speed of propagation as the former (nth approximation), since two wave systems of the first and the $(n+1)$th order can be regarded as independent of each other in the $(n+1)$th approximation.

Hence, for the type of waves considered here, wave systems whose wavelength is a whole number n times as large as that of a system with the same speed of propagation will not satisfy the conditions in the nth approximation, and therefore cannot exist.

It is assumed (G. G. Stokes, *On the highest wave of uniform propagation*, Proc. Cambridge Phil. Soc. **A** (1883) 361) that, if gravity acts alone, waves of all possible wavelengths will actually exist. (The formulas are convergent when the amplitudes lie within a certain finite range.) It is shown here that, if both gravity and surface tension act, this range will be exactly zero for an infinite number of wavelengths and at least very small (since $g - nT_1 k^2$ is very nearly zero) for wavelengths close to these values.

It is seen that the circumstance mentioned can occur only for waves whose wavelengths are larger than that corresponding to the minimum velocity.

V. SECOND ROYAL SOCIETY PAPER

ON THE DETERMINATION OF THE TENSION
OF A RECENTLY FORMED WATER-SURFACE

(Proc. Roy. Soc. **A84** (1910) 395 *)

* [See Introduction, sect. 7.]

[*Reprinted from the* PROCEEDINGS OF THE ROYAL SOCIETY, A. Vol. 84]

On the Determination of the Tension of a recently formed Water-Surface.

By N. BOHR, Copenhagen.

(Communicated by Lord Rayleigh, O.M., F.R.S. Received August 22,— Read November 10, 1910.)

As an addition to my paper, published in the ' Philosophical Transactions,'[*] on the determination of the surface-tension of water, I desire to set forth the following remarks concerning the problem of the value of the tension of a recently formed water-surface, and the circumstances which are of importance for the determination of this tension.

Prof. P. Lenard has, in a paper lately published,[†] determined the surface-tension of a recently formed water-surface by means of the vibration of falling drops, and has for this tension found values which are considerably greater than those found by other methods.   From this, as well as from the results of experiments published in a former paper,[‡] he concludes that a recently formed water-surface has a very great tension, which, however,

---

[*] 'Phil. Trans.,' A, 1909, vol. 209, p. 281.
[†] 'Sitzungsber. d. Heidelberger Akad. d. Wiss.,' Math.-nat. Kl. Jahrg. 1910, Abh. 18.
[‡] ' Ann. d. Phys. u. Chem.,' 1887, vol. 30, p. 209.

in the course of a very short time (fraction of a second) decreases considerably. He remarks that this result is in agreement with experiments published in my paper mentioned above.   I shall, however, in the following try to explain the reasons why I cannot agree in these conclusions.

The determination of the surface-tension published in my paper cited above was carried out by the method of jet-vibrations, the theoretical foundation of which method, as well as of the method used by Prof. Lenard, is due to Lord Rayleigh.*

As to the closer investigation of the vibration of the jets, especially with regard to the satisfaction of the suppositions made in the theoretical treatment of the phenomenon, a great number of vibrations, commencing just at the orifice and in the most stable jets extending to a distance of more than 45 cm. from this (the velocity of the jets was about 425 cm./sec.), were examined by my experiments.   These measurements showed that the wave-length was not the same everywhere, but that, advancing from the orifice, it increased in the beginning rather rapidly and thereupon more slowly until finally from a distance of about 25 cm. from the orifice and as far as the stability of the jet allowed the measurings, the wave-length became practically perfectly constant.   (See the tables, *loc. cit.*, pp. 310—312.)

This result consequently showed the existence of certain irregularities of the phenomenon, which arise in the formation of the jet, and which are rapidly (in about 0·06 sec.) extinguished (*loc. cit.*, p. 309).

These irregularities might partly be thought to originate from possible variations of the value of the surface-tension in the time immediately after the formation of the surface, partly from irregularities of a more mechanical (hydrodynamical) character (*loc. cit.*, p. 299).   Since the last-mentioned irregularities, as explained in my former paper, must decrease rapidly on removal from the issue of the jet, the result of the experiments showed that the surface-tension, in every case from about 0·06 second after formation of the surface, and as long as it was possible to investigate the tension by the used method, was sensibly constant.†   This constant value was considered as

---

* 'Roy. Soc. Proc.,' 1879, vol. 39, p. 71.

† Prof. Lenard remarks in his last paper (*loc. cit.*, p. 4) that the mechanical irregularities certainly must decrease, but cannot, even far from the orifice, completely disappear, on account of the resistance of the air against the movement of the jet. As, however, the effect of the air resistance removing from the orifice very rapidly will become constant, we see that an influence of this resistance on the phenomenon will not affect the above conclusion of the constancy of the surface-tension, but it can only cause an alteration of the value found for this constant tension. As to the question of the magnitude of the influence of the air resistance, I mention here an unpublished experiment made during my first investigation. Around the jet, at a distance of about 10 cm. from the orifice, was placed a large and carefully worked iris diaphragm, so

the sought value for the surface-tension, and it was in every case the only one which could be compared with values found by other methods, in which the investigated surfaces always have been much older than 0·06 second.

Concerning the question of a possible variation of the surface-tension during the time from the formation of the surface until some 0·06 second later, it seems to me that my experiments do not give any reason to conclude an existence of such a variation, there being, as we shall see, no objection in explaining the found variation in the wave-length by help of the velocity-differences between concentric parts of the jet produced by friction during its formation, by which the central parts receive a greater velocity than the parts nearer the surface. These velocity-differences decrease, removing from the issue, on account of the viscosity in the jet; the mean velocity of the jet keeping constant, this effects that the velocity of the outer parts increases at the same time as the velocity of the central parts decreases. That the wave-lengths are shorter close to the orifice than at a greater distance from this has always seemed to me to be a natural consequence of the velocity of the surface (the outer parts) here being smaller and the waves in question being surface-waves (the velocity of the vibrating liquid-particles is becoming smaller removing from the surface, and is vanishing in the axis of the jet). In his above-cited paper (*loc. cit.*, p. 4), Prof. Lenard, however, is of the opinion that the inner mixture—produced during the vibrating motion, on account of the mutual displacing of the concentric parts of the jet—will effect an apparent increase of the mass, and a thereby resulting prolongation of the time of vibration and increase of the wave-lengths.

For the closer examination of this question, I have therefore made the following direct calculation of the wave-lengths, under the assumption that the different concentric parts of the jet are moving with different velocities.

that the jet just passed through the centre of the diaphragm. This was at first open, so that a free space of 5 cm. (the opening of the diaphragm was 10 cm.) surrounded the jet. Thereupon the diaphragm was closed, so that the free space between the jet and the diaphragm was not more than some 0·2 mm., and at the same time a wave-summit of the jet at a distance of 30 cm. from the orifice was fixed in a telescope by help of reflection in the surface of the jet (*loc. cit.*, p. 305). It was then observed that the mentioned summit during the closing of the diaphragm was displaced only very little (less than 0·1 mm.). This simple experiment was repeated several times with exactly the same result. As such a closing of the diaphragm must increase the resistance of the air to a very considerable degree, completely stopping the mass of air set in motion by the jet (the jet produces a sensible blast), the experiment, in my opinion, shows very distinctly that the air resistance cannot have any appreciable influence on the results. As will be shown in the following, an air resistance would besides introduce a correction of the value of the surface-tension, the sign of which would be opposite to that supposed by Prof. Lenard.

The general equations of motion of an incompressible non-viscous fluid unaffected by extraneous forces, are

$$\rho \frac{\mathrm{D}u}{\mathrm{D}t} = -\frac{\partial p}{\partial x}, \qquad \rho \frac{\mathrm{D}v}{\mathrm{D}t} = -\frac{\partial p}{\partial y}, \qquad \rho \frac{\mathrm{D}w}{\mathrm{D}t} = -\frac{\partial p}{\partial z}, \tag{1}$$

and

$$\frac{\partial u}{\partial x} + \frac{\partial v}{\partial y} + \frac{\partial w}{\partial z} = 0, \tag{2}$$

in which u, v, w are the components of the velocity, p the pressure, ρ the density, and

$$\frac{\mathrm{D}}{\mathrm{D}t} = \frac{\partial}{\partial t} + u\frac{\partial}{\partial x} + v\frac{\partial}{\partial y} + w\frac{\partial}{\partial z}.$$

In the problem in question the motion will be steady. Putting $w = \mathrm{W} + \omega$, and supposing that u, v, and ω are so small that products of them, and quantities of the same order of magnitude, can be neglected in the calculations, we get from the equations (1)

$$\rho \mathrm{W}\frac{\partial u}{\partial z} = -\frac{\partial p}{\partial x}, \qquad \rho \mathrm{W}\frac{\partial v}{\partial z} = -\frac{dp}{dy}, \qquad \rho\left(u\frac{\partial \mathrm{W}}{\partial x} + v\frac{\partial \mathrm{W}}{\partial y} + w\frac{\partial w}{\partial z}\right) = -\frac{\partial p}{\partial z}. \tag{3}$$

Introducing polar co-ordinates r and ϑ ($x = r\cos\vartheta$, $y = r\sin\vartheta$), and the radial and tangential velocity α and β, we get by help of the relations

$$u = \alpha\cos\vartheta - \beta\sin\vartheta, \qquad v = \alpha\sin\vartheta + \beta\cos\vartheta,$$

from (3), assuming W to be a function of r only.

$$\rho\mathrm{W}\frac{\partial\alpha}{\partial z} = -\frac{\partial p}{\partial r}, \qquad \rho\mathrm{W}\frac{\partial\beta}{\partial z} = -\frac{1}{r}\frac{\partial p}{\partial\vartheta}, \qquad \rho\left(\alpha\frac{d\mathrm{W}}{dr} + \mathrm{W}\frac{\partial\omega}{\partial z}\right) = -\frac{\partial p}{\partial z}, \tag{4}$$

and from (2)

$$\frac{\partial\alpha}{\partial r} + \frac{\alpha}{r} + \frac{1}{r}\frac{\partial\beta}{\partial\vartheta} + \frac{\partial\omega}{\partial z} = 0. \tag{5}$$

Now, supposing that p, α, β, and ω have the form $f(r)\,e^{in\vartheta + ikz}$, we get from (4) and (5)

$$\frac{\partial^2 p}{\partial r^2} + \frac{\partial p}{\partial r}\left(\frac{1}{r} - \frac{2}{\mathrm{W}}\frac{d\mathrm{W}}{dr}\right) - p\left(\frac{n^2}{r^2} + k^2\right) = 0. \tag{6}$$

In the case W is constant, the solution of (6), subject to the condition to be imposed when $r = 0$, is

$$p_0 = \mathrm{A}\mathrm{J}_n(ikr)\,e^{in\vartheta + ikz}, \tag{7}$$

in which J_n is the symbol of the Bessel's function of nth order.

Putting

$$p = p_0 e^{\int_0^r \psi(r)\,dr}, \tag{8}$$

we get from (6)

$$\frac{d\psi}{dr} + \psi^2 + \psi\left(\frac{1}{r} + \frac{2}{p_0}\frac{\partial p_0}{\partial r} - \frac{2}{\mathrm{W}}\frac{d\mathrm{W}}{dr}\right) - \frac{2}{p_0\mathrm{W}}\frac{\partial p_0}{\partial r}\frac{d\mathrm{W}}{dr} = 0. \tag{9}$$

We will now suppose that $W = c + \sigma$, in which the constant c is the mean velocity of the jet, and σ a quantity small compared with c. In this case ψ is small, and neglecting terms of the same order of magnitude as $(\sigma/c)^2$, we get from (9)

$$\frac{d\psi}{dr} + \psi\left(\frac{1}{r} + \frac{2}{p_0}\frac{\partial p_0}{\partial r}\right) - \frac{2}{cp_0}\frac{\partial p_0}{\partial r}\frac{d\sigma}{dr} = 0. \tag{10}$$

In the experiments the numerical value of ikr will be a very small quantity—the wave-length large in comparison to the diameter of the jet—in order not to complicate the formulæ, we will therefore in the calculation of ψ only use the first term of the expression for $J_n(ikr)$. This gives $\frac{1}{p_0}\frac{\partial p_0}{\partial r} = \frac{n}{r}$, and the solution of (8) becomes

$$\psi = \frac{2n}{c}r^{-(2n+1)}\left\{\int_0^r \frac{d\sigma}{dr}r^{2n}dr + \mathcal{C}\right\}.$$

The motion being finite for $r = 0$, we have $\mathcal{C} = 0$.

Integrating by part we get

$$\psi = \frac{2n}{cr}\sigma - \frac{4n^2}{cr^{2n+1}}\int_0^r \sigma r^{2n-1}dr. \tag{11}$$

Let us suppose that the equation of the surface is

$$r - a = \zeta = Be^{in\vartheta + ikz}.$$

The general surface condition gives

$$\frac{D}{Dt}(r - a - \zeta) = \left(\alpha\frac{\partial}{\partial r} + \frac{\beta}{r}\frac{\partial}{\partial\vartheta} + w\frac{\partial}{\partial z}\right)(r - a - \zeta) = 0,$$

whence we get, neglecting quantities of the same order of magnitude, as by the equations (3)

$$\alpha - W\frac{\partial\zeta}{\partial z} = 0, \qquad \zeta = -\frac{i}{Wk}\alpha.$$

In the same manner we get further, if the principal radii of curvature are R_1 and R_2,

$$\frac{1}{R_1} + \frac{1}{R_2} = \frac{1}{a} - \frac{\zeta}{a^2} - \frac{1}{a^2}\frac{\partial^2\zeta}{\partial\vartheta^2} - \frac{\partial^2\zeta}{\partial z^2} = \frac{1}{a} - \alpha\frac{i(n^2 - 1 + k^2a^2)}{a^2Wk}.$$

Calling the surface-tension T, the dynamical surface-condition will be

$$T\left(\frac{1}{R_1} + \frac{1}{R_2}\right) - p = \text{const.}$$

From this we get, with the same approximation as before and using (4)

$$\left[T\frac{(n^2 - 1 - k^2a^2)}{\rho a^2k^2W^2}\frac{\partial p}{\partial r} - p\right]_{r=a} = 0. \tag{12}$$

From (12) we get, using (7) and (8),

$$k^2 = T\frac{iakJ'_n(iak)}{\rho a^3 J_n(iak)}(n^2 - 1 + a^2k^2)\left\{\frac{1}{W^2}\left(1 + \psi\frac{p_0}{\partial p_0/\partial r}\right)\right\}_{r=a}. \tag{13}$$

$b\ 2$

[85]

From (13) we get, by help of (11) and with the same approximation as used in the calculation of ψ,

$$k = \mathrm{T} \frac{iak\mathrm{J}'_n(iak)}{\rho c^2 a^3 \mathrm{J}_n(iak)} (n^2 - 1 + a^2 k^2) \left[1 - \frac{4n}{ca^{2n}} \int_0^a \sigma r^{2n-1} dr \right]. \tag{14}$$

This equation is, except the last term, the solution given by Lord Rayleigh. We therefore see that the effect of the velocity-differences between concentrical parts of the jet consists in an exchange in the formula for the wave-length λ ($\lambda = 2\pi/k$), of the mean velocity of the jet c by an " effective mean velocity,"

$$c' = c + \frac{2n}{a^{2n}} \int_0^a \sigma r^{2n-1} dr. \tag{15}$$

We see from (15) that the greater n is, the nearer the effective mean velocity will be to the velocity of the surface, which is explained by the fact that the greater n (the number of waves on the circumference of a section of the jet) is, the more rapidly the velocity of the vibrating liquid particles will decrease, moving from the surface towards the axis of the jet.

It can now be shown that c' will be smaller than c, if the velocity of the jet—which will be the case in the experiments—is greatest in the middle and continually decreases approaching the surface ; c being the mean velocity of the jet, we have $\int_0^a \sigma r\, dr = 0$, and in the case in question further $\int_0^r \sigma r\, dr > 0$, where $a > r > 0$. From this we get for $n \geqq 2$ (in the experiments $n = 2$).

$$c' - c = \frac{2n}{a^{2n}} \int_0^a \sigma r^{2n-1} dr = -\frac{2n(2n-2)}{a^{2n}} \int_0^a \left[\int_0^r \sigma r\, dr \right] r^{2n-3} dr < 0. \tag{16}$$

After having seen that the velocity-differences in question will produce a variation of the wave-length in the same direction as found by the experiment, we shall further see how the decrement of the variation of the wave-length can also be approximately explained by the manner in which the velocity-differences will decrease. In order to show this, we shall use the four experiments quoted in the table, *loc. cit.*, p. 310. In the table below is quoted the mean radius of the jet a, the velocity v (calculated from the mean radius and the discharge), and under the indication D_2/D_1, the difference between the wave-length, measured between the wave-summits IV and V, and the constant value to which the wave-lengths were tending, divided by the corresponding difference between the wave-length, measured between the summits II and III, and the mentioned constant value (in these differences are introduced the small corrections for the curvature of the jet and for the wave-amplitudes mentioned in the table, *loc. cit.*, p. 311) ; under the

---

* [The left member of eq. (14) should be k^2 rather than $\overline{k}$.]

indication l is quoted the difference between the mean value of the readings of the summits IV and V and the mean values of the readings II and III. Further, under the indication α is quoted the ratio between the variations of the wave-length in two places corresponding to a time-interval of $1/100$ sec. (calculated on the assumption that the variations decrease after an exponential law). Under the indication α' is finally quoted the ratio between the velocity differences in the jet in two places corresponding to a time-interval of $1/100$ sec., calculated from the theoretical formula, *loc. cit.*, p. 298, *l.* 2. f. b.

	I.	II.	III.	IV.
a	0·06755	0·07554	0·07595	0·08010
v	426	428	426	429
$\dfrac{D_2}{D_1}$	0·54	0·53	0·53	0·50
l	4·30	5·16	5·16	5·59
$\alpha = \left(\dfrac{D_2}{D_1}\right)^{\frac{v}{100\,l}}$	0·54	0·59	0·59	0·59
$\alpha' = e^{-\frac{\mu}{\rho}\left(\frac{\pi 1\cdot2197}{\alpha}\right)^2 \frac{1}{100}}$	0·67	0·72	0·73	0·75

As will be seen, the calculated and the found values for the decrement of the variation of the wave-lengths agree with regard to the order of magnitude, and more was not to be expected from such an approximate calculation. It is thus not justified to expect that the distribution of the velocity in jet-sections so close to the orifice could be completely expressed by the first term in the general formula on p. 298.

After having now seen that my experiments do not give any reasons for the conclusion of the existence of a variation in the surface-tension during the first time after the formation of the surface, we shall proceed to mention the values for the surface-tension of a recently formed water-surface found by Prof. Lenard by his investigations of the vibrations of failing drops.

The surfaces there investigated must, in my opinion, be considered as much older than the surfaces investigated by my experiments, on account of the length of the time used for the formation of the drops. Prof. Lenard remarks in his last paper [*loc. cit.*, p. 11, note (18)], that this time—amounting from 0·17 to 1·05 sec. in his first paper, and from 0·6 to 0·9 sec. in his last—will contribute only a very little to the age of the surface of the drops, new surface continually being formed during this time. This circumstance does

[87]

not, however, seem to me sufficiently to justify the neglecting of this, in this connection, very long time. I should rather be inclined to agree with the opinion set forth in his former paper, according to which the time for the formation of the drops is considered a measure for the age of the surface (*loc. cit.*, p. 233).

A comparison between the experiments of Prof. Lenard (*ibid.*, p. 236) and of Lord Rayleigh[*] on the surface-tension of a solution of soap seems also distinctly to show that the time of formation of the drops has a great influence on the condition of the surface. By the experiments of Lord Rayleigh with vibrating jets, the surface-tension of a solution of soap, 1/100 second after the formation of the surface, was thus found to be very near to that of pure water; while Prof. Lenard by experiments with vibrating drops (time of formation greater than $\frac{1}{4}$ sec.) finds the surface-tension of a soap solution of corresponding strength (1 : 1000) less than half that of water and rather near the stationary value of the surface-tension of a soap solution.

It appears, from the preceding, that the high values of the tension of a recently formed water-surface and the rapid decreasing of this value, which Prof. Lenard has found by his experiments, are not in agreement with the result of my previous experiments, because the tension of water-surfaces of lower age than those investigated by Prof. Lenard have been found much smaller and perfectly constant within the time interval (from 0·06 to 0·11 second after the formation of the surface) during which the method allowed the determination of the tension.

The cause of the great deviation between the results found by the method of drop-vibration and those found by the method of jet-vibration must, in my opinion, be sought in the circumstance that sufficient regard as to the influence of irregularities of mechanical character, arising from the disengagement of the drops, is scarcely taken by the drop method. The investigation of the influence of such irregularities seems also much more difficult by vibrating drops than by vibrating jets, the investigation by the latter being very much facilitated by the perfectly steady character of the phenomenon.

Concluding these remarks, I might call attention to the good agreement between the value of the tension of a water-surface 0·06 second old (73·23 dyne/cm. at 12° C.), found in my paper, and values of the tension of a water-surface found by statical methods[†] (Volkmann[‡] 73·72 dyne/cm. at

[*] 'Roy. Soc. Proc.,' 1890, vol. 47, p. 281.

[†] As to the result found by other methods, I might refer to the discussion in my former paper (*loc. cit.*, pp. 314—317).

[‡] 'Wied. Ann.,' 1895, vol. 56, p. 457.

12° C., Eötvös* 73·06).   This agreement seems to show that the tension of
a water-surface already only 0·06 second after the formation of the surface
(and according to what is discussed in the present paper probably much
earlier) has assumed the constant value which the tension, if contaminations
are kept away, will retain during a very long time.

* 'Math. es Természettud.,' 1885, vol. 3, p. 54 (Budapest).

HARRISON AND SONS, Printers in Ordinary to His late Majesty, St. Martin's Lane.

PART II

ELECTRON THEORY
OF METALS

INTRODUCTION

by

J. RUD NIELSEN

Niels Bohr's work on surface tension was published in a world language and became known immediately to physicists working in the field. On the other hand, because of the failure of his many efforts to get his doctor's dissertation published in a foreign language, his more original and fundamental work on the electron theory of metals has remained largely unknown. Thus, his important result, found already in his M.S. examination paper, that classical statistical physics cannot explain the magnetic properties of matter, is often attributed to Miss van Leeuwen who discovered it independently some ten years later*.

1. M. SC. EXAMINATION PAPER (1909)

When Bohr's first paper on surface tension appeared in the spring of 1909, he was still a student working toward the master's degree. The examination for this degree consisted of several parts, including the writing of a paper on an assigned topic, for which six weeks were allowed. The topic assigned was, generally, in a field in which the student had shown special interest, and to which he had given special attention in his studies. Thus, this "store opgave" ("big problem") corresponded roughly to a master's thesis.

On May 4, 1909, Bohr, who had gone to a parsonage in Vissenbjerg (on the island of Funen) to prepare himself for the examination, wrote to his brother Harald** (who, although two years younger, had obtained the degree of *magister scientiarum* in mathematics in April 1909 and had gone to Göttingen to work with the eminent mathematician E. Landau):

"... I shall leave from here in a week, and then I shall start on the big problem. I am very anxious to learn how it will be ..."

---

* J. H. van Leeuwen, Dissertation, Leiden 1919; J. de Physique (6) **2** (1921) 361.
** The letters from Niels to Harald Bohr, together with a few from Niels to his mother, throw considerable light on his early career. They are reproduced in original and in translation as Part III of this volume.

The problem set by Professor C. Christiansen was the following: "Give an account of the application of the electron theory to explain the physical properties of metals".

Bohr was evidently pleased with the problem. On April 26 he had written to his brother:

"... At the moment I am wildly enthusiastic about Lorentz' (Leiden) electron theory ..."

On June 9 he writes:

"... with regard to the examination problem, it is going tolerably, but I must work hard in order not to have to hurry toward the end ..."

He handed in the paper (handwritten by his mother) on June 28. An English translation of it is given on pp. [131]–[161]. On July 1 he wrote to Harald:

"... Now I have fortunately finished all the writing; that is awfully nice, even though I cannot say as a certain magister [i.e., Harald] that I am fully satisfied with the result. In fact, the problem was so very broad that I, with my fluent pen, had to be content with treating only a few fragments of it. But I hope that the examiners will let it pass, for I think I have included a couple of small matters that are not dealt with elsewhere. However, these things are mostly of a negative nature (you know I have the bad habit of believing that I can find mistakes made by others). On the more positive side, I believe that I have given a hint of the reason for the fact, perhaps less well known to you, that alloys do not conduct electricity as well as the pure metals of which they are composed. I am now anxious to learn what Christiansen will say to the whole thing ..."

In the next letter to his brother, dated July 4, Bohr indicates that he plans to continue work on the electron theory of metals. He writes:

"... I don't think that he [Christiansen] will make difficulties. With regard to the paper itself, you got only so brief information in the last letter; however, it is not so easy to explain it in a letter ... I believe that I got something out of it, and I am thinking of writing a little about it some time next autumn."

However, before he could resume work on the electron theory of metals, he had to complete the examinations for the master's degree. He went again into the country in order to prepare himself. On November 7, 1909, he wrote to his brother (who had returned to Copenhagen), asking him to go to a certain official at the university and request that certain dates be fixed for the written and oral examinations; two days later he wrote and asked Harald to send him Professor Zeuthen's lecture notes on the theory of equations. He passed the examinations satisfactorily shortly afterwards and received the degree of *mag.scient*.

2. THE DOCTOR'S DISSERTATION (1911)

When he had obtained the master's degree, Bohr could resume his work on the electron theory of metals. During the work on the six-weeks examination paper, he had studied critically most of the rather extensive literature on the subject, and had found errors or shortcomings in much of the previous work. An exception was the work of Lorentz, but this was based on rather special assumptions about the interaction between the free electrons and the atoms in a metal. Bohr now decided to try to develop a rigorous electron theory of metals, based on as general assumptions as possible.

Some indications of the progress of the work may be gathered from his letters to Harald. On June 25, 1910, he congratulates his brother who, after receiving the Ph.D. degree, had just been appointed "docent" in mathematics in the University of Copenhagen, and adds:

". . . I have taken a few days off, but I shall now work really hard again . . ."

The next day he writes, in part:

". . . I hope you are not too disgusted by the fact that I am so badly able to conceal that my envy is soon growing over the house tops, but this is not so strange after I have spent four months speculating about a silly question about some silly electrons and have succeeded only in writing circa fourteen more or less divergent rough drafts. Nevertheless, I am afraid that I haven't been able either to conceal from the Docent's clever eyes that my mood hasn't been so bad after all; and, again, that is not so strange, for I believe that I have finally licked the silly question about those silly electrons; in any case, I feel this time at least "a round" more certain than the other times. The solution I am referring to is half statistical and half direct and does not involve problems of probability. It will only take up a couple of lines and is so simple that no one, however much he tries, will be able to understand that it has presented any difficulty (unless, what seems inconceivable, another poor idiot should have been sitting and wasting his time on the same silly questions) . . ."

Bohr then mentions that he has got a new portfolio for all his papers and drafts and continues:

"... honestly, I must confess that I don't know if I am most happy over your appointment, over the good behaviour of my electrons at the moment, or over this portfolio; probably, the only answer is that sensations, like cognition, must be arranged in planes that cannot be compared ..."*

On July 5 he writes to his brother:

"... I hope that the writing will go tolerably well for me, but at the moment it is at a standstill. You know that, for certain stupid persons, a little reaction sets in as soon as they feel that at the moment no doubts exist ..."

Three weeks later, he writes a long letter in which he describes a tragi-comical happening ("suddenly my little home-made edifice collapsed"). He knew that he had made an error of calculation, but could not find it. His friend, the mathematician N. E. Nørlund, whose sister, Margrethe, he married in 1912, helped him to track down the error.

"... The result was again that the two constants, which I badly wanted to show were different, actually turned òut to be equal; ... as you will understand, these many adversities haven't exactly furthered the writing and the study of the literature ..."

To escape the distractions of Copenhagen, Bohr again retreated to the parsonage in Vissenbjerg. On November 24 he writes from there to his brother:

"... That you haven't heard from me sooner ... is because the writing has not been going so well ... But now I think that it is beginning to go; only a little bit, you understand, but, of course, I am not spoiled in this respect ..."

However, on January 2, 1911, he is more optimistic:

"... It is going wonderfully well for me; in fact, I finished the two chapters by New Year ... Now I shall start on the introduction ..."

The dissertation was completed, printed, and on April 12, 1911, accepted by the faculty to be defended for the doctor of philosophy degree. Bohr's father had died shortly after the completion of the dissertation, and Bohr dedicated the book to his memory.

The public defense took place on May 13. The mathematician, Professor Poul

---

* This is an allusion to Bohr's early epistemological thoughts, which included an analogy between the ambiguities of language and the representation of a multivalued function on a Riemann surface. See L. Rosenfeld, *Niels Bohr's Contribution to Epistemology*, Physics Today **16**, No. 10 (1963) 47.

Heegaard, and the physicist, Professor C. Christiansen, were the official "opponents". For his opening and closing statements, Bohr had prepared the following notes:

Statements at thesis
defense
Danish text
p. [124]–[125]

Honourable and learned professors and doctors,
ladies and gentlemen!

The dissertation which the faculty of mathematics and natural sciences has permitted me to defend today deals with the electron theory of metals, i.e., with the theory that seeks to explain the peculiar properties of metals by assuming the presence of small electrically charged particles in the interior of the metals.

The foundation of such a theory was laid in pioneering papers by such investigators as Riecke, Drude, J. J. Thomson and H. A. Lorentz. The theory developed by these scientists agrees on many fundamental points remarkably well with experience; thus, this is true of Drude's calculation of the thermal and electric conductivities of metals and also of the theoretical derivation given by Lorentz of the general laws holding for the thermoelectric effects and the phenomocena nnected with heat radiation. Nevertheless, on many essential points there is no agreement between experience and the theoretical results.

In the present work an attempt is made to treat the electron theory of metals on the basis of more general assumptions than those of the authors mentioned, the main purpose being to investigate which results of the theory depend essentially on the special assumptions made, and which results will remain valid when more general assumptions are made. In other words, the aim has been to investigate on what points the electron theory of metals may possess a flexibility of such a kind that a closer agreement with experience can be attained.

I hope that not many misprints have crept into the dissertation to hamper the reading. However, I should like to point out a few that I have found on reading it after it was printed.

On p. 23, line 6, "equation (3)" should be changed to read "equation (4)".

On p. 41, in the right member of equation (26) $(d/dx)N_R(a + \varepsilon\varphi))$ should be changed to $(d/dx)(N_R(a - \varepsilon\varphi))$. The same remark holds for equation (27) on p. 42.

Equation (12b) on p. 85 has the factor $\Gamma^2(2n/(n-1))$. This should be changed to $\Gamma^{-2}(2n/(n-1))$, the minus sign in the exponents having fallen out in the final printing.

[There followed the official "oppositions" of Prof. Poul Heegaard and Prof. C. Christiansen. After thanking them for their "interesting and instructive oppositions", N. Bohr concluded as follows:]

I should like also to express my thanks to Professor Christiansen for the never failing interest he has shown me during my years of study and for the encouragement to scientific work that he has always given me.

In conclusion, permit me to express my best wishes for a rich and fruitful future for the University of Copenhagen.

On May 14 the Copenhagen newspaper POLITIKEN reported the event as follows (in English translation):

Newspaper account of defense

"Yesterday the late Professor Bohr's other son defended his dissertation, Studies on the Electron Theory of Metals, for the degree of doctor of philosophy.

It was the 25-year old mag.scient. Niels Bohr, who after only one hour and a half was able to leave the University as a Dr.phil. Professor Heegaard was his first opponent. He dealt with the linguistic side of the thesis and had nothing but praise for the erudite treatise. Professor Christiansen continued with a more specialized opposition, but it can be called such only in the most figurative sense of the word.

Professor Christiansen spoke in his usual pleasant way, told little anecdotes, and went so far in his respect for Niels Bohr's work as to regret that the book had not been published in a foreign language. Here in Denmark there is hardly anyone well enough informed about the electron theory of metals to be able to judge a dissertation on this subject.

Dr. Bohr, a pale and modest young man, did not take much part in the proceedings, whose short duration is a record. The small Auditorium 3 was filled to overflowing, and people were standing far out in the corridor."

The article included a drawing, reproduced here, of Bohr at the lectern defending his dissertation.

The doctor's dissertation is reproduced in facsimile on pp. [165]–[290] and an English translation of it is given on pp. [291]–[395].

When he had sent copies of his dissertation to various friends, relatives and Danish scientists, Bohr received a number of letters of congratulation. Perhaps the most interesting of these is one from the physical chemist, Niels Bjerrum*, who was then working with Nernst in Berlin. Bjerrum became a close friend of Bohr. His letter, dated April 27, 1911, reads as follows:

---

* Niels Bjerrum, 1879–1958, was docent in chemistry at the University of Copenhagen from 1912 to 1914 and professor at the Danish Agricultural College from 1914 to 1949. From 1939 to 1946 he served as president of the Agricultural College. Noted for pioneering work on the application of thermodynamics to chemical problems, he was one of the first to apply quantum ideas to molecular structure and band spectra.

Niels Bohrs Disputats.

1. Dr. Bohr. 2. Prof. Chrisiansen. 3. Prof. Heegaard.

Det er kun et Aars Tid siden afdøde Professor Bohrs ene Søn, Harald, erhvervede sig den filosofiske Doktorgrad. I Gaar er den lidt yngre Broder, Magister Niels Bohr, fulgt i hans lære Fodspor.

Meningen havde været, at Disputatsen skulde være foregaaet i Anneget; men paa Grund af det amerikanske Professorbesøg var Handlingen i Stedet for tleven henlagt til et bestemt Auditorium paa selve Universitetet. Dette havde til Følge, at Tilhørerne maatte staa langt ud paa Gangen, og det var for saa vidt af det gode, at Opponenterne temmelig hurtig fik sagt, hvad de havde paa Hjerte.

Metallernes Elektrontheori, der er Emne for Nils Bohrs Afhandling, er jo ogsaa et ret kredilt Emne. og den første af Opponenterne,

Professor Heegaard, dvælede i en Kritik af den sproglige Fremstilling væsentlig ved Udenværkerne. Mere ind paa det reel'e kom den anden Opponent, Professor Christiansen, der havde megen Anerkendelse tilovers for Doktorandens Arbejde.

Prof. Christiansen mindede om, at efter H. C Ørsted var Lorentz den Danske, der havde lært bedst hjemme paa det videnskabelige Omraade. Talen var om, og Professoren havde altid plejet at tage ud til Lorentz, naar han i herhen hørende Spørgsmaal vilde have Besked. Siden Lorenz' Tid havde vi ikke havt nogen Fagmand paa det omhandlede Felt, og Opponenten sluttede derfor med at udtale sin Glæde over, at dette Savn nu med Niels Bohr var afhjulpet.

Report of the defense of Niels Bohr's thesis published in the Copenhagen newspaper *Dagbladet*.

Dr. Bohr.

Niels Bohr defending his thesis

(*Politiken*, 14 May 1911)

Dear Mr. Bohr, M.Sc.

Letter from N. Bjerrum
to Niels Bohr (27 Apr 11)
Danish text p. [125]

Many thanks for sending me your dissertation. So far, I have only managed to dip a little into it; however, I have seen that it treats very penetratingly the now so modern electron theory of metals. Down here in Berlin, with Planck, Rubens, and Nernst, electron conceptions are about the only ones that capture the interest. Since, for this reason, I have lately become more familiar with these conceptions, your work interests me all the more. If only the electrons really are in thermal equilibrium with the metal atoms! For, if so, because of the low specific heat of metals, there must be alarmingly few of them. But it is to be hoped that the future will soon clear up this difficulty.

With my heartfelt congratulations to your so early acquired status as a Dr. phil. and with my kindest regards,

yours very sincerely,
Niels Bjerrum

3. EARLY REACTIONS TO THE DISSERTATION FROM CONTINENTAL PHYSICISTS (1911–1912)

Bohr sent copies of his dissertation to the foreign physicists who had contributed to the electron theory of metals, and to some others. Although the dissertation was written in Danish, several of the Continental physicists were able to understand it to some extent, and some correspondence with Bohr resulted, which is reproduced on pp. [397]–[409].

Bohr had also sent a copy of his dissertation to Carl W. Oseen who in 1909, at the age of 30, had become professor in the University of Uppsala, Sweden*. Oseen, who had no language difficulty in reading the dissertation, acknowledged the receipt of it by the following letter:

Källberget, Leksand   5 July 1911

Letter from C. W. Oseen
to Niels Bohr (5 July 11)
Swedish text p. [125]

Dear Doctor,

I have let a rather long time pass without expressing my thanks for your book "On the Electron Theory of Metals". I wanted first to study it more thoroughly than was possible near the end of a semester. Since I have now had an opportunity to do so, I should like to express my pleasure that such a solid work of mathematical physics has appeared in one of our Nordic countries. I wish

---

* Carl Wilhelm Oseen (1879–1944), Ph.D. 1903, docent 1902–1904, associate professor 1904–1909 in the University of Lund; professor of mechanics and mathematical physics in the University of Uppsala 1909–1933; director of the Nobel Institute 1933–1944. Oseen made pioneering contributions to hydrodynamics, in particular by taking the viscosity of liquids into account and developing methods for calculating the resistance to bodies moving in liquids. He also contributed to the theory of anisotropic liquids, electromagnetism, the quantum theory and the theory of relativity.

especially to express my pleasure over the criticism to which you have subjected the flood of papers on the subject in question. Partly with this particular literature in mind, I have long been of the opinion that what mathematical physics is most in need of at present is criticism.

I hope to meet you personally at the congress in Copenhagen.

With highest regards and many thanks
C. W. Oseen.

The congress referred to was the Scandinavian Mathematical Congress held later in the summer of 1911 in Copenhagen. Oseen and Harald Bohr were among the speakers, and a close and lasting friendship was formed on this occasion between Oseen and the Bohr brothers.

In a paper on the electron theory of dispersion and absorption by metals (Ann. d. Phys. **38** (1912) 731), another Swedish physicist, D. Enskog*, refers to Bohr's dissertation as follows:

"Die erste mit befriedigender Strenge durchgeführte Elektronentheorie der Metalle ist die von H. A. Lorentz. . .

Spätere Verfasser haben—mit Ausgangspunkt in Lorentz' Theorie oder sonst—die Elektronentheorie weiter entwickelt. So haben J. J. Thomson, J. H. Jeans und H. A. Wilson die elektrische Stromstärke bei nach einer Sinusfunktion oszillierenden Feldstärke berechnet. Die Ergebnisse dieser Arbeiten sind zum Teil unrichtig. Sie sind korrigiert von Ishiwara (Jun Ishiwara, Proc. Tokyo Math.-Phys. Soc. (2) **6**, p. 56 bis 65, 1911) und Bohr (Niels Bohr, Studie⟨r⟩ over Metallernes El. theorie (Diss.) Kopenhagen 1911). Ishiwaras Abhandlung enthält eine ausführliche Besprechung u.a. der von H. A. Wilson angewandten Methoden, und in der von Bohr findet sich eine sehr reichhaltige und wertvolle Diskussion der auf dem Gebiete erschienenen Arbeiten.

Ishiwara steht auf dem Boden der Lorentzschen Grundannahmen. Bohr verallgemeinert die Theorie insofern, als er ein beliebiges Wechselwirkungsgesetz zulässt, die Geschwindigkeit der Metallmoleküle nicht von vornherein gleich Null setzt, und die Wechselwirkungen der Elektronen, von denen auch keine speziellen Voraussetzungen gemacht sind, in Betrachtung zieht. Die Metallmoleküle müssen jedoch die Voraussetzung erfüllen, dass ihre "Eigenschaften im Mittel in allen Richtungen gleich sind", auch wenn äussere Kräfte wirken. Das bedeutet z.B., dass ihre Geschwindigkeitsverteilung eine völlig symmetrische ist, was eine nicht kleine Beschränkung der Allgemeinheit heisst. Die Beschränkung scheint freilich hier nicht unzulässig zu sein, wie auch nicht die Lorentzsche,

---

* David Enskog (1884–1947), Ph.D. Uppsala 1917, lecturer in Gävle 1918, professor of mathematics and mechanics at the Tekniska Högskolan in Stockholm 1930. He made fundamental contributions to the kinetic theory of gases and to the theory of radioactivity.

dass man von ihren Geschwindigkeiten ganz absehen kann. Andererseits werden unter gewissen Bedingungen die Bohrschen Voraussetzungen auch von den Kräften erfüllt, die zwischen einem bewegten Elektron und einem Dipol oder einem kleinen Magneten wirken, obgleich diese Kräfte ja nicht zentral sind.— Bohr erhält als Grundgleichung eine Integralgleichung, die sich in verschiedenen Fällen nach Fredholms und anderen Methoden behandeln lässt. Die allgemeinen Resultate enthalten gewisse, nicht analytisch formulierte Funktionen, die in einigen bestimmten Fällen berechnet sind. In anderen können aus ihr qualitative Schlüsse gezogen werden."

4. DISCUSSIONS WITH BRITISH PHYSICISTS

EARLY EFFORTS TO GET THE DISSERTATION PUBLISHED IN ENGLISH (1911–1913)

When he had obtained the doctor's degree the 25-year old Niels Bohr received a fellowship from the Carlsberg Foundation for a year's study in England. Although he undoubtedly had sent copies of his dissertation also to some of the British physicists, there is no evidence that any of them had attempted to decipher it and acquaint themselves with its contents. In order to be able to acquaint the British physicists with his work on the electron theory of metals, and in the hope of getting it published in an English journal, Bohr prepared an English translation of his dissertation before leaving Denmark. He was helped by a friend, Carl Christian Lautrup, who had spent some time in England. However, Lautrup knew very little physics, and Bohr's knowledge of English composition left much to be desired (in those days German was the principal foreign language taught in the Danish schools, French and English having less prominent places in the curriculum). The result was, therefore, not very readable and contained such blunders as rendering "electric charge" as "electric loading".

In a letter to Oseen, dated September 6, 1911, Bohr writes*:

"... At the moment nothing is in my mind except that silly translation, which I labour on with all my might in order to get off [to Cambridge] as soon as possible ..."

He was finally able to leave in late September, but had to spend the first days in Cambridge writing in formulas.

Now began his lengthy efforts to obtain the reaction of the British physicists to the results of his dissertation and to get it published in the English language. To begin with he was full of optimism. Thus, on September 29 he wrote to his brother:

* The complete Danish text and English translation of this letter is given on p. [126].

"Oh Harald! Things are going so well for me. I have just been talking to J. J. Thomson and have explained to him, as well as I could, my ideas about radiation, magnetism, etc. If you only knew what it meant to me to talk to such a man . . . I do believe that he thought there was some sense to what I said. He is now going to read the book . . . I am also glad that I have published those papers [on surface tension] in the Royal Society. But, above all, I am so unspeakably happy and thankful that my paper [i.e., the translation of the dissertation] is finished . . . and that I could give it to Thomson . . ."

Bohr first explored the possibility of getting the dissertation published by the Royal Society. However, in a letter dated October 16, 1911, J. Larmor informed him as follows:

Letter from J. Larmor to Niels Bohr (16 Oct 11)

"It is possible that the Publication Committee of the Royal Society might accept for publication in their Proceedings an abstract of *new matter* of length 5 or 6 pages if the whole is not printed elsewhere. But the rules strictly exclude expository or controversial matter, in fact everything except contributions to new knowledge—unless in special cases when it relates to mistakes in their own publications. Thus, I think there is no prospect of their accepting your interesting dissertation if it were presented."

Bohr also found out very soon that it was difficult to get Thomson's attention. On October 23 he wrote to Harald:

". . . Thomson has so far not been as easy to deal with as I thought the first day. He is an excellent man, incredibly clever and full of imagination (you should hear one of his elementary lectures) and extremely friendly; but he is so immensely busy with so many things, and he is so absorbed in his work that it is very difficult to get to talk to him. He has not yet had time to read my paper, and I do not know yet if he will accept my criticism. He has only talked to me about it a few times for a couple of minutes, and only about a single point, namely about my criticism of his calculation of the absorption of heat rays. You may remember that I pointed out that, in his calculation of the absorption (contrary to the emission), he does not take into account the time taken by the collisions and, therefore, finds a value for the ratio between emission and absorption that is of the wrong order of magnitude for small periods of vibration. Thomson first said that he couldn't see that the duration of the collisions could have such a large influence upon the absorption. I tried to explain it to him and gave him the next day a calculation of a very simple example . . . which showed it very clearly. Since then I have only talked with him about it for a moment, about a week ago, and I believe that he feels that my calculation is correct; however, I am not sure but that he thinks that a mechanical model can be found which will explain the law of heat radiation on the basis of the ordinary laws of electromagnetism, something that is obviously impossible, as I have shown indirectly, and as it has, moreover, later been proved directly by McLaren (see below).

As I have written before, Thomson is extremely nice to talk to, and I have been so glad to do it each time; but the trouble is that there is no definite time ⟨when he is available⟩, and one has to disturb him while he is working (he has very little peace); in spite of this, he is very friendly; but when you have talked with him for a moment he gets to think about one of his own things, and then he leaves you in the midst of a sentence (they say that he would walk away from the King, and that means more in England than in Denmark), and then you have the impression that he forgets all about you until the next time you dare to disturb him . . . I have a couple of times tried to talk to Jeans after his lecture (the last time I began with something he had said in the lecture which I didn't think was quite right); he was very friendly, but is very reticent, and every time he said that he would wait ⟨to discuss the dissertation⟩ until my paper has appeared in English.

About the publication of the paper, I don't know anything definite yet; I have talked to Thomson about it a couple of times (I first wanted to wait until he had read it), once two weeks ago and again last Saturday; he has promised to find out if there is a possibility of getting it out in the Transactions of the Cambridge Philosophical Society. It is evidently not possible to get it published by the Royal Society. I have talked with Larmor . . . and he believes that would be impossible, not because it has been published in Danish, but because it contains criticism of the work of others, and the Royal Society considers it an inviolable rule not to accept criticism that does not originate in its own publications.

It would be fine if I could get it out in the Cambridge Transactions; there it would appear without delay, and there begins to be haste about it; since the Danish publication there have already appeared two long papers in Phil. Mag. on the same subject, of course without ⟨the authors'⟩ knowledge of my paper. There is not much sense to one of them which is based on Wilson's results; but the other (by McLaren) is excellent and gives a result that is more general than one of mine. He has calculated the ratio between absorption and emission for all periods of vibration by considering the case in which the electrons are assumed to move independently of one another in a stationary field of force: a result that is hardly of more than critical interest, since, as I mention, it is only possible to obtain agreement with experiments for long periods of vibration . . ."

Bohr now tells about the lectures he attends and complains that the work in the laboratory (on the advice of Thomson, he worked on a problem dealing with positive rays) leaves him too little time to read; then he continues:

". . . besides, for the present I have a good deal of work with my own paper; I have met an extremely nice young man (Mr. Owen) at the laboratory, and he has been kind enough to offer to look it over for me; he does it very thoroughly, but has had so little time that we have almost come to a standstill, and, if we don't soon get going again, I must try to get it done in some other way . . .''

A week later Bohr writes to his mother:

"Things are going very well; not that I have heard from Thomson yet, but I am in excellent spirits and have many plans . . . Friday I go to Manchester to visit Lorain Smith [a colleague and friend of his father] . . .''

In Manchester Bohr was introduced to Rutherford. He was greatly impressed and inquired about the possibility of spending the next term in Rutherford's laboratory. On January 28, 1912, he writes to his mother and brother:

"A little greeting to let you know that I got a very kind letter from Rutherford today. He writes that it would suit him well that I come to Manchester next term . . .''

Meanwhile, Bohr sought contact with McLaren, in whose paper he had been so much interested. He visited him in Birmingham and later wrote to him; this letter is reproduced on pp. [432]–[434], together with a letter to Oseen on similar topics.

Thomson apparently never read Bohr's dissertation. In March 1912 Bohr went to Manchester and enrolled in an experimental course on radioactivity.

On November 13, 1911, Bohr had spoken on the electron theory of metals at a meeting of the Cambridge Philosophical Society. The announcement of this meeting, and the notes he had prepared for the lecture, are reproduced on pp. [412]–[419]. At the same time he had submitted the translation of his disser-tation in the hope of getting it published by the Society. On February 5, 1912, he wrote to his brother:

". . . I haven't written about the Philosophical Society, for the judging committee hadn't reported before the last meeting; but there is another meeting on February 12, and then I hope to hear from them. I would be glad to get my paper published soon . . .''

In early May he finally received the following letter:

Letter from E. L. Barnes to Niels Bohr (7 May 12)

Trinity College, Cambridge
May 7, 1912

Dear Dr. Bohr

The Council of the Cambridge Philosophical Society yesterday considered the reports of the referees on the paper which you presented on November 13th last. The expense of printing so long a paper is one which the Society cannot undertake. They understand, however, that without materially reducing the value of the paper you could cut it down to one-half its length. If you care to attempt this task, and to send it to us again, I think that it would be accepted.

I will send the paper to you when you inform me of your present address.

Sincerely yours,
E. L. Barnes

Bohr immediately replied:

Manchester, May 8, 1912

Letter from Niels Bohr
to E. L. Barnes (8 May 12)

Dear Dr. Barnes:

Thank you for your kind letter of May 7, in which you inform ⟨me⟩ about the decision of the Council of the Cambridge Philosophical Society regarding my paper. I shall consider the proposal stated in it.

Yours faithfully,
N. Bohr

My address is Hulme Hall, Victoria Park, Manchester.

On May 19 Bohr writes to his brother:

"You have heard what has happened to my paper ... I didn't know quite what to do with it and had, therefore, (in order not to make too much unnecessary fuss) first thought a little about seeing if I could easily shorten it myself, and if I would like that. To make the matter quite clear to you, I am sending you copies of the letter of the secretary and of my answer. I do not yet know definitely what to do. I haven't had time yet to make a serious effort to shorten it, but I believe now that that is what must be done; in fact, I am not sure that it won't be better (easier to read) thereby; and it would require too much work (and work that I dare not undertake now) to make it so complete that there would be sense in trying to get somebody to take it as a book. It is clear to me that what has to be done must be done very soon ..."

On May 27 he writes again to Harald:

"... With regard to the paper, I haven't had any time yet, and I am afraid that I won't have it very soon if I am to do some real work in the laboratory. I am not quite sure either what I want to do with it. I am thinking a little about the possibility of getting it out in complete form. I shall let you know about it soon, when I have found if it is possible. Today I would really like to have it published in full, for I believe that I have found out something, which, if it is true (and, as far as I can see, nobody can claim the opposite before more experiments are made

[107]

(some that I am thinking of doing next year together with Owen, if no one has done them in the meantime)), will remove all the main objections that can be raised (and have lately been raised) against an electron theory of the kind I have treated; and, indeed, if that is the case, the value of my work will be a little different from what it is now considered to be.

Dear Harald, you know how easily I can be wrong, and it is perhaps also silly to tell about such things so soon, but I wanted so much to talk to you tonight, for here I have no one who is really interested in such things. If there is some sense in it, I shall see how soon I can get it written up and get it out, but I have hardly time to really concentrate on such things and get the necessary literature read when I am at the laboratory all day, which is absolutely necessary. You also ask about the work in the laboratory. It is really going quite well. Unfortunately, I must say right off that I am not yet sure how much will come of what Rutherford has put me on . . . But, however it goes, I learn a lot every day, since it is real work, and since Rutherford is such an excellent man and takes a real interest in the work of all the people working with him . . ."

May 28, 1912

". . . I forgot to drop the letter in the mail box yesterday and open it now to tell you that I believe that I have already had to change my ideas a little. I still think that what I wrote had occurred to me is perhaps of no small general significance (if it should turn out to be in agreement with experiments) and that it can explain certain difficulties of a general nature in the electron theory of metals (such as the fact, incomprehensible on the basis of simple considerations, that the Thomson effect (as you may remember) apparently is of the wrong order of magnitude, and also that the specific heat of metals is not larger at low temperatures (a difficulty that I believe you have heard of)); but whether it will be possible to explain the more special matters, which depend on the particular conditions under which the electrons move in the metal, is something else again. Thus, I am inclined to believe that it may not be possible to explain the high values of the electric conductivity of the highly conducting metals from the special assumptions I used in section 4, i.e., without considering the forces between the electrons which, so to speak, make them move together through the metal. This morning I read a very interesting paper about it by Stark (a very well known man)*. (He doesn't know that the ideas he uses to explain the difference between the alloys and the pure metals are, for the most part, the same that I have used). He gives

---

* [J. Stark (Jahrb. d. Rad. u. El. **9** (1912) 188) assumes that each atom in a metal crystal releases a valence electron and that these electrons form a regular lattice holding the positive atoms together. An electron in a lattice can be displaced only on certain surfaces ("Schubflächen") and only when many other electrons are simultaneously displaced.]

interesting hints of an explanation of the electric conductivity, but he writes that he doesn't see how it can be applied to explain the heat conductivity; but this I have an idea of, and I am already thinking of perhaps trying to write a little about it.

Now I shall not bother you any more with all my nonsense; I must now let a little time pass and settle down with all these various things. For my part, I have no idea of how much I can accomplish this year; it will depend on so many external circumstances, and also on what others find, or have found, to write about on the same subject. I only feel that I am perhaps beginning to get back into the field again.

It has been so nice to write about all this to you; but you will understand that none of it is certain."

In the Niels Bohr Archive is found a one-page note concerning the paper by Stark. It is reproduced in Danish and English on pp. [437]–[438].

In his letter to his brother, dated June 12, Bohr for the first time mentions Rutherford's atom model and writes that he has been doing some theoretical work in connection with it. Then he writes:

"... with regard to the things I wrote about last time, I still believe that they may be of importance (if they are right), but I shall not have time to think of publishing them in the short time I have left here ⟨in Manchester⟩, and I have my work in the laboratory. With regard to my dissertation, as I have mentioned, I am now making a last attempt to get it published here, and if I don't succeed I must publish it myself (I am so glad and thankful that I am in a ⟨financial⟩ position to do so) if no one else will have it (and there is little prospect that anyone will, especially now that Oseen is not well); for I am quite determined to get it out in full form, and that very soon. If possible, I am thinking of trying soon to treat the electron theory from a somewhat different angle, perhaps more in accord with the real conditions (corresponding to Stark's ideas that I wrote about last time), but then I must first have the old things off my hands ..."

In the next letter to his brother, Bohr mentions that "perhaps I have found out a little about the structure of atoms". Toward the end of his four-month stay in Manchester, and after his return to Copenhagen, where in September 1912 he became scientific assistant to Professor Martin Knudsen*, he had become pre-

* Martin Knudsen (1871–1949) was docent in physics at the University of Copenhagen from 1901 until 1912 when he succeeded C. Christiansen as professor. Noted for a series of investigations of gases at low pressure which provided quantitative experimental verifications of the kinetic theory of gases, he also carried out hydrographic research and designed instruments for oceanographic explorations.

occupied with this work, which a year later led to the well-known epoch-making results. However, his efforts to get the dissertation published in a world language continued, at least sporadically, and for some years he occasionally expressed a desire to resume work on the electron theory of metals.

The major difficulty in the way of getting the dissertation published in English was evidently that of finding the time and help to prepare an acceptable translation. For some time Bohr thought that E. A. Owen might be able to provide the needed help by revising the English translation which he brought with him to Cambridge, but, although Owen began to do so, the work apparently did not progress very far and was never completed.

On July 1, 1912, Owen writes to Bohr from Bangor:

Extracts from letters
from E. A. Owen to
Niels Bohr (1912-1914)

"Bring all your troubles with you, and I shall do my best to help you. How is the electron theory of the elements getting along? I am so interested to hear the development & I believe it will be a jolly good thing when it is done . . ."

Some time after Bohr's return to Copenhagen, on February 23, 1913, Owen writes a long letter in which he refers to the problem of the translation:

". . . I had leisure this past week to read the translation of your dissertation on the electron theory of metals. If you remember, I still have the first two chapters here. I hope to let you have them all corrected in a short time & I am very sorry to have been so long over it."

On November 28, 1913, Owen thanks for reprints of Bohr's first two papers "On the Constitution of Atoms and Molecules" (Phil. Mag. **26** (1913) 1, 476) and continues:

". . . I should like to be able to read Danish so that I could plod through your dissertation. When are you going to bring it out in book form? You really ought to do so before very long. I should certainly advise you to take the matter up seriously and decide to bring it out in book form as soon as possible—in English and not in German. If I could be of any help to you I should be glad to give you my service . . ."

Almost a year later, on November 11, 1914, he writes again:

". . . I am so glad to have your book with me, translated into English. I have many times during the past few months wished it by me . . ."

Although these last two letters might seem to indicate that the revision of the English translation had been completed, this is not confirmed by any evidence in the Niels Bohr Archive. In later correspondence with Owen the dissertation is not mentioned.

5. NOTE ON THE ELECTRON THEORY OF THERMOELECTRIC PHENOMENA (1912)

In February 1912, O. W. Richardson, then at Princeton University, published a paper "The Electron Theory of Contact Electromotive Force and Thermo-electricity" (Phil. Mag. **23** (1912) 263). In it, he obtained expressions for the Peltier and Thomson coefficients by considering a cyclic process, in which electrons pass through a circuit consisting in part of metals, and applying the first and second laws of thermodynamics.

In the letter to his brother of February 2, of which a paragraph has already been quoted, Bohr refers to Richardson's paper as follows:

". . . There has appeared a paper by Richardson in the last issue of Phil. Mag., which came Saturday, dealing with the electron theory of thermoelectricity. It was a very interesting paper, but he had not at all got hold of the same as I. Thus, Richardson believes to have given a general derivation of the Peltier and Thomson effects, but his formulas are just Lorentz', and not at all my much more general formulas. It was not so easy to find what was wrong, and I would probably never have done it if I had not known my own work. I have written a little about it to Phil. Mag. . . . I believe it is quite good (Owen has been kind enough to help me), and I should be pleased if they would publish it without any fuss; I would then have had great pleasure of Richardson's paper which, as I said before, is really excellent . . ."

Bohr submitted his note to the Phil. Mag. on February 5, 1912. Having received no word from the editor by February 21, he wrote and asked if the note had been received and if it would be published in the next issue of the Phil. Mag. On February 26 he received the following reply:

Letter from W. Francis to Niels Bohr (26 Feb 12)

Printing Office
Red Lion Court, Fleet Street, E.C.
26 February 1912

Dear Sir,

We have another paper in hand by Professor Richardson which will appear shortly in which he refers to your Dissertation "Studier over Metallernes Elektro-nentheorie" which he had not seen previously. We think it will be advisable to postpone any discussion until this has appeared and you have had an opportunity of seeing it.

Yours sincerely
W. Francis

Dr. N. Bohr.

Expressing his disagreement with a firmness that was not characteristic of him, Bohr replied:

[111]

10 Eltisley Avenue
Cambridge
1 March, 1912

W. Francis, Esq.

Dear Sir,

I beg to thank you for your letter of 26 Feb. I do however not understand why you decline to publish my note. I would therefore be very much obliged to you, if you would kindly let me know, whether you received Prof. Richardson's second paper—mentioned in your letter, and in which he discusses my results—before you received my note; for, otherwise I cannot understand, why you decline to publish my note before or at least at the same time as Prof. Richardson's paper.

Yours faithfully
N. Bohr

On March 25 Francis sent Bohr a rough proof of Richardson's paper with the remark that "His reference to the priority of your work (see p. 594) will perhaps render your note unnecessary".

Bohr, now in Manchester, replied:

Hulme Hall
Victoria Park
Manchester
27 March, 1912

W. Francis, Esq.
Dear Sir,

I thank you for your kind letter of March 25, and for the opportunity you have given me to see Prof. Richardson's paper. I see however that you have misunderstood the intention of my note. It was not at all a question of priority, it was only an attempt to show and explain the disagreement between Prof. Richardson's calculations and mine. Since this disagreement still exists to the same extent (Prof. Richardson's reference to my paper deals with quite other questions), I am sending you my note again with a short post script, and shall be very much obliged to you if you will kindly publish it in the next issue of the Phil. Mag., or otherwise as soon as it is possible.

Yours very truly
Niels Bohr

The note was now accepted for publication in Phil. Mag. On May 16, Bohr thanks for the receipt of the proof and, characteristically, adds:

"I am sorry to have corrected so much in it, but I was afraid that some of the sentences was not easy enough to understand. I shall be glad for a second proof."

Bohr's note was finally published in the June issue (Phil. Mag. **23** (1912) 984). It is reprinted on pp. [440]–[442]. Richardson's second paper, in which he generalized his derivations by dropping the assumption that all the free electrons in a metal have the same potential energy, had appeared a month earlier (Phil. Mag. **23** (1912) 594).

In November 1912 Richardson published a third paper on the electron theory of thermoelectric and thermionic effects (Phil. Mag. **24** (1912) 737). The second section of this paper is intended to meet Bohr's criticism. Richardson begins by saying:

"The difference between my formulæ and those obtained by Dr. Bohr does not appear to arise from any inherent defect in the method which I employed, but from the fact that I neglected to include effects arising from a possible difference in the rate of transference of kinetic energy by an electric current in different materials. My calculations may easily be extended so as to include these effects in the following manner."

Modifying the circuit again, he obtains somewhat more general formulas. He points out that these agree with those obtained by Bohr when the volume of the regions within which the forces acting on the electrons are appreciable is small compared with the volume in which these forces are negligible. He finally concludes:

"The agreement in this case seems strongly to support the validity of the thermodynamic method, and to be subversive of the view that the assumption of reversibility is an arbitrary one."

Bohr did not agree with this conclusion and began to write another note. However, preoccupation with the work on atomic structure prevented him from completing and publishing it. In the Niels Bohr Archive is found an unfinished very rough draft, in his mother's handwriting, evidently dictated to her. It is reproduced here on pp. [443]–[449].

6. LECTURES ON THE ELECTRON THEORY OF METALS (1914)

In 1913 Bohr was appointed "docent" at the University of Copenhagen. His chief duty was to lecture to medical students. However, in the spring of 1914 he gave a series of lectures on the electron theory of metals. On March 3 he wrote to Oseen*:

* The complete Danish text and English translation of this letter is given pp. [127]–[129].

"... In a few days I begin to give a series of lectures on the electron theory of metals. I am looking forward to getting back to this subject and to work on it again; but it has indeed changed extraordinarily much in recent years, and ... the problems have become so extremely complicated. I have several new ideas that I shall try to work out; but I begin the lectures without having any idea about what I shall end up with ..."

The notes which Bohr prepared for these lectures are given, in English translation, on pp. [445]–[471]. They do not reveal clearly what his new ideas were, and he never found time to work them out. In the fall of 1914 he went to the University of Manchester as reader in theoretical physics, replacing Charles G. Darwin who had been called into military service. He remained in Manchester until the autumn of 1916, when a professorship in theoretical physics was created for him in the University of Copenhagen.

In a letter from Manchester, dated March 2, 1915, Bohr asks his brother to talk to a young Danish physicist, Hugo Fricke, who had asked him to suggest a topic for a doctor's dissertation, and writes:

"... Apart from such purely mathematical disciplines as potential theory, hydrodynamics, etc., in which I am not sufficiently well versed at the moment myself, and in which most of the problems that are straightforward and not too difficult are already solved, there is literally nothing that is so well founded that it is possible to say anything definite about it. Thus, for example, the electron theory of metals was an excellent field just a few years ago, but at the moment we know neither how to begin with it nor how to end, since we apparently have no basis whatsoever ... I thought for a while to advise him [Fricke] to try to work out the suggestions I made in my dissertation about magnetism, but there everything is, if possible, even more wild and loose, and everything depends on flair and experience. I hope some time to get back to such things myself, and I should only be too glad if we could help him, but, with me sitting here [in Manchester] and he there [in Copenhagen], it is hopeless even to think about it ..."

7. FIRST ATTEMPT TO GET THE DISSERTATION PUBLISHED IN AMERICA (1914)

At the time when he was planning to lecture on the electron theory of metals, Bohr was encouraged by O. W. Richardson to publish his dissertation in America. On February 10, 1914, he wrote from Copenhagen to Richardson (draft in the handwriting of Mrs. Bohr):

Letter from Niels Bohr
to O. W. Richardson
(12 Feb 14)

"I am so very ashamed not to have written about your kind proposal as to the publication of my paper on the electron theory, but I have been so occupied till now that I have had no time to go through the paper. I intend, however, to do it within the next weeks and shall then let you know about it . . ."

Richardson replied on April 4:

". . . I had a lot of correspondence with Dr. Woodward, the director of the Carnegie Institution in Washington. He agreed to take the question up with Dr. Walcott, the Director of the Smithsonian Institution. That is the shape at which things were when I left America at Christmas. I have not heard from either of them since but I hoped that you had been more fortunate and that one of the two would by now have written to tell you that they were willing to publish it. I am surprised to find them so shy of publishing work which is not American. I thought that they would jump at the chance of getting your paper. The only tangible results of the negotiations so far seems to be that Dr. Woodward has borrowed my copy of your dissertation and has not returned it . . ."

Letter from
O. W. Richardson
to Niels Bohr (4 Apr 14)

On April 9 Bohr wrote (draft in his own handwriting):

"Dear Prof. Richardson,

Letter from Niels Bohr
to O. W. Richardson
(9 Apr 14)

I thank you very much for your kind letter and for all the trouble you have had with the publication of my dissertation. I am this term lecturing on the electron theory of metals and have again started working on this subject. It seems to me to be possible to explain the experimental value of the ratio between the thermal and the electrical conductivities, even if the independent motions of the free electron⟨s⟩ is abandoned, and a view is taken similar to the theory of Stark of electric conduction. I intend to write a paper about it, and I may also make some smaller papers out of my dissertation. As soon as I have decided anything I shall let you know.

I was of course very interested in your paper on the photoelectric effect in the March Phil. Mag.*, and I also read with great pleasure the interesting paper of Langmuir in the Phys. Zeitschr.**

With kindest regards
Yours sincerely
N. Bohr

Bohr did not find time to write the contemplated papers.

* [O. W. Richardson, Phil. Mag. **27** (1914) 476.]
** [I. Langmuir, Phys. Zeitschr. **15** (1914) 348, 516.]

8. CORRESPONDENCE WITH G. H. LIVENS AND O. W. RICHARDSON (1915)

Early in 1915 there appeared a paper by G. H. Livens (Phil. Mag. **29** (1915) 171) which gave rise to a correspondence on the electron theory of metals, reproduced on pp. [473]–[480].

In the autumn of 1915, there occurred a discussion with O. W. Richardson, which gave rise to an extensive correspondence, reproduced on pp. [481]–[491]. A brief account of this discussion will be given here.

On September 23, 1915, Richardson wrote to Bohr, who was now in Manchester:

"... In the Berichte der D.P.G. that I am returning there is an article by W. Schottky [Verh. d. D. Phys. Ges. (1915) 109] attacking some of my results. It seems to me to be mostly nonsense but as others may not think so I have decided to publish a reply to it. I enclose a copy of the reply and should be glad to have your opinion of it, if you have time..."

Bohr replied on September 29 in a letter (draft in Mrs. Bohr's handwriting) which began as follows:

"I have taken very great interest in reading your and Schottky's papers. I must say at once, that so far as I have been able to form an opinion I am afraid that I cannot agree with you in all your conclusions. While I agree that some of Schottky's calculations do not seem correct, I am not convinced that some of his arguments do not touch upon real difficulties in your calculations. It appears to me that reasons may be given in support of Schottky's opinion that the Thomson effect cannot have a simple connection with w ..."

The quantity w is, according to Richardson, the heat of vaporization of an electron, or the change in energy accompanying the transference of an electron from the metal to the surrounding enclosure. Bohr expresses his agreement with Schottky's suggestion that this quantity w may not have the same meaning in two fundamental equations applied by Richardson: in one equation w is simply the difference in the potential energy of a *free* electron inside and outside the metal surface, in the other equation w is the total energy to be transferred to the metal in order to liberate an electron. Bohr suggests that a surface layer, which would greatly affect w, could not be expected to have any effect on the Peltier and Thomson coefficients. He indicates reasons for the fact that Schottky's expressions for these coefficients do not satisfy Kelvin's thermodynamic relations and refers to the calculations of them in his dissertation. He encloses a copy of the English translation of the dissertation and writes:

"... I hope you will not be too horrified by the English translation, which was made before I came to England. It is not my intention to trouble you with reading all the old stuff, I thought only that you might be interested to see the calculations made above ..."

On October 9 Richardson replied. He thanks Bohr for his letter and for the English translation of the dissertation, and writes:

"... The English may not be uniformly good but at any rate I can understand it which is more than I managed to do with the original Danish ..."

Richardson goes into considerable detail to refute Bohr's argument that the Thomson effect cannot depend in a simple way upon w. He admits that there is a small difference between the w's in his two fundamental equations and gives expressions for the Thomson and Peltier coefficients in which the value of w corresponds to a displacement under equilibrium conditions, as in the second of his equations.

However, Bohr was not convinced, and on October 16 he wrote another letter to Richardson. He repeats in different form his criticism of Richardson's calculations and his suggestions for the failure of Schottky's expressions for the Thomson and Peltier coefficients to satisfy the Kelvin relations. He emphasizes again that he does not understand the meaning of one of Richardson's fundamental equations:

"... it appears to me that the fact that any liberation of electrons leave⟨s⟩ the metal charged, while v in (5) refer⟨s⟩ to a state where the metal is uncharged, disturbs essentially the analogy with the evaporation of a liquid and makes the correctness of the simple application of the entropy principle doubtful ..."

No reply to this letter is extant. Richardson did not publish his proposed rejoinder to Schottky. In some of his later papers he uses different symbols for the w's entering in his two fundamental equations and discusses the relation between them and the correction required in the expression for the Thomson coefficient (see, e.g., Proc. Roy. Soc. **A105** (1924) 387).

9. LAST ATTEMPT TO GET THE DISSERTATION PUBLISHED IN AMERICA (1920)

Five years after the discussion with Richardson, and undoubtedly as a result of his having received the English translation, a prospect developed again of getting Bohr's dissertation published in America. On April 28, 1920, E. P. Adams wrote from Princeton:

Letter from E. P. Adams
to Niels Bohr (28 Apr 20)
"The National Research Council is desirous of publishing a translation of your memoir on the Electron Theory of Metallic Conduction for the reason that it is fundamental in this field and so as to make it more widely available. I remember Richardson saying that you had a translation of it which you have never published, and I wonder whether you would be willing to let us have a copy of it which we could publish. It might be that you would prefer to revise the original and publish a shorter paper. This would be wholly agreeable.

While I do not believe in a free electron theory of metallic conduction, I do think it important to have as much knowledge as possible about what such a theory will accomplish, and your paper seems to be the most complete investigation that there has been; and your methods should be more widely known."

Bohr replied on June 12, 1920. (He now had a secretary, and a carbon copy of the typed letter is available.) After expressing his pleasure and asking Adams to convey his thanks to the Council, he wrote:

Letter from Niels Bohr
to E. P. Adams
(12 June 20)
". . . Concerning the question of possible alterations in the memoir I may say, that I quite agree with you in your view as regards the difficulty in believing for the moment in a theory of metallic conduction, based on assumptions as those underlying the calculation in the paper in question; and I understand, that it is also the view of the National Research Council, that the eventual interest of a publication essentially lies in the method of treatment used. I should therefore prefer to publish a translation without any alterations and perhaps in a short preface briefly to mention and discuss the problems, which have arisen since the original publication and the new points of view, suggested by them, and I shall be very glad to hear from you about the view of the Council and yourself as regards this point.

The copy of the translation, which I have lent to Richardson, is the only one, I possess, and I should like to have a look at the language, which may be very bad, the translation being made before my first visit to England, I am writing to Richardson with the same post in order to have it back for a perusal, before I send it to you . . ."

Bohr's letter to Richardson, of the same date, reads in part:

Letter from Niels Bohr
to O. W. Richardson
(12 June 20)
"A few days ago I received from Professor Adams in Princeton a very kind letter with the information that the National Res. Council was willing to publish a translation of my old paper on the electron theory of metallic conduction. Although the publication may not have much interest at present, I was very glad and thankful for the information about the decision of the Council, which I no doubt principally owe to your kind interest in the paper, Adams telling me

in his letter that he had spoken about the paper with you. As regards the translation itself I possess no other copy than that, I sent to you when I was in England, and at ⟨your⟩ convenience I should therefore be glad to have it back in order to look through it again before sending it to Adams . . .''

Richardson replied from London on June 19, in part:

"I am glad to hear that the U.S. National Research Council are going to publish a translation of your calculations on the electron theory of metallic conduction . . . I will forward you the copy of your translation on my return home . . .''

Letter from
O. W. Richardson
to Niels Bohr
(19 June 20)

Before having received the translation, Adams wrote to Bohr on July 8, 1920:

". . . I think there will be very little delay in publishing ⟨it⟩ after it is received.

Letter from E. P. Adams
to Niels Bohr (8 July 20)

We would be very glad to have you add a preface, as you suggest, to discuss the new problems and points of view that have arisen since the original was published. And I quite agree with you as to the advisability of publishing the original memoir in full.

If I can be of any assistance as far as the language is concerned, do not hesitate to ask me.''

This was evidently Bohr's last attempt to get his dissertation published in a foreign language. It also failed. The tasks of revising the old English translation and writing a preface to it were too heavy for Bohr at a time when he was preoccupied with the building of an Institute of Theoretical Physics in Copenhagen and with the development of his atomic theory, in particular with the classification of the electron orbits in all the chemical elements and the explanation of the periodic system. There is no evidence in his files of even a first draft of the intended preface, nor of any work on the translation.

10. A BOOK BY PAUL SUTER ON THE ELECTRON THEORY OF METALS (1920)

At about this time a book *Die Elektronentheorie der Metalle mit besonderer Berücksichtigung der Theorie von Bohr und der galvanomagnetischen und thermomagnetischen Erscheinungen* by Dr. Paul Suter* was published in Switzerland (Paul Haupt, Akad. Buchhandlung vorm. Max Drechsel, Bern, 1920).

* Born July 24, 1881, in Kölliken, Kanton Aargau, Switzerland, Paul Suter taught at Kulm, Aargau, from 1901 to 1910. He entered the University of Bern in 1910 and from 1912 to 1916 served as assistant at the physical institute. At the suggestion of Professor P. Gruner, he wrote a prize-winning essay on the electron theory of metals, which was expanded into a doctor's dissertation (Bern 1917). From 1916 to 1952 he taught physics at the Städtisches Gymnasium of Bern. Since his retirement he lives in Schwarzenburg.

This book was published without Bohr's knowledge. His attention was called to it by the following letter:

Letter from F. Hoffmann
to Niels Bohr (14 Apr 21)

14/4 1921.

Sehr geehrter Herr Prof.! Durch die Arbeit von P. SUTER "Die Elektronentheorie der Metalle" (Bern 1920) aufs neue aufmerksam gemacht auf Ihre wertvolle Inaug. Diss. Kopenhagen 1911, habe ich den Wunsch, Ihre Darlegungen genauer zu studieren. Da die Arbeit hier nur schwer zugänglich ist, würden Sie mich zu grossem Dank verpflichten, wenn Sie mir noch ein Exemplar davon zukommen lassen könnten.

Zu Gegendiensten jederzeit gern bereit, zeichne ich mit dem Ausdruck vorzüglicher Hochachtung.

Ihr sehr ergebener
Fr. Hoffmann

Physikal. Technischen Reichsanstalt III
Charlottenburg, Werner Linnensstr. 8/12

Bohr replied (undated rough draft in the handwriting of H. A. Kramers):

Letter from Niels Bohr
to F. Hoffmann (undated)

Sehr geehrter Herr Professor!

Ich habe eben ein Exemplar meiner Kopenhagener Dissertation an Sie geschickt. Ich sollte sehr dankbar sein, wenn Sie mir mit ein Paar Worte erzählen könnten, ob die erwähnte mir unbekannte Arbeit von Herrn Suter eine Dissertation oder ein gewöhnliches Buch ist, damit ich versuchen kann, diese Arbeit mir zu verschaffen.

Mit vorzüglicher Hochachtung
N. Bohr

There is no record of any further correspondence with Hoffmann, nor, strange as it might seem, of any correspondence whatever between Suter and Bohr.

Suter's 114-page book consists of three parts. In the first part is given a survey of the historical development of the electron theory of metals from the early work of Weber to the more recent attempts by Jaffé, Wien, and others, to apply the quantum theory to account for the properties of metals. The second part deals with the derivation of the general statistical equations for the motion of the free electrons in metals. This part is essentially a translation of the first three sections of Chapter I of Bohr's dissertation, which is reproduced almost paragraph for paragraph. Most of the digressions in the dissertation (usually printed with small type) are omitted, and intermediate steps in the derivation of several of Bohr's formulas are given, making the presentation very readable.

However, Suter commits a serious error on pp. 26–27. In deriving the change in the collective momentum of the electrons caused by an external electric field, he is evidently unable to understand Bohr's rather subtle argument and employs instead a procedure first used by Debye (and criticized by Bohr in a footnote on p. 18 of his dissertation), with the result that in some of his formulas the term containing the charge of the electron has the wrong sign. Since Bohr denotes this charge by ε, and Suter by $-e$, their formulas become identical in appearance, and Suter is apparently unaware of the discrepancy.

Suter does not treat any of the problems in which the dimensions of the metal atoms are not assumed to be negligible compared to their mutual separations (Bohr, Chapter I § 4), nor does he discuss any of the problems treated in Chapters II and III of Bohr's dissertation.

The third, and longest, part of Suter's book is devoted to the galvano- and thermomagnetic phenomena. The theory of Gans, based on the work of Lorentz, is first discussed. Several of the formulas derived here are found also in Bohr's examination paper of 1909. Then Chapter IV § 2 of Bohr's dissertation is presented. Again e is identified with ε, without comment, and some intermediate steps in the derivation of the formulas are added. The theories of Livens and Corbino are next discussed, and the formulas derived in the various theories are collected in a table. Experimental data by other investigators are then presented in 15 tables and compared with the predictions of the theories. Finally, a 9-page bibliography is given, which the author believes to be essentially complete (up to 1918) as far as theoretical papers are concerned.

11. LAST CORRESPONDENCE ON THE ELECTRON THEORY OF METALS (1928)

By the time he learned about Suter's book, Bohr was no longer actively concerned with the electron theory of metals. Some seven years later occurred what is believed to be his last correspondence on this subject. He received the following letter:

Mundip House, Headington Hill, Oxford,
England, 19th February 1928

Letter from
W. Hume-Rothery
to Niels Bohr
(19 Feb 28)

Dear Sir,

I hope you will excuse my taking the liberty of writing to you about the following matter, but Dr. N. V. Sidgwick of Lincoln College Oxford, whom I think you know, has advised me to write to you.

I am working on the properties of metallic substances, and in connection with this work, I am writing a book on metals and the various theories of their ultimate structure. In one section of this book I am discussing the electronic theories, and the different ways in which the fundamental equations for the conductivity are deduced. I have found many references to a paper by yourself called

"Studier over Metallernes Elektrontheori" (Copenhagen, 1911).

This paper is, I think, in Danish, and unfortunately I do not read this language. In one of his most interesting papers in the Philosophical Magazine, Mr. G. H. Livens refers to an English translation of this paper, and I am writing to ask whether you can by any chance tell me where an English translation is to be found? I am also able to read French and German if by any chance you have published the paper in either of these languages. If you happen to know of any one in England who has an English copy which might be lent me for a short time I should be most grateful, as I think that I am right in saying that the paper is not to be found in any of the English Libraries.

I am sending you a copy of a recent paper of my own in the hope that it may interest you.

With apologies for troubling you,

Yours very Truly
W. Hume-Rothery

Professor Niels Bohr, F.R.S.

Bohr replied:

February 28. 1928

Letter from Niels Bohr
to W. Hume-Rothery
(28 Feb 28)

Dear Dr. Hume-Rothery,

I thank you for your kind letter and the reprint of your most interesting paper. It is a very long time indeed since I have occupied myself actively with the metallic problem, and although I once took great trouble in preparing an English translation, which was never published, I am sorry to say that I do not know for the moment where any of the few copies are to be found. In the meanwhile an abstract of some of the calculations was published in German by Dr. Paul Suter "Die Elektronentheorie der Metalle", Bern 1920.

Nowadays the old theories based on the classical mechanics can hardly make claim of actual physical interest. Indeed they are left quite behind by the recent fundamental work of Sommerfeld which has just been published in Zeitschrift für Physik*. Although not yet complete, Sommerfeld's work surely means a

* [A. Sommerfeld, Z. Phys. **47** (1928) 1.]

decisive step as regards the adequate quantum theoretical treatment of the metallic problem.

With kind regards, also to Dr. Sidgwick,

Yours sincerely,
N. Bohr

Many years afterwards, when once reminded of the electron theory of metals, Bohr remarked: "I really did a lot of hard work on that subject".

DANISH AND SWEDISH TEXTS OF QUOTED DOCUMENTS

Statements at thesis defense

13 Maj 1911.

Højtærede og højlærde Professorer og Doktorer,
højtærede Forsamling.

Den Afhandling, som jeg af det mathematisk-naturvidenskabelige Fakultet har faaet Tilladelse til i Dag at forsvare for den filosofiske Doktorgrad, omhandler Metallernes Elektrontheori, d.v.s. den Theori der søger at forklare Metallernes særlige Egenskaber derigennem, at den antager Tilstedeværelsen af smaa elektrisk ladede, bevægelige Partikler – de saakaldte fri Elektroner – i Metallernes Indre.

Grundlaget for en saadan Theori er givet gennem banebrydende Arbejder af Forskere som Riecke, Drude, J. J. Thomsen og H. A. Lorentz. Den af disse Forskere udviklede Theori stemmer paa en Række fundamentale Punkter udmærket overens med Erfaringen; dette gælder saaledes Drudes Beregning af Metallernes Varmeledningsevne og deres Ledningsevne for Elektricitet, samt den af Lorentz givne theoretiske Udledning af de almindelige Love, der gælder for de thermoelektriske Fænomener og de til Varmestraalingen knyttede Fænomener. Der er imidlertid paa mange væsentlige Punkter ikke Overensstemmelse mellem Erfaringen og Theoriens Resultater.

I den foreliggende Afhandling er forsøgt at give en Behandling af Metallernes Elektrontheori ud fra mere almindelige Antagelser end de, de nævnte Forskere har lagt til Grund, for derigennem at undersøge hvilke Resultater, der er særlig knyttede til de benyttede specielle Antagelser og hvilke Resultater, der vil gælde uforandret overfor mere almindelige Antagelser; med andre Ord at undersøge paa hvilke Punkter i Metallernes Elektronteori, der er Variationsmuligheder til Stede, og Variationsmuligheder af en saadan Art, at man derigennem kan opnaa en bedre Overenstemmelse med Erfaringen.

Jeg haaber, at der i Afhandlingen ikke ,har indsneget sig ret mange Trykfejl, der har kunnet virke forstyrrende under Læsningen. Jeg vil imidlertid gerne gøre opmærksom paa nogle enkelte, som jeg har fundet ved en senere Gennemlæsning. Paa Side 23 Linje 6 foroven staar "man faar derfor ved Hjælp af Ligningen (3)", der skulde staa "ved Hjælp af Ligningen (4)".

Side 41 Ligning (26) staar paa højre Side af Lighedstegnet "$(d/dx)(N_R(a + \varepsilon\varphi))$"
der skulde staa $(d/dx)(N_R(a - \varepsilon\varphi))$"; samme Bemærkning gælder om Ligning (27)
Side 42.

Side 85 Ligning (12b) staar "$\Gamma^2(2n/(n - 1))$", der skulde staa "$\Gamma^{-2}(2n/(n - 1))$".
Minustegnet i Eksponenten er faldet bort under Rentrykningen.

Jeg vil gerne takke den højtærede Opponent for hans velvillige og lærerige
Opposition.

Jeg vil gerne takke den højtærede Opponent for hans interessante og belærende
Opposition. Jeg vil gerne hertil knytte en Tak til Professor Christiansen for den
aldrig svigtende Interesse, han under min Studietid stedse har vist mig og for den
Opmuntring til videnskabeligt Arbejde, han altid har ydet mig.

Til Slut vil jeg gerne udtale Ønsket om en rig og frugtbar Fremtid for Køben-
havns Universitet.

Letter from N. Bjerrum to Niels Bohr (27 April 11)

27-4-1911

Kære Hr. Magister Bohr.

Mange Tak for Tilsendelsen af Deres Disputats. Jeg har endnu kun faaet
kigget i den, men har dog set, at der højst indtrængende behandles den nu saa
moderne Elektronteori for Metallerne. Hernede i Berlin hos Planck, Rubens og
Nernst er Elektronforestillingerne snart de eneste, der fanger Interessen. Da jeg
derved i den senere Tid er ført mere ind paa Elektronforestillinger, interesserer
Deres Arbejde mig saa meget desmere. Naar nu bare Elektronerne virkelig er i
Varmeligevægt med Metalatomerne! Thi i saa Fald maa der jo paa Grund af Metallernes ringe Varmefylde være uhyggelig faa af dem. Men denne Vanskelighed vil
Fremtiden forhaabentlig snart klare. Med min hjærteligste Lykønskning til Deres
saa tidlig erhvervede Værdighed som Dr. phil. og en venlig Hilsen fra Deres heng.

Niels Bjerrum

Letter from C. W. Oseen to Niels Bohr (5 July 11)

Herr Doktor!

Jag har låtit ganska lång tid gå utan att ha uttalat mitt tack för doktorns arbete:
Studier over metallernes elektronteori. Jag ville först ägna avhandlingen ett mera
noggrannt studium än vad som är möjligt under slutet av en termin. Sedan jag
nu haft tillfälle till ett sådant studium, ber jag att få uttala min glädje att ett så
gediget matematiskt-fysiskt arbete sett dagen i ett av våra nordiska länder. Alldeles
särskilt ber jag att få uttala min glädje över den kritik, som doktorn givit av
avhandlingsströmmen på det ifrågavarande området. Delvis med tanke på just denna

litteratur har jag ganska länge varit av den meningen, att vad den mat. fysiken nu mäst behöver är kritik.

Jag hoppas få träffa doktorn personligen på kongressen i Köpenhamn.

Med utmärkt högaktning
och med tacksamhet
C. W. Oseen

Källberget, Leksand den 5. Juli 1911.

Letter from Niels Bohr to C. W. Oseen (6 Sept 11)

København 6-9-11.

Kære Ven,

Jeg ved ikke hvordan jeg skal takke Dig for Dit Brev og for al Din Venlighed imod mig; jeg kan ikke sige hvor glad og taknemlig jeg er over Din Interesse for mit Arbejde. Jeg glæder mig saa meget til at vi i Fremtiden med hverandre skal af- handle de Sager, der interesserer os begge, og at jeg kan ty til Dit store Kendskab. I Øjeblikket har jeg intet i mit Hovede uden den dumme Oversættelse, som jeg holder paa med af alle Kræfter for saa snart som muligt at komme af Sted; og sender derfor kun denne Gang de venligste Hilsner fra min Forlovede, min Broder og mig selv. Jeg skriver fra England saa snart jeg har noget at berette.

Din hengivne
Niels Bohr

Translation

Copenhagen 6-9-11

Dear friend:

I don't know how to thank you for your letter and for all your kindness to me; I cannot say how glad and grateful I am for your interest in my work. I am looking forward very much to future discussions with you of the matters which interest both of us, and to taking advantage of your great knowledge. At the moment nothing is in my mind except that silly translation which I labour on with all my might, in order to get off [to Cambridge] as soon as possible; hence, for now I only send the kindest regards from my fiancée, my brother and myself. I shall write from England as soon as I have something to tell.

Sincerely yours,
Niels Bohr

Letter from Niels Bohr to C. W. Oseen (3 Mar 14)

96 Øster Søgade, København,
3-3-1914.

Kære Ven!

Jeg er meget skamfuld over saa længe ikke at have svaret paa Dit Brev der glædede og interesserede mig saa meget. Jeg har meget ofte tænkt paa at skrive, men jeg har haft saa overmaade travlt at jeg ikke har faaet Tid.

Lige i Aften har jeg faaet "Annalen" med Din Afhandling, og har straks studeret den saa godt jeg kunde. Jeg kan ikke sige hvor interessant jeg finder den og hvor imponeret jeg er over den Lethed hvormed Du behandler et saa vanskeligt mathematisk Problem. Resultatet forekommer mig overordentlig betydningsfuld, og Kritikken af Planck's Teori uafviselig. Jeg vilde saa gerne engang rigtig høre Din Mening om de Antagelser hvorpaa jeg har bygget. Saa vidt jeg kan se behøver de ikke at være i Modstrid med Antagelsen af Maxwell's Ligninger i det tomme Rum, da jeg jo netop i Modsætning til Planck antager at Emission og Absorption følges fuldkommen ad. For at opnaa indbyrdes Sammenhæng forekommer det mig nødvendig at bryde langt skarpere med den sædvanlige Mekanik end Planck vil gøre, og for Eks. at antage at Systemet i de stationære Tilstande hverken emitterer eller absorberer.

Uden at forandre væsentligt paa mit Synspunkt, har jeg søgt at gøre mine Betragtninger noget klarere ved at bytte om paa Ordningen af mine Argumenter. Den Tankegang som jeg har fulgt vil Du kunne se af den lille Afhandling som jeg sender indlagt (det er blot et halv-populært Foredrag som jeg holdt lige før Jul i den fysiske Forening i København). Ved nøjagtig at følge den Tankegang som er fremsat deri er det muligt at udsige i hvert Tilfælde noget om Stark-effekten. Jeg har skrevet en Afhandling derom og om Zeemaneffekten, som jeg haaber vil udkomme i Phil. Mag. om nogle Dage; det er dog kun et Forsøg paa at vise hvorledes mine Forestillinger muligt tilbyder et Grundlag for Udviklingen af en Teori for disse Fænomener. Jeg tænker at Starks Opdagelse vil føre til en stor Udvidelse af vort Kendskab til Straalingsproblemet, og det synes jo ogsaa at et større Antal Fysikere arbejder saavel teoretisk som eksperimentelt derpaa. Sammen med Dr. Hansen har jeg begyndt at gøre nogle Forsøg der skulde prøve Rigtigheden af mine Forestillinger paa enkelte Punkter; men vi har hidtil ingen Fremgang haft, idet vi begge kun har haft meget liden Tid og kun smaa Penge-midler og ingen Assistance. Til min Stilling hører der nemlig intet Laboratorium, thi det Professor Knudsen havde, da han var Docent, blev lagt over til Professoratet, da jeg blev ansat. Jeg har kun det Hverv at undervise de medicinske Studerende i Fysik og har intet at gøre med den videnskabelige Undervisning i Fysik, og har saaledes ikke Mulighed for at faa Elever eller Assistance. Jeg arbejder derfor paa at faa oprettet en Lærerpost i Theoretisk Fysik (Du husker maaske at der efter min Doktordisputats var Tale om en saadan) men der er vist ikke meget Haab om at det vil lykkes idet Fakultetet stadig ikke ønsker Oprettelsen af en saadan Stilling.

Jeg begynder om nogle Dage at holde en Række Forelæsninger over Metallernes Elektronteori. Jeg glæder mig meget til at komme tilbage til dette Emne og til at arbejde derpaa igen, men det har jo rigtignok forandret sig overordentligt i de sidste Aar og intet synes at staa fast og hele Spørgsmaalene or blevet saa uhyre indviklede. Jeg har forskellige nye Forestillinger, som jeg vil prøve at udarbejde, men jeg begynder Forelæsningerne uden at ane med hvilket Resultat jeg vil slutte.

Jeg haaber saa meget snart at træffe Dig igen og rigtig at tale med Dig om det altsammen. Dersom jeg faar Tid, kommer jeg maaske et lille Besøg til Sverige.

Med de venligste Hilsner fra min Hustru og

Din Ven
Niels Bohr

Translation

96 Øster Søgade, Copenhagen
3-3-1914

Dear friend:

I am very ashamed for not having answered earlier your letter which pleased and interested me so much. I have often thought of writing, but I have been so extremely busy that I haven't found the time.

Just tonight I got the "Annalen" with your paper, and I have immediately studied it as well as I could. I cannot tell you how interesting I find it, and how impressed I am by the ease with which you treat such a difficult mathematical problem. The result appears extremely significant to me, and your criticism of Planck's theory cannot be refuted. I would like once to really hear your opinion of the assumptions I have built on. As far as I can see, they are not necessarily contrary to the assumption that Maxwell's equations are valid in empty space, since—unlike Planck—I assume that emission and absorption go together quite closely. To obtain internal correlation it seems to me necessary to break with the usual mechanics much more radically than Planck is willing to do, and, e.g., to assume that a system neither emits nor absorbs in its stationary states.

Without changing my viewpoint essentially, I have tried to make my considerations somewhat clearer by interchanging the order of my arguments. The idea I have followed will appear to you from the little paper I am enclosing (it is merely a semi-popular lecture which I gave just before Christmas at the Physical

Society of Copenhagen). By following closely the train of ideas set forth in it, it is possible, in any case, to say something about the Stark effect. I have written a paper about it and about the Zeeman effect which I hope will appear in the Phil. Mag. in a few days; however, it is only an attempt to show how my considerations may offer a basis for developing a theory of these phenomena. I believe that Stark's discovery will lead to a great extension of our understanding of the problem of radiation, and, in fact, it seems that a large number of physicists are working on this effect, both theoretically and experimentally. Together with Dr. Hansen, I have started some experiments which should test the validity of my conceptions on certain points, but so far we have not made any progress; for both of us have had very little time, and we have had very little money and no assistance. In fact there is no laboratory attached to my position; for the laboratory which Professor Knudsen had when he was assistant professor was transferred to the professorship when I was appointed. My only job is to teach the medical students, and I have nothing to do with the scientific teaching of physics and, hence, no possibility of getting students or assistance. I, therefore, work at getting a teaching position in theoretical physics established (you may remember that there was talk about it after my dissertation), but there seems to be little hope that this will happen, since the faculty still doesn't want such a position.

In a few days I begin to give a series of lectures on the electron theory of metals. I am looking forward to getting back to this subject and to work on it again; but it has indeed changed extraordinarily much in recent years, and nothing seems to be certain, and the problems have become so extremely complicated. I have several new ideas that I shall try to work out; but I begin the lectures without having any idea about what I shall end up with.

I hope very much to meet you again soon and really talk to you about it all. If I find time I may make a little visit to Sweden.

With kindest regards from my wife and

your friend
Niels Bohr

I. M.Sc. EXAMINATION PAPER

(TRANSLATION *)

Problem:

Give an account of the application of the electron theory to explain the physical properties of metals

Paper submitted by N. Bohr, 28 June, 1909

* [See Introduction, sect. 1.]

[131]

INTRODUCTION

Among the substances which conduct electricity, the metals occupy a special position, not only because of their high conductivity but also in that the passage of electricity through them is not accompanied by observable chemical reactions, as in the case of most other good conductors.

According to the generally accepted views on the spatial distribution of electricity, this is explained by the ability of certain small electrically charged particles (the electrons) to move from one chemical atom to another in the interior of metals. In fact, attempts have been made to explain all the various properties that differentiate metals from electrolytes and insulators by the motion of such "free electrons" in the interior of the metals. Apart from the electric conductivity, the most important of these properties are the high thermal conductivity, whose value for the different metals is markedly related to their electric conductivity, the thermoelectric properties, the galvanomagnetic and thermomagnetic effects, and finally the strong absorption and emission of light and heat rays.

To explain these phenomena, two essentially different theories have been proposed which can be differentiated by the role ascribed to the molecules of a metal. According to the former theory, which has been developed especially by Riecke, Drude and Lorentz, the electrons are assumed to move perfectly freely in approximately straight paths among the molecules which play no other important role than to provide the geometrical limits to the electron paths. According to the second theory, which has been proposed by J. J. Thomson, electrons are assumed to escape from a metal molecule only under special circumstances and to move immediately to another molecule and be captured by it. Hence, in this theory the molecules of a metal play a major role in the explanation of the various phenomena. However, while the former theory has reached a high degree of mathematical completeness, especially through the work of Lorentz, the latter theory has only been briefly sketched. In what follows we shall, therefore, only deal with the former theory.

GENERAL THEORY OF THE MOTION OF ELECTRONS IN A METAL

It will be assumed that electrons move perfectly freely among the molecules of a metal much as gas molecules are thought to move in the pores of a porous medium. That only negligibly few electrons move out into the space surrounding the metal at ordinary temperatures is explained by the assumption that the molecules exert strong attractive forces upon them. While these forces are believed to cancel one another in the interior of the metal, they will have a resultant at the surface directed into the metal; and to escape the electrons must, therefore, possess a certain kinetic energy.

In the interior of a metal the electrons are thought to collide continually, partly with the metallic molecules and partly with one another; in these collisions their velocities undergo sudden changes both in magnitude and in direction. In the absence of external forces, the paths of the electrons are assumed to be made up entirely of small straight segments. In this manner the electrons are thought to partake in the general thermal motion in the interior of the metal, and the high heat conductivity of the metal is explained by their motions.

With regard to the kind of electrons in question, different views have been expressed. In order to explain the observations, the early authors, Riecke and Drude, considered it necessary to assume the simultaneous presence of both negative and positive electrons. J. J. Thomson is the first who, on fundamental grounds, assumes that only a single kind of free electrons is present in metals (*Rapp. du Congrès* (Paris 1900), vol **3**, 138). He bases his assumption on the circumstance that while the presence of a definite kind of negative electron, with a mass more than a thousand times smaller than the masses of chemical atoms, has been demonstrated in many widely different ways, it has never been possible to observe positive electricity except in combination with masses of the same order of magnitude as those of chemical atoms. Moreover, Lorentz (Proc. Acad. Amsterdam **7** (1905) 684) has later shown that a close analysis of the assumption of a stationary distribution of positive as well as negative free electrons in a non-homogeneous piece of metal encounters considerable difficulties of a fundamental nature. In what follows we shall, therefore, assume the presence of only one kind of free electrons and that these are the negatively charged particles commonly known as electrons.

Before entering into the theoretical discussion, we shall briefly mention the assumptions on which the different authors have based their calculations*. In the first place, all of them assume that the dimensions of electrons and also of metallic molecules (or the ranges within which they exert appreciable forces upon each other) are small compared with the distance traversed by electrons between successive collisions. It is assumed further that the number of collisions among the electrons themselves is negligible compared with the number of collisions between

---

* Very recently, Jeans (Phil. Mag. June 1909) has made an attempt to develop the electron theory of metals without introducing any special assumptions. Whether he has succeeded to any great extent is not yet clear, however, since so far (only the first part of Jeans' paper has appeared) only such calculations have been published which treat a single one of the questions to be discussed in the following, and since these calculations, moreover, do not seem to be quite correct, as will be explained in detail later.

electrons and molecules*. With respect to these collisions, Riecke (Ann. d. Phys. **66** (1898) 357) assumes (in any case, these assumptions correspond to his calculations) that the molecules, regarded as immovable, reflect the electrons in a direction that is perfectly independent of the direction of their motion before the collision. Moreover, the electrons are reflected with a definite velocity which (apart from a small temperature correction held undetermined in all calculations) is assumed to be proportional to the square root of the absolute temperature of the metal at the place in question.

As mentioned above, Drude's assumptions about the collisions are not given explicitly; however, if his calculations are to be consistent, they appear to be the same as Riecke's, with the addition that the kinetic energy of the reflected electrons is assumed to be equal to the average kinetic energy of a gas molecule at the same temperature (Ann. d. Phys. **1** (1900) 571). According to the kinetic theory of gases, this added assumption about the electronic kinetic energy, with the help of which Drude succeeded in calculating theoretically the ratio between the electric and thermal conductivities of metals in good agreement with experience, is necessary if the electrons are to move freely and be in thermal equilibrium with their surroundings. In close connection with Drude's idea, Lorentz (Proc. Acad. Amsterdam **7** (1905) 438) has finally extended the theory by taking into account that, in order to be in perfect heat equilibrium, the free electrons in a metal not subject to external influence must be assumed to have velocities of all magnitudes, distributed according to Maxwell's law. The treatment of Lorentz, which will be taken as a basis in what follows, differs from those of the above mentioned authors also in its application of the so-called statistical method used in the kinetic theory of gases. This has the advantage that the theoretical calculations can be carried out more rigorously, while at the same time becoming more lucid.

---

* Riecke (Jahrb. d. Rad. u. El. **3** (1906) 26) believes that the discrepancy between his and Drude's results arises from the fact that Drude only takes collisions among the electrons into account, while he himself only considers collisions between electrons and metallic molecules. To this it may be pointed out, however, that Drude's calculations only seem to be justified if the collisions mentioned (but not definitely specified) are regarded as collisions between electrons and molecules. The effective mean free path occurring in Drude's formulas would otherwise not be the same in the expressions for the thermal and electric conductivities. In fact, since the motion of the center of mass remains unchanged in a collision between two like electrons, such collisions can well influence the heat conductivity but not the electric conductivity (Lorentz, Proc. Acad. Amsterdam **7** (1905) 449). As will be discussed in more detail, the small differences in the numerical coefficients occurring in the expressions for the ratio between the electric and thermal conductivities calculated by the two authors appear to arise from the fact that the theoretical calculations by both authors are not in all respects so rigorous that these numerical coefficients can be said to be perfectly accurately determined from the assumptions made.

To keep account of the motion of the electrons, two Cartesian coordinate systems are used. In one of these the position of an electron at a given time is specified in the usual manner by coordinates (x, y, z). In the other coordinate system the velocity of each electron is represented by a point whose coordinates (ξ, η, ζ) are put equal to the components along the axes of the former system of the velocity of the electron at the given instant. The number of electrons found at time t in the volume element dS and having velocities in the velocity element $d\lambda$ (which number is assumed to be very large) will be designated by

$$f(x, y, z, \xi, \eta, \zeta, t)dS\,d\lambda,$$

where (x, y, z) is a point in the element dS and (ξ, η, ζ) a point in $d\lambda$.

To find an equation to determine f, we consider the change of f with time. It will change for several reasons; for one thing, electrons will enter and leave the volume element dS as a result of their motion; in addition, velocity points will enter and leave $d\lambda$ because of the action of external forces and as a result of collisions between electrons and molecules in the metal. Now, the fundamental formula developed in the kinetic theory of gases (Boltzmann, *Gastheorie*, vol. **1**, p. 114) which takes all of these circumstances into account may be written

$$\frac{\partial f}{\partial t} + \xi \frac{\partial f}{\partial x} + \eta \frac{\partial f}{\partial y} + \zeta \frac{\partial f}{\partial z} + X \frac{\partial f}{\partial \xi} + Y \frac{\partial f}{\partial \eta} + Z \frac{\partial f}{\partial \zeta} = Q.$$

Here X, Y, Z denote the external forces acting upon unit mass, while $Q\,dS\,d\lambda\,dt$ represents the increment in the number of electrons whose velocities are in $d\lambda$ resulting from the collisions occurring in the volume element dS in the time dt.

Lorentz now calculates Q from the assumption that during collisions the electrons and the molecules of the metal act upon each other as elastic spheres, one of which (the molecule) is immovable. Lorentz assumes further that the electrons in a piece of metal not subject to external influence will have the same velocities as the molecules of an ideal gas in thermal equilibrium at the same temperature T. This means that the velocity distribution is expressed by the equation $f = A_T e^{-hr^2}$ where r is the absolute value of the velocity of the electrons $(r^2 = \xi^2 + \eta^2 + \zeta^2)$. The constants A_T and h are related as follows to the number of electrons in unit volume, N_T, the average of the square of the velocity, u^2, and the absolute temperature, T,

$$N_T = A_T \left(\frac{\pi}{h}\right)^{\frac{3}{2}}, \qquad u^2 = \frac{3}{2h}, \qquad \tfrac{1}{2}mu^2 = \alpha T, \qquad h = \frac{3m}{4\alpha T},$$

where α is a constant which is the same for all gases (and m is the mass of an electron).

When the temperature and condition of the metal vary from point to point, Lorentz assumes that the velocity distribution deviates only very little from that stated above (T then denotes the absolute temperature at the place in question), and he shows then how this deviation can be found with the aid of the above-mentioned fundamental formula.

However, before we enter upon these calculations, we shall consider more closely the assumptions made by Lorentz. In fact, his picture seems to suffer from certain, at least formal, flaws in that the law according to which the electrons and the molecules of the metal are assumed to interact does not produce any thermal equilibrium. Thus, one might ask how the electrons come to have different velocities at different places in the metal depending on the temperature there; for, they collide only with molecules and, according to the law on which all the calculations are based, these can neither give energy to the electrons nor take energy away from them. Hence, it may be of interest to show that the Lorentz theory can be developed without any change in the calculations, in case of the applications to be discussed here, even if a somewhat different picture, not suffering from these flaws, is taken as a basis.

It will be assumed that the molecules of the metal have such a complex structure (are systems of many particles) that not only the direction of motion of the electrons but also their absolute velocity after a collision is entirely independent of their motion before collision; the velocities with which electrons are reflected is assumed to depend only on the internal state of the molecule, i.e., on its temperature, defined as the temperature of the metal at the place in question. In a homogeneous piece of metal, not subject to external influences, it seems necessary, if the electrons move freely in thermal equilibrium with the molecules, to assume with Lorentz that their velocities are distributed according to Maxwell's law. In the time dt every molecule in such a metal will be hit by electrons of all possible velocities, and these velocities will be distributed according to this perfectly definite law; in the same time interval the molecule will, if the condition is stationary, reflect electrons with the same velocity distribution. According to what is assumed above, if the condition is not homogeneous, the molecule will reflect the electrons that hit it with exactly the same velocity distribution as in the former case, if only the temperature is the same at the place considered. (If the molecule thereby gives up more energy than it receives, that only means that the temperature of the metal decreases at this place.)

We shall now show how Q can be determined from this assumption. Since the molecules of the metal do not move, the mean free path between successive collisions, here called l, will be independent of the electron velocity. The number of electrons with velocities in dλ colliding with molecules in the volume element dS in the time dt will be

$$(r/l)f\,\mathrm{d}S\,\mathrm{d}\lambda\,\mathrm{d}t.$$

If $f = Ae^{-hr^2}$, the total number of collisions P occurring in the time $\mathrm{d}t$ in the volume element $\mathrm{d}S$ will be

$$P = \mathrm{d}S\,\mathrm{d}t\int\frac{r}{l}Ae^{-hr^2}\mathrm{d}\lambda = \mathrm{d}S\,\mathrm{d}t\,\frac{A}{l}\int_0^\infty re^{-hr^2}4\pi r^2\,\mathrm{d}r = \frac{2\pi A}{h^2 l}\,\mathrm{d}S\,\mathrm{d}t.$$

The number of electrons with velocities in $\mathrm{d}\lambda$ which in the same time collide with and, hence, are reflected from the molecules in $\mathrm{d}S$ are

$$(r/l)Ae^{-hr^2}\mathrm{d}S\,\mathrm{d}\lambda\,\mathrm{d}t = P(h^2/2\pi)re^{-hr^2}\mathrm{d}\lambda.$$

In case of a different velocity distribution, one has

$$P = \mathrm{d}S\,\mathrm{d}t\int\frac{r}{l}f\,\mathrm{d}\lambda.$$

Now, Lorentz assumes further that the number of electrons in unit volume of a homogeneous piece of metal is a definite function of the temperature, the function denoted by N_T above. We must assume, therefore, that the molecules emit or absorb electrons when their temperature changes. This can be expressed simply by putting the number of electrons ejected by the molecules equal to

$$P + \frac{\partial N_T}{\partial T}\frac{\partial T}{\partial t}\,\mathrm{d}S\,\mathrm{d}t = \left(\frac{1}{l}\int rf\,\mathrm{d}\lambda + \frac{\partial N_T}{\partial T}\frac{\partial T}{\partial t}\right)\mathrm{d}S\,\mathrm{d}t.$$

Thus we obtain*

$$Q\,\mathrm{d}S\,\mathrm{d}t = \left\{\left(\frac{1}{l}\int rf\,\mathrm{d}\lambda + \frac{\partial N_T}{\partial T}\frac{\partial T}{\partial t}\right)\frac{h^2}{2\pi}re^{-hr^2} - \frac{r}{l}f\right\}\mathrm{d}S\,\mathrm{d}\lambda\,\mathrm{d}t$$

* [There is evidently an error here. Bohr apparently assumed that the electrons reflected or ejected by the molecules have the Maxwellian velocity distribution. The factor to the parenthesis in the right member of this equation should, therefore, be

$$Ae^{-hr^2}\left/\int Ae^{-hr^2}\mathrm{d}\lambda\right. = \left(\frac{h}{\pi}\right)^{\frac{3}{2}}e^{-hr^2},$$

rather than $(h^2/2\pi)re^{-hr^2}$, and this leads to

$$f = \frac{2}{(\pi h)^{\frac{1}{2}}r}Ae^{-hr^2} + \frac{\xi l}{r}\left(2hAX - \frac{\partial A}{\partial x} + r^2 A\frac{\partial h}{\partial x}\right)e^{-hr^2},$$

rather than the expression given in the text and found by Lorentz on the assumption that the collisions are as those of elastic spheres. The error has undoubtedly arisen because Bohr has overlooked the fact that the expression $P(h^2/2\pi)re^{-hr^2}\mathrm{d}\lambda$ for the number of electrons with velocities in $\mathrm{d}\lambda$ which collide in the volume element $\mathrm{d}S$ in the time $\mathrm{d}t$ is valid only when the velocity distribution before collisions is Maxwellian.]

[137]

The fundamental formula, therefore, becomes

$$\frac{\partial f}{\partial t} + \xi \frac{\partial f}{\partial x} + \eta \frac{\partial f}{\partial y} + \zeta \frac{\partial f}{\partial z} + X \frac{\partial f}{\partial \xi} + Y \frac{\partial f}{\partial \eta} + Z \frac{\partial f}{\partial \zeta}$$
$$= \frac{r}{l} \left\{ \left(\int r f \, d\lambda + l \frac{\partial N_T}{\partial T} \frac{\partial T}{\partial t} \right) \frac{h^2}{2\pi} e^{-hr^2} - f \right\},$$

where the integration on the right is to be carried out over all velocity points*.

This equation seems to be derived rigorously from certain assumptions, among which the most crucial are that the electrons move perfectly freely and that the external forces do not influence the internal state of the molecules of the metal. The justification of these two assumptions may be debatable, but, if they are retained, the results appear to be the same, no matter what additional assumptions are adopted**.

---

* The expression for Q which follows from the picture of Lorentz can be written

$$Q = \frac{r}{l} \left(\frac{1}{4\pi} \int f \, d\omega - f \right),$$

where $d\omega$ is an element of solid angle and where the integration is to be carried out over a sphere of radius r. The correctness of this expression, which differs from that used by Lorentz only by its somewhat simpler form, can be seen in the following way. Since, as mentioned above, the speed of the electrons remains unchanged after collision, while their direction of motion is entirely independent of the direction before collision, the effect of collisions will be that electrons whose velocity points before collision lie in a spherical shell with radii r and $r + dr$, and whose number is

$$\frac{r}{l} \int f r^2 \, dr \, d\omega \cdot dS \, dt = \frac{r^3}{l} \int f \, d\omega \cdot dS \, dr \, dt,$$

after collision will have their velocity points evenly distributed over this same spherical shell. Since the volume of the shell is $4\pi r^2 \, dr$, the number of electrons whose velocity points are brought into $d\lambda$ in the collisions will be

$$\frac{r}{l} \frac{1}{4\pi} \int f \, d\omega \cdot dS \, d\lambda \, dt.$$

** However, we must mention here an attempt to extend the theory of Lorentz made by Gruner (Ber. d. D. phys. Ges. **10** (1908) 509) who apparently retains the fundamental assumptions discussed but arrives at a somewhat different result. In fact, Gruner attempts to explain the observed deviations from Lorentz' theory by assuming that the electrons under certain circumstances can be bound temporarily by molecules of the metal and only be liberated again by impact of other electrons. However, as is seen immediately, these circumstances are included in Lorentz' theory (at least in the form in which it has been presented here), for no assumption has been made as to how long the individual electron stays inside a molecule, provided only that the average number of free electrons is kept constant. The different result is, therefore, caused entirely by certain additional assumptions of a mechanical nature, introduced to keep account of the effects of the collisions. Gruner divides the molecules into two kinds, those that have bound no electrons (positive mole-

From the equation thus derived one can now determine f. In the applications which here will be discussed first, the condition will be stationary; moreover, the condition of the metal will be assumed to be the same everywhere in any plane perpendicular to the x-axis, and the electrons are acted on only by forces parallel to this axis. We now put $f = Ae^{-hr^2} + \varphi$, where φ is assumed to be very small compared to the first term, as will be shown to be permissible. Hence, when f is substituted in the fundamental formula, the terms on the left side arising from φ can be neglected compared to those arising from Ae^{-hr^2}. The latter terms become

$$\xi \left(-2hAX + \frac{\partial A}{\partial x} - r^2 A \frac{\partial h}{\partial x} \right) e^{-hr^2}.$$

With respect to the terms on the right, it follows immediately from the derivation of the formula that the terms arising from Ae^{-hr^2} will vanish and that only terms coming from φ will remain. Lorentz now shows that the equation is satisfied by $\varphi = \xi \psi(r)$. In fact, if this expression is substituted on the right side of the equation, the integral vanishes because of the symmetry, and only $-(r/l)\varphi$ remains. Thus we obtain

$$f = Ae^{-hr^2} + l \left(2hAX - \frac{\partial A}{\partial x} + r^2 A \frac{\partial h}{\partial x} \right) \frac{\xi}{r} e^{-hr^2}.$$

Lorentz now investigates the conditions that φ actually is small compared to Ae^{-hr^2}, as assumed. It is seen that this is the case if the quantities $2lXh$, $lA^{-1}(\partial A/\partial X)$, and $lu^2(\partial h/\partial x) = \frac{3}{2}lh^{-1}(\partial h/\partial x)$, are small. The first quantity will be small if

---

cules) and those that have bound one electron (neutral molecules). Next, Gruner assumes that, in collisions with positive molecules, all electrons with a speed greater than a certain definite speed G will be reflected according to the law assumed by Lorentz (collisions between elastic spheres), while all electrons with speed lower than G will be bound, thus making the molecule neutral. In collisions with neutral molecules, electrons with speed less than G will be reflected according to the law of Lorentz, while in collisions with electrons of speed higher than G electrons will be liberated. All this can be fully subsumed in the picture used here. However, a difference exists in regard to the assumptions made about the velocities of the impinging and liberated electrons in collisions of the last-mentioned kind. In fact, Gruner assumes that the impinging electron continues in its path with unchanged speed and direction, while the liberated electron is ejected in such a manner that, on the average, just as many electrons are liberated with each partic- ular speed and direction as are bound of electrons with the same speed and direction. As may be seen immediately, the consequence of these assumptions, which appear very improbable, are that the influence of those collisions which, according to the basic idea, must be regarded as most important, i.e., collisions of slow electrons with positive molecules and fast electrons with negative molecules, is just the same as if these collisions did not occur at all (for neither the speed nor the direction of the electrons are changed). If the author had assumed (as he does for the other, less violent, collisions) that electrons after these particularly violent collisions are sent out equally in all directions, all his formulas would have been exactly the same as those of Lorentz.

the energy acquired by electrons on traversing their free path under the influence of the external forces is small compared to their kinetic energy $\frac{1}{2}mu^2 = 3m/4h$. The last two quantities will be small if the condition of the metal varies only little between two points a distance l apart. In the applications to be made of the theory, these conditions can always be assumed satisfied, as long as a homogeneous piece of metal is considered. If one is dealing with pieces of different metals in contact, as in the case of the thermoelectric effects, the equations can also be applied if it is assumed that the properties of the metals do not vary discontinuously from one metal to the other, but that such a gradual transition occurs that the properties vary very little over the extremely small distance specified by the mean free path of the electrons.

Now, the quantities to be determined are the amounts of electricity i and heat W carried by the motion of the electrons in unit time through a unit area perpendicular to the x-axis. If the electrons have the charge e and mass m, these quantities are

$$i = e \int \xi f \, d\lambda \quad \text{and} \quad W = \frac{1}{2}m \int \xi r^2 f \, d\lambda.$$

In these integrals terms arising from the term Ae^{-hr^2} in the expression for f will vanish because of the symmetry. The terms arising from φ will contain a factor ξ^2. However, since the average value of ξ^2 over a sphere of radius r is $\frac{1}{3}r^2$, the integrations can be performed by replacing ξ^2 by $\frac{1}{3}r^2$ and $d\lambda$ by $4\pi r^2 dr$. Carrying out the calculations, one obtains

$$i = \tfrac{2}{3}\pi e l \left(\frac{1}{h} 2AX - \frac{1}{h^2} \frac{\partial A}{\partial x} + \frac{2}{h^3} A \frac{\partial h}{\partial x} \right)$$

and

$$W = \tfrac{2}{3}\pi m l \left(\frac{1}{h^2} 2AX - \frac{1}{h^3} \frac{\partial A}{\partial x} + \frac{3}{h^4} A \frac{\partial h}{\partial x} \right).$$

The acting forces X are assumed to arise in part from the electric field E produced by free electric charges as well outside as inside the metal; also, the electrons are assumed to possess a potential V relative to the molecules of the metal arising from attractive forces from these. However, since these forces, as already mentioned, are believed to cancel in the interior of a homogeneous piece of metal, V is constant in such a metal. On the other hand, if the metal is not homogeneous, V is assumed to have a definite value at each place determined by the nature and temperature of the metal at the place in question. The force acting on unit mass will, therefore, be

$$X = \frac{e}{m} E - \frac{1}{m} \frac{\partial V}{\partial x}.$$

[140]

With the velocity distribution obtained, the number of electrons per unit volume is

$$N = \int f \, d\lambda = \left(\frac{\pi}{h}\right)^{\frac{3}{2}} A.$$

The relation between the electric field and the free charge gives

$$\frac{\partial E}{\partial x} = 4\pi e(N - N_T) = 4\pi e \left(\frac{\pi}{h}\right)^{\frac{3}{2}} (A - A_T).$$

If such values as must be assumed for e and N (e = ca. 3×10^{-10} absolute electrostatic units; N = ca. 10^{23}) are substituted, it is seen that in all ordinary cases (when surface charges can be neglected) $N - N_T$ can be considered negligible compared to N_T; hence, we shall assume in what follows that $N = N_T$, i.e., that everywhere N and A depend only on the nature of the metal and the temperature.

If the above expression for X is substituted in the expressions found for i and W, the result is

$$i = \tfrac{2}{3}\pi e l \left[\frac{2}{h} \left(\frac{e}{m} E - \frac{1}{m} \frac{\partial V}{\partial x} \right) A - \frac{1}{h^2} \frac{\partial A}{\partial x} + \frac{2}{h^2} A \frac{\partial h}{\partial x} \right] \tag{a}$$

$$W = \tfrac{2}{3}\pi m l \left[\frac{2}{h^2} \left(\frac{e}{m} E - \frac{1}{m} \frac{\partial V}{\partial x} \right) A - \frac{1}{h^3} \frac{\partial A}{\partial x} + \frac{3}{h^4} A \frac{\partial h}{\partial x} \right]. \tag{b}$$

Now, Lorentz has shown that almost all known facts connected with the conduction of electricity and heat by metals and their thermoelectric properties can be explained, at least qualitatively, from these equations. Lorentz (loc. cit. p. 444) is, therefore, inclined to ascribe general significance to these equations, independently of the basic assumptions from which they were derived, provided only that an appropriate value, dependent upon the nature and temperature of the metal, is assigned to l. However, if we should attempt to go a step further and introduce in all the preceding calculations the assumption that l depends on the absolute velocity of the electron, we should arrive at the following equations

$$i = \tfrac{2}{3}\pi e \left[\frac{2L_1}{h} \left(\frac{e}{m} E - \frac{1}{m} \frac{\partial V}{\partial x} \right) A - \frac{L_1}{h^2} \frac{\partial A}{\partial x} + \frac{L_2}{h^3} A \frac{\partial h}{\partial x} \right]$$

and

$$W = \tfrac{1}{3}\pi m \left[\frac{2L_2}{h^2} \left(\frac{e}{m} E - \frac{1}{m} \frac{\partial V}{\partial x} \right) A - \frac{L_2}{h^3} \frac{\partial A}{\partial x} + \frac{L_3}{h^4} A \frac{\partial h}{\partial x} \right],$$

where, the mean free path being denoted by $L(hr^2)$,

$$L_1 = \int_0^\infty x L(x) e^{-x} \, dx, \qquad L_2 = \int_0^\infty x^2 L(x) e^{-x} \, dx, \quad \text{and} \quad L_3 = \int_0^\infty x^3 L(x) e^{-x} \, dx.$$

(It may be difficult to ascribe any definite mechanical meaning to this assumption, but it seems likely that it would be equivalent to dropping the assumption that the electrons move quite freely and permitting arbitrary forces between molecules and electrons.)

These equations* might command some interest; for, even when L_1, L_2 and L_3 denote arbitrary functions of the temperature, it can be shown that, like the equations of Lorentz, they satisfy the thermodynamical conditions that the thermoelectric effects are believed to obey. We shall how proceed to discuss applications of the equations (a) and (b).

CONDUCTION OF ELECTRICITY AND HEAT

Let a homogeneous piece of metal, kept at a constant temperature at all of its points, be subjected to a constant electric field E. It will then be traversed by a current whose intensity per unit area i can be found directly from equation (a). Since the derivatives of V, A and h vanish in this case, we obtain

$$ i = \frac{4\pi l A e^2}{3hm} E. $$

Thus, the electric conductivity is $\sigma = (4\pi l A e^2/3hm)$, an expression which, with the aid of the relations on p. [135], can be written

$$ \sigma = \sqrt{\frac{2}{3\pi}} \frac{e^2 N l u}{\alpha T}. $$

According to equations (a) and (b), there is a very close connection between the electric current and the heat motion; thus, as seen from equation (b), in the case mentioned, i.e., isothermal electric conduction, there will occur, in addition to the electric current, a flow of heat in the opposite direction (since e is negative here). Hence, to define the heat conductivity unambiguously, one must state the conditions under which the conduction of heat takes place. In most determinations of the heat conductivity, the experiments are carried out in such a manner that the metal rod investigated is surrounded by electric insulators, at least at one end. In such a rod a kind of electric equilibrium will very rapidly be set up such that the electric current vanishes. In that case $i = 0$, and if we subtract equation (a) multiplied by m/eh from equation (b), we obtain

---

* These equations contain as a special case those found by Gruner. This can also be understood from the preceding discussion; for Gruner's theory differs from Lorentz' only in that a definite average mean free path is arbitrarily assumed for the electrons whose speed is larger than a certain speed G and another for those whose speed is smaller than G (also, the influence of certain collisions is ignored in the calculations).

$$W = \tfrac{2}{3}\pi m l \, \frac{A}{h^4} \frac{\partial h}{\partial x} \, .$$

With the aid of the relations given on p. [135], this expression can be written

$$W = \frac{8}{9} \sqrt{\frac{2}{3\pi}} \, \alpha N l u \, \frac{\partial T}{\partial x} \, .$$

The heat conductivity k is, therefore,

$$k = \frac{8}{9} \sqrt{\frac{2}{3\pi}} \, \alpha N l u.$$

From the expressions found for σ and k, one obtains*

* Riecke and Drude have found

$$\frac{k}{\sigma} = \frac{3}{2} \left(\frac{\alpha}{e}\right)^2 T \quad \text{and} \quad \frac{k}{\sigma} = \frac{4}{3} \left(\frac{\alpha}{e}\right)^2 T,$$

respectively, values which agree better with experience than that calculated by Lorentz (Riecke, Jahrb. d. Rad. u. El. 3 (1906) 33). However, since the calculations of Riecke and Drude were not carried out quite rigorously, it would perhaps be of interest to investigate whether the agreement mentioned is due to the manner of calculation or is a consequence of the basic assumptions of these authors (see p. [133]). To carry out these calculations directly would be very difficult, since the function f in this case will be discontinuous; however, if it is assumed, in accord with the results of Riecke and Drude, that the expressions for the electric and thermal conductivities can be written in the form

$$\sigma_c = K_1 Nl(c\sqrt{T})^{-1} \quad \text{and} \quad k_c = K_2 Nl(c\sqrt{T})^3,$$

where $c\sqrt{T}$ denotes the speed with which the electrons are emitted from the molecules of the metal, it is possible to determine the constants K_1 and K_2 from Lorentz' results in the following way.

According to the theory of Lorentz, the number of electrons with speed between r and $r+\mathrm{d}r$ emitted per second by the molecules in unit volume is

$$\frac{r}{l} A e^{-hr^2} 4\pi r^2 \mathrm{d}r = 4\pi^{-\frac{1}{2}} \frac{N}{l} e^{-hr^2} h^{\frac{3}{2}} r^3 \mathrm{d}r.$$

Putting $r = c\sqrt{T}$, this becomes

$$L_c \mathrm{d}c = 4\pi^{-\frac{1}{2}} \frac{N}{l} e^{-hTc^2} h^{\frac{3}{2}} T^2 c^3 \mathrm{d}c.$$

Since the electrons are not supposed to collide with one another, electrons emitted with different speeds can be regarded as independent. Now, since, according to the Riecke-Drude theory, the number of electrons emitted per second by the molecules in unit volume with speed $c\sqrt{T}$ is $(c\sqrt{T}/l)N$, and since the ratio between this number and the number $Lc\mathrm{d}c$ is the same at different places in the metal (and $hT = 3m/4\alpha$), one obtains

[143]

$$\frac{k}{\sigma} = \frac{8}{9}\left(\frac{\alpha}{e}\right)^2 T.$$

This equation expresses the Wiedemann-Franz law, which states that the ratio between the conductivities for heat and electricity is the same for all metals, and also the law of Lorenz stating that this ratio is proportional to the absolute temperature. The values of the quantities α and e entering in this equation are known only with rather low accuracy. However, as Reinganum (Ann. d. Phys. **2** (1900) 398) has pointed out, the ratio α/e can be determined with great accuracy from known quantities. Thus, if the number of molecules in a cubic centimeter of a gas (e.g. hydrogen) at the temperature T and the pressure p dynes per cm^2 is called N, one has $\frac{2}{3}N\alpha T = p$. Moreover, if E is the amount of electricity that must pass through an electrolyte to liberate one cubic centimeter of hydrogen at the same temperature and pressure, then $E = 2Ne$, and, hence, $\alpha/e = 3p/ET$. In this manner we get (since $T = 273°$, $p = 1.0133 \times 10^6 = 1$ atm, $E = 0.385 \times 10^{-10}$ absolute electrostatic units) $\alpha/e = 0.428 \times 10^{-6}$ and from this $k/\sigma = 0.163 \times 10^{-12} T$.

$$\sigma = \int_0^\infty \frac{\sigma_c}{(cT^{\frac{1}{2}}/l)N} L_c dc = 2\pi^{-\frac{1}{2}} K_1 Nlh^{\frac{1}{2}} = \sqrt{\frac{2}{3\pi}} \frac{e^2 Nlu}{\alpha T},$$

$$K_1 = \frac{e^2 u^2}{3\alpha T}, \qquad K_2 = \frac{\alpha}{3u^2},$$

$$k = \int_0^\infty \frac{k_c}{(cT^{\frac{1}{2}}/l)N} L_c dc = 4\pi^{-\frac{1}{2}} K_2 Nlh^{-\frac{3}{2}} = \frac{8}{9}\sqrt{\frac{2}{3\pi}} \alpha Nlu.$$

Substituting the values for K_1 and K_2 in the expressions for σ_c and k_c and putting $c\sqrt{T} = u$, we find

$$[\sigma] = \frac{e^2 Nlu}{3\alpha T} \quad \text{and} \quad [k] = \tfrac{1}{3}\alpha Nlu.$$

The formula for the electric conductivity thus obtained agrees with Riecke's but not with Drude's; on the other hand, the expression for the thermal conductivity agrees with Drude's but not with Riecke's formula. For the ratio between the two conductivities we find $[k]/[\sigma] = (\alpha/e)^2 T$, which does not agree much better with experience than the ratio found by Lorentz.

It may be mentioned here that Lorentz (Proc. Acad. Amsterdam **5** (1903) 666) has calculated the emission of heat from a metal under the assumption that all electrons move with the same speed. Next, calculating the absorption on the basis of Drude's expression for the electric conductivity, he found a radiation formula which, for the waves in question, agrees perfectly with that found by Planck in an entirely different way. However, since Drude's expression is not correct, it is seen that the assumption that all electrons move with the same speed will not lead to Planck's formula. Hence, it may be of interest to point out that if we calculate the emissivity on the basis of Maxwell's distribution law (which can be done very simply from the results of Lorentz), and calculate the absorption with the aid of the Lorentz formula for the electric conductivity derived in the text, we find again Planck's radiation law.

The data available to compare this result with experience are found mainly in investigations of L. Lorenz (Wied. Ann. **13** (1881) 598) and Jäger and Diesselhorst (Berl. Ber. (1889) 726). The results of these authors are listed in the *table* below. Although the agreement in order of magnitude between the calculated and observed values is remarkably good, it is seen, nevertheless, that there are considerable discrepancies. One might think of several ways to explain this. In the first place, heat might be conducted directly between molecules of a metal, for also non-metals conduct heat to some extent. This would explain that the observed values are higher than the calculated, and in particular the fact that the poorer metallic conductors have the highest values of k/σ. However, Koenigsberger (Phys. Zeitschr. **8** (1907) 237), who made calculations on this assumption, found that, if an appropriate "nonmetallic heat conductivity" of the same order of magnitude as that of the insulators is ascribed to metals, it might be possible to assume that the ratio k/σ would be the same for the various metals, but its value would still be

	Lorenz		Jäger and Diesselhorst	
	$\left(\dfrac{k}{\sigma}\right)_{0°} \times 10^{10}$	$\left(\dfrac{k}{\sigma}\right)_{100°} \Big/ \left(\dfrac{k}{\sigma}\right)_{0°}$	$\left(\dfrac{k}{\sigma}\right)_{18°} \times 10^{10}$	$\left(\dfrac{k}{\sigma}\right)_{100°} \Big/ \left(\dfrac{k}{\sigma}\right)_{18°}$
Calculated (Lorentz)	0.445	1.366	0.474	1.282
Aluminium	0.714	1.307	0.707	1.327
Magnesium	0.717	1.398		
Copper	0.735	1.358	0.750	1.297
Silver			0.762	1.284
Nickel			0.777	1.296
Gold			0.808	1.272
Zinc			0.760	1.290
Cadmium	0.755	1.315	0.784	1.282
Lead	0.759	1.304	0.794	1.308
Tin	0.763	1.334	0.817	1.259
Platinum			0.845	1.345
Palladium			0.828	1.349
Iron	0.749	1.530*	0.911	1.326
Bismuth	0.887	1.372	1.069	1.120
Antimony	0.939	1.294		
Brass (red)	0.729	1.360	0.841	1.262
Brass (yellow)	0.749	1.428		
German silver	0.867	1.314		
Manganin			1.017	1.214
Constantan			1.229	1.184

* [This should evidently be 1.350.]

considerably higher than that calculated by Lorentz; in fact, it would be ca. 0.70×10^{-10} at 18° C. (Koenigsberger surmises that this discrepancy is caused by erroneous assumptions of Lorentz, and among such he mentions that the mean free path is taken to be independent of temperature. However, to this it should be pointed out that the calculations of Lorentz show that, whether or not the mean free path depends on temperature, it does not occur at all in this expression for k/σ.) In this connection, it may be mentioned that Reinganum (Phys. Zeitschr. 7 (1906) 787) has pointed out that there seems to be a definite correlation between the value of k/σ and the atomic weight and magnetic properties of metals.

Finally it may be mentioned that if the general formulas on p. [141] are applied, the result is

$$\frac{k}{\sigma} = \frac{4(L_1 L_3 - L_2^2)}{9 L_1^2} \left(\frac{\alpha}{e}\right)^2 T,$$

an expression which, if no definite assumption is made about the dependence of the mean free path upon the speed of the electrons, gives a very wide latitude for the value of k/σ and its dependence on temperature for the individual metals. However, if it is assumed that the effective mean free path increases with the speed of the electrons (the opposite seems in any case to be inconceivable), then this formula will always give a higher value for k/σ than that of Lorentz. Thus, for example, if the mean free path were directly proportional to the speed, $l = rf(T)$, the result would be

$$\frac{k}{\sigma} = \frac{10}{9} \left(\frac{\alpha}{e}\right)^2 T, \qquad \left(\frac{k}{\sigma}\right)_{18°} = 0.59 \times 10^{-10};$$

if it were assumed proportional to the square of the speed, $l = r^2 f(T)$, one would find

$$\frac{k}{\sigma} = \frac{4}{3} \left(\frac{\alpha}{e}\right)^2 T, \qquad \left(\frac{k}{\sigma}\right)_{18°} = 0.71 \times 10^{-10}.$$

If the effective mean free path were assumed proportional to the nth power of the speed of the electrons, $l = r^n f(T)$, n an integer, the result would be

$$\frac{k}{\sigma} = \frac{2(4+n)}{9} \left(\frac{\alpha}{e}\right)^2 T.$$

As may be seen from the table, alloys do not deviate appreciably more from the Wiedemann-Franz and Lorenz laws than do pure metals. On the other hand, if the electric conductivity is considered alone, there is a considerable difference between alloys and pure metals; for one thing, the conductivity of an alloy is often much lower than those of the metals of which it is composed; moreover, it varies in quite a different manner with temperature. While the conductivity of a pure metal quite generally is inversely proportional to the absolute temperature, and thus is extremely high at very low temperatures, the conductivity of many alloys varies very little over the temperature ranges in which they have been investigated. To explain these circumstances, Lord Rayleigh (Nature **54** (1896) 145)* has advanced a theory according to which there is supposed to be an essential difference between the electric conduction in alloys and pure metals, in that alloys cannot be regarded as physically homogeneous but may consist of a mixture of crystals of the pure metals of which they are composed. Thus, in alloys there should occur, in addition to the ordinary electric resistance, an apparent resistance —a thermoelectric polarization—arising from temperature differences produced by Peltier effects at the interfaces between the crystals. However, this theory does not explain the peculiar temperature independence of the conductivity and does not seem to agree with experience**. Thus, if the resistance were caused by a polarization, it would be expected not to be the same for stationary and for rapidly alternating currents. However, Willows (Phys. Zeitschr. **8** (1907) 173) who investigated this question was unable to find any such difference. Although it might be thought that the frequency used was too small (up to 1000 cycles per second), the experiments of Hagen and Rubens (Ann. d. Phys. **11** (1903) 873 and Ber. d. D. phys. Ges. (1904) 128) on the absorption of heat rays by metals seem to show definitely that the conduction of electricity takes place in the same manner in alloys and pure metals. In fact, these authors have shown that, if the conductivity of metals is calculated from their absorption (emission) of heat rays by means of Maxwell's theory of light, values are found that agree perfectly with those determined with the use of stationary currents. This is just as true for alloys as for pure metals. According to Guertler (Jahrb. d. Rad. u. El. **5** (1908) 17), the conditions under discussion are particularly found in alloys of those metals which are capable of forming mixed crystals in all ratios, i.e., just in those alloys that must be regarded as physically homogeneous. However, it seems possible to explain those conditions on the basis of the ordinary electron theory of metals†. As mentioned

---

* L. Lorenz (Wied. Ann. **13** (1881) 600), by the way, has carried out exactly the same calculations as Lord Rayleigh. However, he was not particularly concerned with alloys, but used his calculations to give a very interesting justification for the Lorenz law for metals in general.

** This does not seem to be the opinion of Lord Rayleigh. However, since the thermal conductivity of pure metals does not depend much on temperature, the apparent resistance should, according to his calculations, be directly proportional to the absolute temperature, as is the resistance of pure metals (Rayleigh, loc. cit. p. 154; Lorenz, loc. cit. p. 601).

† Schenck (Phys. Zeitschr. **8** (1907) 234) has attempted to explain these circumstances by assuming that in alloys, unlike in pure metals, the molecules possess a mobility similar to that heretofore ascribed to the electrons. However (Guertler, loc. cit. p. 72), this seems hardly compatible with the extremely slow diffusion of molecules in alloys. Moreover, it is hard to see how this

above, it must be assumed that the atoms of a metal exert strong attractive forces upon the electrons. If all molecules are alike, these forces will in large measure cancel one another in the interior of the metal, and the electrons will become free to a certain degree. On the other hand, if the metal consists of a mixture of molecules which attract electrons with different forces, these forces will not be able to cancel one another to nearly the same extent, and the electrons will be less free, i.e., have a shorter mean free path more dependent upon the speed of the electrons.

From such a view, which seems able to explain the extremely sharp decrease in the conductivity when a very small amount of another metal is added, the close connection between the electric and thermal conductivities of the alloys can be immediately understood (Schulze (Ann. d. Phys. 9 (1902) 555) has shown that closely analogous irregularities are found in the curves giving the dependence of the electric and thermal conductivities upon the ratio of mixing.)

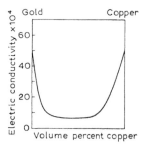

The figure shows a typical case of the dependence of the conductivity upon the ratio of mixing (Guertler, loc. cit. p. 47)

It may be remarked further that from such a view one understands that the ratio k/σ is higher for alloys than for pure metals; for, according to what was stated on p. [146], this is a consequence of the assumption that in alloys the mean free path is more dependent on the speed of the electrons. This great dependence of the mean free path upon the speed can also explain the remarkably low* value of the conductivity of alloys at very low temperatures. (If one would assume that the dissociation of the individual molecules, i.e., the number of free electrons, and their other conditions vary with temperature in the same manner in alloys and pure metals, and if one would assume further that the effective mean free path in the latter is proportional to the square of the speed of the electrons (corresponding to the value of k/σ), then it would seem that the mean free path in alloys with

---

assumption can explain the low conductivity and the peculiar temperature coefficients. (Schenck refers to the fact that the viscosity of two gases—here molecules and electrons—in certain cases can be higher than in either gas—here electrons; however, it seems very difficult to understand how this can help to explain the phenomena.)

* [The manuscript has "high", but this is evidently an error.]

temperature-independent electric conductivity might be expected to be proportional to the fourth power of the speed – for, in pure metals the conductivity is inversely proportional to the temperature, and the square of the speed is directly proportional to the temperature. This would give $k/\sigma = (16/9)(\alpha/e)^2 T$ or $(k/\alpha)_{18°} = 0.95 \times 10^{-10}*$, a value approaching those found for manganin and constantan. The difference can perhaps be explained by a "non-metallic" heat conductivity (see p. [145]) which must be assumed to play a great role in such poorly conducting metals.)

However, in searching for a fuller understanding of the phenomena just mentioned, we must recognize that the electron theory in its present form does not give any explanation of the dependence upon temperature of the conductivity of pure metals. Thus, according to the formula derived on p. [142], the electric conductivity is proportional to $Nl T^{-\frac{1}{2}}$; hence, to explain that it is actually found proportional to T^{-1}, we must assume that the quantity Nl is inversely proportional to the square root of the absolute temperature. (If the mean free path were not regarded as constant, but taken to increase with the velocity of the electrons, it would be necessary to make a corresponding assumption.) However, from the ordinary electron theory one would rather be inclined to assume that Nl would increase with temperature, partly because of an increasing dissociation of the molecules of the metal, and partly because more space becomes available for the free motions of the electrons. This, and the somewhat analogous phenomenon that the electric conductivity decreases markedly on melting for such metals as lead, tin and zinc which expand on melting, has, therefore, been used by J. J. Thomson (*Korpuskular-theorie der Materie*, 1908) to attack the whole theory of free electrons. (J. J. Thomson points out that the decrease in the conductivity in this case cannot be ascribed to a corresponding decrease in the number of free electrons, since this would give rise to a Peltier effect between the solid and the liquid metal of a magnitude that is entirely inconsistent with experience.) However, the phenomena mentioned appear comprehensible to a certain degree on the basis of considerations similar to those advanced in connection with alloys. In fact, it seems that the forces acting in the interior of a metal can cancel more nearly the closer the molecules are to one another. Thus, while in a metal vapor (e.g., mercury vapor which conducts electricity very poorly (J. J. Thomson, loc. cit. p. 48)) we are dealing with separately moving molecules which hold their electrons with great force, we might think that in case of the opposite extreme, the solid metal, the molecules are so closely pressed together that for the most loosely bound electrons (and the metals are distinguished in many ways by rather loosely bound electrons) it is hardly possible to decide to which molecule they belong. From such a point

---

* [See the results given on p. [145].]

of view, it also appears understandable that the conductivity of pure metals increases with increasing external pressure. (The only exception is bismuth, a fact which, however, can probably be ascribed to special circumstances, e.g., crystal structure, since bismuth also in other ways deviates from the other metals; thus, among other things, solid bismuth is the only metal which exhibited marked irregularities in Hagen and Rubens' experiments with heat rays of long wavelength (Ber. d. D. phys. Ges. (1904) 128).) That alloys exhibit a quite different behaviour when compressed (the change in conductivity is much smaller, and not always an increase) is not impossible to understand either, since the conditions must be essentially different when the molecules are not all alike.

In conclusion, it may be pointed out that, even if the electrons are not regarded as perfectly free, there is still a very great difference between the electron theory treated here and the theory developed by J. J. Thomson. This difference can perhaps be characterized by saying that in the present theory the electrons are essentially free, while in Thomson's theory they are essentially bound. Thus, the influence of external forces is explained here as an action upon the electrons themselves, while in Thomson's theory it is described as an action (e.g., a rotation) upon each molecule as a unit.

CONTACT AND THERMOELECTRICITY

It will be briefly discussed here how Lorentz has shown that a complete theory of thermo- and contact-electricity can be derived from equations (a) and (b) on p. [141]. We consider first a thermoelectric chain consisting of two different metals kept at different temperatures through which no current passes (open chain). Putting $E = -\partial\varphi/\partial x$ and substituting $h = (3m/4\alpha T)$, we get in this case

$$\frac{\partial\varphi}{\partial x} = -\frac{1}{e}\frac{\partial V}{\partial x} - \frac{4\alpha}{3e}\frac{\partial T}{\partial x} - \frac{2\alpha}{3e}T\frac{\partial \ln A}{\partial x},$$

whence

$$\varphi_Q - \varphi_P = -\frac{1}{e}(V_Q - V_P) - \frac{4\alpha}{3e}(T_Q - T_P) - \frac{2\alpha}{3e}\int_P^Q T\frac{\partial \ln A}{\partial x}\,\mathrm{d}x.$$

This equation permits several conclusions.

(i) If the temperature is the same throughout the chain, one obtains

$$\varphi_Q - \varphi_P = -\frac{1}{e}(V_Q - V_P) - \frac{2\alpha}{3e}T(\ln A_Q - \ln A_P),$$

an equation which states that the potential difference depends only on the nature of the metals at the end points and that it vanishes if these consist of the same metal. (This is the law expressed in Volta's tension series.)

(ii) If the chain consists of the same metal throughout, the potential difference will depend only on the temperatures at the end points (since in this case A is a function of the temperature only), and the potential difference will be zero if the end points have the same temperature.

(iii) When the chain consists of several metals, and the temperature is different at different places, the potential difference will depend only on the temperatures at the places of contact between different metals, provided that the end points of the chain consist of the same metal and have the same temperature. (The temperature is assumed to be uniform over the short spaces where a transition from one metal to another takes place.) Thus, if the chain consists of two metals I and II, arranged as shown in the figure, and if the temperature at the junction between

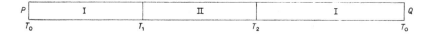

I and II is T_1 and at that between II and I is T_2, while the temperature of the end points, assumed to consist of the metal I, is $T_0 = T_Q = T_P$, we obtain after partial integration

$$F = \varphi_Q - \varphi_P$$
$$= \frac{2\alpha}{3e} \int_P^Q \ln A \frac{\partial T}{\partial x} \, dx = \frac{2\alpha}{3e} \left[\int_{T_0}^{T_1} \ln A_I \, dT + \int_{T_1}^{T_2} \ln A_{II} \, dT + \int_{T_2}^{T_0} \ln A_I \, dT \right]$$

or

$$F = \frac{2\alpha}{3e} \int_{T_1}^{T_2} \ln \left(\frac{A_{II}}{A_I} \right) dT = \frac{2\alpha}{3e} \int_{T_1}^{T_2} \ln \left(\frac{N_{II}}{N_I} \right) dT.$$

We shall now investigate the heat effect that takes place when an electric current passes through such a thermoelectric chain. The temperature is assumed to be kept constant at every point by addition or subtraction of heat from the outside. (Such an addition of heat is strictly not possible if the temperature, as assumed, is to be the same everywhere in planes perpendicular to the x-axis; however, if the chain consists of a thin wire, and if the current is not too strong, the necessary amount of heat can be added without perceptible variation of the temperature over a cross section.) For simplicity, the chain is assumed to have unit cross section. We shall now consider a small element of the chain lying between the two planes determined by coordinates x and $x+dx$. The amount of heat q that must be subtracted from such an element in unit time to keep the temperature constant will be equal to the difference between the energy imparted to the electrons by the external forces and the energy conducted away in the direction of the chain due to the motion of the electrons.

Thus, we obtain

$$q = \left(iE - \frac{\partial}{\partial x}\left(W + \frac{i}{e} V \right) \right) dx.$$

Using equations (a) and (b), eliminating E and V, and substituting the expressions found for the electric and thermal conductivities, we now get

$$q = \left(\frac{i^2}{\sigma} + i\frac{2\alpha}{3e}\frac{\partial \ln A}{\partial x} - \frac{\partial}{\partial x}\left(k\frac{\partial T}{\partial x} \right) \right) dx.$$

While the first term, which is proportional to the square of the current, and the last term, which is independent of the current, represent the Joule heat and the heat carried away by ordinary conduction, respectively, the middle term, which is proportional to the current and changes its sign when the direction of the current is reversed, represents the so-called Peltier and Thomson effects. To investigate these more closely, we shall consider two cases.

We shall first consider the junction between two metals I and II and assume that the temperature is kept at a fixed value at that part of the chain. On carrying out the integration, we find that the heat arising from the term in question and being developed at the junction between the two metals is

$$i\frac{2\alpha}{3e} T \ln\left(\frac{A_{II}}{A_I} \right).$$

Hence, if the "Peltier coefficient" (measured as heat absorption) is called $\Pi_{I, II}$, we obtain

$$\Pi_{I, II} = \frac{2\alpha}{3e} T \ln\left(\frac{A_I}{A_{II}} \right) = \frac{2\alpha}{3e} T \ln\left(\frac{N_I}{N_{II}} \right).$$

On the other hand, if we next consider a part of the chain over which the metal is homogeneous, the heat arising from the term in question, and developed in an element of the chain within which the temperature varies from T to $T + dT$, will be

$$i\frac{2\alpha}{3e} T \frac{d \ln A}{dT} dT.$$

Thus, we obtain for the "Thomson coefficient" μ

$$\mu = -\frac{2\alpha}{3e} T \frac{d \ln A}{dT}.$$

It is seen that the expressions for F, Π and μ satisfy the relations

$$\mu_{II} - \mu_I = T\frac{\partial}{\partial T}\left(\frac{\Pi_{I, II}}{T} \right) \quad \text{and} \quad F = -\int_{T_1}^{T_2} \frac{\Pi_{I, II}}{T} dT$$

derived by Lord Kelvin from thermodynamic considerations.

It may be pointed out here that, if we had applied the equations given on p. [141] in the preceding calculations, we should have found

$$F = \frac{2\alpha}{3e} \int_{T_1}^{T_2} \left[\ln\left(\frac{A_{II}}{A_I}\right) + \left(\frac{L_2}{L_1}\right)_I - \left(\frac{L_2}{L_1}\right)_{II} \right] dT,$$

$$\Pi_{I, II} = \frac{2\alpha}{3e} T \left[\ln\left(\frac{A_{II}}{A_I}\right) - \left(\frac{L_2}{L_1}\right)_I + \left(\frac{L_2}{L_1}\right)_{II} \right],$$

$$\mu = -\frac{2\alpha}{3e} T \frac{\partial}{\partial T} \left(\ln A - \frac{L_2}{L_1} \right).$$

As may be seen, these expressions also satisfy the relations mentioned above*.

ELECTRIC CONDUCTION UNDER PERIODIC FORCES

Heretofore, we have only considered the electric conductivity under stationary conditions; in the theory in question this means that the electric field is assumed to vary very little in a time that is long compared to the time it takes electrons to traverse the free path between collisions. However, this condition cannot be assumed satisfied by the extremely rapidly varying electric fields that one is dealing with in the electromagnetic theory of light. However, we shall now show how this case can be treated on the basis of the fundamental formula derived on p. [138], in much the same way as the cases so far discussed.

We assume that the temperature is constant and that electric forces act only in the direction of the x-axis. The formula then becomes

$$\frac{\partial f}{\partial t} + X \frac{\partial f}{\partial \xi} + \xi \frac{\partial f}{\partial x} + \eta \frac{\partial f}{\partial y} + \zeta \frac{\partial f}{\partial z} = \frac{r}{l} \left[\int rf \, d\lambda \cdot \frac{h^2}{2\pi} e^{-hr^2} - f \right].$$

If we assume that the light consists of plane waves moving in the direction of the z-axis, $\partial f/\partial x$ and $\partial f/\partial y$ will vanish; also $\zeta(\partial f/\partial z) = (\zeta/v)/(\partial f/\partial t)$ (where v is the speed of light) and, since the speed of the electrons is extremely much smaller than that of light, this expression will be negligible compared to $\partial f/\partial t$. We shall now assume that $f = A e^{-hr^2} + \varphi$, where A and h are constants and φ is very small compared to $A e^{-hr^2}$.

Substituting this in the formula and putting $X = (e/m) E \cos' pt$, we obtain

* If L_2/L_1 is put equal to U, these equations are formally identical with those derived by Gruner (In Gruner's formulas (19) and (26), loc. cit. p. 533, two signs have been interchanged by a typographical error.)

$$\frac{\partial \varphi}{\partial t} - \xi 2hAe^{-hr^2}\frac{e}{m}E\cos pt = \frac{r}{l}\left[\int r\varphi\, d\lambda \cdot \frac{h^2}{2\pi}e^{-hr^2} - \varphi\right].$$

If we try to satisfy this equation by $\varphi = \xi\psi(r)$, we get, since the integral vanishes because of symmetry,

$$\frac{\partial \psi}{\partial t} - 2hAe^{-hr^2}\frac{e}{m}E\cos pt = -\frac{r}{l}\psi;$$

this is an ordinary linear differential equation of the first order in ψ and t with the solution

$$\psi = \frac{2hA(e/m)El}{r(1+p^2l^2/r^2)}e^{-hr^2}\left(\cos pt + p\frac{l}{r}\sin pt\right) + Ce^{-(r/l)t}.$$

Since ψ is assumed to be periodic with respect to time, $C = 0$ and

$$f = Ae^{-hr^2} + \xi\frac{2hA(e/m)El}{r(1+p^2l^2/r^2)}e^{-hr^2}\left(\cos pt + p\frac{l}{r}\sin pt\right).$$

The electric current through unit area then is

$$i = e\int \xi f\, d\lambda = e\int \xi^2\psi\, d\lambda = \frac{4\pi e}{3}\int_0^\infty r^4\psi\, dr$$

$$= \frac{8\pi e^2 ElAh}{3m}\left[\int_0^\infty \frac{r^3e^{-hr^2}dr}{(1+p^2l^2/r^2)}\cos pt + \int_0^\infty \frac{plr^2e^{-hr^2}dr}{(1+p^2l^2/r^2)}\sin pt\right].$$

The work done in unit volume in the time dt will be $iE\cos pt$. If we now seek the average energy given off in unit time, we get, since the average of $\cos x \sin x$ is zero and the average of $\cos^2 x$ is $\frac{1}{2}$,

$$U = \frac{4\pi e^2 E^2 lhA}{3m}\int_0^\infty \frac{r^3e^{-hr^2}dr}{(1+p^2l^2/r^2)} = \frac{2\pi e E^2 lA}{3mh}\int_0^\infty \frac{x^2e^{-x}dx}{x+p^2l^2h}$$

$$= \frac{2\pi e^2 E^2 lA}{3mh}\left[\int_0^\infty (x - p^2l^2h)e^{-x}dx + \int_0^\infty \frac{(p^2l^2h)^2e^{-x}dx}{x+p^2l^2h}\right]$$

$$= \frac{2\pi e^2 E^2 lA}{3mh}\left[1 - p^2l^2h + (p^2l^2h)^2\int_0^\infty \frac{e^{-x}dx}{x+p^2l^2h}\right].$$

If now p^2l^2h is very small, the term containing the integral, whose order of magnitude is $(p^2l^2h)^2\ln(p^2l^2h)$, can be regarded as negligible compared to p^2l^2h, and we get

$$U = \frac{2\pi e^2 E^2 lA}{3mh}(1 - p^2l^2h).$$

Ascribing a definite conductivity σ_1 to the metal, one would have $U = \frac{1}{2}\sigma_1 E^2$. Moreover, since the conductivity of a metal under stationary conditions is

$$\sigma = \frac{4\pi e^2 l A}{3mh} = \frac{4e^2 l N}{3m}\left(\frac{h}{\pi}\right)^{\frac{1}{4}},$$

we get

$$\sigma_1 = \sigma\left(1 - \sigma^2\frac{9\pi p^2 m^2}{16N^2 e^4}\right).$$

J. J. Thomson (Phil. Mag. **14** (1907) 224) has given the following formula (converted to our notations) for the relation between the conductivities for stationary and periodic electric fields

$$\sigma_1 = \sigma\left[\sin\frac{pm\sigma}{Ne^2}\bigg/\frac{pm\sigma}{Ne^2}\right] = \sigma\left(1 - \sigma^2\frac{p^2 m^2}{3N^2 e^4} + \cdots\right).$$

As may be seen, the term giving the dependence of the conductivity upon the frequency is much smaller (about six times smaller) than in the formula derived above. However, quite apart from the circumstance that Thomson assumes that all electrons are ejected with the same speed, there seems to be essential shortcomings in his calculations (Jeans, Phil. Mag., June 1909, p. 778), in that the times in which the electrons traverse their free path are put equal, quite independently of the speed acquired by them under the influence of the field.

Jeans, moreover, has given the following formula (loc. cit. p. 778)

$$\sigma_1 = \frac{\sigma}{1 + \sigma^2 p^2 m^2/N^2 e^4} = \sigma\left(1 - \sigma^2\frac{p^2 m^2}{N^2 e^4} + \cdots\right).$$

In his opinion, this is derived without making any special assumptions about the nature of the collisions between electrons and molecules of the metal; in fact, it is supposed to be valid even if the electrons do not have any "free path". However, this does not seem to be correct. In the first place, this formula is not identical with the formula we have derived above on the basis of special assumptions that are included in those of Jeans; furthermore, the error in Jeans' derivation seems very clear. In fact, Jeans assumes that the loss in the total momentum of the electrons in a given direction caused by collisions with molecules of the metal is equal to the average velocity of the electrons in this direction multiplied by a constant that depends only on the nature of electrons and molecules, and on the temperature. He bases this on the assumption that the deviations from the normal (Maxwellian) law of distribution caused by the external forces are so small that the average velocity u_0 of the electrons in a given direction is very small compared to u, the average speed of the molecules, so that quantities of the same order of magnitude as $(u_0/u)^2$ can be neglected and it can be assumed, in particular, that all quantities multiplied by u_0 depend only upon the normal distribution. While this is correct in general, it will not be true for changes in the total momentum of the electrons in a given direction; for, since the normal distribution contributes nothing to it, this quantity will depend only upon the deviation from the normal distribution, and this deviation will be different for stationary and periodic external forces.

The formulas derived above for the electric and thermal conductivities and the thermoelectric effects cannot serve to determine the quantities N and l which enter in the formulas. (Thus only the product Nl occurs in the electric and thermal conductivities, and in the Thomson and Peltier coefficients occur only the ratios between the number of free electrons in one and the same metal at different temperatures and in two different metals at the same temperature.)

However, as pointed out by J. J. Thomson, one can determine an upper limit for N from the specific heat of metals. In fact, to heat a piece of metal one must at least add the heat corresponding to the increase in the kinetic energy of the free electrons. Thus, if the specific heat is denoted by q and the density by d, one has $N < qd/\alpha$, since the energy of the free electrons in unit volume is $N\alpha T$. For such metals as copper and silver, this gives $N < 2 \times 10^{23}$.

Now, in the formula for the dependence of the conductivity upon frequency, the number N is the only unknown quantity, and one might, therefore, consider applying this formula to determine N. However, if we attempt to calculate N from Hagen and Rubens' experiments (Ann. d. Phys. **11** (1903) 484, Table 2) on the absorption by metals of heat rays of wavelength 4 microns, we find that for almost all the pure metals investigated N will exceed the upper limit mentioned above; moreover, for all the alloys investigated and for one of the pure metals (silver), the formula cannot be satisfied for any N, since the apparent conductivity for periodic fields is higher than the stationary conductivity. The latter circumstance shows that in these rapid electric vibrations metals convert energy also in other ways than by collisions of the free electrons, e.g., by vibration of electrons bound in molecules. That such an absorption of energy plays a greater role in alloys than in pure metals can be understood as a result of the lower conductivity of alloys, if it is assumed that the individual molecules absorb just as much energy in alloys as in the pure metals involved. (It would seem that it might be possible from such an assumption to determine how large a fraction this molecular absorption of energy is of the entire absorption; however, it turns out that the data of Hagen and Rubens are not regular enough for such a calculation.)

We shall here mention still another application of the equations derived on p. [141], namely to the electrostatic problem. While it is assumed in the ordinary electrostatics that the electricity is distributed in an infinitely thin layer on the surface of good conductors, such a distribution is not compatible with the theory of free electrons, for these would diffuse from such a surface layer into the interior of the metal. Putting $i = 0$ and assuming the temperature to be constant, we obtain

$$2hA \frac{e}{m} E - \frac{\partial A}{\partial x} = 0,$$

whence

$$\frac{3e}{2\alpha T} NE - \frac{\partial N}{\partial x} = 0 \quad \text{and} \quad \frac{\partial E}{\partial x} = 4\pi e(N - N_T).$$

J. J. Thomson (*Die Korpuskulartheorie der Materie* (1908), p. 80) has shown that one can estimate from those equations how rapidly the free electricity $e(N - N_T)$ decreases on going from the surface into the metal. If it is assumed that the excess of free electricity is small compared to the entire charge of the free electrons in unit volume, the first equation can be written

$$\frac{3e}{2\alpha T} N_T E - \frac{\partial N}{\partial x} = 0.$$

Combining this with the second equation, and putting $e(N - N_T) = \xi$, we get $\partial^2 \xi / \partial x^2 = (6\pi e^2 / \alpha T) N_T \xi$, which is satisfied by $\xi = Be^{-px}$, where $p^2 = (6\pi e^2 / \alpha T) N_T$. If now the charge per unit surface area is Q, we have

$$Q = \int_0^\infty \xi \, dx = B/p,$$

and, therefore, $\xi = Qpe^{-px}$. Since $\partial E / \partial x = 4\pi \xi$, $E = -4\pi Qe^{-px}$, and if the potential difference between the surface and the interior of the metal is called V, we finally obtain

$$V = -\int_0^\infty E \, dx = 4\pi Q/p.$$

The highest possible value for N (ca. 10^{23}) mentioned above, which corresponds to the most rapid decrease of the free electricity and the smallest possibly potential difference, thus gives, according to Thomson, $p =$ ca. 10^8. It is seen that such a value for p corresponds to an extremely thin layer of essentially free electricity. With respect to the potential difference between the surface and the interior of a metal, on the other hand, Thomson calculates its value to be ca. 3×10^{-3} volts for the largest possible charge that one can give a metal surface; hence, he believes that it might not be impossible to observe this potential difference, for example in the interior of a charged sphere whose external surface was given the potential zero. (However, it would seem that there may be fundamental difficulties in this, since, according to the theory discussed here, it does not seem possible to give the surface of a metal, but only its interior, a definite potential.)

[157]

Closely connected with these considerations are the experiments carried out by Bose (Phys. Zeitschr. 7 (1906) 373, 462) and Pohl (Phys. Zeitschr. 7 (1906) 500) to investigate whether the electric conductivity in a very thin metal plate changes its value when the number of free electrons in the plate is artificially increased or decreased by giving it a negative or a positive charge. Contrary to expectation, no change was observed. However, J. J. Thomson (loc. cit. p. 81) has pointed out that the reason for this is probably that the metal plate cannot be regarded as smooth in the ordinary sense, but rather must be extremely uneven compared to molecular dimensions, with the result that the added electrons accumulate in points and elevations and are thus not able to participate in the conduction of electricity.

THE INFLUENCE OF A MAGNETIC FIELD UPON THE MOTION OF FREE ELECTRONS

We shall investigate first the influence of a magnetic field alone upon a piece of metal in which free electrons move. Under the influence of such a field, the paths of the electrons will be curved, and the electrons will move in helical paths whose axes are parallel to the axis of the field. J. J. Thomson (*Rapp. du Congrès de Paris* (1900), vol. **3**, p. 148) believes that this curvature of the path will produce a magnetic field in the opposite direction of the external field and that the piece of metal containing the free electrons will, therefore, act as a diamagnetic body. Thomson mentions that on this basis he has calculated the diamagnetism that the free motions of the electrons will impart to bismuth (in which metal the free paths of the electrons in his opinion are particularly long) and that the result obtained, although much too large, nevertheless, does not deviate more from the observed diamagnetism than can be ascribed to uncertainties in the numerical values used. However, this view, which seems to be generally accepted*, does not appear correct to me. In fact, it seems that the free electrons can exert no magnetic effects at all; for the circumstance that the paths are curved is not sufficient in itself, since a particle moving in a curved path (provided its speed is not comparable with that of light) at every instant will produce the same magnetic effects as if it moved in a straight path with the same speed and direction. Now, consider a small volume element in the interior of a piece of metal which is subjected to the influence of a constant magnetic field. Because of the symmetry, it will be traversed equally frequently by electrons in all directions that lie symmetrical with respect to the axis of the field. Such a volume element will, therefore, not exert any magnetic

---

* Thus, Langevin (Ann. Chim. et Phys. **5** (1905) 90) believes that this may explain the fact that the diamagnetism of bismuth varies greatly with temperature, in contrast to the diamagnetism of other substances.

influence, and, hence, the whole piece of metal will not do so either. (The justification of the last conclusion is based on the assumption that the electrons are perfectly free and that the metal, therefore, can be divided by mathematical surfaces in volume elements that are all alike. On the other hand, in the case of the electrons assumed to move in closed orbits inside the individual chemical atoms, and used in Langevin's theory to explain the magnetic effects, it would only be possible to divide the body in equal elements if each element contains a whole number of atoms.)

On the other hand, when an electric field acts at the same time as the magnetic field, or when the metal does not have the same temperature everywhere, such that the symmetry with respect to the axis of the magnetic field is disturbed, then the influence of the field upon the motion of the electrons may manifest itself and give rise to the so-called galvanomagnetic and thermomagnetic effects.

The theoretical treatment of these phenomena has been very different with the different authors. Thus, the earliest authors, Riecke and Drude, assumed that all electrons under the influence of the electric field have a certain average velocity in a definite direction, and they calculated the influence of the magnetic field in a straightforward manner, as if this velocity were equally distributed among the electrons. Later, however, doubt has been thrown on the permissibility of such a view, and van Everdingen (Arch. Néerl. **6** (1901) 274) has even carried out calculations to show that, when the motions of the individual electrons are taken into account, no Hall effect at all is obtained*. However, very recently Gans (Ann. d. Phys. **20** (1906) 293) has given a complete theoretical treatment of the phenomena in question on the basis of Lorentz' theory and has obtained results which, at least in the case of the Hall effect, deviate from those of Riecke and Drude only by a small difference in the numerical constants that enter.

We shall here only discuss a single case treated by Gans, namely the calculation of the galvanomagnetic effects in a piece of metal whose temperature is kept the same everywhere. Assuming the condition to be stationary, the fundamental formula given on p. [138] becomes in this case

$$X \frac{\partial f}{\partial \xi} + Y \frac{\partial f}{\partial \eta} + Z \frac{\partial f}{\partial \zeta} = \frac{r}{l} \left[\int rf \, d\lambda \cdot \frac{h^2}{2\pi} e^{-hr^2} - f \right].$$

If we now assume that a magnetic field H acts in the direction of the z-axis, and that the electric field has components E_x and E_y in the direction of the x- and y-axes, the forces upon unit mass will be

---

* Shown by Gans' calculations to be incorrect, this result arises from the fact that van Everdingen in his calculations (loc. cit. p. 297) introduces an auxiliary quantity p which he takes to be constant; however, this quantity is not completely independent of the direction of motion of the electrons.

$$X = \frac{e}{m} E_x + \frac{e}{m} H\eta, \qquad Y = \frac{e}{m} E_y - \frac{e}{m} H\xi, \qquad Z = 0.$$

Now, putting $f = Ae^{-hr^2} + \varphi$, we get

$$-(\xi E_x + \eta E_y)2 \frac{e}{m} hAe^{-hr^2} + \frac{e}{m} \left(E_x \frac{\partial\varphi}{\partial\xi} + E_y \frac{\partial\varphi}{\partial\eta} \right)$$

$$+ \frac{e}{m} H \left(\eta \frac{\partial\varphi}{\partial\xi} - \xi \frac{\partial\varphi}{\partial\eta} \right) = \frac{r}{l} \left[\int r\varphi \, d\lambda \cdot \frac{h^2}{2\pi} e^{-hr^2} - \varphi \right].$$

Since φ is small compared to Ae^{-hr^2}, the second term will be small compared to the first, and if we attempt to satisfy the equation by

$$\xi\psi_1(r) + \eta\psi_2(r),$$

we obtain, since the integral vanishes,

$$\xi \left(2 \frac{e}{m} E_x hAe^{-hr^2} + \frac{e}{m} H\psi_2 - \frac{r}{l}\psi_1 \right) + \eta \left(2 \frac{e}{m} E_y hAe^{-hr^2} - \frac{e}{m} H\psi_1 - \frac{r}{l}\psi_2 \right) = 0,$$

which gives

$$\psi_1 = \left(E_x + \frac{e}{m} H \frac{l}{r} E_y \right) 2 \frac{l}{r} \frac{e}{m} hAe^{-hr^2} \left/ \left[1 + \left(\frac{e}{m} H \frac{l}{r} \right)^2 \right] \right.$$

$$\psi_2 = \left(E_y - \frac{e}{m} H \frac{l}{r} E_x \right) 2 \frac{l}{r} \frac{e}{m} hAe^{-hr^2} \left/ \left[1 + \left(\frac{e}{m} H \frac{l}{r} \right)^2 \right] \right. .$$

For the electric current densities in the directions of the x- and y-axes, we now get

$$i_x = e \int \xi f \, d\lambda \quad \text{and} \quad i_y = \int \eta f \, d\lambda.$$

These integrals cannot be evaluated completely; however, if we assume that $(e/m)H(l/u)$ is small and neglect quantities of higher order than $((e/m)H(l/u))^2$, we get

$$i_x = \frac{4\pi e^2 Al}{3mh} \left[E_x \left(1 - \frac{e^2}{m^2} l^2 hH^2 \right) + E_y \frac{\sqrt{\pi h}}{2} l \frac{e}{m} H \right]$$

$$i_y = \frac{4\pi e^2 Al}{3mh} \left[E_y \left(1 + \frac{e^2}{m^2} l^2 hH^2 \right) - E_x \frac{\sqrt{\pi h}}{2} l \frac{e}{m} H \right].$$

If it is now assumed that the piece of metal in question is traversed by an electric current in the direction of the x-axis, there will arise, under the influence of the magnetic field, a potential difference between any two points corresponding to the same x but different y (Hall effect). The corresponding electric field E_y is found from the equations derived by putting $i_y = 0$; neglecting quantities of higher order than $(e/m)H(l/u)$, we obtain

$$E_y = \tfrac{1}{2}\sqrt{\pi h}l\,\frac{e}{m}\,HE_x = \tfrac{3}{8}\pi^{-\frac{1}{2}}h^{\frac{3}{2}}A^{-1}e^{-1}Hi_x = \frac{3\pi H}{8Ne}\,i_x.$$

Substitution of this expression for E_y into that for i_x gives

$$i_x = \frac{4\pi e A l}{3mh}\,E_x\left(1 - \frac{4-\pi}{4}\,hl^2\,\frac{e^2}{m^2}\,H^2\right)$$

$$= E_x\sigma\left(1 - \frac{3(4-\pi)l^2e^2}{16\alpha Tm}\,H^2\right).$$

It is seen that the effects of the field consist in producing a transverse potential difference with a definite direction relative to the electric current and the magnetic field, and in causing a decrease in the electric conductivity.

However, the calculated results do not agree with those found experimentally; in fact, there are metals, e.g., bismuth, possessing a Hall effect in the same direction as that calculated, and for which the electric conductivity decreases in a magnetic field, but also other metals, such as iron, which have a Hall effect in the opposite direction, and whose conductivity is increased in a magnetic field.

For the thermomagnetic effects, quite similar discrepancies between theory and experience are found. Hence, it appears that at least a part of the effects in question are not caused by the motions of free electrons but must arise from motions of electrons bound in the molecules of the metal.

II. THE DOCTOR'S DISSERTATION

(TEXT AND TRANSLATION*)

* [See Introduction, sect. 2.]

STUDIER OVER
METALLERNES ELEKTRONTHEORI

STUDIER OVER
METALLERNES ELEKTRONTHEORI

AFHANDLING FOR DEN FILOSOFISKE

DOKTORGRAD

AF

NIELS BOHR

KØBENHAVN

I KOMMISSION HOS V. THANING & APPEL

Trykt hos J. Jørgensen & Co. (M. A. Hannover)

1911

Denne Afhandling er af det mathematisk-naturviden-skabelige Fakultet antagen til at forsvares for den filosofiske Doktorgrad.

København, den 12. April 1911.

Elis Strömgren,
d. A. Dekanus.

I DEN DYBESTE TAKNEMMELIGHED TILEGNET

MINDET OM MIN FADER

INDHOLDSFORTEGNELSE.

INDLEDNING.

Blandt de Legemer, der leder Elektriciteten, indtager Metallerne en særlig Plads, ikke blot paa Grund af deres store Ledningsevne, men ogsaa fordi Elektricitetens Gennemgang igennem dem, i Modsætning til de fleste andre gode Ledere, ikke er forbunden med paaviselige kemiske Omsætninger. Dette Forhold forklares i Overensstemmelse med den almindelig antagne Opfattelse derved, at visse smaa Partikler — Elektroner —, ladede med Elektricitet, i Metallernes Indre kan bevæge sig fra det ene kemiske Atom til det andet.

Den første nærmere Udarbejdelse af en saadan Forestilling om Metallerne skyldes *W. Weber*[1]). Efter *Weber's* Theori bestaar hvert enkelt Metalmolekyle af et System af elektrisk ladede Partikler, der bevæger sig i krumlinede Baner omkring hverandre; disse Systemer antages imidlertid ikke at være stabile, men med korte Mellemrum vil nogle af Partiklerne forlade Systemet for derefter i tilnærmelsesvis retlinede Baner at bevæge sig igennem Metallet, indtil de atter indfanges af andre Metalmolekyler; i disse vil de saa for en Tid deltage i de intra-molekylære Bevægelser, for derefter atter at sendes ud o. s. v. Idet *Weber* endvidere antager, at de omhandlede Partiklers Bevægelsesenergi er identisk med en væsentlig Del af Metallets Varmeenergi, gives der ved *Weber's* Theori en Mulighed for en Forklaring af den nære Forbindelse, der er mellem Metallernes Ledningsevne for Varme og for Elektricitet, samt for en mekanisk Forstaaelse af den af *F. Kohlrausch*[2]) opstillede Medføringstheori for de thermoelektriske Fænomener, hvilken sidste Theori forklarer disse Fænomener ved at antage, at der

---

[1]) Se f. Eks. *W. Weber:* Pogg. Ann., Bd. 156, p. 1, 1875.
[2]) *F. Kohlrausch:* Nachr. d. Kgl. Ges. d. Wiss. zu Göttingen, 1874, p. 65.

1

til enhver elektrisk Strøm er knyttet en bestemt Varmestrøm, der er proportional med den elektriske Strøm og afhængig af Metallets Natur og Temperatur, og at der omvendt til enhver Varmestrøm svarer en bestemt elektrisk Strøm.

Udfra et Grundlag, der svarer til de her omtalte, af *Weber* benyttede Forestillinger om Forholdene i Metallernes Indre, har *E. Riecke*[1] senere udarbejdet en udførlig Theori for Metallerne, i hvilken der udledes bestemte Udtryk saavel for Metallernes Ledningsevne for Varme og Elektricitet som for de thermoelektriske Fænomener, samt for de galvano- og thermomagnetiske Fænomener. I disse Udtryk indgaar Værdierne for Antallet af Partikler, der udsendes i Tidsenheden af Metalmolekylerne i en Volumenenhed, for Partiklernes elektriske Ladninger, deres Masser og de Hastigheder, hvormed de udsendes, samt endelig for de Vejlængder, som de i Middel tilbagelægger, før de atter indfanges af et andet Metalmolekyle; idet der imidlertid ikke paa Forhaand gøres bestemte Antagelser om de absolute Værdier af nogen af de omtalte Størrelser, tillader *Riecke's* Theori kun forholdsvis ringe Sammenligning med Erfaringerne.

Et meget væsentligt Fremskridt i Metallernes Elektrontheori skyldes *P. Drude*[2], der bragte de i den kinetiske Lufttheori udledte Hovedresultater til Anvendelse paa de ovenfor omtalte fri Partiklers Bevægelse i Metallernes Indre. I den kinetiske Lufttheori bevises f. Eks., at en Samling Partikler, der befinder sig i mekanisk Varmeligevægt med deres Omgivelser, maa have saadanne Hastigheder, at deres kinetiske Energi, hidrørende fra deres fremadskridende Bevægelse, i Middel er lig den kinetiske Energi, hidrørende fra den fremadskridende Bevægelse af et Molekyle af en hvilkensomhelst Luftart, der befinder sig ved samme Temperatur. Ved Hjælp af denne Sætning om Partiklernes Middelenergi og ved at antage, at Partiklernes elektriske Ladninger har samme Størrelse som Ladningen af en engyldig Ion i en Elektrolyt, har *Drude* beregnet Forholdet mellem Metallernes Varme- og Elektricitetsledningsevne i meget nær Overensstemmelse med de eksperimentelt fundne Værdier[3].

Drude's Beregninger er imidlertid ikke fuldkommen nøjagtig gennemførte; han antager blandt andet for Simpelheds Skyld, at alle Partiklerne har den samme absolute Hastighed, medens man efter Ud-

---

[1] *E. Riecke*: Wied. Ann., Bd. 66, pp. 353, 545 og 1199, 1898.
[2] *P. Drude*: Ann. d. Phys., Bd. 1, p. 566 og Bd. 3, p. 369, 1900.
[3] Se *M. Reinganum*: Ann. d. Phys., Bd. 2, p. 398, 1900.

viklingerne i den kinetiske Lufttheori nødvendigvis maa antage, at Partiklerne, dersom de befinder sig i en mekanisk Varmeligevægt som den af *Drude* betragtede, maa have forskellige absolute Hastig- heder, fordelte efter den saakaldte *Maxwell'ske* Hastighedsfordelingslov. En fuldkommen strængt gennemført Theori udfra bestemt angivne Forudsætninger er først givet af *H. A. Lorentz*[1].

Medens *Riecke* og *Drude* antager den samtidige Tilstedeværelse i Metallerne af flere forskellige Slags, baade positive og negative, frit bevægelige Partikler, antager *Lorentz* kun Tilstedeværelsen af en enkelt Slags fri Partikler, ens i alle Metaller. En saadan Antagelse er først indført i Metallernes Elektrontheori af *J. J. Thomson*[2]. Denne Forfatter, der i saa høj Grad har bidraget til det eksperimentelle Grundlag for Elektrontheorien, begrunder sin Antagelse paa den Om- stændighed, at medens man paa mange, vidt forskellige Maader har kon- stateret Tilstedeværelsen af en ganske bestemt Slags negativt ladede Partikler, de saakaldte Elektroner, hvis Masser er overordentlig mange Gange mindre end de kemiske Atomers Masser, er det aldrig lykkedes at iagttage positiv Elektricitet uden i Forbindelse med Masser af samme Størrelsesorden som de kemiske Atomers. *Lorentz*[3] har yder- mere paavist, at der indtræder Vanskeligheder af principiel Art ved at tænke sig flere Slags frit bevægelige Partikler i stationær Fordeling i et Metalstykke, hvis Tilstand — Temperatur og kemiske Sammen- sætning — ikke overalt er den samme.

Til Grund for *Lorentz's* Theori lægges følgende mekaniske Billede. I Metallernes Indre tænkes at befinde sig baade Metalmolekyler og fri Elektroner. Molekylernes og Elektronernes Dimensioner — de Om- raader, indenfor hvilke de paavirker hinanden kendeligt — antages at være forsvindende smaa i Forhold til Middelværdien af deres ind- byrdes Afstande; de tænkes derfor kun at paavirke hinanden ved adskilte Sammenstød, under hvilke sidste de antages at virke paa hin- anden som absolut haarde elastiske Kugler. Saavel Elektronernes Dimensioner som deres Masser tænkes endvidere at være saa smaa i Forhold til Metalmolekylernes, at man dels kan se bort fra Sammen- stødene mellem de fri Elektroner indbyrdes i Forhold til Sammen- stødene mellem Elektronerne og Metalmolekylerne, dels kan betragte Metalmolekylerne som ubevægelige i Forhold til Elektronerne.

---

[1] *H. A. Lorentz*: Proc. Acad. Amsterdam, vol. 7, pp. 438, 585 og 684, 1905.
[2] *J. J. Thomson*: Rapp. d. Congrès d. Physique, Paris 1900, Tom. 3, p. 138.
[3] *H. A. Lorentz*: Proc. Acad. Amsterdam, vol. 7, p. 684, 1905. Jahrb. d. Rad. u. Elek., Bd. 4, p. 125, 1907.

1*

Udfra disse Antagelser har *Lorentz* ikke alene beregnet Udtryk for et Metals Elektricitets- og Varmeledningsevne og for de thermo-elektriske Fænomener, men ogsaa for et Metals Udstraalingsevne og Absorptionsevne for Varmestraaler med store Svingningstider. Som nogle af de interessanteste Resultater af *Lorentz's* Theori kan, foruden den ovenfor omtalte, af *Drude* først paaviste, tilnærmelsesvise Over-ensstemmelse mellem den beregnede Værdi og de eksperimentelt fundne Værdier for Forholdet mellem Metallernes Ledningsevne for Varme og Elektricitet, nævnes den fuldkomne Overensstemmelse mel-lem de udfra *Lorentz's* Theori beregnede Resultater for saavel de thermoelektriske Fænomener som de med Varmestraalingen forbundne Fænomener og de ad thermodynamisk Vej, af henholdsvis Lord *Kelvin* og *Planck* opstillede Theorier for disse Fænomener. Denne sidste Overensstemmelse — der kun fremkommer ved en fuldkommen stræng Gennemførelse af Beregningerne, saaledes først naar der tages Hensyn til den *Maxwell'ske* Hastighedsfordelingslov — er saameget mere interessant, som den, paa Grund af særlige Forhold ved Opstil-lingen af de to sidstnævnte Theorier, ikke bestemt paa Forhaand kunde ventes.

Medens *Lorentz's* Theori er meget fuldkommen i mathematisk Henseende, er de fysiske Antagelser, hvorpaa Theorien bygger, imid-lertid af en saadan Art, at de næppe engang tilnærmesvis kan tænkes opfyldt ved de virkelige Metaller. Det er derfor af Interesse udfra andre og mere almindelige Forudsætninger at undersøge, hvorledes Metallernes Elektrontheori vil forme sig, og da navnlig at se, hvorledes det vil gaa med den omtalte Overensstemmelse med de thermodynamiske Theorier.

Et Forsøg paa at gennemføre en saadan Theori er for Varme-straalingsproblemets Vedkommende gjort af *J. H. Jeans* [1]. *Jeans's* Be-regninger er imidlertid i deres Almindelighed ikke rigtige [2], idet der, som vi skal se, benyttes Antagelser, der kun vil være opfyldt i ganske specielle Tilfælde, saaledes ikke engang ved det ovenfor omtalte, af *Lorentz* behandlede Tilfælde.

Det Maal, jeg har sat mig ved denne Afhandling, er at forsøge at gennemføre Beregningerne for de forskellige Fænomener, der for-klares ved Tilstedeværelsen af fri Elektroner i Metallerne, i saa stor

---

[1] *J. H. Jeans*: Phil. Mag., vol. 17, p. 773 og vol. 18, p. 209, 1909.

[2] *H. A. Wilson* (Phil. Mag., vol. 20, p. 835. 1910) har i en nylig offentliggjort Afhand-ling gjort opmærksom paa Urigtigheden af *Jeans's* Beregninger og forsøgt at gennemføre en rigtigere Beregning for et enkelt specielt Tilfældes Vedkommende. (Som vi i det følgende skal se, er *Wilson's* Beregning imidlertid heller ikke rigtig.)

Almindelighed som muligt, med Bevarelsen af de Hovedsynspunkter, der er lagt til Grund i *Lorentz's* Theori.

Vi skal da i det følgende antage, at der i ethvert Metalstykke findes et vist, af Metallets Natur og Temperatur afhængigt. Antal fri Elektroner, hvilke Elektroner skal antages at være ens i alle Metaller. ↑

↗ Vi skal dernæst endvidere antage, at der i et homogent Metalstykke, der overalt har samme Temperatur, og som ikke paavirkes af ydre Kræfter, vil være fuldkommen mekanisk Varme-Ligevægt mellem de fri Elektroner og Metalmolekylerne, — ved mekanisk Varme-Ligevægt skal her forstaas en saadan statistisk Ligevægt som den, der vil fremkomme, dersom Metalmolekylerne og Elektronerne paavirker hinanden med Kræfter af samme Art som de i den sædvanlige Mekanik betragtede Kræfter, d. v. s. dersom Bevægelserne tilfredsstiller de *Hamilton'ske* Ligninger. — Den omtalte Antagelse er ikke paa Forhaand selvfølgelig, idet man maa antage, at der i Naturen ogsaa findes Kræfter af ganske anden Art end de almindelige mekaniske Kræfter; medens man nemlig paa den ene Side har opnaaet overordentlig store Resultater i den kinetiske Lufttheori ved at antage, at Kræfterne mellem de enkelte Molekyler er af almindelig mekanisk Art, er der paa den anden Side mange af Legemernes Egenskaber, det ikke er muligt at forklare, dersom man antager, at de Kræfter, der virker indenfor de enkelte Molekyler (der efter den almindelig antagne Opfattelse bestaar af Systemer, i hvilke indgaar et stort Antal »bundne« Elektroner), er af en saadan Art. Foruden forskellige almindelig kendte Eksempler herpaa, f. Eks. Beregningen af Legemernes Varmefylde og Beregningen af Varmestraalingsloven for korte Svingningstider, skal vi i det følgende ogsaa se et yderligere Eksempel herpaa, nemlig ved Omtalen af Legemernes magnetiske Forhold.

Foruden de ovenfor omtalte almindelige Antagelser, skal vi i det følgende yderligere antage, at Metalmolekylernes Egenskaber i Middel er ens i alle Retninger, og at denne Isotropi vil bestaa uafhængig af de ydre Kræfters Tilstedeværelse, saaledes at de ydre Kræfters Virkning forklares ved deres direkte Indvirkning paa de fri Elektroners Bevægelser. Dette, der danner en Hovedantagelse i *Drude's* og *Lorentz's* Theorier, skiller den her givne Behandling meget væsentligt fra saadanne Theorier for Metallerne som dem, der er opstillede af *W. Sutherland*[1]) og *J. J. Thomson*[2]). Efter disse sidste Theorier, i hvilke man ikke antager en saadan stadig Tilstedeværelse

---

[1]) *W. Sutherland:* Phil. Mag., vol. 7, p. 423, 1904.
[2]) *J. J. Thomson:* The Corpuscular Theory of Matter, London 1907, p. 86.

6

i Metallet af fri Elektroner som i *Lorentz's* Theori, forklares nemlig de ydre Kræfters Virkning gennem deres Indvirkning paa de enkelte Metalmolekyler med deres tilhørende Elektroner, betragtet som Helheder, og Elektronernes Bevægelse gennem Metallet fremkommer ved, at de, paa Grund af hele Molekylets Bevægelse, udsendes som fri Elektroner i Retninger, der i Middel er afhængige af de ydre Kræfters Retninger.

Undersøgelsen i det følgende falder i to adskilte Tilfælde, alt efter som vi antager, at der, ligesom ved det af *Lorentz* behandlede specielle Tilfælde, finder adskilte Sammenstød Sted mellem Elektronerne og Metalmolekylerne, eller vi antager, at Molekylerne befinder sig saa tæt inde paa hverandre, at Elektronerne under en stor Del af deres Bevægelse er underkastet stærke Kræfter fra Molekylernes Side. I det første af disse Tilfælde har det været muligt for visse Problemers Vedkommende at gennemføre Undersøgelsen i fuldkommen Almindelighed udfra de ovenfor omtalte Hovedantagelser. I det andet Tilfælde derimod, hvor Behandlingen er langt vanskeligere, har det ikke været muligt at gennemføre Undersøgelsen i saa stor Almindelighed; det er dog lykkedes at gennemføre Behandlingen under saadanne Antagelser, der synes tilnærmelsesvis at kunne svare til Forholdene i de virkelige Metaller.

Det har ved denne Lejlighed været Hovedhensigten at drage de videst mulige Konsekvenser af en Theori for Metallerne, grundet paa de ovenfor omtalte almindelige Antagelser. Henvisninger til eksperimentelt fundne Resultater er derfor i denne Afhandling kun medtaget i saadanne Tilfælde, hvor de er af særlig Betydning for de omhandlede theoretiske Synspunkter.

Ved Afslutningen af dette Arbejde vil jeg bede min Lærer ved Universitetet, Hr. Professor *Christiansen*, modtage min bedste Tak for hans værdifulde Vejledning under mine Studier og for den velvillige Interesse, han stedse har vist mig.

København, April 1911.

Niels Bohr.

KAPITEL I.

OPSTILLINGEN AF BETINGELSESLIGNINGERNE FOR ELEKTRONERNES SAMLEDE BEVÆGELSE GENNEM METALLET.

----

§ 1.

Elektrontheoriens almindelige Forudsætninger om Tilstanden i Metallernes Indre.

I det Indre af et Metalstykke tænkes en stor Mængde fri Elektroner at bevæge sig afsted med store Hastigheder i alle mulige Retninger. Disse Elektroner tænkes uafbrudt at forandre deres Bevægelsesretning og Hastighed paa Grund af de Kræfter, hvormed de paavirkes fra hverandre og fra Metalmolekylernes Side. I et homogent Metalstykke, der overalt har samme Temperatur, og som er upaavirket af ydre Kræfter, vil Elektronernes Hastighed i Middel være den samme i alle Retninger. Virker der derimod en ydre elektrisk Kraft, vil Elektronernes Bevægelse forandres, idet alle de enkelte Elektroners Baner vil paavirkes i Kraftens Retning. Idet imidlertid de ydre Kræfter, der her vil være Tale om, vil være meget smaa i Sammenligning med de store Kræfter, hvormed Elektronerne paavirkes fra Metalmolekylernes Side, vil de ydre Kræfter kun have meget ringe Indflydelse paa Karakteren af de enkelte Elektroners Baner, og Elektronernes Hastighedsfordeling vil kun afvige meget lidt fra den normale, d. v. s. den, der vil være til Stede, naar der ingen ydre Kræfter virker. Som Følge af den ydre Krafts Indvirkning vil der dog i Middel være flere Elektroner, der bevæger sig i Kraftens Retning, end i den modsatte, og idet Elektronerne medfører elektriske Ladninger, vil der derfor fremkomme en Strømning af Elektricitet i Kraftens Retning. Idet Elektronerne er i Besiddelse af kinetisk Energi hidrørende fra deres Bevægelse, vil den ydre Kraft, idet den frembringer en Strømning af Elektronerne, ogsaa give Anledning til en Energistrømning (Varmestrømning) gennem Metallet.

Man kunde tænke sig, at der under Indvirkningen af en elektrisk Kraft vilde kunne indtræde en Ligevægt — svarende til den Ligevægt, der opstaar i en Beholder med en Luftart, der er underkastet Tyngdekraftens Paavirkning —, idet Elektronernes Koncentration under Kraftens Indflydelse blev større paa Steder med lavere Potential, indtil den Strømning af Elektroner, der skyldes den elektriske Kraft, vilde holdes i Ligevægt af den Strømning, der skyldtes Elektronernes Diffusion i modsat Retning. En saadan Ligevægt kan imidlertid ikke opstaa i det Indre af et homogent, ensartet opvarmet Metalstykke, idet Koncentrationsforskelle som de omtalte, paa Grund af Elektronernes overordentlig store Antal, vilde give Anledning til Dannelsen af meget store fri elektriske Ladninger, og disse igen til ydre elektriske Kræfter modsat rettede de oprindelige Kræfter og overordentlig mange Gange større end disse. I det Indre af et Metalstykke vil der derfor ikke kunne være elektrisk Ligevægt samtidig med, at der virker ydre elektriske Kræfter, og Elektronernes Koncentration vil kun kunne afvige overordentlig lidt fra den normale; kun i den allernærmeste Nærhed af Overfladen vil der indtræde særlige Forhold paa Grund af de her virkende store Kræfter, der forhindrer Elektronerne i at forlade Metallet.

Dersom Temperaturen i et Metalstykke ikke overalt er den samme, vil Middelværdien af Elektronernes absolute Hastigheder de forskellige Steder være forskellig, idet Hastighederne vil være større, hvor Temperaturen er højere, og der vil derfor finde en Vandring Sted af hurtige Elektroner i en Retning fra højere Temperatur til lavere, og af langsommere den modsatte Vej. En Temperaturforskel vil derfor ogsaa i saadanne Tilfælde, hvor der ingen elektrisk Strøm gaar gennem Metallet, fremkalde en Strømning af Energi fra Steder med højere Temperatur til Steder med lavere, idet de hurtigere Elektroner vil medføre mere kinetisk Energi end de langsommere. Endvidere vil en Temperaturforskel kunne give Anledning til en elektrisk Strøm, selv om der ikke virker ydre elektriske Kræfter, idet blandt andet Elektronerne med de forskellige absolute Hastigheder vil bevæge sig ulige hurtigt igennem Metallet. Paa lignende Maade som i det Tilfælde, hvor der virker en ydre Kraft paa Elektronerne, vil ogsaa her Elektronernes Bevægelser paa de enkelte Steder afvige meget lidt fra deres normale Bevægelser, idet de Afstande, indenfor hvilke Elektronernes Middel-Hastigheder vil variere kendeligt paa Grund af Temperaturens Variation, vil være meget store i Sammenligning med saadanne Afstande, indenfor hvilke de enkelte Elektroners Hastigheder under deres Bevægelse vil have forandret sig meget stærkt paa Grund af de Kræfter, hvormed de paavirkes fra Metalmolekylernes og de andre Elektroners Side.

I det Tilfælde endelig, hvor det betragtede Metalstykke ikke er homogent, d. v. s. hvor dets kemiske Sammensætning ikke er den samme overalt, skal det vel antages, at Middelværdien af Elektronernes absolute Hastigheder paa de forskellige Steder er den samme, dersom Temperaturen er den samme; men i dette Tilfælde vil Antallet af fri Elektroner paa Volumenenheden i Almindelighed ikke være det samme paa de forskellige Steder. Der vil derfor ogsaa i dette Tilfælde finde en Vandring af Elektroner Sted og en deraf resulterende Elektricitets- og Energistrømning. — Et lignende Forhold som det sidst betragtede vil i Virkeligheden ogsaa finde Sted i et homogent Metalstykke, i hvilket Temperaturen ikke overalt er den samme, idet det skal antages, at ikke alene Elektronernes absolute Hastigheder, men ogsaa deres Antal i Volumenenheden vil variere med Temperaturen. — Ligesom i det forrige Tilfælde skal vi ogsaa her antage, at Metallets Tilstand varierer saaledes, at Metallets Egenskaber (her f. Eks. Antallet af fri Elektroner i Volumenenheden) kun forandrer sig meget lidt indenfor de meget smaa Afstande, indenfor hvilke de enkelte Elektroners Hastigheder under deres Bevægelse allerede har forandret sig meget stærkt.

Vi skal nu i dette Kapitel søge Ligninger til Bestemmelse af saavel den Elektricitets- som den Energimængde, der ved Elektronernes Bevægelser føres gennem et Fladeelement i det Indre af Metallet, dersom der hersker en given ydre elektrisk Kraft, og dersom Metallets Temperatur og kemiske Sammensætning i dets enkelte Punkter er given. I de næste Kapitler skal vi derefter ved Hjælp af disse Ligninger behandle Spørgsmaalene om Metallernes Elektricitets- og Varmeledning, de thermoelektriske Fænomener o. s. v.

Ved Opstillingen af de omtalte Ligninger skal vi anvende den i den kinetiske Lufttheori almindelig benyttede, saakaldte statistiske Methode. Denne Methode er først blevet anvendt paa Elektrontheoriens Problemer af *H. A. Lorentz*[1]), der har vist, hvorledes man ved Hjælp af den paa forholdsvis simpel Maade kan undersøge Virkningerne af Elektronernes indviklede Bevægelse i Metallet.

Til at holde Regnskab med Elektronernes Bevægelser skal vi benytte to retvinklede 3-dimensionale Koordinatsystemer. I det ene Koordinatsystem angives en Elektrons Plads i det betragtede Øjeblik paa sædvanlig Maade med Koordinater (x, y, z). I det andet Koordinatsystem angives enhver Elektrons Hastighed ved et Punkt, hvis Koordinater, der betegnes med (ξ, η, ζ), sættes lig med Komposanterne

---

[1]) *Lorentz*: Proc. Acad. Amsterdam, vol. 7, p. 440, 1905.

efter Akserne i det første System af Elektronens Hastighed i det betragtede Øjeblik. Det Antal Elektroner, der til Tiden t befinder sig i Volumenelementet dv, samtidig med at deres Hastighedspunkter befinder sig i Volumenelementet (Hastighedselementet) $d\sigma$, skal nu betegnes ved

$$f(x, y, z, \xi, \eta, \zeta, t)\, d\sigma dv$$

(eller paa Steder, hvor det ikke kan misforstaas, simpelthen ved $fd\sigma dv$), hvor (x, y, z) betegner et Punkt i Elementet dv og (ξ, η, ζ) et Punkt i Elementet $d\sigma$. — Elementerne dv og $d\sigma$ skal i Almindelighed antages saa store, at det omhandlede Antal Elektroner er meget stort; dersom man imidlertid, som det vil være Tilfældet ved nogle af de betragtede Problemer, ikke kan gøre Elementerne saa store, at denne Betingelse er opfyldt, skal man blot ved det omtalte Antal forstaa Middelværdien af Antallet af Elektroner i de betragtede Elementer, taget over en vis Tid. —

Vi skal nu først betragte Fordelingen af de fri Elektroner i et homogent Metalstykke, der er upaavirket af ydre Kræfter, og i hvilket Temperaturen overalt er den samme.

At de omhandlede Elektroner er »fri« af Metalmolekylerne, skal ikke betyde, at de under deres Bevægelse ikke paavirkes af Kræfter fra Molekylernes Side, men blot at de under Vekselvirkningen med Metalmolekylerne optræder som selvstændige mekaniske Systemer, paa hvilke man kan anvende den statistiske Mekaniks Love.

Det maa her bemærkes, at det ikke vil være tilladt at gøre en saadan Antagelse om alle de Elektroner, der befinder sig i Metallet, nemlig ikke om dem, der tænkes »bundne« i stort Antal indenfor hvert enkelt Metalmolekyle; thi Konsekvenserne heraf vilde være, at Metallernes Varmefylde maatte være langt større end den, der findes ved Forsøgene. Det er derfor nødvendigt i mekanisk Henseende bestemt at skelne mellem de fri Elektroner og de i Molekylerne bundne Elektroner, hvilke sidste synes at være afskaarne fra Paavirkning fra andre Molekyler eller Elektroner paa en Maade, der ikke svarer til noget, man kender fra de almindelige mekaniske Systemer.

Det følger nu af Udviklingerne i den statistiske Mekanik, at de fri Elektroner i et Stykke Metal i den ovenfor beskrevne Tilstand vil indtage en Fordeling, der udtrykkes ved [1])

$$\left. \begin{aligned} f = A \cdot \int e^{-\frac{U}{kT}} \cdot dx_2\, dy_2\, dz_2\, d\xi_2\, d\eta_2\, d\zeta_2\, dx_3\, dy_3 \cdots \\ dx_N\, dy_N\, dz_N\, d\xi_N\, d\eta_N\, d\zeta_N\, dq_1\, dq_2 \cdots dq_n\, dp_1 \cdots dp_n, \end{aligned} \right\} \quad (1)$$

---

[1]) Se *P. Debye*: Ann. d. Phys., Bd. 33, p. 455, 1910.

hvor N er Antallet af fri Elektroner i en Volumenenhed af Metallet, og hvor $q_1, q_2 \cdots q_n$ er et Antal generaliserede Stedskoordinater og $p_1, p_2 \cdots p_n$ de dertil svarende generaliserede Bevægelsesmængdekomposanter $\left(p_r = \dfrac{\delta E_p}{\delta \dot{q}_r}, \text{ hvor } \dot{q}_r = \dfrac{dq_r}{dt} \text{ og } E_p \text{ er Molekylernes samlede kinetiske Energi}\right)$, tjenende til Bestemmelse af Metalmolekylernes øjeblikkelige Bevægelsestilstand, samt U den samlede Energi (kinetisk og potentiel) af hele Systemet, bestaaende af alle Elektronerne og Metalmolekylerne i en Volumenenhed af Metallet, svarende til bestemte Værdier $x, x_2, x_3, \cdots, x_N, y, y_2, \cdots \cdots, \eta_N, \zeta, \zeta_2, \cdots, \zeta_N$ for de N Elektroners Steds- og Hastigheds-Koordinater og bestemte Værdier for Størrelserne $q_1, q_2, \cdots, q_n, p_1, \cdots, p_n$. Endvidere er T den absolute Temperatur, k den i en Luftarts Tilstandsligning

$$pv = k \cdot NT$$

(hvor p er Trykket, v Volumenet og N Antallet af Molekyler) indgaaende universelle Konstant, og A en Konstant afhængig af Metallets Natur og Temperatur. Integrationen i Ligningen (1) tænkes udført over alle Værdier for Hastighedskoordinaterne og over alle Stedskoordinater indenfor en Volumenenhed af Metallet.

Dersom Elektronerne, saaledes som ved de Anvendelser, vi i det følgende skal gøre af Ligningen (1), tænkes at bevæge sig uafhængig af hverandre — d. v. s. dersom den potentielle Energi, der hidrører fra deres indbyrdes Paavirkning, kun under en forsvindende ringe Del af Elektronernes Bevægelser har en kendelig Værdi — i stationære Kraftfelter, faar man af (1), idet i dette Tilfælde

$$U = \left[\tfrac{1}{2}mr^2 + P(x, y, z)\right] + \left[\tfrac{1}{2}mr_2^2 + P(x_2, y_2, z_2)\right] + \cdots$$
$$+ \left[\tfrac{1}{2}mr_N^2 + P(x_N, y_N, z_N)\right] + Q(q_1, q_2, \cdots, q_n, p_1, \cdots p_n),$$

hvor m er en Elektrons Masse, og $r = \sqrt{\xi^2 + \eta^2 + \zeta^2}$ dens absolute Hastighed[1]), samt $P(x, y, z)$ en Elektrons potentielle Energi i Punktet (x, y, z),

---

[1]) Vi forudsætter her, ligesom overalt i det følgende, at Elektronernes Hastigheder vil være meget ringe i Sammenligning med Lysets Hastighed (Elektronernes Middelhastighed vil ved almindelig Temperatur være c. $\frac{1}{1000}$ af Lysets Hastighed (se f. Eks. *J. J. Thomson*: The Corpuscular Theory of Matter, p. 52)), og at derfor den Energi, der stammer fra en Elektrons Bevægelse, vil kunne skrives $\tfrac{1}{2}mr^2$, hvor m er en Konstant, uafhængig af Elektronens Hastighed.

$$f = K \cdot e^{-\frac{1}{kT}\left(\frac{m}{2}r^2 + P\right)}, \tag{2}$$

hvor K er en Konstant, der er afhængig af Metallets Natur og Temperatur.

Ligningen (1) — og den deri indeholdte Ligning (2) — giver blandt andet Udtryk for, at Elektronerne i ethvert nok saa lille Volumenelement af det betragtede Metalstykke i Middel bevæger sig i lige stort Antal i alle Retninger, samt at Middelværdien af de enkelte Elektroners kinetiske Energi — hvilken Middelværdi vil have samme Værdi indenfor hvert nok saa lille Volumenelement af Metallet — er lig Middelværdien af den kinetiske Energi, svarende til den fremadskridende Bevægelse af et Molekyle af en Luftart ved samme Temperatur.

Ved Beviserne for Udtryk som Ligning (1) for den statistiske Fordeling af en Samling Partikler forudsættes det i Almindelighed, at Partiklerne kun paavirkes af Kræfter, der er uafhængige af Partiklernes Hastighedskoordinater og kun afhængige af deres Stedskoordinater; man kan imidlertid meget simpelt bevise, at Ligningen (1) ogsaa vil gælde for Elektronernes Fordeling under Antagelsen af Tilstedeværelsen af saadanne Kræfter som de, der paavirker Elektronerne under deres Bevægelse i et magnetisk Felt[1]). Idet nu disse Kræfter staar vinkelret

---

[1]) Ligningen (1) udledes nemlig ved Hjælp af almindelige statistiske Betragtninger (se *Debye*: loc. cit., p. 445—455) udfra den Sætning (se f. Eks. *Gibbs*: Elementary Principles in statistical Mechanics, New York 1902, p. 3—11), at en Samling af ensartede mekaniske Systemer, af hvilke hvert enkelt System bestemmes ved de n generaliserede Stedskoordinater $q_1, q_2, \ldots, q_n$ og de dertil svarende n generaliserede Bevægelsesmængdekomposanter $p_1, p_2, \ldots, p_n$, vil forandre sig saaledes med Tiden: at, dersom man tænker sig de enkelte Systemers Bevægelsestilstand angivet ved Punkter i et $2n$-dimensionalt »retvinklet« Koordinatsystem ved Hjælp af Koordinaterne $q_1, q_2, \ldots, q_n, p_1, p_2, \ldots, p_n$ og betragter det Antal Systemer, hvis tilsvarende Punkter til Tiden t (Systemernes samlede Antal skal tænkes at være saa stort, at de til ethvert Tidspunkt kan betragtes som værende kontinuert fordelt over alle mulige Værdier af de $2n$ Koordinater) befinder sig indenfor et lille »Volumenelement« dv ($dv = dq_1 dq_2 \ldots dq_n dp_1 \ldots dp_n$), vil disse Systemers Punkter til et vilkaarligt senere Tidspunkt t_1 befinde sig indenfor et Volumenelement dv_1 af samme Størrelse som dv.

Denne Sætning vil være rigtig, dersom der for de omhandlede Systemer gælder følgende Betingelsesligning (se *Gibbs*: loc. cit., p. 11, Note):

$$\frac{\delta \dot{q}_1}{\delta q_1} + \frac{\delta \dot{p}_1}{\delta p_1} + \frac{\delta \dot{q}_2}{\delta q_2} + \frac{\delta \dot{p}_2}{\delta p_2} + \ldots + \frac{\delta \dot{q}_n}{\delta q_n} + \frac{\delta \dot{p}_n}{\delta p_n} = 0 \tag{1'}$$

(hvor ligesom ovenfor $\dot{q}_1 = \dfrac{dq_1}{dt}$ o. s. v.).

13

paa Elektronernes Bevægelsesretning, vil de intet Arbejde udføre, og Elektronerne vil derfor ingen potentiel Energi besidde i Forhold til de magnetiske Felter. Man ser derfor, at magnetiske Kræfter ikke vil have nogen Indflydelse paa Elektronernes statistiske Fordeling, dersom der er statistisk Ligevægt til Stede. (Se nærmere Kap. IV.)

Vi har i det foregaaende i mekanisk Henseende opfattet Elektronerne som Punkter, hvis hele Bevægelse fuldstændigt bestemmes ved de 6 Koordinater x, y, z, ξ, η og ζ; opfattes Elektronerne derimod som mekaniske Systemer med en vis Udstrækning, maa man for fuldstændig at bestemme Bevægelsen ogsaa tage Hensyn til Drejninger af Systemerne omkring deres Tyngdepunkter, og indføre tilsvarende Koordinater til Bestemmelse af denne sidste Art af Bevægelse. — Det maa dog her bemærkes, at Ligning (2) stadig vil gælde, dersom man blot betragter den fremadskridende Bevægelse, idet den statistiske Fordeling af denne Be-

Ligningen (1') følger, dersom de Kræfter, hvormed Systemets Partikler paavirkes, er uafhængige af Partiklernes Hastigheder, umiddelbart udfra de *Hamilton'ske* Ligninger (*Gibbs*: loc. cit., p. 4)

$$\dot{q}_r = \frac{\delta E_p}{\delta p_r} \quad \text{og} \quad \dot{p}_r = -\frac{\delta E_p}{\delta q_r} + F_r, \tag{2'}$$

hvor E_p er Systemets kinetiske Energi og $F_1, \ldots, F_n$ er de generaliserede Kraftkomposanter (i det omhandlede Tilfælde Funktioner af $q_1, q_2, \ldots, q_n$).

Dersom nu Partiklerne ogsaa paavirkes med saadanne Kræfter, som Elektroner, der bevæger sig i et magnetisk Felt, er underkastede, vil $F_1, \ldots, F_n$ ikke mere være uafhængige af de generaliserede Hastighedskomposanter $\dot{q}_1, \dot{q}_2, \ldots, \dot{q}_n$, men vil være lineære Funktioner af disse Størrelser. Idet de Kræfter, der hidrører fra det magnetiske Felt, staar vinkelret paa de enkelte Partiklers øjeblikkelige Bevægelsesretning, vil de imidlertid ikke kunne udføre noget Arbejde, og de vil derfor forsvinde ud af Udtrykket

$$dA = F_1 dq_1 + F_2 dq_2 + \ldots + F_n dq_n = (F_1 \dot{q}_1 + F_2 \dot{q}_2 + \ldots + F_n \dot{q}_n) dt = B dt$$

for det i Tiden dt af de Systemet paavirkende Kræfter udførte Arbejde. B vil derfor ogsaa i dette Tilfælde være en linenær Funktion af de generaliserede Hastighedskomposanter $\dot{q}_1, \dot{q}_2, \ldots, \dot{q}_n$, og man vil derfor have

$$\frac{\delta^2 B}{\delta \dot{q}_r \delta \dot{q}_s} = 0 \quad \left(\begin{matrix} r = 1, 2, \ldots, n \\ s = 1, 2, \ldots, n \end{matrix} \right).$$

Dette giver

$$\frac{\delta F_r}{\delta \dot{q}_s} + \frac{\delta F_s}{\delta \dot{q}_r} = 0. \tag{3'}$$

Vi har nu ved Hjælp af (2') og (3')

$$\sum_{r=1}^{r=n} \frac{\delta F_r}{\delta p_r} = \sum_{r=1}^{r=n} \sum_{s=1}^{s=n} \frac{\delta F_r}{\delta \dot{q}_s} \frac{\delta \dot{q}_s}{\delta p_r} = \sum_{r=1}^{r=n} \sum_{s=1}^{s=n} \frac{\delta F_r}{\delta \dot{q}_s} \frac{\delta^2 E_p}{\delta p_r \delta p_s} = 0. \tag{4'}$$

Af (2') og (4') følger nu (1'). q. e. d.

vægelse og af den drejende Bevægelse er uafhængige af hinanden. —
Efter den statistiske Mekanik vil nu den kinetiske Energi, der svarer til
en saadan Drejning, dersom der skal være statistisk Ligevægt, i Middel
være ligesaa stor som den, der svarer til den fremadskridende Bevægelse.
Man kan imidlertid næppe tilskrive Elektronerne en saadan Bevægelsesenergi, hidrørende fra Drejninger omkring deres Tyngdepunkter. Det er
nemlig, som bekendt, ved Forsøg over Forholdet mellem Luftarternes
Varmefylde ved konstant Tryk og ved konstant Rumfang vist, at Molekylerne af saadanne Luftarter, hvis Molekyler antages at bestaa af et
enkelt Atom — hvilke Molekyler dog sikkert maa antages at danne
langt mere udstrakte Systemer end de enkelte fri Elektroner —, ikke
besidder en saadan Energi.

Dersom der virker ydre Kræfter, og Metallets Temperatur og
kemiske Sammensætning ikke er ens i alle Punkter, vil Hastighedsfordelingen ikke mere udtrykkes ved Ligning (1), men ved Ligningen

$$f = f_0 + \psi, \tag{3}$$

hvor f_0 betegner den samme, af Metallets Natur og Temperatur i det
betragtede Punkt afhængige, Funktion af x, y, z, ξ, η og ζ, som udtrykkes ved højre Side af Ligningen (1), medens ψ efter de i det
foregaaende omtalte Forudsætninger vil være en Størrelse, der kun er
meget lille i Sammenligning med det første Led.

Ved alle de følgende Beregninger skal vi nu se bort fra saadanne
Led, der forholder sig til de betydende Led som ψ til f; derved faar
hele Theorien en meget simpel Karakter, og de søgte Udtryk for
Elektricitets- og Energistrømningen bliver lineært afhængige af Størrelsen af de ydre Kræfter, Temperaturvariationerne o. s. v. Disse
Resultater, der saaledes direkte udledes udfra Karakteren af det hele
Billede, som Elektrontheorien lægger til Grund, svarer fuldstændig til,
hvad man finder ved Forsøgene. Man finder f Eks. saaledes den
elektriske Strøm, saa nøje som man kan maale, fuldkommen proportional med den elektriske Kraft indenfor de meget vide Grænser,
indenfor hvilke Forsøgene er blevne anstillede [1]).

Idet vi nu gaar over til de nærmere Beregninger, skal vi betragte
to forskellige Tilfælde, nemlig først det simplere Tilfælde, ved hvilket
de fri Elektroner under langt den største Del af deres Bevægelse ikke
paavirkes af Kræfter fra Metalmolekylernes og de andre Elektroners
Side, og ved hvilket hele Vekselvirkningen mellem Elektronerne ind-

---

[1]) Se f. Eks. *E. Lecher:* Sitzungsber. d. Wiener. Akad. d. Wiss., math.-nat. Kl., Bd. 116
Abt. II a, p. 49, 1907.

byrdes og mellem disse og Metalmolekylerne saaledes sker ved adskilte Sammenstød, samt derefter det mere komplicerede Tilfælde, ved
hvilket Elektronerne under en stor Del af deres Bevægelse antages at
være paavirket af stærke Kræfter fra Metalmolekylernes Side.

---

§ 2.

Opstilling af Betingelsesligningerne for Elektronernes samlede Bevægelse i det Tilfælde, hvor man antager, at der finder adskilte Sammenstød Sted.

Det skal her antages, at de lineære Dimensioner af de
Omraader, indenfor hvilke Elektronerne og Metalmolekylerne paavirker hinanden med en kendelig Kraft, — Omraader, indenfor hvilke der skal siges at finde Sammenstød Sted, —
er meget smaa i Forhold til de Veje, som Elektronerne i
Middel gennemløber mellem to Sammenstød.

Foruden den nævnte Antagelse skal vi foreløbig kun forudsætte,
at Metalmolekylernes Egenskaber er ens i alle Retninger, og at denne
Symmetri skal bestaa paa hvert enkelt Sted, uafhængig af, om der
virker ydre Kræfter, eller om Temperaturen ikke overalt i Metallet er
den samme. Denne Antagelse er efter det i Indledningen omtalte
karakteristisk for den Art af Elektrontheori, vi her beskæftiger os
med, i hvilken Theori Metallernes Egenskaber forklares ved de ydre
Kræfters Indflydelse paa de fri Elektroner selv og ikke paa Metalmolekylerne.

Bortset fra de omtalte Antagelser skal Metalmolekylerne og Elektronerne kunne tænkes at paavirke hinanden paa ganske vilkaarlig
Maade. Metalmolekylerne behøver saaledes ikke alle at være ens eller
symmetrisk byggede o. s. v.

Paa Grund af den ovenfor omtalte Antagelse om Forholdet mellem
Elektronernes fri Vejlængder og Metalmolekylernes og Elektronernes
Dimensioner behøver vi her ved Beregning af Virkningen af Elektronernes Bevægelse (Elektricitets- og Energistrømningen) kun at betragte Elektronerne, naar de ikke befinder sig i Sammenstød. Af
Ligningen (2) Side 12 faar vi i det her omhandlede Tilfælde — idet P
er 0 (konstant) overalt, undtagen i Omraader, der er forsvindende

smaa i Forhold til hele Metallets Volumen — for Elektronernes Fordeling i et homogent, af ydre Kræfter upaavirket Metalstykke, der overalt har samme Temperatur,

$$f = K \cdot e^{-\frac{m}{2kT} r^2}.\tag{4}$$

Kaldes Antallet af fri Elektroner i Volumenenheden for N, faar
vi følgende Relation imellem N og K (idet man integrerer over alle
Hastighedselementer):

$$N = \int K \cdot e^{-\frac{m}{2kT} r^2} \cdot d\sigma = K \int_0^\infty e^{-\frac{m}{2kT} r^2} \cdot 4\pi r^2 dr = K \cdot \left(\frac{2\pi kT}{m}\right)^{\frac{3}{2}}.\tag{5}$$

Dersom der ikke er Ligevægt — d. v. s. dersom vi ikke længer
antager, at der ingen ydre Kræfter virker, samt at Metallets Temperatur
og kemiske Sammensætning er ens i alle Punkter — vil Fordelingen
efter Ligningen (3) Side 14 være

$$f = K \cdot e^{-\frac{m}{2kT} r^2} + \psi,\tag{6}$$

hvor ψ som omtalt vil være meget lille i Forhold til det første Led.

De Størrelser, det gælder om at bestemme, er den Elektricitetsmængde og den Energimængde der ved Elektronernes Bevægelse i
Tidsenheden føres gennem et Fladeelement i det Indre af Metallet,
hvilket Fladeelement vi her for Simpelheds Skyld skal vælge vinkelret
paa X-aksen. Disse Størrelser vil kunne bestemmes, dersom man for
nhver Værdi af den absolute Hastighed kender den samlede Bevægelsesmængde efter X-aksen, som indehaves af de Elektroner, der har
denne absolute Hastighed.

Den Bevægelsesmængde efter X-aksen, der indehaves af de Elektroner, der befinder sig i Volumenelementet dV — der her skal tænkes
saa stort, at det indeholder et stort Antal Metalmolekyler —, og hvis
Hastighedspunkter befinder sig indenfor en Kugleskal omkring Be
gyndelsespunktet med Radier r og $r + dr$, vil vi betegne ved $G_x(r)\, dr\, dV$.
For at søge en Ligning til Bestemmelse af $G_x(r)$ skal vi nu betragte
denne Størrelses Forandring med Hensyn til Tiden.

Den Tilvækst, som $G_x(r)\, dr\, dV$ faar i Tiden dt, vil dels skyldes
Elektronernes Bevægelse, ved hvilken der vil gaa Elektroner og dermed
Bevægelsesmængde ind og ud af det betragtede Volumenelement dV,
samt de ydre Kræfters accelererende Virksomhed, ved hvilken der
vil gaa Elektroner ind og ud af det betragtede Hastighedsomraade,

dels skyldes de Sammenstød mellem Metalmolekylerne og Elektronerne og mellem disse sidste indbyrdes, der finder Sted indenfor det betragtede Volumenelement.

Ved Beregningen af den Tilvækst i den betragtede Bevægelses-mængde, der skyldes de to førstnævnte Aarsager, vil det — idet vi som tidligere omtalt (Side 14) overalt bortkaster Led, der forholder sig til de betydende Led som ψ til f — være tilladt kun at tage Hensyn til Virkningen paa de Elektroner, der til det givne Tidspunkt hører til en Fordeling, der udtrykkes ved det første Led paa højre Side af Lighedstegnet i Ligning (6), idet der nemlig i Udtrykket for den søgte Tilvækst vil indgaa Størrelser, der afhænger af det omhandlede Led. — (Dersom de ydre Kræfter, der virker paa Elektronerne, stammer fra et magnetisk Felt, vil den Tilvækst i Bevægelsesmængden, som disse Kræfter frembringer paa en Fordeling, der udtrykkes ved det første Led i Formlen (6), dog være 0, og man maa derfor ved Under-søgelsen over Indflydelsen af et magnetisk Felt (*Hall-Effekt*) tage Hensyn til Led, der stammer fra ψ. Vi skal imidlertid i det følgende foreløbig kun beskæftige os med ydre elektriske Kræfter (hvilke Kræfter i Omegnen af det betragtede Punkt skal antages at have et Potential) og skal senere i Kapitel IV omtale Indflydelsen af magne-tiske Kræfter.) — Den Tilvækst i Bevægelsesmængden, der skyldes Sammenstødene, vil til Gengæld kun afhænge af ψ, det vil sige af Afvigelserne fra den normale Fordeling; at denne Tilvækst kan blive af samme Størrelsesorden som den først omtalte Tilvækst, skyldes den store Virkning af Kræfterne mellem Elektronerne og Metalmolekylerne (de overordentlig hyppige Sammenstød) i Forhold til Virkningen af de ydre Kræfter.

Ved Beregningen af den Tilvækst i Bevægelsesmængden, der skyldes de forskellige Aarsagers Indvirkning paa de Elektroner, der paa det betragtede Tidspunkt hører til en Fordeling, der udtrykkes ved

$$f_0 = K \cdot e^{-\frac{m}{2kT} r^2},$$

skal vi nu tænke os denne Fordeling delt i to, $f_0 = f_1 + f_2$, af hvilke vi skal sætte

$$f_1 = K_0 \cdot e^{-\frac{1}{kT_0}\left(\frac{m}{2} r^2 + \varepsilon\varphi\right)}, \tag{7}$$

hvor K_0 og T_0 betegner Værdierne af K og T i et Punkt (x_0, y_0, z_0), der tænkes beliggende indenfor det betragtede Volumenelement dV, og hvor ε betegner en Elektrons Ladning, samt φ de ydre Kræfters

Potential. Idet nu φ tænkes valgt saaledes, at det er o i Punktet (x_0, y_0, z_0), vil Fordelingen f_2 i et Punkt $(x, y, z) = (x_0 + x_1, y_0 + y_1, z_0 + z_1)$, der ligger meget nær ved Punktet (x_0, y_0, z_0), med Bortkastning af smaa Størrelser af samme Orden som Produkter af x_1, y_1 og z_1, kunne skrives

$$f_2 = K \cdot e^{-\frac{m}{2kT} r^2} \cdot \left[x_1 \left(\frac{\varepsilon}{kT} \frac{\delta \varphi}{\delta x} + \frac{1}{K} \frac{\delta K}{\delta x} + \frac{m r^2}{2kT^2} \frac{\delta T}{\delta x} \right) + y_1 \left(\frac{\varepsilon}{kT} \frac{\delta \varphi}{\delta y} + \cdots \right) + z_1 \left(\cdots \right) \right]. \tag{8}$$

Man beviser nu i den statistiske Mekanik, at en Fordeling, der udtrykkes ved f_1 (sammenlign Ligning (2) Side 12), vil være stationær, dersom φ er uafhængig af Tiden. Dette vil her betyde, at den Forandring i Tiden dt i Elektronernes Fordeling — og dermed ogsaa i Bevægelsesmængdens —, der skyldes Elektronernes Bevægelse, for de Elektroners Vedkommende, der i det betragtede Øjeblik hører til Fordelingen f_1, akkurat vil ophæves af de ydre Kræfters Virksomhed; og vi skal derfor kun undersøge den Tilvækst i Bevægelsesmængden i det betragtede Volumenelement og Hastighedsomraade, der skyldes Fordelingen f_2 [1]. Ved Beregningen heraf kan vi imidlertid se bort fra de ydre Kræfters Virksomhed, da saavel f_2 som dens Variationer efter Hastighedskoordinaterne ξ, η og ζ vil være uendelig smaa indenfor det betragtede Volumenelement dV; den Tilvækst i Bevægelsesmængden, der skyldes Elektronernes Bevægelse, vil derimod være af endelig Størrelse, idet f_2's Variationer efter Koordinaterne x, y og z er endelige. Da Fordelingen f_2 i ethvert enkelt Punkt (x, y, z) er uafhængig af Elektronernes Bevægelsesretning (kun afhængig af r), vil den Be-

---

[1] *P. Debye* (loc. cit., p. 476), der har forsøgt at beregne Metallernes Elektricitets- og Varmeledningsevne udfra de samme fysiske Forudsætninger som *Lorentz*, men ved Hjælp af en noget ændret Behandlingsmaade, gaar ved sine Beregninger (loc. cit., p. 477) ud fra, at Elektronernes Fordeling, dersom der virker ydre Kræfter, kun vil afvige meget lidt fra den stationære Fordeling (7) i Modstrid med, at Elektrontheorien, saaledes som vi har set, jo netop maa gaa ud fra, at Fordelingen kun afviger meget lidt fra den Fordeling (4), der vil være til Stede i et af ydre Kræfter upaavirket Metalstykke. Ved Beregningen af Elektronernes Bevægelse mellem Metalmolekylerne (loc. cit., p. 478--480) tager *Debye* endvidere ikke Hensyn til de ydre Kræfters Virkning — dersom der toges Hensyn dertil, vilde den af ham antagne Fordeling, som ovenfor omtalt, ikke give Anledning til nogen samlet Bevægelse af Elektronerne —. Af disse Grunde finder *Debye* (se loc. cit., p. 481, Lign. (75)), hvad han ikke selv har bemærket, en Værdi for Elektricitetsledningsevnen, der vel i Størrelse er lig den af *Lorentz* fundne, men som har modsat (urigtigt) Fortegn, d. v. s. at Elektriciteten efter *Debye's* Beregninger vil bevæge sig i modsat Retning af den elektriske Kraft.

vægelsesmængde, der ved Elektronernes Bevægelse føres gennem et Fladeelement, have en Retning vinkelret paa Elementet. Den Bevægelsesmængde efter X-aksen, der med Elektronerne i det betragtede Hastighedsomraade føres gennem et Fladeelement $dS = dydz$ vinkelret paa X-aksen i Tiden dt, vil nu være

$$dydzdt \int m\xi \cdot \xi f_2 \, d\sigma = \frac{4\pi}{3} mr^4 f_2 \, drdydzdt,$$

hvor Integrationen er udført over et Hastighedsomraade, der begrænses af to Kugleflader omkring Begyndelsespunktet med Radierne r og $r + dr$. Den søgte Tilvækst i Bevægelsesmængden i Volumenelementet $dV = dxdydz$ bliver derfor

$$- dx \frac{\delta}{\delta x} \left(\frac{4\pi}{3} mr^4 f_2 \, drdydzdt \right) = - \frac{4\pi}{3} mr^4 \frac{\delta f_2}{\delta x} \, dr \, dV dt.$$

Idet vi nu benytter Ligningen (8) og bortkaster uendelig smaa Størrelser af højere Orden med Hensyn til x_1, y_1 og z_1, faar vi derfor for den Tilvækst i Bevægelsesmængden $G_x(r) \, dr \, dV$, der skyldes Elektronernes Bevægelse og de ydre Kræfters Virksomhed i Tiden dt [1]),

$$- \frac{4\pi}{3} mK \left(\frac{\varepsilon}{kT} \frac{\delta \varphi}{\delta x} + \frac{1}{K} \frac{\delta K}{\delta x} + \frac{mr^2}{2kT^2} \frac{\delta T}{\delta x} \right) r^4 e^{-\frac{mr^2}{2kT}} \, dr \, dV dt. \quad (9)$$

Vi skal nu undersøge den Tilvækst i den betragtede Bevægelsesmængde, der skyldes Sammenstødene mellem Elektronerne og Metalmolekylerne og mellem Elektronerne indbyrdes. Paa Grund af den Antagelse, vi har gjort Side 15 om Forholdet mellem de af Elektro-

---

[1]) *H. A. Wilson* (loc. cit., p. 836) gaar ved sine i det foregaaende (Side 4, Note 2) omtalte Beregninger ud fra, at den Tilvækst i Bevægelsesmængden $G_x(r) \, dr \, dV$, der skyldes de ydre Kræfter, vil være lig $- \varepsilon \frac{\delta \varphi}{\delta x} dNdt$, hvor dN er Antallet af Elektroner, der befinder sig i Volumenelementet dV, og hvis absolute Hastigheder ligger mellem r og $r + dr$. Idet $dN = 4\pi r^2 Ke^{-\frac{mr^2}{2kT}} dr \, dV$, bliver det af *Wilson* benyttede Udtryk for Tilvæksten i Bevægelsesmængden $G_x(r) \, dr \, dV$ lig $- 4\pi K\varepsilon \frac{\delta \varphi}{\delta x} r^2 e^{-\frac{mr^2}{2kT}} dr \, dV dt$. Dette Udtryk er imidlertid, som man ser ved Sammenligning med det ovenstaaende Udtryk (9) ikke rigtigt. Fejlen hidrører fra, at den søgte Tilvækst sættes lig Tilvæksten i Bevægelsesmængde i Tiden dt af de Elektroner, der i det betragtede Øjeblik hører til den omhandlede Gruppe, uden at der tages Hensyn til, at der ved de ydre Kræfters Virksomhed i Tiden dt vil gaa Elektroner ud og ind af denne Gruppe.

2*

nerne mellem to Sammenstød i Middel gennemløbne fri Vejlængder og Elektronernes og Metalmolekylernes Dimensioner, vil i det her betragtede Tilfælde Sandsynligheden for, at en Elektron indenfor en vis lille Tid *dt* — der skal antages at være stor i Forhold til saadanne Tider, der medgaar til de enkelte Sammenstød, men lille i Forhold til de Tider, som Elektronerne i Middel bruger om at tilbagelægge deres fri Vejlængder mellem to Sammenstød — paa en eller anden bestemt Maade vil støde sammen med et Metalmolekyle eller en anden Elektron, kun være afhængig af Elektronens Hastighed, d. v. s. uafhængig af den Bane, Elektronen har beskrevet siden forrige Sammenstød.

Vi skal her først betragte Virkningen af Sammenstødene mellem Elektronerne og Metalmolekylerne. Da Molekylernes Egenskaber i Middel er ens i alle Retninger, vil Fordelingen af Hastighedspunkterne for Elektroner, der før Sammenstødet har en bestemt Hastighed og Retning, efter et Sammenstød i Middel være symmetrisk med Hensyn til Elektronernes Retning før Stødet. Den Bevægelsesmængde, der i Tiden *dt* ved Sammenstød i Volumenelementet *dV* af Elektroner, hvis Hastighedspunkter før Stødet befinder sig i et lille Element *dσ'* omkring Punktet (ξ', η', ζ'), bringes ind i et Hastighedsomraade, der begrænses af to Kugleflader omkring Begyndelsespunktet med Radierne *r* og *r + dr*, vil derfor have samme Retning som Radiusvektor til Punktet (ξ', η', ζ'), og i Størrelse (idet man sætter $\sqrt{(\xi')^2 + (\eta')^2 + (\zeta')^2} = \rho$, og ved $Q(\rho, r)$ betegner en Funktion af ρ og r) kunne udtrykkes ved

$$m\rho \cdot f(\xi', \eta', \zeta')\, d\sigma' \cdot Q(\rho, r)\, dr\, dV\, dt.$$

Komposanten efter *X*-aksen af denne Bevægelsesmængde vil derfor være

$$m\xi' \cdot f(\xi', \eta', \zeta')\, d\sigma' \cdot Q(\rho, r)\, dr\, dV\, dt,$$

og den hele Bevægelsesmængde efter *X*-aksen, der bringes ind i det betragtede Hastighedsomraade ved Sammenstød af Elektroner, der før Stødet har Hastighedspunkter i et Omraade, der begrænses af to Kugleflader med Radier ρ og $\rho + d\rho$, vil derfor være

$$G_x(\rho)\, d\rho \cdot Q(\rho, r)\, dr\, dV\, dt.$$

Den samlede Bevægelsesmængde efter *X*-aksen, der ved Sammenstødene bringes ind i Hastighedsomraadet mellem Kuglefladerne med Radier *r* og *r + dr*, er derfor

$$dr\, dV\, dt \cdot \int_0^\infty G_x(\rho)\, Q(\rho, r)\, d\rho.$$

Paa ganske tilsvarende Maade slutter man, at den Bevægelsesmængde efter X-axen, der ved Sammenstødene i Volumenelementet dV
føres ud af Hastighedsomraadet mellem Kuglefladerne med Radier r
og $r + dr$, vil kunne udtrykkes ved

$$dr\, dV\, dt \cdot G_x(r)\, F(r),$$

hvor $F(r)$ betegner en Funktion af r alene. Den Tilvækst i Bevægelsesmængden $G_x(r)\, dr\, dV$, der skyldes Sammenstødene mellem Elektronerne og Metalmolekylerne i Tiden dt, bliver altsaa [1])

$$-\left[\left(G_x(r)\, F(r) - \int_0^\infty G_x(\rho)\, Q(\rho, r)\, d\rho\right]\, dr\, dV\, dt. \tag{10}$$

Uden nærmere Antagelser om Elektronernes og Metalmolekylernes
Art kan man ikke fuldstændig bestemme Funktionerne $F(r)$ og $Q(\rho, r)$.
Vi kan imidlertid vise, at $Q(\rho, r)$ altid maa opfylde en vis Betingelse,
dersom Vekselvirkningen mellem Metalmolekylerne og de fri Elektroner,
som her antaget, foregaar efter de almindelige mekaniske Love.

Det Antal Elektroner, der i Middel i Tiden dt ved Sammenstød
i Volumenelementet dV mellem saadanne Elektroner, hvis Hastigheds-

---

[1]) *J. H. Jeans* (Phil. Mag., vol. 17, p. 775, 1909) gaar i sit, i det foregaaende (Side 4) omtalte, Forsøg paa at give en almindelig Theori for de med Metallernes Absorption og
Emission af Varmestraaler forbundne Fænomener ved Opstillingen af den Betingelsesligning, der lægges til Grund for Beregningerne, ud fra den Antagelse, at det
samlede Tab af Bevægelsesmængde i en bestemt Retning, som Elektronerne lider
i Tidsenheden ved Sammenstødene med Metalmolekylerne, kan sættes lig Elektronernes samlede Bevægelsesmængde i den omhandlede Retning, multipliceret med
en Konstant, der kun er afhængig af Metallets Natur og Temperatur. Denne Antagelse er imidlertid i Almindelighed ikke rigtig (smlg. *H. A. Wilson:* loc. cit.,
p. 836). Det omhandlede Tab af Bevægelsesmængde efter X-aksen udtrykkes nemlig
med vore Betegnelser efter Udtrykket (10) ved

$$\int_0^\infty \left[G_x(r)\, F(r) \doteq \int_0^\infty G(\varrho)\, Q(\varrho, r)\, d\varrho \right]\, dr, \qquad *$$

hvilket Udtryk — idet Elektronernes Bevægelsesmængde ved de forskellige Problemer ikke fordeler sig paa samme Maade mellem Elektronerne med de forskellige absolute Hastigheder — i Almindelighed ikke vil være proportional med den
samlede Bevægelsesmængde efter X-aksen

$$\int_0^\infty G_x(r)\, dr.$$

Jeans's Beregningsmaade giver derfor kun rigtige Resultater i enkelte specielle
Tilfælde, saaledes f. Eks. naar $F(r) = $ Konstant, og $Q(\varrho, r) = 0$ (se Side 33).

* $[G(\rho)$ should be $G_x(\rho).]$

punkter før Sammenstødet befinder sig i et lille Volumenelement $d\sigma'$ omkring Punktet (ξ', η', ζ'), og Metalmolekylerne bringes ind i et lille Hastighedsomraade $d\sigma$ omkring Punktet (ξ, η, ζ), vil vi betegne ved

$$\lambda(\xi', \eta', \zeta', \xi, \eta, \zeta) \cdot f(\xi', \eta', \zeta') \, d\sigma \, d\sigma' d V dt. \qquad (11)$$

Funktionen λ, der er uafhængig af Elektronernes Fordeling f, vil, da Metalmolekylernes Egenskaber antages i Middel at være ens i alle Retninger, kun afhænge af den relative Stilling af Punkterne (ξ', η', ζ'), (ξ, η, ζ) og Begyndelsespunktet. $\lambda(\xi', \eta', \zeta', \xi, \eta, \zeta)$ kan derfor skrives $\chi(\rho, r, \vartheta)$, hvor ρ og r er Radivektorerne henholdsvis til Punktet (ξ', η', ζ') og til Punktet (ξ, η, ζ), og hvor ϑ er Vinklen imellem disse Radivektorer. Vi skal nu vise, at Funktionen $\chi(\rho, r, \vartheta)$ maa opfylde en vis Betingelse, hvilket kan indses ved at betragte det Tilfælde, hvor der ingen ydre Kræfter virker, og Temperaturen overalt er den samme, og hvor derfor Elektronernes stationære Fordeling vil svare til Formlen (4) Side 16.

Tænker vi os en bestemt Fordeling af de enkelte Elektroners og Metalmolekylers Steds- og Hastighedskoordinater — (idet de enkelte Metalmolekyler ikke antages at være simple Partikler, hvis Egenskaber er ens i alle Retninger, vil der i Almindelighed behøves mere end 6 Koordinater til fuldstændig at karakterisere et Molekyles Tilstand) —, og betragter vi dernæst en Fordeling med de samme Stedskoordinater og med Hastighedskoordinaterne lige store med modsat Tegn, vil, idet de enkelte Sammenstød, da de antages at foregaa efter de almindelige mekaniske Love, vil kunne gaa for sig i modsat Retning, hele Bevægelsen i Metallet ved den sidste Fordeling være akkurat modsat den ved den første Fordeling (d. v. s. akkurat de samme Tilstande vil følge efter hinanden med samme Mellemrum, blot i omvendt Rækkefølge). Idet nu saadanne to »modsatte« Fordelinger af de enkelte Elektroners og Metalmolekylers Steds- og Hastighedskoordinater i et homogent Metalstykke, der er upaavirket af ydre Kræfter, og hvis Temperatur overalt er den samme, vil være lige sandsynlige[1]), vil i det betragtede Tilfælde det Antal Elektroner, der angives ved Udtrykket (11), i Middel være lig det Antal Elektroner, hvis Hastighedspunkter i samme Tid

---

[1]) Dette følger umiddelbart ud fra den Maade, hvorpaa en Fordelings Sandsynlighed defineres i den statistiske Mekanik (jvnfr. f. Eks. *Jeans*: Dynamical Theory of Gases, Cambridge 1904, Kap. 3—5). (Uden en saadan gennemført rationel Behandling af Begrebet en Fordelings Sandsynlighed, som den, der findes paa det citerede Sted, kunde en Betragtning, som den i Texten staaende, paa Forhaand forekomme usikker; se f. Eks. *Boltzmann*: Vorlesungen über Gastheorie, Leipzig 1896, I, p. 42—45).

ved Sammenstødene bringes fra et Element $d\sigma_1$, symmetrisk med $d\sigma$ med Hensyn til Begyndelsespunktet, til et Element $d\sigma'_1$, symmetrisk med $d\sigma'$. Dette Antal er med Benyttelse af de ovenfor indførte Betegnelser lig

$$\chi(r, \rho, \vartheta) \cdot f(-\xi, -\eta, -\zeta)\, d\sigma'_1 d\sigma_1\, dV dt,$$

og man faar derfor ved Hjælp af Ligningen (3) *

$$\chi(\rho, r, \vartheta) \cdot e^{-\frac{m}{2kT}\rho^2} = \chi(r, \rho, \vartheta)\; e^{-\frac{m}{2kT}r^2},$$

hvoraf følger

$$\chi(\rho, r, \vartheta) = e^{-\frac{m}{2kT}r^2} \cdot s(\rho, r, \vartheta), \qquad (12)$$

hvor $s(\rho, r, \vartheta)$ er en symmetrisk Funktion af ρ og r.

Den Bevægelsesmængde, der i Tiden dt, ved Sammenstødene mellem Elektronerne og Metalmolekylerne i Volumenelementet dV, af Elektroner, hvis Hastighedspunkter før Stødet befinder sig i et lille Element $d\sigma'$ omkring Punktet (ξ', η', ζ'), bringes ind i et Hastighedsomraade, der begrænses af to Kugleflader omkring Begyndelsespunktet med Radierne r og $r + dr$, — (og hvis Retning, efter hvad der tidligere er sagt, falder sammen med Retningen af Radiusvektoren til (ξ', η', ζ'), medens dens Størrelse, med de i det foregaaende (Side 20) indførte Betegnelser, udtrykkes ved $m\rho \cdot f(\xi', \eta', \zeta')\, d\sigma' \cdot Q(\rho, r)\, dr\, dV dt$) — vil nu være

$$dV dt \cdot f(\xi', \eta', \zeta')\, d\sigma' \cdot \int_{\varphi=0}^{\varphi=2\pi} \int_{\vartheta=0}^{\vartheta=\pi} mr \cos\vartheta \cdot \chi(\rho, r, \vartheta) \cdot r^2 \sin\vartheta\, dr\, d\vartheta\, d\varphi =$$

$$dV dt \cdot f(\xi', \eta', \zeta')\, d\sigma' \cdot 2\pi mr^3 e^{-\frac{m}{2kT}r^2} dr \cdot \int_0^\pi s(\rho, r, \vartheta) \cos\vartheta \sin\vartheta\, d\vartheta.$$

Man faar altsaa

$$Q(\rho, r) = r^4 e^{-\frac{m}{2kT}r^2} \cdot S(\rho, r), \qquad (13)$$

hvor

$$S(\rho, r) = \frac{2\pi}{r\rho} \int_0^\pi s(\rho, r, \vartheta) \cos\vartheta \sin\vartheta\, d\vartheta$$

er en symmetrisk Funktion af ρ og r.

* ["Ligningen (3)" should read "Ligningen (4)".]

Vi skal nu gaa over til at betragte Virkningen af Sammenstødene mellem Elektronerne indbyrdes. Idet Hastighederne af de Elektroner, som den enkelte Elektron støder sammen med, ikke er ens fordelt i alle Retninger, vil her, i Modsætning til hvad der fandt Sted i det ovenfor betragtede Tilfælde, Elektronens Hastighed efter Stødet i Middel ikke være fordelt symmetrisk med Hensyn til Hastigheden før Stødet. Er imidlertid Afvigelserne fra den normale Fordeling, saaledes som det her antages at være Tilfældet, meget ringe, kan man dog, som vi skal se, beregne Virkningen af Sammenstødene paa ganske tilsvarende Maade som ovenfor.

Vi skal tænke os Elektronerne til ethvert Tidspunkt paa en eller anden Maade delt i to Grupper, hvoraf den ene A skal svare til den normale Fordeling $f = K \cdot e^{-\frac{m}{2kT}r^2}$, medens den anden B, der vil være meget lille i Forhold til A, svarer til en Fordeling $f = \psi$ (se Ligningen (6) Side 16). Ved Beregningen af Virkningerne af Sammenstødene skal vi nu betragte dels Sammenstød imellem Elektroner af samme Gruppe dels Sammenstød mellem Elektroner af de to forskellige Grupper. Vi har nu:

1) Virkningen af de indbyrdes Sammenstød mellem Elektronerne i Gruppen A vil, da Hastighedsfordelingen er den normale, ikke have nogen Indflydelse paa denne Fordeling til Følge.

2) Virkningen af Sammenstødene af Elektronerne i Gruppen B med Elektronerne i Gruppen A vil paa Grund af Gruppen A's Symmetri kunne beregnes paa ganske tilsvarende Maade som den, vi benyttede ved Sammenstødene mellem Elektronerne og Metalmolekylerne. Opfattes nemlig Elektronerne i Gruppen B ved Sammenstødene som de stødende Elektroner, vil Hastighedspunkterne efter Stødet, af saavel den stødte som den stødende Elektron, i Middel være symmetrisk fordelt med Hensyn til den stødende Elektrons Hastighedsretning før Stødet. De betragtede Sammenstød giver derfor en Tilvækst til Bevægelsesmængden $G_x(r)\,dr\,dV$, der kan skrives paa ganske samme Form som Udtrykket (10) Side 21.

3) Virkningen af de indbyrdes Sammenstød mellem Elektronerne i Gruppen B vil være forsvindende paa Grund af disse Sammenstøds forholdsvis overordentlig ringe Antal.

Vi kan nu endvidere vise, hvorledes Funktionen $Q(\rho, r)$ ogsaa i det her omhandlede Tilfælde vil opfylde den Betingelse, som angives ved Ligning (13). Af det ovenstaaende fremgaar det, at Virkningen af de indbyrdes Sammenstød mellem de fri Elektroner vil være den samme som Virkningen af Sammenstødene mellem disse Elektroner — (som vi her skal opfatte som stødende Partikler) — og en Samling Elektroner, hvis Fordeling udtrykkes ved $f_0 = K \cdot e^{-\frac{m}{2kT}r^2}$.

Det samlede Antal Elektroner, der i Tiden dt i Volumenelementet dV ved saadanne Sammenstød, ved hvilke den stødende Elektron før Stødet befinder sig i et lille Hastighedsomraade $d\sigma'$, bringes ind i et lille Hastighedsomraade $d\sigma$, skal vi her i Lighed med i det forrige Tilfælde betegne ved

$$\chi_1\,(\rho,\,r,\,\vartheta)\cdot f\,(\xi',\,\eta',\,\zeta')\,d\sigma'd\sigma\,dVdt.$$

Ved Beregningen af $\chi_1(\rho,\,r,\,\vartheta)$, der kun er afhængig af Fordelingen f_0, maa vi her, foruden til saadanne Sammenstød, ved hvilke den stødende Elektron bringes fra Hastighedsomraadet $d\sigma'$ til Hastighedsomraadet $d\sigma$, ogsaa tage Hensyn til saadanne Sammenstød, ved hvilke den stødte Elektron bringes ind i Hastighedsomraadet $d\sigma$, samt endelig til saadanne, ved hvilke den stødte Elektron før Sammenstødet befinder sig i dette lille Hastighedsomraade og derfor ved Stødet bringes ud deraf. Den første Art af Sammenstød svarer ganske til den, der blev betragtet ovenfor ved Beregningen af Virkningen af Sammenstødene mellem de fri Elektroner og Metalmolekylerne; disse Sammenstød vil derfor give et Bidrag til $\chi_1(\rho,\,r,\,\vartheta)$ af samme Form som Udtrykket paa højre Side af Ligningen (12) Side 23. Samme Resultat vil man ogsaa faa ved den anden Art af Sammenstød; ogsaa ved saadanne Sammenstød vil nemlig de »modsatte« Sammenstød give tilsvarende Bidrag til $\chi_1(r,\,\rho,\,\vartheta)$, som de »direkte« Sammenstød til $\chi_1(\rho,\,r,\,\vartheta)$. Dette sidste vil vel ikke være opfyldt ved den tredje Art af Sammenstød; i dette Tilfælde indser man imidlertid umiddelbart, at det Bidrag, som de omhandlede Sammenstød vil give til Størrelsen $\chi_1(\rho,\,r,\,\vartheta)$, vil forholde sig til det Bidrag, som den samme Art af Sammenstød vil give til $\chi_1(r,\,\rho,\,\vartheta)$, som

$$f_0\,(\xi,\,\eta,\,\zeta) : f_0\,(\xi',\,\eta',\,\zeta') = e^{-\frac{m}{2kT}r^2} : e^{-\frac{m}{2kT}\rho^2}.$$

Vi faar derfor, at Størrelsen $\chi_1(\rho,\,r,\,\vartheta)$ kan skrives paa samme Form som den, der udtrykkes ved højre Side af Ligningen (12) Side 23, og at derfor ogsaa i Tilfælde af Sammenstød mellem de fri Elektroner indbyrdes $Q(\rho,\,r)$ vil have den Form, der udtrykkes ved Ligning (13).

Betegner vi nu den Tilvækst, som $G_x(r)$ modtager i Tiden dt, med $\dfrac{dG_x(r)}{dt}\,dt$, faar vi ved Hjælp af Udtrykkene (9) og (10)

$$\left[\frac{dG_x(r)}{dt}\right] = -\frac{4\pi}{3}\,mK\left(\frac{\varepsilon}{kT}\frac{\delta\varphi}{\delta x} + \frac{1}{K}\frac{\delta K}{\delta x} + \frac{mr^2}{2kT^2}\frac{\delta T}{\delta x}\right)r^4 e^{-\frac{m}{2kT}r^2}$$
$$- G_x(r)\,F(r) + \int_0^\infty G_x(\rho)\,Q(\rho, r)\,d\rho, \qquad (14)$$

hvor Parenthesen om Leddet paa venstre Side af Lighedstegnet skal betyde, at Differentialkvotienten her er at opfatte som en Middelværdi for denne Størrelse under de forhaandenværende Omstændigheder — d. v. s. for givne Værdier for de ydre Kræfter, Temperaturens Variation o. s. v., samt for Størrelserne $G_x(\rho)$ —.

Vi skal her straks omtale Betydningen af Leddet paa venstre Side af Ligning (14) ved de forskellige Anvendelser, vi i det følgende skal gøre af denne Ligning

Ved de fleste Anvendelser søger man blot en Middelværdi af $G_x(r)$ under givne ydre Omstændigheder, d. v. s. for givne Værdier af de ydre Kræfter, Temperaturens Variation o. s. v. Ved saadanne Anvendelser vil, idet Størrelserne $F(r)$ og $Q(\rho, r)$ er meget store (dette er en Følge af den i det foregaaende omtalte Antagelse om, at de enkelte Elektroners oprindelige Bevægelsesmængde i en bestemt Retning meget hurtigt vil mistes ved Sammenstødene), Leddet paa venstre Side være meget ringe i Sammenligning med de sidste Led paa højre Side, undtagen i saadanne særlige Tilfælde, hvor $G_x(r)$ forandrer sig overordentlig hurtigt med Tiden, d. v. s. saa hurtigt, at den har forandret sig mærkeligt indenfor de meget korte Tidsrum, i hvilke Elektronerne mister Størstedelen af deres oprindelige Bevægelsesmængde. Dette sidste vil finde Sted, dersom det første Led paa højre Side forandrer sig meget hurtigt med Tiden; af de i dette Led indgaaende Størrelser vil kun de ydre Kræfter kunne forandre sig meget hurtigt, hvilket f. Eks. skal antages at være Tilfældet ved Beregningen af Metallernes Absorption for Varmestraaler. — (Paa Grund af den Maade, hvorpaa Ligning (13) er udledet, vil den dog, som vi senere (Kap. III) nærmere skal omtale, kun kunne benyttes til Beregning af Absorptionen for Varmestraaler med saa store Svingningstider, at de ydre Kræfter kun vil forandre sig meget lidt indenfor Tider af samme Størrelsesorden som de Tider, der medgaar til de enkelte Sammenstød; Tider, der her er antagne at være meget smaa i Forhold til de Tider, indenfor hvilke de enkelte Elektroner i Middel mister en betydelig Del af deres oprindelige Bevægelsesmængde i en bestemt Retning.) — I alle andre Tilfælde af den her omhandlede

Art vil man derimod kunne se bort fra Leddet paa venstre Side af Ligning (14) og altsaa sætte denne Side af Ligningen lig 0. Saada.ne Problemer som de sidst omtalte, ved hvilke Tilstanden vel antages at variere med Tiden, men dog saa langsomt, at de enkelte Tilstande kan behandles som Ligevægtstilstande, vil vi, med en almindelig anvendt Betegnelse, i det følgende kalde »quasi-stationære«.

Foruden til saadanne Anvendelser, hvor man søger en Middel-værdi af $G_x(r)$ for givne ydre Omstændigheder, skal Ligningen (14) ogsaa benyttes til Undersøgelse af de smaa, meget hurtige Variationer i $G_x(r)$, der skyldes de enkelte Elektroners uregelmæssige Bevægelse i Metallet, og som f. Eks. antages at give Anledning til Metallernes Udsendelse af Varmestraaler. Ved saadanne Anvendelser vil Leddet paa venstre Side spille en Hovedrolle, idet det angiver Middelværdien for $\dfrac{dG_x(r)}{dt}$ for en given Fordeling af Elektronernes samlede Bevægelses-mængde blandt Elektronerne med de forskellige absolute Hastigheder.

§ 3.

Specielle Eksempler paa Beregningernes Gennemførelse i Tilfælde, hvor man antager, at der finder adskilte Sammenstød Sted.

Ligningen (14) er den fuldt almindelige Ligning til Bestemmelse af Elektronernes samlede Bevægelsesmængde i det Tilfælde, at der finder adskilte Sammenstød Sted. Den tillader, sammenholdt med Ligningen (13), som vi skal se, allerede udfra sin Form at undersøge Sammenhængen mellem visse af Metallernes Egenskaber. For nær-mere at bestemme Størrelserne $F(r)$ og $Q(\rho, r)$, hvad der er nødven-digt for at naa kvantitative Resultater, maa man imidlertid gøre spe-cielle Antagelser om Virkningen af Sammenstødene mellem Elektronerne og Metalmolekylerne og mellem Elektronerne indbyrdes. Før vi gaar videre, skal vi derfor her omtale nogle Eksempler paa Indførelsen af saadanne Antagelser.

Efter *H. A. Lorentz's* Theori[1]) beregnes, som omtalt i det fore-gaaende (Side 3), Virkningen af Sammenstødene udfra den Antagelse,

---

[1]) *Lorentz*: Proc. Acad. Amsterdam, vol. 7, p. 439, 1905.

at Elektronerne og Metalmolekylerne paavirker hinanden som absolut haarde elastiske Kugler. Saavel Elektronernes Dimensioner som deres Masser tænkes endvidere at være saa smaa i Forhold til Metalmolekylernes, at man dels kan se bort fra Sammenstødene mellem de fri Elektroner indbyrdes i Forhold til Sammenstødene mellem Elektronerne og Metalmolekylerne dels kan betragte Metalmolekylerne som ubevægelige i Forhold til Elektronerne. Som man let kan overbevise sig om, vil Virkningen af et Sammenstød under de omtalte Omstændigheder være den, at Elektronen forlader Metalmolekylet med uforandret absolut Hastighed, medens dens Bevægelsesretning efter Stødet vil være fuldkommen uafhængig af Retningen før Stødet, d. v. s. jævnt fordelt i alle Rumvinkler [1]); Elektronen vil derfor ved et Sammenstød i Middel miste hele sin oprindelige Bevægelsesmængde i en bestemt Retning.

Idet Elektronernes absolute Hastighed ikke forandres ved Sammenstødene, vil $Q(r, \rho)$ i det omhandlede Tilfælde være lig o, medens $F(r)$ kan beregnes paa følgende Maade. Da Metalmolekylerne er antagne at være ubevægelige, vil de fri Vejlængder, som Elektronerne gennemløber mellem to paa hinanden følgende Sammenstød, være uafhængige af Elektronernes Hastighed. Kaldes Middelvejlængden nu l, bliver det Antal Sammenstød, som de Elektroner, hvis absolute Hastighed ligger mellem r og $r + dr$, lider i Tiden dt i Volumenelementet dV, — (idet Antallet af de betragtede Elektroner i Volumenelementet dV betegnes ved $N(r)\,dr\,dV$) — lig

$$\frac{r\,dt}{l} \cdot N(r)\,dr\,dV.$$

Den Bevægelsesmængde efter X-aksen, som en Elektron med absolut Hastighed r i Middel besidder — og efter det i det foregaaende (Side 20) omtalte altsaa ogsaa umiddelbart før et Sammenstød —, er nu $\dfrac{G_x(r)}{N(r)}$, og den samlede Bevægelsesmængde efter X-aksen, der mistes ved de omtalte Sammenstød, er derfor

$$\frac{r}{l} \cdot G_x(r)\,dr\,dV\,dt.$$

Efter Udtrykket (10) Side 21 faar man altsaa

$$F(r) = \frac{r}{l}. \tag{15}$$

P. Gruner [2]) har gjort Forsøg paa at udvide Lorentz's Theori ved i Beregningerne at indføre, hvad han kalder Metalmolekylernes »Ioni-

---

[1]) Se *Maxwell*: Scientific Papers, vol. I, p. 379.
[2]) *Gruner*: Verh. d. Deutsch. Phys. Ges., Bd. 10, p. 509, 1908.

sation« og »Elektronbinding«. *Gruner* antager saaledes, at de fri Elektroner under særlige Omstændigheder kan bindes midlertidigt af Metalmolekylerne og først igen frigøres ved Stød af andre fri Elektroner. I den omhandlede Theori skelnes derfor mellem 2 Slags Metalmolekyler: de, der ikke har bundet Elektroner (positive Molekyler), og de, der har bundet saadanne (neutrale Molekyler). I Tilknytning hertil gøres forskellige Antagelser om Virkningerne af Sammenstødene mellem Elektronerne og Metalmolekylerne; det antages saaledes, at Elektroner, der støder sammen med positive Molekyler, bindes af disse, dersom deres absolute Hastighed før Stødet er mindre end en vis bestemt Hastighed G, medens det omvendt antages, at der ved Sammenstød mellem Elektroner med Hastighed større end G og neutrale Molekyler frigøres en Elektron. Endvidere gøres forskellige yderligere Antagelser om Hastigheden efter Stødet saavel af de stødende som af de frigjorte Elektroner. Disse sidste specielle Antagelser er imidlertid af en saadan Art, at de giver Anledning til alvorlige Indvendinger. I Erkendelsen heraf og for dog at bevare det væsentlige i Theorien har *Gruner*[1]) senere forandret denne saaledes, at der ikke længer benyttes særlige Antagelser om, hvorledes Ionisationen og Elektronbindingen foregaar, men saaledes, at det eneste Hensyn, der i Beregningerne tages til, at en saadan finder Sted, sker derigennem, at man antager den samtidige Tilstedeværelse af saavel positive som neutrale Molekyler. *Gruner* antager nu, at alle Elektronerne ved Sammenstød med positive Molekyler bliver kastede tilbage som fra elastiske Kugler, medens han ved Sammenstød mellem Elektronerne og neutrale Molekyler — hvilke sidste ikke tænkes at virke saa stærkt paa Elektronerne som de positive — antager, at kun de Elektroner, hvis Hastighed er mindre end G, kastes tilbage som fra elastiske Kugler, medens derimod de Elektroner, hvis Hastighed er større, slet ikke paavirkes.

Som man umiddelbart indser, bliver Forskellen mellem *Lorentz's* og *Gruner's* Theorier kun den, at Middelvejlængden i den sidstnævnte Theori ikke er den samme for alle Elektroner, men har en Værdi, l_1, for Elektroner, hvis Hastighed er mindre end G, en anden, l_2, for Elektroner, hvis Hastighed er større end G; $(l_2 > l_1)$. Vi faar derfor i dette Tilfælde

$$F(r) = \frac{r}{l_1} \text{ for } r < G \text{ og } F(r) = \frac{r}{l_2} \text{ for } r > G. \tag{16}$$

Det maa dog bemærkes, at *Gruner's* Antagelse kun vil medføre nogen væsentlig Forandring i *Lorentz's* Beregninger, dersom man yder-

---

[1]) *Gruner*: Phys. Zeitschr., Bd. 10, p. 48, 1909.

ligere antager, at Størrelsen G ikke er meget forskellig fra Elektronernes Middelhastighed ved den paagældende Temperatur.

Gruner's Theori synes, navnlig i den sidste Fremstilling, at maatte betragtes som et Forsøg paa, ved Hjælp af en første grov Tilnærmelse, at undersøge, hvilken Indflydelse det vil have paa Theoriens Resultater, dersöm Sammenstødenes Karakter antages for afhængig af Elektronernes Hastighed. For nærmere at kunne undersøge dette Spørgsmaal, skal vi nu gaa over til at betragte Molekylerne som Kraftcentrer, der udøver tiltrækkende eller frastødende Kræfter paa Elektronerne.

Vi skal her tænke os Molekylerne som faste Kraftcentrer, der frastøder eller tiltrækker Elektronerne med Kræfter, der forholder sig omvendt som n^{te} Potens af Afstanden. De Omraader, indenfor hvilke Molekylerne udøver kendelige Virkninger paa Elektronernes Bevægelser, skal vedblivende antages for smaa i Sammenligning med Molekylernes indbyrdes Afstande. Idet imidlertid de omtalte Omraader i dette Tilfælde ikke er skarpt begrænsede, kan man ikke her paa samme Maade som i de forrige Tilfælde tale om en Middelvejlængde. Vi maa her anvende en Behandlingsmaade, der svarer til den af *Maxwell*[1]) benyttede, og som af ham er blevet fuldstændig gennemført overfor Problemerne i den kinetiske Lufttheori i det Tilfælde, hvor Kræfterne forholder sig omvendt som den 5te Potens af Afstanden, hvilket Tilfælde som bekendt frembyder særlig simple Forhold. Ved det Problem, vi her skal behandle, hvor vi betragter Metalmolekylerne som ubevægelige og ser bort fra Virkningerne af Sammenstødene mellem Elektronerne indbyrdes, vil hele Behandlingen imidlertid være langt simplere end ved det af *Maxwell* behandlede Problem, og det vil her være ligesaa let at beregne Resultaterne for en vilkaarlig Potens (der dog maa antages at være større end 2) af Afstanden som for den 5te.

Idet Elektronernes absolute Hastighed ikke forandres ved et »Sammenstød« af den her omhandlede Art, hvor de fra Metalmolekylerne udgaaende Kræfter har et fast Potential, vil den Størrelse, som det i dette Tilfælde udelukkende kommer an paa at bestemme, være den Vinkel, som Elektronernes Bevægelsesretning før Sammenstødet danner med Retningen efter Stødet. Idet vi tænker os en Plan lagt gennem Molekylet, vinkelret paa Elektronens Retning før Sammenstødet, kan den omtalte Vinkel, som man umiddelbart indser, kun

---

[1]) *Maxwell*: Phil. Mag. (4), vol. 35, pp. 129 og 185, 1868.

afhænge af Afstanden a fra Molekylet til Projektionen af Elektronen
før Sammenstødet paa denne Plan, samt Elektronens absolute Hastighed r. Det kan nu vises, at dersom Kraften paa Enhed af en
Elektrons Masse skrives $\mu \cdot \rho^{-n}$, hvor ρ er Afstanden fra Elektronen til Molekylet, vil den søgte Vinkel ϑ være en Funktion af
$a \cdot r^{\frac{2}{n-1}} \cdot \mu^{-\frac{1}{n-1}}$ [1]$)$, og altsaa $\vartheta = \varphi\left(a \cdot r^{\frac{2}{n-1}}\right)$.

Dersom den Kraft, hvormed Molekylerne paavirker Elektronerne, er en
Tiltrækning, frembyder der sig dog en særlig Vanskelighed, idet (for $n > 3$)
alle de Elektroner, for hvilke $a \cdot r^{\frac{2}{n-1}}$ er mindre end en bestemt endelig
Størrelse, vil bøjes saaledes ud af deres Baner, at de træffer selve Kraftcentret og her ankommer med uendelig stor Hastighed. *Boltzmann* [2]$)$ har
imidlertid vist, hvorledes man kan omgaa denne Vanskelighed ved at antage, at den Lov, hvorefter Kraftcentret tiltrækker Elektronerne i Centrets
umiddelbare Nærhed (d. v. s. i en Afstand, der er forsvindende i Forhold
til de allerede smaa Afstande, hvor Stødet kan opfattes som begyndt), ikke
udtrykkes ved Formlen $\mu \cdot \rho^{-n}$, men ved en saadan, at Hastigheden forbliver
endelig. En saadan Antagelse vil nemlig ikke medføre nogen Forandring i
den Omstændighed, at ϑ kan udtrykkes som en Funktion af $a \cdot r^{\frac{2}{n-1}}$.

Vi skal nu beregne den samlede Bevægelsesmængde efter en bestemt Retning, som Elektronerne mister ved Sammenstød med et af
de omhandlede Molekyler. Det Antal Elektroner med Hastighedspunkter
indenfor et lille Element $d\sigma$ omkring Punktet (ξ, η, ζ), der i Tiden dt

---

[1]$)$ *Maxwell:* loc. cit., p. 143. Det nøjagtige Udtryk for ϑ er

$$\vartheta = \pi - 2 \int_0^{x_0} \frac{dx}{\sqrt{1 - x^2 - \frac{2}{n-1}\left(\frac{x}{\alpha}\right)^{n-1}}},$$

hvor $\alpha = a \cdot r^{\frac{2}{n-1}} \cdot \mu^{-\frac{1}{n-1}}$, og hvor x_0 betegner den mindste positive Rod i Ligningen

$$1 - x^2 - \frac{2}{n-1}\left(\frac{x}{\alpha}\right)^{n-1} = 0.$$

Det i Teksten anførte Resultat kan iøvrigt umiddelbart indses ved en Dimensionsbetragtning. Man har nemlig (idet L betegner en »Længde« og T en
»Tid«) $\mu \cdot \rho^{-n} \sim L \cdot T^{-2}$ og derfor $\mu \sim L^{1+n} \cdot T^{-2}$, endvidere er $r \sim L \cdot T^{-1}$
og $a \sim L$; idet nu ϑ skal være et dimensionsløst Tal, maa ϑ derfor nødvendigt
være en Funktion af $\mu \cdot r^{-2} \cdot a^{-(n-1)}$, eller, hvad der udsiger det samme, af
$a \cdot r^{\frac{2}{n-1}} \cdot \mu^{-\frac{1}{n-1}}$ (jvnfr. *Jeans:* Dynamical Theory of Gases, p. 275, hvor der
findes en lignende Dimensionsberegning for et tilsvarende Problem).

[2]$)$ *Boltzmann:* Sitzungsb. d. Wiener Akad. d. Wiss., math.-nat. Kl., Bd. 89, Abt. II,
p. 720, 1884. Se ogsaa *P. Czermak:* ibid., p. 723.

»rammer« det betragtede Molekyle saaledes, at Afstanden a ligger mellem a og $a + da$, er lig

$$f(\xi, \eta, \zeta)\, d\sigma \cdot r dt \quad 2\pi a da.$$

Ved disse Elektroners Sammenstød mistes en Bevægelsesmængde, hvis Retning falder sammen med Retningen af Radiusvektoren til Punktet (ξ, η, ζ), og hvis Størrelse er

$$m r \,(1 - \cos \vartheta) \cdot 2\pi a r f(\xi, \eta, \zeta)\, da d\sigma dt.$$

Idet vi nu integrerer over alle a fra o til ∞ — dette vil i Virkeligheden være det samme som kun at integrere over meget smaa a, idet $(1 - \cos \vartheta)$ kun afviger overmaade lidt fra o, dersom a ikke er meget lille (Integrationen vil dog, som en nærmere Betragtning viser, kun være tilladelig, dersom $n > 2$) —, faar vi, idet

$$\int_0^\infty (1 - \cos \vartheta) \cdot 2\pi a da = \int_0^\infty \left(1 - \cos\left(\varphi\left(a r^{\frac{2}{n-1}}\right)\right)\right) 2\pi a da$$

$$= r^{-\frac{4}{n-1}} \int_0^\infty \left(1 - \cos\left(\varphi\left(x\right)\right)\right) 2\pi x dx = r^{-\frac{4}{n-1}} \cdot c,$$

den samlede Bevægelsesmængde, som de Elektroner, hvis Hastighedspunkter før Stødet befinder sig indenfor Elementet $d\sigma$, mister i Tiden dt ved Sammenstød med det betragtede Molekyle, lig

$$m c r^{\frac{2n-6}{n-1}} f(\xi, \eta, \zeta)\, d\sigma dt.$$

Projektionen paa X-aksen af denne Bevægelsesmængde er lig

$$c r^{\frac{n-5}{n-1}} \cdot m \xi f(\xi, \eta, \zeta)\, d\sigma dt.$$

Integreres dernæst over alle Hastighedspunkter mellem to Kugleflader med Radierne r og $r + dr$, findes den Bevægelsesmængde efter X-aksen, som Elektroner med absolute Hastigheder mellem r og $r + dr$ mister ved Sammenstød med det betragtede Metalmolekyle, lig

$$c r^{\frac{n-5}{n-1}} \cdot G_x(r)\, dr dt.$$

Summeres nu over alle Metalmolekylerne i Volumenelementet dV, og antages det, at de alle paavirker Elektronerne med Kræfter, der aftager som samme Potens af Afstanden, men ellers ikke behøver at være lige store, faar vi den samlede Bevægelsesmængde efter X-aksen, som de betragtede Elektroner mister i Tiden dt, lig

$$C \cdot r^{\frac{n-5}{n-1}} G_x(r)\, dr\, dV dt,$$

og efter Udtrykket (9) altsaa

$$F(r) = C \cdot r^{\frac{n-5}{n-1}}. \tag{17}$$

Dersom man vilde antage, at der fandtes Molekyler, der paavirkede Elektronerne med Kræfter, der forholdt sig som forskellige Potenser af Afstanden, vilde man faa

$$F(r) = \Sigma C_n \cdot r^{\frac{n-5}{n-1}}.$$

(Baade i dette og i det foregaaende Tilfælde maa Metalmolekylerne antages at være saa godt »blandede«, at Antallet af de forskellige Slags Molekyler kan regnes for at være lige stort indenfor hvert Volumenelement, hvis Dimensioner kan sammenlignes med Elektronernes fri Vejlængder.)

Dersom vi i Lign. (17) sætter $n = \infty$, faar vi $F(r) = C \cdot r$, hvilket, dersom vi sætter $C = \frac{1}{l}$, er den samme Værdi, som vi ovenfor (Side 28 Lign. (15)) fandt ved Omtalen af *Lorentz's* Theori. Dette er ogsaa, hvad vi kunde vente, thi dersom vi skriver Udtrykket for Kraften pr. Enhed af Masse, hvormed Molekylerne paavirker Elektronerne, paa Formen $\mu \cdot \rho^{-n} = d \cdot \left(\frac{b}{\rho}\right)^n$, ser vi, at Kraften, dersom n er meget stor, vil være overordentlig lille, dersom $\rho > b$, og overordentlig stor, dersom $\rho < b$, og dette er netop det Forløb, som Kraften vil have, dersom Molekylerne og Elektronerne er haarde elastiske Kugler, hvis Radiers Sum er lig b (jvnfr. *Jeans*: Dynamical Theory of Gases, p. 276). — Ved Sammenligning mellem Ligningerne (15) og (17) ser vi, at Forskellen mellem de to omhandlede Tilfælde kan udtrykkes ved at sige, at den »effektive fri Middelvejlængde« ($r : F(r)$) i det sidste Tilfælde vokser med voksende r, medens den i det første Tilfælde er konstant. (Den effektive fri Middelvejlængde vil aldrig aftage med voksende r, hvilke Kræfter der end antages at virke mellem Molekylerne og Elektronerne.) —

Dersom vi sætter $n = 5$, faar vi $F(r) = C$. I dette Tilfælde vil derfor det samlede Tab af Bevægelsesmængde i en bestemt Retning, som Elektronerne lider i Tidsenheden ved Sammenstødene, være proportionalt med Elektronernes samlede Bevægelsesmængde i den omhandlede Retning (se Noten Side 21), og Beregningernes Udførelse vil for mange Problemers Vedkommende derfor være betydelig simplere end i de fleste andre Tilfælde.

3

Paa ganske tilsvarende Maade som ved de Tilfælde, vi her har betragtet, kan vi beregne Funktionen $F(r)$ i det Tilfælde, hvor Metalmolekylerne antages at være Systemer sammensatte af to elektrisk ladede Partikler, hvis Ladninger har samme absolute Størrelse men modsat Fortegn[1]), samt i det Tilfælde, Metalmolekylerne antages at være smaa Magneter[2]). — Dimensionerne af disse Systemer og Magneter, d. v. s. Afstanden henholdsvis mellem de ladede Partikler og mellem Magnetpolerne, skal antages at være forsvindende smaa i Forhold til de Afstande, indenfor hvilke Metalmolekylerne allerede udøver en kendelig Virkning paa Elektronernes Bevægelse, hvilke sidste Afstande her, ligesom i det forrige Tilfælde, skal antages at være smaa i Forhold til Molekylernes indbyrdes Afstande. —

I de her omhandlede Tilfælde vil, ligesom i det forrige Tilfælde, Elektronernes absolute Hastighed ikke forandres ved »Sammenstødet«. Vinklen ϑ vil derimod, foruden af Elektronens absolute Hastighed r og af Afstanden a (se Side 31), ogsaa afhænge af Vinklen τ mellem Molekylets Akse og en Plan vinkelret paa Elektronens Bevægelsesretning før Sammenstødet, samt af Vinklen v mellem Projektionen af denne Akse paa en saadan Plan og Forbindelseslinien mellem Projektionerne af Metalmolekylet og af Elektronen før Stødet paa den samme Plan. Den Kraft paa Enhed af Masse, hvormed et Metalmolekyle paavirker en Elektron, dersom Molekylet antages at være et elektrisk System som det ovenfor omtalte, vil i Størrelse være $\mu \cdot \dfrac{1}{\rho^3} \cdot \psi(\vartheta')$, hvor ρ er Radiusvektoren fra Molekylet til Elektronen og ϑ' Vinklen mellem denne Radiusvektor og Molekylets Akse. I det Tilfælde, Metalmolekylet antages at være en Magnet, vil Kraften være $\mu \cdot \dfrac{r}{\rho^3} \cdot \psi(\vartheta', \tau', v')$, hvor r som sædvanlig betegner Elektronens absolute Hastighed (r er i det omhandlede Tilfælde konstant under Sammenstødet), og hvor ρ og ϑ' har samme Betydning som i det forrige Tilfælde, medens τ' betegner Vinklen mellem Elek-

---

[1]) Antagelsen af Tilstedeværelsen af saadanne Systemer (»electric-doublets«) i Legemernes Indre er benyttet af *J. J. Thomson* (Phil. Mag., vol. 20, p. 238, 1910) til at forklare forskellige af Legemernes optiske Forhold. Det kan iøvrigt bemærkes, at det er lykkedes *J. H. Jeans* (Phil. Mag., vol. 20, p. 380, 1910) fuldstændig at gennemføre den mathematiske Beregning af en Elektrons Bane i en saadan »doublet«s Felt.

[2]) Udførlige Beregninger over en Elektrons Bevægelse i en Elementarmagnets Felt er foretagne af *C. Störmer* (Arch. d. Sciences phys. e natur. (4), Tom. 24, pp. 5, 113, 221 og 317, 1907) gennem numerisk Integration af de til dette Problem hørende Differentialligninger. Jvnfr. ogsaa *P. Gruner* (Jahrb. d. Rad. u. Elek., Bd. 6, p. 149, 1909).

tronens øjeblikkelige Hastighed og Radiusvektoren fra Molekylet til Elektronen, og v' Vinklen mellem Elektronens Hastighed og en Plan gennem Molekylets Akse og Elektronens øjeblikkelige Plads. Ved en Dimensionsbetragtning, der ganske svarer til den ovenfor (i Noten Side 31) udførte, finder man nu i de to Tilfælde for Vinklen ϑ henholdsvis

$$\vartheta = \varphi_1 \left((a \cdot r), \tau, v\right) \quad \text{og} \quad \vartheta = \varphi_2 \left((a \cdot r^{\frac{1}{2}}), \tau, v\right).$$

Ved nu at gaa frem paa ganske samme Maade som ovenfor og tage Middeltal over Vinklerne τ og v udfra den Antagelse, at Molekylernes Akser er ligelig fordelte i alle mulige Retninger, vil man finde i de to Tilfælde henholdsvis

$$F(r) = C \cdot r^{-1} \quad \text{og} \quad F(r) = C.$$

I det Tilfælde, Metalmolekylerne antages at være »elektriske Dobbeltsystemer«, faar man altsaa ganske samme Form for $F(r)$, som i det Tilfælde Molekylerne antages at være Kraftcentrer, der paavirker Elektronerne med Kræfter, der forholder sig omvendt som 3^{dje} Potens af Afstanden; medens man i det Tilfælde, Molekylerne antages at være smaa Magneter, faar samme Form for $F(r)$ som ved Betragtning af Kraftcentrer, der paavirker Elektronerne omvendt som 5te Potens af Afstanden.

Alle de her behandlede Tilfælde har været saadanne, ved hvilke det antages, at Elektronernes absolute Hastigheder ikke forandres ved Sammenstødene, og ved hvilke derfor $Q(\rho, r) = 0$; man kunde ogsaa udtrykke dette ved at sige, at der i de behandlede Tilfælde ved Sammenstødene omsættes Bevægelsesmængde og ikke Energi. Dersom imidlertid denne Antagelse var strængt opfyldt, vilde Sammenstødene ikke saaledes, som det i det foregaaende er antaget, kunne bevirke, at Elektronernes Fordeling, naar der virker ydre Kræfter, og Temperaturen ikke er konstant, kun afviger meget lidt fra den normale Fordeling $f = K \cdot e^{-\frac{m}{2kT} r^2}$, men kun, at Fordelingen indenfor de enkelte Grupper med samme absolute Hastighed paa meget lidt nær vil være symmetrisk med Hensyn til Hastighedskoordinaternes Begyndelsespunkt. Vi maa derfor, for at gøre det mekaniske Billede fuldstændigt, antage, at en Elektron ved Sammenstødene ogsaa under Omstændigheder kan forandre sin absolute Hastighed, saa at den enkelte Elektron under sin Bevægelse vil have forandret sin kinetiske Energi meget stærkt indenfor saadanne smaa Afstande, indenfor hvilke Temperaturen og Elektronernes potentielle Energi, hidrørende fra de ydre Kræfter,

3*

kun varierer overordentlig lidt; vi skal blot i de ovenfor behandlede Tilfælde antage, at de Tider, indenfor hvilke den enkelte Elektrons Bevægelsesmængde under Elektronens Bevægelse i Middel forandrer sig meget stærkt, vil være smaa i Forhold til saadanne Tider, indenfor hvilke dens kinetiske Energi i Middel har forandret sig mærkeligt.

I de Tilfælde, ved hvilke Elektronernes absolute Hastigheder forandres ved Sammenstødene — Tilfælde, ved hvilke Omsætningen at kinetisk Energi ikke er forsvindende i Forhold til Omsætningen af Bevægelsesmængde — er Beregningen af Funktionerne $F(r)$ og $Q(r, \rho)$ i Almindelighed langt vanskeligere end i de forrige Tilfælde. Jeg har gennemført Beregningen af $Q(\rho, r)$ i et af de allersimpleste Eksempler, nemlig for Sammenstød mellem de fri Elektroner indbyrdes i det Tilfælde, at disse antages at paavirke hinanden som absolut haarde elastiske Kugler; i dette Tilfælde kan endvidere Funktionen $F(r)$ findes ved Hjælp af *Maxwell's* Beregninger af Middelvejlængden mellem to paa hinanden følgende Sammenstød i en Luftart, hvis Molekyler antages at være elastiske Kugler (se f. Eks. *Jeans*: Dynamical Theory of Gases, p. 231). Da Beregningerne er temmelig langvarige, og vi ikke skal gøre nogen Brug af dem i det følgende, skal jeg her blot meddele Resultatet.

$$
\begin{aligned}
F(r) &= \frac{1}{l}\left(\frac{kT}{m\pi}\right)^{\frac{1}{2}}\left\{e^{-\frac{m}{2kT}r^2} + \left(\frac{mr}{kT} + \frac{1}{r}\right)\int_0^r e^{-\frac{m}{2kT}z^2}\,dz\right\}, \\
Q(\rho, r) &= \frac{1}{l}\left(\frac{kT}{m\pi}\right)^{\frac{1}{2}}\frac{r}{\rho^3}e^{-\frac{m}{2kT}r^2}\left\{8x + \left(\frac{m}{kT}\right)^2\left(\frac{1}{3}x^3y^2 - \frac{1}{15}x^5\right)\right. \\
&\quad\left. + \left(\frac{4m}{kT}x^2 - 8\right)e^{\frac{m}{2kT}x^2}\int_0^x e^{-\frac{m}{2kT}z^2}\,dz\right\},
\end{aligned}
\tag{18}
$$

hvor $x = r$, $y = \rho$, for $\rho > r$, medens $x = \rho$, $y = r$, for $\rho < r$.

I disse Udtryk betegner l Middelvejlængden mellem to paa hinanden følgende Sammenstød, beregnet paa *Maxwell's* Maade (se *Jeans*: loc. cit., p. 234).

Man ser, hvorledes $Q(\rho, r)$ tilfredsstiller den gennem Ligningen (13) Side 23 udtrykte Betingelse, som vi har vist, at enhver Funktion $Q(r, \rho)$ maa tilfredsstille.

§ 4.

Opstilling af Betingelsesligningerne for Elektronernes samlede Bevægelse i det Tilfælde, hvor Metalmolekylernes Dimensioner ikke antages at være forsvindende i Forhold til deres indbyrdes Afstande.

Vi skal nu betragte det Tilfælde, hvor Metalmolekylerne tænkes at være saa nær paa hinanden, at Elektronerne under en stor Del af deres Bevægelse er underkastede Virkningerne af de fra Molekylerne udgaaende Kræfter. Behandlingen af dette Tilfælde vil i Almindelighed være langt vanskeligere end af det Tilfælde, hvor vi betragtede adskilte Sammenstød. At Forholdene i det forrige Tilfælde var simplere, hidrører fra, at Sandsynligheden for Sammenstød, og derfor ogsaa for at modtage og afgive Bevægelsesmængde til Metalmolekylerne, kun var afhængig af Elektronernes øjeblikkelige Hastighed og ikke af deres Baner; man kunde derfor uden større Vanskelighed opstille en Relation mellem den i Tidsenheden til Metalmolekylerne afgivne Bevægelsesmængde og den Bevægelsesmængde, som Elektronerne besad i det betragtede Øjeblik. I det her omhandlede Tilfælde, hvor Vekselvirkningen mellem Metalmolekylerne og Elektronerne antages at foregaa paa en saa langt mere indviklet Maade, synes det derimod ikke muligt at opstille en almindelig (blot formel) Relation imellem Elektronernes Bevægelsesmængde og den i Tidsenheden til Metalmolekylerne afgivne Bevægelsesmængde. Vi skal derfor her kun gennemføre Behandlingen i et særlig simpelt Tilfælde, der imidlertid, dersom man gaar ud fra Rigtigheden af de i Indledningen omtalte Hovedantagelser, synes at maatte frembyde betydelige Lighedspunkter med Forholdene i de virkelige Metaller.

Vi skal antage, at der ikke finder nogen Vekselvirkning Sted mellem de fri Elektroner indbyrdes, samt at de Kræfter, hvormed Metalmolekylerne paavirker Elektronerne, danner et stationært elektromagnetisk Felt; endvidere skal vi ved Beregningerne antage, at de ydre Kræfter ligeledes er stationære.

Vi skal dog ikke antage, at disse Betingelser er strængt opfyldte; i et saadant stationært Felt vil der nemlig ikke finde nogen udjævnende Omsætning af Energi Sted (d. v. s. den enkelte Elektrons kinetiske Energi vil til ethvert Tidspunkt være bestemt udfra Begyndelsesenergien og Elektronens øjeblikkelige Plads (x, y, z)), og dersom man derfor ikke antager, at Elektronerne har Lejlighed til paa anden Maade at omsætte Energi enten med hverandre indbyrdes eller med Metalmolekylerne, vil der ikke være tilstrækkelig Aarsag til at bevirke, at Elektronernes Hastigheds- fordeling, dersom der virker ydre Kræfter eller Temperaturen ikke er konstant, kun vil afvige meget lidt fra den normale Fordeling (smlg. det analoge Tilfælde Side 35). En saadan udjævnende Energiomsætning kan dels tænkes at foregaa ved Vekselvirkning (Sammenstød) mellem de fri Elektroner, der befinder sig i det stationære Felt, dels ved Variationer i Feltet; vi skal imidlertid herom blot antage, at Virkningen af Sammen- stødene mellem Elektronerne indbyrdes i Middel vil være meget ringe i Sammenligning med Vekselvirkningen mellem Metalmolekylerne og Elek- tronerne, samt at Molekylernes Kraftfelter — og ligeledes de ydre Kræfter — kun vil forandre sig meget lidt indenfor saadanne smaa Tidsrum, indenfor hvilke de enkelte Elektroner under deres Bevægelse i Middel vil have mistet næsten hele deres oprindelige Bevægelsesmængde i en bestemt Retning.

Disse Antagelser synes nu, som ovenfor berørt, med Tilnærmelse at maatte være opfyldte ved de virkelige Metaller. For det første maa man nemlig antage, at Vekselvirkningen mellem Elektronerne og Metal- molekylerne vil være langt større end Vekselvirkningen mellem de fri Elektroner indbyrdes paa Grund af disse sidstes ringe »Størrelse« i For- hold til Molekylerne (de enkelte Molekyler antages jo nemlig at indeholde et meget stort Antal bundne Elektroner og synes saaledes at maatte opfattes som store Systemer i Forhold til de enkelte Elektroner), og idet Antallet af fri Elektroner i det højeste kan være af samme Stør- relsesorden som Molekylernes Antal (Antagelsen af et væsentlig større Antal fri Elektroner vilde nemlig ikke være forenelig med de eksperi- mentelt fundne Værdier for Metallernes Varmefylde[1]). For det andet maa Bevægelserne af et Molekyle, betragtet som en Helhed, i Middel være mange Gange langsommere end de enkelte Elektroners Bevægelse paa Grund af Elektronernes ringe Masse i Forhold til Molekylernes (et Metalmolekyles Masse er c. 10^5 Gange saa stor som en Elektrons); og idet der endvidere kun synes at kunne være en yderst ringe direkte Vekselvirkning mellem de fri Elektroner og de enkelte i Mole- kylerne bundne Elektroner (man maatte ellers vente en Fordeling af Energien imellem dem, der svarede til den statistiske Mekaniks Love (se Side 10)), maa Molekylernes Kraftfelter antages at kunne betragtes som stationære overfor Elektronernes Bevægelser.

Vi skal nu betragte de fri Elektroners af hverandre uafhængige Bevægelse i et saadant stationært Kraftfelt. Idet det elektriske Poten- tial i et Punkt (x, y, z) i det Indre af Metallet, hidrørende fra Metal- molekylernes Kraftfelter, betegnes ved λ, vil Elektronernes Fordeling i

---

[1] Se f. Eks. *J. H. Jeans:* Phil. Mag., vol. 17, p. 793, 1909.

et af ydre Kræfter upaavirket homogent Metalstykke, der overalt har samme Temperatur, efter Ligning (2) Side 12 i dette Tilfælde udtrykkes ved

$$f = K \cdot e^{-\frac{1}{kT}\left(\frac{m}{2}r^2 + \varepsilon\lambda\right)}. \tag{19}$$

Kaldes Antallet af fri Elektroner i Volumenenheden for N, har man

$$\left.\begin{aligned} N &= \iint K \cdot e^{-\frac{1}{kT}\left(\frac{m}{2}r^2 + \varepsilon\lambda\right)} d\sigma dv = K \int e^{-\frac{m}{2kT}r^2} d\sigma \cdot \int e^{-\frac{\varepsilon\lambda}{kT}} dv \\ &= K \cdot \left(\frac{2\pi kT}{m}\right)^{\frac{3}{2}} \cdot \int e^{-\frac{\varepsilon\lambda}{kT}} dv, \end{aligned}\right\} \tag{20}$$

hvor Integrationen med Hensyn til $d\sigma$ er udført over alle Hastighedselementer, medens Integralet med Hensyn til dv tænkes udført over en Volumenenhed.

Dersom der virker ydre Kræfter, eller Metallet ikke er homogent, eller ikke overalt har samme Temperatur, vil Ligevægten blive forstyrret, og Elektronernes statistiske Fordeling blive en anden. Efter det i Begyndelsen af dette Kapitel omtalte vil man dog altid kunne sætte

$$f = K \cdot e^{-\frac{1}{kT}\left(\frac{m}{2}r^2 + \varepsilon\lambda\right)} + \psi, \tag{21}$$

hvor K er en Funktion af Metallets Tilstand — Natur og Temperatur — i det betragtede Punkt, medens ψ er en Størrelse, der er meget lille i Forhold til det første Led.

Ved Undersøgelsen af Elektronernes samlede Bevægelse skal vi ikke her, som i det Tilfælde, hvor vi betragtede adskilte Sammenstød, betragte de Elektroner under et, der har samme absolute Hastighed, men derimod de, der har samme Sum $\frac{m}{2}r^2 + \varepsilon\lambda = a$ af kinetisk Energi og potentiel Energi i Forhold til Metalmolekylerne.

Komposanten efter X-aksen af den Bevægelsesmængde, der indehaves af de Elektroner, hvis Energi ligger mellem a og $a + da$, og som befinder sig i Volumenelementet dV — der skal tænkes saa stort, at det indeholder et meget stort Antal Metalmolekyler — skal vi betegne ved $G_x(a)\,da\,dV$.

Ved Beregningen af $G_x(a)$ skal vi nu paa tilsvarende Maade som i det forrige Tilfælde (se Side 17) tænke os den Fordeling, der svarer til det første Led i Ligningen (21), delt i to, f_1 og f_2, af hvilke vi skal sætte

$$f_1 = K_0 \cdot e^{-\frac{1}{kT_0}\left(\frac{m}{2}r^2+\varepsilon\lambda+\varepsilon\varphi\right)}, \tag{22}$$

hvor K_0 og T_0 betegner Værdierne for K og T i et Punkt (x_0, y_0, z_0), beliggende indenfor det betragtede Volumenelement dV, og hvor φ betegner de ydre Kræfters Potential. Idet φ tænkes valgt saaledes, at det er o i Punktet (x_0, y_0, z_0), vil Fordelingen f_2 i et Punkt $(x, y, z) = (x_0 + x_1, y_0 + y_1, z_0 + z_1)$, der ligger meget nær ved Punktet (x_0, y_0, z_0), med Bortkastning af smaa Størrelser af samme Orden·som Produkter af x_1, y_1 og z_1, kunne skrives

$$\left.\begin{aligned} f_2 = K \cdot e^{-\frac{1}{kT}\left(\frac{m}{2}r^2+\varepsilon\lambda\right)} &\left[x_1\left(\frac{\varepsilon}{kT}\frac{\delta\varphi}{\delta x} + \frac{1}{K}\frac{\delta K}{\delta x} + \frac{1}{kT^2}\left(\frac{m}{2}r^2 + \varepsilon\lambda\right)\frac{\delta T}{\delta x}\right) \right. \\ &\left. + y_1\left(\frac{\varepsilon}{kT}\frac{\delta\varphi}{\delta y} + \cdots\right) + z_1\left(\cdots\right) \right] \end{aligned}\right\} \tag{23}$$

Idet nu saavel λ som φ her antages at være konstante Funktioner af x, y og z, vil Fordelingen f_1 være stationær (smlg. Side 18); og de Elektroner, der til et vist Tidspunkt hører til denne Fordeling, vil derfor vedblive at svare til en Fordeling, der udtrykkes ved (22). De Elektroner, der til et vist Tidspunkt hører til Restfordelingen $f_R = f_2 + \psi$, vil derfor ogsaa vedblivende høre til denne sidste Fordeling. Idet nu endvidere Fordelingen f_1, der i ethvert Punkt (x, y, z) er symmetrisk med Hensyn til Hastighedskoordinaternes Begyndelsespunkt, ikke vil give Anledning til nogensomhelst samlet Bevægelse af Elektronerne, behøver vi ved Beregningen af Elektronernes samlede Bevægelse derfor kun at betragte Restfordelingen. Vi skal nu her enkeltvis betragte de forskellige Aarsager, der bevirker Elektronernes samlede Bevægelse.

Vi skal først betragte det Tilfælde, hvor der ikke virker ydre Kræfter, og hvor Elektronernes samlede Bevægelse derfor kan betragtes som hidrørende fra en Art »fri Diffusion« fra saadanne Steder i Metallet, hvor deres Koncentration er større, til saadanne, hvor den er lavere. I dette Tilfælde vil for hver enkelt Elektron Summen a af dens kinetiske Energi og dens potentielle Energi med Hensyn til Metalmolekylerne være konstant.

Antallet af Elektroner i Volumenelementet dV, hvis Energi ligger mellem a og $a + da$, skal vi betegne ved $N(a)\,da\,dV$. Det Antal Elektroner med Energi mellem a og $a + da$, der i Tidsenheden gaar igennem et Fladeelement dS vinkelret paa X-aksen, hvis Dimensioner skal antages store i Forhold til de enkelte Metalmolekylers, og som tænkes at befinde sig i Nærheden af Punktet (x_0, y_0, z_0), — hvilket

Antal efter de indførte Betegnelser vil være lig $\frac{1}{m} G_x(a)\,da\,dS$ — vil nu i dette Tilfælde kunne skrives

$$\frac{1}{m} G_x(a)\,da\,dS = - D(a) \cdot \frac{dN(a)}{dx}\,da \cdot dS, \qquad (24)$$

hvor »Diffusionskvotienten« $D(a)$ vil være en Funktion af a, kun af-hængig af Metallets Natur og Temperatur i det betragtede Punkt. Idet nu Antallet af de Elektroner i Volumenelementet dV, der hører til Restfordelingen f_R, og hvis Energi ligger imellem a og $a + da$, beteg-nes ved $N_R(a)\,da\,dV$, vil vi i det her betragtede Tilfælde, hvor Antallet af Elektroner i Volumenenheden, hørende til Fordelingen f_1, er konstant paa de forskellige Steder af Metallet, i Stedet for Ligningen (24) kunne skrive

$$\frac{1}{m} G_x(a)\,da\,dS = - D(a)\frac{dN_R(a)}{dx}\,da\,dS. \qquad (25)$$

I det Tilfælde, hvor der virker ydre Kræfter — og vi skal hertil medregne saadanne Tilfælde, i hvilke de fra Metalmolekylerne udgaa-ende Kræfter, paa Grund af Inhomogenitet i Metallet, i Middel giver Anledning til en Kraft i en bestemt Retning —, fremkommer Elektro-nernes samlede Bevægelse derved, at alle de enkelte Elektroners Baner paavirkes i Kraftens Retning. Da de ydre Kræfter antages at være overordentlig smaa i Sammenligning med de fra Metalmolekylerne ud-gaaende Kræfter, vil de enkelte Elektroners Bevægelse i dette Tilfælde kun afvige meget lidt fra Bevægelsen i det ovenfor betragtede Tilfælde, og en mærkelig samlet Bevægelse af Elektronerne i en bestemt Ret-ning fremkommer kun paa Grund af Elektronernes overordentlig store Antal. For de Elektroners Vedkommende, der hører til Fordelingen f_1, og hvis Antal i Volumenenheden her i Modsætning til i det forrige Tilfælde ikke er konstant paa de forskellige Steder i Metallet, vil nu, som ovenfor omtalt, Elektronernes samlede Bevægelse, hidrørende fra de ydre Kræfters Indvirkning og fra Koncentrationsforskelle, ophæve hinanden. For de Elektroners Vedkommende, der hører til Restfordelingen, kan man imidlertid paa Grund af det ringe Antal Elektroner indenfor det betragtede Volumenelement dV, der hører til denne Fordeling, se bort fra de ydre Kræfters Indvirkning paa Elektronernes Bevægelse, og vi vil derfor i dette Tilfælde — hvor $a + \varepsilon\varphi$ vil være konstant under de enkelte Elektroners Bevægelser —, paa tilsvarende Maade som i det ovenfor betragtede Tilfælde, kunne skrive

$$\frac{1}{m} G_x(a)\,da\,dS = - D(a)\frac{dN_R(a + \varepsilon\varphi)}{dx}\,da\,dS. \qquad (26) \qquad *$$

---

* $[N_R(a+\varepsilon\varphi)$ should read $N_R(a-\varepsilon\varphi).]$

Ved Hjælp af Ligningen (23) faar vi nu, idet vi bortkaster uendelig smaa Størrelser af højere Orden med Hensyn til x_1, y_1 og z_1 og tager i Betragtning, at det Antal Elektroner i Volumenelementet dV, der svarer til Fordelingen ψ, vil være meget ringe i Sammenligning med det hele Antal Elektroner (dette gælder ogsaa om Variationerne af de omhandlede Antal med Hensyn til Koordinaterne x, y og z):

$$* \qquad \frac{dN_R(a+\varepsilon\varphi)}{dx}\,da\,dV = \frac{dN_x(a)}{dx}\,da\,dV$$

$$= \left(\frac{\varepsilon}{kT}\frac{\delta\varphi}{\delta x} + \frac{1}{K}\frac{\delta K}{\delta x} + \frac{a}{kT^2}\frac{\delta T}{\delta x}\right)N(a)\,da\,dV. \qquad (27)$$

Ved Hjælp af Ligningerne (25) resp. (26) og (27) faar vi derfor i alle Tilfælde følgende Ligning til Bestemmelse af $G_x(a)$:

$$G_x(a) = -m\left(\frac{\varepsilon}{kT}\frac{\delta\varphi}{\delta x} + \frac{1}{K}\frac{\delta K}{\delta x} + \frac{a}{kT^2}\frac{\delta T}{\delta x}\right)D(a)N(a). \qquad (28)$$

For nærmere at belyse de her anførte Beregninger skal vi kort omtale, hvorledes Beregningerne vilde have formet sig, dersom vi, i Stedet for at anvende den ovenfor givne Behandlingsmaade, ogsaa i det her om-handlede Tilfælde havde forsøgt at benytte samme Fremgangsmaade som i det Tilfælde, hvor vi betragtede adskilte Sammenstød, d. v. s. dersom vi havde søgt en Ligning til Bestemmelse af $G_x(a)$ ved at undersøge de Tilvækster, som Bevægelsesmængden $G_x(a)\,da\,dV$ modtager i Tiden dt, dels paa Grund af Elektronernes øjeblikkelige Bevægelse og de ydre Kræfters Indvirkning, dels paa Grund af Vekselvirkningen mellem Metalmolekylerne og Elektronerne (smlg. Side 16).

Beregningen af den Tilvækst, der skyldes de førstnævnte Aarsager, — ved hvilken Beregning man kun behøver at tage Hensyn til de Elektroner, der i det betragtede Øjeblik hører til en Fordeling, der udtrykkes ved det første Led i Ligningen (21) (smlg. Side 17), — vil her kunne udføres paa tilsvarende Maade som i det Tilfælde, hvor vi betragtede adskilte Sammenstød, idet vi her blot udfører Beregningen over et lille Volumenelement dv, hvis Dimensioner er forsvindende i Forhold til Metal-molekylernes Dimensioner (d. v. s. saa smaa, at man kan betragte λ som konstant indenfor det betragtede Volumenelement), og dernæst integrerer over Volumenelementet dV. Man vil paa denne Maade finde et Udtryk for den søgte Tilvækst i Bevægelsesmængden $G_x(a)\,da\,dV$ lig Udtrykket paa højre Side af Ligningen (28) multipliceret med $R(a)\,da\,dVdt$, hvor $R(a)$ betegner en Funktion af a, afhængig af Metallets Natur og Temperatur. (Et saadant Udtryk vil, som man kan vise ved Benyttelse af Ligningen (1) Side 10, gælde i Almindelighed — d. v. s. ogsaa dersom man ikke ser bort fra Vekselvirkningen mellem de fri Elektroner indbyrdes og ikke betragter Metalmolekylernes Kraftfelter som stationære —, idet da a blot

* $[N_R(a+\varepsilon\varphi)$ should read $N_R(a-\varepsilon\varphi).]$

skal betegne Summen af en Elektrons kinetiske Energi og dens øjeblikke-lige potentielle Energi i Forhold til Metalmolekylerne og de andre Elek-troner, d. v. s. den Energi, som Systemet af Elektroner og Molekyler vilde miste, dersom man, efter pludselig at have standset alle Partiklernes Bevægelser uden at forandre deres indbyrdes Stillinger, førte den paa-gældende Elektron ud af Systemet og uendelig langt bort.)

Den Tilvækst i den omhandlede Bevægelsesmængde, der skyldes Vekselvirkningen mellem Metalmolekylerne og Elektronerne, — hvilken Tilvækst kun vil hidhøre fra de Elektroner, der i det betragtede Øjeblik hører til Fordelingen ψ (smlg. Side 17) — eller, som man ogsaa kan ud-trykke det, den Tilvækst, der skyldes Metalmolekylernes »Modstand« imod Elektronernes samlede Bevægelse i X-aksens Retning, vil imidlertid være langt vanskeligere at bestemme. Den omhandlede Modstand hidrører nemlig fra, at Elektronerne under deres Bevægelse vil være til Stede i større Antal paa saadanne Steder, hvor de fra Molekylerne hidrørende Kræfter er rettede imod Bevægelsen, end paa saadanne Steder, hvor Kræfterne har samme Retning som denne — eller anderledes udtrykt »Elek-tronerne hober sig paa Grund af Bevægelsen op foran Metalmolekylerne« —; og denne Elektronernes »Ophobning« (der·udtrykkes gennem Fordelingen ψ) vil her, i Modsætning til i det Tilfælde, hvor vi betragtede adskilte Sammenstød, i Almindelighed være forskellig alt efter den Maade, hvorpaa Elektronernes samlede Bevægelse er frembragt.

Ved de Betragtninger, vi ovenfor (Side 40—41) har benyttet, har vi imidlertid indirekte bevist, at i det Tilfælde, man antager, at Elektronerne bevæger sig uafhængigt af hverandre, og at Metalmolekylernes Kraftfelter er stationære, vil den omhandlede Ophobning i alle de betragtede Tilfælde, for hver enkelt Gruppe af Elektroner med samme samlede Energi, kun afhænge af den samlede Bevægelsesmængde efter X-aksen, som den om-handlede Gruppe besidder, idet vi har vist, at Bevægelsen af de Elek-troner, der hører til Fordelingen $f_2 + \psi$, hvad enten Elektronernes samlede Bevægelse skyldes Temperaturforskelle, eller den skyldes stationær ydre Kræfter, kan betragtes som en »fri Diffusion« (Fordelingen f_2 medfører, lige saa lidt som Fordelingen f_1, nogen Ophobning »foran« Molekylerne).

Den sidst søgte Tilvækst i Bevægelsesmængden $G_x(a)\,da\,dV$ vil derfor i det omhandlede Tilfælde kunne udtrykkes ved $F(a) \cdot G_x(a)\,da\,dVdt$, hvor $F(a)$ er en Funktion af a, afhængig af Metallets Natur og Temperatur; og sættes nu Summen af de fundne Tilvækster i den betragtede Bevægelses-mængde lig o, faar vi derfor, som man maatte vente, en Ligning til Bestemmelse af $G_x(a)$ af samme Form som Ligning (28). I de mere almindelige Tilfælde synes det imidlertid meget vanskeligt nærmere at undersøge Elektronernes omtalte Ophobning, og derfor meget vanskeligt ad denne Vej at opstille Ligninger til Bestemmelse af Elektronernes samlede Bevægelse.

Angaaende Anvendelserne af Ligning (28) maa vi bemærke, at det ved Udledningen af Ligningen er forudsat, at de ydre Kræfter er stationære. Ligningen kan derfor kun benyttes overfor stationære (eller quasi-stationære) Problemer (smlg. Side 27).

Som et enkelt Eksempel paa de i denne Paragraf omtalte Beregninger skal vi her betragte det Tilfælde, hvor Metalmolekylerne og Elektronerne antages at paavirke hinanden som absolut haarde elastiske Legemer. Metalmolekylerne skal antages at kunne være af ganske vilkaarlig Form og at kunne have ganske vilkaarlige indbyrdes Afstande. — (Denne Antagelse skiller dette Eksempel fra de Eksempler, som vi betragtede i § 3 af dette Kapitel, og ved hvilke vi antog, at Metalmolekylernes Dimensioner var forsvindende i Forhold til Molekylernes indbyrdes Afstande.) — Vi skal her kun antage, at Metalmolekylernes Dimensioner og Masser er meget store i Forhold til Elektronernes, saa at man dels kan se bort fra Virkningen af Sammenstødene mellem Elektronerne indbyrdes i Forhold til Virkningen af Sammenstødene mellem Elektronerne og Metalmolekylerne, dels kan regne Metalmolekylerne for at være ubevægelige indenfor saadanne smaa Tidsrum, i hvilke de enkelte Elektroner i Middel mister næsten hele deres oprindelige Bevægelsesmængde i en bestemt Retning.

I det her betragtede Tilfælde vil Grupperne af Elektroner med samme Energi svare til Grupper af Elektroner med samme absolute Hastighed. Tænker vi os nu en Elektron, der upaavirket af ydre Kræfter bevæger sig mellem Metalmolekylerne, vil dens Bane efter de Antagelser, vi har gjort om Molekylerne, være ganske uafhængig af Elektronens absolute Hastighed. Idet endvidere den Vejstrækning, som en Elektron vil gennemløbe i et bestemt Tidsrum, vil være proportional med dens absolute Hastighed, vil Diffusionskoefficienten $D(a)$ i dette Tilfælde være ligefrem proportional med Elektronernes absolute Hastighed. Indfører vi nu dette i Ligningen (28) Side 42 og sætter $a = \frac{1}{2}mr^2$, samt benytter, at λ i det her omhandlede Tilfælde er o i visse Omraader (nemlig udenfor Metalmolekylerne) og uendelig stor i de øvrige Omraader (indenfor Metalmolekylerne), finder vi en Ligning til Bestemmelse af Bevægelsesmængden, der ganske svarer til den Ligning, vi fandt ved Omtalen af *H. A. Lorentz's* Theori (d. v. s. den Ligning, som man faar ved at indføre den ved Ligningen (15) Side 28 angivne Værdi for $F(r)$ i Ligningen (14) Side 26 og sætte $Q(\rho, r)=o$), blot med den Forskel, at der i Stedet for Middelvejlængden l indgaar en Konstant med en anden fysisk Betydning.

Heraf vil f. Eks. følge, at Forholdet mellem Elektricitets- og Varmeledningsevnen vil blive det samme i alle Tilfælde, hvor man antager, at Metalmolekylerne paavirker Elektronerne som haarde Legemer (eller med Kræfter der varierer overordentlig hurtigt med Afstanden (smlg. Side 33)), uanset Molekylernes Form og indbyrdes Afstande, og lig den Værdi, som *Lorentz* har beregnet udfra den Antagelse, at Molekylerne er haarde elastiske Kugler, hvis Radier er meget smaa i Forhold til Molekylernes indbyrdes Afstande (se nærmere i næste Kapitel).

Vi skal her til Slut bemærke, at i alle andre Tilfælde vil $D(a)$ vokse stærkere med a end i det ovenfor betragtede Tilfælde, hidrørende fra, at de hurtige Elektroner ikke alene kommer hurtigere frem ad de samme Baner, men ogsaa lettere bryder sig Vej end de langsommere (undtagen netop i det Tilfælde, at Metalmolekylerne er absolut haarde Legemer). Betydningen heraf vil blive omtalt i det næste Kapitel.

KAPITEL II.

BEHANDLINGEN AF STATIONÆRE PROBLEMER.

§ 1.

Udledning af Udtryk for Elektricitets- og Energibevægelsen gennem Metallet.

Vi skal nu ved Hjælp af de i forrige Kapitel udledte Ligninger beregne saavel den Elektricitetsmængde som den Energimængde, der ved Elektronernes Bevægelse i Tidsenheden føres gennem en Fladeenhed i det Indre af et Metalstykke. Vi skal ved alle Beregninger i dette Kapitel antage, at de ydre Forhold — d. v. s. de ydre Kræfter og Temperaturen i Metallets forskellige Dele — er stationære eller i det mindste quasi-stationære (se Side 27).

Vi skal her først betragte det Tilfælde, hvor der antages at finde adskilte Sammenstød Sted. Betegner vi den Elektricitetsmængde, der med Elektronerne i Tiden dt føres gennem et Fladeelement dS vinkelret paa X-aksen, med $i_x dSdt$, og paa tilsvarende Maade den kinetiske Energi, der i samme Tid føres igennem det samme Fladeelement med $W_x dSdt$, har vi, med Benyttelse af de samme Betegnelser som i forrige Kapitel,

$$i_x = \frac{\varepsilon}{m} \int_0^\infty G_x(r)\, dr \qquad (1\,a)$$

$$W_x = \frac{1}{2} \int_0^\infty r^2\, G_x(r)\, dr. \qquad (2\,a)$$

Til Bestemmelse af $G_x(r)$ har vi, idet vi under de her omhandlede Omstændigheder kan sætte den højre Side af Ligning (14) Side 26 lig 0,

$$-\frac{4\pi}{3}\,mK\left(\frac{\varepsilon}{kT}\,\frac{\delta\varphi}{\delta x}+\frac{1}{K}\,\frac{\delta K}{\delta x}+\frac{mr^2}{2kT^2}\,\frac{\delta T}{\delta x}\right)r^4\,e^{-\frac{m}{2kT}r^2}$$
$$-G_\mathrm{x}(r)\,F(r)+\int_0^\infty G_\mathrm{x}(\rho)\,Q(\rho,r)\,d\rho=0.\qquad\Bigg\}\ (3\,\mathrm{a})$$

Denne Ligning er en Integralligning af den *Fredholm'ske* Type[1]),

$$\Psi(r)=\psi(r)+\int_a^b\Psi(\rho)\cdot\tau(r,\rho)\,d\rho,$$

hvis Løsning kan skrives paa Formen

$$\Psi(r)=\psi(r)+\int_a^b\psi(\rho)\cdot\pi(r,\rho)\,d\rho,$$

hvor Funktionen $\pi(r,\rho)$ kun er afhængig af $\tau(r,\rho)$ (d. v. s. uafhængig af $\psi(r)$). Af det *Fredholm'ske* Udtryk for $\pi(r,\rho)$ fremgaar det endvidere umiddelbart, at $\pi(r,\rho)$ vil være en symmetrisk Funktion af r og ρ, dersom $\tau(r,\rho)$ er en symmetrisk Funktion af disse Variable.

Sætter vi nu $G_\mathrm{x}(r)=\Psi(r)\cdot g(r)$, hvor $g(r)=(F(r))^{-\frac{1}{2}}\cdot r^2 e^{-\frac{mr^2}{4kT}}$, faar vi af Ligningen (3 a), med Benyttelse af Ligningen (13) Side 23 for $Q(\rho,r)$,

$$\Psi(r)=-\frac{4\pi}{3}\,mK\left(\frac{\varepsilon}{kT}\,\frac{\delta\varphi}{\delta x}+\frac{1}{K}\,\frac{\delta K}{\delta x}+\frac{mr^2}{2kT^2}\,\frac{\delta T}{\delta x}\right)g(r)$$
$$+\int_0^\infty\Psi(\rho)\cdot g(\rho)\,g(r)\,S(\rho,r)\,d\rho.$$

Idet nu den Funktion $g(\rho)\,g(r)\,S(\rho,r)$, der svarer til Funktionen $\tau(r,\rho)$ i den *Fredholm'ske* Ligning, her er symmetrisk med Hensyn til r og ρ, faar vi efter det ovenfor omtalte

---

[1]) Se f. Eks. *M. Bôcher*: Introduction to the Study of Integral Equations, Cambridge 1909, p. 29—38. (I den ovenstaaende Ligning (3 a) er Integralet taget mellem uendelige Grænser, medens det ved Beviset for den *Fredholm'ske* Løsning forudsættes, at Integralet er taget mellem endelige Grænser. Dette medfører dog her ingen Vanskelighed, idet det allerede af den fysiske Betydning af Ligningen (3 a) fremgaar, at det ikke vil gøre nogen væsentlig Forandring hverken for Ligningens Løsning eller for de Anvendelser, vi skal gøre af denne, om Integralet tages mellem Grænserne 0 og ∞ eller mellem 2 Grænser, af hvilke den ene er meget lille og den anden meget stor i Forhold til Elektronernes Middelhastighed ved den paagældende Temperatur.)

$$\Psi(r) = -\frac{4\pi}{3}\,mK\left(\frac{\varepsilon}{kT}\frac{\delta\varphi}{\delta x} + \frac{1}{K}\frac{\delta K}{\delta x} + \frac{mr^2}{2kT^2}\frac{\delta T}{\delta x}\right)g(r)$$

$$-\int_0^\infty \frac{4\pi}{3}\,mK\left(\frac{\varepsilon}{kT}\frac{\delta\varphi}{\delta x} + \frac{1}{K}\frac{\delta K}{\delta x} + \frac{m\rho^2}{2kT^2}\frac{\delta T}{\delta x}\right)g(\rho)\cdot S(\rho, r)\,d\rho,$$

hvor $S(\rho, r)$ betegner en symmetrisk Funktion af ρ og r, kun af-hængig af $g(r)$ og $S(\rho, r)$. Af Ligningerne (1 a) og (2 a) faar vi nu

$$i_x = -A_1\left(\frac{\delta\varphi}{\delta x} + \frac{kT}{\varepsilon K}\frac{\delta K}{\delta x}\right) - A_2\frac{\delta T}{\delta x} \qquad (4\,a)$$

og

$$W_x = -A_2 T\left(\frac{\delta\varphi}{\delta x} + \frac{kT}{\varepsilon K}\frac{\delta K}{\delta x}\right) - A_3\frac{\delta T}{\delta x}, \qquad (5\,a)$$

hvor

$$A_1 = \frac{4\pi\varepsilon^2 K}{3kT}\left\{\int_0^\infty r^4\big(F(r)\big)^{-1}e^{-\frac{mr^2}{2kT}}dr + \int_0^\infty g(r)\left[\int_0^\infty g(\rho)\,S(\rho, r)\,d\rho\right]dr\right\},$$

$$A_2 = \frac{2\pi\varepsilon mK}{3kT^2}\left\{\int_0^\infty r^6\big(F(r)\big)^{-1}e^{-\frac{mr^2}{2kT}}dr + \int_0^\infty g(r)\left[\int_0^\infty \rho^2 g(\rho)\,S(\rho, r)\,d\rho\right]dr\right\} =$$

$$\frac{2\pi\varepsilon mK}{3kT^2}\left\{\int_0^\infty r^6\big(F(r)\big)^{-1}e^{-\frac{mr^2}{2kT}}dr + \int_0^\infty r^2 g(r)\left[\int_0^\infty g(\rho)\,S(\rho, r)\cdot d\rho\right]dr\right\}$$

og

$$A_3 = \frac{\pi m^2 K}{3kT^2}\left\{\int_0^\infty r^8\big(F(r)\big)^{-1}e^{-\frac{mr^2}{2kT}}dr + \int_0^\infty r^2 g(r)\left[\int_0^\infty \rho^2 g(\rho)\,S(\rho, r)\,d\rho\right]dr\right\}.$$

Ligningerne (4 a) og (5 a) vil gælde i alle Tilfælde, i hvilke det antages, at der finder adskilte Sammenstød Sted; de indeholder derfor som specielle Tilfælde de af *Lorentz*[1]) og *Gruner*[2]) givne Ligninger. Den Omstændighed, at den samme »Konstant« (d. v. s. Funktion af Me-tallets Natur og Temperatur i det betragtede Punkt) A_2 indgaar i Udtryk-kene for i_x og W_x, viser, at der i alle de omhandlede Tilfælde bestaar en særegen ensartet Forbindelse mellem Elektricitetens og Varmens Bevægelse gennem Metallet. Den fysiske Betydning heraf vil nærmere blive omtalt ved Behandlingen af de thermoelektriske Fænomener.

[1]) *H. A. Lorentz*: Proc. Acad. Amsterdam, vol. 7, p. 447, 1905.
[2]) *P. Gruner*: Verh. d. Deutsch. Phys. Ges., Bd. 10, p. 524, 1908.

Som et enkelt Eksempel paa en Bestemmelse af Størrelserne A_1, A_2 og A_3 skal vi her betragte det simple Tilfælde, hvor Metalmolekylerne antages at være faste Kraftcentrer, der paavirker Elektronerne med Kræfter, der forholder sig omvendt som den n^{te} Potens af Afstanden, og hvor man ser bort fra Vekselvirkningen mellem Elektronerne indbyrdes. I dette Tilfælde har vi (se Side 33) $F(r) = C \cdot r^{\frac{n-5}{n-1}}$ og $Q(\rho, r) = 0$ (og altsaa $S(\rho, r) = 0$). Vi faar derfor

$$
\left.
\begin{aligned}
A_1 &= \frac{4\pi\varepsilon^2 K}{3kT} \frac{1}{C} \int_0^\infty r^{\frac{3n+1}{n-1}} e^{-\frac{mr^2}{2kT}} dr = \frac{4\pi\varepsilon^2 K}{3mC} \left(\frac{2kT}{m}\right)^{\frac{n+1}{n-1}} \Gamma\left(\frac{2n}{n-1}\right) \\[2mm]
A_2 &= \frac{2\pi\varepsilon m K}{3kT^2} \frac{1}{C} \int_0^\infty r^{\frac{5n-1}{n-1}} e^{-\frac{mr^2}{2kT}} dr = \frac{2\pi\varepsilon K}{3TC} \left(\frac{2kT}{m}\right)^{\frac{2n}{n-1}} \Gamma\left(\frac{3n-1}{n-1}\right) \\[2mm]
&= A_1 \cdot \frac{2n}{n-1} \frac{k}{\varepsilon} \\[4mm]
A_3 &= \frac{\pi m^2 K}{3kT^2} \frac{1}{C} \int_0^\infty r^{\frac{7n-3}{n-1}} e^{-\frac{mr^2}{2kT}} dr = \frac{\pi m K}{3TC} \left(\frac{2kT}{m}\right)^{\frac{3n-1}{n-1}} \Gamma\left(\frac{4n-2}{n-1}\right) \\[2mm]
&= A_1 \cdot \frac{2n(3n-1)}{(n-1)^2} \frac{k^2}{\varepsilon^2} T.
\end{aligned}
\right\} \quad (6)
$$

Vi skal nu betragte det Tilfælde, hvor Metalmolekylerne antages at være saa nær inde paa hverandre, at Elektronerne under en stor Del af deres Bevægelse er underkastede Virkningerne af de fra Molekylerne udgaaende Kræfter. I Modsætning til, hvad der fandt Sted ved det ovenfor behandlede simplere Tilfælde — adskilte Sammenstød —, kan vi her, som omtalt i forrige Kapitel, ikke angive en almindelig Ligning til Bestemmelse af Elektronernes Bevægelsesmængde, og vi kan derfor ikke her angive den fuldt almindelige Form for Udtrykkene for Elektricitets- og Energistrømningen. Vi har kun udledt Betingelsesligninger for Bevægelsesmængden under visse indskrænkende Antagelser, nemlig at Metalmolekylernes Kraftfelter kan betragtes som stationære, og at Vekselvirkningen mellem de fri Elektroner indbyrdes kan regnes for forsvindende i Sammenligning med Virkningen mellem Metalmolekylerne og Elektronerne De følgende Beregninger omhandler derfor kun dette sidste Tilfælde.

4

Vi skal nu her søge den Elektricitetsmængde og den Sum af kinetisk Energi og potentiel Energi i Forhold til Metalmolekylerne, der ved Elektronernes Bevægelse i Tiden dt føres gennem et Flade-element dS vinkelret paa X-aksen. Idet disse Størrelser paa tilsvarende Maade som i det forrige Tilfælde betegnes henholdsvis ved $i_x dS dt$ og $W_x dS dt$, faar man med de Betegnelser, vi har benyttet i forrige Kapitel,

$$i_x = \frac{\varepsilon}{m} \int_0^\infty G_x(a)\, da \qquad (1\,b)$$

og

$$W_x = \frac{1}{m} \int_0^\infty a\, G_x(a)\, da. \qquad (2\,b)$$

Til Bestemmelse af $G_x(a)$ har vi (Ligning (28) Side 42)

$$G_x(a) = - m \left(\frac{\varepsilon}{kT} \frac{\delta \varphi}{\delta x} + \frac{1}{K} \frac{\delta K}{\delta x} + \frac{a}{kT^2} \frac{\delta T}{\delta x} \right) D(a)\, N(a). \qquad (3\,b)$$

Ved Hjælp af (1 b) og (2 b) faar vi nu

$$i_x = - A_1 \left(\frac{\delta \varphi}{\delta x} + \frac{kT}{\varepsilon K} \frac{\delta K}{\delta x} \right) - A_2 \frac{\delta T}{\delta x} \qquad (4\,b)$$

$$W_x = - A_2 T \left(\frac{\delta \varphi}{\delta x} + \frac{kT}{\varepsilon K} \frac{\delta K}{\delta x} \right) - A_3 \frac{\delta T}{\delta x}, \qquad (5\,b)$$

hvor

$$A_1 = \frac{\varepsilon^2}{kT} \int_0^\infty D(a)\, N(a)\, da, \quad A_2 = \frac{\varepsilon}{kT^2} \int_0^\infty a\, D(a)\, N(a)\, da,$$

$$A_3 = \frac{1}{kT^2} \int_0^\infty a^2\, D(a)\, N(a)\, da. \qquad \left. \right\} \quad (7)$$

Som man ser, har Ligningerne (4 b) og (5 b) ganske samme Form som Ligningerne (4 a) og (5 a), og der gælder derfor de samme Bemærkninger om dem, som vi har gjort ved de sidstnævnte Ligninger.

Ligningerne (4) og (5) bestemmer fuldstændig Elektricitets- og Varmebevægelsen i Metallet. Idet vi ikke har gjort nogen særlig Antagelse om X-aksens Retning, viser Ligningerne, at Elektricitets-

og Energistrømningen gennem et Fladeelement kun afhænger af Variationerne i de ydre Kræfters Potential, Temperaturen o. s. v. i Retning
vinkelret paa Elementet. Vi skal nu i de følgende Paragrafer udfra
disse Ligninger undersøge Metallernes Elektricitets- og Varmeledning
og de thermo-elektriske Fænomener.

§ 2.

Elektricitets- og Varmeledning.

Tænker vi os et homogent Metalstykke, der i alle sine Punkter
har samme Temperatur, udsat for Virkningen af en konstant elektrisk
Kraft $E = -\dfrac{\partial \varphi}{\partial x}$, vil det gennemløbes af en e l e k t r i s k S t r ø m, hvis
Styrke pr. Fladeenhed i umiddelbart kan findes ved Hjælp af Lig-
ningerne (4). Idet Differentialkvotienterne af K og T i dette Tilfælde
er o, faar vi

$$i = A_1 \cdot E.$$

Idet den e l e k t r i s k e L e d n i n g s e v n e betegnes ved σ, faar vi altsaa

$$\sigma = A_1. \tag{8}$$

Dersom vi specielt antager, at der finder adskilte Sammenstød
Sted, og at Kræfterne mellem Metalmolekylerne og Elektronerne
under disse Sammenstød forholder sig omvendt som n^{te} Potens af
Afstanden, faar vi ved Hjælp af Ligningerne (6) Side 49 og Ligningen
(5) Side 16

$$\sigma = \frac{4N\varepsilon^2}{3\sqrt{\pi}\,mC} \left(\frac{m}{2kT}\right)^{\frac{n-5}{2(n-1)}} \Gamma\left(\frac{2n}{n-1}\right). \tag{9}$$

Dersom vi lader n vokse mod uendelig, hvilket vil være ens-
betydende med at antage, at Molekylerne og Elektronerne paavirker
hinanden som absolut haarde elastiske Kugler (se Side 33), faar vi,
idet vi sætter $C = \dfrac{1}{l}$,

$$\sigma = \tfrac{4}{3}\, lN\varepsilon^2 (2\pi mkT)^{-\frac{1}{2}},$$

hvilket er den Værdi for σ, som *Lorentz* [1]) har fundet.

---

[1]) *H. A. Lorentz:* Proc. Acad. Amsterdam, vol. 7, p. 448, 1905. (I *Lorentz's* Afhandling
benyttes andre Betegnelser.)

4*

Ved Elektronernes Bevægelse føres der foruden Elektricitet tillige
Energi — Varme — igennem Metallet. Det er imidlertid i Almindelighed ikke muligt at angive en bestemt Værdi for den Energimængde,
der ved Elektronernes Bevægelse føres igennem et vist Fladeelement,
idet Elektronernes Energi, der er sammensat dels af kinetisk Energi,
dels af potentiel Energi i Forhold til Metalmolekylerne og i Forhold
til fri Elektricitetsmængder (ydre Kraftfelter), samt endelig af »indre«
Energi, kun kan angives paa nær en arbitrær Konstant. Kun i et
enkelt Tilfælde, nemlig hvor det samlede Antal Elektroner, der gaar
gennem Fladeelementet, er o, vil den omtalte Ubestemthed bortfalde.
I dette Tilfælde vil den Energimængde, der i Tidsenheden føres gennem en Fladeenhed vinkelret paa X-aksen, være angivet ved den i det
foregaaende omtalte Størrelse W_x. Da i dette Tilfælde den elektriske
Strøm er o, faar man, idet man eliminerer $\dfrac{\delta\varphi}{\delta x}$ af Ligningen $i_x = 0$ og
Ligningen for W_x (Ligningerne (4) og (5)),

$$W_x = -\frac{A_3 A_1 - T A_2^2}{A_1}\frac{\delta T}{\delta x}.$$

Energistrømningen i dette eneste fuldkommen behandlelige Tilfælde skal vi nu definere som Varmeledning, og faar altsaa, idet
Varmeledningsevnen betegnes med γ,

$$\gamma = \frac{A_3 A_1 - T A_2^2}{A_1}. \tag{10}$$

Vi ser, at vi paa denne Maade har faaet Varmeledningsevnen bestemt
som en Størrelse, der kun er afhængig af Metallets Tilstand — Natur
og Temperatur — paa det betragtede Sted; det er endvidere den
Størrelse, der svarer til den, der er fundet ved de eksperimentelle
Undersøgelser over Varmeledningsevnen. Ved de fleste af disse
Undersøgelser — som f. Eks. *L. Lorenz's* [1] — iagttages Temperaturens Variation i en Metalstang, der er omgivet af elektriske Isola-
torer. I en saadan Metalstang vil der meget hurtigt indtræde en
elektrisk Ligevægt, saaledes at der ingen elektrisk Strøm gaar, og
Forholdene i Stangen vil derfor fuldstændig svare til Forholdene i det
ovenfor omtalte Tilfælde. Ved andre Undersøgelser — som f. Eks.
Jaeger og *Diesselhorst's* [2] — iagttages Fordelingen af Temperaturen

---

[1] *L. Lorenz*: Wied. Ann., Bd. 13, p. 422, 1881.

[2] *W. Jaeger* u. *H. Diesselhorst*: Wiss. Abh. d. phys.-tech. Reichsanstalt, Berlin,
Bd. 3, p. 269, 1900.

og af det elektriske Potential i en Metalstang, der gennemløbes af en elektrisk Strøm, hvorefter Varmeledningsevnen bestemmes ved Hjælp af en af *F. Kohlrausch*[1]) angivet Relation imellem Temperaturen og Potentialet i Stangens forskellige Punkter. Som vi skal se ved Omtalen af de thermo-elektriske Fænomener, kommer man, ved Hjælp af Ligningerne (4) og (5) og ved at definere Varmeledningsevnen som oventor, til et Udtryk for Varmeudviklingen i et Volumenelement af Metallet, der svarer til det, som er lagt til Grund for Opstillingen af den *Kohlrausch'ske* Relation.

Dersom vi atter specielt betragter adskilte Sammenstød og antager, at Kræfterne mellem Metalmolekylerne og Elektronerne forholder sig omvendt som n^{te} Potens af Afstanden, faar vi ved Hjælp af Ligningerne (10), (6), (8) og (9)

$$\gamma = A_1 \cdot \frac{2n}{n-1} \frac{k^2}{\varepsilon^2} T = \frac{8nNk^2T}{3\sqrt{\pi}\,(n-1)\,mC} \left(\frac{m}{2kT}\right)^{\frac{n-5}{2(n-1)}} \Gamma\!\left(\frac{2n}{n-1}\right). \quad (11)$$

Dersom vi lader n vokse imod ∞ og sætter $C = \frac{1}{l}$, faar vi

$$\gamma = \tfrac{8}{3} lNk^2T (2\pi mkT)^{-\frac{1}{2}},$$

hvilket er *Lorentz's* Værdi for γ.

For Forholdet mellem Ledningsevnerne for Varme og Elektricitet faar vi nu af (8) og (10)

$$\varkappa = \frac{\gamma}{\sigma} = \frac{A_3 A_1 - TA_2^2}{A_1^2}. \quad (12)$$

I det ovenfor betragtede specielle Tilfælde faar vi af Ligningerne (9) og (11)

$$\varkappa = \frac{2n}{n-1} \frac{k^2}{\varepsilon^2} T \quad (13)$$

og heraf for n lig ∞ den Værdi, som *Lorentz* har fundet[2]),

$$\varkappa = 2 \frac{k^2}{\varepsilon^2} T. \quad (14)$$

---

[1]) *F. Kohlrausch*: Sitzungsber. d. Berliner Akad. d. Wiss. 1899, p. 711.

[2]) *H. A. Lorentz*: Proc. Acad. Amsterdam, vol. 7, p. 449, 1905.
Vi skal her i Forbindelse med *Lorentz's* Værdi for $\varkappa$ omtale de Værdier, som *E. Riecke* (Wied. Ann., Bd. 66, pp. 353, 545 og 1199, 1898; Ann. d. Phys., Bd. 2, p. 835,

Medens de fundne Udtryk for den elektriske Ledningsevne og for Varmeledningsevnen indeholder de paa Forhaand ukendte eller i det mindste meget usikkert kendte Størrelser N og l (C), er dette,

---

1900; Jahrb. d. Rad. u. Elek., Bd. 3, p. 24, 1906; Phys. Zeitschr., Bd. 10, p. 508, 1909) og *P. Drude* (Ann. d. Phys., Bd. 1, p. 566 og Bd. 3, p. 369, 1900) tidligere har beregnet for denne Størrelse, nemlig med vore Betegnelser henholdsvis

$$\varkappa = \frac{27}{8} \frac{k^2}{\varepsilon^2} T \left(1 + \frac{2}{3} \frac{T}{N} \frac{dN}{dT} \right) \quad \text{og} \quad \varkappa = 3 \frac{k^2}{\varepsilon^2} T;$$

disse Værdier stemmer nemlig bedre overens med Erfaringerne end den af *Lorentz* beregnede (se f. Eks. *Riecke*: Phys. Zeitschr. Bd. 10, p. 513—514, 1909).

De nævnte Forfattere gør samme Antagelser om Virkningerne af Sammenstødene mellem Metalmolekylerne og Elektronerne, som *Lorentz*, d. v. s. de antager, at en Elektron ved et Sammenstød i Middel mister hele sin oprindelige Bevægelsesmængde i en bestemt Retning; de antager ogsaa, at de fri Elektroner befinder sig i Varmeligevægt med Metalmolekylerne — d. v. s. de antager, at Middelværdien af en Elektrons kinetiske Energi er lig Middelværdien af den kinetiske Energi af den fremadskridende Bevægelse af et Luftmolekyle ved samme Temperatur —; de antager derimod i Modsætning til *Lorentz*, at alle Elektronerne har samme absolute Hastighed. Efter den kinetiske Lufttheori er imidlertid Tilstedeværelsen af den *Maxwell'ske* Hastighedsfordeling saa nøje knyttet til Tilstedeværelsen af en Varmeligevægt i den ovenfor omtalte Forstand, at det synes, at man næppe er berettiget til at tillægge den nævnte bedre Overensstemmelse nogen Vægt. Vi kan iøvrigt vise, at denne bedre Overensstemmelse for den væsentligste Del ikke hidrører fra den omtalte Antagelse, men fra den ikke fuldt nøjagtige Maade, hvorpaa Beregningerne er udførte.

For Elektricitetsledningsevnen finder *Riecke* ved at gennemføre Beregningerne af de enkelte Elektroners Baner $\sigma = \frac{2}{3} l N \varepsilon^2 (3mkT)^{-\frac{1}{2}}$. *Drude* finder derimod $\sigma = \frac{1}{2} l N \varepsilon^2 (3mkT)^{-\frac{1}{2}}$, en Værdi, der, som man ser, er $\frac{3}{4}$ saa stor som *Riecke's*. Denne Forskel kommer af, at *Drude* beregner Elektronernes Middelhastighed i en bestemt Retning ved Hjælp af de Middelhastigheder, hvormed de enkelte Elektroners fri Baner mellem to Sammenstød gennemløbes, uden at tage Hensyn til, at de enkelte Baner gennemløbes i forskellige Tider, alt eftersom Elektronerne bevæger sig med eller imod den ydre elektriske Kraft.

For Varmeledningsevnen finder *Riecke* og *Drude* henholdsvis følgende Værdier:

$$\gamma = \frac{9}{4} l N k^2 T (3mkT)^{-\frac{1}{2}} \left(1 + \frac{2}{3} \frac{T}{N} \frac{dN}{dT} \right) \quad \text{og} \quad \gamma = \frac{3}{2} l N k^2 T (3mkT)^{-\frac{1}{2}}.$$

Forskellen mellem disse to Værdier hidrører fra, at *Riecke* og *Drude* definerer Varmeledningsevnen forskellig; medens *Drude*, ligesom vi ovenfor, definerer Varmeledningen som Energibevægelse uden elektrisk Strøm, definerer *Riecke* Varmeledningen som Energibevægelse uden ydre elektrisk Kraft. (De Korrektionsled hidrørende fra Forandringerne i de fri Elektroners Antal med Temperaturen, som indkommer i *Riecke's* Formel, stammer saaledes fra Led, der svarer til Leddet med $\frac{\delta K}{\delta x}$ i Ligning (5) (i Teksten), hvilket Led ved den af *Riecke* benyttede Definition fra Varmeledningen kommer til at spille en Rolle.) *Riecke's* Værdi for Varmeledningsevnen vil derfor, efter hvad vi ovenfor i Teksten har omtalt, ikke kunne benyttes til Sammenligning med de eksperimentelt fundne Værdier for denne Størrelse.

Vi ser altsaa, at den Værdi, man faar for Forholdet mellem Elektricitets- og

som vi ser, ikke Tilfældet med Udtrykkene (13) og (14) for Forholdet mellem Ledningsevnerne. Disse Udtryk giver derfor Lejlighed til at sammenligne Theorien med de eksperimentelt fundne Resultater.

De i Udtrykkene (13) og (14) indgaaende Størrelser ε (Elementarkvantum for Elektricitet) og k (den absolute Luftkonstant henført til et enkelt Molekyle) kendes vel kun med en temmelig ringe Nøjagtighed; men som *M. Reinganum*[1]) har gjort opmærksom paa, kan Forholdet $k/ε$ bestemmes med stor Nøjagtighed ved Hjælp af velkendte Størrelser. Kaldes saaledes Antallet af Molekyler i en cm³ af en Luftart (f. Eks. Brint) ved den absolute Temperatur T og Trykket p for N, har man (se Side 11) $p = kNT$; er endvidere den Elektricitetsmængde, der maa ledes igennem en Elektrolyt for at udskille en cm³ Brint ved samme Tryk og Temperatur, lig E, har man $E = 2εN$ og altsaa $k/ε = 2p/ET$. Vi faar paa denne Maade (idet, for $T=273$ og $p=1,013 \cdot 10^6$ (1 Atm.), $E=0,259 \; 10^{10}$ (abs. elektrostatisk Maal)) $k/ε = 0,287 \cdot 10^{-6}$.

Tabellen Side 56 indeholder en Oversigt over Resultaterne af *Jaeger* og *Diesselhorst's* Forsøg[2]); Værdierne for σ, γ og ϰ er angivne ved 18°, og ved $\dfrac{T}{σ}\dfrac{Δσ}{ΔT}$, $\dfrac{T}{γ}\dfrac{Δγ}{ΔT}$ og $\dfrac{T}{ϰ}\dfrac{Δϰ}{ΔT}$ skal forstaas $\dfrac{T_{18^0}}{σ_{18^0}}\dfrac{σ_{100^0} - σ_{18^0}}{T_{100^0} - T_{18^0}}$ og de analoge Udtryk.

Varmeledningsevnen, udfra det Grundlag, hvorpaa *Riecke's* og *Drude's* Theorier er opbyggede, bliver (Forholdet mellem *Drude's* Værdi for γ og *Riecke's* for σ)

$$ϰ = \frac{9}{4}\frac{k^2}{ε^2} T,$$

en Værdi, der kun afviger meget lidt fra *Lorentz's*, og som ikke stemmer væsentlig bedre overens med Erfaringerne end denne.

Vi skal endvidere her ved Omtalen af *Lorentz's* Beregninger omtale et Forsøg, som *G. Jäger* (Sitzungsber. d. Wiener Akad. d. Wiss., math.-nat. Kl., Bd. 117, Abt. II a, p. 869, 1908) har gjort paa, udfra de samme fysiske Antagelser som *Lorentz*, at beregne Forholdet mellem Elektricitets- og Varmeledningsevnen uden Benyttelse af den statistiske Fremgangsmaade, men ved Betragtning af selve Elektronernes Baner. *Jäger* finder en Værdi for σ, der er ¾ Gange saa stor som *Lorentz's*, hvilket hidrører fra, at han beregner denne Størrelse paa tilsvarende Maade som *Drude* (smlg. ovenfor) At *Jäger* desuagtet finder samme Værdi for ϰ som *Lorentz*, hidrører fra, at der i hans Beregninger af Varmeledningsevnen er indkommet en Regnefejl, hvorved ogsaa Værdien for denne Størrelse bliver ¾ Gange for stor. Det antages nemlig (loc. cit., p. 873), at de Elektroner, der gaar igennem et Fladeelement, i Middel har haft deres sidste Sammenstød i en Afstand af ½l fra Fladen (l er Middelvejlængden), medens denne Afstand i Virkeligheden er ⅔l. (Fejlen er indkommet ved, at det i Beregningerne af den omtalte Middelafstand antages, at der gaar ligemange Elektroner gennem Fladeelementet i alle mulige Retninger, medens Antallet af Elektroner, der gaar gennem Fladen, i Virkeligheden er proportional med Cosinus af Vinklen mellem Elektronernes Retning og Normalen til Fladen.)

[1]) *M. Reinganum:* Ann. d. Phys., Bd. 2, p. 398, 1900.
[2]) *W. Jaeger* und *H. Diesselhorst:* loc. cit., p. 424. [Tallene er beregnede ved Hjælp af Slutningstabellen. I de Tilfælde, hvor der er undersøgt flere Prøver af det samme Metal, er kun benyttet de Værdier, der svarer til de Metalprøver, der indeholder de færreste Urenheder.]

	$\sigma \cdot 10^{-17},$	$-\dfrac{T}{\sigma}\dfrac{\Delta\sigma}{\Delta T}$	$\gamma \cdot 10^{-7},$	$\dfrac{T}{\sigma}\dfrac{\Delta\gamma}{\Delta T}$	$\varkappa \cdot 10^{10},$	$\dfrac{T}{\varkappa}\dfrac{\Delta\varkappa}{\Delta T}$
Sølv	5,53	0,84	4,21	$-$ 0,05	0,762	1,03
Kobber..............	5,06	0,86	3,76	$-$ 0,06	0,742	1,07
Guld	3,72	0,79	2,93	$+$ 0,01	0,788	1,03
Aluminium...........	2,84	0,82	2,01	$+$ 0,09	0,707	1,18
Zink	1,486	0,84	1,110	$-$ 0,04	0,747	1,05
Kadmium	1,182	0,88	0,927	$-$ 0,11	0,784	1,02
Palladium	0,840	0,78	0,704	$+$ 0,20	0,838	1,26
Platin	0,832	0,81	0,696	$-$ 0,15	0,837	1,25
Nikkel	0,765	0,89	0,594	$-$ 0,09	0,777	1,07
Tin	0,745	0,93	0,608	$-$ 0,23	0,817	0,94
Jern................	0,699	0,97	0,636	$-$ 0,12	0,911	1,18
Staal	0,452	0,98	0,453	$-$ 0,02	1,003	0,98
Bly	0,436	0,88	0,346	$-$ 0,05	0,794	1,11
Vismut	0,0756	0,91	0,081	$-$ 0,60	1,071	0,43
Messing .... (85,7 Cu; 7,15 Zn; 6,39 Sn; 0,58 Ni)	0,710	0,22	0,597	$+$ 0,67	0,841	0,95
Manganin .... (84 Cu; 4 Ni; 12 Mn.)	0,214	0,00	0,217	$-$ 0,77	1,016	0,76
Konstantan........... (60 Cu; 40 Ni)	0,184	0,00	0,224	$-$ 0,66	1,229	0,67

De Værdier, der svarer til *Lorentz's* Theori (Ligning (14)), er

$$\varkappa = 0,479 \cdot 10^{-10} \quad \text{og} \quad \frac{T}{\varkappa}\frac{\Delta\varkappa}{\Delta T} = 1.$$

Man ser af Tabellen, at de forskellige Metaller, tiltrods for at deres Ledningsevner for Varme og Elektricitet er meget forskellige, dog har omtrent samme Forhold mellem Ledningsevnerne — *Wiedemann-Franz's* Lov —, samt at dette Forhold for de enkelte Metaller varierer omtrent som den absolute Temperatur — *Lorenz's* Lov —.

Som man ser, er der endvidere, hvad Størrelsesordnen angaar, god Overensstemmelse mellem den beregnede Værdi for $\varkappa$ og de eksperimentelt fundne.

Denne Overensstemmelse, der som omtalt i det foregaaende først er paavist af *Drude*, er et af Elektrontheoriens smukkeste Resultater og synes bestemt at vise, at saavel Elektricitets- som Varmebevægelsen i Metallerne maa føres af Partikler, der har elektriske Ladninger af samme Størrelsesorden som Ionerne i en Elektrolyt, og som befinder sig i mekanisk Varmeligevægt med Metalmolekylerne, det vil sige i Middel

* [$(T/\sigma)\Delta\gamma/\Delta T$ should read $(T/\gamma)\Delta\gamma/\Delta T$.]

har en kinetisk Energi, hidrørende fra deres fremadskridende Bevægelse, der er lig Middelenergien af den fremadskridende Bevægelse af et Luftmolekyle ved samme Temperatur (se *Reinganum:* loc. cit, p. 403).

Som man ser af Tabellen paa forrige Side, er der dog betydelige Afvigelser mellem Værdierne for ϰ for de enkelte Metaller, ligesom endvidere alle de eksperimentelt fundne Værdier for ϰ ikke er saa lidt større end den af *Lorentz* beregnede Værdi.

Før vi fortsætter med denne Sammenligning mellem Theorien og Erfaringen, maa vi her først omtale et Forhold, der kunde tænkes at bevirke, at de eksperimentelt fundne Resultater ikke umiddelbart kan sammenlignes med den her behandlede Theori. Man kan nemlig tænke sig, at Varmeledningen i Metallerne for en Del ogsaa kan foregaa paa anden Maade end ved de fri Elektroners Bevægelse, idet jo ogsaa de Legemer, der kun i yderst ringe Grad leder Elektriciteten, besidder en vis, omend ringe Varmeledningsevne. Tilstedeværelsen af en saadan ikke-metallisk Varmeledningsevne vilde bevirke, at Forholdet ϰ var større end det, der svarer til det her beregnede. Efter Beregning af *J. Koenigsberger*[1]) vil imidlertid en saadan Varmeledningsevne af samme Størrelsesorden som den, man finder ved Isolatorerne, kun have meget ringe Indflydelse paa Værdien for ϰ hos de bedst ledende Metaller, men vilde vise sig tydeligt ved de slettest ledende Metaller og vilde kunne forklare den Omstændighed, at man for de slettest ledende Metaller gennemgaaende finder de største Værdier for ϰ. *Koenigsberger* tænker sig derfor Muligheden af, at Forholdet mellem den metalliske Varmeledningsevne og Elektricitetsledningsevnen i Virkeligheden er det samme hos alle de rene Metaller — (hos Legeringerne synes Forholdet at maatte være noget større) — og lig c. 0,70 · 10⁻¹⁰ ved 18⁰.

De ovenfor omtalte Afvigelser er imidlertid ikke andet end, hvad man kunde vente efter vore foregaaende Beregninger. Ligning (13) Side 53 viser saaledes, at Forholdet mellem Varme- og Elektricitetsledningsevnen til en vis Grad er afhængigt af de specielle Antagelser, som man indfører om de mellem Metalmolekylerne og Elektronerne virkende Kræfter; ligesom denne Ligning viser, at dette Forholds Værdi i de omhandlede Tilfælde[2]) altid vil være større end den af *Lorentz*

---

[1]) *J. Koenigsberger*: Phys. Zeitschr., Bd. 8, p. 237, 1907. (Se ogsaa *M. Reinganum*: Phys. Zeitschr., Bd. 10, pp. 355 og 645, 1909 og Bd. 11, p. 673, 1910, hvilken sidste Forfatter viser, at den Del af Varmeledningsevnen, der kan hidrøre fra indre Varmestraaling, for Metallernes Vedkommende vil være forsvindende i Forhold til den iagttagne Varmeledningsevne.)

[2]) Værdien for ϰ vil blive større end *Lorentz's* Værdi i alle saadanne Tilfælde, hvor den »effektive fri Middelvejlængde« (se Side 33) i det forholdsvis lille Interval omkring Elektronernes Middelhastighed, indenfor hvilket næsten alle Elektronernes absolute Hastigheder ligger, vokser paa regelmæssig Maade med voksende r (d. v. s.

beregnede Værdi. (Det kan saaledes bemærkes, at dersom vi i Ligningen (13) sætter $n = 3$ — svarende til de af *J. J. Thomson* betragtede »doublets« (se Side 34) —, faar vi en Værdi for x, der stemmer udmærket overens med de eksperimentelt fundne Værdier for de bedst ledende Metaller, og som falder meget nær sammen med den af *Koenigsberger* (se ovenfor) angivne hypothetiske Værdi for x for alle rene Metaller.)

Efter hvad vi har omtalt i Slutningen af forrige Kapitel (Side 44), vil man endvidere ogsaa i det Tilfælde, hvor Metalmolekylernes Dimensioner ikke antages at være forsvindende i Forhold til deres indbyrdes Afstande, dersom man ser bort fra Vekselvirkningen mellem de fri Elektroner indbyrdes og betragter Metalmolekylernes Kraftfelter som stationære, genfinde *Lorentz's* Værdi for x, dersom Metalmolekylerne antages at være haarde elastiske Legemer (eller, hvad der udtrykker det samme, dersom Kræfterne mellem Metalmolekylerne og Elektronerne antages at variere overordentlig hurtigt med Afstanden), medens Værdien for x i andre Tilfælde af den her omhandlede Art (jvnfr. Slutbemærkningen Side 45) i Lighed med i de ovenfor betragtede Tilfælde (smlg. Note 2, Side 57) i Almindelighed vil være større end *Lorentz's*.

Dersom man derimod vilde antage, at Virkningen af Sammenstødene mellem de fri Elektroner indbyrdes ikke var ringe i Forhold til Vekselvirkningen mellem Metalmolekylerne og Elektronerne, vil Værdien for x kunne blive betydelig mindre end den af *Lorentz* fundne. Dette hidrører fra, at Sammenstødene mellem de fri Elektroner indbyrdes vil kunne bevirke en betydelig Nedsættelse af Varmeledningsevnen, idet de vil virke hindrende paa de enkelte Elektroners fri Bevægelser fra Sted til Sted i Metallet paa lignende Maade som Sammenstødene med Metalmolekylerne, medens de kun vil have yderst ringe Indflydelse paa Elektricitetsledningsevnen, idet de stødende Elektroners samlede Bevægelsesmængde i en be-

---

f. Eks. dersom den effektive fri Middelvejlængde i det paagældende Interval med tilstrækkelig Tilnærmelse kan skrives paa Formen $a + b(r - r_0)$, hvor r_0 betegner Elektronernes Middelhastighed, og a og b er positive Konstanter). I saadanne særlige Tilfælde, hvor den effektive fri Middelvejlængde antages at vokse meget uregelmæssigt med voksende r, vil imidlertid Værdien for x under Omstændigheder kunne blive mindre end den af *Lorentz* beregnede Værdi. Dette sidste vil f. Eks. være Tilfældet ved *Gruner's* Theori (se Side 29), i hvilken Theori Værdien for x findes mindre end *Lorentz's*, dersom Forholdet mellem den indførte »kritiske« Hastighed G og Elektronernes Middelhastighed ved den paagældende Temperatur samt Forholdet mellem de to Middelvejlængder l_1 og l_2 ligger mellem visse bestemte Grænser (se *Gruner*: Phys. Zeitschr., Bd. 10, p. 50, 1909).

stemt Retning ikke vil forandres ved Sammenstødene [1]). Hele Virkningen paa Elektricitetsledningsevnen vil kun hidrøre fra, at der ved de betragtede Sammenstød omsættes Bevægelsesmængde mellem de forskellige Grupper af Elektroner med forskellige absolute Hastigheder, og at disse forskellige Grupper ulige hurtigt vil miste deres Bevægelsesmængde ved Sammenstød med Metalmolekylerne; i saadanne særlige Tilfælde, hvor det sidste ikke vil finde Sted — f. Eks. dersom man betragter adskilte Sammenstød og antager, at Metalmolekylerne paavirker Elektronerne med Kræfter, der forholder sig omvendt som 5te Potens af Afstanden, (se Side 33) — vil derfor de betragtede Sammenstød mellem Elektronerne indbyrdes ingensomhelst Virkning kunne have paa Elektricitetsledningsevnen.

Vi maa endelig her bemærke, at man efter de foregaaende Beregninger ikke paa Forhaand kan vente nogen systematisk Afhængighed mellem Afvigelserne fra den *Wiedemann-Franz'ske* Lov og Afvigelserne fra den *Lorenz'ske* Lov. Ligning (13) viser nemlig, at ϰ vel er afhængig af den Maade, hvorpaa Kræfterne mellem Metalmolekylerne og Elektronerne varierer med Afstanden, men at dette Forhold, dersom man betragter adskilte Sammenstød og antager, at de omtalte Kræfter forholder sig omvendt som en eller anden Potens af Afstanden, for hvert enkelt Metal vil være proportionalt med den absolute Temperatur; og Afvigelser fra den *Lorenz'ske* Lov vil derfor i saadanne Tilfælde kun kunne fremkomme, dersom man antager, at selve de Kræfter, hvormed Metalmolekylerne paavirker Elektronerne, varierer med Temperaturen (f. Eks. som Følge af Forandringer indenfor selve Molekylerne).

Medens Elektrontheorien i hvert Tilfælde tilnærmelsesvis tillader at bestemme Forholdet mellem Metallernes Elektricitets- og Varmeledningsevne, idet dette Forhold, som vi har set, kun i temmelig ringe Grad er afhængigt af de specielle Antagelser om de Kræfter, der virker mellem Metalmolekylerne og Elektronerne, stiller Sagen sig ganske anderledes, naar det drejer sig om de e n k e l t e L e d n i n g s e v n e r. De i det foregaaende beregnede Udtryk for Ledningsevnerne tillader,

---

[1]) Jvnfr. *H. A. Lorentz*: Proc. Acad. Amsterdam, vol. 7, p. 449, 1905. (Vi maa i Sammenhæng med det her anførte omtale et Forsøg, som *P. Debye* (loc. cit., p. 484) har gjort paa at beregne Virkningen af Sammenstød mellem de fri Elektroner indbyrdes paa Elektricitets- og Varmeledningsevnen. *Debye* finder som Resultat af sine Beregninger, at Forholdet mellem Ledningsevnerne, i Modsætning til hvad vi ovenfor har omtalt, med Tilnærmelse ikke vil forandres ved de omhandlede Sammenstød. Dette hidrører imidlertid fra, at *Debye* antager, at en Elektron i Middel vil miste hele sin Bevægelsesmængde i en bestemt Retning ved et Sammenstød; dette sidste er efter det ovenfor omtalte ikke rigtigt, idet netop det karakteristiske for Sammenstød mellem fri Elektroner indbyrdes er, at Elektronernes samlede Bevægelsesmængde — og dermed deres Bidrag til den elektriske Strøm — ikke forandres ved Sammenstødet.)

som ovenfor bemærket, ikke nogen direkte Sammenligning med Erfaringerne paa Grund af de i dem indgaaende ubekendte Størrelser; heller ikke den Maade, hvorpaa Ledningsevnerne varierer med Temperaturen, kan forklares uden at indføre særlige Antagelser.

Det maa i denne Sammenhæng bemærkes, at de i det foregaaende ud fra specielle simple Antagelser udledte Formler for Elektricitetsledningsevnen ikke, dersom man antager, at de i Formlerne indgaaende Konstanter N og C, resp. l, er uafhængige af Temperaturen, kan bringes i Overensstemmelse med den Maade, hvorpaa Metallernes Elektricitetsledningsevne eksperimentelt findes at variere med Temperaturen. — (Ligning (9) viser saaledes, at Elektricitetsledningsevnen, dersom N og C ikke antages at variere med Temperaturen, i alle de betragtede Tilfælde (d. v. s. for alle $n > 2$) højst vil aftage ligesaa hurtigt som $T^{-\frac{1}{2}}$, medens derimod Ledningsevnen for de fleste rene Metaller aftager tilnærmelsesvis som T^{-1}.) — Dette er imidlertid næppe andet end, hvad man paa Forhaand kunde vente, idet Forholdene for Elektronernes Bevægelser i Metallerne sikkert maa antages i Virkeligheden at være meget indviklede og paa meget indviklet Maade at kunne forandre sig med Temperaturen [1]).

De eksperimentelle Resultater er ogsaa her langt mere uensartede end i det ovenfor omtalte Tilfælde; navnlig er der her en langt større Forskel paa Legeringernes og de rene Metallers Forhold. Mange Legeringers Ledningsevne er saaledes langt ringere end de rene Metallers,

---

[1]) *J. Koenigsberger* (Jahrb. d. Rad. u. Elek., Bd. 4, p. 158, 1907) og *J. Koenigsberger* og *K. Schilling* (Ann. d. Phys., Bd. 32, p. 179, 1910) har forsøgt udfra Elektrontheorien at forklare den Maade, hvorpaa de faste Stoffers Elektricitetsledningsevne varierer med Temperaturen. De nævnte Forfattere betragter Tilstedeværelsen af de fri Elektroner i Metallet som et Resultat af en Dissociationsproces af Metalmolekylerne og antager derfor, at den Maade, hvorpaa Antallet N af fri Elektroner i Volumenenheden forandrer sig med Temperaturen, kan beregnes ud fra almindelige thermodynamiske Betragtninger. Udfra en saadan Antagelse kan man vel forklare, at Elektricitetsledningsevnen hos næsten alle undersøgte Stoffer besidder et Maksimum for en bestemt Temperatur (hos de rene Metaller er denne Temperatur dog i hvert Tilfælde meget lav); men for at opnaa en nærmere Overensstemmelse med Erfaringerne maa de nævnte Forfattere (se f. Eks. Ann. d. Phys., loc. cit., p. 218) yderligere indføre passende Antagelser om den fri Middelvejlængdes Forandring med Temperaturen — (de nævnte Forfattere bygger paa et Udtryk for Elektricitetsledningsevnen, der svarer til *Lorentz's* Theori; Forholdene vilde iøvrigt i den omhandlede Henseende ikke blive bedre, dersom man gik ud fra et Udtryk for σ svarende til Ligningen (9), idet man da maatte gøre lignende Antagelser om Forandringen af C med Temperaturen) —, Antagelser, for hvilke de ikke giver nogen theoretisk Begrundelse.

hvoraf de er sammensatte, og forandrer sig paa ganske anden Maade med Temperaturen (smlg. Tabellen Side 56) [1].

Vi maa her omtale, at man har forsøgt at forklare disse sidste Forhold ved at opstille en Theori, efter hvilken der skulde være en væsentlig Forskel imellem Elektricitetsledningen igennem Legeringerne og de rene Metaller, hidrørende fra, at de første ikke skulde kunne betragtes som fysisk homogene, men som bestaaende af smaa Dele (Krystaller) af de rene Metaller, hvoraf de er sammensatte. I Legeringerne skulde der derfor foruden den almindelige elektriske Modstand optræde en tilsyneladende Modstand, nemlig en thermoelektrisk Polarisation, hidrørende fra Temperaturforskelle, frembragte ved *Peltier*-Effekt i Berøringsfladerne mellem de omtalte Smaadele. En saadan Theori er uafhængig af hinanden opstillet af *Lorenz, Ostwald, Rayleigh* og *Liebenow* [2]. Senere Undersøgelser synes imidlertid bestemt at vise, at Legeringernes Forhold ikke kan forklares ved Antagelsen af en fysisk Inhomogenitet af den omhandlede Art [3].

Før vi forlader disse Spørgsmaal, skal vi her omtale, at Legeringernes Forhold synes at kunne forklares ud fra Elektrontheorien [4] uden

---

[1]) Angaaende de eksperimentelle Resultater vedrørende Legeringernes Ledningsevne for Elektricitet se f. Eks. *W. Guertler*: Jahrb. d. Rad. u. Elek., Bd. 5, p. 17, 1908.

[2]) *L. Lorenz*: Wied. Ann., Bd. 13, p. 600, 1881; *W. Ostwald*: Zeitschr. f. physik. Chem., Bd. 11, p. 520, 1893; Lord *Rayleigh*: Nature, vol. 54, p. 154, 1896; *C. Liebenow*: Zeitschr. f. Elektrochem., Bd. 4, p. 201, 1897.

[3]) Se f. Eks. *K. Baedeker*: Die elektrischen Erscheinungen in metallischen Leitern, Braunschweig 1911, p. 47—48.

[4]) Vi maa her omtale et Forsøg, som *R. Schenck* (Ann. d. Phys., Bd. 32, p. 261, 1910) har gjort paa at forklare Forskellen mellem Legeringerne og de rene Metaller udfra Elektrontheorien. I *Schenck's* Theori opfattes Legeringerne som »faste Opløsninger« af det ene Metal i det andet, og det opløste Metals Molekyler antages i Middel at besidde en ligesaa stor Bevægelsesenergi som et Luftmolekyle ved samme Temperatur. Idet Molekylernes Bevægelser antages at være underkastede meget store »Gnidningsmodstande«, vil de dog kun meget langsomt bevæge sig fra Sted til Sted og derfor ikke selv tage direkte Del i Varme- eller Elektricitetsledningen; deres Bevægelser antages imidlertid at øve Indflydelse paa de fri Elektroners Bevægelser. For det første antages Molekylernes Bevægelse at forøge »Gnidningsmodstanden« mod Elektronernes Bevægelse. For endvidere at forklare, at Legeringerne udviser en større Værdi for Forholdet mellem Varme- og Elektricitetsledningsevnen end de rene Metaller, antager *Schenck* (loc. cit. p. 272), »at Molekylerne er i Stand til at afgive deres kinetiske Energi til de fri Elektroner ved Stød og saaledes at bidrage til Varmeudligningen«. Dette udtrykkes nærmere saaledes (loc. cit., p. 273): »Den kinetiske Energi, som Elektronerne i Legeringerne har at transportere, er større end i de rene Metaller; til deres egen kommer endnu den, som de har optaget ved Sammenstødene med de opløste Molekyler. Den af Elektronerne i Legeringen overførte samlede kinetiske Energi er

$$i = \frac{N_e - N_\mu}{N_e}$$ Gange saa stor som den i det rene Metal«. (N_e er de fri Elektroners Antal, N_μ Antallet af de opløste Molekyler i Rumfangsenheden.) I Overensstem-

at indføre nogen særlig Forskel mellem dem og de rene Metaller, dersom man, hvad der ikke forekommer mig usandsynligt, kan antage, at den samtidige Tilstedeværelse af uensartede Molekyler i Legeringerne, vil bevirke, at Kraftfelterne i det Indre af Metallet i disse er stærkere end i de rene Metaller, hvor Kræfterne kan tænkes i højere Grad at op-hæve hinanden. Dette vilde nemlig ikke alene kunne forklare Lege-ringernes ringe Ledningsevne som en Følge af den større Modstand mod Elektronernes samlede Bevægelse; men idet de omhandlede Kraftfelter vilde have større Indflydelse paa de langsomme Elektroners Bevægelse end paa de hurtigeres, vilde Varmeledningen nedsættes i

---

melse med det her anførte sætter derfor *Schenck* i Formlerne for Elektricitets- og Varmeledningsevnen (*Drudes* Formler benyttes) *ik* i Stedet for *k* ($\frac{3}{2}kT$ betegner et Luftmolekyles Middelenergi ved Temperaturen T, hidrørende fra dettes fremad-skridende Bevægelse) og finder derfor Forholdet mellem Varme- og Elektricitets-ledningsevnen i^2 Gange saa stort for Legeringerne som for de rene Metaller.

De forskellige Antagelser, der gøres i den omhandlede Theori, synes imid-lertid ikke berettigede. For det første antages kun det opløste Stofs Molekyler og ikke Opløsningsmidlets Molekyler at være i Besiddelse af Bevægelsesenergi; dette synes ganske utænkeligt, — en hel anden Sag er det jo, at man ved mange Spørgsmaal, f. Eks. det opløste Stofs osmotiske Tryk, Diffusion o. s. v., ikke be-høver at tage direkte Hensyn til Bevægelsen af Opløsningsmidlets Molekyler —; tilskriver man imidlertid ogsaa Opløsningsmidlets Molekyler Bevægelsesenergi, falder, som man umiddelbart indser, hele den i *Schenck's* Theori omhandlede For-skel mellem Legeringerne og de rene Metaller fuldstændig bort. Rent bortset fra dette, synes imidlertid heller ikke de Slutninger, *Schenck* drager af de opløste Mole-kylers Bevægelse at være berettigede. Saaledes vil vel Sandsynligheden for Sammenstød mellem Elektronerne og Metalmolekylerne, og dermed de ovenfor omtalte ›Gnid-ningsmodstande‹, være større, naar Molekylerne tænkes at være i Bevægelse, end naar de tænkes at være i Hvile, men Forøgelsen vil i det her omhandlede Tilfælde paa Grund af Elektronernes store Hastigheder i Forhold til Metalmole-kylernes være ganske forsvindende. (Man beviser saaledes i den kinetiske Luft-theori (se f. Eks. *Jeans*: Dynamical Theory of Gases, p. 234), at Sandsynligheden for Sammenstød mellem to forskellige Slags Luftmolekyler med Masserne m_1 og m_2 vil være $K \cdot \sqrt{1 + \dfrac{m_1}{m_2}}$, hvor K betyder Sandsynligheden for Sammenstød, naar Molekylerne med Masse m_2 tænkes ubevægelige. Ved de almindelige Metaller vil Forholdet mellem Elektronernes Masser og Metalmolekylernes Masser være c. 10^{-5} og Kvadratroden derfor ikke afvige mærkeligt fra 1.) Hvad endvidere *Schenck's* Antagelser om de fri Elektroners Energioverføring angaar, synes dette efter min Mening ikke at kunne opfattes anderledes, end at de fri Elektroners Middelenergi i *Schenck's* Theori sættes lig med $i \cdot \frac{3}{2}kT$. (Det synes nemlig ikke berettiget at skelne mellem Elektronernes ›egen‹ Energi og ›den‹, som de har optaget ved Sammenstødene med Metalmolekylerne‹.) Dette vil imidlertid være i Modstrid med de Antagelser, der theoretisk begrundes i den kinetiske Lufttheori, og som danner Grundlaget for Metallernes Elektrontheori, efter hvilke Middelenergien af et Molekyle af en hvilken som helst Slags, der befinder sig i mekanisk Varmelige-vægt med Molekyler af samme eller andre Slags, er lig $\frac{3}{2}kT$.

forholdsvis ringere Grad end Elektricitetsledningen; og endelig maatte Temperaturkoefficienten (regnet med Fortegn) for Elektricitetslednings-evnen ventes at være større end for de rene Metaller, idet den effek-tive Middelvejlængde (forudsat, at Molekylernes Kraftfelter antages at være uafhængige af Temperaturen) vilde stige stærkere med Tempe-raturen end hos disse (smlg. Forandringerne med aftagende n i Vær-dierne for $\frac{\gamma}{\sigma}$ og $\frac{1}{\sigma}\frac{d\sigma}{dT}$ givne ved Ligningerne (13) og (9)).

---

§ 3.

Thermoelektriske Fænomener.

Vi har i det foregaaende betragtet to særlige Tilfælde — Elektri-citets- og Varmeledning — af henholdsvis Elektricitets- og Energi-bevægelse i et homogent Metalstykke. Vi skal nu udfra Ligningerne (4) og (5) betragte de mere almindelige Tilfælde af Elektricitets- og Varmebevægelse i et Metalstykke, og vi skal se, hvorledes de om-handlede Ligninger fører til en fuldstændig Theori for de saakaldte thermoelektriske Fænomener. Ved Behandlingen af disse Spørgs-maal skal vi i Hovedsagen følge den elegante Fremstillingsmaade, der er angivet af *H. A. Lorentz* [1].

Ligningerne (4) og (5) kan, idet man indfører de i det foregaaende ved σ og γ betegnede Størrelser (se Ligningerne (8) og (10)) og sætter $A_2 = \mu\frac{k}{\varepsilon}A_1$, skrives paa Formen

$$i_x = -\sigma\left(\frac{\partial\varphi}{\partial x} + \frac{k}{\varepsilon}\frac{T}{K}\frac{\partial K}{\partial x} + \mu\frac{k}{\varepsilon}\frac{\partial T}{\partial x}\right) \tag{15}$$

og

$$W_x = \mu\frac{k}{\varepsilon}Ti_x - \gamma\frac{\partial T}{\partial x}, \tag{16}$$

hvor Størrelsen K efter Ligning (20) Side 39 bestemmes ved

$$K = N\left(\frac{m}{2\pi kT}\right)^{\frac{3}{2}}\cdot\left[\int e^{-\frac{\varepsilon\lambda}{kT}}dv\right]^{-1} \tag{17}$$

(hvilket, dersom man betragter »adskilte Sammenstød« ($\lambda = 0$), giver $K = N\left(\frac{m}{2\pi kT}\right)^{\frac{3}{2}}$ (smlg. Lign. (5) Side 16)).

Ligningerne (15) og (16) bestemmer fuldstændig Elektricitets- og Varmebevægelsen i et Metalstykke, i hvilket Metallets Natur og Tem-

---

[1] *H. A. Lorentz:* Proc. Acad. Amsterdam, vol. 7, pp. 451 og 585, 1905.

peratur er ens i alle Punkter af et Plan vinkelret paa X-aksen; i de
mere almindelige Tilfælde, hvor Metallets Natur og Temperatur varierer
paa vilkaarlig Maade, maa man ogsaa benytte de til (15) og (16) svarende Ligninger, der faas ved at ombytte x med henholdsvis y og z.
Vi skal imidlertid i det følgende for Simpelheds Skyld, da det ikke
medfører nogen væsentlig Forandring i Resultaternes Karakter, antage,
at det Metalstykke, vi betragter, har Form af en tynd Traad — en
thermoelektrisk Kæde —, hvis Tværsnits Dimensioner overalt er meget
smaa i Forhold til Krumningsradien, og i hvilken Metallets Natur og
Temperatur antages at være meget nær ens i alle Punkter af hvert
enkelt Tværsnit. I et saadant Tilfælde vil man, som man umiddelbart
indser, kunne nøjes med at benytte Ligningerne (15) og (16), dersom blot
x i disse Ligninger antages at betegne Buelængden paa en Kurve lagt
igennem Tyngdepunkterne af Kædens Tværsnit regnet ud fra et fast
Punkt paa denne Kurve.

Vi skal nu først betragte en **aaben thermoelektrisk Kæde**,
det vil sige en Kæde, i hvilken der ingen elektrisk Strøm gaar. I dette
Tilfælde faar vi af Ligningen (15), idet vi sætter $i_x = 0$,

$$\frac{d\varphi}{dx} = -\frac{k}{\varepsilon}\left(\frac{T}{K}\frac{dK}{dx} + \mu\frac{dT}{dx}\right), \qquad (18)$$

en Ligning, ved Hjælp af hvilken man kan bestemme Potentialdifferensen mellem to hvilkesomhelst Punkter af Kæden.

Det maa dog bemærkes, at de her omhandlede Potentialdifferenser
er Differenser mellem Potentialet i Punkter i det Indre af Metallet. Disse
Potentialdifferenser er derfor ikke direkte tilgængelige for Maalinger, idet
det maa antages, at der finder et betydeligt Potentialfald Sted i umiddelbar
Nærhed af Metallets Overflade. (Et saadant Potentialfald iagttages saaledes
ved Undersøgelser over Metallernes Udsendelse af Elektroner ved høje
Temperaturer og over den Varmemængde, der frigøres ved Metallernes
»Absorption« af fri Elektroner (se f. Eks. *Richardson* og *Cooke:* Phil. Mag.,
vol. 20, p. 173, 1910).) De nedenstaaende Beregninger kan derfor ikke
benyttes til direkte Sammenligning med Erfaringerne, undtagen i saadanne
Tilfælde, hvor det gælder Potentialdifferensen mellem to Punkter af Kæden,
der bestaar af samme Metal, og som har samme Temperatur.

Ligningen (18) tillader at drage forskellige Slutninger.

1) Dersom **Temperaturen overalt i Kæden er konstant**,
faar man ved Integration

$$\varphi_P - \varphi_Q = -\frac{k}{\varepsilon}\,T\log\frac{K_P}{K_Q}, \qquad (19)$$

hvilken Ligning udsiger, at Potentialdifferensen mellem to Punkter, *P* og *Q*, af Kæden kun vil være afhængig af Metallets Natur i disse Punkter og vil være o, dersom denne er den samme. — Heraf følger, at Metallerne med Hensyn til de betragtede Potentialdifferenser for hver Temperatur vil kunne lade sig ordne i en saakaldt Spændingsrække. —

Af Ligning (18) faas i det her omhandlede Tilfælde ved Benyttelse af (17)

$$\frac{d\varphi}{dx} = -\frac{k}{\varepsilon}\frac{T}{N}\frac{dN}{dx} - \int\frac{\partial\lambda}{\partial x}e^{-\frac{\varepsilon\lambda}{kT}}dv : \int e^{-\frac{\varepsilon\lambda}{kT}}dv. \tag{20}$$

Da Antallet af Elektroner i Volumenelementet *dv* efter Ligning (19) Side 39, dersom der er statistisk Ligevægt, vil være proportionalt med $e^{-\frac{\varepsilon\lambda}{kT}}$, vil i et saadant Tilfælde det sidste Led i Ligning (20) være lig Middelværdien af $\frac{\partial\lambda}{\partial x}$ beregnet over alle Elektronerne; idet vi betegner en saadan Middelværdi med $\overline{\frac{\partial\lambda}{\partial x}}$, faar vi af Ligning (20)

$$-N\varepsilon\left(\frac{d\varphi}{dx} + \overline{\frac{\partial\lambda}{\partial x}}\right) - kT\frac{dN}{dx} = \mathrm{o},$$

hvilken Ligning umiddelbart udtrykker Betingelsen for, at de elektriske Kræfter, der virker paa Elektronerne, holder Ligevægt med Forandringerne i Elektronernes Tryk, $p = k \cdot NT$ (se Side 11); i det her betragtede Tilfælde, hvor der dels er elektrisk Ligevægt, og dels Temperaturen er konstant, vil nemlig Elektronerne ikke besidde nogen ordnet Bevægelse i en bestemt Retning, og der vil derfor ikke fremkomme nogen »Ophobning af Elektronerne foran Metalmolekylerne« (se Side 43, smlg. iøvrigt det følgende Tilfælde).

2) Dersom Kæden overalt bestaar af samme Metal, vil, idet i dette Tilfælde *K* og μ er Funktioner af Temperaturen alene, Udtrykket for *d*φ være et eksakt Differential med Hensyn til *dT*, og Potentialdifferensen mellem to Punkter af Kæden vil derfor kun afhænge af Temperaturen i disse Punkter og være o, dersom Temperaturen er den samme.

Ligning (18) tillader i dette Tilfælde ikke en saa simpel Tydning som i det forrige Tilfælde, hvilket hidrører fra, at der i dette Tilfælde vil gaa en Varmestrøm igennem Kæden, og der derfor vil fremkomme en Modstand mod Elektronernes ordnede Bevægelse fra Metalmole-

5

kylernes Side. Dette ses simplest, dersom man betragter et Metalstykke, i hvilket det antages, at der finder adskilte Sammenstød Sted. Dersom man antager, at de Kræfter, hvormed Metalmolekylerne paavirker Elektronerne, aftager omvendt som n^{te} Potens af Afstanden, faar man i dette Tilfælde ved Hjælp af Ligningerne (6) Side 49 $\mu = \dfrac{2n}{n-1}$; af (18) faas nu ved Hjælp af (17), idet vi sætter $\lambda = 0$,

$$\frac{d\varphi}{dx} = -\frac{k}{\varepsilon}\left(\frac{T}{N}\frac{dN}{dT} + \frac{n+3}{2(n-1)}\right)\frac{dT}{dx}. \tag{21}$$

Dersom $n = \infty$, bliver det sidste Led indenfor Parenthesen lig $\frac{1}{2}$, hvilket svarer til Koefficienten til det paagældende Led i *Lorentz's* Udtryk for $\dfrac{d\varphi}{dx}$. Dersom $n = 5$, bliver det omhandlede Led lig 1, og dette svarer til et Udtryk for $\dfrac{d\varphi}{dx}$, som *G. Jäger*[1]) har udledt ved at gaa ud fra, at der skulde være Ligevægt mellem de ydre elektriske Kræfter og Elektronernes Tryk; denne sidste Overensstemmelse er ogsaa let forstaaelig, idet vi Side 33 har omtalt, at den i Tidsenheden til Metalmolekylerne afgivne Bevægelsesmængde — og følgelig den resulterende Kraft, hvormed Molekylerne modsætter sig Elektronernes ordnede Bevægelse —, dersom $n = 5$, er proportional med Elektronernes samlede Bevægelsesmængde, og denne er i dette Tilfælde lig 0, da der ingen elektrisk Strøm gaar.

3) Dersom Kæden bestaar af flere forskellige Metaller, og Temperaturen er forskellig paa de forskellige Steder, vil Potentialdifferensen mellem Kædens Endepunkter, dersom disse bestaar af samme Metal og·har samme Temperatur, kun afhænge af Temperaturen i Berøringsstederne mellem de forskellige Metaller (Temperaturen antages her at være konstant paa de korte Stykker, hvor den gradvise Overgang mellem Metallerne tænkes at finde Sted). Rigtigheden af denne Sætning fremgaar umiddelbart af det, der er sagt ved Omtalen af de to ovenfor behandlede Tilfælde. Vi skal nu her beregne Størrelsen af

---

[1]) *G. Jäger*: Sitzungsber. d. Wiener Akad. d. Wiss., math.-nat. Kl., Bd. 117, Abt. II a, p. 859, 1908. [*Jäger* begrunder (loc. cit., p. 859, Note 1) ikke sin Fremgangsmaade udfra en saadan Betragtning som den i Teksten staaende, men derudfra, at han antager, at man kan se bort fra Sammenstødene mellem Elektronerne og Metalmolekylerne i Forhold til Sammenstødene mellem Elektronerne indbyrdes, en Antagelse, der imidlertid, efter hvad der er omtalt i det foregaaende (Side 58), ikke kan forenes med de eksperimentelt fundne Værdier for Forholdet mellem Varme- og Elektricitetsledningsevnen, idet Værdien af dette Forhold, beregnet udfra en saadan Antagelse, vil blive mange Gange for lille.]

Potentialdifferensen mellem Endepunkterne af en Kæde, der bestaar af to Metaller P og Q, idet vi antager, at Temperaturen i Berøringsstedet mellem P og Q er T_1 og mellem Q og P er T_2, samt at Temperaturen i Endepunkterne x_1 og x_2, der antages at bestaa af Metallet P, er lig T_0. Af Ligning (18) faar vi, idet vi betegner den omhandlede Potentialdifferens — den elektromotoriske Kraft i Kæden — med F,

$$F = \varphi_{x_2} - \varphi_{x_1} = - \frac{k}{\varepsilon} \int\limits_{x_1}^{x_2} \left(\frac{T}{K} \frac{dK}{dx} + \mu \frac{dT}{dx} \right) dx.$$

Ved partiel Integration af det første Led under Integraltegnet faar vi

$$F = \frac{k}{\varepsilon} \int\limits_{x_1}^{x_2} (\log K - \mu) \frac{dT}{dx} dx = \frac{k}{\varepsilon} \int\limits_{T_0}^{T_1} (\log K_P - \mu_P) \, dT$$

$$+ \frac{k}{\varepsilon} \int\limits_{T_1}^{T_2} (\log K_Q - \mu_Q) \, dT + \frac{k}{\varepsilon} \int\limits_{T_2}^{T_0} (\log K_P - \mu_P) \, dT,$$

og heraf

$$F = \frac{k}{\varepsilon} \int\limits_{T_1}^{T_2} \left(\log \left(\frac{K_Q}{K_P} \right) - (\mu_Q - \mu_P) \right) dT. \tag{22}$$

Dette Udtryk adskiller sig fra det af *Lorentz*[1]) givne Udtryk ved Tilføjelsen af det sidste Led under Integraltegnet; dette Led forsvinder nemlig af *Lorentz's* Beregninger, fordi μ efter hans Theori har samme Værdi for alle Metaller, nemlig $\mu = 2$ (se Side 66). I *Gruner's*[2]) Theori (se Side 29) vil derimod μ være forskellig for to Metaller, dersom den i denne Theori indgaaende kritiske Hastighed G, eller Forholdet mellem de to Middelvejlængder l_1 og l_2, er forskellig i de omhandlede Metaller. Det af *Gruner* givne Udtryk for F har derfor samme Form som det i (22) givne almindelige Udtryk.

Vi skal nu undersøge den V a r m e t o n i n g, d e r f i n d e r S t e d, n a a r d e r g a a r e n e l e k t r i s k S t r ø m g e n n e m e n s a a d a n t h e r m o - e l e k t r i s k K æ d e. Tilstanden skal antages at være stationær, idet Temperaturen tænkes at blive holdt konstant i ethvert Punkt af Kæden, ved at der udefra tilføres eller bortledes Varme. — Idet Kæden antages

---

[1]) *H. A. Lorentz*: Proc. Acad. Amsterdam, vol. 7, p. 453, 1905.

[2]) *P. Gruner*: Phys. Zeitschr., Bd. 10, p. 50, 1909.

at bestaa af en tynd Traad, vil dette kunne lade sig gøre, uden at Temperaturen i et enkelt Tværsnit varierer mærkeligt. —

Vi skal nu betragte et lille Element af Kæden beliggende mellem to Planer, der bestemmes ved Koordinaterne x og $x + dx$, og vi skal søge den Varmemængde dQ, der i Tidsenheden maa ledes bort fra et saadant Element for at holde Temperaturen konstant.

Denne Varmemængde vil være lig Differensen mellem den Energi, der tilføres Elektronerne i Elementet ved de ydre Kræfters Virksomhed, og den Energi, der ved Elektronernes Bevægelse føres bort i Kædens Længderetning. Idet Tilstanden antages at være stationær, vil der føres ligesaa mange Elektroner ud af som ind i det betragtede Element af Kæden, og den sidst omtalte Energimængde vil derfor kunne bestemmes ved Hjælp af Udtrykket for W_x, der angiver Summen af den kinetiske Energi og den potentielle Energi i Forhold til Metalmolekylerne, der i Tidsenheden ved Elektronernes Bevægelse føres igennem en Fladeenhed af Kædens Tværsnit (jvnfr. Side 52); vi har derfor, idet Arealet af Kædens Tværsnit betegnes ved ω,

$$dQ = -\left(\omega\, i_x \frac{d\varphi}{dx} + \frac{d(\omega W_x)}{dx}\right) dx.$$

Dette giver ved Hjælp af Ligningerne (15) og (16), idet vi borteliminerer $\frac{\delta\varphi}{\delta x}$ og bemærker, at $\omega i_x = i$ i det omhandlede Tilfælde er uafhængig af x,

$$dQ = \left(\frac{i^2}{\omega\sigma} + i\frac{k}{\varepsilon} T \frac{d(\log K - \mu)}{dx} - \frac{d\left(\omega k \frac{dT}{dx}\right)}{dx}\right) dx. \qquad (23)$$

Det første Led i Udtrykket for dQ, der er proportionalt med Strømstyrkens Kvadrat, angiver den *Joule'ske* Varme, og det sidste Led, der er uafhængigt af den elektriske Strøm, den Varmeudvikling, der skyldes den almindelige Varmeledning; endelig giver det mellemste Led, der svarer til en Varmeudvikling

$$dQ_R = i\frac{k}{\varepsilon} T \frac{d(\log K - \mu)}{dx}, \qquad (24)$$

der er proportional med Strømstyrken, og som forandrer sit Fortegn, naar Strømmens Retning forandres, Udtryk for de saakaldte *Peltier-* og *Thomson*-Effekter. For at undersøge denne sidste Varmetoning nærmere skal vi her betragte to forskellige Tilfælde.

Vi skal først betragte et Stykke af Kæden, indenfor hvilket Temperaturen er konstant, og i hvilket der finder en Overgang Sted fra Metallet P til Metallet Q. Ved Hjælp af Udtrykket (24) finder vi ved

* [The last term in the parenthesis should read d$(\omega\gamma(dT/dx)/dx$.]

Integration, at den Varmemængde, der udvikles i Tidsenheden i den betragtede Del af Kæden, og som angives ved det omhandlede Led i Formlen (23), vil være lig

$$i\,\frac{k}{\varepsilon}\,T\left(\log\frac{K_Q}{K_P}-(\mu_Q-\mu_P)\right).$$

Idet vi nu ved *Peltier*-Effekten $\pi_{P,Q}$ skal forstaa den Varmemængde, der absorberes ved Berøringsstedet mellem to Metaller P og Q, dersom der gaar en Elektricitetsenhed gennem Berøringsstedet fra P til Q, faar vi

$$\pi_{P,Q}=\frac{k}{\varepsilon}\,T\left(\log\frac{K_P}{K_Q}-(\mu_P-\mu_Q)\right). \tag{25}$$

Betragter vi dernæst et Stykke af Kæden, indenfor hvilket Metallet er det samme, faar vi af Udtrykket (24), idet K og μ i dette Tilfælde vil være Funktionen af T alene, at den Varmemængde, der angives ved det omhandlede Led i (23), og som udvikles i Tidsenheden i et lille Stykke af Kæden, indenfor hvilket Temperaturen forandrer sig fra T til $T+dT$, vil være

$$i\,\frac{k}{\varepsilon}\,T\,\frac{d(\log K-\mu)}{dT}\,dT.$$

Idet vi nu ved *Thomson*-Effekten ρ skal forstaa den Varmemængde, der absorberes i et Stykke af en Metalstang, i hvilket Temperaturen varierer 1^0, naar der gaar en Elektricitetsenhed gennem Stangen i Retning fra lavere Temperatur til højere, faar vi

$$\rho=-\frac{k}{\varepsilon}\,T\,\frac{d(\log K-\mu)}{dT}. \tag{26}$$

Dersom vi antager, at der finder adskilte Sammenstød Sted, og at Molekylernes Kraftfelter — Loven for Sammenstødene — er ens i alle Metaller og ved alle Temperaturer, vil Udtrykkene for F, π og ρ, idet vi indsætter Værdien for K, given ved Ligningen (17) Side 63 ($\lambda = 0$), og bemærker at μ i dette Tilfælde er konstant, blive

$$\left.\begin{aligned}F &=\frac{k}{\varepsilon}\int_{T_1}^{T_2}\log\left(\frac{N_Q}{N_P}\right)dT,\quad \pi_{P,Q}=\frac{k}{\varepsilon}\,T\log\left(\frac{N_P}{N_Q}\right),\\[2mm] \rho &=-\frac{k}{\varepsilon}\,T\,\frac{d\log N}{dT}+\frac{3}{2}\,\frac{k}{\varepsilon},\end{aligned}\right\} \tag{27}$$

hvilket er de Værdier, som *Lorentz* har fundet. Idet $\tfrac{3}{2}kT$ er Middelværdien af en Elektrons kinetiske Energi ved Temperaturen T, vil det sidste Led i Udtrykket for *Thomson*-Effekten $\tfrac{3}{2}k/\varepsilon$ være den

Varmemængde, der skal tilføres en Samling fri Elektroner, hvis samlede elektriske Ladning er en Elektricitetsenhed, naar Temperaturen skal stige en Grad.

De her udledte Udtryk for den thermoelektriske Kraft og *Peltier*- og *Thomson*-Effekterne tillader ikke en nøjere direkte Sammenligning med Erfaringen paa Grund af de i dem indgaaende ubekendte Størrelser K og μ. — Vi skal derimod i det følgende betragte den indbyrdes Sammenhæng mellem de tre omhandlede Størrelser. — Vi skal her kun omtale den Omstændighed, at Udtrykket for *Thomson*-Effekten indeholder et Led, nemlig det ovenfor betragtede Led $\frac{3}{2}k/\varepsilon$, der er betydelig større end de iagttagne Værdier for *Thomson*-Effekten [1]. Dersom man antager adskilte Sammenstød og samme Lov for Sammenstødene i alle Metaller og ved alle Temperaturer (se Lign. (27)), maa man derfor, for at kunne forklare *Thomson*-Effekten i Overensstemmelse med Erfaringen, antage, at de fri Elektroners Antal i Volumenenheden i alle Metaller varierer omtrentlig paa samme Maade med Temperaturen, nemlig som $T^{\frac{3}{2}}$; kun i dette Tilfælde vil nemlig de to Led i Udtrykket for *Thomson*-Effekten omtrentlig hæve hinanden [2]. Som det fremgaar af Ligningen (26), vil der imidlertid yderligere indkomme Led i Udtrykket for *Thomson*-Effekten, dersom μ varierer med Temperaturen, d. v. s. dersom Metalmolekylernes Kraftfelter forandrer sig med Temperaturen. Det kan i denne Sammenhæng bemærkes, at Antagelsen af en saadan Forandring, efter hvad vi har omtalt i forrige Paragraf (se Noten Side 60), er nødvendig for at forklare den elektriske Ledningsevnes iagttagne Afhængighed af Temperaturen.

---

[1] Se f. Eks. *J. Kunz*: Phil. Mag., vol. 16. p. 781, 1908. [*Kunz* mener, at man kan forklare den omhandlede tilsyneladende Uoverensstemmelse mellem Theorien og Erfaringerne ved at antage, at Elektroner, der bevæger sig gennem Metallet i Retning fra lavere Temperatur til højere, ikke vil absorbere den samme Energi fra Metallet, dersom de bevæger sig under Indflydelse af en ydre elektrisk Kraft, som naar de diffunderer frit som Følge af Temperaturforskelle. En saadan Antagelse vil imidlertid stride imod de Hovedprinciper, der lægges til Grund for Metallernes Elektrontheori — og hvis Rigtighed ogsaa antages af *Kunz* selv paa det citerede Sted —, nemlig at de ydre Kræfter kun vil have en meget ringe Indflydelse paa de enkelte Elektroners Bevægelse.]

[2] Se f. Eks. *J. J. Thomson*: The Corpuscular Theory of Matter, p. 79. [Paa det citerede Sted angives dog, hidrørende fra den ikke fuldt nøjagtige Beregningsmaade (den kinetiske Energi, der under en Luftarts Bevægelse føres igennem en Flade i en bestemt Retning, er nemlig ikke lig Molekylernes Middelenergi $\frac{3}{2}kT$ multipliceret med det Antal Molekyler, der gaar gennem Fladen, men er, som en nærmere Beregning viser, $\frac{5}{3}$ Gange saa stor), at N maa forholde sig som $T^{\frac{1}{2}}$ i Stedet for som $T^{\frac{3}{2}}$.]

Vi maa her bemærke, at de fundne Udtryk for π og ρ vel afhænger af Metalmolekylernes Kraftfelter, for saa vidt som disse bestemmer de fri Elektroners Bevægelser paa de enkelte Steder i Metallet, men ikke er direkte afhængige af Forandringen fra Sted til Sted i Metallet af Middelværdien af Elektronernes Potential i Forhold til Metalmolekylerne, og at der derfor f. Eks. ingen umiddelbar Sammenhæng er imellem Værdien af *Peltier*-Effekten og de i det foregaaende omtalte Spændingsdifferenser mellem to Metaller i Berøring. Betragter man saaledes to Metalstykker P og Q, i hvis Indre alle Forhold antages at være ens, undtagen at Elektronernes potentielle Energi i hvert Punkt i det Indre af P er λ_0 større end den potentielle Energi i det tilsvarende Punkt i Q, vil Spændingsforskellen mellem de to Metaller, der angives ved (19), som det ses ved Betragtning af Udtrykket (17) for K, være $\varphi_P - \varphi_Q = -\lambda_0$. Derimod vil i et saadant Tilfælde *Peltier*-Effekten være o, idet Værdien af $(\log K - \mu)$, som det ses ved Betragtningen af Ligningerne (7) og (17), vil være uafhængig af en saadan konstant Forskel i Elektronernes potentielle Energi. Dette er ogsaa, hvad man paa Forhaand kunde vente, idet de omhandlede Spændingsdifferenser i et saadant Tilfælde netop holder Ligevægt med de fra Metalmolekylerne udgaaende Kræfter, saa at Elektronerne i det Indre af Metallet vil bevæge sig, som om Metalstykket var homogent[1]).

Vi skal nu gaa over til at omtale de i det foregaaende udledte Resultaters Stilling til den af Lord *Kelvin*[2]) opstillede t h e r m o d y n a-m i s k e T h e o r i for de thermoelektriske Forhold. I denne Theori betragtes Elektricitetens Gennemgang gennem en sluttet thermoelektrisk Kæde som en reversibel Kredsproces, ved hvilken der kan omsættes Varme til Arbejde og omvendt, — de Varmemængder, der her er Tale om, er dem, der udvikles eller absorberes ved *Peltier*- og *Thomson*-Effekterne, og det omhandlede Arbejde er det, som den thermoelektriske Kraft udfører under Elektricitetens Bevægelse i Kæden —. Ved at anvende Varmetheoriens anden Hovedsætning paa denne Kredsproces faar man følgende Betingelse:

$$\int \frac{dQ_R}{T} = 0, \qquad (28)$$

hvor Integrationen tænkes udført over hele Kæden.

Ved Hjælp af denne Ligning og ved at anvende Energisætningen finder man følgende af Lord *Kelvin* opstillede Betingelser (loc. cit., p. 135):

$$\rho_Q - \rho_P - T\frac{d\left(\dfrac{\pi_{P,Q}}{T}\right)}{dT} = 0 \quad \text{og} \quad E_{P,Q} - \frac{\pi_{P,Q}}{T} = 0. \qquad (29)$$

[1]) Se f. Eks. *R. Clausius:* Pogg. Ann., Bd. 90, p. 520, 1853.
[2]) *W. Thomson*: Trans. Roy. Soc., Edinburgh 1854, p. 123.

hvor $E_{P,Q}$ betegner den thermoelektriske Kraft i en Kæde, der bestaar af to Metaller, og hvor det ene Berøringssted er en Grad varmere end det andet; den omhandlede Kraft er regnet positiv i Retning fra Metallet P igennem det varmere Lodningssted til Metallet Q.

Disse Betingelser har som bekendt vist sig, i hvert Tilfælde med Tilnærmelse, at stemme overens med de eksperimentelt fundne Resultater [1]).

Den her omtalte Udledning af Ligningerne (28) og (29) er imidlertid, som Lord *Kelvin* allerede omtaler (loc. cit., p. 128), og som navnlig *Boltzmann* [2]) har fremhævet, ikke strengt begrundet, idet den her omhandlede reversible Kredsproces nødvendigt er forbunden med irreversible Processer, nemlig Udviklingen af den *Joule'ske* Varme og Varmeledningen, og det paa en saadan Maade, at de Varmetoninger, der skyldes de sidstnævnte Processer, ikke paa nogen Maade samtidig kan gøres forsvindende smaa i Sammenligning med den thermoelektriske Varmetoning. Som *Boltzmann* har vist, vil Varmetheoriens anden Hovedsætning i Virkeligheden altid være tilfredsstillet, naar Størrelserne E, π og ρ — i Stedet for Betingelserne (28) — blot opfylder følgende Betingelser (loc. cit., p. 1281):

og

$$\left.\begin{array}{c} \left| \rho_Q - \rho_P - T \cdot \dfrac{d\left(\dfrac{\pi_{P,Q}}{T}\right)}{dT} \right| \leqq \dfrac{2}{\sqrt{T}} \left(\sqrt{\left(\dfrac{\gamma}{\sigma}\right)_P} + \sqrt{\left(\dfrac{\gamma}{\sigma}\right)_Q} \right) \\[4ex] \left| E_{P,Q} - \dfrac{\pi_{P,Q}}{T} \right| \leqq \dfrac{2}{\sqrt{T}} \left(\sqrt{\left(\dfrac{\gamma}{\sigma}\right)_P} + \sqrt{\left(\dfrac{\gamma}{\sigma}\right)_Q} \right). \end{array}\right\} \quad (30)$$

Ligningen (28) og de deraf følgende Betingelser (29) kan derfor ikke begrundes ad ren thermodynamisk Vej.

Som vi ser, tilfredsstiller det i det foregaaende (Side 68) udfra Elektrontheorien udledte Udtryk (24) for den thermoelektriske Varmetoning dQ_R Ligningen (28); de fundne Udtryk for F, π og ρ vil derfor ogsaa tilfredsstille Betingelserne (29). Denne Overensstemmelse mellem Elektrontheoriens og den thermodynamiske Theoris Resultater er for et fuldkommen strengt gennemregnet Eksempels Vedkommende først paavist af *H. A. Lorentz* [3]). Som vi har set, finder den omtalte

---

[1]) Se f. Eks. *Baedeker*: Die elektrischen Erscheinungen in metallischen Leitern, p. 81—86.

[2]) *Boltzmann*: Sitzungsber. d. Wiener Acad. d. Wiss., math.-nat. Kl., Bd. 96, Abt. II, p. 1258, 1887.

[3]) *H. A. Lorentz*: Proc. Acad Amsterdam, vol. 7, p. 589, 1905; se ogsaa Jahrb. d. Rad. u. Elek., Bd. 2, p. 377, 1905. [Dersom man ikke tager Hensyn til Tilstedeværelsen af den *Maxwell'ske* Hastighedsfordeling, vil man ikke komme til en saadan Overensstemmelse (se f. Eks. *Riecke*: Wied. Ann., Bd. 66, p. 1200, 1898).]

Overensstemmelse Sted i alle de her betragtede Tilfælde; nemlig, forudsat Opfyldelsen af de i Indledningen omtalte Hovedantagelser, i alle Tilfælde, hvor der finder adskilte Sammenstød Sted, og endvidere i det Tilfælde, hvor Elektronernes Bevægelser kan betragtes som foregaaende uafhængige af hverandre i et stationært Kraftfelt, et Tilfælde, der, som omtalt i Kapitel I Side 38, kan antages tilnærmelsesvis at svare til de virkelige Forhold i Metallerne.

Den her omhandlede Overensstemmelse mellem Elektrontheorien og Lord *Kelvin's* Theori for de thermoelektriske Fænomener er saameget mere interessant, som den ikke paa Forhaand kunde ventes, idet Elektrontheorien, der tænker sig saavel Elektricitetens som Varmens Bevægelse knyttet til de samme Partikler, tilsyneladende ikke giver nogen Berettigelse til ved Behandlingen formelt at skille de reversible Processer fra de irreversible [1].

---

[1] *J. Kunz* (Phil. Mag., vol. 16, p. 767, 1908) forsøger at forklare den omtalte Overensstemmelse, idet han beregner de thermoelektriske Varmetoninger ved Hjælp af det Arbejde, som de elektriske Kræfter, der svarer til de Spændingsforskelle, der efter Elektrontheorien er til Stede mellem to Metaller i Berøring og mellem de forskellige Punkter af et homogent Metalstykke, der ikke overalt har samme Temperatur, udfører under Elektricitetens Gennemgang gennem Kæden. *Kunz* mener nu (loc. cit., p. 779), at man kan betragte de omhandlede Varmetoninger som hidrørende fra en reversibel Kredsproces, idet man ved Beregningen af de omtalte Spændingsdifferenser ikke behøver at tage direkte Hensyn til Varmeledningen. Hertil er imidlertid at bemærke, at en Betragtning som den af *Kunz* anvendte kun viser, at de omhandlede Varmetoninger er formelt reversible i den Forstand, at de skifter Fortegn, naar Strømmens Retning forandres, men ikke, at de omhandlede Varmetoninger er at betragte som hidrørende fra en af irreversible Processer uafhængig reversibel Kredsproces, paa hvilken man for sig alene kan anvende Varmetheoriens anden Hovedsætning. Endvidere kan anføres, at den af *Kunz* benyttede Beregningsmaade for de thermoelektriske Varmetoninger kun giver rigtigt Resultat under saadanne simple Antagelser som dem, der lægges til Grund ved *Kunz's* Betragtninger. *Kunz* antager nemlig ligesom *Drude*, at alle Elektronerne har samme Hastighed, og at en elektrisk Strøm derfor altid vil medføre en bestemt Mængde Energi, kun afhængig af Temperaturen; tager man imidlertid Hensyn til Elektronernes forskellige Hastigheder, vil man finde, at den elektriske Strøm i højere eller mindre Grad vil føres fortrinsvis af de hurtigere Elektroner, alt efter de Antagelser, man gør om Kræfterne mellem Metalmolekylerne og Elektronerne, og man maa derfor ved Beregningen af Varmetoningen tage Hensyn til den forskellige Mængde af Energi, den elektriske Strøm fører med sig i de forskellige Metaller og ved de forskellige Temperaturer. En nærmere Beregning viser endvidere, at i saadanne mere almindelige Tilfælde vil den Varmetoning, der hidrører alene fra det til Spændingsdifferenserne knyttede Arbejde, ikke tilfredsstille Betingelsen (28); men denne Ligning vil kun tilfredsstilles af Summen af denne Varmetoning og den, der hidrører fra den med Strømmen førte Energi.

Man har forsøgt at forklare den omhandlede Overensstemmelse ved at sammenligne Elektronernes Gennemgang gennem en thermo-elektrisk Kæde med en Kredsproces, ved hvilken en Luftart føres gennem en Række Tilstande, der i Tryk og Temperatur svarer til Elektronernes Tryk og Metallets Temperatur paa de forskellige Steder af Kæden, og ved Hjælp af en saadan Betragtning at beregne de ther-moelektriske Varmetoninger og den elektromotoriske Kraft i Kæden udfra de Varmemængder, som en Luftart vil modtage og afgive, og det Arbejde, den vil kunne udføre under den omhandlede Kredsproces[1]). Ved en Betragtning som den nævnte kommer man til Udtryk for F, π og ρ, der svarer til dem, som *Lorentz* har fundet. At de Resultater, man finder ad en saadan Vej, maatte være i Overensstemmelse med Lord *Kelvin's* Theori, er paa Forhaand indlysende, idet Resultaterne netop er beregnede ved Betragtning af en Kredsproces, ved hvilken en Luft-art føres gennem en Række Ligevægtstilstande, altsaa en almindelig reversibel thermodynamisk Kredsproces. En saadan Betragtning kan imidlertid ikke opfattes som andet end som en Illustration til Lord *Kelvin's* Theori, idet man, da der uden nærmere Begrundelse ses bort fra Elektricitets- og Varmeledningen, ikke paa Forhaand kan være sikker paa, at de beregnede Varmetoninger stemmer overens med dem, der virkelig finder Sted i Kæden. En Beregning udfra et mere almindeligt Grundlag end det, hvorpaa *Lorentz* bygger, viser ogsaa, at en saadan Overensstemmelse i Almindelighed ikke vil være til Stede, idet der, som vi har set, i Udtrykkene for de omhandlede Varmetoninger foruden saadanne Størrelser (som f. Eks. Elektronernes

---

[1]) Smlg. *J. J. Thomson*: The Corpuscular Theory of Matter, p. 73; se ogsaa *K. Bae-deker*: Phys. Zeitschr., Bd. 11, p. 809, 1910. [I den sidste Afhandling gøres der et yderligere Forsøg paa at belyse Elektrontheoriens Behandling af de thermoelektriske Fænomener, ved Hjælp af de udfra Eksperimenter over Metallernes Udsending af Elek-troner ved høj Temperatur dragne Slutninger, efter hvilke der i et Hulrum dannet i et homogent Metalstykke, dersom der skal være Ligevægt til Stede, maa befinde sig en ganske bestemt Mængde fri Elektroner pr. Volumenenhed, udøvende et bestemt Tryk, afhængigt af Metallets Natur og Temperatur. *Baedeker* betragter nu en reversibel Kreds-proces, ved hvilken en Samling Elektroner føres gennem en Række Tilstande, der i Tryk og Temperatur svarer til dem, de vilde være til Stede i Hulrum dannede paa de forskellige Steder i Kæden. Dersom Elektronernes Koncentration i de be-tragtede Hulrum antages at være proportional med de fri Elektroners Koncentration i selve Metallet, giver en saadan Beregningsmaade samme Resultat som den oven for i Teksten omtalte, og de Bemærkninger, der er knyttede til denne sidste, gælder derfor ogsaa for *Baedeker's* Theori. Er den omtalte Betingelse vedrørende Elek-tronernes Koncentration derimod ikke opfyldt (se *Baedeker*: loc. cit., p. 810), vil der ikke være nogen simpel Forbindelse mellem de af *Baedeker* beregnede Varme-toninger og de i Kæden optrædende *Peltier-* og *Thomson*-Effekter.]

Antal i Volumenenheden), der karakteriserer de omhandlede Ligevægtstilstande, ogsaa indgaar Størrelser (som den i det foregaaende ved μ
betegnede), der karakteriserer den Maade, hvorpaa Elektronernes samlede Bevægelsesmængde vil fordele sig mellem Elektronerne med de
forskellige absolute Hastigheder, dersom Ligevægten forstyrres, Størrelser, der ifølge Sagens Natur er udelukkede fra at kunne komme til
at indgaa i Resultater, der er udledte udfra en Betragtning som den
ovennævnte.

Forskellen mellem det her behandlede Problem og de sædvanlige
(strengt begrundede) Anvendelser af Varmetheoriens anden Hovedsætning viser sig ogsaa tydeligt ved Forskellen mellem den i det foregaaende udfra Elektrontheorien givne Udledning af Betingelsen (28) og
de sædvanlige Verifikationer af thermodynamiske Resultater ved Hjælp
af kinetiske Theorier. Medens man saaledes ved de sidstnævnte Problemer udelukkende betragter statistiske Ligevægtstilstande og disses
Forskydninger, betragter vi derimod her systematiske Afvigelser fra
den statistiske Ligevægt, nemlig Elektronernes ordnede Bevægelse
mellem Metallets Molekyler; og Rigtigheden af Ligningen (28) udledes
her direkte udfra den særlige Form, som Ligningerne til Bestemmelse
af Elektronernes samlede Bevægelsesmængde i en bestemt Retning
antager — d. v. s. Sammenhængen mellem Koefficienterne til $\frac{\delta \varphi}{\delta x}$ og
til $\frac{\delta T}{\delta x}$ samt, i de Tilfælde (adskilte Sammenstød) hvor vi har antaget
en »Energiomsætning« (se Side 35) mellem Metalmolekylerne og Elek-
tronerne eller mellem de sidste indbyrdes, den særlige Form for Funktionen $Q(\rho, r)$ given ved Ligning (13) Side 23 —, og den deraf udledte
Form for Udtrykkene for Elektricitets- og Energistrømningen gennem
Metallet given ved Ligningerne (15) og (16)[1].

---

[1] Det kan her bemærkes, at *W. Voigt* (se f. Eks. Wied. Ann., Bd. 67, p. 717, 1899)
har vist, at dersom den af Lord *Kelvin* opstillede Betingelse (28) skal være opfyldt, maa Ligningerne for Elektricitets- og Varmebevægelsen i et isotropt Medium
nødvendigvis have en Form, der svarer til den, der angives ved Ligningerne (15)
og (16).

KAPITEL III.

BEHANDLING AF IKKE-STATIONÆRE PROBLEMER.

Metallernes Absorption og Emission af Varmestraaler.

Kirchhoff har som bekendt, ud fra den Antagelse, at der blandt Legemer, der er fuldstændig afskaarne fra ydre Paavirkninger, vil opstaa en Temperaturligevægt, der ikke vil forstyrres ved Legemernes gensidige Varmestraaling, paavist, at der i et Hulrum omgivet af Legemer i Temperaturligevægt pr. Volumenenhed vil være en ganske bestemt, kun af Temperaturen afhængig, Mængde straalende Energi til Stede, fordelt paa en ganske bestemt Maade mellem Straaler med forskellige Svingningstider; samt at der vil bestaa et bestemt Forhold mellem et hvilketsomhelst Legemes Absorptionsevne og dets Emissionsevne for Varmestraaler, kun afhængigt af Temperaturen og de paagældende Straalers Svingningstid, hvilket Forhold vil staa i en meget simpel Relation til den ovenfor omtalte Energimængde. Det har derfor været af stor Interesse at søge ad theoretisk Vej at bestemme et eller andet Legemes Absorptions- og Emissionsevne for Varmestraaler, idet man ad denne Vej vil kunne naa til en theoretisk Bestemmelse af den ovenfor omtalte Energimængde og dens Fordeling paa Varmestraaler af de forskellige Svingningstider eller, som man ofte udtrykker det, til en theoretisk Bestemmelse af Varmestraalingsloven. I Metallernes Elektrontheori behandler man nu imidlertid netop saadanne Legemer, ved hvilke en Bestemmelse som den omtalte er mulig. De herhen hørende Beregninger er ved Betragtning af et Metalstykke, for hvilket der gøres særlig simple Antagelser om Vekselvirkningen mellem Molekylerne og Elektronerne, udført af *H. A. Lorentz*[1]) for saadanne Varmestraalers Ved-

---

[1]) *H. A. Lorentz*: Proc. Acad. Amsterdam, vol. 5, p. 666, 1903; se ogsaa *H. A. Lorentz*: Theory of Electrons, Leipzig 1909, Kap. II.

kommende, hvis Svingningstider er store i Forhold til de meget smaa Tider, indenfor hvilke de enkelte Elektroner mister næsten hele deres oprindelige Bevægelsesmængde i en bestemt Retning. *Lorentz's* Beregninger er i meget god Overensstemmelse med eksperimentelle Undersøgelser over Varmestraaler med Svingningstider som de nævnte, en Overensstemmelse, der maa anses for et af Elektrontheoriens interessanteste Resultater.

J. J. Thomson[1]) har senere forsøgt at udstrække Beregningerne til ogsaa at omfatte Straaler med kortere Svingningstider. *Thomson* antager ligesom *Lorentz*, at der finder adskilte Sammenstød Sted, og at en Elektron ved et Sammenstød i Middel mister hele sin oprindelige Bevægelsesmængde i en bestemt Retning; medens imidlertid Beregningerne for Straaler med store Svingningstider kun er afhængige af »Resultatet« for Sammenstødene, viser *Thomson* derimod, hvorledes Beregningen af Emissionsevnen for Straaler med Svingningstider, der er af samme Størrelsesorden som Varigheden af de enkelte Sammenstød, er i højeste Grad afhængig af de Antagelser, man gør om Elektronernes Bevægelse under de enkelte Sammenstød. Den Ligning, som *Thomson* opstiller til Beregning af selve Varmestraalingsloven, og som benyttes til at beregne Udtryk for denne Lov udfra forskellige specielle Antagelser om Elektronernes Bevægelse under Sammenstødene[2]), er imidlertid efter min Mening ikke rigtig, idet der ikke ved Beregningerne af Absorptionsevnen, saaledes som ved Beregningen af Emissionsevnen, er taget tilstrækkeligt Hensyn til Elektronernes Bevægelse under de enkelte Sammenstød. Vi skal i det følgende komme tilbage til dette Spørgsmaal.

J. H. Jeans[3]) har forsøgt at gennemføre Beregningerne af Forholdet mellem et Metalstykkes Absorption og Emission af Varmestraaler uden at gøre specielle Antagelser om Metallets Natur. *Jeans's* Beregninger vil imidlertid, som omtalt i det foregaaende, ikke gælde i Almindelighed, men kun under visse meget specielle Antagelser; Antagelser, der f. Eks. ikke vil være opfyldt ved det af *Lorentz* betragtede Tilfælde. *H. A. Wilson*[4]), der har gjort opmærksom paa Mangler ved *Jeans's* Beregninger, har forsøgt at gennemføre Beregningerne under de samme specielle Antagelser som *Lorentz*, men overfor Varmestraaler med kortere Svingningstider end de af *Lorentz* under-

---

[1]) *J. J. Thomson*: Phil. Mag., vol. 14, p. 217, 1907.
[2]) *J. J. Thomson*: Phil. Mag., vol. 14, p. 225—231, 1907 og vol. 20, p. 238, 1910.
[3]) *J. H. Jeans*: Phil. Mag., vol. 17, p. 773, 1909.
[4]) *H. A. Wilson*: Phil. Mag., vol. 20, p. 835, 1910.

søgte, nemlig for Svingningstider, der kun antages at være store i Forhold til de Tider, der medgaar til de enkelte Sammenstød. *Wilson's* Resultat er imidlertid, som berørt i det foregaaende, og som vi her nærmere skal se, ikke rigtigt.

Vi skal nu i dette Kapitel vise, hvorledes Beregningerne af Metallernes Absorption og Emission af Varmestraaler med »store« Svingningstider kan gennemføres i alle saadanne Tilfælde, som vi har undersøgt i Kapitel I. I de Tilfælde, hvor vi antager, at der finder adskilte Sammenstød Sted, og i hvilke Beregningerne kan udføres uden specielle Forudsætninger, vil Undersøgelsen kunne udstrækkes til Varmestraaler af alle Svingningstider, der blot er store i Forhold til Tiden for et enkelt Sammenstød; i de andre Tilfælde kun til Varmestraaler, hvis Svingningstider er store i Sammenligning med de Tider, indenfor hvilke de enkelte Elektroner i Middel mister næsten hele deres oprindelige Bevægelsesmængde i en bestemt Retning.

---

§ 1.

Absorption af Varmestraaler.

Dersom et Metalstykke er udsat for Virkningen af en konstant ydre elektrisk Kraft, vil de fri Elektroner faa en Middelbevægelse i Kraftens Retning, og Elektronerne vil derfor under deres Bevægelser i Middel modtage kinetisk Energi under Kraftens Indvirkning — hvilket vi ogsaa kan udtrykke ved at sige, at Elektronerne vil absorbere Energi fra det ydre elektriske Felt —; denne Energi vil ved Vekselvirkningen (Sammenstødene) med Metalmolekylerne omsættes til Varme, det vil sige til kinetisk Energi fordelt paa tilfældige uregelmæssige Bevægelser. Ganske tilsvarende Forhold finder nu Sted, dersom et Metalstykke udsættes for Varmestraaler, hvilke efter den elektromagnetiske Lysteori jo bestaar af hurtigt varierende elektriske og magnetiske Felter.

Vi skal her først betragte saadanne Varmestraaler, hvis Svingningstider er store i Forhold til de meget smaa Tider, indenfor hvilke de enkelte Elektroner i Middel mister Størstedelen af deres Bevægelsesmængde i en bestemt Retning. I dette Tilfælde vil Tilstanden i hvert enkelt Øjeblik kunne betragtes som stationær — »quasi-stationær« (se Side 27) —.

Vi skal nu betegne Komposanterne efter Koordinatakserne af den ydre elektriske Kraft i et Punkt af Metallet til Tiden t med E_x, E_y og E_z. Den elektriske Strøm gennem et Fladeelement dS vinkelret paa X-aksen vil da i det her betragtede Tilfælde være $i_x dS = \sigma E_x dS$, hvor σ betegner den almindelige elektriske Ledningsmodstand; og ganske lignende Udtryk vil gælde for den elektriske Strømning efter Y- og Z-aksen. Den Energi, der absorberes i et Volumenelement dV af Metallet dt, vil derfor være   **

$$(i_x E_x + i_y E_y + i_z E_z)\, dV dt = \sigma \cdot (E_x^2 + E_y^2 + E_z^2)\, dV dt.$$

Og den Energi, der absorberes i Tidsenheden i en Volumenenhed af Metallet, vil være

$$\sigma \cdot (\overline{E_x^2} + \overline{E_y^2} + \overline{E_z^2}),$$

hvor de vandrette Streger skal betegne, at Størrelserne er beregnede som Middelværdier saavel over Volumen som over Tid.

Den Energimængde, der i Middel paa Grund af Straalingen vil være til Stede i en Volumenenhed af det tomme Rum mellem Molekylerne og Elektronerne — denne Energi er ikke den hele Energi, der paa Grund af Straalingen vil være til Stede i Metallet; der vil nemlig ogsaa være Energi til Stede hidrørende fra de fri Elektroners ordnede Bevægelse (og eventuelt ogsaa fra Medsvingninger af de indenfor Molekylerne bundne Elektroner) —, vil være

$$2 \cdot \frac{1}{8\pi} (\overline{E_x^2} + \overline{E_y^2} + \overline{E_z^2}),$$

hvor Faktoren 2 hidrører fra, at den magnetiske Energi ved elektromagnetiske Svingninger som de her betragtede i Middel vil være lige saa stor som den elektriske Energi. (Vi har ved Beregningen af Absorptionen ikke taget Hensyn til Indvirkningen af de magnetiske Kræfter; thi idet Elektronernes Hastigheder vil være meget smaa i Sammenligning med Lysets Hastighed (smlg. Noten Side 11), vil Virkningen paa Elektronernes Bevægelser af de magnetiske Kræfter i det her omhandlede Tilfælde være forsvindende i Sammenligning med Virkningen af de elektriske Kræfter.)

Idet vi nu ved Absorptionskoefficienten α skal betegne den Energimængde, der absorberes i en Volumenenhed af Metallet, dersom den straalende elektromagnetiske Energi pr. Volumenenhed (beregnet paa den ovenfor omtalte Maade) er lig 1, faar vi

$$\alpha = 4\pi \cdot \sigma. \tag{1}$$

---

* ["Ledningsmodstand" should read "Ledningsevne".]

** [The words "i Tiden" should be inserted between "Metallet" and "dt".]

Man kan i Almindelighed ikke ved Hjælp af de her anførte Beregninger direkte finde den Energi, der absorberes i et Metalstykke, der er udsat for Varmestraaler; thi de fri Elektroners ordnede Bevægelse (og de i Molekylerne bundne Elektroners Medsvingninger) vil virke tilbage paa de ydre elektromagnetiske Felter, idet de foruden til Absorption vil give Anledning til de Fænomener, der betegnes ved Tilbagekastning og Brydning. Dersom man imidlertid, som ved de Anvendelser, vi i det følgende skal gøre af Ligning (1), betragter et Metalstykke, der er saa lille, at man kan se bort fra Tilbagevirkningen af dets Elektroner paa de ydre Felter — (Metalstykket skal dog antages at indeholde et meget stort Antal Elektroner og Molekyler, saa at man kan benytte de i de foregaaende Kapitler udførte Beregninger for den elektriske Ledningsmodstand o. s. v.; denne sidste Fordring er imidlertid forenelig med den ovenstaaende Fordring, idet Tilbagevirkningen af de enkelte Elektroner paa de ydre Felter er ganske overordentlig ringe, hvilket f. Eks. for de fri Elektroners Vedkommende hidrører fra, at de enkelte Elektroner paa Grund af deres store Hastigheder og hyppige Sammenstød kun i meget ringe Grad kan følge de ydre Kræfters Indflydelse) —, vil man ved direkte Anvendelse af Ligningen (1) finde, at den Energimængde, der absorberes i Metalstykket, hvis Volumen vi skal betegne ved ΔV, dersom det udsættes for et Bundt af Varmestraaler, for hvilket den i Volumenenheden tilstedeværende Energi er A — og altsaa den igennem en Fladeenhed vinkelret paa Bundtet pr. Sek. gaaende Energi er cA, hvor c er Lysets Hastighed — vil være

$$\alpha\, A\, \Delta V.$$

Forsøg af *Hagen* og *Rubens*[1]) over Metallernes Refleksion og Emission af Varmestraaler viser nu, at Metallernes Absorption af Varmestraaler med store Svingningstider (d. v. s. større end 10^{-13} Sek.) virkelig kan beregnes i Overensstemmelse med Erfaringen udfra Metallernes elektriske Ledningsevne for stationære Strømme, det vil sige efter Ligning (1). Forsøgene viser imidlertid endvidere, at for kortere Svingningstider vil den elektriske Ledningsevne beregnet paa denne Maade ikke stemme overens med den, der findes for stationære Strømme.

Vi skal nu gaa over til at beregne Absorptionen af Varmestraaler med saadanne kortere Svingningstider. Vi kan her, som ovenfor berørt, kun gennemføre Behandlingen i det Tilfælde, hvor vi antager,

---

[1]) *E. Hagen* og *H. Rubens*: Ann. d. Phys., Bd. 11, p. 873, 1903.

at der finder adskilte Sammenstød Sted, idet vi i de andre Tilfælde ved Opstillingen af Betingelsesligningerne i Kapitel I kun har behandlet stationære Tilstande.

Idet vi her skal antage, at Temperaturen er konstant og Metallet
homogent, vil Ligningen til Bestemmelse af Elektronernes samlede Bevægelse, dersom der finder adskilte Sammenstød Sted, være (Lign. (14)
Side 26)

$$\left[\frac{dG_x(r)}{dt}\right] = -\frac{4\pi m\varepsilon K}{3kT}\frac{\delta\varphi}{\delta x}r^4 e^{-\frac{mr^2}{2kT}} - G_x(r)F(r) + \int_0^\infty G_x(\rho)\,Q(\rho,r)\,d\rho. \quad (2)$$

Denne Ligning er ikke umiddelbart at opfatte som en almindelig
Differentialligning, idet Leddet paa venstre Side af Lighedstegnet (se
Side 26) kun angiver Middelværdien af $\frac{dG_x(r)}{dt}$ for givne Værdier af $\frac{\delta\varphi}{\delta x}$
og af Størrelserne $G_x(\rho)$. Ved de Anvendelser, vi her skal gøre, kan vi
imidlertid betragte Ligningen som en almindelig Differentialligning,
idet vi her kun søger en Middelværdi af $G_x(r)$, og idet der ikke vil
være nogen mærkelig systematisk Forbindelse mellem Værdierne af
$\frac{dG_x(r)}{dt}$ og forudgaaende Tilstande (undtagen naturligvis, hvad der ud-
siges gennem selve Lign. (2)). At der ikke findes en saadan Forbindelse,
hidrører fra, at Sandsynligheden for Sammenstød (se Side 20) er antaget
at være uafhængig af Elektronernes Baner; dette sidste gælder dog strengt
taget kun, dersom den Tid, de enkelte Sammenstød tager, antages at
være uendelig kort; antager man derimod, at Sammenstødene i Middel
varer en vis lille Tid ν — (den Tid, Elektronerne bruger om at gennemløbe Strækninger af samme Størrelsesorden som Elektronernes eller
Molekylernes Dimensioner) —, vil der være systematiske Forbindelser
mellem $\frac{dG_x(r)}{dt}$ og forudgaaende Tilstande, der har fundet Sted indenfor Tider, der kan sammenlignes med ν. Ligningen (2) kan derfor
kun benyttes som Differentialligning, dersom dt kan betragtes som
stor i Sammenligning med ν, hvilket vil sige, dersom $\frac{\delta\varphi}{\delta x}$ — og dermed $G_x(r)$ beregnet som Middelværdi — ikke forandrer sig mærkeligt
indenfor Tidsrum af samme Størrelsesorden som ν. De følgende Beregninger vil derfor kun gælde for Absorptionen af saadanne Varmestraaler, hvis Svingningstider er store i Sammenligning med de Tider,
der medgaar til de enkelte Sammenstød; Tider, der imidlertid her skal
antages at være meget smaa i Sammenligning med de Tider, inden-

6

for hvilke Elektronerne i Middel mister Størstedelen af deres oprindelige Bevægelsesmængde i en bestemt Retning.

Vi skal nu antage, at der virker en elektrisk Kraft efter X-aksen,
der udtrykkes ved

$$E_x = - \frac{\delta \varphi}{\delta x} = E \cos pt.$$

Betingelsen for, at Ligningen (2) vil være tilfredsstillet af en Løsning
af Formen

$$G_x(r) = \psi_1(r) \cos pt + \psi_2(r) \sin pt \tag{3}$$

(man kan let vise, at Lign. (2) kun kan have én periodisk Løsning),
vil være

$$p\psi_2(r) = \frac{4\pi m \varepsilon K}{3kT} E r^4 e^{-\frac{mr^2}{2kT}} - \psi_1(r) F(r) + \int_0^\infty \psi_1(\rho) Q(\rho, r) d\rho \tag{4}$$

og

$$-p\psi_1(r) = -\psi_2(r) F(r) + \int_0^\infty \psi_2(\rho) Q(\rho, r) d\rho. \tag{5}$$

Disse to sammenhørende Integralligninger vil have en Løsning,
der kan skrives paa Formen (dette indses lettest ved at løse den
sidste med Hensyn til $\psi_2(r)$ ved Hjælp af *Fredholm's* Løsning (se
Side 47) og indsætte i den første og løse denne paa samme Maade):

$$\psi_1(r) = \frac{4\pi m \varepsilon K}{3kT} E \left(r^4 e^{-\frac{mr^2}{2kT}} \cdot F_1(r) + \int_0^\infty \rho^4 e^{-\frac{mr^2}{2kT}} \cdot Q_1(\rho, r) d\rho \right), \tag{6}$$

hvor $F_1(r)$ og $Q_1(\rho, r)$ kun afhænger af $F(r)$, $Q(\rho, r)$ og p (d. v. s. ikke
afhænger af det første Led paa højre Side af Lighedstegnet i Lign. (4)),
samt et dertil ganske svarende Udtryk for $\psi_2(r)$, i hvilket vi blot skal
tænke os Indeks 1 ombyttet med Indeks 2.

Idet vi nu sætter

$$P_1(r) = F_1(r) + \int_0^\infty Q_1(r, \rho) d\rho, \tag{7}$$

samt

$$\sigma_p = \frac{4\pi \varepsilon^2 K}{3kT} \cdot \int_0^\infty r^4 e^{-\frac{mr^2}{2kT}} P_1(r) dr, \tag{8}$$

hvor σ_p vil være en Størrelse, der saavel er afhængig af Metallets Natur og Temperatur som af p, og betegner det tilsvarende Udtryk for Indeks 2 med δ_p, faar vi ved Hjælp af (3), (6), (7) og (8) for den elektriske Strøm gennem en Fladeenhed vinkelret paa X-aksen

$$i_x = \frac{\varepsilon}{m} \int_0^\infty G_x(r)\, dr = E \cdot (\sigma_p \cos pt + \delta_p \sin pt), \qquad (9)$$

og for den Energi, der i Tidsenheden absorberes i en Volumenenhed af Metallet,

$$\frac{p}{2\pi} \int_0^{\frac{2\pi}{p}} i_x E \cos pt\, dt = \tfrac{1}{2} E^2 \cdot \sigma_p.$$

Idet nu den Energi, der i Middel vil være til Stede i en Volumenenhed af Metallet (se Side 79), vil være $\frac{1}{4\pi} \cdot \overline{E_x^2} = \frac{1}{8\pi} \cdot E^2$, faar man for Absorptionskoefficienten (se Side 79) α_p for Varmestraaler med Svingningstid $\frac{2\pi}{p}$.

$$\alpha_p = 4\pi \cdot \sigma_p. \qquad (10)$$

Dette Udtryk, der med den ovenfor (Side 81) omtalte Indskrænkning for p gælder i alle Tilfælde, hvor man betragter adskilte Sammenstød, skal vi benytte ved den senere Sammenligning med Emissionsevnen.

I Analogi med de Betragtninger, der har ført os til Ligningen (1) Side 79, kan man udtrykke Ligningen (10) ved at sige, at Metallet overfor en elektrisk Kraft, der varierer som $\cos pt$, besidder en »effektiv elektrisk Ledningsevne« σ_p forskellig fra (mindre end) Ledningsevnen σ_0 for stationære Strømme. For saadanne hurtigt varierende elektriske Kræfter fremkommer der endvidere, som det fremgaar af Ligning (9) en Faceforskel mellem den elektriske Strøm og den elektriske Kraft, der med * de benyttede Betegnelser udtrykkes ved $\operatorname{arctg} \frac{\delta_p}{\sigma_p}$. Denne Faceforskel, der * fremkommer paa Grund af de fri Elektroners Inerti, viser sig som en Forsinkelse af Strømmen i Forhold til den elektriske Kraft; Optræ- delsen af en saadan Faceforskel vil have samme Virkning paa Bereg- * ningen af Metallernes optiske Forhold som Tilstedeværelsen af en (effektiv) negativ Dielektricitetskonstant.

6*

* ["Face-" should read "Fase-".]

Vi skal nu her nærmere betragte Udtrykkene for σ_p og δ_p i det simple Tilfælde, hvor Metalmolekylerne antages at være faste Kraft-centrer, der paavirker Elektronerne med Kræfter, der forholder sig omvendt som n^{te} Potens af Afstanden, og i hvilket man ser bort fra Sammenstødene mellem Elektronerne indbyrdes. I dette Tilfælde har man (se Side 33)

$$F(r) = C \cdot r^{\frac{n-5}{n-1}} \quad \text{og} \quad Q(\rho, r) = 0.$$

Man faar derfor

* $$F_1(r) = \left(F(r) + \frac{p^2}{F(r)}\right)^{-1} = \frac{\frac{1}{C}r^{-\frac{n-5}{n-1}}}{1 - \frac{p^2}{C^2}r^{-2\frac{n-5}{n-1}}} \quad \text{og} \quad Q_1(\rho, r) = 0,$$

og deraf ved Hjælp af (7) og (8)

$$\sigma_p = \frac{4\pi\varepsilon^2 K}{3kTC} \cdot \int_0^\infty \frac{r^{\frac{3n+1}{n-1}} e^{-\frac{mr^2}{2kT}}}{1 + \frac{p^2}{C^2}r^{-2\frac{n-5}{n-1}}} dr. \tag{11}$$

Dette Udtryk adskiller sig fra Udtrykket for σ for stationære Strømme (Side 49 Lign. (6), Udtrykket for A_1) ved Tilføjelsen af det sidste Led i Nævneren under Integraltegnet. Integrationen i Udtrykket (11) kan i Almindelighed ikke udføres fuldstændigt; men dersom vi antager, at p er saa lille, at Nævneren for alle Værdier af r, der afgiver et væsentligt Bidrag til Integralet (vi antager $n > 2$), er meget nær 1, kan man med Tilnærmelse sætte

** $$\sigma_p = \frac{4\pi\varepsilon^2 K}{3kTC} \cdot \int_0^\infty \left(r^{\frac{3n+1}{n-1}} - \frac{p^2}{C^2}r^{\frac{n+11}{n-1}}\right)e^{-\frac{mr^2}{2kT}} dr$$

$$= \frac{4\pi\varepsilon^2 K}{3kTC}\left(\frac{1}{2}\left(\frac{2kT}{m}\right)^{\frac{2n}{n-1}}\Gamma\left(\frac{2n}{n-1}\right) - \frac{p^2}{C^2}\frac{1}{2}\left(\frac{2kT}{m}\right)^{\frac{n+5}{n-1}}\Gamma\left(\frac{n+5}{n-1}\right)\right);$$

dette giver

$$\sigma_p = \sigma_0\left(1 - \frac{p^2}{C^2}\left(\frac{m}{2kT}\right)^{\frac{n-5}{n-1}} \cdot \Gamma\left(\frac{n+5}{n-1}\right)\Gamma^{-1}\left(\frac{2n}{n-1}\right)\right).$$

Ved Hjælp af Ligningen (9) Side 51 kan dette skrives

$$\sigma_p = \sigma_0\left(1 - \sigma_0^2\frac{p^2 m^2}{N^2\varepsilon^4}\frac{9\pi}{16}\Gamma\left(\frac{n+5}{n-1}\right)\Gamma^{-3}\left(\frac{2n}{n-1}\right)\right). \tag{12a}$$

* [The denominator should be $1 + (p^2/C^2)r^{-(n-5)/(n-1)}$.]

** [The exponent of r in the second term between the brackets should be $(n+1)/(n-1)$.]

Paa tilsvarende Maade faas

$$\delta_p = \frac{4\pi\varepsilon^2 K}{3kTC} \cdot \int_0^\infty \frac{\frac{p}{C} r^{\frac{2n+6}{n-1}} e^{-\frac{mr^2}{2kT}}}{1 + \frac{p^2}{C^2} r^{-2\frac{n-5}{n-1}}} \, dr,$$

og med samme Tilnærmelse som ovenfor faas heraf

$$\delta_p = \sigma_0^2 \frac{pm}{N\varepsilon^2} \frac{3\sqrt{\pi}}{4} \Gamma\left(\frac{3n+5}{2(n-1)}\right) \Gamma^2\left(\frac{2n}{n-1}\right). \qquad (12b) \qquad *$$

Vi skal nu betragte nogle specielle Tilfælde svarende til forskellige Værdier af n.

1) For $n = \infty$ faar vi af (12a) og (12b)

$$\sigma_p = \sigma_0\left(1 - \sigma_0^2 \frac{p^2m^2}{N^2\varepsilon^4} \frac{9\pi}{16}\right) \quad \text{og} \quad \delta_p = \sigma_0^2 \frac{pm}{N\varepsilon^2} \frac{3\pi}{8}, \qquad (13)$$

hvilket er de Udtryk, der vil gælde, dersom Metalmolekylerne antages at være **haarde elastiske Kugler**, eller anderledes udtrykt, dersom Elektronerne i Middel mister hele deres Bevægelsesmængde i en bestemt Retning ved et enkelt Sammenstød.

J. J. Thomson[1]) har som omtalt i det foregaaende beregnet et Udtryk for σ_p udfra en Antagelse om Virkningerne af Sammenstødene, der svarer til de her omhandlede. Den Værdi, *Thomson* har fundet, nemlig (med vore Betegnelser)

$$\sigma_p = \sigma_0 \cdot \left[\frac{\sin\frac{\sigma_0 pm}{N\varepsilon^2}}{\frac{\sigma_0 pm}{N\varepsilon^2}}\right]^2 = \sigma_0\left(1 - \frac{1}{3}\frac{\sigma_0^2 p^2 m}{N^2\varepsilon^4} \cdot \cdot\right), \qquad (13')$$

adskiller sig imidlertid, som man ser, meget væsentligt fra den ovenfor beregnede. (Det Led, der angiver Ledningsevnens Afhængighed af Svingningstallet, er saaledes c. 5 Gange mindre end det tilsvarende Led i Lign. (13)). Denne Forskel hidrører dels derfra, at *Thomson* antager, at alle Elektronerne har samme absolute Hastighed, dels fra, at de Tider, i hvilke Elektronerne tilbagelægger deres fri Vejlængder, ved *Thomson's* Beregninger sættes lige store, uden at der tages Hensyn til de Hastigheder, Elektronerne opnaar under de ydre Kræfters Indvirkning[2]).

Udtrykket (13') benyttes af *Thomson* til Beregning af et Metals Absorption for Straaler af alle mulige Svingningstider. Udtrykket er imidlertid — bortset fra de omtalte Indvendinger, der ikke berører Størrelsesordenen af Svingningstidens Indflydelse paa Absorptionskoefficienten — efter min

---

[1]) *J. J. Thomson*: Phil. Mag., vol. 14, p. 225, 1907.
[2]) Jvnfr. *J. H. Jeans*: Phil. Mag., vol. 17, p. 778, 1909.

* [The last factor in the expression for δ_p should read $\Gamma^{-2}(2n/(n-1))$.]

Mening paa Grund af den Maade, hvorpaa det er udledt, ikke anvendeligt (smlg. Side 81) overfor Straaler, hvis Svingningstider ikke er korte i Sammenligning med de Tider, der medgaar til de enkelte Sammenstød (se Side 77).

H. A. Wilson[1]), der, som omtalt Side 77, ogsaa har beregnet σ_p udfra den ovennævnte Antagelse om Sammenstødenes Art, har ligeledes fundet et Resultat, der ikke falder sammen med Udtrykket (13). Wilson har beregnet sin Værdi for σ_p paa to forskellige Maader; af disse svarer den ene til den her benyttede Beregningsmaade, men den Ligning, der i Wilsons Beregninger svarer til Ligningen (2), er, som det er forklaret Side 19, ikke rigtig; ved den anden Beregningsmaade betragter Wilson ligesom J. J. Thomson derimod de enkelte Elektroners Bevægelser og undersøger, hvor stor en Energi der i Middel absorberes af en Elektron under Tilbagelæggelsen af en enkelt fri Vejlængde mellem to paa hinanden følgende Sammenstød. Ved denne Beregning er der imidlertid ligesom ved Thomson's ikke taget Hensyn til, at den Tid, der medgaar til at gennemløbe en saadan fri Vejlængde, i Middel vil være større, naar Elektronen bevæger sig imod den ydre elektriske Kraft, end naar den bevæger sig i Kraftens Retning. Wilson's Resultater saavel for σ_0 — Værdien for σ_0 findes saaledes $\frac{3}{2}$ Gange saa stor som Lorentz's Værdi — som for σ_p er derfor ikke rigtige.

2) Dersom vi i Ligningerne (12a) og (12b) sætter $n = 5$, faar vi

$$\sigma_p = \sigma_0\left(1 - \sigma_0^2 \frac{p^2 m^2}{N^2 \varepsilon^4}\right) \quad \text{og} \quad \delta_p = \sigma_0^2 \frac{pm}{N\varepsilon^2}. \tag{14}$$

I dette Tilfælde, der svarer til den Antagelse, at Metalmolekylerne er smaa Magneter (se Side 35), vil, som man kan vise, Talkoefficienterne i Korrektionsleddet paa σ og i Udtrykket for δ_p (saaledes som det specielt ses ved Sammenligning med det foregaaende og det efterfølgende Eksempel) være mindre end i alle andre Tilfælde, hvor man betragter adskilte Sammenstød og ser bort fra Vekselvirkningen mellem Elektronerne indbyrdes.

Den ved Lign. (14) givne Værdi for σ_p svarer til den Værdi for σ_p, som J. H. Jeans[2]) har beregnet, nemlig

$$\sigma_p = \sigma_0\left(1 + \sigma_0^2 \frac{p^2 m^2}{N^2 \varepsilon^4}\right)^{-1}, \tag{14'}$$

et Udtryk, der efter Jeans's Mening skulde gælde i fuld Almindelighed ogsaa uden Antagelse af adskilte Sammenstød. Grunden til, at vi netop genfinder Jeans's Udtryk ved at sætte $n = 5$, er forklaret i det foregaaende (se Side 33).

---

[1]) H. A. Wilson: Phil. Mag., vol. 20, p. 835, 1910.
[2]) J. H. Jeans: Phil. Mag., vol. 17, p. 778, 1909.

* [σ should be σ_p.]

3) Dersom vi endelig sætter $n = 3$, faar vi

$$\sigma_p = \sigma_0\left(1 - \sigma_0\frac{p^2m^2}{N^2\varepsilon^4}\frac{27\pi}{64}\right) \quad \text{og} \quad \delta_p = \sigma_0^2\frac{pm}{N\varepsilon^2}\frac{45\pi}{128}, \qquad (15) \quad *$$

Udtryk, der vil gælde, dersom Metalmolekylerne antages at være saadanne elektriske Dobbeltsystemer som dem, vi har betragtet Side 34.

Beregner man nu Refleksionskoefficienten for en Metalflade for vinkelret indfaldende Varmestraaler udfra Antagelsen af en Sammenhæng mellem den elektriske Kraft og den elektriske Strøm som den, der udtrykkes ved Ligning (9), finder man [1]), idet Refleksionskoefficienten betegnes ved R, og idet man ved Beregningerne bortkaster Led, der forholder sig til de betydende Led som Kvadratet paa Størrelsen δ_p/σ_p, og ser bort fra en mulig Virkning af de bundne Elektroners Medsvingninger, idet en saadan Virkning af samme Størrelsesorden som den, man finder hos Isolatorerne, paa Grund af Metallernes store Ledningsevne i dette Tilfælde vil være forsvindende i Forhold til de betydende Led (se *Drude*: loc. cit., p. 941), følgende Ligning:

$$\left(\frac{1+R}{1-R}\right)^2 = \frac{2\pi}{p}\cdot\sigma_p\left(1 + \frac{\delta_p}{\sigma_p}\right).$$

Idet Koefficienten til $\frac{2\pi}{p}$ paa Ligningens højre Side for uendelig langsomme Svingninger er lig σ_0, viser Ligningen, at der i Formlen for Refleksionskoefficienten for Varmestraaler med kortere Svingningstider i Stedet for den almindelige elektriske Ledningsevne σ_0 indgaar en effektiv Ledningsevne $\sigma_p\left(1 + \frac{\delta_p}{\sigma_p}\right)$. Idet nu for tilstrækkelig store Svingningstider Korrektionsleddet i Udtrykket for σ_p efter de i det foregaaende udførte Beregninger er af samme Størrelsesorden som $\left(\frac{\delta_p}{\sigma_p}\right)^2$, maa man derfor efter den her omhandlede Theori (Antagelsen af adskilte Sammenstød) vente en Forøgelse ($\delta_p > 0$) af den tilsyneladende Ledningsevne for aftagende Svingningstider.

Sammenligner man nu de her udledte Resultater med *Hagen's* og

---

[1]) Se *P. Drude*: Ann. d. Phys., Bd. 14, p. 940, 1904. *Drude's* Resultat svarer til det nedenfor i Teksten givne Udtryk, dersom man for σ_p og δ_p indsætter de i det foregaaende, for det Tilfælde, at Metalmolekylerne antages at paavirke Elektronerne med Kræfter, der forholder sig omvendt som 5te Potens af Afstanden, fundne Udtryk (Lign. (14)); dette hidrører fra, at Modstanden mod Elektronernes Bevægelse hos *Drude* sættes proportional med Elektronernes Hastighed (smlg. det foregaaende, Side 33).

* [Inside parenthesis: σ_0 should be σ_0^2.]

Rubens's [1]) ovenfor omtalte Forsøg over Metallernes Refleksionsevne for Varmestraaler, ser man imidlertid, at der for de fleste Metaller (blandt de rene Metaller er kun Sølv undtaget) finder en Aftagen af den tilsyneladende Ledningsevne Sted for aftagende Svingningstider [2]).

Denne Uoverensstemmelse mellem Theorien og Erfaringen hidrører efter min Mening fra, at Antagelsen om Tilstedeværelsen af adskilte Sammenstød mellem Elektronerne og Metalmolekylerne i Metallernes Indre ikke er berettiget.

Dersom der nemlig finder adskilte Sammenstød Sted, vil en ydre Kraft væsentlig frembringe en Forandring i Elektronernes Bevægelse, men ikke i deres rumlige Fordeling (en saadan Forandring vil kun indtræde i den allernærmeste Nærhed af Metalmolekylerne); er Metalmolekylernes Dimensioner — de Omraader, indenfor hvilke de paavirker Elektronerne kendeligt — derimod ikke forsvindende i Forhold til hele Metallets Volumen, vil den Ophobning af Elektronerne foran Metalmolekylerne, der finder Sted under Elektronernes samlede Bevægelse (se Side 43), i væsentlig Grad kunne forandre Elektronernes rumlige Fordeling, og der vil derfor i dette Tilfælde foruden den Virkning, der skyldes Elektronernes Enerti og som betragtet for sig vil frembringe en effektiv negativ Dielektricitetskonstant, ogsaa være en Virkning hidrørende fra Elektronernes Forskydning i Metalmolekylernes Kraftfelter (ikke at forveksle med Metalmolekylernes Polarisation, hidrørende fra de bundne Elektroners Bevægelse, hvilken, ved Sving-

---

[1]) *E. Hagen* u. *H. Rubens:* Ann. d. Phys., Bd. 11, p. 884, 1903.

[2]) Jvnfr. *Drude*: loc. cit., p. 944. [Den her omhandlede Aftagen af den tilsyneladende Ledningsevne er af flere senere Forfattere (se f. Eks. *J. H. Jeans* Phil. Mag., vol. 17, p. 779, 1909) opfattet som værende i Overensstemmelse med en Elektrontheori af den her behandlede Art, idet den er blevet betragtet som hidrørende fra en Aftagen af den effektive Ledningsevne beregnet ved Hjælp af Absorptionskoefficienten (smlg. ovenfor Side 83). Man har ad denne Vej, ved Hjælp af saadanne Udtryk for σ_p som de i det foregaaende omtalte (Lign. (13')—(14')), søgt at beregne Antallet N af fri Elektroner i Volumenenheden af de forskellige Metaller; dette Antal er nemlig den eneste paa Forhaand ubekendte Størrelse, der indgaar i de omhandlede Formler. De Værdier for N, man paa denne Maade har beregnet, er for de fleste Metallers Vedkommende større end saadanne Værdier, der lader sig forene med de eksperimentelt fundne Værdier for Metallernes Varmefylde (se *Jeans'*: loc. cit., p. 780). Den her omtalte Brug af de omhandlede Formler er imidlertid, som vi ovenfor har set, ikke tilladelig, idet den Indflydelse paa den tilsyneladende Ledningsevne, beregnet udfra Metallernes Refleksionskoefficienter, der hidrører fra den effektive Dielektricitetskonstant (se Side 83), er af modsat Tegn og af lavere Størrelsesorden end den, der hidrører fra Forandringen i den tilsyneladende Ledningsevne, beregnet ud fra Absorptionskoefficienterne.]

ningstider som de her omhandlede, vil være af yderst ringe Indflydelse (se forrige Side)), der betragtet for sig vil frembringe en positiv Dielektricitetskonstant.

---

§ 2.

Emission af Varmestraaler.

Dersom en Elektron har en retlinet Bevægelse med konstant Hastighed, vil den ikke udsende Energi; i et saadant Tilfælde vil nemlig det Elektronen omgivende elektromagnetiske Felt, som man kan udtrykke det, føres med Elektronen under dennes Bevægelse.

Anderledes forholder det sig derimod, dersom Elektronens Hastighed forandres enten i Størrelse eller Retning; i dette Tilfælde vil nemlig Elektronen udsende Energi, hvilket man kan udtrykke ved at sige, at en Del af Elektronens elektromagnetiske Felt vil løsrive sig fra Elektronen og forplante sig bort i det omgivende Rum i alle mulige Retninger med Lysets Hastighed. Idet vi her antager, at Elektronernes absolute Hastighed vil være meget ringe i Forhold til Lysets Hastighed c, vil den Energi, der paa denne Maade udsendes af en Elektron i Tiden dt være lig[1])

$$\frac{2\varepsilon^2}{3c^3} j^2 dt, \tag{16}$$

hvor j betegner den absolute Værdi for Elektronens Acceleration. Denne Energimængde er beregnet som den Energimængde, der vil sendes ud igennem en lukket Flade, hvis Afstand fra Elektronen er stor i Sammenligning med saadanne Afstande, indenfor hvilke de Kræfter, der stammer fra Elektronens medførte Felt, har en mærkelig Værdi.

Et Stykke Metal, i hvilket de fri Elektroners Bevægelse paa Grund af Sammenstødene — Paavirkningen fra Metalmolekylernes Kraftfelter — er underkastet store, hurtigt paa hinanden følgende Forandringer, vil derfor være Stedet for en Udsendelse af Energi. Vi skal nu nærmere undersøge Størrelsen af denne Energiudsendelse og den Maade, hvorpaa den fordeler sig paa Straaler med forskellige Svingningstider. Som omtalt i det foregaaende har *Lorentz* behandlet

---

[1]) Se f. Eks. *H. A. Lorentz*: Theory of Electrons, p. 52, Lign. (82). [I den citerede Formel er anvendt en anden Enhed for Elektricitetsmængder end den, vi her benytter (abs. elektrostatisk Maal), hvilket bevirker, at Udtrykket for den omhandlede Energimængde er divideret med 4π.]

dette Spørgsmaal i det Tilfælde, hvor man ser bort fra Vekselvirkningen mellem Elektronerne indbyrdes og antager, at Metalmolekylerne
paavirker Elektronerne som haarde elastiske Kugler, samt hvor det
drejer sig om at bestemme den udstraalede Energis Fordeling paa
Straaler med Svingningstider, der er meget store i Sammenligning
med de Tider, indenfor hvilke Elektronerne mister næsten hele deres
oprindelige Bevægelsesmængde i en bestemt Retning. Ved *Lorentz's*
Behandling følges de enkelte Elektroners Bevægelser; dette vilde
imidlertid blive meget indviklet i de mere almindelige Tilfælde, vi her
skal betragte. (At *Lorentz's* Tilfælde er simplere, hidrører fra, at Elektronernes Bevægelsesretning efter et Sammenstød, i hans Theori, antages i Middel at være fuldkommen uafhængig af Bevægelsesretningen
før Sammenstødet.) Vi skal derfor her betragte Elektronernes samlede
Bevægelse i en bestemt Retning under ét, idet vi hermed følger den
Vej, der er angivet af *Jeans.*

Vi skal nu undersøge den Energi, der udsendes fra et lille Volumenelement af Metallet ΔV, hvilket Element skal antages at indeholde et
meget stort Antal Metalmolekyler og Elektroner og samtidig være
saa lille, at den Energi, der udsendes af de enkelte Elektroner, ikke
absorberes i kendelig Grad i Elementet selv (smlg. Side 80). For at beregne den søgte Energimængde skal vi tænke os en enkelt Elektron,
hvis Bevægelsesmængde i hvert enkelt Øjeblik i Størrelse og Retning
skal være lig den samlede Bevægelsesmængde af Elektronerne i Volumenelementet ΔV, og paa denne Elektron anvende Formlen. (16).

[Om en saadan Anvendelse af Formlen (16) maa vi dog bemærke
følgende. Formlen er kun udledt for en punktformig Ladnings Bevægelse,
og den er som ovenfor omtalt udledt ved Integration over den Energi,
der vil udsendes gennem de enkelte Elementer af en lukket Flade, der
omgiver Ladningen. Dersom vi nu søger den Energi, der sendes gennem
et saadant Fladeelement paa Grund af Elektronernes samlede Bevægelse,
kan vi vel betragte denne som samlet i en punktformig Ladning, men
denne Ladnings Bevægelse vil ikke være Middel af Elektronernes samtidige
Bevægelser, men derimod Middel af Elektronernes Bevægelser i Tidspunkter
valgte saaledes for hver enkelt Elektron, at Lysstraaler udgaaende fra Elektronerne i disse Tidspunkter samtidig vil naa det betragtede Fladeelement.
Idet vi nu imidlertid antager, at Elektronerne har Hastigheder, der er overordentlig mange Gange mindre end Lyset, vil enhver Bevægelse blandt
Elektronerne forplante sig overordentlig mange Gange langsommere end
Lyset, og der vil derfor i Middel være meget nær de samme systematiske
Forbindelser mellem de enkelte Elektroners Bevægelser[1]), hvad enten vi

---

[1]) *J. H. Jeans* (Phil. Mag., vol. 18, p. 213, 1909) udtaler den Mening, at man kan beregne
et Metalstykkes Energiudstraaling som en Sum af Straalingerne fra de enkelte
Elektroner, betragtet hver for sig, da der, efter den statistiske Mekanik, i Middel

betragter deres samtidige Bevægelser i almindelig Forstand, eller vi betragter Bevægelser, der er »effektivt samtidige« i Forhold til det betragtede Fladeelement i den ovenomtalte Forstand. Vi vil derfor i Middel faa den rigtige Værdi for Energistrømmen gennem Fladeelementet, dersom vi beregner den ud fra Betragtning af en enkelt punktformig Ladning, der er i Besiddelse af en Bevægelsesmængde lig Summen af de enkelte Elektroners samtidige Bevægelsesmængder. Idet det samme gælder for hvert enkelt Element af den ovenfor omtalte lukkede Flade — for hvert enkelt Element vil der naturligvis være en forskellig »effektiv Samtidighed« —, vil det gælde for hele Energistrømmen gennem Fladen.]

Kaldes nu Komposanterne efter Koordinatakserne af Elektronernes samlede Bevægelsesmængde i ΔV for g_x, g_y og g_z, faar man af Formlen (16) som Udtryk for den Energi, der i Middel udsendes i Tidsenheden,

$$\mathcal{J} = \frac{2\varepsilon^2}{3m^2c^3} \cdot \frac{1}{\mathfrak{T}} \int_0^{\mathfrak{T}} \left(\left(\frac{dg_x}{dt}\right)^2 + \left(\frac{dg_y}{dt}\right)^2 + \left(\frac{dg_z}{dt}\right)^2 \right) dt,$$

hvor $\mathfrak{T}$ betegner et Tidsrum, der skal antages at være overordentlig stort i Forhold til alle i Betragtning kommende Svingningstider. Da der i Middel ikke vil være nogen Afhængighed mellem Bevægelsesmængderne efter de tre Akser (i de i Kapitel I opstillede Ligninger til Bestemmelse af Elektronernes samlede Bevægelsesmængde efter X-aksen indgaar nemlig kun Elektronernes Bevægelsesmængde i den omhandlede Akses Retning), vil man kunne betragte den Udstraaling, der svarer til hvert enkelt af Leddene, for sig, og man vil derfor paa Grund af Symmetrien faa

ikke vil være nogensomhelst systematisk Forbindelse mellem to forskellige Elektroners samtidige Hastighedskoordinater. En saadan Beregningsmaade vil dog kun give rigtigt Resultat, dersom man antager, at man kan se bort fra Vekselvirkningen mellem Elektronerne indbyrdes, idet der, i Tilfælde af en saadan Vekselvirkning, vel ikke i Middel vil være nogen systematisk Forbindelse mellem de forskellige Elektroners samtidige Hastighedskoordinater, men derimod imellem deres samtidige Accelerationer. Ved et Sammenstød mellem to Elektroner vil der saaledes ikke udsendes nogen Energi, idet Elektronernes samlede Bevægelsesmængde ikke forandres, og saadanne Sammenstød vil derfor kun have indirekte Virkning paa Metalstykkets Energiudstraaling (smlg. det analoge Spørgsmaal (se det foregaaende, Side 58) vedrørende de omhandlede Sammenstøds Indvirkning paa Elektricitetsledningsevnen, og derigennem paa Absorptionsevnen); havde man derimod beregnet Energiudstraalingen som en Sum af Straalingerne fra de enkelte Elektroner betragtet hver for sig — d. v. s. beregnet ud fra de enkelte Elektroners Accelerationer ved Hjælp af Formlen (16) —, vilde man have fundet, at de betragtede Sammenstød vilde have en (i Forhold til deres Antal) betydelig direkte Indflydelse paa Metalstykkets Energiudstraaling.

$$\mathfrak{F} = \frac{2\varepsilon^2}{m^2c^3} \cdot \frac{1}{\mathfrak{T}} \int\limits_0^{\mathfrak{T}} \left(\frac{dg_x}{dt}\right)^2 dt. \qquad (17)$$

For at undersøge, hvorledes denne Energi vil fordele sig paa Straaler med forskellige Svingningstider, skal vi skrive $\dfrac{dg_x}{dt}$ som en Sum af harmoniske Svingninger ved Hjælp af et *Fourier'sk* Integral gældende fra Tiden o til Tiden $\mathfrak{T}$. Vi vil saaledes kunne skrive

$$\frac{dg_x}{dt} = \frac{1}{\pi} \int\limits_0^\infty (A_p \cos pt + B_p \sin pt)\, dp, \qquad (18)$$

hvor

$$A_p = \int\limits_0^{\mathfrak{T}} \frac{dg_x}{dt} \cos pt\, dt \quad \text{og} \quad B_p = \int\limits_0^{\mathfrak{T}} \frac{dg_x}{dt} \sin pt\, dt. \qquad (19)$$

Som Lord *Rayleigh*[1]) har bevist, følger ud af (18) og (19)

$$\int\limits_0^{\mathfrak{T}} \left(\frac{dg_x}{dt}\right)^2 dt = \frac{1}{\pi} \int\limits_0^\infty (A_p^2 + B_p^2)\, dp.$$

Medens Størrelserne A_p og B_p vil variere overordentlig stærkt og uregelmæssigt med p og $\mathfrak{T}$, vil, som vi skal vise, Størrelsen $(A_p^2 + B_p^2) : \mathfrak{T}$, dersom $\mathfrak{T}$ er tilstrækkelig stor, nærme sig en bestemt Grænseværdi, der vil variere regelmæssigt med p.

Af (17) faar vi nu, at den i Tidsenheden udstraalede Energi kan skrives

$$\mathfrak{F} = \frac{2\varepsilon^2}{\pi m^2 c^3} \cdot \frac{1}{\mathfrak{T}} \int\limits_0^\infty (A_p^2 + B_p^2)\, dp, \qquad (20)$$

i hvilket Udtryk Koefficienten til dp vil være den Energi, der svarer til Straaler med Svingningstal pr. Sek. mellem $\dfrac{1}{2\pi} p$ og $\dfrac{1}{2\pi}(p + dp)$.

Af (19) faas ved partiel Integration

$$A_p = \left| g_x \cos pt \right|_0^{\mathfrak{T}} + p \int\limits_0^{\mathfrak{T}} g_x \sin pt\, dt.$$

---

[1]) Lord *Rayleigh*: Phil. Mag (5), vol. **27**, p. 465, 1889.

Idet det første Led paa højre Side af denne Ligning ikke vil vokse (d. v. s. ikke vil oscillere med tiltagende Amplitude) med voksende $\mathfrak{T}$, vil man ved Beregningen af $(A_\mathfrak{p}^2 + B_\mathfrak{p}^2)$ kunne se bort fra dette Led i Sammenligning med det sidste Led. Vi faar derfor, idet man ligeledes ser bort fra det tilsvarende Led i Udtrykket for $B_\mathfrak{p}$,

$$A_\mathfrak{p} = p \int_0^\mathfrak{T} g_\mathrm{x} \sin ptdt \qquad \text{og} \qquad B_\mathfrak{p} = -p \int_0^\mathfrak{T} g_\mathrm{x} \cos ptdt. \qquad (21)$$

Af (21) faar vi nu, idet vi betegner Værdien af g_x til Tiden t med $g_\mathrm{x}(t)$,

$$A_\mathfrak{p}^2 = p^2 \int_0^\mathfrak{T}\int_0^\mathfrak{T} g_\mathrm{x}(t_1) g_\mathrm{x}(t_2) \sin pt_1 \sin pt_2 \, dt_1 \, dt_2$$

og

$$B_\mathfrak{p}^2 = p^2 \int_0^\mathfrak{T}\int_0^\mathfrak{T} g_\mathrm{x}(t_1) g_\mathrm{x}(t_2) \cos pt_1 \cos pt_2 \, dt_1 \, dt_2.$$

Dette giver

$$A_\mathfrak{p}^2 + B_\mathfrak{p}^2 = p^2 \int_0^\mathfrak{T}\int_0^\mathfrak{T} g_\mathrm{x}(t_1) g_\mathrm{x}(t_2) \cos p(t_2 - t_1) \, dt_1 \, dt_2,$$

hvilket kan skrives

$$A_\mathfrak{p}^2 + B_\mathfrak{p}^2 = 2p^2 \int_0^\mathfrak{T} g_\mathrm{x}(t_1) \int_{t_1}^\mathfrak{T} g_\mathrm{x}(t_2) \cos p(t_2 - t_1) \, dt_2 \, dt_1 \ ^1). \qquad (22)$$

---

[1]) Ligningen (22) er den samme Ligning som *Jeans* (Phil. Mag., vol. 17, p. 789, 1909) lægger til Grund for sine Beregninger. Han beregner derudfra $A_\mathfrak{p}^2 + B_\mathfrak{p}^2$ ved at benytte en Ligning af Formen (med vore Betegnelser):

$$\left[\frac{dg_\mathrm{x}}{dt}\right] = -\alpha \cdot g_\mathrm{x}, \qquad (1')$$

hvor Leddet paa venstre Side af Lighedstegnet skal betegne Middelværdien af Differentialkvotienten for en given Værdi af g_x. Af Ligningen (1') udleder *Jeans*

$$g_\mathrm{x}(t_2) = g_\mathrm{x}(t_1) e^{-\alpha(t_2 - t_1)} + j, \qquad (2')$$

hvor j betegner en Størrelse, hvis Middelværdi er 0. Ved at indsætte Udtrykket (2') i Ligningen (22) og udføre Integrationen med Hensyn til t_2 finder *Jeans* sluttelig (loc. cit., p. 790)

$$A_\mathfrak{p}^2 + B_\mathfrak{p}^2 = \frac{2\alpha p^2}{\alpha^2 + p^2} \cdot \int_0^\mathfrak{T} (g_\mathrm{x}(t_1))^2 \, dt_1, \qquad (3')$$

ved Hjælp af hvilken Ligning $A_\mathfrak{p}^2 + A_\mathfrak{p}^2$ simpelt lader sig bestemme.

Disse Beregninger er imidlertid i Almindelighed ikke rigtige (smlg. *H. A. Wilson*

Ved Beregningen af $A_p^2 + B_p^2$ udfra Ligningen (22), skal vi nu først betragte det Tilfælde, hvor vi antager, at der finder adskilte Sammenstød Sted.

I dette Tilfælde har vi med de i det foregaaende anvendte Betegnelser

$$g_x = \Delta V \cdot \int_0^\infty G_x(r)\, dr, \qquad (23)$$

hvor $G_x(r)$, idet vi her tænker os, at vi betragter et Metalstykke, der ikke paavirkes af ydre Kræfter, og som overalt har samme Temperatur, bestemmes ved Ligningen (Lign. (14) Side 26)

$$\left[\frac{dG_x(r)}{dt}\right]_{G_x(\rho)} = -G_x(r)\, F(r) + \int_0^\infty G_x(\rho)\, Q(\rho, r)\, dr, \qquad (24)$$

hvor Leddet paa Ligningens venstre Side skal betegne Middelværdien af $\frac{dG_x(r)}{dt}$ taget for alle mulige Maader, hvorpaa Bevægelsesmængden kan være fordelt mellem de enkelte Elektroner, saaledes at Størrelserne $G_x(\rho)$ antager de Værdier, der er tænkt indsatte i Ligningens højre Side.

Dersom vi nu multiplicerer Leddet paa venstre Side af Ligningen (24) med $\cos p\,(t_2 - t_1)$ og integrerer med Hensyn til t_2 fra t_1 til $\mathfrak{T}$, og dernæst tager Middelværdien af dette Integral for alle mulige Maader, hvorpaa Tilstanden ved Tiden t_1 er betegnet ved bestemte Værdier af $G_x(\rho)$, faar vi, idet vi benytter tilsvarende Betegnelser som ovenfor og kalder den søgte Middelværdi for a,

$$a = \left[\int_{t_1}^{\mathfrak{T}} \left[\frac{dG_x(r)}{dt}\right]_{(G_x(\rho))_{t_2}} \cdot \cos p\,(t_2 - t_1)\, dt_2\right]_{(G_x(\rho))_{t_1}}. \qquad (25)$$

Idet der nu i dette Tilfælde, hvor vi betragter adskilte Sammenstød, i Middel ikke er nogen systematisk Forbindelse mellem $\frac{dG_x(r)}{dt}$ og for-

---

loc. cit., p. 836); som vi har set i første Kapitel, vil nemlig den Hastighed, hvormed g_x aftager, være afhængig af den Maade, hvorpaa Bevægelsesmængden er fordelt mellem Elektronerne med de forskellige absolute Hastigheder; og denne Fordeling vil paa systematisk Maade forandre sig, medens g_x aftager, idet Elektronerne med de forskellige absolute Hastigheder vil miste deres Bevægelsesmængde i en bestemt Retning med forskellig Hurtighed.

udgaaende Tilstande (smlg. Side 81) — vi antager her, ligesom ved Beregningen af Absorptionen, at $\cos pt$ kun forandrer sig overordentlig lidt i de Tider, der medgaar til et enkelt Sammenstød —, vil man have

$$a = \left[\int_{t_1}^{\mathfrak{T}} \frac{dG_x(r)}{dt_2} \cos p(t_2 - t_1)\, dt_2 \right]_{(G_x(\rho))_{t_1}} . \qquad (26)$$

Af (26) faas ved partiel Integration

$$a = \left[(G_x(r))_{\mathfrak{T}} \cdot \cos p\, (\mathfrak{T} - t_1) - (G_x(r))_{t_1} + p \int_{t_1}^{\mathfrak{T}} G_x(r) \sin p(t_2 - t_1)\, dt_2 \right]_{(G_x(\rho))_{t_1}} .$$

Idet nu Middelværdien af det første Led indenfor Parenthesen, dersom $\mathfrak{T}$ er tilstrækkelig stor, vil være 0, faar vi, idet vi sætter

$$\xi_2(r) = \left[\int_{t_1}^{\mathfrak{T}} G_x(r) \sin p(t_2 - t_1)\, dt_2 \right]_{(G_x(\rho))_{t_1}} , \qquad (27)$$

$$a = - (G_x(r))_{t_1} + p\, \xi_2(r). \qquad (28)$$

Idet vi nu endvidere sætter

$$b = \left[\int_{t_1}^{\mathfrak{T}} \left[\frac{dG_x(r)}{dt} \right]_{(G_x(\rho))_{t_1}} \cdot \sin p\,(t_2 - t_1)\, dt_2 \right]_{(G_x(\rho))_{t_1}} \qquad (29) \qquad *$$

og

$$\xi_1(r) = \left[\int_{t_1}^{\mathfrak{T}} G_x(r) \cos p\,(t_2 - t_1)\, dt_2 \right]_{(G_x(\rho))_{t_1}} , \qquad (30)$$

faar vi ved at gaa frem paa ganske samme Maade som ovenfor

$$b = -p\, \xi_1(r). \qquad (31)$$

Dersom vi nu multiplicerer begge Sider af Ligningen (24) med henholdsvis $\cos p\,(t_2 - t_1)$ og $\sin p\,(t_2 - t_1)$ og integrerer med Hensyn til t_2 fra t_1 til $\mathfrak{T}$ og dernæst tager Middelværdien af Integralerne for alle mulige Maader, hvorpaa Bevægelsesmængden kan være fordelt mellem de enkelte Elektroner, saaledes at Tilstanden ved Tiden t_1 er

---

* [Subscript to the first factor of integrand should be $(G_x(\rho))_{t_2}$ rather than $(G_x(\rho))_{t_1}$.]

betegnet ved givne Værdier for $G_x(\rho)$, faar vi ved Hjælp af Ligningerne (25), (27), (28), (29), (30) og (31)

$$- (G_x(r))_{t_1} + p\,\xi_2(r) = - \xi_1(r)\,F(r) + \int_0^\infty \xi_1(\rho)\,Q(\rho, r)\,d\rho \qquad (32)$$

og

$$- p\,\xi_1(r) = - \xi_2(r)\,F(r) + \int_0^\infty \xi_2(\rho)\,Q(\rho, r)\,d\rho. \qquad (33)$$

Idet Ligningerne (32) og (33) til Bestemmelse af $\xi_1(r)$ og $\xi_2(r)$, som man ser, ganske svarer til Ligningerne (4) og (5) Side 82 til Bestemmelse af $\psi_1(r)$ og $\psi_2(r)$, faar man her paa ganske samme Maade som ved disse Ligninger

$$\xi_1(r) = (G_x(r))_{t_1} \cdot F_1(r) + \int_0^\infty (G_x(\rho))_{t_1} \cdot Q_1(\rho, r)\,d\rho, \qquad (34)$$

hvor $F_1(r)$ og $Q_1(\rho, r)$ betegner de samme Funktioner som i Ligningen (6) Side 82. Ved Hjælp af Ligningerne (23), (30) og (34) samt Ligningen (7) Side 82 faar vi nu

$$\left[\int_{t_1}^{\mathfrak{X}} g_x(t_2)\cos p\,(t_2 - t_1)\,dt_2 \right]_{(G_x(\rho))_{t_1}} = \Delta V \cdot \int_0^\infty (G_x(r))_{t_1}\,P_1(r)\,dr. \qquad (35)$$

Idet vi nu antager, at Tiden $\mathfrak{X}$ er saa lang, at $g_x(t)$ indenfor denne Tid ikke alene antager alle mulige Værdier, som den kan antage, men antager enhver af disse saa mange Gange, at man kan regne, at den har haft denne Værdi, med Bevægelsesmængden fordelt paa de enkelte Elektroner paa alle mulige Maader, vil man kunne indsætte Udtrykket (35) i Ligningen (22) i Stedet for Integralet i denne Ligning med Hensyn til t_2; vi faar derfor

$$A_p^2 + B_p^2 = 2p^2 \cdot (\Delta V)^2 \cdot \int_0^{\mathfrak{X}} \left(\int_0^\infty G_x(r)\,dr \cdot \int_0^\infty G_x(r)\,P_1(r)\,dr \right)_{t_1} dt_1. \qquad (36)$$

Som man ser, har vi nu opnaaet, at Størrelsen under Integraltegnet er henført til samme Tidspunkt, og derved er Opgaven blevet overordentlig meget simplificeret.

Vi har hidtil, for at lette Oversigten, i vore Beregninger benyttet Integraler til Udtrykkene for Elektronernes samlede Bevægelsesmængde o. s. v.; dette er strengt taget ikke rigtigt, idet der kun er et endeligt Antal Elektroner til Stede i ΔV; men man indser let, at det ingen væsentlig Forandring vilde have medført, dersom man havde erstattet Integralerne med Summer over de enkelte Elektroner. Ved den nærmere Beregning af Udtrykket (36) stiller det sig imidlertid ganske anderledes, idet det Integral, der indgaar i dette Udtryk, er i højeste Grad afhængigt af Antallet af Elektroner. Vi skal derfor i Stedet for Integralerne med Hensyn til r i Udtrykket (36) indføre Summer over de ΔN Elektroner i Volumenelementet ΔV.

Vi faar, idet vi betegner en enkelt Elektrons Hastighed med (ξ, η, ζ),

$$A_p^2 + B_p^2 = 2p^2 m^2 \cdot \int_0^{\tau} \left(\sum_1^{\Delta N} \xi_n \cdot \sum_1^{\Delta N} \xi_n P_1(r_n) \right)_{t_1} dt_1. \tag{37}$$

Man beviser nu imidlertid i den statistiske Mekanik, at der, hvilke Kræfter der end antages at virke mellem Elektronerne, i Middel ikke vil bestaa nogen systematisk Afhængighed mellem de enkelte Elektrons samtidige Hastighedskoordinater — (der vil derimod være systematiske Forbindelser mellem Elektronernes samtidige Accelerationer (se Noten Side 90); dette er der imidlertid ved de foregaaende Beregninger taget Hensyn til, nemlig gennem Benyttelsen af Ligningen (24)) —. Leddene med $\xi_n \cdot \xi_m P(r_m)$, hvor n og m er forskellige, vil derfor intet Bidrag give til Integralet i (37), og vi faar

$$A_p^2 + B_p^2 = 2p^2 m^2 \cdot \int_0^{\tau} \left(\sum_1^{\Delta N} \xi_n^2 P_1(r_n) \right) \cdot dt. \tag{38}$$

Idet nu Middelværdien af $\xi_n^2 P_1(r_n)$ paa Grund af Symmetrien (idet $r^2 = \xi^2 + \eta^2 + \zeta^2$) vil være lig $\frac{1}{3}$ af Middelværdien af $r_n^2 P_1(r_n)$, og idet endvidere Elektronernes Hastighedsfordeling i Middel udtrykkes ved Ligningen (4) Side 16, faar man

$$\left. \begin{aligned} A_p^2 + B_p^2 &= 2p^2 m^2 \cdot \mathfrak{T} \ \Delta V \int_0^{\infty} \frac{r^2}{3} P_1(r) \cdot K e^{-\frac{m r^2}{2kT}} \cdot 4\pi r^2 dr \\ &= \frac{2m^2 kT}{\varepsilon^2} \sigma_p p^2 \mathfrak{T} \Delta V, \end{aligned} \right\} \tag{39}$$

hvor σ_p betegner den ved Ligningen (8) Side 82 definerede Størrelse.

7

Af Ligningerne (20) og (39) faar vi nu, idet den Energi, der i Tidsenheden udstraales fra et lille Metalstykke med Volumen ΔV fordelt paa Straaler med Svingningstal mellem $\frac{1}{2\pi} p$ og $\frac{1}{2\pi}(p + dp)$, betegnes ved $\beta_p dp \Delta V$,

$$\beta_p = \frac{4kT}{\pi c^3} \sigma_p p^2. \tag{40}$$

Indsætter vi i (40) den Værdi for σ_p, der er givet ved Ligningen (11) Side 84, og sættes henholdsvis $n = \infty$ og $n = 5$, faar vi de Værdier, der er givne for β_p af henholdsvis *Wilson*[1]) og *Jeans*[2]).

Vi skal nu betragte det Tilfælde, hvor Metalmolekylernes Dimensioner ikke antages at være forsvindende i Forhold til deres indbyrdes Afstande. I dette Tilfælde vil man ikke kunne anvende en Fremgangsmaade, der svarer til den, vi har benyttet i det ovenfor omhandlede Tilfælde, idet der her i Middel vil være systematiske Forbindelser mellem den Hastighed, hvormed Bevægelsesmængden aftager, og tidligere Tilstande indenfor Tidsrum, der er af samme Størrelsesorden som de Tider, indenfor hvilke de enkelte Elektroner i Middel mister Størstedelen af deres oprindelige Bevægelsesmængde i en bestemt Retning[3]). (Et Tidsrum af denne Størrelsesorden skal vi i det følgende for Korteds Skyld betegne ved τ.)

Vi skal her kun betragte det Tilfælde (se Side 37), i hvilket Metalmolekylernes Kraftfelter antages at være stationære, og hvor man ser bort fra Vekselvirkningen mellem Elektronerne indbyrdes; og vi skal endvidere kun gennemføre Beregningerne for saadanne Varmestraalers Vedkommende, hvis Svingningstider er store i Forhold til den ovenfor omtalte Størrelse τ.

Idet Elektronerne her antages at bevæge sig uafhængigt af hverandre, vil der ikke være systematiske Forbindelser mellem deres samtidige Bevægelser; og Udstraalingen fra et lille Metalstykke ΔV vil

---

[1]) *H. A. Wilson*: loc. cit., p. 842.

[2]) *J. H. Jeans*: Phil. Mag., vol. 17, p. 790, 1909.

[3]) *Jeans* benytter samme Fremgangsmaade i alle Tilfælde, han bemærker blot (loc. cit., p. 776), at Størrelsen *dt* i den Ligning, der svarer til Ligning (1') i Noten Side 93. i det her omhandlede Tilfælde skal betragtes som stor i Forhold til saadanne Tidsrum, indenfor hvilke g_x forandrer sig meget stærkt. Bortset fra de Mangler ved *Jeans's* Beregninger, som vi ovenfor har omtalt, og som naturligvis ogsaa vil være til Stede i dette mere almindelige Tilfælde, synes den nævnte Antagelse om *dt* ikke at kunne bringes i Overensstemmelse med Benyttelsen af Ligningen (1') som Differentialligning (nemlig til at udlede (2')), ved hvilken Benyttelse man jo netop nødvendigvis maa gaa ud fra, at g_x kun forandrer sig meget lidt i Tiden *dt*.

derfor kunne beregnes som Summen af Udstraalingerne fra de enkelte Elektroner betragtede hver for sig. Idet Antallet af Elektroner i Metalstykket betegnes ved ΔN, faar vi i dette Tilfælde paa ganske samme Maade som ovenfor, svarende til Ligningerne (20) og (22), for den Mængde Energi, der udstraales i en Tidsenhed,

$$\mathcal{F} = \frac{2\varepsilon^2}{\pi c^3} \cdot \frac{1}{\mathfrak{T}} \int\limits_0^\infty \left\{ \sum_1^{\Delta N} (A_\mathrm{p}^2 + B_\mathrm{p}^2)_\mathfrak{n} \right\} dp, \tag{41}$$

hvor

$$(A_\mathrm{p}^2 + B_\mathrm{p}^2)_\mathfrak{n} = p^2 \int\limits_0^\mathfrak{T} \int\limits_0^\mathfrak{T} \xi_\mathfrak{n}(t_1)\, \xi_\mathfrak{n}(t_2) \cos p\,(t_2 - t_1)\, dt_1 dt_2, \tag{42}$$

i hvilken Ligning $\xi_\mathfrak{n}(t)$ betegner en enkelt Elektrons Hastighed efter X-aksen til Tiden t.

Da vi nu her kun skal betragte Svingningstider, der er store i Sammenligning med τ, kan $\cos p\,(t_2 - t_1)$ sættes lig 1 for alle saadanne Værdier af t_1 og t_2, ved hvilke der i Middel vil være nogen kendelig systematisk Forbindelse imellem $\xi_\mathfrak{n}(t_1)$ og $\xi_\mathfrak{n}(t_2)$; man vil imidlertid indse, at man, ogsaa for de øvrige Værdier af t_1 og t_2, i Integralet i (42) kan erstatte $\cos p\,(t_2 - t_1)$ med 1; thi disse sidste Værdier vil, paa Grund af Uafhængigheden mellem $\xi_\mathfrak{n}(t_1)$ og $\xi_\mathfrak{n}(t_2)$, ikke i noget af de to Tilfælde i Middel give noget kendeligt Bidrag til Værdien af Integralet i (42). Vi faar da af (42)

$$(A_\mathrm{p}^2 + B_\mathrm{p}^2)_\mathfrak{n} = p^2 \left(\int\limits_0^\mathfrak{T} \xi_\mathfrak{n}(t)\, dt \right)^2, \tag{43}$$

i hvilket Udtryk Integralet paa højre Side vil være lig Projektionen paa X-aksen af det Stykke, den paagældende Elektron har bevæget sig i Tiden $\mathfrak{T}$.

Idet vi antager, at der ikke finder nogen Vekselvirkning Sted mellem de fri Elektroner indbyrdes, og at Metalmolekylernes Kraftfelter er stationære, vil Summen a af en enkelt Elektrons kinetiske Energi og dens potentielle Energi i Forhold til Metalmolekylerne være konstant under Elektronens Bevægelse, og vi har i 1ste Kapitel omtalt, hvorledes de enkelte Elektroners Bevægelse kan betragtes som en »fri Diffusion« med Diffusionskoefficienten $D(a)$ (se Side 41). — Ved Beregningerne i 1ste Kapitel har vi, for at bringe Forbindelse til Veje mellem det her omhandlede Tilfælde og Forholdene i de virkelige

7*

Metaller, kun antaget, at a i Middel er mærkelig konstant indenfor Tider af samme Størrelsesorden som τ (se Side 38); vi skal imidlertid her for Simpelheds Skyld antage, at a vil være konstant under hele den enkelte Elektrons Bevægelse i den lange Tid $\mathfrak{T}$; dette vil nemlig af Grunde, der ganske svarer til, hvad der blev omtalt, da vi ovenfor erstattede $\cos p\,(t_2 - t_1)$ med 1, ikke have nogen kendelig Indvirkning paa Værdien af Udtrykket for $\Sigma\,(A_\mathrm{p}^2 + B_\mathrm{p}^2)_\mathrm{n}$. —

Bestemmelsen af Størrelsen $\Sigma\,(A_\mathrm{p}^2 + B_\mathrm{p}^2)_\mathrm{n}$ er nu ved Ligningen (43) bleven henført til Spørgsmaalet om Fordelingen til Tiden $\mathfrak{T}$ af et stort Antal uafhængigt af hinanden diffunderende Partikler, der til Tiden $t = 0$ befinder sig i en bestemt Plan (f. Eks. YZ-planen). Dette Spørgsmaal kan, som *A. Einstein*[1]) har vist, behandles paa en meget simpel Maade. Partiklernes Fordeling vil nemlig bestemmes ved Ligningen

$$\frac{\delta f}{\delta t} = D \cdot \frac{\delta^2 f}{\delta x^2},$$

hvor $f(x, t)\ dx$ betegner Antallet af Partikler til Tiden t mellem to parallele Planer i Afstandene x og $x + dx$ fra den faste Plan, og hvor Diffusionskoefficienten D svarer til den ovenfor omtalte Størrelse $D(a)$. Løsningen af denne Ligning under den Betingelse, at $f(x, t) = 0$ for $t = 0$ og $x \lessgtr 0$, — hvilken Løsning kendes fra *Fourier's* Theori for Varmeledningen i et isotropt Medium — vil være

$$f(x, t) = C t^{-\frac{1}{2}} e^{-\frac{x^2}{4 D t}},$$

og Middelværdien af Kvadratet paa Partiklernes Afstande fra den faste Plan til Tiden $\mathfrak{T}$, vil være (*Einstein:* loc. cit., p. 559)

$$\int_0^\infty x^2 f(x, \mathfrak{T})\, dx : \int_0^\infty f(x, \mathfrak{T})\, dx = 2 D \mathfrak{T}.$$

Af Ligningen (43) faar vi derfor nu

$$\sum_1^{\Delta N}{}'(A_\mathrm{p}^2 + B_\mathrm{p}^2)_\mathrm{n} = 2 p^2 \mathfrak{T} \sum_1^{\Delta N}{}' D(a),$$

og idet vi ligesom i første Kapitel skal betegne Antallet af Elektroner i Volumenenheden, hvis Energi ligger imellem a og $a + da$, ved $N(a)\, da$, faar vi

---

[1]) *A. Einstein*: Ann. d. Phys., Bd. 17, p. 556, 1905.

$$\sum_{1}^{\Delta N}{}'(A_p^2 + B_p^2)_n = 2p^2 \mathfrak{T} \cdot \Delta V \int_0^\infty D(a) \, N(a) \, da. \qquad (44)$$

Idet endvidere Metallets elektriske Ledningsevne σ vil være (se Lign. (7) Side 50 og Lign. (8) Side 51)

$$\sigma = \frac{\varepsilon^2}{kT} \int_0^\infty D(a) \, N(a) \, da,$$

faar vi sluttelig med Benyttelse af de samme Betegnelser som paa Side 98 ved Hjælp af Ligningerne (41) og (44)

$$\beta_p = \frac{4kT}{\pi c^3} \sigma p^2. \qquad (45)$$

Vi skal nu benytte de i det foregaaende udførte Beregninger til Bestemmelse af Varmestraalingsloven for Straaler med store Svingningstider. Tænker vi os et lille Metalstykke i Ligevægt med den elektromagnetiske Straaling i det omgivende tomme Rum, maa den Energi fordelt paa Straaler med bestemte Svingningstider, som Metallet absorberer i Tidsenheden, i Middel være lig den Energi, som det udsender i den samme Tid fordelt paa Straaler med de samme Svingningstider. Vi faar derfor, idet vi kalder den Energi, der i Ligevægtstilstanden findes i en Volumenenhed af det omgivende tomme Rum fordelt paa Straaler med Svingningstal mellem $\frac{1}{2\pi} p$ og $\frac{1}{2\pi}(p + dp)$, for $E(p) \, dp$, og idet vi betegner Metalstykkets Volumen ved ΔV,

$$E(p) \, dp \cdot \alpha_p \Delta V = \beta_p \, dp \, \Delta V,$$

hvoraf faas

$$E(p) = \frac{\beta_p}{\alpha_p}.$$

Idet vi nu indsætter de fundne Værdier for α_p og β_p enten fra Ligningerne (10) og (40) eller fra Ligningerne (1) og (45), faar vi i begge Tilfælde

$$E(p) = \frac{kT}{\pi^2 c^3} \cdot p^2. \qquad (46)$$

Som det fremgaar af, hvad der er omtalt ved Bestemmelserne af α_p og β_p (se Side 86 og 98), er Ligningen (46) den samme som den af *Jeans* udledte, men forskellig fra den af *Wilson* fundne; dette sidste hidrører efter *Wilsons*

Mening fra, at der. ved hans Beregninger ikke er taget Hensyn til Sammenstødene mellem Elektronerne indbyrdes (loc. cit., p. 844), men som vi
har set, er Ligningen, i hvert Tilfælde saa længe man antager, at der
finder adskilte Sammenstød Sted, fuldkommen uafhængig af den Indflydelse,
man tilskriver Vekselvirkningen mellem Elektronerne indbyrdes i Forhold
til Vekselvirkningen mellem Elektronerne og Metalmolekylerne.

Dersom vi i Stedet for p indfører Bølgelængden λ, der staar i
Forbindelse med p gennem Ligningen $\frac{1}{2\pi} p = \frac{c}{\lambda}$, og betegner Energien i Volumenenheden fordelt paa Straaler med Bølgebredder mellem
λ og $\lambda + d\lambda$ med $E(\lambda) d\lambda$, faar vi af (46)

$$E(\lambda) = 8\pi k T \cdot \lambda^{-4}. \tag{47}$$

Denne Formel, hvis Rigtighed vi her har bevist for alle de i
Kapitel I behandlede Tilfælde, er den samme som den af *Lorentz*
udledte, og er som omtalt i det foregaaende i udmærket Overensstemmelse med Eksperimenterne over Varmestraaler med store Bølgelængder.

For nærmere at kunne undersøge Betydningen af Formlen (47)
maa vi her omtale et Forsøg, der er gjort af Lord *Rayleigh*[1]) og
Jeans[2]) paa at beregne Varmestraalingsloven for Straaler med alle
mulige Svingningstider umiddelbart ud fra selve de til Grund liggende
Antagelser, hvorpaa den elektromagnetiske Theori bygger. De nævnte
Forfattere har nemlig forsøgt ved Hjælp af den statistiske Mekaniks
almindelige Behandlingsmetoder at undersøge Betingelserne for Ligevægt mellem den elektromagnetiske Straaling og et hvilketsomhelst
Elektronsystem, der befinder sig i et Rum, der er fuldstændig afskaaret
fra ydre Paavirkninger. Resultatet af disse Undersøgelser er nu, at
Betingelsen for Ligevægt er, at den straalende Energi i Volumenenheden
udtrykkes ved Formlen (47), resp. (46), for Straaler med alle mulige
Svingningstider. Dette vil imidlertid, som man indser, medføre, at
hele Energien paa Volumenenhed bliver uendelig stor, hvilket betyder,
at der slet ikke kan opnaas en Ligevægt, men at hele Energien efterhaanden vil gaa over paa Straalingen og her mere og mere koncen-
treres paa Straaler med kortere og kortere Svingningstider, ja til Slut
vil hele Energien være koncentreret paa Straaler med uendelig korte
Svingningstider.

---

[1]) Lord *Rayleigh*: Phil. Mag. (5), vol. 49, p. 539, 1900. Nature, vol. 72, pp. 54 og
243, 1905.

[2]) *J. H. Jeans*: Phil. Mag. (6), vol. 10, p. 91, 1905 og vol. 17, p. 229, 1909; se ogsaa
H. A. Lorentz: Nuovo Cimento (5), Tom. 16, p. 5, 1908.

Paa Grundlag af disse Resultater har *Jeans* fremsat den Anskuelse, at der ved Forsøgene over Varmestraalingen slet ikke er Tale om nogen virkelig Ligevægt, men kun om en Tilstand, der forskyder sig meget langsomt paa Grund af de almindelige Legemers ringe Absorptions- og Emissionsevne for Straaler med meget korte Svingningstider. Dette er imidlertid, som det fremgaar af den Diskussion[1]), der har været ført herom, ikke tilstrækkeligt til at forklare de ved Eksperimenterne fundne Fænomener. Det synes derfor, som om det er udelukket at forklare Varmestraalingsloven i Overensstemmelse med Erfaringen, dersom man vil fastholde de Grundantagelser, hvorpaa den elektromagnetiske Theori bygger[2]). Dette hidrører antagelig fra, at den elektromagnetiske Theori ikke er i Overensstemmelse med de virkelige Forhold, og at den kun kan give rigtige Resultater, dersom den anvendes overfor et stort Antal Elektroner (som ved de almindelige Legemer), eller til at bestemme Middelbevægelsen af en enkelt Elektron i forholdsvis lange Tidsrum (saaledes som ved Beregningen af Katodestraalernes Bevægelse), men ikke kan benyttes til at undersøge en enkelt Elektrons Bevægelse indenfor korte Tidsrum. Vi skal ikke her, da det ligger ganske udenfor de Spørgsmaal, som behandles i denne Afhandling, komme ind paa en Omtale af de Forsøg, der er gjort paa at indføre principielle Forandringer i den elektromagnetiske Theori; Forsøg, der synes at skulle føre til meget interessante Resultater.

Man kunde nu spørge om, hvorledes de Beregninger, som vi har udført i dette Kapitel, stiller sig til det her omtalte. Vi maa da først og fremmest gøre opmærksom paa, at vi slet ikke har forsøgt at finde en Ligevægtstilstand i den Forstand som Lord *Rayleigh* og *Jeans*. Vi er nemlig stiltiende gaaet ud fra, at der i Overensstemmelse med Erfaringen vil eksistere en Ligevægtstilstand, ved hvilken den elektromagnetiske Energi i Volumenenheden vil være en forholdsvis lille Størrelse, nemlig saa lille, at de elektromagnetiske Felter hidrørende fra Straalingen kun vil have en yderst ringe Indflydelse paa de fri Elektroners Bevægelser i Forhold til de Kræfter, hvormed Elektronerne paavirkes af hinanden og af Metalmolekylerne[3]). — (Der-

---

[1]) Se f. Eks. *O. Lummer* og *E. Pringsheim*: Phys. Zeitschr., Bd. 9, p. 449, 1908; og *H. A. Lorentz*: Phys. Zeitschr., Bd. 9, p. 562, 1908.

[2]) Se *A. Einstein*: Phys. Zeitschr., Bd. 10, pp. 185 og 817, 1909; se ogsaa *M. Planck*: Ann. d. Phys., Bd. 31, p. 758, 1910; og *A. Einstein* u. *L. Hopf*: Ann. d. Phys., Bd. 33, p. 1105, 1910.

[3]) Naar flere Forfattere (*J. J. Thomson*: Phil. Mag., vol. 14, p. 217, 1907 og vol. 20, p. 238, 1910 og *J. Kunz*: Physical. Review, vol. 28, p. 313, 1909; smlg. ogsaa *J. H. Jeans*: Phil. Mag., vol, 18, p. 217, 1909 og vol. 20, p. 642, 1910) er af den

som man derimod vilde søge den Ligevægtstilstand, der vilde opstaa i et fra ydre Paavirkning afskaaret Rum, dersom de elektromagnetiske Grundantagelser var fuldstændig rigtige, maatte man, som *Jeans*[1]) har gjort opmærksom paa, nødvendigvis tage Hensyn til de omtalte elektromagnetiske Felter, der efterhaanden, som den straalende Energi voksede, vilde blive af langt større Indflydelse paa de enkelte Elektroners Bevægelser end Metalmolekylernes Kraftfelter.) — Idet vi endvidere kun har behandlet saadanne Varmestraaler, hvis Svingningstider er store i Forhold til Varigheden af de enkelte elektromagnetiske »Processer« (Sammenstødene), synes det, som om det ikke er uberettiget paa Forhaand at vente, at de i dette Kapitel udførte Beregninger har samme Stilling til og samme Krav paa Overensstemmelse med Erfaringen som den øvrige Metallernes Elektrontheori. — Dette synes navnlig at maatte gælde for den Udledning, vi har givet i det sidst behandlede Tilfælde, idet vi her ikke har gjort Brug af nogen Antagelse om, at der overhovedet finder pludselige Sammenstød Sted.

Mening, at man udfra Elektrontheorien, ved en Beregningsmaade som den omtalte, ved Hjælp af passende Antagelser om Virkningen mellem Metalmolekylerne og Elektronerne kan forklare i hvert Tilfælde Karakteren af den eksperimentelt fundne Straalingslov for alle Svingningstider — f. Eks. den meget hurtige Aftagen af Energien fordelt paa Straaler med meget korte Svingningstider —, synes dette dog ikke rigtigt. Ved alle de nævnte Undersøgelser er der nemlig ikke taget tilstrækkeligt Hensyn til den overordentlig hurtige Aftagen af Absorptionsevnen for Straaler med meget korte Svingningstider, men kun til Emissionsevnens Aftagen (smlg. Side 77). Dersom man imidlertid ogsaa tager Hensyn til Absorptionsevnens Aftagen, vil man efter min Mening i alle Tilfælde nødvendigvis komme til en Straalingslov, der ikke lader sig forene med Erfaringerne; thi det modsatte Resultat vilde være i Strid med Lord *Rayleigh's* og *Jeans's* Beregninger. Førtes man nemlig, udfra Forudantagelser som de ovenfor nævnte om Tilstedeværelsen af en Ligevægt som den eksperimentelt fundne, til en Straalingslov, der for alle Svingningstider lod sig forene med disse Antagelser, vilde denne svare til en virkelig Ligevægt og ikke (smlg. ovenfor ved Omtalen af *Jeans's* Theori) til en tilsyneladende Ligevægt.

[1]) *J. H. Jeans*: Phil. Mag., vol. 18, p. 215, 1909.

KAPITEL IV.

ET MAGNETISK FELTS INDVIRKNING PAA DE FRI ELEKTRONERS BEVÆGELSE.

---

§ 1.

Tilfælde, hvor der er statistisk Ligevægt til Stede.

Vi skal her først undersøge, hvad Virkning et ydre magnetisk Felt vil have paa et Metalstykke, der befinder sig saavel i elektrisk Ligevægt som i Temperaturligevægt.

Under det magnetiske Felts Indvirkning vil de fri Elektroners Baner blive krummede i en bestemt Retning i Forhold til Feltets Akse — dersom man antager, at der finder adskilte Sammenstød Sted, vil Elektronernes Baner mellem Sammenstødene bestaa af smaa Stykker af Skruelinier, hvis Akser er parallele med Feltets Akse —. Man kunde nu tænke sig, at denne Elektronernes Bevægelse vilde frembringe ydre magnetiske Virkninger. En saadan Antagelse vil imidlertid, som vi skal se, ikke være rigtig. Krumningen alene af Banerne vil nemlig ikke frembringe magnetiske Virkninger, idet den magnetiske Kraft, der stammer fra en Elektrons Bevægelse, kun vil være afhængig af Elektronens Sted og Hastighed i det betragtede Øjeblik (vi antager her ligesom i det foregaaende, at Elektronernes Hastighed kan regnes for forsvindende i Sammenligning med Lyset) og ikke af Elektronens Acceleration. Den samlede magnetiske Virkning af de fri Elektroners Bevægelser kan derfor bestemmes ud fra den statistiske Fordeling af Elektronernes Steds- og Hastigheds-Koordinater til enhver given Tid. Som vi nu imidlertid har omtalt i 1ste Kapitel Side 13, vil denne Fordeling, dersom der er statistisk Ligevægt til Stede, være ganske uafhængig af en ydre magnetisk Kraft. Hvad enten der virker en magnetisk Kraft eller ej, vil Elektronernes Hastigheder i ethvert nok saa lille Volumenelement af Metallet i Middel være ligelig fordelt i alle Retninger, og der vil derfor ikke udgaa nogen magnetisk Virkning fra et saadant Element og derfor heller ikke fra hele Metalstykket.

Dersom man ikke behandler Spørgsmaalet statistisk, men derimod betragter de enkelte Elektroners Bevægelser, finder man, at enhver Elektron, der bevæger sig helt i det Indre af Metallet, paa Grund af dens Banes Krumning under det ydre magnetiske Felts Indvirkning, under sin Bevægelse i Middel vil frembringe en magnetisk Kraft modsat rettet den ydre magnetiske Kraft. Flere Forfattere [1]) har derfor ment, at et Metalstykke, hvori der bevæger sig fri Elektroner, vil virke som et diamagnetisk Legeme. Dette er imidlertid, som vi ovenfor har set, ikke rigtigt, hvilket hidrører fra, at man ved en Betragtningsmaade som den sidst omtalte, hvor de enkelte Elektroners Baner betragtes hver for sig, maa tage særligt Hensyn til Forholdene i den umiddelbare Nærhed af Metallets Overflade.

For nærmere at vise dette skal vi her betragte Forholdene ved et specielt, særlig simpelt Eksempel.

Vi skal betragte en Samling Elektroner, der bevæger sig fuldkommen uafhængig af hverandre (ligesom Molekylerne i en stærkt fortyndet Luftart) i en lukket Beholder, og vi skal endvidere for Simpelheds Skyld yderligere antage, at Elektronerne, naar de rammer Beholderens Vægge, vil kastes tilbage som elastiske Kugler fra en glat Flade. Vi skal nu undersøge Virkningen af en ydre magnetisk Kraft paa en saadan Samling Elektroner.

Vi skal først for dette specielle Eksempel eftervise, hvad vi ovenfor har bevist i Almindelighed, nemlig at Elektronernes Hastighed i ethvert nok saa lille Volumenelement af Beholderen vil være ligelig fordelt i alle Retninger, dersom der er statistisk Ligevægt til Stede. Ovenstaaende Figur skal forestille et lille Stykke af Beholderen i Nærheden af Væggen. Denne, der tænkes at staa vinkelret paa Figurens Plan, er afbildet ved Linien ab, og Beholderens Indre tænkes at ligge tilhøjre for denne Linie. Den magnetiske Kraft tænkes endvidere at staa vinkelret paa Figurens Plan. Paa Figuren er medtaget Banerne for et Antal Elektroner, der alle tænkes at have samme absolute Hastighed og bevæge sig i Baner, der ligger i

---

[1]) *J. J. Thomson*: Rapp. d. Congrès d. Physique, Paris 1900, Tom. 3, p. 148 og *W. Voigt*: Ann. d. Phys., Bd. 9, p. 130 (Note 1), 1902; se ogsaa *P. Langevin*: Ann. d. Chim. et d. Phys., Tom. 5, p. 90, 1905. [Naar *Z. Thullie* (Prace mat.-fiz. Warszava, Tom. 19, p. 207, 1908) i Modsætning til de nævnte Forfattere mener at kunne slutte, at de fri Elektroners Virkning ikke vil være diamagnetisk men paramagnetisk, synes dette at skyldes en Fortegnsfejl ved *Thullie's* Beregninger; smlg. loc. cit. Lign. (1) p. 209 og Lign. (4) p 210.]

Figurens Plan. Vi skal nu her for Simpelheds Skyld kun betragte saadanne Elektroner som dem, hvis Baner er medtagne paa Figuren. At der vil være statistisk Ligevægt blandt en saadan Samling Elektroner, dersom deres Hastigheder i Omegnen af ethvert Punkt er ligelig fordelt i alle Retninger i Figurens Plan, følger, dersom man betragter et Punkt i en saadan Afstand fra Væggen, at en Elektron, der begynder sin Bane i Punktet, paa Grund af den magnetiske Krafts afbøjende Virkning ikke vil kunne naa Væggen, umiddelbart udfra Symmetrien med Hensyn til den magnetiske Krafts Retning; og at det samme ogsaa vil gælde om et Punkt, der ligger i nok saa stor Nærhed af Beholderens Væg, følger af, at Elektronerne (efter den ovenfor omtalte Antagelse om Tilbagekastningen fra Væggen) vil kastes saaledes tilbage, at deres Baner i Middel danner den direkte Fortsættelse af de Baner, Elektronerne vilde gennemløbe, dersom Væggen ikke var til Stede. — Ved ganske lignende Betragtninger kan man indse, at en tilsvarende statistisk Ligevægt vil gælde for alle Elektronerne, d. v. s. ogsaa, naar man medtager saadanne, der ikke ligger i Figurens Plan, eller som befinder sig i Nærheden af et Stykke af Væggen, der ikke er parallelt med den ydre magnetiske Kraft. — Bevægelsen af de betragtede Elektroner vil derfor ikke give Anledning til nogen ydre magnetisk Virkning.

Vi skal nu gaa over til at undersøge de enkelte Elektroners Baner i det her omhandlede Eksempel. Betragter vi igen saadanne Elektroner, der er afbildede paa Figuren, vil man indse, at en Elektron, der under sin Bevægelse ikke kommer i Berøring med Beholderens Væg, idet den beskriver en lukket Bane, i Middel vil frembringe en magnetisk Kraft, der vil have modsat Retning af den ydre magnetiske Kraft. Følger man derimod de Elektroner, der under Gennemløbningen af deres Baner kommer i Berøring med Væggen, vil man se, at disse, som det fremgaar af Figuren, under det magnetiske Felts Indvirkning faar en Slags krybende Bevægelse langs med Væggen; og disse Elektroner vil derfor, betragtet for sig, frembringe et magnetisk Felt i samme Retning som det ydre Felt, og som vil være ligesaa stort og modsat rettet det, som frembringes af Elektronerne i det Indre.

Vi har saaledes set, at et Metalstykke, i hvilket der er saavel elektrisk Ligevægt som Temperaturligevægt, ikke vil besidde nogensomhelst magnetiske Egenskaber hidrørende fra de fri Elektroners Tilstedeværelse. (Det maa dog bemærkes, at der i et Metalstykke, der udsættes for en variabel magnetisk Kraft, vil fremkomme en Strømning af Elektronerne — de saakaldte *Foucault'ske* Strømme — frembragt af de elektriske Kræfter, der efter den elektromagnetiske Theori er uløseligt knyttede til det magnetiske Felts Variation; en Strømning, der vil frembringe en magnetisk Virkning, der vil modvirke Feltets Variation. Ganske tilsvarende Forhold vil naturligvis finde Sted, naar et Metalstykke pludselig udsættes for Virkningerne af et magnetisk Felt [1]); men i et saadant Til-

---

[1]) Som *Langevin* (loc. cit., p. 82—93) har paavist (smlg. ogsaa *Z. Thullie*: Anz. d. Akad. d. Wiss. Krakau, math.-nat. Kl., 1907, p. 749), vil et Forhold, der ganske svarer

fælde vil den omhandlede Strømning af Elektronerne, naar det magnetiske Felt er blevet konstant, meget hurtigt forsvinde uden at efterlade nogen blivende Virkning paa Elektronernes samlede Bevægelse.)

---

§ 2.

De galvano- og thermomagnetiske Fænomener.

Vi har hidtil antaget, at der i det betragtede Metalstykke var saavel elektrisk Ligevægt som Temperaturligevægt; er dette imidlertid ikke Tilfældet, vil Elektronerne i Middel antage en ordnet Bevægelse

til det her omhandlede, ogsaa finde Sted, dersom man betragter de »bundne« Elektroner, der antages at bevæge sig i lukkede Baner i Metalmolekylernes Indre; idet der ogsaa her, dersom der pludselig opstaar en magnetisk Kraft, vil frembringes en Bevægelse blandt Elektronerne, der vil fremkalde en magnetisk Kraft modsat rettet den ydre magnetiske Kraft. — Naar *W. Voigt* (loc. cit., p. 125) og *J. J. Thomson* (Phil. Mag., vol. 6, p. 688, 1903) ikke har fundet en saadan Virkning hos en Samling Elektroner, der bevæger sig i lukkede Baner under Paavirkning af en Centralkraft (Tiltrækning), hidrører dette, som *Langevin* (loc. cit., p. 83) har gjort opmærksom paa, fra, at de nævnte Forfattere ikke har taget tilstrækkeligt Hensyn til de elektriske Kræfter, der er knyttede til Forandringerne i det magnetiske Felt. [Det kan her bemærkes, at sammenholdt med det ovenfor (i Teksten) omtalte, danner *Voigt's* og *Thomson's* Beregninger en Illustration til den statistiske Ligevægtsfordelings Uafhængighed af Tilstedeværelsen af magnetiske Kræfter (i det omhandlede Tilfælde, hvor Elektroner bevæger sig i lukkede Baner, vil der ikke fremkomme nogen »Randvirkning«; smlg. ovenfor).] — *Langevin* har ud fra det her omtalte søgt at forklare Legemernes diamagnetiske Egenskaber, idet han antager, at en saadan en Gang opstaaet Bevægelse vil holde sig uforandret, saalænge det magnetiske Felt holder sig konstant. Dette sidste vil dog, efter hvad vi har omtalt i første Kapitel (Side 13), ikke være Tilfældet, dersom Elektronernes Bevægelser følger de almindelige mekaniske Love, og dersom der finder blot den ringeste Energiomsætning Sted mellem de enkelte Elektroner indenfor de forskellige Molekyler (Tilstedeværelsen af en vis Energiomsætning mellem de paagældende Elektroner maa man efter den omhandlede Theori antage, idet de betragtede magnetiske Virkninger jo netop forklares ved Elektronernes Paavirkning af ydre Kræfter, d. v. s. Kræfter, der stammer fra udenfor Molekylet værende Elektroner), saaledes at der til Slut vil fremkomme en mekanisk statistisk Ligevægt.

Det følger af det ovenstaaende, at det ikke er muligt udfra Elektrontheorien at forklare Legemernes magnetiske Egenskaber, dersom man ikke tager Hensyn til saadanne Virkninger, der, som f. Eks. Energiudstraalingen (Smlg. *W. Voigt*: loc. cit., p. 146 og *J. J. Thomson*: loc. cit., p. 689; se ogsaa *R. Gans*: Nach. d. Kgl. Ges. d. Wiss. Göttingen, math.-phys. Kl., 1910, p. 213—230. — En Energiudstraaling vil dog kun kunne forklare de paramagnetiske, ikke de diamagnetiske Fænomener. —), vil kunne bevirke, at der ikke blandt de indenfor Molekylerne bundne Elektroner indstiller sig en saadan mekanisk statistisk Ligevægt som den, vi har antaget for de fri Elektroners Vedkommende. De her omtalte Forhold synes at staa i den nøjeste Forbindelse med de i Slutningen af forrige Kapitel omtalte Forhold.

i bestemte Retninger, og Indvirkningen af det magnetiske Felt paa de enkelte Elektroners Bevægelser vil derfor kunne vise sig og vil give Anledning til de saakaldte galvanomagnetiske og thermomagnetiske Fænomener.

Vi skal nu søge Betingelsesligninger for Elektronernes samlede Bevægelse, dersom der foruden saadanne ydre Kræfter som dem, der blev betragtet i Kapitel I, ogsaa virker ydre magnetiske Kræfter. Vi skal her kun betragte det Tilfælde, i hvilket det antages, at der finder adskilte Sammenstød Sted. Ligesom i første Kapitel skal vi søge den samlede Bevægelsesmængde, der indehaves af de Elektroner, der befinder sig i Volumenelementet dV og hvis absolute Hastigheder ligger mellem r og $r + dr$. Komposanterne efter Koordinatakserne af denne Bevægelsesmængde skal vi betegne ved $G_x(r)\,dr\,dV$, $G_y(r)\,dr\,dV$ og $G_z(r)\,dr\,dV$.

Vi har i Kapitel I (Side 16—26) beregnet den Tilvækst, som den omhandlede Bevægelsesmængde vil faa i Tiden dt paa Grund af de ydre elektriske Kræfter og paa Grund af Elektronernes Bevægelser, samt paa Grund af Sammenstødene mellem Metalmolekylerne og Elektronerne og mellem disse sidste indbyrdes; vi skal nu her betragte den Tilvækst, der skyldes de magnetiske Kræfter. Denne sidste Tilvækst er meget simpel at beregne; thi da de Kræfter, der stammer fra det magnetiske Felt, staar vinkelret paa de enkelte Elektroners Bevægelsesretninger, vil de ikke forandre Elektronernes absolute Hastigheder, og de vil derfor ikke bevirke, at der gaar Elektroner ud eller ind af den betragtede Gruppe (smlg. Noten Side 19); hele den søgte Tilvækst i den omhandlede Bevægelsesmængde vil derfor være lig Summen af de enkelte Elektroners Tilvækst i Bevægelsesmængde i Tiden dt.

Under det magnetiske Felts Indvirkning vil de enkelte Elektroner paavirkes af Kræfter, hvis Komposanter efter Koordinatakserne, dersom Komposanterne af den magnetiske Kraft (maalt i abs. magnetisk Maal) betegnes ved H_x, H_y og H_z, vil være (vi benytter her et Koordinatsystem, i hvilket Z-aksens positive Retning er valgt saaledes, at Omløbsretningen i XY-planen set fra positive z har modsat Retning af Viserne paa et Uhr)

$$M_x = \frac{\varepsilon}{c} H_z \cdot \eta - \frac{\varepsilon}{c} H_y \cdot \zeta, \quad M_y = \frac{\varepsilon}{c} H_x \cdot \zeta - \frac{\varepsilon}{c} H_z \cdot \xi \quad \text{og} \quad M_z = \frac{\varepsilon}{c} H_y \cdot \xi - \frac{\varepsilon}{c} H_x \cdot \eta,$$

og den Tilvækst, som Bevægelsesmængden $G_x(r)\,dr\,dV$ faar i Tiden dt hidrørende fra de magnetiske Kræfter, bliver derfor

$$\Sigma M_x dt = \left(\frac{\varepsilon}{c} H_z \, \Sigma\eta - \frac{\varepsilon}{c} H_y \cdot \Sigma\zeta \right) dt = \left(\frac{\varepsilon}{cm} H_z \cdot G_y(r) - \frac{\varepsilon}{cm} H_y \cdot G_z(r) \right) dr\,dV\,dt.$$

Ved Hjælp af de ovenfor omtalte Beregninger i Kapitel I faar vi nu som Betingelsesligninger for Elektronernes samlede Bevægelse, naar man ogsaa tager Hensyn til ydre magnetiske Kræfter,

* $$\left[\frac{dG_x(r)}{dt}\right] = -\frac{4\pi}{3}mK\left(\frac{\varepsilon}{kT}\frac{\delta\varphi}{\delta x} + \frac{1}{K}\frac{\delta K}{\delta x} + \frac{mr^2}{2kT^2}\frac{\delta T}{\delta x}\right)r^4 e^{-\frac{mr^2}{kT}} - G_x(r)\,F(r)$$
$$+ \int_0^\infty G_x(\rho)\,Q(\rho,r)\,d\rho + \frac{\varepsilon}{cm}H_z G_y(r) - \frac{\varepsilon}{cm}H_y\,G_z(r), \tag{1}$$

og de to tilsvarende Ligninger, der faas ved cyklisk Ombytning af x, y og z.

Ligningen (1) og de tilsvarende Ligninger tillader under givne Forhold at bestemme Størrelserne $G_x(r)$, $G_y(r)$ og $G_z(r)$, ved Hjælp af hvilke man atter kan beregne den elektriske Strøm og den Varmestrøm, der under givne Forhold til ethvert Tidspunkt vil gaa igennem et Fladeelement i det Indre af Metallet; af Udtrykkene for disse Størrelser vil man dernæst ved en Behandling, der ganske svarer til den, vi har gennemført overfor de thermoelektriske Fænomener, atter kunne beregne de forskellige galvano og thermomagnetiske Virkninger.

Vi skal her kun behandle et enkelt af de omtalte Fænomener, nemlig de galvanomagnetiske Virkninger, der fremkommer under stationære Forhold i et homogent Metalstykke, der overalt har samme Temperatur, og som er udsat for Indvirkningen af et homogent magnetisk Felt; vi skal endvidere kun betragte det simple Tilfælde, i hvilket Metalmolekylerne antages at være faste Kraftcentrer, der paavirker Elektronerne med Kræfter, der forholder sig omvendt som n^{te} Potens af Afstanden, og hvor man ser bort fra Sammenstødene mellem Elektronerne indbyrdes.

Idet vi antager, at den magnetiske Krafts Retning er parallel med Z-aksen, faar vi i det omhandlede Tilfælde til Bestemmelse af $G_x(r)$ og $G_y(r)$, idet her $Q(\rho,r) = 0$, Ligningerne

$$-\frac{4\pi m\varepsilon K}{3kT}\frac{\delta\varphi}{\delta x}r^4 e^{-\frac{mr^2}{2kT}} - G_x(r)\cdot F(r) + \frac{\varepsilon}{cm}H_z\cdot G_y(r) = 0$$
og
$$-\frac{4\pi m\varepsilon K}{3kT}\frac{\delta\varphi}{\delta y}r^4 e^{-\frac{mr^2}{2kT}} - G_y(r)\cdot F(r) - \frac{\varepsilon}{cm}H_z\cdot G_x(r) = 0. \tag{2}$$

Af Ligningerne (2) faar man

$$G_x(r) = -\frac{4\pi m\varepsilon K}{3kT}\frac{\dfrac{\delta\varphi}{\delta x}F(r) + \dfrac{\delta\varphi}{\delta y}\dfrac{\varepsilon}{cm}H_z}{(F(r))^2 + \left(\dfrac{\varepsilon}{cm}H_z\right)^2}r^4 e^{-\frac{mr^2}{2kT}}, \tag{3}$$

* [The exponential factor should read $e^{-mr^2/2kT}$.]

[280]

og et ganske tilsvarende Udtryk for $G_y(r)$, der faas af (3) ved at ombytte x med y og y med $-x$. For den elektriske Strøm gennem en Fladeenhed henholdsvis vinkelret paa X-aksen og paa Y-aksen har vi nu

$$i_x = \frac{\varepsilon}{m}\int_0^\infty G_x(r)\,dr \quad \text{og} \quad i_y = \frac{\varepsilon}{m}\int_0^\infty G_y(r)\,dr\,;$$

dette giver ved Hjælp af (3) og den tilsvarende Ligning

$$i_x = -\sigma_1\frac{\delta\varphi}{\delta x} - \delta_1\frac{\delta\varphi}{\delta y} \quad \text{og} \quad i_y = -\sigma_1\frac{\delta\varphi}{\delta y} + \delta_1\frac{\delta\varphi}{\delta x}, \tag{4}$$

hvor σ_1 og δ_1 bestemmes ved Ligningerne

$$\sigma_1 = \frac{4\pi\varepsilon^2 K}{3kT}\int_0^\infty \frac{F(r)\,r^4 e^{-\frac{mr^2}{2kT}}}{(F(r))^2+\left(\frac{\varepsilon}{cm}H_z\right)^2}\,dr \quad \text{og} \quad \delta_1 = \frac{4\pi\varepsilon^2 K}{3kT}\int_0^\infty \frac{\frac{\varepsilon}{cm}H_z\,r^4 e^{-\frac{mr^2}{2kT}}}{(F(r))^2+\left(\frac{\varepsilon}{cm}H_z\right)^2}\,dr. \tag{5}$$

Ved Beregningen af disse Udtryk skal vi antage, at $\frac{\varepsilon}{cm}H_z\,(F(r))^{-1}$ for alle Værdier af r, der giver et væsentligt Bidrag til Integralerne, vil være en meget lille Størrelse. — Idet $\frac{\varepsilon}{cm}H_z$, som man kan vise ved Hjælp af de ovenstaaende Udtryk for M_x og M_y, vil være den Vinkel, som Projektionen af en Elektrons Hastighed paa XY-planen vil have drejet sig i Tidsenheden under Indflydelse af den magnetiske Kraft, og idet $(F(r))^{-1}$ er lig den Tid τ, i hvilken den oprindelige Bevægelsesmængde i en bestemt Retning af en Elektron med absolut Hastighed r paa Grund af Sammenstødene med Metalmolekylerne i Middel vil aftage til $\frac{1}{e}$ af sin Værdi, vil den fysiske Betydning af den nævnte Antagelse være, at den magnetiske Kraft kun vil have ringe Indflydelse paa de enkelte Elektroners Baner i Sammenligning med de Virkninger, der stammer fra Metalmolekylerne. — Idet vi nu, svarende til det her omhandlede Tilfælde, sætter $F(r) = C \cdot r^{\frac{n-5}{n-1}}$ (se Side 33), faar vi ganske tilsvarende til Udtrykkene for σ_p og δ_p Side 84 og 85, blot med Ombytning af p med $\frac{\varepsilon}{cm}H_z$,

$$\left.\begin{aligned}
\sigma_1 &= \sigma\left(1 - \sigma^2\frac{H_z^2}{N^2\varepsilon^2 c^2}\frac{9\pi}{16}\Gamma\left(\frac{n+5}{n-1}\right)\Gamma^{-3}\left(\frac{2n}{n-1}\right)\right) \\
\delta_1 &= \sigma^2\frac{H_z}{N\varepsilon c}\frac{3\sqrt\pi}{4}\Gamma\left(\frac{3n+5}{2(n-1)}\right)\Gamma^{-2}\left(\frac{2n}{n-1}\right).
\end{aligned}\right\} \tag{6}$$

Man bemærker her den fuldkomne formelle Analogi, der er til Stede mellem Beregningerne af Metallernes Absorption for Varmestraaler og Beregningerne af Metallernes galvanomagnetiske Egenskaber. Den fysiske Betydning af denne Analogi, der vil være til Stede i alle Tilfælde, hvor man betragter adskilte Sammenstød, kan udtrykkes ved, at den Face-forskel, der indtræder mellem den elektriske Kraft og den elektriske Strøm for hurtigt svingende elektriske Felter, og som (se Side 84) udtrykkes ved arc tg δ_p/σ_p, vil være lig den Vinkel, som den elektriske Strøm danner med den elektriske Kraft under Indvirkning af en magnetisk Kraft, der staar vinkelret paa den elektriske Kraft, og som (se Ligning (4)) udtrykkes ved arc tg δ_1/σ_1, dersom Svingningstiden $2\pi/p$ for det elektriske Felt er lig den Tid $2\pi/\dfrac{\varepsilon}{cm} H_z$, som en Elektron bruger om at gennemløbe en Om-drejning i sin skruelinieformede Bane omkring en Akse parallel med den magnetiske Kraft.

Vi skal nu beregne de galvanomagnetiske Virkninger ved Hjælp af Ligningerne (4). Tænker vi os, at det betragtede Metalstykke gen-nemløbes af en elektrisk Strøm i X-aksens Retning, vil der under det magnetiske Felts Indvirkning fremkomme en Potential-Forskel mellem to Punkter af Metallet, der svarer til samme x, men til forskelligt y — *Hall*-Effekt. — Den hertil svarende elektriske Kraft E_y kan findes ved at sætte Udtrykket for i_y lig 0; dette giver

$$E_y = -\frac{\partial\varphi}{\partial y} = -\frac{\delta_1}{\sigma_1}\frac{\partial\varphi}{\partial x}.$$

Indsættes dette i Udtrykket for i_x, faar man

$$i_x = -\sigma_1\left(1 + \left(\frac{\delta_1}{\sigma_1}\right)^2\right)\frac{\partial\varphi}{\partial x} \quad \text{og} \quad E_y = \frac{\delta_1}{\sigma_1^2 + \delta_1^2} i_x. \tag{7}$$

Vi ser af Ligningerne (7), hvorledes det magnetiske Felt foruden den omtalte *Hall*-Effekt ogsaa bevirker en Forandring i den elek-triske Ledningsevne σ, idet denne efter Lign. (7) under Indvirkning af den magnetiske Kraft bliver

$$\sigma_H = \sigma_1\left(1 + \left(\frac{\delta_1}{\sigma_1}\right)^2\right). \tag{8}$$

Indsætter vi nu Værdierne for σ_1 og δ_1, givne ved Ligningerne (6), i (7) og (8), faar vi med den ovenfor benyttede Tilnærmelse

$$E_y = i_x \frac{H_z}{N\varepsilon c} \frac{3\sqrt{\pi}}{4} \Gamma\left(\frac{3n+5}{2(n-1)}\right) \Gamma^{-2}\left(\frac{2n}{n-1}\right)$$

og

$$\sigma_H = \sigma\left(1 - \sigma^2 \frac{H_z^2}{N^2\varepsilon^2 c^2} \frac{9\pi}{16}\left(\Gamma\left(\frac{n+5}{n-1}\right)\Gamma\left(\frac{2n}{n-1}\right) - \Gamma^2\left(\frac{3n+5}{2(n-1)}\right)\right)\Gamma^{-4}\left(\frac{2n}{n-1}\right)\right). \tag{9}$$

* ["Face-" should read "Fase-".]

Vi skal endnu her undersøge den Energimængde W_y, der med Elektronerne føres igennem en Fladeenhed vinkelret paa Y-aksen. Idet vi har

$$W_y = \frac{1}{2} \int_0^\infty r^2\, G_y(r)\, dr,$$

faar vi af Ligningen (3), idet vi indfører $F(r) = C \cdot r^{\frac{n-5}{n-1}}$ og bortkaster smaa Størrelser af højere Orden med Hensyn til $\frac{\varepsilon}{cm} H_z\, (F(r))^{-1}$, med Benyttelse af Ligningerne (5)

$$W_y = -\frac{2n}{n-1}\frac{k}{\varepsilon} T \sigma_1 \frac{\partial \varphi}{\partial y} + \frac{3n+5}{2(n-1)}\frac{k}{\varepsilon} T \sigma_1 \frac{\partial \varphi}{\partial x};$$

idet $i_y = 0$, faar vi ved Hjælp af (4)

$$W_y = \frac{n-5}{2(n-1)}\frac{k}{\varepsilon} T \sigma_1 E_y.$$

Vi har hidtil antaget, at Temperaturen overalt i Metalstykket er den samme; dersom imidlertid Metalstykket er afskaaret fra ydre Varmetilførsel, vil der fremkomme Temperaturforskelle i det, indtil den Varmeledning, som disse vil fremkalde, holder Ligevægt med den oven for omtalte Energistrømning. Med den samme Tilnærmelse, som vi har benyttet ved Beregningen af W_y, vil Varmeledningsevnen γ være uafhængig af den magnetiske Kraft, og idet man endvidere med samme Tilnærmelse kan sætte σ_1 lig σ, faar vi for Temperaturvariationen efter Y-aksen i det her omhandlede Tilfælde

$$\frac{\partial T}{\partial y} = \frac{W_y}{\gamma} = \frac{n-5}{2(n-1)}\frac{k}{\varepsilon} T \frac{\sigma}{\gamma} E_y$$

og heraf ved Benyttelse af Ligningen (13) Side 53

$$\frac{\partial T}{\partial y} = \frac{n-5}{4n}\frac{\varepsilon}{k} E_y. \tag{10}$$

[Angaaende Ligningen (10) maa vi bemærke, at E_y betegner den ovenfor omhandlede elektriske Kraft efter Y-aksen, der vil fremkomme i et Metalstykke, i hvilket Temperaturen overalt er konstant. Dersom denne sidste Antagelse, saaledes som i det her betragtede Tilfælde, ikke er opfyldt, vil den elektriske Kraft efter Y-aksen blive forandret paa Grund af de Spændingsforskelle, der fremkommer som Følge af Temperaturdifferenserne, og som vi har omtalt i Kapitel II ved Behandlingen af de thermoelektriske Forhold. Idet disse Spændingsforskelle med første Til-

8

nærmelse vil være uafhængige af den magnetiske Kraft, faar vi, ved Hjælp af Ligning (21) Side 66, for den elektriske Kraft efter Y-aksen i det her omhandlede Tilfælde

$$E'_y = E_y + \frac{k}{\varepsilon}\left(\frac{T}{N}\frac{dN}{dT} + \frac{n+3}{2(n-1)}\right)\frac{\delta T}{\delta y},$$

og heraf ved Hjælp af Lign. (10)

$$E'_y = E_y\left(\frac{n-5}{4n}\frac{T}{N}\frac{dN}{dT} + \frac{9n^2-10n-15}{8n(n-1)}\right). \qquad (11)$$

Den her beregnede Spændingsdifferens er dog ikke (smlg. Side 64) saaledes som den ovenfor betragtede Størrelse E_y direkte tilgængelig for Maalinger.]

Vi skal nu anvende de her udviklede Formler til Undersøgelse af nogle specielle Tilfælde.

1) Dersom vi i Ligningerne (9) og (10) sætter $n = \infty$, faar vi gældende for det Tilfælde, at Metalmolekylerne antages at være **elastiske Kugler**,

$$E_y = i_x\frac{3\pi}{8}\frac{H_z}{N\varepsilon c}, \quad \sigma_H = \sigma\left(1 - \frac{9\pi(4-\pi)}{64}\frac{\sigma_2 H_z^2}{N^2\varepsilon^2 c^2}\right) \text{ og } \frac{\delta T}{\delta y} = i_x\frac{3\pi}{32}\frac{H_z}{Nkc}. \quad (12)$$

Disse Formler (saavel som den Værdi for den »adiabatiske *Hall*-Effekt«, som vi faar ved i (11) at sætte $n = \infty$) er beregnede af *R. Gans*[1]), der har givet en meget fuldstændig Behandling af de galvano- og thermomagnetiske Fænomener ud fra de samme Antagelser om Virkningerne af Sammenstødene mellem Metalmolekylerne og Elektronerne som dem, der er lagt til Grund i *H. A. Lorentz's* Theori.

Idet vi her er gaaet ud fra, at der kun findes én bestemt Slags Elektroner (med negativ elektrisk Ladning ε), viser den første af Ligningerne (12), at, dersom den elektriske Strøm har samme Retning som X-aksens positive Retning, og dersom den magnetiske Kraft har samme Retning som Z-aksens positive Retning, vil der fremkomme en elektrisk Kraft efter Y-aksen, der vil have samme Retning som sidstnævnte Akses negative Retning. Endvidere viser Ligningerne (12) en Formindskelse af den elektriske Ledningsevne ($\sigma_H < \sigma$) under Indflydelse af den magnetiske Kraft.

De her omtalte Resultater stemmer imidlertid ikke overens med de eksperimentelt fundne[2]), idet man saavel finder Metaller, der som Vismut besidder en *Hall*-Effekt, hvis Retning stemmer overens med

[1]) *R. Gans*: Ann. d. Phys., Bd. 20, p. 293, 1906.
[2]) Angaaende Literaturen over de eksperimentelle Undersøgelser vedrørende disse Spørgsmaal se f. Eks. *H. Zahn*: Jahrb. d. Rad. u. Elek., Bd. 5, p. 166, 1908.

* [In the parenthesis σ_2 should be σ^2.]

den ovenfor beregnede, og hvis elektriske Ledningsevne aftager i et magnetisk Felt, som andre, der som Jern besidder en *Hall*-Effekt med modsat Retning af den ovenfor beregnede, og som viser en Forøgelse af den elektriske Ledningsevne i et magnetisk Felt.

Det maa her fremhæves, at den omtalte Uoverensstemmelse mellem de beregnede og de eksperimentelt fundne Resultater ikke er en Følge af de her gjorte specielle Antagelser (de *Lorentz'ske*), idet de ovenfor beregnede Resultater (*Hall*-Effektens Fortegn og den elektriske Ledningsevnes Formindskelse ($\sigma_H \leq \sigma$)) vil gælde i alle Tilfælde, der falder ind under de i Indledningen omtalte Hovedantagelser, d. v. s. i alle Tilfælde, hvor man antager, at Metalmolekylernes Egenskaber i Middel er ens i alle Retninger, ogsaa naar der virker ydre Kræfter, saaledes at Virkningen af disse sidste Kræfter kan beregnes ud fra deres direkte Indvirkning paa de fri Elektroners Bevægelser. For at forklare de galvanomagnetiske Virkninger i Overensstemmelse med Erfaringen ved Hjælp af en Theori, der bygger paa Forestillingen om ensartede Partiklers (Elektroners) Bevægelse i Metallernes Indre, tvinges man derfor, i hvert Tilfælde for visse Metallers Vedkommende, til at antage, at de ydre magnetiske Kræfter ikke blot direkte paavirker de fri Elektroners Bevægelser, men ogsaa indirekte derigennem, at de udøver en polariserende Virkning paa Metalmolekylerne. En saadan Antagelse synes saa meget mere naturlig, som der bestaar en meget nøje Sammenhæng mellem Metallernes magnetiske Egenskaber, der jo forklares ved at antage en saadan polariserende Virkning, og de her omhandlede Forhold; Størrelsen af *Hall*-Effekten er saaledes for de ferro-magnetiske Metallers Vedkommende ikke proportional med den magnetiske Kraft, men med Magnetiseringen.

Bortset fra den her omtalte Uoverensstemmelse mellem Theorien og Erfaringen, der viser sig ved, at den observerede *Hall*-Effekt ikke altid har samme Fortegn som den beregnede, frembyder der sig for en Theori, der bygger paa de specielle Antagelser, som *Lorentz* har lagt til Grund, yderligere en principiel Vanskelighed ved Forklaringen af den galvanomagnetiske transversale Temperaturdifferens. Denne Temperaturdifferens har nemlig ved alle de Metaller, for hvilke *Hall*-Effektens Fortegn stemmer med det ovenfor beregnede, og som viser en Aftagen af den elektriske Ledningsevne i et magnetisk Felt — og hvor altsaa disse Virkninger i hvert Tilfælde kvalitativt kan forklares ved en Elektrontheori af den her omhandlede Art —, et Fortegn, der er modsat det, der er givet ved Ligningerne (12)[1].

---

[1] Se *Gans*: loc. cit., p. 307--308; jvnfr. ogsaa *Gruner*: Arch. d. sciences phys. et nat. (4), Tom. 28, p. 607—608, 1909.

8*

som haarde Legemer, men derimod med Kræfter, der varierer paa passende Maade med Afstanden.

2) Dersom vi sætter $n = 5$ — hvilket vil svare til den Antagelse at Metalmolekylerne paavirker Elektronerne som Elementarmagneter (se Side 33) —, faar vi af Ligningerne (9) og (10)

$$E_y = \frac{i_x H_z}{N \varepsilon c}, \quad \sigma_H = \sigma \quad \text{og} \quad \frac{dT}{dy} = 0. \tag{13}$$

Disse Formler er dem, som man vilde finde, dersom man tænkte sig, at alle Elektronerne havde samme Hastighed efter X-aksen (ide de f. Eks. tænktes at bevæge sig i et kontinuert Medium med indre Gnidning); idet $N\varepsilon$ angiver den elektriske Ladning af de fri Elektrone i Volumenenheden, vil nemlig i et saadant Tilfælde $\frac{i_x}{N\varepsilon}$ være Elektro nernes fælles Hastighed; idet endvidere den elektriske Kraft E_y i e saadant Tilfælde for hver enkelt Elektrons Vedkommende fuldstændig vil ophæve de afbøjende Virkninger af den magnetiske Kraft, vil Til stedeværelsen af et magnetisk Felt slet ingen Indflydelse have paa Ledningsevnen eller give Anledning til nogen Temperaturvariation. — At en saadan Beregningsmaade vil føre til et rigtigt Resultat i de Tilfælde, hvor $n = 5$, hidrører fra, at Elektronernes samlede Bevægelses mængde efter X-aksen i dette Tilfælde vil være ligelig fordelt mellen Elektronerne med de forskellige absolute Værdier for deres relative Hastighed i Forhold til Middelhastigheden efter X-aksen. —

Det kan her bemærkes, at i alle andre Tilfælde, hvor man be tragter adskilte Sammenstød og ikke tager Hensyn til Vekselvirkninge mellem Elektronerne indbyrdes, vil, saaledes som det specielt ses ve det foregaaende og det efterfølgende Eksempel, E_y være større en den i dette specielle Tilfælde fundne Værdi $\frac{i_x H_z}{N\varepsilon c}$, samt at Formind skelsen af den elektriske Ledningsevne her antager sin mindste Værdi nemlig 0.

3) Dersom vi sætter $n = 3$ — hvilket svarer til at antage at Metalmolekylerne er elektriske Dobbeltsystemer som dem der er omtalt i Kapitel I Side 34 —, faar vi af Ligningerne (9 og (10)

$$E_y = i_x \frac{45\pi}{128} \frac{H_z}{N\varepsilon c}, \quad \sigma_H = \sigma \left(1 - \frac{27\pi (256 - 75\pi)}{16384} \frac{\sigma^2 H^2}{N^2 \varepsilon^2 c^2} \right)$$

og

$$\frac{dT}{dy} = -i_x \frac{15\pi}{256} \frac{H_z}{Nkc}.$$

(14)

*

Vi ser, at i dette Tilfælde har Værdien for den galvanomagnetiske transversale Temperaturdifferens modsat Fortegn af den udfra de *Lorentz'ske* Antagelser beregnede Værdi, altsaa (smlg. ovenfor) en Værdi, der i Fortegn stemmer overens med de eksperimentelt fundne. Vi ser saaledes gennem dette Eksempel, hvorledes man, for ved de her behandlede Fænomener at bringe Overensstemmelse med Erfaringen til Veje, maa antage, at Vekselvirkningen mellem Metalmolekylerne og Elektronerne ikke foregaar som mellem haarde Kugler, men at Paavirkningen fra Metalmolekylernes Side derimod sker gennem langsommere varierende Kraftfelter; dette er i udmærket Overensstemmelse med, hvad vi i det foregaaende har set ved Omtalen af Forholdet mellem Elektricitets- og Varmeledningsevnen, hvor vi for at bringe Overensstemmelse til Veje mellem de beregnede og de eksperimentelt fundne Værdier netop ogsaa maatte antage Tilstedeværelsen af saadanne Kræfter mellem Metalmolekylerne og Elektronerne (f. Eks. fandtes for de rene Metallers Vedkommende udmærket Overensstemmelse mellem den beregnede og den eksperimentelt fundne Værdi for Forholdet mellem Ledningsevnerne ved at sætte $n = 3$ (se Side 58)).

Det maa sluttelig bemærkes, at man gennem tilsvarende Betragtninger som dem, vi benyttede i § 4 i 1ste Kapitel, kan indse, at den ovenstaaende Paavisning af Nødvendigheden af at antage, at Metalmolekylerne paavirker Elektronerne gennem kontinuerte Kraftfelter, ikke alene vil gælde, naar man antager, at der finder adskilte Sammenstød Sted, men ogsaa for saadanne Tilfælde, hvor Metalmolekylernes Dimensioner ikke antages at være forsvindende i Forhold til deres indbyrdes Afstande. Man kan nemlig vise, at dersom man antager, at man kan se bort fra Vekselvirkningen mellem Elektronerne indbyrdes i Forhold til Vekselvirkningen mellem disse sidste og Metalmolekylerne og betragte Metalmolekylernes Kraftfelter som stationære overfor Elektronernes Bevægelser — Antagelser, der som omtalt Side 38 med Tilnærmelse kan ventes opfyldte ved de virkelige Metaller —, vil Forholdet $\frac{\delta T}{\delta y} : E_y$, dersom man antager, at Molekylerne er haarde elastiske Legemer, være lig det,

* [In the copy of the dissertation selected for reproduction, a parenthesis wrongly printed between the two fractional factors has been erased. Moreover, H^2 should be H_z^2.]

der angives ved Ligningerne (12), nemlig $\varepsilon/4k$ (Udtrykkene for de en-kelte Størrelser vil derimod i Almindelighed ikke være de samme), medens dette Forhold, dersom man antager Tilstedeværelsen af pas-sende Kræfter mellem Molekylerne og Elektronerne, vil kunne faa samme Fortegn som det, der angives ved Ligningerne (14) (smlg. det analoge Tilfælde ved Omtalen af Forholdet mellem Ledningsevnerne for Varme og Elektricitet (Side 58)).

THESES.

Ved Hjælp af den i denne Afhandling udledte Ligning (14) Side 26 og saadanne Udtryk for $F(r)$ og $Q(\rho, r)$ som dem, der er givne ved Ligningerne (18) Side 36, kan man med enhver ønsket Tilnærmelse numerisk beregne Forholdet mellem Varmeledningskoefficienten og Koefficienten for Interdiffusion i sig selv for en Luftart, hvis Molekyler antages at paavirke hinanden som haarde elastiske Kugler, en Beregning, som det ikke har været muligt at udføre ved Hjælp af de Metoder, der sædvanlig anvendes overfor disse Problemer.

Det synes, som om Værdien for en Vandoverflades Spænding kun i meget ringe Grad forandrer sig i den første Tid efter Overfladens Dannelse, og at derfor Værdien for den omhandlede Spænding — i Modsætning til en almindelig udbredt Mening — kan lade sig bestemme med stor Nøjagtighed ved Hjælp af statiske Metoder.

Medens der paa Overfladen af en usammentrykkelig Vædske uden indre Gnidning, der tænkes underkastet enten alene Tyngdekraften eller alene Virkningen af en Overfladespænding, svarende til enhver Værdi for Bølgelængden, vil kunne eksistere rent periodiske Bølger, der under deres fremadskridende Bevægelse bevarer deres Form uforandret, er dette, som man kan vise, derimod ikke Tilfældet, dersom Vædsken tænkes underkastet saavel Tyngdekraftens som en Overfladespændings samtidige Indvirkning, idet der da, svarende til visse Værdier (i uendeligt Antal) for Bølgelængden, ikke vil kunne eksistere saadanne under deres Fremadskriden uforandrede Bølger. Dette staar i nøje Forbindelse med den Omstændighed, at medens Bølgernes Forplantningshastighed i de to første Tilfælde vil henholdsvis stadig vokse og stadig aftage med Bølgelængden, vil der i det sidste Tilfælde eksistere en Minimumshastighed, svarende til en bestemt Værdi for Bølgelængden.

Poincaré (Thermodynamique, 2ième édition, Paris 1908, p. 384) udtaler den Mening, at Lord *Kelvin's* Theori for de thermoelektriske Fænomener kan begrundes strengt ud fra de thermodynamiske Principper alene, idet Problemet kan behandles paa en saadan Maade, at man kan se bort fra Varmeledningen. Dette er imidlertid i Modstrid med Resultatet af *Boltzmann's* Undersøgelser (Sitzungsb. d. Wiener Akad. d. Wiss., math.-nat. Kl., Bd. 96, Abt. II, p. 1258, 1887). Urigtigheden i *Poincaré's* Bevisførelse hidrører fra, at han i sine Beregninger (loc. cit., p. 385—386) ikke tager Hensyn til den Entropitilvækst, der finder Sted ved Varmeledningen, og som hidrører fra, at den Varmemængde, der ledes igennem et Element af en thermoelektrisk Kæde, ledes ind i dette ved højere Temperatur og ud deraf ved lavere.

Som berørt i Afhandlingen har man opstillet en Theori for at forklare forskellige af Legeringernes Forhold, efter hvilken Legeringerne, i Modsætning til de rene Metaller, ikke skulde kunne betragtes som fysisk homogene, og derfor en væsentlig Del af Legeringernes elektriske Modstandsevne skulde hidrøre fra en thermoelektrisk Polarisation. En saadan Theori synes som omtalt uholdbar. Naar imidlertid *Lederer* (Sitzungsb. d. Wiener Akad. d. Wiss., math.-nat. Kl., Bd. 117, Abt. II a, p. 311, 1908) mener at have modbevist Theorien derigennem, at han eksperimentelt har paavist, at den i en Del af et elektrisk Kredsløb udviklede Varme kan beregnes i Overensstemmelse med Erfaringerne ved Hjælp af *Joule's* Lov udfra Styrken af den elektriske Strøm og Størrelsen af den elektriske Modstand, hvad enten den betragtede Del af Kredsløbet er dannet af en Legering eller et Stykke rent Metal, er dette dog ikke rigtigt; en Overensstemmelse som den omtalte maatte nemlig paa Grund af Energisætningen være til Stede, hvad enten Legeringernes Modstand skyldtes en thermoelektrisk Polarisation eller ej.

Det synes ikke muligt, paa Elektrontheoriens nuværende Standpunkt, udfra denne Theori at forklare Legemernes magnetiske Egenskaber.

STUDIES ON THE ELECTRON THEORY OF METALS

DISSERTATION FOR THE DEGREE OF DOCTOR OF PHILOSOPHY

by

NIELS BOHR

Translated by

J. RUD NIELSEN

TRANSLATOR'S PREFACE

When his dissertation *Studier over Metallernes Elektrontheori* had been printed and defended in the spring of 1911, Bohr had a copy bound in half-leather wih blank sheets between consecutive pages. On these sheets he afterwards indicated, in ink or pencil, a number of intended or considered revisions of the text. Some sentences or paragraphs were to be rewritten, a few paragraphs were to be added, and others were marked to be deleted. These annotations were, for the most part, made during the summer of 1911 when Bohr, with the help of a friend, Carl Christian Lautrup, prepared an English translation to take along to Cambridge in September of that year. Most of the changes indicated in ink were incorporated in this translation, and a number of other changes of argument or emphasis were made. However, as mentioned in the Introduction (p. [103]), the translation was linguistically very imperfect.

The translation given here conforms for the most part to the original Danish text, with the revisions indicated in ink in the leather-bound copy. The pencilled notes, many of which are now illegible, have generally been ignored. When the minor changes made in the 1911 English version appear to be intended as improvements, rather than caused by language difficulties, they have been considered.

The aim has been to prepare a translation such as Bohr would have liked to have in 1911. Care has been taken not to introduce concepts or terms developed since that time. However, the word "Metalmolekyle", which Bohr uses throughout except in the first paragraph of his dissertation, has been translated as "metal atom", or simply "atom", rather than as "metal molecule". In the Danish text the formulae have a separate numeration in each chapter; it was found more convenient, in the translation, to number them consecutively throughout, since this makes references to formulae somewhat simpler.

Places where appreciable changes from the original Danish text occur have been marked by asterisks, and the nature of these changes is indicated in footnotes*.

---

* In these footnotes the Danish text will be referred to as I, while the notes written in ink on the blank sheets in the leather-bound copy of the dissertation will be collectively denoted by II, and the English translation prepared by Bohr and Lautrup in 1911 will be referred to as III. The asterisks in the present English translation are placed at the end of paragraphs that have been revised, or at places where paragraphs have been deleted or new paragraphs inserted.

This dissertation is accepted by the faculty of mathematical and natural sciences to be defended for the degree of doctor of philosophy.

Copenhagen, April 12, 1911.

Elis Strömgren, Dean

DEDICATED IN DEEPEST GRATITUDE TO THE
MEMORY OF MY FATHER

LIST OF CONTENTS

INTRODUCTION

Among the substances which conduct electricity, metals occupy a special position, not only because of their high conductivity but also in that the passage of electricity through them is not accompanied by observable transport of chemical matter, as in the case of most other good conductors. According to the generally accepted view, this is explained by assuming that certain small electrically charged particles – the electrons – are able to move from one chemical atom to another in the interior of metals.

The first detailed treatment of such an idea about the conditions in metals is due to W. Weber[1]. According to Weber's theory, each metal atom is a system of electrically charged particles moving in curved paths around one another. These systems, however, are not assumed to be stable, but at short intervals some of the particles will leave the system and move through the metal in approximately straight paths, until they are captured again by other metal atoms; in these, they will then for a time take part in intra-atomic motions, thereafter to be sent out again, etc. As Weber, furthermore, assumes that the kinetic energy of these particles forms an essential part of the thermal energy content of the metal, his theory offers a possibility of explaining the close connection existing between the conductivities of metals for heat and electricity. It may also provide a mechanical basis for the transport theory of the thermoelectric phenomena proposed by F. Kohlrausch[2], according to which these phenomena are explained on the assumption that any electric current is accompanied by a certain heat flow, proportional to the current and dependent on the nature and temperature of the metal, and, conversely, that any flow of heat is accompanied by a certain electric current.

From assumptions corresponding to Weber's ideas about the conditions in the interior of metals, E. Riecke[3] has later worked out a detailed theory of metallic conduction in which explicit expressions are deduced for the conductivities for electricity and heat, as well as for the thermoelectric and galvano- and thermomagnetic coefficients. These expressions involve the number of particles emitted in unit time by the metallic atoms present in unit volume, the electric charges of these particles, their masses, the velocities with which they are emitted, and, finally, the mean free path they travel before being captured by another atom; however, since no assumptions about the values of these quantities are introduced, Riecke's theory permits only to a rather small extent a comparison with experimental results.

[1] See, for instance, W. Weber: Pogg. Ann. **156** (1875) 1.
[2] F. Kohlrausch: Nachr. d. Kgl. Ges. d. Wiss. zu Göttingen (1874) 65.
[3] E. Riecke: Wied. Ann. **66** (1898) 353, 545, 1199.

A major advance in the electron theory of metals is due to P. Drude[1] who applied the principal results of the kinetic theory of gases to the motions of the free electric particles in metals. Thus, in the kinetic theory of gases it is shown that a collection of particles in mechanical equilibrium with the surroundings will have such velocities that the translatory kinetic energy of a particle is equal, on the average, to the mean translatory kinetic energy of a molecule of any gas at the same temperature. Using this theorem about the average energy of the particles, and assuming that their electric charges are equal to that of a monovalent ion in an electrolyte, Drude has calculated the ratio between the thermal and electric conductivities of metals in close agreement with the experimental values[2].

Drude's calculations, however, are not perfectly rigorous. Thus, he assumes, for simplicity, that all the particles have the same absolute velocity, while it must be assumed, according to the kinetic theory of gases, that particles of an assemblage in mechanical and thermal equilibrium, as envisaged by Drude, will have different absolute velocities, distributed in accordance with the so-called Maxwell's distribution law. An exact theory, based on clearly stated assumptions, was first given by H. A. Lorentz[3].

While Riecke and Drude assume the simultaneous presence in metals of different kinds of free particles, positive as well as negative, Lorentz assumes the existence of only a single type of free particle, the same in all metals. Such an assumption was first introduced in the electron theory of metals by J. J. Thomson[4]. This author, who has contributed so much to the experimental foundation of the electron theory, bases his assumption on the circumstance that, while the existence of a distinct kind of negatively charged particles—the so-called electrons, whose masses are very small compared to the masses of the chemical atoms—has been demonstrated in a wide variety of ways, positive electricity has only been observed associated with masses of the same order of magnitude as those of the chemical atoms. Moreover, Lorentz has shown that difficulties of a fundamental nature arise if a stationary distribution of several kinds of free particles is assumed to exist in a piece of metal whose state, i. e., temperature and chemical composition, is not the same at all points[5].

Lorentz' theory is based on the following mechanical picture. In the interior of metals both atoms and free electrons are assumed to be present. The dimensions of the atoms and the electrons, i.e., the ranges within which they affect each other appreciably, are assumed to be very small compared to their average mutual

[1] P. Drude: Ann. d. Phys. 1 (1900) 566; 3 (1900) 369.
[2] See M. Reinganum: Ann. d. Phys. 2 (1900) 398.
[3] H. A. Lorentz: Proc. Acad. Amsterdam 7 (1905) 438, 585, 684.
[4] J. J. Thomson: *Rapp. du Congrès de Physique* (Paris 1900), vol. 3, p. 138.
[5] H. A. Lorentz: Proc. Acad. Amsterdam 7 (1905) 684; Jahrb. d. Rad. u. El. 4 (1907) 125.

distances; thus, they are thought to interact only in separate collisions, in which they behave as hard elastic spheres. Moreover, the dimensions and masses of electrons are thought to be so much smaller than those of atoms that collisions among the free electrons can be neglected compared to collisions between electrons and atoms, and that the atoms can be regarded as immovable in comparison to the electrons.

From these assumptions, Lorentz has not only calculated expressions for the conductivities of a metal for electricity and heat, and for the thermoelectric phenomena, but also for the emissivity and absorptivity of heat rays with long periods of vibration. As some of the most interesting results of Lorentz' theory, in addition to the approximate agreement of the calculated and observed values for the ratio between the thermal and electrical conductivities of metals, first found by Drude, may be mentioned the perfect agreement of the results for the thermoelectric phenomena, and those connected with heat radiation, with the thermodynamical theories for these phenomena given by Lord Kelvin and Planck, respectively. This agreement, found only when the calculations are rigorous, and when Maxwell's distribution law is taken into account, is the more remarkable, as it could not be expected beforehand, on account of the doubtful assumptions underlying these latter theories.

However, while Lorentz' theory is mathematically very perfect, the physical assumptions on which it is based can hardly be expected to be valid, even approximately, for actual metals. Moreover, on many essential points the agreement between the theory and the experimental results is unsatisfactory. It would, therefore, be of interest to develop the electron theory of metals from more general assumptions, and to investigate which results of the theory are connected with the special assumptions, and which results remain unchanged when more general assumptions are adopted. In particular, it would be of interest to see what happens to the agreement with the thermodynamical theories.

An attempt to work out such a theory for the problems of heat radiation has been made by J. H. Jeans[1]. As we shall see, however, his calculations are not correct in general, since he uses assumptions which are valid only in certain very special cases, and not even in that treated by Lorentz[2].

The aim of the present work is to attempt to carry out the calculations for the various phenomena that are explained by the presence of free electrons in metals

[1] J. H. Jeans: Phil Mag. **17** (1909) 773; **18** (1909) 209.
[2] H. A. Wilson, in a recently published paper (Phil Mag. **20** (1910) 835), has called attention to the incorrectness of Jeans' calculations and has attempted to carry out a more correct calculation for a single special case. (However, as we shall see in the following, Wilson's calculations are also incorrect.)

in as great generality as possible, while retaining the fundamental points of view underlying the theory of Lorentz.

Thus, in the following we shall assume *that free electrons are present in any piece of metal, their number depending on the nature and temperature of the metal, but their kind being the same in all metals.*

We shall further assume *that mechanical heat equilibrium will exist between the free electrons and the atoms in a homogeneous piece of metal of uniform temperature and not subjected to external forces.* By mechanical heat equilibrium we shall here understand such a dynamical statistical equilibrium which will occur if the atoms and electrons act upon each other with forces of the same kind as those considered in ordinary mechanics, i.e., if the motions satisfy Hamilton's equations.

This assumption cannot be considered obvious beforehand; in fact, experiments on the specific heat of bodies show that thermal equilibrium as that mentioned cannot occur everywhere, in particular, not among the electrons supposed to be bound in great numbers inside the individual atoms. Furthermore, it must be assumed that the Maxwell-Lorentz equations for the electro-magnetic phenomena are not strictly satisfied, for Lord Rayleigh's and Jeans' investigations on heat radiation have shown (see Chapter III) that, in an electromagnetic system for which these equations hold rigorously, statistical equilibrium in which the individual electrons possess kinetic energy cannot exist. Without entering further into these questions, we shall simply assume that among free electrons (contrary to those bound in atoms) having velocities small compared to the speed of light a statistical equilibrium as that envisaged above will occur. We shall assume also that the motion of these electrons can be calculated, on the average, as the motion of material particles with constant masses, subject to forces obeying the ordinary electro-magnetic laws*.

In addition to these general assumptions, we shall suppose in the following that *the properties of the individual atoms of the metal are, on the average, the same in all directions, and that this isotropy will remain, independently of the presence of external forces*, so that the effect of the external forces is explained by their direct influence upon the motions of the free electrons. This assumption, which is fundamental in the theories of Drude and Lorentz, makes the treatment given here differ very essentially from theories of metals such as those proposed by W. Sutherland[1] and J. J. Thomson[2]. In fact, according to these theories, which do not assume the continual presence of free electrons in the metal, as does Lorentz'

---

* [This paragraph from II replaces the following paragraph of I: "Den omtalte Antagelse . . . Legemernes magnetiske Forhold".]

[1] W. Sutherland: Phil. Mag. **7** (1904) 423.
[2] J. J. Thomson: *The corpuscular theory of matter* (London 1907), p. 86.

theory, the effect of the external forces is explained by their influence upon the individual atoms with their bound electrons, considered as integral systems, and a motion of the electrons through the metal takes place, because, as a result of the motion of the entire atom, free electrons are emitted in directions that depend, on the average, upon the directions of the external forces.

In the following investigation two distinct cases are considered. Either we assume, as in the special case treated by Lorentz, that the motions of the electrons consist of free paths and collisions with the atoms of the metal, or we suppose that the atoms are so close to one another that the electrons are subject to strong forces from them during a large part of their motion. In the first of these cases it has been possible to carry out the investigation of certain problems in perfect generality, on the basis of the above-mentioned fundamental assumptions. In the second case, on the other hand, the treatment is much more difficult, and it has not been possible to carry out the investigation in as great generality; it has been possible, nevertheless, to treat the problems by introducing assumptions that appear to correspond approximately to the actual conditions in metals.

The main purpose of the present work has been to draw the widest possible consequences of a theory of metals based on the general assumptions stated above. References to experimental results are given, therefore, only in cases where they are of special importance to the theoretical viewpoints under consideration.

At the conclusion of this work, I wish to ask my teacher at the University, Professor Christiansen, to accept my best thanks for his valuable guidance during my studies and for the kind interest he has always shown me.

Copenhagen, April, 1911.
Niels Bohr.

CHAPTER I

DERIVATION OF EQUATIONS FOR THE COLLECTIVE MOTION OF THE ELECTRONS IN A METAL

§ 1

The General Assumptions of the Electron Theory about the Conditions in the Interior of Metals

In the interior of a piece of metal we shall assume the existence of a large number of free electrons moving with high velocities in all directions. These electrons are thought to change their direction of motion and speed continually, on account of the forces exerted upon them by the atoms and by other electrons. In a homogeneous piece of metal of uniform temperature, not exposed to external forces, the velocities of the electrons will be evenly distributed in all directions. In the presence of an external force, on the other hand, the motion of the electrons will be altered, the path of each electron being affected in the direction of the force. However, since the external forces to be considered are very small compared to the large forces exerted on the electrons by the atoms of the metal, they will only have a very small influence upon the character of the individual electron paths, and the distribution of velocities among the electrons will, at each place, deviate very little from the normal distribution, i.e., from the distribution that would exist in the absence of external forces. Also, the external forces cannot produce appreciable changes in the concentration of electrons in the interior of the metal (such as the concentration gradient produced by gravity in a vessel containing a gas), for changes in concentration large enough to have an appreciable effect upon the motions of the electrons would, because of the enormous number of electrons, create large free electric charges, giving rise to forces tending to destroy the differences in concentration that would be very large compared to the original external forces*.

As a consequence of the external force, however, more electrons, on the average, will move in the direction of the force than in the opposite direction and, since the electrons carry electric charges, a current of electricity will appear in the direction of the force. Moreover, since the electrons possess kinetic energy because of their random motion, the external force, in producing a flow of electrons, will also cause a flow of energy, or heat, through the metal**.

If the temperature of a piece of metal is not the same everywhere, the mean value

---

* [In accordance with II, the preceding sentence has been added.]
** [In accordance with II, a paragraph, printed in small type in I, has been deleted.]

of the speed of the electrons will vary from place to place, being greater where the temperature is higher, and a flow of fast electrons in the direction of decreasing temperature, and of slow electrons in the opposite direction, will take place. Since the faster electrons carry more kinetic energy than the slower electrons, a difference in temperature will, also when no electric current passes through the metal, produce a flow of energy from places of higher to places of lower temperature. Furthermore, a difference in temperature may produce an electric current even in the absence of external forces, partly because electrons of different speeds travel at different rates through the metal, and partly because the number of free electrons per unit volume must be assumed to depend on the temperature*. As in the presence of external forces, the motions of the electrons will, also in this case, deviate very little from their normal motions, since the distances over which the mean speed and concentration of the electrons vary appreciably as a result of the temperature variation will be very large compared to the distances within which the velocities of the individual electrons are altered greatly by the forces exerted on them by the atoms of the metal and the other electrons**.

Finally, in case the metal is not homogeneous, i.e., when its chemical composition is not the same at all points, we shall again assume that the mean speed of the electrons is the same everywhere, provided the temperature is the same; however, in this case the number of free electrons per unit volume will generally not be the same at different places. Hence, also in this case, a directed flow of electrons will occur, resulting in currents of electricity and energy***. As in the former case, we shall also here assume that the state of the metal, e. g., the number of free electrons per unit volume, varies only very little over the small distance within which the motions of the individual electrons have undergone very large changes.

In this chapter we shall now derive equations for determining the amounts of electricity and energy carried by the motions of electrons through a surface element in the interior of the metal when a given external force is applied and the temperature and chemical composition are known at every point. With the aid of these equations, we shall then, in the next chapter, discuss the questions of the electric and thermal conductivities of metals, the thermoelectric phenomena, etc.

In setting up these equations, we shall use the so-called statistical method generally employed in the kinetic theory of gases. This method was first applied to problems of the electron theory by H. A. Lorentz[1], who showed how it is possible

---

* [To conform to II, the clause "and partly because . . ." has been added.]

** [In accordance with II, the words "and concentration" have been added to the preceding sentence.]

*** [To conform to II, the sentence "Et lignende Forhold . . . variere med Temperaturen" has been deleted.]

[1] H. A. Lorentz: Proc. Acad. Amsterdam **7** (1905) 440.

to investigate the effects of the complicated motions of the electrons in a metal by means of this method in a rigorous and rather simple manner.

To describe the motions of the electrons, we shall use two rectangular 3-dimensional coordinate systems. In one of these, the position of an electron at the time considered is given in the usual way by coordinates (x, y, z). In the other coordinate system, the velocity of any electron is represented by a point having coordinates (ξ, η, ζ) which are equal to the components of the velocity of the electron at the given instant along the axes of the former coordinate system. The number of electrons present at time t in the volume element dv, and having velocities in the velocity element $d\sigma$, will be denoted by

$$f(x, y, z, \xi, \eta, \zeta, t)d\sigma\, dv$$

(or, when misunderstanding cannot arise, simply by $f\,d\sigma\,dv$), where (x, y, z) indicates a point in dv and (ξ, η, ζ) a point in the velocity element $d\sigma$ *.

We shall usually assume in the following that the elements dv and $d\sigma$ are so large that the number of electrons in $d\sigma\, dv$ is very large; if, however, as will be the case in some of the problems considered, it is not possible to choose the elements so large that this condition is satisfied, we shall by the number of electrons in the elements simply understand the average number taken over a certain short time interval.

To begin with, we shall consider the distribution of free electrons in a homogeneous piece of metal on which no external forces act, and in which the temperature is everywhere the same.

That these electrons are "free" is not to mean that they are not acted on by forces from the atoms of the metal, but merely that, in their interaction with the atoms, they behave as separate mechanical systems, to which the laws of statistical mechanics can be applied.

It should be pointed out here that it would not be permissible to make such an assumption about all the electrons present in the metal, namely not about those supposed to be bound in large numbers inside each individual atom; for, if that were not true, the specific heat of metals would be much higher than found by experiments. Hence, it is necessary, from a mechanical point of view, to distinguish sharply between free electrons and electrons bound in the atoms, the latter electrons being apparently cut off from the influence of other atoms or electrons in a manner that does not correspond to anything known for ordinary mechanical systems.

From the theorems of statistical mechanics, it now follows that the distribution of the free electrons in a piece of metal, in the state specified above, will be expressed by[1]

* [Beginning here, three and a half pages have been crossed out in II and replaced by partly illegible pencilled notes. The present translation conforms to I and to III.]

[1] See P. Debye: Ann. d. Phys. **33** (1910) 455.

$$f = A \int e^{-U/kT} dx_2 \, dy_2 \, dz_2 \, d\xi_2 \, d\eta_2 \, d\zeta_2 \ldots dx_N \, dy_N \, dz_N \, d\xi_N \, d\eta_N \, d\zeta_N \, dq_1 \, dq_2 \ldots$$
$$dq_n \, dp_1 \ldots dp_n, \quad (1)$$

where N is the number of free electrons in a unit volume of the metal, where $q_1, q_2, \cdots, q_n$ are a number of generalized positional coordinates and $p_1, p_2, \cdots, p_n$ the corresponding generalized momenta ($p_1 = \partial E_p / \partial \dot{q}_r$, where $\dot{q}_r = \partial q_r / \partial t$, and E_p is the total kinetic energy of the atoms) which determine the instantaneous state of motion of the atoms, and where U is the total energy of the entire system consisting of all the electrons and atoms in a unit volume of the metal in the state specified by given positional and velocity coordinates

$$x, x_2, x_3, \ldots, x_N, y, y_2, \ldots, \eta_N, \zeta, \zeta_2, \ldots \zeta_N$$

of the N electrons and the parameters $q_1, q_2, \cdots, q_n, p_1, \cdots, p_n$ for the atoms. Furthermore, T is the absolute temperature, and k is the universal constant which enters in the equation of state for an ideal gas

$$pv = kNT$$

(where p is the pressure, v the volume and N the number of molecules). Finally, A is a constant dependent upon the nature and temperature of the metal. The integrations in eq. (1) are to be carried out over all values of the velocity components and, for each positional coordinate, over a unit volume of the metal.

If, as in the applications to be made of eq. (1) in the following, the electrons are assumed to move independently of each other, i.e., if their mutual potential energy has an appreciable value only during a negligible part of their motion, we get from (1), in a stationary field of force,

$$f = K e^{-(\frac{1}{2}mr^2 + P)/kT}, \quad (2)$$

where K is a constant dependent on the nature and temperature of the metal, m is the mass of an electron, $r = \sqrt{\xi^2 + \eta^2 + \zeta^2}$ is its speed[1], and $P(x, y, z)$ its potential energy at the point (x, y, z), This is true, since in this case

$$U = [\tfrac{1}{2}mr^2 + P(x, y, z)] + [\tfrac{1}{2}mr_2^2 + P(x_2, y_2, z_2)] + \ldots + [\tfrac{1}{2}mr_N^2 + P(x_N, y_N, z_N)]$$
$$+ Q(q_1, q_2, \ldots, q_n, p_1 \cdots p_n).$$

Eq. (1)—and eq. (2) which is a special case of (1)—state, for one thing, that the electrons in any arbitrarily small volume element of the metal move equally in all directions, and, further, that the average value of their kinetic energy is the same

[1] It is assumed here, as in all what follows, that the velocities of the electrons are negligibly small compared to the speed of light. (At ordinary temperature the average speed of the electrons will be ca. $\frac{1}{3000}$ of the speed of light (see, e.g., J. J. Thomson: *The Corpuscular Theory of Matter*, p. 52)). Hence, the kinetic energy of an electron can be written as $\frac{1}{2}mr^2$, where m is a constant independent of the speed of the electron.

in any small volume element, being equal to the average translatory kinetic energy of a molecule of a gas at the same temperature.

In the derivation of relations such as eq. (1) for the statistical distribution of a collection of particles, it is usually assumed that the particles are acted on only by forces that are independent of the velocities of the particles and depend only on their positional coordinates; however, it can easily be shown that eq. (1) will hold also for the distribution of electrons in the presence of such forces as act upon the electrons when they move in a magnetic field[1]. However, being perpendicular to the direction of motion of the electrons, these forces will perform no work, and the electrons will, therefore, possess no potential energy relative to the magnetic field. Thus, it is seen that, in case of statistical equilibrium, the magnetic forces will have no influence upon the statistical distribution of the electrons. (See further Chapter IV.)

[1] In fact, eq. (1) is derived by general statistical considerations (see, e.g., P. Debye: loc. cit. p. 445–455) from the theorem (see, e.g., W. Gibbs: *Elementary Principles in Statistical Mechanics* (New York 1902), p. 3–11) that a collection of identical mechanical systems, of which each individual system is determined by n generalized positional coordinates $q_1, q_2, \ldots, q_n$ and the corresponding n momenta $p_1, p_2, \ldots, p_n$ will vary with time in such a way that, if we specify the state of motion of a single system by a point in a $2n$-dimensional orthogonal coordinate system having the coordinates $q_1, q_2, \ldots, q_n, p_1, p_2, \ldots, p_n$ and consider those systems whose representative points are found at a time t within a small "volume element" dv ($dv = dq_1, dq_2, \ldots, dq_n. dp_1 dp_2, \ldots dp_n$), the representative points of these same systems will at an arbitrary later time t_1 be found within a volume element dv_1 of the same magnitude as dv. (The total number of systems is assumed to be so large that their representative points at any time may be regarded as being continuously distributed in the $2n$-dimensional space.)

This theorem will hold if the systems in question satisfy the following condition (see Gibbs: loc. cit. p. 11, Note)

$$\frac{\partial \dot{q}_1}{\partial q_1} + \frac{\partial \dot{p}_1}{\partial p_1} + \frac{\partial \dot{q}_2}{\partial q_2} + \frac{\partial \dot{p}_2}{\partial p_2} + \ldots + \frac{\partial \dot{q}_n}{\partial q_n} + \frac{\partial \dot{p}_n}{\partial p_n} = 0, \tag{1'}$$

where, as above, $\dot{q}_1 = \partial q_1/\partial t$, etc.

If the forces acting on the particles of the system are independent of their velocities, eq. (1') follows directly from Hamilton's equations (Gibbs: loc. cit. p. 4)

$$\dot{q}_r = \frac{\partial E_p}{\partial p_r} \quad \text{and} \quad \dot{p}_r = -\frac{\partial E_p}{\partial q_r} + F_r, \tag{2'}$$

where E_p is the kinetic energy of the system, and $F_1, F_2, \ldots, F_n$ are the generalized components of force (in the case in question, these are functions of $q_1, q_2, \ldots, q_n$).

If the particles are affected also by such forces as act on electrons moving in a magnetic field, $F_1, F_2, \ldots, F_n$ will no more be independent of the generalized velocity components $\dot{q}_1, \dot{q}_2, \ldots, \dot{q}_n$ but will be linear functions of these quantities. Being perpendicular to the instantaneous direction of motion of the individual particles, the forces arising from the magnetic field will perform no work, however, and will consequently disappear from the expression

$$dA = F_1 dq_1 + F_2 dq_2 + \ldots + F_n dq_n = (F_1 \dot{q}_1 + F_2 \dot{q}_2 + \ldots + F_n \dot{q}_n)dt = B dt$$

We shall here consider the electrons as mass points whose entire motion is determined by 6 coordinates x, y, z, ξ, η, ζ. On the other hand, if the electrons were regarded as mechanical systems with a certain extension in space, we must, in order to completely determine their motions, take into consideration also the rotation of the electrons about their centres of mass and introduce suitable coordinates to describe this type of motion. According to statistical mechanics, the kinetic energy corresponding to such a rotation will, in case of statistical equilibrium, have the same mean value as that corresponding to the translatory motion. However, it is hardly possible to ascribe such a kinetic energy of rotation to the electrons; for, as is well known from the value of the ratio between the specific heats at constant pressure and at constant volume, not even the molecules of monatomic gases possess such an energy, although they must be regarded as much more extended systems than the individual free electrons*.

We shall express the velocity distribution of the free electrons in a homogeneous piece of metal, not exposed to external forces and of uniform temperature, by

$$f = f_0 .$$

In the presence of external forces, and when the temperature and chemical composition are not the same at all points, the distribution at a given point will, according to the above-mentioned assumptions, be expressed by

$$f = f_0 + \psi, \tag{3}$$

where f_0 is the function of x, y, z, ξ, η, ζ that would pertain to a homogeneous force-free metal having the temperature and composition of the given point, while ψ is a function that is very small compared to f_0.

In all of the following calculations, we shall now neglect terms which, as compared to the significant terms, are of the order of magnitude as ψ to f; the theory thereby

---

* [Three changes have been introduced in this paragraph from II: (a) The first sentence is replaced by one from II, expressing the same idea. (b) The sentence "Det maa dog . . . uafhængige af hinanden" has been deleted. (c) The words "not even" are added.]

---

for the work performed in the time interval dt by the forces acting on the system. Hence, B will, also in this case, be a linear function of the generalized velocity components $\dot{q}_1, \dot{q}_2, \ldots, \dot{q}_n$, and therefore

$$\frac{\partial^2 B}{\partial \dot{q}_r \partial \dot{q}_s} = 0 \qquad (r = 1, 2, \ldots, n; \ s = 1, 2, \ldots, n).$$

This gives

$$\frac{\partial F_r}{\partial \dot{q}_s} + \frac{\partial F_s}{\partial \dot{q}_r} = 0 \tag{3'}$$

We now get, with the aid of (2') and (3'),

$$\sum_{r=1}^{r=n} \frac{\partial F_r}{\partial p_r} = \sum_{r=1}^{r=n} \sum_{s=1}^{s=n} \frac{\partial F_r}{\partial \dot{q}_s} \frac{\partial \dot{q}_s}{\partial p_r} = \sum_{r=1}^{r=n} \sum_{s=1}^{s=n} \frac{\partial F_r}{\partial \dot{q}_s} \frac{\partial^2 E_p}{\partial p_r \partial p_s} = 0. \tag{4'}$$

From (2') and (4') we finally obtain (1'), q.e.d.

assumes a rather simple character, and the desired expressions for the transfer of electricity and energy will depend linearly upon the external forces, the variations in temperature, etc. These results, deduced directly from the very nature of the picture on which the electron theory is based, correspond exactly to what is found experimentally. Thus, as accurately as can be measured, the electric current is found to be proportional to the electric field over the very wide range of the experiments performed[1].

Proceeding now to detailed calculations, we shall consider two different cases; first, the simpler case in which the free electrons during the greater part, by far, of their motion are unaffected by appreciable forces from the atoms of the metal or from other electrons, so that the whole interaction between the electrons and the atoms, and among the electrons mutually, takes place in distinct, separate collisions. Next, we shall consider the more complicated case in which the electrons, during a large part of their motion, are assumed to be acted on by strong forces from the atoms of the metal.

§ 2

Derivation of Equations for the Collective Motion of the Electrons in the Case in which Separate Collisions are Assumed to Occur

We shall assume that *the linear dimensions of the regions inside which the electrons and the metal atoms interact appreciably*, i.e., the regions where collisions are said to be taking place, *are very small compared to the average distance travelled by the electrons between collisions.*

For the present, we shall not make any special assumptions about the forces between the atoms and the electrons; we shall only assume that the properties of the individual metal atoms are the same in all directions, and that this symmetry exists at every point in the metal, also in the presence of external forces and when the temperature is not uniform. According to what was said in the Introduction, this assumption is characteristic of the kind of electron theory with which we are here dealing, in which the properties of metals are explained by the action of the external forces upon the electrons themselves, and not upon the atoms*.

On account of the above-mentioned assumption about the ratio between the mean free path of the electrons and the dimensions of the metal atoms and the electrons, we need, in calculating the effect of the collective motion of the electrons (flow of electricity and energy), only consider the electrons when they are not suffering

[1] See, e.g., E. Lecher: Sitzungsber. d. Wiener Akad. d. Wiss., math.-nat. Kl. **116**, Abt. IIa (1907) 49.

* [In accordance with II and III, the following short paragraph has been deleted: "Bortset fra . . . symmetrisk byggede o.s.v.".]

collisions. Since the electrons, as stated in the Introduction, are assumed to be in mechanical and thermal equilibrium with their surroundings, the kinetic theory of gases gives for the distribution of the free electrons in a homogeneous piece of metal, of uniform temperature and unaffected by external forces,

$$f = Ke^{-\frac{1}{2}mr^2/kT}, \tag{4}$$

where m is the mass of an electron and $r = \sqrt{\xi^2 + \eta^2 + \zeta^2}$ its speed; T is the absolute temperature, K is a constant dependent upon the nature and temperature of the metal, and k is the universal constant in the equation of state of an ideal gas, $pv = NkT$, p denoting the pressure, v the volume, and N the number of molecules.

Denoting the number of free electrons per unit volume by N, we have

$$N = K \int_0^\infty e^{-\frac{1}{2}mr^2/kT} 4\pi r^2 \, dr = K \left(\frac{2\pi kT}{m}\right)^{\frac{3}{2}}. \quad * \tag{5}$$

If the equilibrium is disturbed, i.e., if external forces are present, or if the metal is not of uniform temperature and chemical composition, we have

$$f = Ke^{-\frac{1}{2}mr^2/kT} + \psi, \tag{6}$$

where ψ, as mentioned above, is small compared to the first term[1].

The quantities to be calculated are the amounts of electricity and energy transferred in unit time by the motion of the electrons through a surface element in the interior of the metal, which, for simplicity, we shall take to be perpendicular to the x-axis. These quantities can be determined if, for every value r of the speed, we know the total momentum in the direction of the x-axis carried by the electrons having this particular speed.

* [The preceding two paragraphs have been revised so as to conform to II. In the first one, that part of the paragraph which begins with "Since the electrons" replaces the sentence "Af Ligningen (2) ... $f = Ke^{-mr^2/2kT}$" in I. In the second one, the two lines "følgende Relation ... Hastighedselementer)" have been deleted, together with the second term in equation (5).]

[1] P. Debye (loc. cit. p. 476), who has attempted to calculate the electric and thermal conductivities of metals from the same physical assumptions as Lorentz, but with a somewhat different procedure, assumes in his calculations (loc. cit. p. 477) that, in the presence of external forces, the distribution of the electrons will differ very little from the stationary distribution $f = Ke^{-\frac{1}{2}(mr^2 + \epsilon\varphi)/kT}$, where ϵ is the electronic charge, and φ is the potential of the external forces. This is at variance with the fact that, according to the electron theory, the distribution of the electrons, as we have seen, will differ only very little from the distribution existing in a piece of metal not acted on by external forces. Moreover, in calculating the motion of the electrons among the metal atoms (loc. cit. p. 478–480), Debye also neglects the influence of the external forces—if he had considered this influence, the distribution assumed by him, being stationary, would not give rise to any collective motion of the electrons at all. For these reasons, Debye (loc. cit. p. 481, eq. (75)) finds a value for the electric conductivity which is equal in magnitude to that found by Lorentz but, what he failed to notice, has the opposite (wrong) sign; i.e., according to Debye's calculations, the electricity should move in the opposite direction of the electric force.

The momentum in the direction of the x-axis of the electrons in the volume element dV—taken so large that it contains a large number of metal atoms,—whose velocity points lie within a spherical shell with radii r and $r + dr$, we shall denote by $G_x(r)dr\,dV$. To obtain an equation for the determination of $G_x(r)$, we shall now consider the variation with time of this quantity.

The increment of $G_x(r)dr\,dV$ in the time dt is due partly to the instantaneous motion of the electrons by which electrons, and with them momentum, enter and leave the volume element dV, partly to the action of the external forces causing electrons to pass into and out of the range of speed considered, and, finally, to collisions between electrons and atoms, and among electrons themselves, occurring in the volume element dV.

In calculating the increment of the momentum in question due to the first two causes, neglecting, as mentioned above, terms of a magnitude relative to the significant terms as ψ to f, we need consider only the influence upon the electrons which at the given time belong to a distribution expressed by the first term on the right hand side of eq. (6)*. (If the external forces acting on the electrons are due to a magnetic field, the increment in the momentum produced by such forces on a distribution given by the first term in eq. (6) will vanish; hence, to investigate the influence of the magnetic field (Hall effect), it is necessary to consider terms arising from ψ. However, for the present, we shall consider only electric external forces (assumed derivable from a potential in the neighborhood of the point in question); later on, in Chapter IV, we shall discuss the influence of magnetic forces.)

On the other hand, the increment in the momentum due to collisions will depend only upon ψ, i.e., on the deviations from the normal distribution**.

To calculate the increment in the momentum $G_x(r)dr\,dV$, due to the influence of the first two causes on the electrons which at a given time are distributed according to the first term on the right side of eq. (6), we shall imagine this distribution to be divided into two parts, f_1 and f_2, of which

$$f_1 = K_0 e^{-(\frac{1}{2}mr^2 + \varepsilon\varphi)/kT_0}, \tag{7}$$

where K_0 and T_0 denote the values of K and T at a point (x_0, y_0, z_0) lying in the volume element dV, and where ε is the charge of an electron and φ is the potential

---

* [On the white sheet facing p. 17, an alternative derivation of eq. (9), p. 19, is given. However, it leads to an error in sign and has not been used in III. The development there is clearer than that given in I; it has, therefore, been taken as basis for the present translation. The paragraph including equations (7) and (8) replaces the corresponding paragraph in I, and the sentence beginning with "Idet vi nu ..." just before eq. (9) on p. 19 in I has been slightly changed to conform to the change in eq. (8).]

** [In accordance with III, the sentence "at denne Tilvækst ... de ydre Kræfter" has been deleted.]

of the external forces. If φ is so chosen that it is zero at the point (x_0, y_0, z_0), the distribution

$$f_2 = K e^{-\frac{1}{2}mr^2/kT} - K_0 e^{-(\frac{1}{2}mr^2 + \varepsilon\varphi)/kT_0} \tag{8}$$

will vanish at this point and be infinitely small at any point within dV.

As seen from eq. (2), f_1 represents a statistical equilibrium distribution, provided φ is independent of the time. This means that, for the electrons which at the given time belong to the distribution f_1, the change in the time dt in the distribution of the electrons—and hence the change in the momentum—due to the instantaneous motion of the electrons will be just cancelled by the influence of the external forces; hence, we need only compute the increment in $G_x(r)dr\,dV$ arising from the distribution f_2 *. Moreover, in doing so, we can neglect the influence of the external forces, since f_2, as well as its derivatives with respect to the velocity components ξ, η and ζ, is infinitely small in the volume element dV. On the other hand, the increment in $G_x(r)$ due to the motion of the electrons is finite, because the derivatives of f_2 with respect to the coordinates x, y and z are finite. Since the distribution f_2 at any point x, y, z is independent of the direction of motion of the electrons (but dependent only on r), the momentum carried through a given surface element will be directed along its normal. Thus the momentum in the x-direction, carried by the electrons in the range of speed considered through the surface element $dS = dy\,dz$ in the time dt, will be

$$dy\,dz\,dt \int m\xi \cdot \xi f_2 \,d\sigma = \tfrac{4}{3}\pi m r^4 f_2 \,dr\,dy\,dz\,dt,$$

the integration having been carried out over a region in velocity space limited by spheres with centres in the origin and of radii r and $r+dr$, respectively. The desired increase in the momentum in the volume element $dV = dx\,dy\,dz$ is, therefore,

$$-dx\,\frac{\partial}{\partial x}\left(\tfrac{4}{3}\pi m r^4 f_2 \,dr\,dy\,dz\,dt\right) = -\tfrac{4}{3}\pi m r^4 \frac{\partial f_2}{\partial x}\,dr\,dV\,dt.$$

Using eq. (8), and noticing that the two terms on the right side are equal at the point (x_0, y_0, z_0) in dV, we get for the increment in $G_x(r)dr\,dV$ due to the motion of the electrons and to the influence of the external forces in the time dt [1]

$$-\tfrac{4}{3}\pi m K \left(\frac{\varepsilon}{kT}\frac{\partial\varphi}{\partial x} + \frac{1}{K}\frac{\partial K}{\partial x} + \frac{mr^2}{2kT^2}\frac{\partial T}{\partial x}\right) r^4 e^{-\frac{1}{2}mr^2/kT}\,dr\,dV\,dt. \tag{9}$$

* [To conform with II, the footnote on p. [188] has been moved to p. [309] following eq. (6).]

[1] H. A. Wilson (loc. cit. p. 836), in his calculations mentioned above (p. [299], footnote), assumes that the increase in the momentum $G_x(r)dr\,dV$ due to the external forces is equal to $-\varepsilon(\partial\varphi/\partial x)dN\,dt$, where dN is the number of electrons in the volume element dV with speeds between r and $r+dr$.

We shall now investigate the increment in the momentum in question arising from the collisions between the electrons and metal atoms and among the electrons themselves. On account of the assumption, mentioned on p. [308], concerning the ratio between the mean free path of the electrons and the dimensions of the electrons and the atoms, the probability that an electron within a certain short time d*t*—large compared to the duration of a collision but small compared to the average time spent by an electron between collisions—will collide in a specified manner with an atom or another electron depends only on the instantaneous velocity of the electron, but not upon the path it has travelled since its last collision.

We shall first consider the effect of collisions between the electrons and the metal atoms. Since the properties of the atoms are, on the average, the same in all directions, the velocity points of the electrons which before a collision have a definite speed and direction will after the collision be distributed symmetrically with respect to this direction. The momentum which in the time d*t* is brought into a region of velocities bounded by two spheres with centres at the origin and radii r and $r+\mathrm{d}r$, as a result of collisions in the volume element $\mathrm{d}V$ of electrons whose velocity points before collision lie in a small element $\mathrm{d}\sigma'$ around (ξ', η', ζ'), will, therefore, have the same direction as the radius vector to (ξ', η', ζ'). Putting $\sqrt{\xi'^2+\eta'^2+\zeta'^2} = \rho$, we can write its magnitude as

$$m\rho f(\xi', \eta', \zeta')\mathrm{d}\sigma'\, Q(\rho, r)\mathrm{d}r\mathrm{d}V\,\mathrm{d}t,$$

where $Q(\rho, r)$ is some function of ρ and r.

The x-component of this momentum is

$$m\xi' f(\xi', \eta', \zeta')\mathrm{d}\sigma'\, Q(\rho, r)\mathrm{d}r\mathrm{d}V\,\mathrm{d}t,$$

and the x-component of the total momentum brought into the region of velocities considered, as a result of collisions of electrons which before collision have velocity points in a region bounded by two spheres with radii ρ and $\rho+\mathrm{d}\rho$, will therefore be

$$G_x(\rho)\mathrm{d}\rho\, Q(\rho, r)\mathrm{d}r\mathrm{d}V\mathrm{d}t.$$

The x-component of the total momentum brought by the collisions into the velocity region bounded by spheres with radii r and $r+\mathrm{d}r$ is, therefore,

$$\mathrm{d}r\,\mathrm{d}V\,\mathrm{d}t\int_0^\infty G_x(\rho)Q(\rho, r)\mathrm{d}\rho.$$

Since $\mathrm{d}N = 4\pi r^2 K\mathrm{e}^{-\frac{1}{2}mr^2/kT}\mathrm{d}r\mathrm{d}V$, the expression used by Wilson for the increment in $G_x(r)\mathrm{d}r\mathrm{d}V$ is equal to $-4\pi K\varepsilon(\partial\varphi/\partial x)r^2\mathrm{e}^{-\frac{1}{2}m/kT}\mathrm{d}r\,\mathrm{d}V\mathrm{d}t$. However, as seen by comparing with eq. (9) above, this is not correct. The error arises from equating the increment to be computed to the increment in the momentum of the electrons which at the time considered belong to the specified speed group, without taking into account that the external forces will cause electrons to leave or enter this group in the time d*t*.

In similar manner we conclude that the x-component of the momentum which, as a result of collisions in dV, leaves the velocity region bounded by the spheres with radii r and $r+dr$ can be written as

$$dr\,dV\,dt\,G_x(r)F(r),$$

where $F(r)$ is a function only of r. The increment in the momentum $G_x(r)dr\,dV$ resulting from collisions between the electrons and the metal atoms in the time dt is, therefore[1],

$$-\left[G_x(r)F(r)-\int_0^\infty G_x(\rho)Q\,(\rho,r)\,d\rho\right]dr\,dV\,dt. \tag{10}$$

Without more detailed assumptions about the forces acting between the electrons and the atoms of the metal, it is not possible to determine the functions $F(r)$ and $Q(\rho,r)$ completely. It can be shown, however, that if these forces obey the ordinary laws of mechanics, as here assumed, $Q(\rho,r)$ must satisfy a certain condition, of which use will be made in the following.

The average number of electrons with velocity points in the small element $d\sigma'$ around (ξ',η',ζ'), which as a result of collisions with atoms in the volume element dV are brought in the time dt into the small velocity element $d\sigma$ around (ξ,η,ζ), we shall denote by

$$\lambda(\xi',\eta',\zeta',\xi,\eta,\zeta)f(\xi',\eta',\zeta')d\sigma\,d\sigma'\,dV\,dt. \tag{11}$$

Since the properties of the metal atoms are assumed to be the same in all directions, the function λ, which is independent of the distribution f of the electrons, will depend only on the relative positions of the points (ξ',η',ζ'), (ξ,η,ζ) and the origin. Hence, $\lambda(\xi',\eta',\zeta',\xi,\eta,\zeta)$ can be written as $\chi(\rho,r,\vartheta)$, where ρ and r are

---

[1] J. H. Jeans (Phil. Mag. **17** (1909) 775), in his attempt mentioned above (p. [399]) to give a general theory of the phenomena connected with the absorption and emission of heat rays by metals, bases his calculations on the assumption that the total loss of momentum in a given direction, suffered by the electrons in unit time due to collisions with atoms, is equal to the total momentum of the electrons in this direction multiplied by a constant that depends only on the nature and temperature of the metal. However, this assumption is incorrect (cf. H. A. Wilson: loc. cit. p. 836). In fact, this loss is in our notations, according to eq. (10),

$$\int_0^\infty\left[G_x(r)F(r)-\int_0^\infty G_x(\rho)Q(\rho,r)d\rho\right]dr,$$

and, since the momentum of the electrons in different problems is not distributed in the same way among electrons with different speeds, this expression will, in general, not be proportional to the x-component of the total momentum,

$$\int_0^\infty G_x(r)dr.$$

Hence, Jeans' method gives the correct results only in certain special cases, e.g., when $F(r)=$ constant and $Q(\rho,r)=0$ (see p. [321]).

the lengths of the radius vectors to the points (ξ', η', ζ') and (ξ, η, ζ), respectively, and where ϑ is the angle between these radius vectors*.

If we imagine a definite distribution of the positional and velocity coordinates of the individual electrons and atoms (if the atoms are not assumed to be simple particles with the same properties in all directions, more than 6 coordinates will generally be needed to specify the state of an atom) and next consider a distribution with the same values of the space coordinates but with velocity components of the same magnitude but of opposite sign, then, since collisions obeying the ordinary laws of mechanics are reversible, the entire motion within the metal for the second distribution will be just the reverse of that for the first distribution, i.e., the same states occur at the same time intervals but in the opposite order. Now, since two such "opposite" distributions of the space and velocity coordinates of the individual electrons and atoms are equally probable in a homogeneous piece of metal of uniform temperature unaffected by external forces, the number of electrons given by (11) will in this case be equal, on the average, to the number of electrons whose velocity points are brought by collisions in the time dt from an element $d\sigma_1$, symmetrical to $d\sigma$ with respect to the origin, into an element $d\sigma_1'$ symmetrical to $d\sigma'$. The latter number, in terms of the notations used above, is

$$\chi(r, \rho, \vartheta)f(-\xi, -\eta, -\zeta)d\sigma_1' \, d\sigma_1 \, dV \, dt,$$

and we, therefore, get, with the aid of eq. (4),

$$\chi(\rho, r, \vartheta)e^{-\frac{1}{2}m\rho^2/kT} = \chi(r, \rho, \vartheta)e^{-\frac{1}{2}mr^2/kT},$$

from which it follows that

$$\chi(\rho, r, \vartheta) = e^{-\frac{1}{2}mr^2/kT}s(\rho, r, \vartheta), \tag{12}$$

where $s(\rho, r, \vartheta)$ is a function symmetrical with respect to ρ and r.

The momentum carried in the time dt into a velocity region bounded by two spheres of radii r and $r+dr$, as a result of collisions in dV of electrons whose velocity points before collisions lie in an element $d\sigma'$ around (ξ', η', ζ'), will now be

$$dV \, dt \, f(\xi', \eta', \zeta')d\sigma' \int_{\varphi=0}^{\varphi=2\pi} \int_{\vartheta=0}^{\vartheta=\pi} mr \cos\vartheta \cdot \chi(\rho, r, \vartheta)r^2 \sin\vartheta \, dr \, d\vartheta \, d\varphi$$

$$= dV \, dt f(\xi', \eta', \zeta')d\sigma' \cdot 2\pi mr^3 e^{-\frac{1}{2}mr^2/kT} \, dr \int_0^\pi s(\rho, r, \vartheta) \cos\vartheta \sin\vartheta \, d\vartheta.$$

But, as stated above, the direction of this momentum will coincide with the radius vector to (ξ', η', ζ'), and its magnitude will be, in the notations introduced on p. [312],

$$m\rho f(\xi', \eta', \zeta')d\sigma' Q(\rho, r)dr \, dV \, dt.$$

* [To conform to II and III the sentence "Vi skal nu . . . Formlen (4) Side 16" has been deleted.]

Consequently,

$$Q(\rho, r) = r^4 e^{-\frac{1}{2}mr^2/kT} S(\rho, r),$$ (13)

where

$$S(\rho, r) = \frac{2\pi}{r\rho} \int_0^\pi s(\rho, r, \vartheta) \cos \vartheta \sin \vartheta \, d\vartheta$$

is a symmetrical function in ρ and r.

We shall now proceed to consider the effect of mutual collisions among the electrons. Since the velocities of the electrons with which a given electron collides are not equally distributed in all directions, its velocity after collision will not, on the average, be distributed symmetrically with respect to its velocity before collision, contrary to what occurred in the case considered above. Nevertheless, if the deviations from the normal distribution are very small, as we shall assume, it is possible to calculate the effect of the collisions in much the same way.

We shall imagine the electrons divided at any time into two groups, A and B, of which A has the normal distribution $f = Ke^{-\frac{1}{2}mr^2/kT}$, while B is very small compared to A and has a distribution $f = \psi$ (see eq. (6)). In computing the effect of collisions, we shall consider separately collisions between electrons belonging to the same group and collisions between electrons belonging to different groups. We now have:

(i) Since the velocity distribution in group A is normal, collisions among electrons of this group will produce no change in the distribution.

(ii) On account of the symmetry of group A, the effect of collisions of electrons of group B with electrons of group A can be calculated in a way quite similar to that used for collisions between electrons and metal atoms. In fact, if the electrons of group B are regarded as the impinging particles in the collisions, the velocity point of an impinging electron after collision, as well as that of the electron hit, will on the average be symmetrically distributed with respect to the direction of the impinging electron before collision. Thus, these collisions produce an increment in the momentum $G_x(r) dr dV$ which can be written in exactly the same form as the expression (10).

(iii) The effect of mutual collisions among the electrons of group B will be negligible, on account of the comparatively very small number of such collisions.

We can further show that the function $Q(\rho, r)$, also in this case, will satisfy the condition given by eq. (13). From this discussion, it follows that the effect of the mutual collisions among the free electrons will be the same as that of collisions between these electrons (which we shall regard as the impinging particles) and a collection of electrons having a distribution $f_0 = Ke^{-\frac{1}{2}mr^2/kT}$.

The total number of electrons brought in the time dt into the small velocity element $d\sigma$, as a result of such collisions in the volume element dV in which the velocity of the impinging electron lies in the element $d\sigma'$, we shall, in analogy with the previous case, denote by

$$\chi_1(\rho, r, \vartheta)f(\xi', \eta', \zeta')d\sigma' d\sigma dV dt.$$

In calculating $\chi_1(\rho, r, \vartheta)$, which depends only on the distribution f_0, we must not only consider the collisions in which the impinging electron is brought from the velocity element $d\sigma'$ into $d\sigma$, but also the collisions in which the struck electron is brought into $d\sigma$, as well as the collisions in which the latter electron is in $d\sigma$ before collision and, hence, is brought out of this velocity element when hit. The first kind of collisions is quite analogous to that considered above when the effect of the collisions between the free electrons and the metal atoms was calculated; these collisions will, therefore, make a contribution to $\chi_1(\rho, r, \vartheta)$ of the same form as the expression on the right side of eq. (12). The same result is obtained for the second kind of collisions; for also among these collisions will the "opposite" collision make a contribution to $\chi_1(r, \rho, \vartheta)$ corresponding to that made by the "direct" collision to $\chi_1(\rho, r, \vartheta)$. This will not be true for the third type of collisions; however, we see immediately that the contributions made by these collisions to $\chi_1(\rho, r, \vartheta)$ and $\chi_1(r, \rho, \vartheta)$, respectively, will be in the ratio

$$f_0(\xi, \eta, \zeta)/f_0(\xi', \eta', \zeta') = e^{-\frac{1}{2}mr^2/kT}/e^{-\frac{1}{2}m\rho^2/kT}.$$

Thus, we see that the quantity $\chi_1(\rho, r, \vartheta)$ can be written in the same form as the right hand side of eq. (12), and hence that, also for the mutual collisions among the free electrons, $Q(\rho, r)$ will have the form given by eq. (13).

Denoting the total increment of $G_x(r)$ in the time dt by $dG_x(r)/dt$, we get, with the aid of expressions (9) and (10),

$$\left[\frac{dG_x(r)}{dt}\right] = -\frac{4}{3}\pi m K \left(\frac{\varepsilon}{kT}\frac{\partial\varphi}{\partial x} + \frac{1}{K}\frac{\partial K}{\partial x} + \frac{mr^2}{2kT^2}\frac{\partial T}{\partial x}\right) r^4 e^{-\frac{1}{2}mr^2/kT}$$
$$-G_x(r)F(r) + \int_0^\infty G_x(\rho)Q(\rho, r)d\rho, \quad (14)$$

where the bracket on the left-hand side indicates that the derivative is to be regarded as an average value under the specified circumstances, i.e., for given values of the external forces, the temperature variation, etc., and the quantities $G_x(\rho)$.

It is strictly not correct to speak of such an average value, for the variation of $G_x(r)$ depends in a systematic manner upon the preceding values of $G_x(\rho)$ over a time interval of the same order of magnitude as the duration of a single collision (see p. [312]). However, we can ignore this circumstance, since, in all applications of (14), we shall assume that dt can be regarded as large compared to this duration, which will be assumed to be very small compared to the time in which an individual electron, on the average, loses the greater part of its original momentum*.

* [To conform to III, the sentence in I beginning with "Vi skal her" has been deleted, and the preceding paragraph from III has been inserted and, as indicated in III, printed in small type.]

In most applications of eq. (14), we shall only seek a mean value of $G_x(r)$ under specified *external conditions*, i. e., for given values of the external forces, the variation in temperature, etc. Since in such applications the values of $F(r)$ and $Q(\rho, r)$ are very large (this follows from the assumption mentioned above that the original momentum of an electron is lost very rapidly in collisions), the term on the left side will be very small compared to the last terms on the right, except in such special cases in which the mean value of $G_x(r)$ varies exceedingly rapidly with time, i.e., so rapidly that it changes appreciably within the very short time intervals in which the individual electrons lose the greater part of their original momentum. This can occur only if the first term on the right varies very rapidly with time; in this term, only the external forces can vary rapidly, as they will be assumed to do in the calculation of the absorption of heat rays by metals.

(However, as will be discussed in detail in Chapter III, because of the manner in which eq. (13) is derived, it can only be applied to calculate the absorption of heat rays of so long periods that the external forces vary very little during individual collisions, i.e., during time intervals assumed to be very short compared to the average time in which an individual electron loses a large part of its original momentum.)

In all other cases of the kind here considered, we can neglect the term on the left side of eq. (14) and put it equal to zero. Problems such as those just mentioned, in which the state varies with time, but so slowly that each momentary state can be regarded as an equilibrium state, we shall in the following denote as "quasi-stationary".

Besides in applications in which it is used to determine the mean value of $G_x(r)$ under given external conditions, eq. (14) will also be used to investigate the small very rapid variations in $G_x(r)$ due to the random motion of the electrons, which are assumed to give rise to the emission of heat rays by the metal. In such applications, the term on the left side will play a major role, giving the average value of $dG_x(r)/dt$ for a specified distribution of the total momentum of the electrons among electrons of different speeds.

§ 3

Special Examples of the Calculations in Cases in which Separate Collisions are Assumed to Take Place

Eq. (14) is a perfectly general equation for determining the collective momentum of the electrons when separate collisions take place. Together with eq. (13), it permits us, as we shall see, merely from its form, to investigate the relation between certain properties of metals. To determine the quantities $F(r)$ and $Q(\rho, r)$, which is necessary to obtain quantitative results, we must, however, make special assumptions about the effects of collisions between electrons and metal atoms, and among the electrons mutually. Before proceeding further, we shall, therefore, mention some examples of the introduction of such assumptions.

In the theory of H. A. Lorentz[1], as mentioned above (p. [298]), the effects of the collisions are calculated from the assumption that the electrons and the metal atoms act upon each other as *perfectly hard elastic spheres*. Moreover, the dimensions of the electrons, as well as their masses, are assumed to be so small compared to those of the metal atoms that mutual collisions among the free electrons can be neglected in comparison with collisions between electrons and atoms, and the metal atoms can be regarded as immovable compared to the electrons. As easily seen, the effect of a collision under these circumstances will be that the electron leaves the atom with unaltered speed, while its direction of motion after the collision is entirely independent of its direction before the collision, i. e., is evenly distributed in all directions in space[2]; the electron will, therefore, in a collision lose all its original momentum in a given direction.

Since the speed of the electrons is not changed in collisions, $Q(r, \rho)$ will in this case be zero, while $F(r)$ can be determined in the following manner. As the atoms are assumed to be immovable, the free paths travelled by the electrons between successive collisions will be independent of their velocity. If the mean free path is denoted by l, the number of collisions, suffered in the time dt by the $N(r)dr\,dV$ electrons in the volume element dV having speeds between r and $r+dr$, will be

$$\frac{r\,dt}{l}\, N(r)dr\,dV.$$

The momentum in the direction of the x-axis which an electron with speed r possesses, on the average,—and, hence, as mentioned above (p. [312]) also immediately before a collision—is now $G_x(r)/N(r)$, and the total x-component of momentum lost in the collisions considered is, therefore,

$$\frac{r}{l}\, G_x(r)dr\,dV\,dt.$$

Hence, from the expression (10),

$$F(r) = \frac{r}{l}\,. \tag{15}$$

P. Gruner[3] has attempted to extend Lorentz' theory by introducing what he calls the "ionization" and "electron binding" of the metal atoms. Thus, Gruner

---

[1] H. A. Lorentz: Proc. Acad. Amsterdam **7** (1905) 439.

[2] See Maxwell: *Scientific Papers*, vol. I, p. 379.

[3] P. Gruner: Verh. d. Deutsch. Phys. Ges. **10** (1908) 509.

assumes that the free electrons, under certain circumstances, are bound temporarily by atoms and are set free again only when the atoms are struck by other free electrons. Hence, his theory distinguishes between two kinds of atoms: those which have bound no electrons (positive atoms) and those which have bound one or more electrons (neutral atoms). In this connection, certain assumptions are made about the collisions between electrons and metal atoms. Thus, it is assumed that electrons colliding with positive atoms are bound by these, if their speed before collision is smaller than a certain definite speed, G, while on the other hand, an electron is set free in collisions between electrons of speed greater than G and neutral atoms. Moreover, several additional assumptions are made about the speed after the collision of the impinging and the liberated electrons. These special assumptions are, however, of such a nature that serious objections can be raised. Admitting this, but seeking to retain the essential part of the theory, Gruner[1] has later modified it, so that no special assumptions are made about the mechanisms of ionization and electron binding, the only regard paid in the calculations to these processes being the assumption of the presence of both positive and neutral metal atoms. Gruner now assumes that all electrons colliding with positive atoms are reflected as from elastic spheres, while in collisions with neutral atoms—which are supposed not to act as strongly upon the electrons—only electrons with speed less than G are reflected as from elastic spheres, while electrons with greater speed are not affected at all.

As may be readily seen, the only difference between Lorentz' and Gruner's theories is that the mean free path in the latter theory is not the same for all electrons, but has one value l_1 for electrons with speeds less than G and another value l_2 for electrons with speeds greater than G ($l_2 > l_1$). Hence, we get in this case

$$F(r) = \frac{r}{l_1} \quad \text{for} \quad r < G \quad \text{and} \quad F(r) = \frac{r}{l_2} \quad \text{for} \quad r > G. \qquad (16)$$

However, it may be remarked that Gruner's assumption will cause an essential change in Lorentz' calculations only if it is further assumed that the quantity G is not very different from the average speed of the electrons at the temperature in question.

It appears that Gruner's theory, especially in its latter form, may be regarded as an attempt to investigate, to a first rough approximation, what the effect in the results of the theory would be, if the nature of the collisions were assumed to depend on the speed of the electrons. To investigate this question more closely, we shall now proceed to consider the atoms as centres of force, attracting or repelling the electrons.

We shall regard the atoms as *fixed centres of force, which repel or attract the*

[1] P. Gruner: Phys. Zeitschr. **10** (1909) 48.

electrons with forces inversely proportional to the n-th power of distance. The range within which the atoms exert an appreciable influence upon the motion of the electrons will again be assumed small compared to the mutual distances of the molecules. However, since this range has no sharp limit, we cannot, as in the previous cases, speak of a mean free path. We might treat the problem in a manner similar to that used by Maxwell[1] in the kinetic theory of gases, and carried through completely by him for the case that the forces vary inversely as the fifth power of the distance, a case known to be particularly simple. Since we regard the metal atoms as immovable and neglect the mutual collisions among the electrons, such a calculation would be much simpler than that of Maxwell, and it would be as easy to obtain the results for an arbitrary power (greater than 2) of the distance as for the fifth power. However, the desired result can be found immediately by a dimensional analysis*.

Since the speed of the electrons is not changed in "collisions" when the forces from the metal atoms have a fixed potential, $Q(\rho, r)$ will vanish, as in Lorentz' theory. We further see that $F(r)$ in this case, in which the probability of a collision depends only on the speed of the electron, must be proportional to the number N of metal atoms in unit volume. If the force per unit mass exerted by an atom upon an electron is denoted by $\mu\rho^{-n}$, where ρ is the distance from the atom to the electron, the quantity $F(r)/N$ will, apart from a dimensionless factor, depend only upon μ and on the speed r of the electrons. As is seen from eq. (14), $F(r)$ has the dimension of a reciprocal time. Denoting a time by T and a length by L, we thus have $F(r)/N \sim L^3 T^{-1}$; moreover, $r \sim L T^{-1}$ and $\mu\rho^{-n} \sim L T^{-2}$ or $\mu \sim L^{n+1} T^{-2}$. It follows that $F(r)/N$ and $\mu^{2/n-1} r^{(n-5)/(n-1)}$ have the same dimension. Thus,

$$F(r) = Cr^{(n-5)/(n-1)}, \tag{17}$$

where C is a constant depending only on the nature and temperature of the metal.

It should be pointed out that a more detailed investigation shows that the stipulation made above, that $n > 2$, is a necessary one, since otherwise the assumption, that an appreciable influence of an atom upon the motion of an electron takes place only within a range that is small compared to the mutual distance between the atoms, cannot be satisfied.

If the atoms exert an attractive force upon the electrons, a special complication arises for $n > 3$, since the electrons then, in a certain finite fraction of the collisions, will reach the centre of force and arrive there with infinite velocity, after which their motion is undefined. However, as indicat-

[1] Maxwell: Phil. Mag. **35** (1868) 129, 185.

* [The text between this and the next asterisk is based on III. That part of I which has been revised begins with "Elektronernes absolute" on p. 30 and ends with the expression $F(r) = \Sigma C_n r^{(n-5)/(n-1)}$ on p. 33.]

ed by Boltzmann[1], this difficulty can be evaded by assuming that the force in the immediate proximity of the centre (i.e., at a distance from it that is small compared to the distance at which the collision is regarded as having commenced) is not given by $\mu\rho^{-n}$ but by an expression making the velocity finite. Such an assumption, involving only the qualitative character of the collisions, will not affect the dimensional analysis and, hence, not the expression for $F(r)$ given above.

Since, according to the assumption made above about the probability of collisions, the contributions of the individual atoms are independent of one another, the above expression for $F(r)$ will be valid also when the atoms are not all identical, provided only that they act upon the electrons with forces varying as the same power of the distance. On the other hand, if we assume the presence of atoms acting upon the electrons with forces varying as different powers of the distance, we get

$$F(r) = \sum C_n r^{(n-5)/(n-1)}.\,*$$

(However, in this as well as in the former case, the metal atoms must be so well "mixed" that the number of each type of atom can be considered the same in every volume element of dimensions comparable to the mean free path of the electrons.)

Putting $n = \infty$ in eq. (17), we get $F(r) = Cr$; with $C = 1/l$, this is the expression found on p. [318], where the theory of Lorentz was discussed. This is what might be expected; for, writing the force per unit mass exerted by the atoms on the electrons in the form $\mu\rho^{-n} = a(b/\rho)^n$, we see that, for very large n, the force will be exceedingly small for $\rho > b$ and exceedingly large for $\rho < b$, i. e., it will behave as if the atoms and electrons were hard elastic spheres with a sum of radii equal to b (see Jeans: *Dynamical Theory of Gases*, p. 276).

Comparing (15) and (17), we see that the difference between the two cases can be expressed by saying that the "effective mean free path" in the latter case increases with increasing r, while in the former case it is constant. (The effective mean free path will never decrease with increasing r, no matter what forces are supposed to act between the atoms and the electrons; see p. [329].)

Putting $n = 5$, we get $F(r) = C$. In this case (see footnote p. [313]) the component in a given direction of the total momentum transferred in collisions from the electrons to the atoms in unit time will be proportional to the corresponding component of the total momentum of the electrons, and in many problems the calculations will, therefore, be considerable simpler than for most other values of n.

In much the same manner as in the cases considered, we can determine the function $F(r)$ when the metal atoms are assumed to be *electric doublets, composed*

---

[1] L. Boltzmann: Sitzungsb. d. Wiener Akad. d. Wiss., math.-nat. Kl. **89**, Abt. 2 (1884) 720. See also P. Czermak: ibid. p. 723.

of two particles with charges of equal magnitude but opposite signs[1], and also when the atoms are assumed to be *elementary magnets*[2].

The dimensions of these doublets or magnets, i.e., the distance between the charges or the magnetic poles, are here considered to be negligibly small compared to the range within which the atoms affect the motion of the electrons appreciably, and this range in turn is assumed to be small compared to the mutual distances between the atoms.

In these cases, as in that considered above, the speed of the electrons is not changed in a "collision", and $Q(\rho, r)$ will therefore vanish also here.

If the metal atom is assumed to be an electric doublet, the magnitude of the force per unit mass exerted by it on an electron will be $(\mu/\rho^3)\psi_1(\vartheta)$, where ρ is the length of the vector from the atom to the electron, and ϑ is the angle between this vector and the axis of the atomic dipole. If the atom is a magnet, the force will be $(\mu r/\rho^3)\psi_2(\vartheta, \tau, v)$, where r as usual denotes the speed of the electron (r is constant during a collision), and ρ and ϑ have the same meanings as in the previous case, while τ denotes the angle between the instantaneous velocity of the electron and the vector from the atom to the electron, and v is the angle between the velocity of the electron and a plane through the axis of the atom and the instantaneous position of the electron. Now, if the axes of the atoms are uniformly distributed in all directions, the dimension of $F(r)$ will, also in these cases, depend only on μ and r (ψ_1 and ψ_2 are assumed to be dimensionless). Thus, we have in these cases $(\mu/\rho^3)\psi_1(\vartheta) \sim LT^{-2}$, $\mu \sim L^4T^{-2}$ and $(\mu r/\rho^3)\psi_2(\vartheta, \tau, v) \sim LT^{-2}$, $\mu \sim L^3T^{-1}$, respectively. This gives, in the same manner as above, $F(r) = Cr^{-1}$ and $F(r) = C$, respectively*.

Hence, if the metal atoms are assumed to be electric doublets, the same form is obtained for $F(r)$ as when they are supposed to be centres of force acting upon the electrons with forces inversely proportional to the third power of the distance; while, if the atoms are assumed to be elementary magnets, the form for $F(r)$ is the same as when the atoms are centres of force affecting the electrons with forces varying inversely as the fifth power of the distance.

[1] The assumption that such electric doublets are present in the interior of bodies has been employed by J. J. Thomson (Phil. Mag. **20** (1910) 238) to explain certain of their optical properties. It may be remarked also that J. H. Jeans (Phil. Mag. **20** (1910) 380) has succeeded in carrying out the complete mathematical calculation of the path of an electron in the field of such a doublet.
[2] Detailed calculations of the motion of an electron in the field of an elementary magnet have been carried out by C. Störmer (Arch. d. Sciences phys. et nat. **24** (1907) 5, 113, 221, 317) by numerical integration of the differential equations pertaining to this problem. See also P. Gruner: Jahrb. d. Rad. u. El. **6** (1909) 149.

* [The two paragraphs "I de her . . . $F(r) = Cr^{-1}$ og $F(r) = C$" of I have been replaced by the preceding two paragraphs which appear in both II and III.]

In all the cases treated here, it has been assumed that the speed of the electron does not change in collisions, and that, therefore, $Q(\rho, r) = 0$; this can also be expressed by saying that, in the cases considered, momentum but not energy is transferred in collisions. However, if this assumption were strictly valid, the collisions could not cause the distribution of the electrons, in the presence of external forces or non-uniform temperature, to deviate only slightly from the normal distribution, $f = Ke^{-\frac{1}{2}mr^2/kT}$; rather, the collisions would only cause the distribution within each group of electrons with the same speed r to be very nearly symmetrical with respect to the origin of the velocity coordinates. To make the mechanical picture complete, we must therefore assume that there is a certain probability that electrons change their speed in collisions, so that the kinetic energy of an electron may change greatly while the electron travels a distance so small that its potential energy arising from the external forces and the temperature vary only very little; in the cases discussed above, we shall only assume that, on the average, the time interval in which the momentum of a moving electron changes greatly is very small compared to the time in which its kinetic energy has changed appreciably.

In cases where the speeds of the electrons are altered in collisions, i.e., cases in which the transfer of kinetic energy is not negligible compared to the transfer of momentum, the determination of the functions $F(r)$ and $Q(\rho, r)$ is, in general, much more difficult than in the former cases. I have carried out the calculation of $Q(\rho, r)$ for one of the simplest examples, namely for *mutual collisions among the free electrons* when these are assumed to act upon one another as hard elastic spheres; moreover, in this case the function $F(r)$ can be obtained from Maxwell's calculation of the mean free path in a gas, the molecules of which are assumed to be elastic spheres (see for example Jeans: *Dynamical Theory of Gases*, p. 231). Since the calculations are rather lengthy, and since no use will be made of them in the following, I shall here only state the result:

$$F(r) = \frac{1}{l}\left(\frac{kT}{\pi m}\right)^{\frac{1}{2}}\left\{e^{-\frac{1}{2}mr^2/kT} + \left(\frac{mr}{kT} + \frac{1}{r}\right)\int_0^r e^{-\frac{1}{2}mz^2/kT}\,dz\right\},$$

$$Q(\rho, r) = \frac{1}{l}\left(\frac{kT}{\pi m}\right)^{\frac{1}{2}}\frac{r}{\rho^3}e^{-\frac{1}{2}mr^2/kT}\left\{8x + \left(\frac{m}{kT}\right)^2(\tfrac{1}{3}x^3 y^2 - \tfrac{1}{15}x^5)\right.$$

$$\left. + \left(\frac{4m}{kT}x^2 - 8\right)e^{-\frac{1}{2}mx^2/kT}\int_0^x e^{-\frac{1}{2}mz^2/kT}\,dz\right\}, \quad (18)$$

where $x = r$, $y = \rho$ for $\rho > r$, while $x = \rho$, $y = r$ for $\rho < r$. In these expressions l denotes the mean free path between successive collisions, calculated in the manner of Maxwell (see Jeans: loc. cit. p. 234).

It is seen that the expression for $Q(\rho, r)$ satisfies the condition expressed by eq.

(13). We can further show that $F(r)$ and $Q(\rho, r)$ satisfy the condition

$$F(r) = \int_0^\infty Q(r, \rho)\mathrm{d}\rho,$$

which is characteristic of the cases in which only mutual collisions among the free electrons are considered; it states that the total momentum in a given direction lost by the electrons in collisions is equal to zero*.

§ 4

Derivation of Equations for the Collective Motion of the Electrons in Case the Dimensions of the Metal Atoms are not Small Compared to their Mutual Distances

We shall now consider the case in which the metal atoms are assumed to be so near one another that the electrons are influenced by forces from the atoms during a large part of their motion. The treatment in this case will, in general, be much more difficult than in the case in which separate collisions are assumed. That the problem is simpler in the latter case is due to the fact that the probability of collisions, and, hence, of the transfer of momentum between electrons and metal atoms, depends only on the instantaneous speed of the electrons and not on their paths; it was possible, therefore, without great difficulty, to establish a relation between the time-rate of transfer of momentum to the atoms, or in other words, the "resistance" of the atoms against the collective motion of the electrons, and the instantaneous total momentum possessed by the electrons. However, such a general relation does not exist in the more complicated case now considered. From a statistical point of view, the "resistance" mentioned arises from the fact that the electrons, during their collective motion, will be present in greater numbers in such places where the forces from the atoms are directed against their motion than in places where the forces are in the same direction, i.e., on account of their collective motion, the electrons, so to speak, pile up in front of the atoms. Contrary to the case in which separate collisions were assumed, this "piling up" of the electrons will, in general, depend upon the way in which the collective motion of the electrons is produced. We shall, therefore, consider only a special case which can be treated in a simple way irrespective of the mechanism of momentum transfer, a case which, however, since it satisfies the fundamental assumptions stated in the Introduction, appears to present some points of similarity to the conditions in real metals**.

* [The preceding paragraph has been slightly expanded by the addition of part of a sentence from II. Version I contains only the first sentence.]
** [That part of the preceding paragraph which begins with "it was possible" has been taken from II. The corresponding part in I begins with "man kunde derfor".]

Thus, we shall assume that *no interaction occurs among the free electrons*, and that *the forces exerted by the metal atoms upon the electrons form a stationary electromagnetic field*; moreover, we shall assume in the calculations that the external forces are stationary.

We shall not assume, however, that these conditions are strictly fulfilled; for in such a stationary field no equalizing transfers of energy will take place (i.e., the kinetic energy of an electron will at each moment be determined by its initial energy and its instantaneous position (x, y, z)), and, if no exchanges of energy among electrons or between electrons and atoms occurred, there would be no sufficient cause to insure that the velocity distribution of the electrons, in the presence of external forces and non-uniform temperature, will differ only slightly from the normal distribution (compare the analogous case, p. [322]). An equalizing transfer of energy may be imagined to take place, partly through interactions (collisions) among the free electrons moving in the stationary field, and partly through fluctuations of the field; however, we shall only assume that the effect of mutual collisions among the electrons is, on the average, very small compared to the interactions between the metal atoms and the electrons, and that the force fields of the atoms — as well as the external forces — change very little during such time intervals in which an electron, on the average, loses almost all its initial momentum in a given direction.

As mentioned above, these assumptions seem to be approximately consistent with the conditions in the real metals. In support of this statement, we may refer, in the first place, to the large mass and great extension of the metal atoms as compared to the electrons (the mass of an atom is ca. 10^5 times that of an electron, and the atoms are supposed to be systems containing a large number of bound electrons); secondly, we may mention the circumstances that the number of free electrons is, at most, of the same order of magnitude as the number of atoms (a larger number being inconsistent with the experimental values of the specific heats of metals[1]) and that the interaction between the free electrons and electrons bound in the atoms must be exceedingly small compared to the interactions of the free electrons with the atoms, regarded as closed systems (the opposite assumption would lead to a distribution of energy among the bound electrons inconsistent with experience, see p. [304])*.

We shall now consider the motion of the free electrons travelling independently of one another in such a stationary field of force. If the electric potential of the field produced by the atoms at the point (x, y, z) in the interior of the metal is denoted by λ, the statistical distribution of the electrons in a piece of metal unaffected by external forces and of uniform temperature will, according to eq. (2), be expressed by

$$f = K\mathrm{e}^{-(\frac{1}{2}mr^2 + \varepsilon\lambda)/kT}. \tag{19}$$

If the number of free electrons per unit volume is denoted by N, we have

[1] See, for example, J. H. Jeans: Phil. Mag. **17** (1909) 793.

* [The preceding paragraph has been revised so as to conform to III. It replaces the paragraph "Disse Antagelser . . . Elektronernes Bevægelse" in the original version.]

$$N = \iint K e^{-(\frac{1}{2}mr^2 + \varepsilon\lambda)/kT} \, d\sigma \, dv$$

$$= K \int e^{-\frac{1}{2}mr^2/kT} \, d\sigma \int e^{-\varepsilon\lambda/kT} \, dv$$

$$= K \left(\frac{2\pi kT}{m}\right)^{\frac{3}{2}} \int e^{-\varepsilon\lambda/kT} \, dv, \tag{20}$$

where the integration with respect to $d\sigma$ is over all possible velocities, while the integration with respect to dv is to be extended over a unit volume.

In the presence of external forces, or if the metal is not homogeneous or of uniform temperature, the equilibrium will be disturbed, and the statistical distribution of the electrons will be altered. However, according to what was stated at the beginning of this chapter, we can always write

$$f = K e^{-(\frac{1}{2}mr^2 + \varepsilon\lambda)/kT} + \psi, \tag{21}$$

where K is a function of the state of the metal—its nature and temperature—at the point in question, while ψ is a quantity that is very small compared to the first term.

In the investigation of the collective motion of the electrons, we shall not, as in the case of separate collisions, consider together the electrons having the same speed, but rather those having the same sum $\frac{1}{2}mr^2 + \varepsilon\lambda = a$ of kinetic energy and potential energy relative to the metal atoms.

The component along the x-axis of the total momentum possessed by the electrons present in the volume element dV—assumed so large that it contains a very large number of metal atoms,—and having energies between a and $a + da$, we shall denote by $G_x(a)da\,dV$.

To calculate $G_x(a)$, we shall now, as in the previous case (see p. [310]), imagine the distribution expressed by the first term on the right side of eq. (21) divided into two parts f_1 and f_2, of which

$$f_1 = K_0 e^{-(\frac{1}{2}mr^2 + \varepsilon\lambda + \varepsilon\varphi)/kT_0}, \tag{22}$$

where K_0 and T_0 denote the values of K and T at a point (x_0, y_0, z_0) in the volume element dV, and where φ is the potential of the external forces chosen in such a way that it vanishes at the point (x_0, y_0, z_0); the distribution f_2 will then be

$$f_2 = K e^{-(\frac{1}{2}mr^2 + \varepsilon\lambda)/kT} - K_0 e^{-(\frac{1}{2}mr^2 + \varepsilon\lambda + \varepsilon\varphi)/kT_0}; \tag{23}$$

it will vanish at (x, y_0, z_0) and be extremely small at any point in the volume element dV *.

* [The preceding paragraph has been revised so as to conform to III. The corresponding paragraph in I is the one that ends with eq. (23), p. 40.]

Since the potential energy $\varepsilon\lambda + \varepsilon\varphi$ of the electrons is here assumed to be a stationary function of x, y and z, the distribution f_1 will be a distribution of statistical equilibrium (compare p. [311]), and the electrons belonging to this distribution at a given time will, therefore, continue to belong to this distribution. The electrons belonging at a given time to the residual distribution

$$f_R = f_2 + \psi$$

will, therefore, also continue to belong to this distribution. Moreover, since the distribution f_1 is symmetrical with respect to the origin of the velocity coordinates at any point (x, y, z) and, hence, does not give rise to any collective motion, or drift, of the electrons, we need, in order to determine this motion, to take into account only the residual distribution f_R. We shall now consider separately the different causes which produce a collective motion of the electrons.

We first consider the case in which no external forces are acting, and where the collective motion of the electrons can, therefore, be regarded as arising from a kind of "free diffusion" from places in the metal where the concentration of electrons is higher to places where it is lower. In this case, the sum of the kinetic energy and the potential energy with respect to the metal atoms will be constant for each electron.

We shall denote the number of electrons in the volume element dV with energy between a and $a + da$ by $N(a)da\,dV$. The excess number of electrons with energies between a and $a + da$ which in unit time pass in the positive direction through a surface element dS near (x_0, y_0, z_0), perpendicular to the x-axis and large compared to the dimensions of the atoms,—which number with the notations introduced equals $(G_x(a)/m)da\,dS$,— can now be written

$$\frac{1}{m} G_x(a)da\,dS = -D(a)\frac{dN(a)}{dx}da\,dS, \tag{24}$$

where the "diffusion coefficient" $D(a)$ is a function of a, dependent only on the nature of the metal and the temperature at the point considered. Denoting the number of electrons in dV belonging to the residual distribution f_R, and having energy between a and $a + da$, by $N_R(a)da\,dV$, and realizing that in the present case the number of electrons per unit volume belonging to the distribution f_1 is the same everywhere in the metal, we can instead of eq. (24) write

$$\frac{1}{m} G_x(a)da\,dS = -D(a)\frac{dN_R(a)}{dx}da\,dS. \tag{25}$$

When external forces act—and we shall here include cases in which the field of the metal atoms due to inhomogeneity produces a resultant force in a certain direction,— the collective motion of the electrons arises from the fact that the paths of all the electrons are affected in the direction of the force. However, since the external forces are assumed to be extremely small compared to the forces from the metal atoms, the motion of the individual electrons will differ very little from the motion in the case considered above, and an appreciable collective motion of the electrons in a certain direction is produced only because of the exceedingly large number of the electrons. As for the electrons belonging to the distribution f_1, of which the number per unit volume is not the same at all places in the metal, as in the former case, the collective motions arising from the external forces and from differences in concentration will completely cancel each other. Moreover, on account of the comparatively small number of electrons in the volume element $\mathrm{d}V$ belonging to the residual distribution, we can neglect the effect of the external forces on the motion of these electrons and, therefore, in this case, where $a + \varepsilon\varphi$ is constant during the motion of each electron, much as in the case considered above, write

$$\frac{1}{m} G_x(a)\mathrm{d}a\,\mathrm{d}S = -D(a)\frac{\mathrm{d}N_\mathrm{R}(a-\varepsilon\varphi)}{\mathrm{d}x}\mathrm{d}a\,\mathrm{d}S. \tag{26}$$

We now have

$$N_\mathrm{R}(a-\varepsilon\varphi)\mathrm{d}a\,\mathrm{d}V = \int\int f_\mathrm{R}\,\mathrm{d}\sigma\,\mathrm{d}v, \tag{27}$$

where the integration with respect to $\mathrm{d}\sigma$ at every point (x, y, z) is to be taken over the part of velocity space that corresponds to a kinetic energy between $a - \varepsilon\lambda - \varepsilon\varphi$ and $a - \varepsilon\lambda - \varepsilon\varphi + \mathrm{d}a$, and where the integration with respect to $\mathrm{d}v$ is over the volume element $\mathrm{d}V$.

In evaluating the right side of eq. (26), we can neglect the distribution ψ, since the number of electrons per unit volume of this distribution is a very small fraction of the whole number of electrons (the same holds for the variations with respect to the coordinates x, y and z of the numbers in question, compare p. [315]). We, therefore, need to consider only the distribution f_2. In differentiating the right side of eq. (27), we can neglect the differentiation of the limits of the integral, since f_2 is infinitesimal within the volume element $\mathrm{d}V$, and we thus get

$$\frac{\mathrm{d}N_\mathrm{R}(a-\varepsilon\varphi)}{\mathrm{d}x}\mathrm{d}a\,\mathrm{d}V = \int\int\frac{\partial f_2}{\partial x}\,\mathrm{d}\sigma\,\mathrm{d}v.$$

Using eq. (23), and noticing that the two terms on the right side are equal at the point (x_0, y_0, z_0) in $\mathrm{d}V$, we get, to the same approximation as above,

$$\frac{\mathrm{d}N_\mathrm{R}(a-\varepsilon\varphi)}{\mathrm{d}x}\mathrm{d}a\,\mathrm{d}V = \left(\frac{\varepsilon}{kT}\frac{\partial\varphi}{\partial x} + \frac{1}{K}\frac{\partial K}{\partial x} + \frac{a}{kT^2}\frac{\partial T}{\partial x}\right)N(a)\mathrm{d}a\,\mathrm{d}V.$$

Substituting this in eq. (26), we obtain the following equation for the determination of $G_x(a)$

$$G_x(a) = -m \left(\frac{\varepsilon}{kT} \frac{\partial \varphi}{\partial x} + \frac{1}{K} \frac{\partial K}{\partial x} + \frac{a}{kT^2} \frac{\partial T}{\partial x} \right) D(a)N(a). \qquad (28)^*$$

With regard to the applications of eq. (28), it should be pointed out that, since it was deduced under the assumption that the external forces are stationary, it can be used only in the treatment of stationary (or quasi-stationary) problems (compare p. [317]).

As a single application of the results obtained in this section, we shall consider the case in which the metal atoms and electrons are assumed to interact as *hard elastic bodies*. We shall not make any special assumptions about the shape or mutual distances of the metal atoms. (This differentiates this example from those considered in § 3 of this chapter, where it was assumed that the dimensions of the metal atoms were small compared to their mutual distances.) We shall only assume that the dimensions and masses of the atoms are very large compared to those of the electrons, so that we can neglect the effect of mutual collisions among the electrons as compared to the effect of collisions between electrons and atoms and can consider the atoms as immovable within such short intervals of time in which the electrons, on the average, lose the greater part of their original momentum in a given direction.

In the case in question, groups of electrons with the same energy will correspond to groups with the same speed. Now, consider an electron moving among the metal atoms unaffected by external forces; according to the assumptions made about the atoms, its path will be entirely independent of its speed. Moreover, since the distance travelled by an electron in a given time is proportional to its speed, the diffusion coefficient $D(a)$ will in this case be directly proportional to the speed of the electrons. Introducing this fact in eq. (28), putting $a = \frac{1}{2}mr^2$, and finally considering that λ in this case vanishes in certain regions of space (i.e., outside the atoms) and is infinitely large in the rest of space (inside the atoms), we obtain an equation for determining the momentum which corresponds exactly to the equation found when we discussed the theory of H. A. Lorentz (i.e., the equation obtained by substituting the expression for $F(r)$ given by eq. (15) into eq. (14), and putting $Q(\rho, r) = 0$), the only difference being that the mean free path l is replaced by a constant with a different physical significance.

It follows from this that the ratio between the electric and thermal conductivities will be the same in all cases in which it is assumed that the metal atoms affect the electrons as hard bodies (i.e., with forces that vary extremely rapidly with the distance (compare p. [321])), irrespective of the shape and mutual distances of the atoms, and, hence, that its value will be the same as that computed by Lorentz on

---

* [The preceding three paragraphs have been revised so as to conform to III. In accordance with II, five paragraphs in small type have been deleted.]

the assumption that the atoms are hard elastic spheres with radii very small compared to their mutual distances (see the next chapter p. [337]).

In conclusion, we may remark that in all other cases $D(a)$ will increase more rapidly with increasing a than in the case considered above; for the faster electrons not only travel faster along the same paths, but also force their way better than the slower electrons (except just in the case that the metal atoms are perfectly hard bodies). The significance of this will be discussed in the next chapter.

CHAPTER II

STATIONARY PROBLEMS

§ 1

Expressions for the Transfer of Electricity and Energy through a Metal

Applying the equations derived in the previous chapter, we shall now calculate the *amounts of electricity and energy transferred in unit time by the motion of electrons through a unit surface in the interior of a piece of metal*. In all calculations in this chapter, we shall assume that the external forces and temperature in the different parts of the metal are *stationary*, or at least quasi-stationary (see p. [317]).

We shall first consider the case in which the motion of the electrons consists of *free paths and separate collisions*. Denoting the amount of electricity transferred with the electrons in the time dt through a surface element dS perpendicular to the x-axis by $i_x dS dt$, and, similarly, denoting the kinetic energy conducted in this time through the same surface element by $W_x dS dt$, we have, with the notations of the previous chapter,

$$i_x = \frac{\varepsilon}{m} \int_0^\infty G_x(r) dr \qquad (29a)$$

and

$$W_x = \frac{1}{2} \int_0^\infty r^2 G_x(r) dr. \qquad (30a)$$

For the determination of $G_x(r)$, we have, since in the present case the right side of eq. (14) can be put equal to zero,

$$-\tfrac{4}{3}\pi m K \left(\frac{\varepsilon}{kT} \frac{\partial \varphi}{\partial x} + \frac{1}{K} \frac{\partial K}{\partial x} + \frac{mr^2}{2kT^2} \frac{\partial T}{\partial x} \right) r^4 e^{-\frac{1}{2}mr^2/kT}$$

$$-G_x(r)F(r) + \int_0^\infty G_x(\rho)Q(\rho, r) d\rho = 0. \quad (31a)$$

This is an integral equation of the Fredholm type[1],

---

[1] See, e.g., M. Bôcher: *Introduction to the Study of Integral Equations* (Cambridge 1909), p. 29–38. In eq. (31a) above, the integral is taken between infinite limits, while in the proof of Fredholm's solution it is assumed that the integral is taken between finite limits. This, however, does not give rise to any difficulty here, since, from the physical meaning of eq. (31a) alone, it

$$\Psi(r) = \psi(r) + \int_a^b \Psi(\rho)\tau(r, \rho)d\rho,$$

the solution of which can be written in the form

$$\Psi(r) = \psi(r) + \int_a^b \psi(\rho)\pi(r, \rho)d\rho,$$

where the function $\pi(r, \rho)$ depends only on $\tau(r, \rho)$ (i.e., is independent of $\psi(r)$). From Fredholm's expression for $\pi(r, \rho)$, it follows, further, that $\pi(r, \rho)$ will be a symmetrical function of r and ρ, provided $\tau(r, \rho)$ is symmetrical in these variables.

Now, putting $G_x(r) = \Psi(r)g(r)$, where $g(r) = F(r)^{-1}r^2 e^{-\frac{1}{2}mr^2/kT}$, and using eq. (13) for $Q(\rho, r)$, we get from (31a)

$$\Psi(r) = -\tfrac{4}{3}\pi m K \left(\frac{\varepsilon}{kT} \frac{\partial\varphi}{\partial x} + \frac{1}{K} \frac{\partial K}{\partial x} + \frac{mr^2}{2kT^2} \frac{\partial T}{\partial x} \right) g(r)$$
$$+ \int_0^\infty \Psi(\rho)g(\rho)g(r)S(\rho, r)d\rho.$$

Since the function $g(\rho)g(r)S(\rho, r)$, which corresponds to the function $\tau(r, \rho)$ in Fredholm's equation, is symmetrical with respect to r and ρ, we get, according to what was stated above,

$$\Psi(r) = -\tfrac{4}{3}\pi m K \left(\frac{\varepsilon}{kT} \frac{\partial\varphi}{\partial x} + \frac{1}{K} \frac{\partial K}{\partial x} + \frac{mr^2}{2kT^2} \frac{\partial T}{\partial x} \right) g(r)$$
$$- \int_0^\infty \tfrac{4}{3}\pi m K \left(\frac{\varepsilon}{kT} \frac{\partial\varphi}{\partial x} + \frac{1}{K} \frac{\partial K}{\partial x} + \frac{m\rho^2}{2kT^2} \frac{\partial T}{\partial x} \right) g(\rho)\mathscr{S}(\rho, r)d\rho,$$

where $\mathscr{S}(\rho, r)$ is a symmetrical function of ρ and r dependent only upon $g(r)$ and $S(\rho, r)$. From eqs. (29a) and (30a), we now obtain

$$i_x = -A_1 \left(\frac{\partial\varphi}{\partial x} + \frac{kT}{\varepsilon K} \frac{\partial K}{\partial x} \right) - A_2 \frac{\partial T}{\partial x} \tag{32a}$$

and

$$W_x = -A_2 T \left(\frac{\partial\varphi}{\partial x} + \frac{kT}{\varepsilon K} \frac{\partial K}{\partial x} \right) - A_3 \frac{\partial T}{\partial x}, \tag{33a}$$

where

$$A_1 = \frac{4\pi\varepsilon^2 K}{3kT} \left\{ \int_0^\infty r^4 F(r)^{-1} e^{-\frac{1}{2}mr^2/kT} \, dr + \int_0^\infty g(r) \left[\int_0^\infty g(\rho)\mathscr{S}(\rho, r)d\rho \right] dr \right\},$$

---

follows that it would make no difference, neither for the solution of the equation, nor for the applications to be made of it, whether the integral is taken between 0 and ∞ or between two limits of which one is very small and the other very large compared to the average speed of the electrons at the temperature in question.

$$A_2 = \frac{2\pi\varepsilon mK}{3kT^2} \left\{ \int_0^\infty r^6 F(r)^{-1} e^{-\frac{1}{2}mr^2/kT} \, dr + \int_0^\infty g(r) \left[\int_0^\infty \rho^2 g(\rho) \mathscr{S}(\rho, r) d\rho \right] dr \right\}$$

$$= \frac{2\pi\varepsilon mK}{3kT^2} \left\{ \int_0^\infty r^6 F(r)^{-1} e^{-\frac{1}{2}mr^2/kT} \, dr + \int_0^\infty r^2 g(r) \left[\int_0^\infty g(\rho) \mathscr{S}(\rho, r) d\rho \right] dr \right\}$$

and

$$A_3 = \frac{\pi m^2 K}{3kT^2} \left\{ \int_0^\infty r^8 F(r)^{-1} e^{-\frac{1}{2}mr^2/kT} \, dr + \int_0^\infty r^2 g(r) \left[\int_0^\infty \rho^2 g(\rho) \mathscr{S}(\rho, r) d\rho \right] dr \right\}.$$

Eqs. (32a) and (33a) will hold in all cases in which separate or distinct collisions are assumed to occur; thus they contain as special cases the equations given by Lorentz[1] and Gruner[2]. The cicumstance that the same "constant" (i.e., function of the nature and temperature of the metal at the point considered) A_2 enters into the expressions for i_x and W_x shows that a peculiar universal connection exists between the flows of electricity and heat through the metal. The physical significance of this will be discussed below when the thermoelectric phenomena are considered.

As a single example of the determination of the quantities A_1, A_2 and A_3, we shall here consider the simple case in which the metal atoms are assumed to be fixed centres of force affecting the electrons with forces inversely proportional to the nth power of the distance, and where the mutual interaction among the electrons is neglected. In this case we have (see p. [321]) $F(r) = Cr^{(n-5)/(n-1)}$ and $Q(\rho, r) = 0$ and hence also $\mathscr{S}(\rho, r) = 0$). We, therefore, obtain

$$A_1 = \frac{4\pi\varepsilon^2 K}{3kT} \frac{1}{C} \int_0^\infty r^{(3n+1)/(n-1)} e^{-\frac{1}{2}mr^2/kT} \, dr = \frac{4\pi\varepsilon^2 K}{3mC} \left(\frac{2kT}{m}\right)^{(n+1)/(n-1)} \Gamma\left(\frac{2n}{n-1}\right),$$

$$A_2 = \frac{2\pi\varepsilon mK}{3kT^2} \frac{1}{C} \int_0^\infty r^{(5n-1)/(n-1)} e^{-\frac{1}{2}mr^2/kT} \, dr = \frac{2\pi\varepsilon K}{3TC} \left(\frac{2kT}{m}\right)^{2n/(n-1)} \Gamma\left(\frac{3n-1}{n-1}\right)$$

$$= A_1 \frac{2n}{n-1} \frac{k}{\varepsilon},$$

$$A_3 = \frac{\pi m^2 K}{3kT^2} \frac{1}{C} \int_0^\infty r^{(7n-3)/(n-1)} e^{-\frac{1}{2}mr^2/kT} \, dr = \frac{\pi mK}{3TC} \left(\frac{2kT}{m}\right)^{(3n-1)/(n-1)} \Gamma\left(\frac{4n-2}{n-1}\right)$$

$$= A_1 \frac{2n(3n-1)}{(n-1)^2} \frac{k^2}{\varepsilon^2} T. \tag{34}$$

We shall now consider the case in which *the metal atoms are assumed to be so*

[1] H. A. Lorentz: Proc. Acad. Amsterdam **7** (1905) 447.
[2] P. Gruner: Verh. d. Deutsch. Phys. Ges. **10** (1908) 524.

near one another that the electrons are influenced by the forces from the atoms during the greater part of their motion. Contrary to what was true in the simpler case —separate collisions—discussed above, we cannot here, as explained in the previous chapter, give a general equation for determining the collective momentum of the electrons, and, hence, we cannot state the general form for the expressions for the flows of electricity and energy. We have only derived equations for determining the momentum of the electrons under certain limiting conditions, namely that the field of force from the metal atoms can be regarded as stationary, and that the mutual interaction among the electrons can be considered very small compared to the interaction between the metal atoms and the electrons. The following calculations, therefore, only apply to this case.

We shall now determine the amount of electricity and the sum of kinetic energy and potential energy relative to the metal atoms which in the time dt are transferred through a surface element dS perpendicular to the x-axis as a result of the motions of the electrons. Denoting these quantities by $i_x dS dt$ and $W_x dS dt$, as in the previous case, and using the notations introduced in Chapter I, we get

$$i_x = \frac{\varepsilon}{m} \int_0^\infty G_x(a) da \qquad (29b)$$

and

$$W_x = \frac{1}{m} \int_0^\infty a G_x(a) da, \qquad (30b)$$

where $G(a)$ is determined by (see eq. (28))

$$G_x(a) = -m \left(\frac{\varepsilon}{kT} \frac{\partial \varphi}{\partial x} + \frac{1}{K} \frac{\partial K}{\partial x} + \frac{a}{kT^2} \frac{\partial T}{\partial x} \right) D(a) N(a). \qquad (31b)$$

From (29b) and (30b) we now get

$$i_x = -A_1 \left(\frac{\partial \varphi}{\partial x} + \frac{kT}{\varepsilon K} \frac{\partial K}{\partial x} \right) - A_2 \frac{\partial T}{\partial x}, \qquad (32b)$$

$$W_x = -A_2 T \left(\frac{\partial \varphi}{\partial x} + \frac{kT}{\varepsilon K} \frac{\partial K}{\partial x} \right) - A_3 \frac{\partial T}{\partial x}, \qquad (33b)$$

where

$$A_1 = \frac{\varepsilon^2}{kT} \int_0^\infty D(a) N(a) da, \qquad A_2 = \frac{\varepsilon}{kT^2} \int_0^\infty a D(a) N(a) da,$$

$$A_3 = \frac{1}{kT^2} \int_0^\infty a^2 D(a) N(a) da. \qquad (35)$$

It is seen that eqs. (32b) and (33b) have exactly the same form as eqs. (32a) and

(33a); hence the remarks that were made about these equations hold also for those obtained here.

Eqs. (32) and (33) determine completely the flow of electricity and heat through the metal. Since we have made no special assumptions about the direction of the *x*-axis, the equations show that the flow of electricity and energy through a surface element depends only on the variation of the potential of the external forces, the temperature, etc., in a direction perpendicular to the element. In the following sections we shall investigate, on the basis of these equations, the electric and thermal conductivities of metals and the thermoelectric phenomena.

§ 2

Conduction of Electricity and Heat

If a homogeneous piece of metal of uniform temperature is subjected to a constant electric field $E = -\partial\varphi/\partial x$, an *electric current* will pass through it, the strength of which per unit area i can be found directly from eqs. (32). Since the derivatives of K and T vanish in this case, we get

$$i = A_1 E.$$

Denoting the *electric conductivity* by σ, we consequently get

$$\sigma = A_1. \tag{36}$$

In particular, if we assume that separate collisions occur and that the forces between metal atoms and electrons are inversely proportional to the *n*th power of the distance, we obtain, with the aid of eqs. (34) and (5),

$$\sigma = \frac{4N\varepsilon^2}{3\sqrt{\pi}mC}\left(\frac{m}{2kT}\right)^{(n-5)/2(n-1)}\Gamma\left(\frac{2n}{n-1}\right). \tag{37}$$

If we let n increase to infinity, which is the same as assuming that the atoms and electrons act upon each other as hard elastic spheres (see p. [321]), we get, putting $C = 1/l$,

$$\sigma = \tfrac{4}{3}lN\varepsilon^2(2\pi mkT)^{-\frac{1}{2}},$$

which is the expression for σ found by Lorentz[1].

By the motion of the electrons, not only electricity but also energy—*heat*—is transferred through the metal. However, it is, in general, not possible to assign a definite value to the amount of energy transferred by the motion of electrons

---

[1] H. A. Lorentz: Proc. Acad. Amsterdam **7** (1905) 448. (In Lorentz' paper different notations are used.)

through a given surface element; for the energy of the electrons, which is composed partly of kinetic energy, partly of potential energy relative to the metal atoms and to quantities of free electricity (external force fields), and partly of "internal" energy, can only be specified to within an arbitrary constant. Only in a single case, namely when the net number of electrons passing through a surface element in a given direction vanishes, will this indeterminacy disappear. In this case the amount of energy transferred in unit time through a unit surface perpendicular to the x-axis will be given by the quantity W_x mentioned above. Since the electric current is zero in this case, we get, after eliminating $\partial\varphi/\partial x$ from the equation $i_x = 0$ and the equation for W_x (eqs. (32) and (33)),

$$W_x = -\frac{A_3 A_1 - A_2^2 T}{A_1}\frac{\partial T}{\partial x}.$$

The flow of energy in this case, which is the only one that can be treated completely, we shall speak of as *conduction of heat*; thus, denoting the thermal conductivity by γ, we get

$$\gamma = \frac{A_3 A_1 - A_2^2 T}{A_1}. \tag{38}$$

As may be seen, in this manner we have determined the thermal conductivity as a quantity dependent only on the state of the metal—its nature and temperature—at the place of interest; moreover, this is the quantity that is measured in the experimental investigations of heat conduction. In most of these investigations —e.g., those of L. Lorenz[1]—the variation of the temperature in a metal bar, surrounded by electric insulators, is observed. In such a metal bar electric equilibrium will rapidly be established, so that no current flows; the conditions in the metal bar will, therefore, conform exactly to the case discussed above. In other investigations—as, for example, those of Jaeger and Diesselhorst[2]—the distribution of temperature and electric potential is observed in a metal rod carrying a current, and the thermal conductivity is determined with the aid of a relation, given by F. Kohlrausch[3], between the temperature and the potential at the various points of the rod. As will be seen in connection with the discussion of the thermoelectric phenomena, from eqs. (32) and (33) and the definition of the thermal conductivity given here, an expression for the development of heat is obtained corresponding to that on which the relation of Kohlrausch is based.

In particular, if we assume separate collisions and that the forces between metal

[1] L. Lorenz: Wied. Ann. **13** (1881) 422.
[2] W. Jaeger u. H. Diesselhorst: Wiss. Abh. d. phys.-tech. Reichsanstalt **3** (1900) 269.
[3] F. Kohlrausch: Sitzungsber. d. Berliner Akad. d. Wiss. (1899) 711.

atoms and electrons are inversely proportional to the nth power of the distance we get from eqs. (38), (34) and (37)

$$\gamma = A_1 \frac{2n}{n-1} \frac{k^2}{\varepsilon^2} T = \frac{8nNk^2T}{3\sqrt{\pi}(n-1)mC} \left(\frac{m}{2kT}\right)^{(n-5)/2(n-1)} \Gamma\left(\frac{2n}{n-1}\right). \quad (39)$$

If we let n increase to infinity and put $C = 1/l$, we get

$$\gamma = \tfrac{8}{3} lNk^2T(2\pi mkT)^{-\frac{1}{2}},$$

which is Lorentz' value.

For the *ratio between the thermal and the electric conductivities* we now get from eqs. (36) and (38)

$$\kappa = \frac{\gamma}{\sigma} = \frac{A_3 A_1 - A_2^2 T}{A_1^2}. \quad (40)$$

In the special case considered above, we get from eqs. (37) and (39)

$$\kappa = \frac{2n}{n-1} \frac{k^2}{\varepsilon^2} T, \quad (41)$$

whence, for $n = \infty$,

$$\kappa = 2 \frac{k^2}{\varepsilon^2} T, \quad (42)$$

which is the expression found by Lorentz[1].

---

[1] H. A. Lorentz: Proc. Acad. Amsterdam **7** (1905) 449.

In connection with Lorentz' value for κ, we shall mention those which E. Riecke (Wied. Ann. **66** (1898) 353, 545, 1199; Ann. d. Phys. **2** (1900) 835; Jahrb d. Rad. u. El. **3** (1906) 24; Phys. Zeitschr. **10** (1909) 508) and P. Drude (Ann. d. Phys. **1** (1900) 566, **3** (1900) 369) have previously calculated for this quantity, in our notations,

$$\kappa = \frac{27}{8} \frac{k^2}{\varepsilon^2} T \left(1 + \frac{2}{3} \frac{T}{N} \frac{dN}{dT}\right) \quad \text{and} \quad \kappa = 3 \frac{k^2}{\varepsilon^2} T,$$

respectively; actually, these expressions are in better agreement with experiments than that obtained by Lorentz (see, e.g., E. Riecke: Phys. Zeitschr. **10** (1909) 513).

These authors make the same assumptions about the effects of collisions between electrons and metal atoms as Lorentz, i.e., they assume that, on the average, an electron loses in a collision all of its original momentum in a given direction. They also assume that the electrons are in mechanical and thermal equilibrium with the metal atoms, i.e., that the mean value of the kinetic energy of an electron is equal to the mean value of the translatory kinetic energy of a molecule of a gas at the same temperature; however, contrary to Lorentz, they assume, for simplicity, that all electrons have the same speed. However, according to the kinetic theory of gases, Maxwell's velocity distribution is so closely connected with the existence of thermal equilibrium that we are scarcely justified in attaching any importance to the better agreement mentioned. Furthermore,

While the expressions found for the electric and thermal conductivities contain the quantities N and l (or C), which are unknown at the outset, or at least very imperfectly known, this is seen not to be the case for the expressions (41) and (42) for the ratio between the conductivities. These expressions, therefore, permit a comparison between the theory and the experimental results.

The quantities ε (the elementary electric charge) and k (the gas constant per molecule), which enter into the expressions (41) and (42), appear to be known only with comparatively low accuracy; however, as M. Reinganum has pointed out[1], the ratio k/ε can be determined with high accuracy from well known quantities. Thus, denoting the number of molecules in one cm^3 of a gas (e.g. hydrogen) at the absolute temperature T and the pressure p by N, we have (see p. [305]) $p = NkT$; if, further, the amount of electricity that must be passed through an electrolyte to separate one cm^3 of hydrogen at the same temperature and pressure is called E, we have $E = 2\varepsilon N$ and, hence, $k/\varepsilon = 2p/ET$. Since, for $T = 273°$ and $p = 1.013$ (1 atm), $E = 0.259 \times 10^{10}$ (absolute electrostatic units), we obtain $k/\varepsilon = 0.287 \times 10^{-6}$.

[1] M. Reinganum: Ann. d. Phys. **2** (1900) 398.

---

it can be shown that this better agreement for the most part arises not from the assumption mentioned, but from the somewhat inexact manner in which the calculations are carried out.

For the electric conductivity, Riecke finds, by calculating the paths of the individual electrons, $\sigma = \frac{2}{3}lN\varepsilon^2(3mkT)^{-\frac{1}{2}}$. Drude, on the other hand, finds $\sigma = \frac{1}{2}lN\varepsilon^2(3mkT)^{-\frac{1}{2}}$, i.e., a value only $\frac{3}{4}$ of that obtained by Riecke. This difference arises from the fact that Drude calculates the average velocity of the electrons in a given direction from their average speed along a free path, without taking into account that the time of travel is different for the different paths, depending on whether the electron moves with or against the external electric field.

For the thermal conductivity, Riecke and Drude find the following expressions

$$\gamma = \tfrac{9}{4}lNk^2T(3mkT)^{-\frac{1}{2}}\left(1 + \frac{2}{3}\frac{T}{N}\frac{dN}{dT}\right)$$

and

$$\gamma = \tfrac{3}{2}lNk^2T(3mkT)^{-\frac{1}{2}},$$

respectively. The difference between these two expressions arises from the fact that Riecke and Drude define the thermal conductivity differently; while Drude, as we have done it above, defines the conduction of heat as flow of energy in the absence of an electric current, Riecke defines it as flow of energy in the absence of an external electric field. (Thus, the correction terms in Riecke's formula involving the variation of the number of free electrons with temperature arise from terms which correspond to $\partial K/\partial x$ in eq. (5) and enter into the calculations when heat conduction is defined as done by Riecke). Hence, as mentioned in the text above, Riecke's value for the thermal conductivity cannot be used for comparison with the experimental values found for this quantity.

Thus, on the assumptions on which the theories of Riecke and Drude are based, the value obtained for the ratio between the thermal and electric conductivities (i.e., the ratio between Drude's value for γ and Riecke's for σ) is found to be

$$\kappa = \frac{9}{4}\frac{k^2}{\varepsilon^2}T,$$

The table on p. [340] gives a summary of results obtained in experiments by Jaeger and Diesselhorst[1]; the values of σ, γ and κ are for 18° C, and under the headings $T\Delta\sigma/\sigma\Delta T$, $T\Delta\gamma/\gamma\Delta T$ and $T\Delta\kappa/\kappa\Delta T$ are given the values of

$$\frac{T_{18°}(\sigma_{100°}-\sigma_{18°})}{\sigma_{18°}(T_{100°}-T_{18°})}$$

and the analogous expressions.

The values which correspond to Lorentz' theory (eq. (42)) are

$$\kappa = 0.479 \times 10^{-10} \quad \text{and} \quad \frac{T}{\kappa}\frac{\Delta\kappa}{\Delta T} = 1.$$

It is seen from the table that the different metals, although their thermal and electric conductivities differ widely, nevertheless have almost the same ratio between the conductivities—the Wiedemann-Franz law,— and that this ratio for the individual metals varies approximately as the absolute temperature—the law of Lorenz. Moreover, as regards order of magnitude, there is good agreement between the calculated values of κ and those found experimentally.

This agreement, which, as mentioned above, was first pointed out by Drude, is one of the most beautiful results of the electron theory, and it seems to show definitely that the flow of heat, as well as of electricity, in the metals is carried by particles having electric charges of the same order of magnitude as the ions in electrolytes and being in thermal equilibrium with the metal atoms, i.e., having translatory kinetic energies equal to those of gas molecules at the same temperature (see Reinganum: loc. cit. p. 403).

[1] W. Jaeger u. H. Diesselhorst: loc. cit. p. 424. (The numerical values are computed from their last table. In cases in which several samples of the same metal were investigated, only data for the metal samples containing the smallest amounts of impurities have been used.)

a value which deviates little from that found by Lorentz and does not agree much better with experiments.

In connection with Lorentz' calculations, we may also mention an attempt made by G. Jaeger (Sitzungsber. d. Wiener Akad. d. Wiss., math-nat. Kl. **117**, Abt. IIa (1908) 869) to calculate the ratio between the thermal and electric conductivities from Lorentz' assumptions, without using statistical methods, but by considering the electron paths themselves. Jäger finds a value for σ which is $\frac{3}{4}$ that of Lorentz; this is due to the fact that he calculates this quantity in a similar manner as Drude (see above). That Jaeger, nevertheless, finds the same value for κ as **Lorentz** arises from the fact that an error has occurred in his calculation of the thermal conductivity, making this $\frac{3}{4}$ times too large. In fact, he assumes (loc. cit. p. 873) that the electrons passing through a surface element have suffered their last collision, on the average, at a distance $\frac{1}{2}l$ from the surface (l is the mean free path), while this distance actually is $\frac{2}{3}l$. (The error has arisen because, in calculating this average distance, he has assumed that the same number of electrons go through the surface element in any direction, while actually the number of electrons passing through the surface is proportional to the cosine of the angle between the direction of motion of the electrons and the normal to the surface.)

	$\sigma \cdot 10^{-17}$	$-T\Delta\sigma/\sigma\Delta T$	$\gamma \cdot 10^{-7}$	$T\Delta\gamma/\gamma\Delta T$	$\kappa \cdot 10^{10}$	$T\Delta\kappa/\kappa\Delta T$
silver	5.53	0.84	4.21	—0.05	0.762	1.03
copper	5.06	0.86	3.76	—0.06	0.742	1.07
gold	3.72	0.79	2.93	+0.01	0.788	1.03
aluminium	2.84	0.82	2.01	+0.09	0.707	1.18
zinc	1.486	0.84	1.110	—0.04	0.747	1.05
cadmium	1.182	0.88	0.927	—0.11	0.784	1.02
palladium	0.840	0.78	0.704	+0.20	0.838	1.26
platinum	0.832	0.81	0.696	—0.15	0.837	1.25
nickel	0.765	0.89	0.594	—0.09	0.777	1.07
tin	0.745	0.93	0.608	—0.23	0.817	0.94
iron	0.699	0.97	0.636	—0.12	0.911	1.18
steel	0.452	0.98	0.453	—0.02	1.003	0.98
lead	0.436	0.88	0.346	—0.05	0.794	1.11
bismuth	0.0756	0.91	0.081	—0.60	1.071	0.43
brass 85.7 Cu, 7.15 Zn, 6.39 Sn, 0.58 Ni	0.710	0.22	0.597	+0.67	0.841	0.95
manganin 84 Cu, 4 Ni, 12 Mn	0.214	0.00	0.217	—0.77	1.016	0.76
constantan 60Cu, 40 Ni	0.184	0.00	0.224	—0.66	1.229	0.67

It is seen from the table, however, that there are appreciable differences among the values of κ for the different metals, and also that these values are considerably larger than that calculated by Lorentz.

Before continuing this comparison between theory and experiments, we must mention a circumstance that might be responsible for the fact that the experimental results cannot be compared directly with the theory here discussed. Thus, it might be thought that the conduction of heat in a metal can take place in a different manner than by the motion of free electrons; for also bodies that conduct electricity to an extremely small extent possess a certain, although very small, thermal conductivity. The presence of such a *non-metallic thermal conductivity* would cause the ratio κ to be larger than that calculated here. However, according to calculations by J. Koenigs-

---

In a recent treatise (Sitzungsber. d. Heidelberger Akad. d. Wiss., math.-nat. Kl. (1911) Abh. 10, p. 16), M. Reinganum expresses the opinion that Lorentz' value for the thermal conductivity γ, and consequently his value for κ, is too small because the electrons crossing a surface element will change their kinetic energy in the time between the passing through the surface element and the next collision, as a result of the electric field which is present in heat conduction and which cancels the electric current that otherwise would accompany the flow of energy. However, in my opinion, this is not correct, for the energy transferred through the metal must be equal to the energy transferred by the electrons through a given surface, irrespective of the later history of these electrons.*

* [The paragraph "In a recent treatise . . . of these electrons" has been added to the footnote to conform to III.]

berger[1], such a thermal conductivity of the order of magnitude as that found in insulators would have only a small influence upon the value of κ for the best conducting metals, but would be plainly noticeable in case of the most poorly conducting metals, and would explain the circumstance that, on the whole, the highest values of κ are found for these. Hence, Koenigsberger suggests the possibility that the ratio between the thermal and electric conductivities actually may be the same for all pure metals and has the value 0.70×10^{-10} at 18 °C. (It seems that the ratio must be somewhat higher in alloys.)

The above-mentioned differences are, however, just what might be expected from our preceding calculations. Thus, eq. (41) shows that *the ratio between the thermal and electric conductivities depends to a certain extent upon the special assumptions introduced about the forces acting between the metal atoms and the electrons.* Moreover, this equation shows that, *in the cases considered[2], the value of this ratio will always be higher than that calculated by Lorentz.* (Thus, it may be pointed out that, if we put $n = 3$ in eq. (41)—corresponding to the "electric doublets" considered by J. J. Thomson (see p. [322])—, we obtain a value for κ which agrees very well with the values found experimentally for the best conducting metals and is very close to the hypothetical value assumed by Koenigsberger (see above) for all the pure metals.)

According to what was said at the end of the previous chapter (see p. [329]), if we neglect the mutual interaction among the free electrons and regard the atomic force field as stationary, we will, *also in the case in which the dimensions of the metal atoms are not negligible compared to their mutual distances, find Lorentz' value for κ, provided we assume the atoms to be hard elastic bodies* (or assume the forces between the atoms and electrons to vary very rapidly with the distance, while, in other cases of the kind considered here (compare the concluding remark, p. [330]), as in the case mentioned above, the value of κ will, in general, *be larger than that of Lorentz[2].*

[1] J. Koenigsberger: Phys. Zeitschr. **8** (1907) 237. (See also M. Reinganum: Phys. Zeitschr. **10** (1909) 355, 645; ibid. **11** (1910) 673, where it is shown that the part of the thermal conductivity which might arise from internal heat radiation will be negligibly small in metals compared to the observed thermal conductivity.)

[2] The value of κ will be higher than Lorentz' value in all cases in which the "effective" mean free path (see p. [321]), corresponding to the comparatively small interval around the average in which the speed of almost all the electrons lies, increases regularly with increasing r (e.g., in all cases where the effective mean free path in this interval can be written, with sufficient accuracy, in the form $a+b(r-r_0)$, where r_0 is the average speed of the electrons, and a and b are constants.) In special cases, however, where the effective mean free path increases in a very irregular manner with r, κ can, under certain circumstances, become smaller than the value calculated by Lorentz. This will be the case, for example, in Gruner's theory (see p. [318]), in which the value of κ comes out smaller than that of Lorentz, provided the ratio between the "critical" speed G and the average speed of the electrons, and also the ratio between the two mean free paths l_1 and l_2, lie between certain definite limits (see P. Gruner: Phys. Zeitschr. **10** (1909) 50).

On the other hand, if the effect of the mutual collisions among the free electrons is not assumed to be small compared to that of the interaction between the metal atoms and the electrons, the value of κ could be considerably smaller than that found by Lorentz. This is due to the fact that collisions among the free electrons may cause a considerable reduction in the thermal conductivity by hindering the motion of the electrons from place to place in the metal, much as do the collisions with the atoms, while they would have little influence upon the electric conductivity, since the collective momentum of the electrons in a given direction is not changed by these collisions[1]. The entire influence upon the electric conductivity arises from the facts that, in these collisions, momentum is interchanged among groups of electrons having different speeds, and that these different groups of electrons will lose their momentum at different rates through collisions with metal atoms; in special cases in which this does not take place — e.g., in the case of separate collisions and forces between atoms and electrons inversely proportional to the fifth power of the distance (see p. [321]) — the mutual collisions among the electrons will, therefore, have no effect upon the electric conductivity.

We must, finally, remark that, according to the preceding calculations, we cannot beforehand expect any systematic relation between the deviations from the Wiedemann-Franz law and the deviations from the law of Lorenz. Thus, while eq. (41) shows that κ depends on the way in which the forces between the metal atoms and the electrons vary with the distance, it also shows that, if we assume separate collisions and forces varying inversely as some power of the distance, this ratio will be proportional to the absolute temperature for any metal; deviations from the law of Lorenz will, therefore, in such cases occur only if it is assumed that the forces with which the metal atoms affect the electrons themselves vary with temperature (e.g., as a result of changes taking place within the atoms).

The electron theory, at any rate approximately, allows us to determine the ratio between the thermal and electric conductivities of metals, since this ratio, as we have seen, depends only to a small degree on special assumptions about the forces acting between metal atoms and electrons; however, the situation is quite different when the question is about the individual conductivities. As already mentioned, the expressions obtained above for the conductivities do not allow any direct comparison with experiments because of the unknown quantities entering into them; nor can the manner in which the conductivities vary with temperature be explained without making special assumptions.

[1] Compare H. A. Lorentz: Proc. Acad. Amsterdam 7 (1905) 449 (In this connection, we may mention an attempt by P. Debye (loc. cit. p. 484) to calculate the influence of collisions among the free electrons upon the electric and thermal conductivities. As a result of his calculations, Debye finds that the ratio between the conductivities, contrary to what we have stated above, is not changed appreciably by these collisions. However, this conclusion arises from the fact that Debye assumes that an electron, on the average, will lose all its momentum in a given direction in a collision; as mentioned above, this is not correct, since it is a characteristic property of the mutual collisions among the free electrons that the total momentum of the electrons — and hence their contribution to the electric current — is not changed in collisions.)

In this connection, it may be pointed out that, if the constants N and C (or l) are assumed to be temperature independent, the formulas for the electric conductivity derived above on the basis of special simple assumptions cannot be brought into accordance with the way in which the electric conductivity of metals is found experimentally to vary with temperature. (Thus, eq. (37) shows that, if N and C are assumed to be independent of temperature, the electric conductivity, in all the cases considered (i.e., for $n > 2$), will at most decrease as $T^{-\frac{1}{2}}$, whereas actually the conductivities of most pure metals decrease approximately as T^{-1} with increasing temperature.) This is, however, scarcely other than what might be expected beforehand, since the conditions under which electrons move in metals must surely be very complicated and change with temperature in a complicated manner[1].

The experimental results are, also here, far more diverse than in the case mentioned above; there is, in particular, a far greater difference in the behavior of the alloys and the pure metals. Thus, the conductivity of many alloys is much lower than those of the pure metals of which they are composed and varies with temperature in quite a different manner (compare the table, p. [340])[2].

We must mention here that attempts have been made to explain these properties of alloys by means of a theory, according to which there should be an essential difference between the conduction of electricity in alloys and in pure metals, arising from the circumstance that the former are not to be regarded as physically homogeneous but as consisting of small parts (crystals) of the pure metals of which they are composed. Hence, in alloys there should, in addition to the ordinary electric resistance, appear an apparent resistance, i.e., a thermoelectric polarization arising from temperature differences produced by a Peltier effect at the surfaces of contact between the small parts mentioned. Such a theory has been proposed independently by Lorenz,

[1] J. Koenigsberger (Jahrb. d. Rad. u. El. **4** (1907) 158) and J. Koenigsberger u. K. Schilling (Ann. d. Phys. **32** (1910) 179) have attempted, on the basis of the electron theory, to explain the manner in which the electric conductivities of solids vary with temperature. They consider the presence of free electrons in the metal as the result of a process of dissociation of the atoms and, hence, assume that the temperature variation of the number N of free electrons can be calculated from general thermodynamic considerations. From such an assumption, it is indeed possible to explain that the electric conductivities of almost all substances investigated have a maximum at a certain temperature (although this temperature must, in any case, be very low for pure metals); but, to obtain a closer agreement with experiments, these authors (see e.g., Ann. d. Phys., loc. cit. p. 218) had to make appropriate additional assumptions about the variation of the mean free path with temperature, assumptions for which they give no theoretical arguments (They base their work on an expression for the electric conductivity corresponding to Lorentz' theory; however, the agreement with experiments would be no better if they had started from an expression for σ corresponding to eq. (37), since that would have necessitated making similar assumptions about the temperature variation of C.)

[2] With regard to the experimental results on the electric conductivity of alloys, see e.g. W. Guertler: Jahrb. d. Rad. u. El. **5** (1908) 17.

Ostwald, Rayleigh and Liebenow[1]. However, later investigations seem to indicate that the properties of alloys at ordinary temperatures cannot be explained by the assumption of such a physical inhomogeneity[2].

Before leaving these questions, we shall mention that it seems possible to explain the properties of alloys by the electron theory without postulating any special difference between them and pure metals, provided we assume, what does not seem improbable to me, that the presence of different kinds of atoms in the alloys has the effect of making the field of force in the interior stronger than in pure metals, where the forces may be supposed to cancel one another to a higher degree[3]. For this would not only explain the low conductivity of alloys as a result of the greater resistance against the collective motion of the electrons; but, since the fields of force in question would exert a greater influence upon the motion of the

---

[1] L. Lorenz: Wied. Ann. **13** (1881) 600; W. Ostwald: Zeitschr. f. Physik. Chem. **11** (1893) 520; Lord Rayleigh: Nature **54** (1896) 154; C. Liebenow: Zeitschr. f. Elektrochem. **4** (1897) 201.

[2] See, e.g., K. Baedeker: *Die elektrischen Erscheinungen in metallischen Leitern* (Braunschweig 1911), p. 47–48. (It may, however, be remarked that an attempt made by E. L. Lederer (Sitzungsber. d. Wiener Akad. d. Wiss., math.-nat. Kl. **117**, Abt. IIa (1908) 311) to disprove this theory by showing experimentally that the heat developed by an electric current in alloys can be calculated from the resistance and the magnitude of the current in exactly the same way as for pure metals is not correct, in my opinion, since this result is a direct consequence of the theorem of the conservation of energy.)

[3] We must mention here an attempt by R. Schenck (Ann. d. Phys. **32** (1910) 261) to explain the differences between alloys and pure metals on the basis of the electron theory. In Schenck's theory, alloys are conceived as "solid solutions" of one metal in another, and the dissolved metal's atoms are assumed to possess, on the average, a kinetic energy equal to that of a gas molecule at the same temperature. As the motion of the atoms is assumed subject to a great "frictional resistance", they can move only very slowly from place to place and, hence, cannot take any direct part in the conduction of heat or electricity; however, their motions are assumed to influence the motions of the free electrons. In the first place, the motion of the atoms is assumed to augment the "friction" hindering the motion of the free electrons. In order to explain that alloys show a higher value for the ratio between the thermal and electric conductivities than pure metals, Schenck (loc. cit. p. 273) further assumes "that atoms are able to transfer their kinetic energy to free electrons in collisions, and thus contribute to the establishment of a heat balance". This is more fully expressed as follows (loc. cit. p. 273) :"The kinetic energy that the electrons have to transport is larger in alloys than in pure metals; to their own energy is added the energy they have acquired by collisions with dissolved atoms. The total kinetic energy transferred by the electrons in the alloys is $i = (N_e - N_\mu)/N_e$ times as large as in the pure metal" (N_e is the number of free electrons, N_μ the number of dissolved atoms in unit volume). In accordance with these statements, Schenck substitutes ik for k in Drude's formula for the thermal conductivity ($\frac{3}{2}kT$ is the average translatory kinetic energy of a gas molecule at the temperature T), and, therefore, finds the ratio between the thermal and electric conductivities i^2 times as high for alloys as for pure metals.

However, the various assumptions made in this theory do not seem justified. In the first place, only the dissolved atoms are assumed to possess kinetic energy but not those of the solvent; this is quite inconceivable,—another thing is that, in many problems, e.g., the osmotic pressure,

slower electrons than upon that of the faster ones, the thermal conductivity would be reduced to a greater extent than the electric conductivity; and, finally, the temperature coefficient (taken with sign) of the electric conductivity would be expected to be higher than for pure metals, since the effective mean free path (assuming that the force fields of the atoms are independent of the temperature) would increase more rapidly with temperature than in pure metals (compare the changes with decreasing n in the values of γ/σ and $\sigma^{-1} \, d\sigma/dT$ given by eqs. (41) and (37)).

§ 3

Thermoelectric Phenomena

In the preceding, we have considered two special cases of transfer of electricity and heat in the interior of a homogeneous piece of metal, i.e., conduction of electricity and heat, respectively. We shall now, with the aid of eqs. (32) and (33), consider the more general cases of the transfer of electricity and heat in a metal, and we shall see how these equations lead to a complete theory of the *thermoelectric phenomena*. In treating these problems, we shall largely follow the elegant manner of presentation given by H. A. Lorentz[1].

[1] H. A. Lorentz: Proc. Acad. Amsterdam **7** (1905) 451, 585.

---

diffusion, etc., of solutes, it is not necessary to take the solvent molecules directly into consideration;—on the other hand, if kinetic energy is ascribed also to the atoms of the solvent metal, it is readily seen that the whole difference, assumed in Schenck's theory to exist between alloys and pure metals, completely disappears. Apart form this, the conclusions which Schenck draws from the motion of the dissolved atoms do not seem justified. Thus, the probability of collisions between electrons and metal atoms, and, hence, the above-mentioned "frictional resistance", will undoubtedly be greater when the atoms are assumed to be in motion than when they are immovable, but, on account of the great speed of the electrons compared to that of the atoms, this increase will be negligibly small. (Thus, in the kinetic theory of gases (see, e.g., Jeans: *Dynamical Theory of Gases*, p. 234) it is proved that the probability of collisions between two kinds of gas molecules of masses m_1 and m_2, respectively, will be $K\sqrt{1+m_1/m_2}$ where K is the probability for collisions when the molecules with mass m_2 are immovable. For the ordinary metals, the ratio between the masses of the electrons and the atoms will be ca. 10^{-5} and the square root will, therefore, deviate very little from one.) Moreover, as regards Schenck's assumptions about the energy transfer of the free electrons, they can, it seems to me, only be taken to mean that, in his theory, the average kinetic energy of a free electron is $i\frac{3}{2}kT$. (For it does not seem justifiable to distinguish between the "own energy" of the electrons and "the energy they have received by collisions with metal atoms".) However, this would be inconsistent with the assumptions which are established in the kinetic theory of gases, and which form the basis for the electron theory of metals, according to which the average translatory kinetic energy of a molecule of any kind whatever, in mechanical heat equilibrium with molecules of the same or another kind, is equal to $\frac{3}{2}kT$.

Substituting the quantities σ and γ defined by eqs. (36) and (38), and putting $A_2 = A_1 \mu k/\varepsilon$, we can write eqs. (32) and (33) in the form

$$i_x = -\sigma \left(\frac{\partial \varphi}{\partial x} + \frac{k}{\varepsilon} \frac{T}{K} \frac{\partial K}{\partial x} + \mu \frac{k}{\varepsilon} \frac{\partial T}{\partial x} \right) \tag{43}$$

and

$$W_x = \mu \frac{k}{\varepsilon} T i_x - \gamma \frac{\partial T}{\partial x}, \tag{44}$$

where the quantity K, according to eq. (20), is determined by

$$K = N \left(\frac{m}{2\pi kT} \right)^{\frac{3}{2}} \left| \int e^{-\varepsilon \lambda/kT} \, dv \right|^{-1} \tag{45}$$

(which, for "separate collisions" ($\lambda = 0$), gives $K = N(m/2\pi kT)^{\frac{3}{2}}$ (compare eq. (5)).

Eqs. (43) and (44) determine completely the transfer of electricity and heat in a piece of metal in which the nature of the metal and the temperature are the same at all points in planes perpendicular to the x-axis; in the general case, in which the nature of the metal and the temperature vary in an arbitrary manner, we must use, also, the equations obtained from (43) and (44) by interchanging x with y and z, respectively. However, since it makes no essential difference in the results, we shall in the following, for simplicity, assume that the piece of metal in question has the form of a thin wire—a thermoelectric circuit—with cross sections of dimensions that are very small compared to the radius of curvature of the wire, and in which the nature of the metal and the temperature are very nearly the same at all points of a cross section. As seen immediately, in such cases it will be sufficient to apply eqs. (43) and (44), provided x is taken to denote the distance along a curve passing through the centres of mass of the cross sections of the wire, measured from a fixed point on this curve.

We shall first consider an *open thermoelectric circuit*, i.e., a circuit in which no electric current flows. Putting $i_x = 0$ in eq. (43), we get in this case

$$\frac{\partial \varphi}{\partial x} = -\frac{k}{\varepsilon} \left(\frac{T}{K} \frac{dK}{dx} + \mu \frac{dT}{dx} \right), \tag{46}$$

an equation from which we can determine the potential difference between any two points in the circuit.

It must be pointed out, however, that this potential difference is that between two points in the interior of the metal. This potential difference is, therefore, not open to direct measurement, since it must be assumed that a considerable potential drop occurs in the immediate proximity of the surface of the metal. (In fact, such a potential drop has been revealed in the investigations of the emission of electrons by metals at high temperatures and of the heat liberated when free electrons are "absorbed" by a metal (see, e.g., Richardson and Cooke: Phil. Mag. **20** (1910) 173). The following calculations cannot, therefore, be used for direct comparison with experiment, except in cases where the question is about a potential difference between two points of the circuit in which the metal and temperature are the same.

From eq. (46), we may draw various conclusions.

(i) *If the circuit is of uniform temperature*, we get by integration

$$\varphi_P - \varphi_Q = -\frac{k}{\varepsilon} T \ln \frac{K_P}{K_Q}, \tag{47}$$

which shows that *the potential difference between two points P and Q of the circuit depends only on the nature of the metal at these points and vanishes if it is the same.* From this it follows that, with regard to these potential differences, the metals can at any temperature be arranged in a so-called tension series.

From eqs. (46) and (45), we get, in the case considered,

$$\frac{d\varphi}{dx} = -\frac{k}{\varepsilon} \frac{T}{N} \frac{dN}{dx} - \int \frac{\partial \lambda}{\partial x} e^{-\varepsilon\lambda/kT} dv \bigg/ \int e^{-\varepsilon\lambda/kT} dv. \tag{48}$$

Since, according to eq. (19), the number of electrons in the volume element dv, in case of statistical equilibrium, is proportional to $e^{-\varepsilon\lambda/kT}$, the last term on the right side of eq. (48) will, in this case, be equal to the value of $\partial\lambda/\partial x$ averaged over all of the electrons; denoting this average value by $\overline{\partial\lambda/\partial x}$, we get from eq. (48)

$$-N\varepsilon \left(\frac{\partial\varphi}{\partial x} + \frac{\overline{\partial\lambda}}{\partial x} \right) - kT \frac{dN}{dx} = 0,$$

which directly expresses the condition that the electric forces acting on the electrons balance the variations in the pressure of the electrons, $p = NkT$ (see p. [305]); for, in the case considered here, in which there is as well electric equilibrium as uniform temperature, the electrons will not have any collective directed motion, and no "piling up" of electrons in front of metal atoms will take place (see p. [324]; compare also the following case).

(ii) *If all parts of the circuit consist of the same metal,* so that K and μ are functions of the temperature only, the expression for $d\varphi$ will be an exact differential with respect to dT, and *the potential difference between two points of the circuit will, therefore, only depend on the temperature at these points, and will vanish if the temperature is the same.*

Eq. (46) does not allow as simple a physical interpretation as in the former case, since, in this case, there will be a flow of heat through the circuit, and, as a consequence of this collective motion of the electrons, there will be a piling up of them in front of the atoms, and the atoms will exert a resultant force of resistance upon the electrons. This is most easily seen, if we consider a piece of metal in which separate collisions are assumed to take place. Assuming that the forces exerted by the metal atoms upon the electrons decrease inversely as the *n*th power of the distance, we get, from eq. (34), $\mu = 2n/(n-1)$; from (46) we get, with the aid of (45), putting $\lambda = 0$,

$$\frac{\mathrm{d}\varphi}{\mathrm{d}x} = -\frac{k}{\varepsilon}\left(\frac{T}{N}\frac{\mathrm{d}N}{\mathrm{d}T} + \frac{n+3}{2(n-1)}\right)\frac{\mathrm{d}T}{\mathrm{d}x}. \tag{49}$$

If $n = \infty$, the last term in the parenthesis will be $\frac{1}{2}$, which is equal to the coefficient to the corresponding term in Lorentz' expression for $\mathrm{d}\varphi/\mathrm{d}x$. If $n = 5$, this term will be equal to 1, and this corresponds to an expression for $\mathrm{d}\varphi/\mathrm{d}x$ derived by G. Jaeger[1] on the assumption that the external electric forces and the electron pressure balance each other; this circumstance is readily understood, since, as pointed out on p. [321], the momentum imparted in unit time to the metal atoms—and hence the resultant force with which the atoms oppose the motion of the electrons,—when $n = 5$, is proportional to the total momentum of the electrons, which vanishes in this case, since there is no electric current.

(iii) *If the circuit is composed of different metals and the temperature is not uniform*, it follows from what has been said about the former two cases that *the potential difference between the ends of the circuit—the electromotive force,—if these consist of the same metal and have the same temperature, will depend only on the temperature at the places of contact between different metals.* (The temperature is assumed to be essentially constant over the very short segments where the gradual transition from one metal to another takes place.) We shall now calculate the potential difference F between the ends of a circuit consisting of two metals P and Q, assuming that the temperature is T_1 at the place of contact between P and Q, and is T_2 between Q and P, and that the temperature at the end points x_1 and x_2, consisting of the metal P, is T_0. From eq. (46), we get for the electromotive force

$$F = \varphi_{x_2} - \varphi_{x_1} = -\frac{k}{\varepsilon}\int_{x_1}^{x_2}\left(\frac{T}{K}\frac{\mathrm{d}K}{\mathrm{d}x} + \mu\frac{\mathrm{d}T}{\mathrm{d}x}\right)\mathrm{d}x.$$

By partial integration of the first term under the integral sign, we get

---

[1] G. Jaeger: Sitzungsber. d. Wiener Akad. d. Wiss., math.-nat. Kl. **117**, Abt IIa (1908) 859. (Jaeger does not base his calculations on the considerations mentioned in the text, but on the assumption that the collisions between electrons and metal atoms can be neglected in comparison with the mutual collisions among the electrons, an assumption which, however, as has been mentioned in the preceding (p. [342]), cannot be reconciled with the experimental values for the ratio between the thermal and electric conductivities, since it would make this ratio much too small.)

$$F = \frac{k}{\varepsilon}\int_{x_1}^{x_2}(\ln K - \mu)\,\frac{\mathrm{d}T}{\mathrm{d}x}\,\mathrm{d}x = \frac{k}{\varepsilon}\int_{T_0}^{T_1}(\ln K_P - \mu_P)\mathrm{d}T$$

$$+ \frac{k}{\varepsilon}\int_{T_1}^{T_2}(\ln K_Q - \mu_Q)\mathrm{d}T + \int_{T_2}^{T_0}(\ln K_P - \mu_P)\mathrm{d}T$$

and from this

$$F = \frac{k}{\varepsilon}\int_{T_1}^{T_2}\left[\ln\left(\frac{K_Q}{K_P}\right) - (\mu_Q - \mu_P)\right]\mathrm{d}T. \tag{50}$$

This expression differs from that given by Lorentz[1] by the addition of the second term under the integral sign; this term vanishes in Lorentz' calculations because μ, according to his theory, has the same value, $\mu = 2$, for all metals (see p. [318]). In Gruner's theory[2], on the other hand, μ will be different for the two metals if the critical speed G, or the ratio between the two mean free paths l_1 and l_2, is different for the two metals. The expression for F given by Gruner has, therefore, the same form as the general expression (50).

We shall now examine *the development of heat taking place when an electric current passes through a thermoelectric circuit.* We shall assume that the state of the circuit is stationary, imagining the temperature to be kept constant at every point in the circuit by conduction of heat to or from the outside. Since the circuit is assumed to consist of a thin wire, this may be done without any appreciable variation of the temperature over a single cross section.

We now consider a small element of the circuit lying between two cross sections, specified by the coordinates x and $x+\mathrm{d}x$, and determine the amount of heat $\mathrm{d}Q$ that must be conducted away from this element in unit time, in order that its temperature may remain constant.

This amount of heat will be equal to the difference between the energy imparted to the electrons in the element by the action of the external forces and that conducted away longitudinally by the collective motion of the electrons. Since the state is assumed to be stationary, as many electrons leave the element as enter it, and the latter amount of energy can, therefore, be determined from the expression for W_x, which represents the sum of the kinetic energy and the potential energy relative to the metal atoms carried in unit time through a unit area of a cross section of the circuit by the collective motion of the electrons (compare p. [335]); denoting the cross sectional area by ω, we, therefore, have

$$\mathrm{d}Q = -\left(\omega i_x\,\frac{\mathrm{d}\varphi}{\mathrm{d}x} + \frac{\mathrm{d}(\omega W_x)}{\mathrm{d}x}\right)\mathrm{d}x.$$

[1] H. A. Lorentz: Proc. Acad. Amsterdam **7** (1905) 453.
[2] P. Gruner: Phys. Zeitschr. **10** (1909) 50.

With the help of eqs. (43) and (44), eliminating $d\varphi/dx$, and noticing that $\omega i_x = i$ is independent of x, we get from this

$$dQ = \left(\frac{i^2}{\omega\sigma} + i \frac{k}{\varepsilon} T \frac{d(\ln K - \mu)}{dx} - \frac{d(\omega\gamma \, dT/dx)}{dx} \right) dx. \tag{51}$$

The first term in this expression for dQ, proportional to the square of the current, represents the Joule heat, and the last term, which is independent of the current, indicates the development of heat due to ordinary heat conduction; the middle term, finally, corresponding to a development of heat

$$dQ_R = i \frac{k}{\varepsilon} T \frac{d(\ln K - \mu)}{dx}, \tag{52}$$

which is proportional to the electric current and changes sign when the current changes direction, represents the so-called Peltier and Thomson effects. To examine this development of heat more closely, we shall consider two different cases.

Consider first a part of the circuit in which the temperature is uniform, and where a junction occurs from a metal P to a metal Q. From the expression (52), we find by integration that the amount of heat developed in unit time in this part of the circuit, given by the middle term in (51), will be equal to

$$i \frac{k}{\varepsilon} T \left(\ln \frac{K_Q}{K_P} - (\mu_Q - \mu_P) \right).$$

Defining the *Peltier coefficient* $\Pi_{P,Q}$ as the amount of heat absorbed at the junction of two metals P and Q when unit of electricity passes through the junction from P to Q, we get

$$\Pi_{P,Q} = \frac{k}{\varepsilon} T \left(\ln \frac{K_P}{K_Q} - (\mu_P - \mu_Q) \right). \tag{53}$$

Considering next a part of the circuit in which the metal is the same, we obtain from eq. (52), since K and μ in this case are functions only of T, that the amount of heat represented by the second term in (51), developed in unit time in a small segment of the circuit in which the temperature varies from T to $T + dT$, will be

$$i \frac{k}{\varepsilon} T \frac{d(\ln K - \mu)}{dT} \, dT.$$

Defining the *Thomson coefficient* ρ as the amount of heat absorbed in a segment of a metal bar, along which the temperature varies by one degree, when one unit of electricity passes through the bar in the direction of increasing temperature, we get

$$\rho = -\frac{k}{\varepsilon} T \frac{d(\ln K - \mu)}{dT}. \tag{54}$$

If we assume that separate collisions take place and that the atomic field of force—the law holding for the collisions—is the same in all metals and at all temperatures, we get from (50), (53) and (54), substituting the value for K given by eq. (45) (with $\lambda = 0$), and noticing that μ is a constant in this case,

$$F = \frac{k}{\varepsilon} \int_{T_1}^{T_2} \ln \left(\frac{N_Q}{N_P} \right) \mathrm{d}T,$$

$$\Pi_{P,Q} = \frac{k}{\varepsilon} T \ln \left(\frac{N_P}{N_Q} \right),$$

$$\rho = -\frac{k}{\varepsilon} T \frac{\mathrm{d} \ln N}{\mathrm{d}T} + \frac{3}{2} \frac{k}{\varepsilon},$$

(55)

which are the expressions found by Lorentz. Since $\frac{3}{2}kT$ is the average kinetic energy of an electron at temperature T, the last term in the expression for the Thomson coefficient, $\frac{3}{2}k/\varepsilon$, represents the amount of heat that must be supplied to a collection of free electrons of total charge equal to one unit of electricity, in order to raise its temperature one degree.

The expressions derived here for the thermoelectric power and the Peltier and Thomson coefficients do not allow a closer direct comparison with experiments, because of the unknown quantities K and μ entering in them. However, in the following we shall consider the relations between these three quantities. We shall here only mention the circumstance that the expression for the Thomson coefficient contains a term, namely that mentioned above, which is much larger than the observed values for the Thomson coefficient[1]. If we assume separate collisions and the same law of collisions for all metals (see eq. (55)), we must, therefore, in order to calculate the Thomson coefficient in accord with experiments, assume that the number of free electrons in unit volume varies in approximately the same way with temperature in all metals, namely as $T^{\frac{3}{2}}$; for only in this case can the two terms in the expression for the Thomson coefficient approximately

---

[1] See, e.g., J. Kunz: Phil. Mag. **16** (1908) 781. (Kunz is of the opinion that this apparent disagreement between theory and experiments can be explained by the assumption that electrons travelling through a metal in the direction of increasing temperature will not absorb the same amount of energy from the metal when they move under the influence of an external force as when they diffuse freely as a result of a temperature gradient. However, such an assumption would be inconsistent with the main principles underlying the electron theory of metals—the correctness of which is also assumed by Kunz in the paper cited—, namely that the external forces have only very little influence upon the motion of the individual electrons.)

cancel each other[1]. However, as seen from eq. (54), additional terms will enter into the expression for the Thomson coefficient if μ varies with temperature. In this connection, we may point out that, according to what we have discussed in the previous section (see the footnote on p. [343]), the assumption of such a variation is necessary to explain the observed temperature dependence of the electric conductivity.

We must remark here that, although the expressions found for Π and ρ depend on the force fields of the metal atoms, in so far as these affect the motion of the free electrons at any particular place in the metal, they do not depend directly on the variation from place to place of the mean value of the potential energy of the electrons relative to the metal atoms, and, consequently, there is, for example, no direct connection between the value of the Peltier coefficient and the afore-mentioned potential differences between two metals in contact. Thus, if we consider two pieces of metal P and Q, in the interior of which all conditions are assumed to be the same, except that the potential energy of the electrons at each point in the interior of P is λ_0 higher than at the corresponding point in Q, the potential difference between the two metals, given by (47), will, as seen from the expression (45) for K, be $\varphi_P - \varphi_Q = -\lambda_0$. On the other hand, in this case the Peltier coefficient vanishes, since the value of $\ln K - \mu$, as seen from eqs. (35) and (45), is independent of a constant difference in the potential energy of the electrons. This is also what might be expected beforehand, since the potential difference at the place of contact, in this case, will just balance the action of the forces from the metal atoms, so that the electrons in the interior of the metal will move as if the metal were homogeneous[2].

We shall now proceed to examine the relation of the results derived above to the *thermodynamic theory* of the thermoelectric phenomena given by Lord Kelvin[3]. In this theory, the passing of electricity through a closed thermoelectric circuit is considered as a reversible cyclic process in which heat can be converted into mechanical work, and vice versa;—the amounts of heat considered are those developed or absorbed in the Peltier and Thomson effects, and the work in question is that performed by the thermoelectric electromotive force as the electricity moves through the circuit. Applying the second law of thermodynamics to this cyclic process, we obtain the following condition

---

[1] See, e.g., J. J. Thomson: *The Corpuscular Theory of Matter*, p. 79. (In this book it is stated, however, that N must vary as $T^{\frac{1}{4}}$ rather than as $T^{\frac{1}{2}}$. This is due to a method of calculation that is not quite correct. The kinetic energy carried by the motion of a gas in a given direction through a surface is not equal to the average energy of a molecule $\frac{3}{2}kT$ multiplied by the number of molecules passing through the surface but, as a more detailed calculation shows, is $\frac{5}{3}$ as large. This is easily seen if we consider a gas flowing in a tube and assume that a certain part of the gas is enclosed between two pistons which move with the gas. The transfer of energy by the gas molecules through a cross section of the tube between the two pistons is equal to the sum of the displacement of the internal energy of the gas ($vN = \frac{3}{2}kT$) and the work done on the piston in front of the gas ($vp = vNkT$)).*

* [In accordance with a note in III, a few lines have been added to this footnote.]

[2] See, e.g., R. Clausius: Pogg. Ann. **90** (1853) 520.

[3] W. Thomson: Trans. Roy. Soc. Edinburgh (1854) 123.

$$\int \frac{\mathrm{d}Q_R}{T} = 0, \tag{56}$$

where the integration is to be extended over the whole circuit.

With the aid of this equation, and applying the law of the conservation of energy, we obtain the following relations, given by Lord Kelvin (loc. cit. p. 135),

$$\rho_Q - \rho_P - T \frac{\mathrm{d}(\Pi_{P,Q}/T)}{\mathrm{d}T} = 0 \quad \text{and} \quad E_{P,Q} - \frac{\Pi_{P,Q}}{T} = 0, \tag{57}$$

where $E_{P,Q}$ denotes the thermoelectric power in a circuit consisting of two metals in which the temperature of one junction is one degree higher than that of the other, $E_{P,Q}$ being counted positive in the direction from the metal P through the hotter junction to the metal Q.

As is known, these relations agree, at any rate approximately, with the results found experimentally[1].

This derivation of eqs. (56) and (57) is, however, not well-founded, as already mentioned by Lord Kelvin (loc. cit. p. 128), and especially emphasized by Boltzmann[2], in that the cyclic process considered necessarily involves irreversible processes, namely the development of Joule heat and heat conduction, and this in such a way that the development of heat due to these processes cannot in any way be made negligible compared to the thermoelectric development of heat.

As shown by Boltzmann, the second law of thermodynamics will always be satisfied if the quantities E, Π and ρ—instead of the relations (57)—merely obey the following conditions (loc. cit. p. 1281)

$$\left| \rho_Q - \rho_P - T \frac{\mathrm{d}(\Pi_{P,Q}/T)}{\mathrm{d}T} \right| \leqq \frac{2}{\sqrt{T}} \left(\sqrt{\left(\frac{\gamma}{\sigma}\right)_P} + \sqrt{\left(\frac{\gamma}{\sigma}\right)_Q} \right)$$

and $\tag{58}$

$$\left| E_{P,Q} - \frac{\Pi_{P,Q}}{T} \right| \leqq \frac{2}{\sqrt{T}} \left(\sqrt{\left(\frac{\gamma}{\sigma}\right)_P} + \sqrt{\left(\frac{\gamma}{\sigma}\right)_Q} \right).$$

Hence, eq. (56), and the relations following from it, cannot be proved by purely thermodynamic considerations[3].

---

[1] See, e.g., Baedeker: *Die elektrischen Erscheinungen in metallischen Leitern*, p. 81–86.

[2] L. Boltzmann: Sitzungsber. d. Wiener Akad. d. Wiss., math.-nat. Kl. **96**, Abt. II (1887) 1258.

[3] H. Poincaré (*Thermodynamique*, 2nd ed. (Paris 1908), p. 384) expresses the opinion that Lord Kelvin's theory of the thermoelectric phenomena can be deduced rigorously from thermodynamical principles, i.e., that the problem can be treated in such a way that the heat conduction can be neglected. This, however,—which is at variance with Boltzmann's calculations—, does not seem correct to me, for, in Poincaré's calculations, no regard is taken of the increase in entropy connected with the heat conduction and originating in the circumstance that the heat transferred through an element of a thermoelectric circuit is supplied to the element at a higher temperature and is removed from it at a lower temperature*.

* [Footnote added in III.]

We see that the expression (52) for the thermoelectric development of heat dQ_R, derived from the electron theory, satisfies eq. (56); the expressions for F, Π and ρ will, therefore, also satisfy the conditions (57). *This agreement between results of the electron theory and the thermodynamic theory* was first proved by H. A. Lorentz for a rigorously treated example[1]. As we have seen, *this agreement exists in all the cases considered here*; namely, provided the fundamental assumptions stated in the Introduction are valid, *in all cases in which separate collisions occur, and also in the case in which the electrons can be considered to move independently of one another in a stationary field of force*, a case which, as mentioned in Chapter I, p. [325], may correspond approximately to the actual conditions in metals.

This agreement between the electron theory and Lord Kelvin's theory for the thermoelectric phenomena is the more remarkable, as it might not be expected beforehand; in fact, the electron theory, which assumes that the transfer of heat as well as electricity are effected by the same particles, apparently offers no justification for a formal differentiation between reversible and irreversible processes[2].

Attempts have been made to explain the agreement in question by comparing

---

[1] H. A. Lorentz: Proc. Acad. Amsterdam **7** (1905) 589; See also Jahrb. d. Rad. u. El. **2** (1905) 377. (If Maxwell's velocity distribution is not taken into account, this agreement will not be obtained (see, e.g., Riecke: Wied. Ann. **66** (1898) 1200).

[2] J. Kunz (Phil. Mag. **16** (1908) 767) has attempted to explain this agreement by calculating the thermoelectric development of heat from the work performed during the passage of electricity through the circuit by the electric forces corresponding to the potential differences which, according to the electron theory, exist between two metals in contact, or between two points in a homogeneous piece of metal of non-uniform temperature. Kunz is of the opinion (loc. cit. p. 770) that this development of heat can be regarded as arising from a reversible cyclic process, since, in calculating these potential differences, it is not necessary to take heat conduction directly into account. To this we may remark, however, that a consideration as that used by Kunz only shows that these developments of heat are formally reversible only in the sense that they change sign when the direction of the current is reversed, but not at all that they arise from a reversible cyclic process independent of irreversible phenomena, to which alone the second law of thermodynamics can be directly applied. Moreover, it may be mentioned that Kunz' calculation of the thermoelectric heat development only gives the correct result under such simple assumptions as those on which Kunz' considerations are based. Kunz assumes, with Drude, that all the electrons have the same speed, and that the electric current always carries with it a definite amount of energy dependent only on the temperature; however, if the different speeds of the electrons are taken into account, it is found that the electric current will be carried to a different degree by electrons of different speeds, depending on the assumptions made about the forces between the metal atoms and the electrons; hence, in the calculation of the heat developed, we must take into account the different amounts of energy carried by the current in different metals and at different temperatures. A closer calculation shows, further, that, in such more general cases, the heat development caused only by the work connected with the potential difference will not satisfy condition (56); rather, this equation will only be satisfied by the sum of this heat development and that arising from the energy carried by the current.

the passage of electricity through a thermoelectric circuit with a cyclic process in which a gas is carried through a series of states whose pressure and temperature correspond to the pressure of the electrons and the temperature of the metal in the different places of the circuit, and, thus, to calculate the thermoelectric heat development and the electromotive force in the circuit from the amounts of heat which the gas would absorb or give up and the work it would be able to perform in such a cyclic process[1].

From such a consideration, expressions for F, Π and ρ are obtained which correspond to those found by Lorentz. That the results thus obtained must be in accord with Lord Kelvin's theory is obvious beforehand, since they are calculated by considering a cyclic process in which a gas is carried through a series of equilibrium states, i.e., an ordinary reversible thermodynamic cyclic process. This consideration, however, can only be regarded as an illustration of Lord Kelvin's theory, for having neglected the conduction of electricity and heat, without any argument in support of such neglect, we cannot be sure beforehand that the calculated developments of heat will agree with those that actually occur in the circuit. A calculation based on more general assumptions than those of Lorentz shows, indeed, that such an agreement will not exist in general; for, as we have seen, in the expressions for the heat developments in question there enter, in addition to quantities (such as the number of free electrons in unit volume) which characterize the equilibrium states mentioned, also quantities (such as that denoted by μ above) which characterize the way the collective motion of the electrons will be distributed among electrons of different speeds when the equilibrium is disturbed, quantities which obviously cannot enter into results derived from a consideration such as that mentioned above.

---

[1] Compare J. J. Thomson: *The Corpuscular Theory of Matter*, p. 73; see also K. Baedeker: Phys. Zeitschr. **11** (1910) 809, Ann. d. Phys. **35** (1911) 75, F. Krüger: Phys. Zeitschr. **11** (1910) 800; **12** (1911) 360. (Baedeker has made a further attempt to elucidate the electron theory treatment of the thermoelectric phenomena, with the aid of conclusions drawn from experiments on the emission of electrons from metals at high temperatures. According to these conclusions, a cavity in a metal, in a state of equilibrium, must contain a definite number of electrons per unit volume, exerting a definite pressure depending on the nature and temperature of the metal. Baedeker now considers a reversible cyclic process in which a collection of free electrons is carried through a series of states having pressures and temperatures corresponding to those that would exist in cavities in the different parts of the circuit. If the concentration of the electrons in these cavities is assumed proportional to the concentration of electrons in the metal itself, such a consideration gives the same result as that mentioned in the text above, or that of Lorentz' theory, and the remarks made there also apply to Baedeker's theory. On the other hand, if this condition about the concentration of electrons is not satisfied (see Baedeker: loc. cit. p. 810), there will be no simple connection between the heat development calculated by Baedeker and the Peltier and Thomson effects appearing in the circuit.)

The difference between the problem discussed here and the ordinary (rigorous) applications of the second law of thermodynamics is also plainly indicated by the difference in the derivation of the relation (56) from the electron theory and the usual verifications of thermodynamic results by means of the kinetic theories. Thus, while, in the latter problems, we exclusively consider statistical equilibrium states and displacements of such states, we consider here only systematic deviations from statistical equilibrium, namely the collective directed motion of the electrons among the atoms of the metal; and the validity of eq. (56) is here deduced directly from the special form of the equations which determine the collective momentum of the electrons in a given direction,—i.e., the relation between the coefficients to $\partial\varphi/\partial x$ and $\partial T/\partial x$, and, in the cases (separate collisions) in which we have assumed an "interchange of energy" (see p. [323]) between metal atoms and electrons or among the electrons mutually, the special form for the function $Q(\rho, r)$, given by eq. (13),—and from the expressions (43) and (44) for the flow of electricity and energy through the metal derived from these equations[1].

---

[1] It may be pointed out that W. Voigt (see, e.g., Wied. Ann. **67** (1899) 717) has shown that, if condition (56) given by Lord Kelvin is to be valid, the equations for the flow of electricity and heat in an isotropic medium must necessarily have a form corresponding to that given by eqs. (43) and (44).

CHAPTER III

NON-STATIONARY PROBLEMS

Absorption and Emission of Heat Rays by Metals

On the assumption that the temperature equilibrium of a collection of bodies that are perfectly shielded from external influences will not be disturbed by mutual heat radiation, Kirchhoff has shown that the ratio between the power of absorption and the power of emission for heat rays is the same for all bodies and depends only on the temperature and on the period of vibration of the heat rays in question; further, that, in a cavity surrounded by bodies in temperature equilibrium, there will be present an amount of radiant energy per unit volume which depends only on the temperature and is distributed among rays of different periods of vibration in a way that can be readily calculated from the above-mentioned ratio between the powers of absorption and emission. It has been of great interest, therefore, to attempt to calculate theoretically the powers of absorption and emission for heat rays of some particular body, since in this way we may arrive at a theoretical determination of the amount of radiant energy density mentioned and its distribution among rays of different periods of vibration, or, as it is usually expressed, at a determination of the law of heat radiation. Now, in the electron theory of metals we are just dealing with bodies for which such a calculation can be carried out. Considering a piece of metal for which especially simple assumptions are made about the interactions between atoms and free electrons, H. A. Lorentz[1] has made such a calculation for heat rays having periods of vibration large compared to the very short times in which individual electrons lose almost all of their original momentum in a given direction. Lorentz' calculations are in very good agreement with experiments, a fact that must be regarded as one of the most interesting results of the electron theory.

J. J. Thomson[2] has later attempted to extend the calculations to rays with shorter periods of vibration. Like Lorentz, Thomson assumes that separate collisions take place and that an electron in a collision, on the average, loses all of its original

---

[1] H. A. Lorentz: Proc. Acad. Amsterdam **5** (1903) 666; see also H. A. Lorentz: *Theory of Electrons* (Leipzig 1909) Chapter II (where the calculations are carried out in a more rigorous way).

[2] J. J. Thomson: Phil. Mag. **14** (1907) 217.

momentum in a given direction; however, while the calculations for rays with long periods of vibration depend only on the "result" of the collisions, Thomson shows that the calculation of the power of emission for rays with periods of the same order of magnitude as the duration of collisions depends, to the highest degree, upon the assumptions made about the motion of the electrons during individual collisions. The final equation obtained by Thomson for the determination of the law of heat radiation, and used by him to calculate expressions for this law from various assumptions about the motions of electrons during collisions[1], is, however, in my opinion, not correct; for in the calculation of the power of absorption, contrary to the calculation of the power of emission, the motion of the individual electrons during collisions is not sufficiently taken into account. We shall return to this question in the following.

J. H. Jeans[2] has attempted to calculate the ratio between the powers of absorption and emission for heat rays by metals, without making any special assumptions about the motions of the free electrons in the metals. However, as previously mentioned, his calculations are not valid in general, but only under certain very special assumptions, which, for example, would not be satisfied in the case considered by Lorentz. H. A. Wilson[3], who has pointed out flaws in Jeans' calculations, has attempted to carry out the calculations under the same special assumptions as Lorentz, but for heat rays of shorter periods of vibrations, i.e., for periods of vibrations that are assumed to be long only compared to the duration of individual collisions. However, as already mentioned, and as will be seen in the following, Wilson's calculations are not correct either.

In this chapter, we shall calculate the ratio between the powers of absorption and emission of metals for heat rays with "long" periods of vibration, under the assumptions used in the previous chapters. In the cases in which we assume that separate collisions take place, in which cases the calculations can be carried out without further special assumptions, the investigation is extended to heat rays with periods of vibration that are assumed to be long only compared to the duration of individual collisions; in other cases, the investigation is limited to rays with periods of vibration long compared to the time in which an electron, on the average, loses the greater part of its original momentum in a given direction.

[1] J. J. Thomson: Phil. Mag. **14** (1907) 225; **20** (1910) 238.
[2] J. H. Jeans: Phil. Mag. **17** (1909) 773.
[3] H. A. Wilson: Phil. Mag. **20** (1910) 835.

§ 1

Absorption of Heat Rays

If a piece of metal is exposed to the influence of a constant external electric field, the free electrons will acquire an average motion in the direction of the electric force, and the electrons will therefore, on the average, receive kinetic energy under the influence of the field,—in other words, the electrons will absorb energy from the external electric field;—through the interaction (collisions) with metal atoms, this energy will be converted into heat, i.e., into kinetic energy distributed over random irregular motions. Now, a very similar situation will occur if a piece of metal is exposed to heat rays; for these, according to the electromagnetic theory of light, consist of rapidly varying electric and magnetic fields.

We shall first consider heat rays with periods of vibration that are large compared to the time intervals within which individual electrons, on the average, lose the greater part of their momentum in a given direction. In this case, the state can at every single instant be regarded as stationary—or rather "quasi-stationary" (see p. [317]).

Denoting the components of the electric field strength at a point in the metal at the time t by E_x, E_y and E_z, the electric current through a surface element dS perpendicular to the x-axis will, in this case, be $i_x dS = \sigma E_x dS$, where σ is the ordinary electric conductivity; and analogous expressions will hold for the electric current in the directions of the y- and z-axes. The energy absorbed in a volume element dV of the metal in the time dt will, therefore, be

$$(i_x E_x + i_y E_y + i_z E_z)dV\,dt = \sigma(E_x^2 + E_y^2 + E_z^2)dV\,dt,$$

and the energy absorbed per unit time in unit volume of the metal will be

$$\sigma(\overline{E_x^2} + \overline{E_y^2} + \overline{E_z^2}),$$

where the bars indicate that the quantities are to be computed as mean values, averaged over space as well as over time.

The average amount of energy present in unit volume of the empty space between atoms and electrons will be

$$2 \times \frac{1}{8\pi}\,(\overline{E_x^2} + \overline{E_y^2} + \overline{E_z^2}),$$

where the factor 2 arises from the fact that in electromagnetic vibrations, as those considered here, the magnetic energy will, on the average, be equal to the electric energy. This energy is not the whole energy present in the metal on account of the radiation; for there will also be energy connected with the collective motion of the free electrons (and perhaps also from resonance of the bound electrons in the atoms). (In calculating the absorption, we have not taken the influence of the magnetic field into account; for, the speed of the electrons being very small compared to the speed of light (see the footnote on p. [305]), its effect upon the motion of the electrons will, in this case, be negligible compared to that of the electric field).

Defining the coefficient of absorption α as the amount of energy absorbed in unit volume of the metal when the radiant energy density (calculated as indicated above) is equal to one, we get

$$\alpha = 4\pi\sigma. \tag{59}$$

From these calculations, it is, in general, not possible to find directly the energy absorbed by a piece of metal exposed to heat rays; for the collective motion of the free electrons (and the resonance of the electrons bound in the atoms) will react upon the external electromagnetic field and, in addition to absorption, give rise to the phenomena known as reflection and refraction. However, if we, as in the applications to be made of eq. (59) in the following, consider a piece of metal that is so small that the reaction of its electrons upon the external field can be neglected, we get directly from eq. (59) that the amount of energy absorbed in the piece of metal of volume ΔV exposed to a beam of heat rays in which the energy per unit volume is A,—so that the energy passing through a unit surface perpendicular to the beam is cA, where c is the speed of light,—will be

$$\alpha A \Delta V.$$

(The piece of metal must, however, contain a very large number of electrons and atoms, in order that we may make use of the expressions obtained in the previous chapters for the electric conductivity, etc. This requirement is compatible with the assumptions made above; for the reaction upon the external field is exceedingly small; as to the free electrons, this is true because, due to their high speeds and frequent collisions, they are able to follow the influence of the external field only to a very small extent.)

Experiments by Hagen and Rubens[1] on the reflection and emission of heat rays by metals show that the absorption of heat rays with long periods of vibration (i.e., longer than 10^{-13} sec) actually can be calculated in agreement with experiments from the electric conductivity for stationary currents. However, the experiments also show that, when the period of vibration is shorter, the electric conductivity, calculated in this way, does not agree with that found for stationary currents.

We shall now proceed to calculate the absorption of heat rays with such shorter

[1] E. Hagen and H. Rubens: Ann. d. Phys. 11 (1903) 873.

periods. As mentioned above, we can only treat the case in which *separate collisions* take place, since, in deriving the equations for the other cases in Chapter I, we only considered stationary states.

If we assume that the metal is homogeneous and of uniform temperature, the equation determining the collective motion of the electrons, in case of separate collisions (eq. (14)), will be

$$\left[\frac{dG_x(r)}{dt}\right] = -\frac{4\pi m\varepsilon K}{3kT}\frac{\partial\varphi}{\partial x}r^4 e^{-\frac{1}{2}mr^2/kT} - G_x(r)F(r) + \int_0^\infty G_x(\rho)Q(\rho, r)d\rho. \tag{60}$$

In this equation, the left member indicates the average value of $dG_x(r)/dt$ for given values of $\partial\varphi/\partial x$ and of the quantities $G_x(\rho)$ (see p. [316]). Seeking an average value of $G_x(r)$ under given external circumstances, and assuming that $\partial\varphi/\partial x$ does not vary appreciably during a single collision—a time interval assumed to be very small compared to the time in which an electron loses the greater part of its original momentum in a given direction,—we can, however, treat eq. (60) as an ordinary differential equation, neglecting the bracket on the left side, and letting $G_x(r)$ denote the desired average value; in fact, this is allowed, since in this case no systematic correlation (except, of course, that expressed by eq. (60) itself) exists between $dG_x(r)/dt$ and preceding states corresponding to appreciably different values for $\partial\varphi/\partial x$ and, hence, for the average of $G_x(r)$. It must be remarked, however, that, for heat rays with periods so short that the above-mentioned conditions are not satisfied, the present method of calculation will not be correct*.

We shall now assume that an electric field expressed by

$$E_x = -\frac{\partial\varphi}{\partial x} = E\cos pt$$

acts in the direction of the x-axis. The conditions that eq. (60), which can readily be shown to have only one periodic solution, will be satisfied by a solution of the form

$$G_x(r) = \psi_1(r)\cos pt + \psi_2(r)\sin pt \tag{61}$$

are

$$p\psi_2(r) = \frac{4\pi m\varepsilon K}{3kT}Er^4 e^{-\frac{1}{2}mr^2/kT} - \psi_1(r)F(r) + \int_0^\infty \psi_1(\rho)Q(\rho, r)d\rho \tag{62}$$

and

$$-p\psi_1(r) = -\psi_2(r)F(r) + \int_0^\infty \psi_2(\rho)Q(\rho, r)d\rho. \tag{63}$$

These two integral equations have a solution which can be written in the form

* [The preceding paragraph has been revised slightly to conform to III.]

$$\psi_1(r) = \frac{4\pi m\varepsilon K}{3kT} E \left(r^4 e^{-\frac{1}{2}mr^2/kT} F_1(r) + \int_0^\infty \rho^4 e^{-\frac{1}{2}mr^2/kT} Q_1(\rho, r)d\rho \right), \qquad (64)$$

where $F_1(r)$ and $Q_1(\rho, r)$ depend only on $F(r)$, $Q(\rho, r)$ and p (i.e., are independent of the first term on the right side of eq. (62)), together with an analogous expression for $\psi_2(r)$, obtained by interchanging indices 1 and 2. This is seen most easily by solving (63) for $\psi_2(r)$ by means of Fredholm's formula (see p. [332]), substituting the result in (62), and solving the equation thus obtained in the same way).

Now, putting

$$P_1(r) = F_1(r) + \int_0^\infty Q_1(r, \rho)d\rho \qquad (65)$$

and

$$\sigma_p = \frac{4\pi\varepsilon^2 K}{3kT} \int_0^\infty r^4 e^{-\frac{1}{2}mr^2/kT} P_1(r)dr, \qquad (66)$$

where σ_p is a quantity depending on the nature and temperature of the metal as well as on p, and, denoting the corresponding quantity for the index 2 by δ_p, we get, with the aid of eqs. (61), (64), (65), and (66), for the electric current through a unit area perpendicular to the x-axis

$$i_x = \frac{\varepsilon}{m} \int_0^\infty G_x(r)dr = E(\sigma_p \cos pt + \delta_p \sin pt) \qquad (67)$$

and for the energy absorbed in unit time by unit volume of the metal

$$\frac{p}{2\pi} \int_0^{2\pi/p} i_x E \cos pt\, dt = \tfrac{1}{2}E^2\sigma_p.$$

Now, since the average amount of energy present in unit volume of the metal is

$$\overline{E_x^2}/4\pi = E^2/8\pi,$$

we obtain for the coefficient of absorption α_p (see p. [360]) for heat rays with the period of vibration $2\pi/p$

$$\alpha_p = 4\pi\sigma_p. \qquad (68)$$

With the above-mentioned (p. [361]) restrictions on p, this expression is valid in all cases in which separate collisions are assumed to occur; we shall make use of it later for comparison with the power of emission.

In analogy to the considerations that led to eq. (59), we can express eq. (68) saying that a metal, in an electric field varying as $\cos pt$, possesses an "effective electric conductivity" σ_p different from (smaller than) the conductivity σ_0 for stationary currents. Moreover, as seen from eq. (69), for such rapidly oscillating electric fields, there will appear a phase difference between the electric current and the field. This phase difference, equal to $\tan^{-1}(\delta_p/\sigma_p)$, arising from the inertia of the free electrons, consists in a retardation of the current relative to the electric field; the appearance of this phase difference will have the same effect on the calculation of the optical properties of metals as an effective dielectric constant.

We shall now examine more closely the expressions for σ_p and δ_p in the simple case in which the metal atoms are assumed to be fixed centres of force acting upon the electrons with forces inversely proportional to the nth power of the distance, and in which the mutual collisions among the electrons can be neglected. In this case,

$$F(r) = Cr^{(n-5)/(n-1)} \quad \text{and} \quad Q(\rho, r) = 0.$$

We therefore get

$$F_1(r) = \left(F(r) + \frac{p^2}{F(r)}\right)^{-1} = \frac{\frac{1}{C} r^{-(n-5)/(n-1)}}{1 + \frac{p^2}{C^2} r^{-2(n-5)/(n-1)}} \quad \text{and} \quad Q_1(\rho, r) = 0,$$

whence, with the aid of (65) and (66),

$$\sigma_p = \frac{4\pi\varepsilon^2 K}{3kTC} \int_0^\infty \frac{r^{(3n+1)/(n-1)} e^{-\frac{1}{2}mr^2/kT}}{1 + \frac{p^2}{C^2} r^{-2(n-5)/(n-1)}} \, dr. \tag{69}$$

This expression differs from that for stationary currents by the addition of the last term in the denominator under the integral sign. The integration of (69) cannot, in general, be carried out exactly; however, assuming that p is so small that the denominator, for all values of r which contribute essentially to the integral (we assume $n > 2$), is very nearly equal to one, we have approximately

$$\sigma_p = \frac{4\pi\varepsilon^2 K}{3kTC} \int_0^\infty \left(r^{(3n+1)/(n-1)} - \frac{p^2}{C^2} r^{(n+1)/(n-1)} \right) e^{-\frac{1}{2}mr^2/kT} \, dr$$

$$= \frac{4\pi\varepsilon^2 K}{3kTC} \left[\frac{1}{2} \left(\frac{2kT}{m}\right)^{2n/(n-1)} \Gamma\left(\frac{2n}{n-1}\right) - \frac{p^2}{C^2} \frac{1}{2} \left(\frac{2kT}{m}\right)^{(n+5)/(n-1)} \Gamma\left(\frac{n+5}{n-1}\right) \right];$$

this gives

$$\sigma_p = \sigma_0 \left[1 - \frac{p^2}{C^2} \left(\frac{m}{2kT}\right)^{(n-5)/(n-1)} \Gamma\left(\frac{n+5}{n-1}\right) \Gamma^{-1}\left(\frac{2n}{n-1}\right) \right].$$

With the aid of eq. (37), this can be written

$$\sigma_p = \sigma_0 \left[1 - \sigma_0^2 \frac{p^2 m^2}{N^2 \varepsilon^4} \frac{9\pi}{16} \Gamma \left(\frac{n+5}{n-1} \right) \Gamma^{-3} \left(\frac{2n}{n-1} \right) \right] . \tag{70}$$

Similarly, we get

$$\delta_p = \frac{4\pi \varepsilon^2 K}{3kTC} \int_0^\infty \frac{\dfrac{p}{C} r^{(2n+6)/(n-1)} e^{-\frac{1}{2}mr^2/kT}}{1 + \dfrac{p^2}{C^2} r^{-2(n-5)/(n-1)}} \, dr,$$

and, to the same approximation as above,

$$\delta_p = \sigma_0^2 \frac{pm}{N\varepsilon^2} \frac{3\sqrt{\pi}}{4} \Gamma \left(\frac{3n+5}{2(n-1)} \right) \Gamma^{-2} \left(\frac{2n}{n-1} \right) . \tag{71}$$

We shall now consider some special cases corresponding to different values of n. (i) For $n = \infty$, we get, from (70) and (71),

$$\sigma_p = \sigma_0 \left(1 - \sigma_0^2 \frac{p^2 m^2}{N^2 \varepsilon^4} \frac{9\pi}{16} \right) \quad \text{and} \quad \delta_p = \sigma_0^2 \frac{pm}{N\varepsilon^2} \frac{3\pi}{8}, \tag{72}$$

which are the expressions holding if the metal atoms are assumed to be *hard elastic spheres*, or, in other words, if, on the average, electrons in a single collision lose all of their original momentum in a given direction.

As mentioned above, J. J. Thomson[1] has derived an expression for σ_p from an assumption about the effects of collisions which corresponds to that made here. However, the expression obtained by Thomson, i.e., in our notations,

$$\sigma_p = \sigma_0 \left[\frac{\sin \dfrac{\sigma_0 pm}{N\varepsilon^2}}{\dfrac{\sigma_0 pm}{N\varepsilon^2}} \right]^2 = \sigma_0 \left(1 - \frac{1}{3} \frac{\sigma_0^2 p^2 m^2}{N^2 \varepsilon^4} + \ldots \right) \tag{72'}$$

differs, as will be seen, very essentially from that given above. (Thus, the term giving the dependence of the conductivity on the frequency is about 5 times smaller than the corresponding term in (72).) This difference is partly due to the circumstance that Thomson assumes that all electrons have the same speed and partly to the fact that, in Thomson's calculations, the times in which the electrons travel through their free paths are considered equal, irrespective of the velocities acquired by the electrons under the influence of the external forces[2].

The expression (72') is used by Thomson to calculate the absorption by metals of rays of all possible periods of vibration. However, apart from the objections mentioned above, which do

[1] J. J. Thomson: Phil. Mag. **14** (1907) 225.

[2] Compare J. H. Jeans: Phil. Mag. **17** (1909) 778.

not affect the order of magnitude of the dependence of the coefficient of absorption on the period, this expression, because of the way it is derived, is, in my opinion, not applicable (compare p. [361]) to rays with periods of vibration that are short compared to the duration of the individual collisions (see p. [358]; see also the footnote p. [360]).

As mentioned on p. [358], H. A. Wilson[1] has also calculated σ_p from the above-mentioned assumption about the nature of the collisions, and has likewise found a result that does not agree with the expression (72). Wilson obtained his value for σ_p in two different ways; one of these corresponds to that used here, but the equation corresponding in his calculations to eq. (60) is, as explained on p. [311], not quite correct. In his other calculation, Wilson, like Thomson, on the other hand, considers the motion of the individual electrons and examines the amount of energy absorbed, on the average, by an electron travelling through a free path between successive collisions. However, in this calculation, as in Thomson's, no account is taken of the fact that the time required to travel through such a free path will be greater when the electron travels against the external electric force than when it moves in the direction of the force. Wilson's results—his value for σ_0 is $\frac{3}{2}$ times that of Lorentz—as well for σ_0 as for σ_p, are, therefore, not correct.

(ii) If, in eqs. (70) and (71), we put $n = 5$, we get

$$\sigma_p = \sigma_0 \left(1 - \sigma_0^2 \frac{p^2 m^2}{N^2 \varepsilon^4}\right) \quad \text{and} \quad \delta_p = \sigma_0^2 \frac{pm}{N\varepsilon^2}. \tag{73}$$

In this case, corresponding to the assumption that the metal atoms are *small magnets* (see p. [322]), it can be shown that the expression for δ_p, and the correction term in σ_p, are smaller than in all other cases in which separate collisions are assumed and the mutual interactions among the electrons are neglected (this is indicated by a comparison with the previous and the following cases).

The value for σ_p, given in eq. (73), corresponds to that obtained by J. H. Jeans[2], i.e.,

$$\sigma_p = \sigma_0 \left(1 + \sigma_0^2 \frac{p^2 m^2}{N^2 \varepsilon^4}\right)^{-1}, \tag{73'}$$

an expression which, in Jeans' opinion, should be of perfectly general validity, also when separate collisions are not assumed. The reason that, when putting $n = 5$, we obtain Jeans' expression is explained in the preceding (see p. [321]).

(iii) Finally, putting $n = 3$, we get

$$\sigma_p = \sigma_0 \left(1 - \sigma_0^2 \frac{p^2 m^2}{N^2 \varepsilon^4} \frac{27\pi}{64}\right) \quad \text{and} \quad \delta_p = \sigma_0^2 \frac{pm}{N\varepsilon^2} \frac{45\pi}{128}, \tag{74}$$

expressions which would be valid if the metal atoms were *electric doublets*, as those considered on p. [322].

Now, calculating the reflection coefficient R of a metal surface for perpendicu-

[1] H. A. Wilson: Phil. Mag. **20** (1910) 835.
[2] J. H. Jeans: Phil. Mag. **17** (1909) 778.

larly incident heat rays with the aid of eq. (67), neglecting terms which are to the significant terms as the square of δ_p/σ_p is to 1, and neglecting also possible resonance in the atoms, which effect would be of the same order of magnitude as in insulators and, hence, because of the high conductivity of the metals, negligible compared to the significant terms (see Drude: loc. cit. p. 941), we find[1] the following equation

$$\left(\frac{1+R}{1-R}\right)^2 = \frac{2\pi}{p}\,\sigma_p\left(1+\frac{\delta_p}{\sigma_p}\right).$$

Since the coefficient to $2\pi/p$ would be equal to σ_0 for infinitely slow vibrations, the equation shows that in the expression for the reflection coefficient for heat rays of shorter periods of vibration there enters, instead of the ordinary electric conductivity σ_0, an effective conductivity $\sigma_p(1+\delta_p/\sigma_p)$. Since, for sufficiently long periods of vibration, the correction term in the expression for σ_p, according to the above calculation, is of the same order of magnitude as $(\delta_p/\sigma_p)^2$, we must, on the present theory (i.e., on the assumption of separate collisions), expect an increase $(\delta_p > 0)$ of the apparent conductivity with decreasing period of vibration.

Comparing the results derived here with Hagen and Rubens' above-mentioned experiments on the reflection of heat rays by metals[2], we see, however, that, for most metals (among the pure metals, silver is the only exception), the apparent conductivity decreases with decreasing period of vibration[3].

This disagreement between theory and experiments is, in my opinion, due to the

---

[1] See P. Drude: Ann. d. Phys. 14 (1904) 940. Drude's result corresponds to that given below in the text, if for σ_p and δ_p we substitute the expressions found above eq. (73) for the case that the metal atoms are assumed to act upon electrons with forces inversely proportional to the fifth power of the distance; this arises from the fact that the resistance against the motion of the electrons in Drude's calculation is considered to be proportional to the speed of the electrons (compare the preceding, p. [321]).

[2] E. Hagen u. H. Rubens: Ann. d. Phys. 11 (1903) 884.

[3] Compare Drude, loc. cit. p. 944 (This decrease in the apparent conductivity has by several recent authors (see, e.g., J. H. Jeans: Phil. Mag. 17 (1909) 779) been regarded as being in accordance with an electron theory of the kind treated here, in that it has been assumed to arise from a decrease in the effective conductivity, calculated from the coefficient of absorption (see p. [362] above). Thus, from expressions for σ_p as those given above (eqs. (72') and (73')), attempts have been made to calculate the number N of free electrons per unit volume in the different metals; this number is, in fact, the only quantity, unknown beforehand, entering into the formulas in question. The values for N, thus obtained, are for most metals larger than values that are compatible with the experimental specific heats of the metals (see Jeans: loc. cit. p. 780). However, as we have seen above, this use of the formulas mentioned is not permissible, since the influence of the effective dielectric constant (see p. [363]) on the apparent conductivity, calculated from the reflection coefficients of metals, has the opposite sign from, and is of a lower order of magnitude than, the influence of the change in the apparent conductivity obtained from the coefficients of absorption.

circumstance that the assumption of separate collisions between atoms and electrons is not valid.

In fact, if separate collisions take place, the external forces will mainly cause a change in the motion of the electrons but not in their distribution in space (such a change will occur only in the immediate neighborhood of the atoms of the metal); on the other hand, if the dimensions of the atoms,—i.e., the ranges within which they affect the electrons appreciably,—are not negligible compared to the total volume of the metal, the piling up of electrons in front of the atoms during their collective motion (see p. [324]) might change their spatial distribution appreciably, and there would then, besides the effect due to the inertia of the electrons, which by itself would produce a negative effective dielectric constant, also be an effect caused by the displacement of the electrons in the force field of the atoms (not to be confused with the polarization of the atoms arising from the bound electrons, which, for periods of vibration considered here, would have exceedingly little effect (see the previous page)), which by itself would produce a positive dielectric constant.

§ 2

Emission of Heat Rays

If an electron moves in a straight line with constant speed, it will not emit energy; for in such a case the electromagnetic field around the electron will, so to speak, be carried with the electron during its motion.

The situation will, however, be different if the velocity of the electron changes either in magnitude or direction; in this case the electron will emit energy, which may be expressed by saying that a part of its electromagnetic field breaks away and is transmitted in all directions in the surrounding space with the speed of light. If the speed of the electron is very small compared to the speed of light c, the energy emitted in the time dt will be[1]

$$\frac{2\varepsilon^2}{3c^3} j^2 \, dt, \tag{76}$$

where j is the magnitude of the acceleration of the electron. The amount of energy is calculated as that which will be transmitted through a closed surface surrounding the electron at a distance from it which is large compared to the range within which the field carried with the electron has an appreciable value.

[1] See, e.g., H. A. Lorentz: *Theory of Electrons*, p. 52, eq. (82). (In this formula, the unit of electric charge is different from that used here (absolute electrostatic unit), with the result that the expression for the energy in question is divided by 4π.)

ELECTRON THEORY OF METALS

A piece of metal, in which the motions of the free electrons, on account of inter-action with the metal atoms, are subjected to large changes in rapid succession, will, therefore, be a source of emission of energy[1]. We shall now examine the magnitude of this emission of energy, and the manner in which it is distributed among rays of different periods of vibration.

As mentioned above, Lorentz has treated the problem of the distribution of the emitted energy among rays with periods of vibration long compared to the time in which electrons lose almost all their momentum in a given direction; he assumed that the metal atoms act upon electrons as hard elastic spheres and neglected the mutual interactions among the electrons. In Lorentz' calculations, the motions of the individual electrons were considered; however, it would be very complicated to do this in the more general cases we shall deal with here. (That Lorentz' case is simpler is due to the fact that in his theory the direction of motion of the electrons after collision is assumed to be, on the average, entirely independent of the direction before collision.) We shall, therefore, consider the collective motion of the electrons as a whole, following the method of calculation indicated by Jeans.

We shall examine the energy emitted from a small volume element of a metal ΔV which is assumed to contain a very large number of metal atoms and electrons and at the same time to be so small that the energy emitted by the individual electrons is not absorbed to an appreciable extent in the volume element itself (see p. [360]). To calculate this amount of energy, we imagine a single electron, the momentum of which at each instant is equal in magnitude and direction to the total momentum of all the electrons in the volume element ΔV, and to this electron we apply the formula (76).

(With regard to such an application of the formula (76), we must, however, make the following remark. The formula is derived for the motion of a point charge and, as mentioned above, it is obtained by integration of the energy passing through the elements of a closed surface surrounding the charge. If we now seek the energy passing through such a surface element as a result of the collective motion of the electrons, we may well regard this as collected into a point charge; however, the motion of this charge will not be the average of the simultaneous motions of the electrons, but rather the average of the motions of the electrons at such an instant for each electron that light rays emitted by the various electrons at these instants will all reach the surface element simultaneously. Nevertheless, if we assume that the velocities of the electrons are exceedingly small compared to the speed of light, each motion among the electrons will be transmitted exceedingly much slower than light, and, on the average, there will be very nearly the same systematic correlations among the motions of the individual electrons, whether we consider their simultaneous motions in the ordinary sense or consider motions that are "effectively simultane-

---

[1] J. J. Thomson: *Rapp. du Congrès de Physique* (Paris 1900), vol. **3**, p. 148*.
* [This footnote is taken from III.]

ous" relative to the surface element considered in the above-mentioned sense[1]. We shall thus, on the average, get the correct value for the flow of energy through the surface element if we calculate it by considering a single point charge having a momentum equal to the sum of the simultaneous momenta of the individual electrons. Since this is true for every element of the closed surface —there will, of course, be a different "effective simultaneity" for each surface element—it will hold for the entire flow of energy through the surface.)

Now, denoting the components of the total momentum of the electrons in ΔV by g_x, g_y and g_z, we get, from formula (76) for the energy emitted, on the average, in unit time, the expression

$$J = \frac{2\varepsilon^2}{3m^2c^3} \frac{1}{\mathscr{T}} \int_0^{\mathscr{T}} \left[\left(\frac{\mathrm{d}g_x}{\mathrm{d}t}\right)^2 + \left(\frac{\mathrm{d}g_y}{\mathrm{d}t}\right)^2 + \left(\frac{\mathrm{d}g_z}{\mathrm{d}t}\right)^2 \right] \mathrm{d}t,$$

where $\mathscr{T}$ is a time interval exceedingly large compared to the periods of vibration under consideration. Since there will be no correlation among the momenta along the three axes (for, in the equations derived in Chapter I for the determination of the total momentum along the x-axis, only the momenta of the electrons in this direction enter), we may consider separately the emission corresponding to each term; we get, on account of the symmetry,

$$J = \frac{2\varepsilon^2}{m^2c^3} \frac{1}{\mathscr{T}} \int_0^{\mathscr{T}} \left(\frac{\mathrm{d}g_x}{\mathrm{d}t}\right)^2 \mathrm{d}t. \tag{77}$$

In order to examine the distribution of this energy among rays with different periods of vibration, we write $\mathrm{d}g_x/\mathrm{d}t$ as a sum of harmonic vibrations, by means of a Fourier integral valid from $t = 0$ to $t = \mathscr{T}$. Thus, we have

$$\frac{\mathrm{d}g_x}{\mathrm{d}t} = \frac{1}{\pi} \int_0^\infty (A_p \cos pt + B_p \sin pt)\mathrm{d}p, \tag{78}$$

---

[1] J. H. Jeans (Phil. Mag. **18** (1909) 213) expresses the opinion that the emission of energy by a piece of metal can be calculated as the sum of the emissions from the individual electrons considered separately, since, according to statistical mechanics, there will, on the average, be no systematic correlations between the simultaneous velocities of two different electrons. However, this method of calculation will only give correct results if the mutual interactions among the electrons can be neglected; for, while these interactions will not, on the average, involve systematic correlations among the simultaneous velocities of the different electrons, they will give rise to correlations among their simultaneous accelerations. Thus, in a collision between two electrons, no energy will be emitted, since the total momentum of the electrons is not altered, and such collisions will, therefore, only have an indirect effect on the energy radiation of a piece of metal (compare the analogous question (see p. [342] above) of the effect of such collisions upon the electric conductivity and, thereby, on the power of absorption); on the other hand, if we had calculated the energy emission as the sum of the radiations from the individual electrons separately, i.e., from the accelerations of the individual electrons, with the aid of the formula (76), we should have found that these collisions would have a considerable direct influence (in proportion to their number) on the energy emission of the metal.

where

$$A_p = \int_0^{\mathscr{T}} \frac{dg_x}{dt} \cos pt\, dt \quad \text{and} \quad B_p = \int_0^{\mathscr{T}} \frac{dg_x}{dt} \sin pt\, dt. \tag{79}$$

From (78) and (79) follows, as proved by Lord Rayleigh[1],

$$\int_0^{\mathscr{T}} \left(\frac{dg_x}{dt}\right)^2 dt = \frac{1}{\pi} \int_0^{\infty} (A_p^2 + B_p^2)\, dp.$$

While the quantities A_p and B_p vary extremely rapidly and irregularly with p and $\mathscr{T}$, the quantity $(A_p^2 + B_p^2)/\mathscr{T}$ will, as we shall see, if $\mathscr{T}$ is sufficiently large, approach a definite limiting value, which will vary in a regular manner with p.

From (77), we now find that the energy emitted in unit time can be written

$$J = \frac{2\varepsilon^2}{\pi m^2 c^3} \frac{1}{\mathscr{T}} \int_0^{\infty} (A_p^2 + B_p^2)\, dp, \tag{80}$$

where the coefficient to dp is the energy carried by rays having frequencies between $p/2\pi$ and $(p+dp)/2\pi$.

From (79) we get, by partial integration,

$$A_p = |g_x \cos pt|_0^{\mathscr{T}} + p \int_0^{\mathscr{T}} g_x \sin pt\, dt.$$

Since the first term on the right side of this equation will not increase (i.e., will not oscillate with increasing amplitude) as $\mathscr{T}$ increases, it can be neglected compared to the last term in the calculation of $(A_p^2 + B_p^2)$. Neglecting, also, the corresponding term in the expression for B_p, we get

$$A_p = p \int_0^{\mathscr{T}} g_x \sin pt\, dt \quad \text{and} \quad B_p = -p \int_0^{\mathscr{T}} g_x \cos pt\, dt. \tag{81}$$

Denoting the value of g_x at the time t by $g_x(t)$, we now get from (81)

$$A_p^2 = p^2 \int_0^{\mathscr{T}} \int_0^{\mathscr{T}} g_x(t_1)g_x(t_2) \sin pt_1 \sin pt_2\, dt_1\, dt_2$$

and

$$B_p^2 = p^2 \int_0^{\mathscr{T}} \int_0^{\mathscr{T}} g_x(t_1)g_x(t_2) \cos pt_1 \cos pt_2\, dt_1\, dt_2,$$

whence

$$A_p^2 + B_p^2 = p^2 \int_0^{\mathscr{T}} \int_0^{\mathscr{T}} g_x(t_1)g_x(t_2) \cos p(t_2 - t_1)\, dt_1\, dt_2,$$

[1] Lord Rayleigh: Phil. Mag. **27** (1889) 465.

which can be written

$$A_p^2 + B_p^2 = 2p^2 \int_0^{\mathscr{T}} g_x(t_1) \int_{t_1}^{\mathscr{T}} g_x(t_2) \cos p(t_2 - t_1) \mathrm{d}t_2 \, \mathrm{d}t_1 .^{[1]} \qquad (82)$$

In calculating $A_p^2 + B_p^2$ from eq. (82), we shall first consider the case in which *separate collisions* are assumed to take place.

In this case we have, with the former notations,

$$g_x = \Delta V \int_0^\infty G_x(r) \mathrm{d}r. \qquad (83)$$

For a piece of metal unaffected by external forces and of uniform temperature, $G_x(r)$ is determined by the equation (see eq. (14))

$$\left[\frac{\mathrm{d}G_x(r)}{\mathrm{d}t} \right]_{G_x(\rho)} = -G_x(r)F(r) + \int_0^\infty G_x(\rho)Q(\rho, r)\mathrm{d}\rho, \qquad (84)$$

where the term on the left side denotes the average value of $\mathrm{d}G_x(r)/\mathrm{d}t$, taken for all the possible ways in which the momentum can be distributed among the individual electrons, such that the quantities $G_x(\rho)$ assume the values supposed to be substituted on the right side of the equation.

If we now multiply the term on the left side of eq. (84) by $\cos p(t_2 - t_1)$, integrate with respect to t_2 from t_1 to $\mathscr{T}$, and take the average of this integral for all possible ways in which the state at the time t_1 is characterized by definite values of $G_x(\rho)$,

---

[1] Eq. (82) is the same as that on which Jeans (Phil. Mag. **17** (1909) 789) bases his calculations. From it, he calculates $A_p^2 + B_p^2$ by using an equation of the form (in our notations)

$$\left[\frac{\mathrm{d}g_x}{\mathrm{d}t} \right] = -\alpha g_x, \qquad (1')$$

where the term on the left side denotes the average value of the differential quotient for a given value of g_x. From eq. (1'), Jeans derives

$$g_x(t_2) = g_x(t_1)\mathrm{e}^{-\alpha(t_2 - t_1)} + j, \qquad (2')$$

where j is a quantity whose average value is zero. Substituting the expression (2') in eq. (82), and integrating with respect to t_2, Jeans finally finds (loc. cit. p. 790)

$$A_p^2 + B_p^2 = \frac{2\alpha p^2}{\alpha^2 + p^2} \int_0^{\mathscr{T}} (g_x(t_1))^2 \, \mathrm{d}t_1 , \qquad (3')$$

by means of which $A_p^2 + B_p^2$ can be readily evaluated.

However, these calculations are, in general, not correct (compare H. A. Wilson: loc. cit. p. 836); for, as we have seen in Chapter I, the rapidity with which g_x decreases will depend on the way in which the momentum is distributed among electrons of different speeds; and this distribution will change in a systematic manner as g_x decreases, since electrons with different speeds lose their momentum in a given direction at a different rate.

we find, using notations corresponding to those employed above, and denoting the desired average value by a,

$$a = \left[\left[\int_{t_1}^{\mathscr{T}} \left[\frac{dG_x(r)}{dt} \right]_{(G_x(\rho))_{t_2}} \cos p(t_2 - t_1)dt_2 \right] \right]_{(G_x(\rho))_{t_1}}. \tag{85}$$

Assuming, as in the calculation of the absorption (see p. [361]), that $\cos pt$ does not vary appreciably within the duration of a single collision, and that, therefore also here, dt can be considered large compared to this duration, no systematic correlation will exist between $dG_x(r)/dt$ and previous states having appreciably different values for $\cos pt$ and $G_x(\rho)$, and we, therefore, have

$$a = \left[\left[\int_{t_1}^{\mathscr{T}} \frac{dG_x(r)}{dt_2} \cos p(t_2 - t_1)dt_2 \right] \right]_{(G_x(\rho))_{t_1}}. \tag{86}$$

From (86) we get by partial integration

$$a = \left[(G_x(r))_{\mathscr{T}} \cos p(\mathscr{T} - t_1) - (G_x(r))_{t_1} + p \int_{t_1}^{\mathscr{T}} G_x(r) \sin p(t_2 - t_1)dt_2 \right]_{(G_x(\rho))_{t_1}}.$$

Since the average value of the first term in the bracket vanishes if $\mathscr{T}$ is sufficiently large, we get, putting

$$\xi_2(r) = \left[\int_{t_1}^{\mathscr{T}} G_x(r) \sin p(t_2 - t_1)dt_2 \right]_{(G_x(\rho))_{t_1}}, \tag{87}$$

$$a = -(G_x(r))_{t_1} + p\xi_2(r). \tag{88}$$

Also, putting

$$b = \left[\left[\int_{t_1}^{\mathscr{T}} \left[\frac{dG_x(r)}{dt} \right]_{(G_x(\rho))_{t_2}} \sin p(t_2 - t_1)dt_2 \right] \right]_{(G_x(\rho))_{t_1}} \tag{89}$$

and

$$\xi_1 = \left[\int_{t_1}^{\mathscr{T}} G_x(r) \cos p(t_2 - t_1)dt_2 \right]_{(G_x(\rho))_{t_1}}, \tag{90}$$

we get, in exactly the same manner as above,

$$b = -p\xi_1(r). \tag{91}$$

If we now multiply both sides of eq. (84) by $\cos p(t_2 - t_1)$ and $\sin p(t_2 - t_1)$, respectively, integrate with respect to t_2 from t_1 to $\mathscr{T}$, and then take the average of the integrals for all possible ways in which the momentum can be distributed among the individual electrons in such a way that the state at the time t_1 is specified by given values of $G_x(\rho)$, we get, with the aid of eqs. (85), (87), (88), (89), (90) and (91),

$$-(G_x(r))_{t_1} + p\xi_2(r) = -\xi_1(r)F(r) + \int_0^\infty \xi_1(\rho)Q(\rho, r)\mathrm{d}\rho \qquad (92)$$

and

$$-p\xi_1(r) = -\xi_2(r)F(r) + \int_0^\infty \xi_2(\rho)Q(\rho, r)\mathrm{d}\rho. \qquad (93)$$

Since eqs. (92) and (93) for the determination of $\xi_1(r)$ and $\xi_2(r)$ correspond perfectly, as will be seen, to eqs. (62) and (63) for the determination of $\psi_1(r)$ and $\psi_2(r)$, we get, in exactly the same way as in the case of these equations,

$$\xi_1(r) = (G_x(r))_{t_1} F_1(r) + \int_0^\infty (G_x(\rho))_{t_1} Q_1(\rho, r)\mathrm{d}\rho, \qquad (94)$$

where $F_1(r)$ and $Q_1(\rho, r)$ are the same functions as occur in eq. (64). With the aid of eqs. (83), (90), (94), and (65), we now get

$$\left[\int_{t_1}^{\mathscr{T}} g_x(t_2) \cos p(t_2 - t_1)\mathrm{d}t_2\right]_{(G_x(\rho))_{t_1}} = \Delta V \int_0^\infty (G_x(r))_{t_1} P_1(r)\mathrm{d}r. \qquad (95)$$

Assuming that the time $\mathscr{T}$ is so long that $g_x(t)$ not only assumes all possible values within this time but assumes each value so many times that we may suppose that it has had each value with the momentum distributed among the individual electrons in all possible ways, we can substitute the expression (95) in eq. (82) instead of the integral with respect to t_2 occurring in this equation; remembering the definition (83) of g_x, we, therefore, get

$$A_p^2 + B_p^2 = 2p^2\Delta V^2 \int_0^{\mathscr{T}} \left(\int_0^\infty G_x(r)\mathrm{d}r \int_0^\infty G_x(r)P_1(r)\mathrm{d}r\right)_{t_1} \mathrm{d}t_1. \qquad (96)$$

As will be seen, we have now succeeded in making the integrand pertain to a single moment of time, and the problem has thereby been very much simplified.

We have, so far, in our calculations, for simplicity, used integrals to express the collective momentum of the electrons; this is strictly not correct, since there is only a finite number of electrons present in ΔV, but it is easily seen that it would have made no essential difference if we had replaced the integrals by sums over the individual electrons. However, in the evaluation of the expression (96), the matter is quite different, as the integral entering here depends to the highest degree on the number of electrons. Hence, instead of the integrals with respect to r in (96), we shall introduce sums over the ΔN electrons in the volume element ΔV.

Denoting the velocity of a single electron by ξ, η, ζ, we get

$$A_p^2 + B_p^2 = 2p^2m^2 \int_0^{\mathcal{T}} \left(\sum_1^{\Delta N} \xi_n \sum_1^{\cdot \Delta N} \xi_n P_1(r_n) \right)_{t_1} dt_1 . \tag{97}$$

Now, according to statistical mechanics, no matter what forces are assumed to act among the electrons, there will be no systematic correlations among the velocity coordinates of individual electrons (there will be, on the other hand, correlations among their simultaneous accelerations (see footnote p. [369]), but this has been taken into account in the preceding calculation, i.e., by the use of eq. (84)). Terms with $\xi_n \xi_m P_1(r_m)$ in which n and m are different will, therefore, not contribute to the value of the integral in (97), and we get

$$A_p^2 + B_p^2 = 2p^2m^2 \int_0^{\mathcal{T}} \left(\sum_1^{\Delta N} \xi_n^2 P_1(r_n) \right) dt. \tag{98}$$

Since, on account of the symmetry ($r^2 = \xi^2 + \eta^2 + \zeta^2$), the average value of $\xi_n^2 P_1(r_n)$ will be $\frac{1}{3}$ times the average value of $r_n^2 P_1(r_n)$, and, since the velocity distribution of the electrons is given by eq. (4), we get from (98)

$$A_p^2 + B_p^2 = 2p^2m^2 \mathcal{T} \Delta V \int_0^\infty \tfrac{1}{3}r^2 P_1(r) K e^{-\frac{1}{2}mr^2/kT} 4\pi r^2 \, dr$$

$$= \frac{2m^2kT}{\varepsilon^2} \sigma_p p^2 \mathcal{T} \Delta V, \tag{99}$$

where σ_p is the quantity defined by eq. (66).

Denoting by $\beta_p \, dp \Delta V$ the energy emitted in unit time by a small piece of metal of volume ΔV, distributed among rays with periods of vibration between $p/2\pi$ and $(p + dp)/2\pi$, we get, from eqs. (80) and (99),

$$\beta_p = \frac{4kT}{\pi c^3} \sigma_p p^2. \tag{100}$$

Substituting in (100) the value for σ_p given by eq. (69), and putting $n = \infty$ and $n = 5$, respectively, we obtain the values for β_p given by Wilson[1] and Jeans[2], respectively.

We shall now consider the case in which *the dimensions of the metal atoms are not assumed to be negligible compared to their mutual distances*. In this case, we cannot use a procedure similar to that used in the case just discussed, for there will now exist systematic correlations between the rapidity with which the momentum decreases and the preceding states within time intervals of the same order of magnitude as the times within which the individual electrons, on the

[1] H. A. Wilson: loc. cit. p. 842.
[2] J. H. Jeans: Phil. Mag. **17** (1909) 790.

average, lose the greater part of their original momentum in a given direction[1]. (Such a time interval we shall in the following for brevity denote by τ.)

We shall here only consider the case (see p. [325]) in which the force fields of the metal atoms are assumed to be stationary, and where the mutual interactions among the electrons are neglected; moreover, we shall only carry out the calculations for heat rays with periods of vibration that are large compared to the above-mentioned time interval τ.

The electrons being assumed to move independently of one another, there will be no systematic correlations among their simultaneous motions; the radiation from a small piece of metal ΔV can, therefore, be calculated as the sum of the emissions by the individual electrons considered separately. Denoting the number of electrons in the piece of metal by ΔN, we get in this case, in exactly the same way as above, for the amount of energy emitted in unit time, corresponding to eqs. (80) and (82),

$$J = \frac{2\varepsilon^2}{\pi c^3} \frac{1}{\mathscr{T}} \int_0^\infty [\sum_1^{\Delta N} (A_p^2 + B_p^2)_n] \mathrm{d}p, \tag{101}$$

where

$$(A_p^2 + B_p^2)_n = p^2 \int_0^{\mathscr{T}} \int_0^{\mathscr{T}} \xi_n(t_1)\xi_n(t_2) \cos p(t_2-t_1)\mathrm{d}t_1\,\mathrm{d}t_2, \tag{102}$$

$\xi_n(t)$ denoting the x-component of the velocity of the nth electron at the time t.

As we shall only consider periods of vibration large compared to τ, $\cos p(t_2-t_1)$ can be put equal to one for all values of t_1 and t_2 for which there is an appreciable correlation between $\xi_n(t_1)$ and $\xi_n(t_2)$; it will be seen, however, that, also for all other values of t_1 and t_2, is it permissible to replace $\cos p(t_2-t_1)$ by 1 in the integral (102); in fact, because of the independence of $\xi_n(t_1)$ and $\xi_n(t_2)$, these other values of t_1 and t_2 will in neither case make any appreciable contribution to the value of the integral (102). We thus get from (102)

$$(A_p^2 + B_p^2)_n = p^2 \left(\int_0^{\mathscr{T}} \xi_n(t)\mathrm{d}t \right)^2, \tag{103}$$

in which equation the integral on the right side is equal to the projection on the x-axis of the path travelled by the nth electron in the time $\mathscr{T}$.

---

[1] Jeans applies the same procedure in all cases; he only remarks (loc. cit. p. 776) that $\mathrm{d}t$ in the equation corresponding to eq. (1') in the footnote p. [371] must be regarded as large compared to time intervals within which g_x varies greatly. Apart from the deficiencies in Jeans' calculations mentioned above, which will, of course, also be present in this more general case, his assumption about $\mathrm{d}t$ seems be to consistent with the use of eq. (1') as a differential equation to derive eq. (2'); in fact, in this application it would be necessary to assume that g_x changes very little in the time $\mathrm{d}t$.

If it is assumed that no mutual interaction occurs among the free electrons, and that the force fields of the metal atoms are stationary, the sum a of the kinetic energy of a single electron and its potential energy relative to the metal atoms will remain constant during its motion; moreover, as pointed out in Chapter I, the motion of the electron can be regarded as a "free diffusion" with a diffusion coefficient $D(a)$ (see p. [327]). To connect this case with the conditions in the real metals, we assumed, in the calculations in Chapter I, only that a is essentially constant, on the average, within time intervals of the same order of magnitude as τ (see p. [325]); however, we shall here assume, for simplicity, that a is constant during the entire motion of the electron in the long time $\mathscr{T}$; for reasons quite similar to those mentioned above when we replaced $\cos p(t_2 - t_1)$ by 1, this will have no appreciable influence upon the value of the expression for $\Sigma(A_p^2 + B_p^2)_n$.

Now, the determination of the quantity $\Sigma(A_p^2 + B_p^2)_n$ has by eq. (103) been reduced to the problem of the distribution at the time $\mathscr{T}$ of a large number of independently diffusing particles located at the time $t = 0$ in a certain plane (e.g., the yz-plane). As Einstein[1] has shown, this question can be treated in a very simple manner. In fact, the distribution of the particles will be determined by the equation

$$\frac{\partial f}{\partial t} = D \frac{\partial^2 f}{\partial x^2},$$

where $f(x, t)\mathrm{d}x$ is the number of particles at the time t between two parallel planes at the distances x and $x + \mathrm{d}x$ from the fixed plane, and where the diffusion coefficient D corresponds to the quantity $D(a)$ mentioned above. The solution of this equation, under the conditions that $f(x, t) = 0$ for $t = 0$ and $x \lessgtr 0$,—a solution known from Fourier's theory of heat conduction in an isotropic medium,—is

$$f(x, t) = C t^{-\frac{1}{2}} \mathrm{e}^{-x^2/4Dt},$$

and the average value of the square of the distance of the particles from the fixed plane at the time $\mathscr{T}$ will be (Einstein: loc. cit. p. 559)

$$\int_0^\infty x^2 f(x, \mathscr{T})\mathrm{d}x \bigg/ \int_0^\infty f(x, \mathscr{T})\mathrm{d}x = 2D\mathscr{T}.$$

Hence, we get from eq. (103)

$$\sum_1^{\Delta N} (A_p^2 + B_p^2)_n = 2p^2 \mathscr{T} \sum_1^{\Delta N} D(a)$$

and, denoting, as in Chapter I, the number of electrons per unit volume having energies between a and $a + \mathrm{d}a$ by $N(a)\mathrm{d}a$, we get

[1] A. Einstein: Ann. d. Phys. **17** (1905) 556.

$$\sum_1^{\Delta N} (A_p^2 + B_p^2)_n = 2p^2 \mathscr{T} \Delta V \int_0^\infty D(a)N(a)\mathrm{d}a. \tag{104}$$

Moreover, since the electric conductivity σ of the metal is (see eqs. (35) and (36))

$$\sigma = \frac{\varepsilon^2}{kT} \int_0^\infty D(a)N(a)\mathrm{d}a,$$

we finally get from eqs. (101) and (104), using the same notations as on p. [374],

$$\beta_p = \frac{4kT}{\pi c^3} \sigma p^2. \tag{105}$$

We shall now apply these calculations to derive *the law of heat radiation for rays with large periods of vibration*. In a small piece of metal in equilibrium with the electromagnetic radiation in the empty space surrounding it, the energy absorbed by the metal in unit time, distributed among rays with certain periods of vibration, must, on the average, be equal to the energy emitted by the metal during the same time, distributed in the same way among rays with the same periods of vibration. Denoting the energy present, in the state of equilibrium, in unit volume of the surrounding empty space with periods of vibration between $2\pi/p$ and $2\pi/(p+\mathrm{d}p)$ by $E(p)\mathrm{d}p$, and denoting the volume of the metal by ΔV, we, therefore, get

$$E(p)\mathrm{d}p\,\alpha_p\,\Delta V = \beta_p\,\mathrm{d}p\,\Delta V,$$

whence

$$E(p) = \frac{\beta_p}{\alpha_p}.$$

Substituting the expressions found for α_p and β_p, either from eqs. (68) and (100) or from eqs. (59) and (105), we get in both cases

$$E(p) = \frac{kT}{\pi^2 c^3} p^2. \tag{106}$$

As follows from what was said in connection with the determination of α_p and β_p (see p. [365] and p. [374]), eq. (106) is identical with that derived by Jeans, but is different from that found by Wilson; in Wilson's opinion, this is due to the fact that the mutual collisions among the electrons were neglected in his calculations (loc. cit. p. 844); however, as we have seen, the equation, in any case, as long as separate collisions are assumed to take place, is entirely independent of the influence ascribed to the mutual interactions among the electrons as compared to the interactions between the electrons and the metal atoms.

Introducing instead of p the wavelenth λ, related to p by the equation $p/2\pi = c/\lambda$, and denoting the energy per unit volume distributed among rays with wavelengths between λ and $\lambda + \mathrm{d}\lambda$ by $E(\lambda)\mathrm{d}\lambda$, we get from (106)

$$E(\lambda) = 8\pi kT\lambda^{-4}. \tag{107}$$

This formula, *the validity of which we have here proved for all the cases considered in Chapter I*, is the same as that derived by Lorentz, and, as mentioned in the preceding, it is in excellent agreement with the experiments on heat rays with long wavelengths.

To examine more closely the significance of formula (107), we must mention attempts made by Lord Rayleigh[1] and Jeans[2] to derive the law of heat radiation for rays of all possible periods of vibration directly from the assumptions on which the electromagnetic theory is based. In fact, these authors have attempted, with the aid of the general principles of statistical mechanics, to investigate the condition for equilibrium between electromagnetic radiation and an arbitrary system of electrons present in a closed space perfectly isolated from external influences. According to the result of these investigations, the condition for equilibrium is that the radiant energy per unit volume be expressed by formulas (106) or (107) for rays with all possible periods of vibration. As is seen, however, this would have the consequence that the total amount of energy per unit volume would become infinite; this would mean that a state of equilibrium could not be reached at all, but that the entire energy would gradually be transferred to the radiation, and more and more be concentrated on rays with shorter and shorter periods of vibration.

On the basis of these results, Jeans has expressed the opinion that we are not dealing with a real state of equilibrium in the experiments on heat radiation, but only with a state that changes very slowly, on account of the low power of absorption and emission of ordinary bodies for rays with very small periods of vibration. However, as appears from the discussion[3] that has taken place on this question, this is hardly sufficient to explain the phenomena found experimentally. Hence, it seems impossible to explain the law of heat radiation if one insists upon retaining the fundamental assumptions underlying the electromagnetic theory[4]. This is presumably due to the circumstance that the electromagnetic theory is not in accordance with the real conditions and can only give correct results when applied to a large number of electrons (as are present in ordinary bodies) or to determine the average motion of a single electron over comparatively long intervals of time (such as in the calculation of the motion of cathode rays) but cannot be used to examine the motion of a single electron within short intervals of time. Since it would take us quite outside the scope of the problems treated in the present work,

[1] Lord Rayleigh: Phil. Mag. **49** (1900) 539; Nature **72** (1905) 54, 243.

[2] J. H. Jeans: Phil. Mag. **10** (1905) 91; **17** (1909) 229; see also H. A. Lorentz: Nuovo Cimento **16** (1908) 5.

[3] See for example O. Lummer u. E. Pringsheim: Phys. Zeitschr. **9** (1908) 449, H. A. Lorentz: Phys. Zeitschr. **9** (1908) 562 and J. H. Jeans: Phys. Zeitschr. **9** (1908) 853.

[4] See A. Einstein: Phys. Zeitschr. **10** (1909) 185, 817, M. Planck: Ann. d. Phys. **31** (1910) 758 and A. Einstein u. L. Hopf: Ann. d. Phys. **33** (1910) 1105.

we shall not enter upon a discussion of the attempts made to introduce fundamental changes in the electromagnetic theory, attempts that seem to be leading to interesting results.

The question may now be asked, how the calculations made in this chapter are related to the problem just mentioned. First of all, we must point out that we have not attempted to determine a state of equilibrium in the same sense as Lord Rayleigh and Jeans. Rather, we have tacitly assumed that, in accordance with experience, a state of equilibrium will exist, in which the electromagnetic energy per unit volume is a comparatively small quantity, i.e., so small that the electromagnetic fields arising from the radiation will only have a very small influence upon the motion of the free electrons, as compared to the forces which the electrons exert upon one another and are acted on by the metal atoms[1]. (If, on the contrary, the problem is to investigate the state of equilibrium which would appear in a closed space isolated from external influences, then we must, as Jeans[2] has pointed out, provided the fundamental electromagnetic assumptions were perfectly valid, necessarily take into account the electromagnetic fields in question, which eventually, as the radiant energy increases, would exert a far greater influence upon the motions of the electrons than the fields of the metal atoms.)—Moreover, as we have treated only heat rays with periods of vibration large compared to the duration of single electromagnetic processes (e.g., collisions), it seems not unreasonable to expect beforehand that the calculations made in this chapter have the same relation to—and the same claims to accordance with—experience as the rest of the electron theory of metals. This seems especially to be true for the derivation given in the last case treated, since no use at all was made of any assumptions about the occurrence of sudden collisions.

[1] Several authors (J. J. Thomson: Phil. Mag. **14** (1907) 217; **20** (1910) 238 and J. Kunz: Phys. Rev. **28** (1909) 313; compare also J. H. Jeans: Phil. Mag. **18** (1909) 217; **20** (1910) 642) are of the opinion that, by a type of calculation, such as that mentioned, based on the ordinary electron theory and suitable assumptions about the interactions between metal atoms and electrons, we can explain, at any rate, the essential character of the radiation law for all periods of vibration, —e.g., the very rapid decrease in the energy distributed among rays with very short periods of vibration. This, however, does not seem to me to be correct; for in the investigations mentioned, sufficient attention has not been paid to the exceedingly rapid decrease of the power of absorption for rays with very short periods of vibration, but only to the decrease in the power of emission (see p. [358]). If the decrease in the power of absorption is also taken into account, we will, in my opinion, in all cases arrive at a law of radiation that is not consistent with the result of experiments; for the opposite would conflict with the calculations of Lord Rayleigh and Jeans. In fact, if, from preconceived assumptions, such as those mentioned above, we were led to a law of radiation that was consistent with these assumptions for all periods of vibration, it would correspond to a real equilibrium and not to an apparent equilibrium (compare the discussion given above of Jeans' theory).

[2] J. H. Jeans: Phil. Mag. **18** (1909) 215.

CHAPTER IV

INFLUENCE OF A MAGNETIC FIELD ON THE MOTION
OF THE FREE ELECTRONS IN A METAL

§ 1

States of Statistical Equilibrium

We shall first examine the influence of a magnetic field on a piece of metal which is in a state of both electric and temperature equilibrium.

Under the influence of the magnetic field, the paths of the free electrons will become curved in a definite direction relative to the axis of the field,—if separate collisions occur the paths of the electrons between collisions will consist of segments of helices with axes parallel to the axis of the field. It might be imagined that this change in the motion of the electrons would produce external magnetic effects. As we shall see, however, such an assumption would not be correct. In fact, the curvature of the paths will not produce any magnetic effects; for the magnetic field arising from the motion of a single electron will depend only on the position and the velocity of the electron at the instant considered, and not on the acceleration of the electron (we assume, as in the preceding, that the speed of the electrons may be regarded as negligible compared to the speed of light). The total magnetic effect of the motion of the free electrons can, therefore, at any moment be determined directly from the statistical distribution of the space and velocity coordinates of the electrons.

To investigate the effect of the magnetic forces upon the statistical distribution of the electrons, we shall consider the alteration they will produce in a given distribution f in the time dt. Denoting the components of the force per unit mass arising from the magnetic field by X_m, Y_m and Z_m, we get for the change ΔN in the number of electrons in the volume element dV having velocity coordinates in the element $d\sigma$ (see p. [304])

$$\Delta N = \left(X_m \frac{\partial f}{\partial \xi} + Y_m \frac{\partial f}{\partial \eta} + Z_m \frac{\partial f}{\partial \zeta} \right) dV \, d\sigma \, dt.$$

If now f is a function of r and x, y, z, as it must be in the state of equilibrium, we get

$$\Delta N = (X_m \xi + Y_m \eta + Z_m \zeta) \frac{1}{r} \frac{\partial f}{\partial r} \, dV \, d\sigma \, dt;$$

since the magnetic force is perpendicular to the direction of motion of the individual electron, the first factor on the right side of this equation is zero, and hence $\Delta N = 0$. Thus, we see that, if a state of equilibrium exists, the presence of magnetic forces will not result in any change in the statistical distribution of the electrons*. Whether a magnetic force is present or not, the velocities of the electrons in any arbitrarily small volume element of the metal will be equally distributed in all directions, and no magnetic effect will issue from such a volume element, and consequently not from the whole piece of metal.

If we do not treat this problem statistically, but rather consider the motions of the individual electrons, we find that each electron moving entirely in the interior of the metal will, on account of the curvature of its path caused by the external magnetic field, produce a magnetic field which, on the average, is in the opposite direction of the external field. From this fact, several authors[1] have concluded that a piece of metal containing free electrons will act as a diamagnetic body. However, as we have seen above, this is not correct. The reason for the error is that, when the problem is treated in the manner just described, i.e., by considering the paths of the individual electrons separately, it is necessary to pay special attention to the conditions in the immediate neighborhood of the surface of the metal.

To make this clear, we shall here consider these circumstances in a special, very simple, example.

* [In accordance with II, the preceding part of this paragraph replaces a sentence: "Som vi nu imidlertid har omtalt . . .".]

[1] J. J. Thomson: *Rapp. du Congrès de Physique* (Paris 1900), vol. 3, p. 148 and W. Voigt: Ann. d. Phys. 9 (1902) 130, Note 1; see also P. Langevin: Ann. d. Chim. et d. Phys. 5 (1905) 90. (Z. Thullie (Prace mat.–fiz. Warszawa 19 (1908) 207), contrary to the authors named, believes that the effect of the free electrons will not be diamagnetic but paramagnetic; however, as far as I can see, this is due to an error of sign in his calculations; compare loc. cit. eq. (1), p. 209 and eq. (4), p. 210.)

Consider a collection of electrons moving independently of one another in a closed container, and assume, for simplicity, that, when the electrons strike the walls of the container, they are reflected like elastic spheres from a hard smooth surface. We shall now examine the effect of an external magnetic field upon such a collection of electrons.

We first show, for this special case, what we have proved in perfect generality above, namely that, in statistical equilibrium, the velocities of the electrons in any arbitrarily small volume element of the container will be equally distributed in all directions. The figure represents a small part of the container near the wall. The wall, assumed to be perpendicular to the plane of the figure, is represented by the line ab, and the interior of the container is imagined to lie to the right of this line. In the figure are drawn the paths of a number of electrons, all of which have the same speed and move in the plane of the figure. For simplicity, we consider only electrons such as those whose paths are included in the figure. That such a collection of electrons will be in statistical equilibrium, if their velocities at every point are equally distributed in all directions in the plane of the figure, follows immediately, for points at such a distance from the wall that electrons starting from these points will be prevented by the action of the magnetic field from reaching the wall, from the symmetry with respect to the magnetic field; that the same is true for points arbitrarily close to the wall is a consequence of the fact that electrons (according to the assumption made) will be reflected from the wall in such a way that their paths, on the average, form direct continuations of the paths they would travel if the wall were not present. Quite similar considerations show that a corresponding statistical equilibrium will exist for all of the electrons, i.e., also when such electrons are included which do not move in the plane of the figure, or which lie near a part of the wall that is not parallel to the external magnetic field. Hence, the motions of the electrons considered will not give rise to any external magnetic effects.

We shall now proceed to examine the paths of the individual electrons in the example considered. Again, if we consider such electrons as are indicated in the figure, we see that an electron, which during its motion does not come in contact with the wall of the container but describes a closed orbit, will, on the average, produce a magnetic field in the opposite direction of the external field. On the other hand, if we follow the motions of the electrons, which along their paths come in contact with the wall, we see from the figure that, under the influence of the external field, they acquire a sort of creeping motion along the wall; these electrons, considered separately, will, therefore, produce a magnetic field in the same direction as the external field, which will be equal and opposite to the field produced by the electrons in the interior.

Thus, we have seen that *a piece of metal in electric and thermal equilibrium will not possess any magnetic properties whatever due to the presence of free electrons.* (It must be pointed out, however, that in a piece of metal exposed to a variable magnetic field there will arise a collective motion of the electrons—the so-called Foucault current—produced by the electric field which, according to the electromagnetic theory, is inextricably connected with the variation of the magnetic field; this current will produce a magnetic effect that will counteract the variation of the magnetic field. Quite similar conditions will, of course, arise when a piece of metal is suddenly exposed to a magnetic field[1], but also in this case will the induced

---

[1] As shown by Langevin (loc. cit. p. 82–93; compare also Z. Thullie: Anz. d. Akad. d. Wiss. Krakau, math.-nat. Kl. (1907) 749), a condition corresponding to that mentioned here will also occur if we consider the "bound" electrons assumed to move in closed orbits in the interior

collective motion of electrons disappear very rapidly after the magnetic field has become constant, without leaving any permanent effect upon the statistical distribution of the velocities of the electrons.)

§ 2

Galvano- and Thermomagnetic Phenomena

We have, so far, assumed that the piece of metal considered is in both electric and thermal equilibrium; however, if this is not the case, the electrons will, on the average, assume a collective motion in a definite direction, and the influence of the magnetic field on the motion will reveal itself, giving rise to the so-called *galvano-magnetic* and *thermomagnetic* phenomena.

We shall now derive the equations for the collective motion of the electrons when, in addition to such external forces as are considered in Chapter I, there also act external magnetic forces. We shall only consider the case in which separate

---

of the metal atoms; if a magnetic field is applied suddenly, it will, also in this case, produce a motion of the electrons which will give rise to an opposing magnetic field. The reason that W. Voigt (loc. cit. p. 125) and J. J. Thomson (Phil Mag. **6** (1903) 688) have not found such an effect in a collection of electrons moving in closed orbits subject to attractive central forces is, as Langevin (loc. cit. p. 83) has pointed out, that these authors have not considered sufficiently the electric field connected with the variation in the magnetic field. (It may be remarked here that, in view of what is stated in the text above, the calculations of Voigt and Thomson illustrate the independence of the statistical equilibrium distribution of the presence of magnetic fields; in the case mentioned in which the electrons move in closed orbits, there will be no "boundary" effect; compare the text above.) Starting from the circumstance mentioned, Langevin has attempted to explain the diamagnetic properties of bodies by assuming that a motion arising in this way will persist without change, as long as the magnetic field remains constant. However, according to what we have discussed in Chapter I (p. [306]), this will not be the case if the motions of the electrons obey the general mechanical laws, and if there occurs any transfer of energy whatever among the electrons in the atoms, such that finally a statistical mechanical equilibrium is reached (the presence of a certain amount of energy exchange among these electrons must be assumed, according to the theory in question; for the magnetic effects considered are explained just by the action of external forces upon these electrons, i.e., by forces arising from electrons outside the atom).

It follows from the above that it is not possible to explain the magnetic properties of bodies on the basis of the electron theory if such effects as, e.g., the emission of energy are neglected, which might have the effect of preventing a statistical mechanical equilibrium among the bound electrons as that assumed for the free electrons. (Compare W. Voigt, loc. cit. p. 146, and J. J. Thomson, loc. cit. p. 689; see also R. Gans: Nachr. d. Kgl. Ges. d. Wiss. Göttingen, math.–phys. Kl. (1910) 213. An emission of energy, however, might perhaps explain the paramagnetic but not the diamagnetic phenomena.) The circumstances mentioned here seem to be intimately connected with those discussed at the end of the last chapter.

collisions occur. As in the first chapter, we shall determine the total momentum possessed by the electrons present in a volume element dV and having speeds between r and $r+dr$. The components along the coordinate axes of this momentum we shall denote by

$$G_x(r)dr\,dV, \qquad G_y(r)dr\,dV \quad \text{and} \quad G_z(r)dr\,dV.$$

In Chapter I (pp. [309]–[316]) we have calculated the increment which this momentum will receive in the time dt due to the action of external forces and to the motions of the electrons, as well as to collisions between metal atoms and electrons, and among the electrons mutually; we shall now calculate the increment caused by the magnetic forces. This is done very easily; for, since the forces arising from the magnetic field are perpendicular to the direction of motion of the individual electrons, they will not change the speed of the electrons and, thus, not cause electrons to leave or enter the speed group considered (compare the note p. [311]); the entire increment in the momentum in question will, therefore, be equal to the sum of the increases in the time dt in the momenta of the individual electrons in the group considered.

Denoting the components of the magnetic field (in absolute magnetic units) by H_x, H_y and H_z, and using a coordinate system in which the direction of rotation in the xy-plane is counterclockwise when viewed from positive z, we have for the components of the force exerted by the magnetic field upon an individual electron

$$M_x = \frac{\varepsilon}{c}H_z\eta - \frac{\varepsilon}{c}H_y\zeta, \qquad M_y = \frac{\varepsilon}{c}H_x\zeta - \frac{\varepsilon}{c}H_z\xi, \quad \text{and} \quad M_z = \frac{\varepsilon}{c}H_y\xi - \frac{\varepsilon}{c}H_x\eta,$$

and the increase in the momentum $G_x(r)dr\,dV$ in the time dt produced by the magnetic forces is, therefore,

$$\sum M_x\,dt = \left(\frac{\varepsilon}{c}H_z\sum\eta - \frac{\varepsilon}{c}H_y\sum\zeta\right)dt = \left(\frac{\varepsilon}{cm}H_z\,G_y(r) - \frac{\varepsilon}{cm}H_y\,G_z(r)\right)dr\,dV\,dt.$$

From the above-mentioned calculations in Chapter I, we now get, as equations for determining the collective motion of the electrons when external magnetic forces are also taken into account,

$$\left[\frac{dG_x(r)}{dt}\right] = -\tfrac{4}{3}\pi m K\left(\frac{\varepsilon}{kT}\frac{\partial\varphi}{\partial x} + \frac{1}{K}\frac{\partial K}{\partial x} + \frac{\tfrac{1}{2}mr^2}{kT^2}\frac{\partial T}{\partial x}\right)r^4 e^{-\tfrac{1}{2}mr^2/kT}$$

$$- G_x(r)F(r) + \int_0^\infty G_x(\rho)Q(\rho,r)d\rho + \frac{\varepsilon}{cm}H_z\,G_y(r) - \frac{\varepsilon}{cm}H_y\,G_z(r) \quad (108)$$

and the two corresponding equations, obtained by cyclic permutation of x, y and z.

Eq. (108), and the corresponding equations, allow us to determine the quantities

$G_x(r)$, $G_y(r)$ and $G_z(r)$ under given circumstances. From these quantities we can calculate the electric current and the flow of heat which, under the given circumstances, will pass through a surface element in the interior of the metal at any given time; by a procedure quite analogous to that employed for the thermoelectric phenomena, we can then calculate the various galvano- and thermomagnetic effects.

We shall treat only a single one of the phenomena mentioned, namely the galvanomagnetic effects which appear, under stationary conditions, in a homogeneous piece of metal of uniform temperature when it is exposed to the influence of a homogeneous magnetic field; moreover, we shall only consider the simple case in which the metal atoms are assumed to be fixed centres of force, acting upon the electrons with forces inversely proportional to the nth power of the distance, and in which the mutual collisions among the electrons are neglected.

Assuming that the magnetic field is parallel to the z-axis, we get in this case, since $Q(\rho, r) = 0$, the following equations for the determination of $G_x(r)$ and $G_y(r)$

$$-\frac{4\pi m\varepsilon K}{3kT}\frac{\partial\varphi}{\partial x}r^4 e^{-\frac{1}{2}mr^2/kT} - G_x(r)F(r) + \frac{\varepsilon}{cm}H_z G_y(r) = 0,$$

$$-\frac{4\pi m\varepsilon K}{3kT}\frac{\partial\varphi}{\partial y}r^4 e^{-\frac{1}{2}mr^2/kT} - G_y(r)F(r) - \frac{\varepsilon}{cm}H_z G_x(r) = 0. \tag{109}$$

From eqs. (109), we get

$$G_x(r) = -\frac{4\pi m\varepsilon K}{3kT}\frac{\dfrac{\partial\varphi}{\partial x}F(r) + \dfrac{\partial\varphi}{\partial y}\dfrac{\varepsilon}{cm}H_z}{F(r)^2 + \left(\dfrac{\varepsilon}{cm}H_z\right)^2}r^4 e^{-\frac{1}{2}mr^2/kT} \tag{110}$$

and an analogous expression for $G_y(r)$, obtained from (110) by interchanging x with y and y with $-x$. For the electric current through unit area perpendicular to the x- and y-axes, respectively, we now have

$$i_x = \frac{\varepsilon}{m}\int_0^\infty G_x(r)\mathrm{d}r \quad\text{and}\quad i_y = \frac{\varepsilon}{m}\int_0^\infty G_y(r)\mathrm{d}r,$$

whence, by (110) and the corresponding equation,

$$i_x = -\sigma_1\frac{\partial\varphi}{\partial x} - \delta_1\frac{\partial\varphi}{\partial y} \quad\text{and}\quad i_y = -\sigma_1\frac{\partial\varphi}{\partial y} + \delta_1\frac{\partial\varphi}{\partial x}, \tag{111}$$

where σ_1 and δ_1 are determined by the equations

$$\sigma_1 = \frac{4\pi\varepsilon^2 K}{3kT} \int_0^\infty \frac{F(r)r^4 e^{-\frac{1}{2}mr^2/kT}}{F(r)^2 + \left(\frac{\varepsilon}{cm}H_z\right)^2}\, dr$$

and (112)

$$\delta_1 = \frac{4\pi\varepsilon^2 K}{3kT} \int_0^\infty \frac{\frac{\varepsilon}{cm}H_z r^4 e^{-\frac{1}{2}mr^2/kT}}{F(r)^2 + \left(\frac{\varepsilon}{cm}H_z\right)^2}\, dr.$$

In the evaluation of these expressions, we shall assume that $(\varepsilon/cm)H_z F(r)^{-1}$ is a very small quantity for all values of r that make an appreciable contribution to the integrals. Since, as can be shown from the above expressions for M_x and M_y, $(\varepsilon/cm)H_z$ is the angle which the projection on the xy-plane of the velocity of an electron will rotate in unit time under the influence of the magnetic force, and since $F(r)^{-1}$ is equal to the time τ in which a component of the original momentum of an electron with speed r will, on the average, decrease to $1/e$ of its value; the physical meaning of this assumption is that the magnetic force will have only a small influence upon the paths of the individual electrons, as compared to the effects arising from the metal atoms. Now, putting $F(r) = Cr^{(n-5)/(n-1)}$ (see p. [321]), as is appropriate in the case considered, we get

$$\sigma_1 = \sigma\left(1 - \sigma^2 \frac{H_z^2}{N^2\varepsilon^2 c^2} \frac{9\pi}{16} \Gamma\left(\frac{n+5}{n-1}\right) \Gamma^{-3}\left(\frac{2n}{n-1}\right)\right)$$

and (113)

$$\delta_1 = \sigma^2 \frac{H_z}{N\varepsilon c} \frac{3\sqrt{\pi}}{4} \Gamma\left(\frac{3n+5}{2(n-1)}\right) \Gamma^{-2}\left(\frac{2n}{n-1}\right),$$

which correspond to the expressions obtained for σ_p and δ_p (pp. [361]–[364]), except that p has been replaced by $(\varepsilon/cm)H_z$.

We notice here the perfect formal analogy existing between the calculation of the absorption of heat rays by metals and the calculation of their galvanomagnetic properties. The physical meaning of this analogy, which will be present in all cases in which separate collisions are considered, may be expressed by saying that the phase difference appearing between the electric field and the electric current for rapidly oscillating electric fields, which is given by $\tan^{-1}(\delta_p/\sigma_p)$ (see p. [363]), will be equal to the angle $\tan^{-1}(\delta_1/\sigma_1)$ between the electric current and the electric field under the influence of a magnetic field perpendicular to the electric field (see eqs. (111)), provided the period of vibration $2\pi/p$ of the oscillating electric field is equal to the time $2\pi/(\varepsilon/cm)H_z$ which it takes an electron to perform one revolution in its helical path around an axis parallel to the magnetic field.

We shall now calculate the galvanomagnetic effects with the aid of eqs. (111). If the piece of metal considered is traversed by an electric current in the direction of the x-axis, there will appear, under the influence of the magnetic field, a potential difference between two points in the metal with the same x-coordinate but with different y-coordinates—*the Hall effect*. The electric field strength E_y corresponding to this potential difference is found by putting the expression for i_y equal to zero; this gives

$$E_y = -\frac{\partial \varphi}{\partial y} = -\frac{\delta_1}{\sigma_1} \frac{\partial \varphi}{\partial x}.$$

Substituting this in the expression for i_x, we get

$$i_x = -\sigma_1 \left(1 + \left(\frac{\delta_1}{\sigma_1}\right)^2\right) \frac{\partial \varphi}{\partial x} \quad \text{and} \quad E_y = \frac{\delta_1}{\sigma_1^2 + \delta_1^2} i_x. \tag{114}$$

We see from eqs. (114) that, in addition to the Hall effect, the magnetic field also produces a *change in the electric conductivity*, which under the influence of the magnetic field becomes

$$\sigma_H = \sigma_1 \left(1 + \left(\frac{\delta_1}{\sigma_1}\right)^2\right). \tag{115}$$

Now, substituting the values for σ_1 and δ_1 given by eqs. (113) into (114) and (115), we get, to the same approximation as above,

$$E_y = i_x \frac{H_z}{N\varepsilon c} \frac{3\sqrt{\pi}}{4} \Gamma\left(\frac{3n+5}{2(n-1)}\right) \Gamma^{-2}\left(\frac{2n}{n-1}\right)$$

and

$$\tag{116}$$

$$\sigma_H = \sigma \left(1 - \sigma^2 \frac{H_z^2}{N^2 \varepsilon^2 c^2} \frac{9\pi}{16} \left(\Gamma\left(\frac{n+5}{n-1}\right) \Gamma\left(\frac{2n}{n-1}\right) - \Gamma^2\left(\frac{3n+5}{2(n-1)}\right) \Gamma^{-4}\left(\frac{2n}{n-1}\right)\right)\right).$$

We shall, finally, examine the amount of energy W_y transferred by the electric current through a unit area perpendicular to the y-axis. Since

$$W_y = \frac{1}{2} \int_0^\infty r^2 G_y(r) \, dr,$$

we obtain from eq. (110), substituting $F(r) = Cr^{(n-5)/(n-1)}$, dropping small quantities of higher order of magnitude with respect to $(\varepsilon/cm)H_z F(r)^{-1}$, and introducing the notations defined in eq. (112),

$$W_y = -\frac{2n}{n-1} \frac{k}{\varepsilon} T\sigma_1 \frac{\partial \varphi}{\partial y} + \frac{3n+5}{2(n-1)} \frac{k}{\varepsilon} T\delta_1 \frac{\partial \varphi}{\partial x};$$

putting $i_y = 0$, we get, with the aid of (111),

$$W_y = \frac{n-5}{2(n-1)} \frac{k}{\varepsilon} T\sigma_1 E_y. \tag{117}$$

So far, we have assumed that the temperature is the same in all parts of the metal; however, if the metal is insulated from transfer of heat from the surroundings, temperature differences will appear in it, until the resulting heat conduction balances the energy flow considered above. To the same approximation as used in calculating W_y, the thermal conductivity γ will be independent of the magnetic field and, since, to this approximation, we can put σ_1 equal to σ, we obtain for the variation of the temperature in the direction of the y-axis, in the case in question,

$$\frac{\partial T}{\partial y} = \frac{W_y}{\gamma} = \frac{n-5}{2(n-1)} \frac{k}{\varepsilon} T \frac{\sigma}{\gamma} E_y,$$

whence, with the aid of eq. (41),

$$\frac{\partial T}{\partial y} = \frac{n-5}{4n} \frac{\varepsilon}{k} E_y. \tag{118}$$

(With regard to eq. (118), we must point out that E_y denotes the y-component of the electric field strength considered above appearing in a piece of metal in which the temperature is uniform. If this condition is not satisfied, as in the case considered, the y-component of the electric field strength will be altered by the potential differences caused by the temperature differences discussed in Chapter II in connection with the thermoelectric effects. Since these potential differences will be independent of the magnetic field to a first approximation, we get, with the aid of eq. (49), for the y-component of the electric field strength in this case

$$E'_y = E_y + \frac{k}{\varepsilon} \left(\frac{T}{N} \frac{dN}{dT} + \frac{n+3}{2(n-1)} \right) \frac{\partial T}{\partial y},$$

whence, with the help of eq. (118),

$$E'_y = E_y \left(\frac{n-5}{4n} \frac{T}{N} \frac{dN}{dT} + \frac{9n^2 - 10n - 15}{8n(n-1)} \right). \tag{119}$$

However, the potential difference calculated here is not open to direct measurement (compare p. [347]), as is the quantity E_y considered above.)

We shall now apply the formulas derived here to investigate some simple cases.

(i) Substituting $n = \infty$ in eqs. (116) and (118), we get, for the case in which the metal atoms are assumed to be hard elastic spheres,

$$E_y = i_x \frac{3\pi}{8} \frac{H_z}{N\varepsilon c}, \qquad \sigma_H = \sigma \left(1 - \frac{9\pi(4-\pi)}{64} \frac{\sigma^2 H_z^2}{N^2 \varepsilon^2 c^2} \right)$$

and

$$\frac{\partial T}{\partial y} = i_x \frac{3\pi}{32} \frac{H_z}{Nkc}. \tag{120}$$

These formulas (as well as the expression for the "adiabatic Hall effect" obtained by putting $n = \infty$ in (119)) have been derived by R. Gans[1], who has given a very complete treatment of the galvanomagnetic and thermomagnetic phenomena, based on the same assumptions about the effects of collisions between metal atoms and electrons as those underlying the theory of H. A. Lorentz.

Since we have assumed that there is only one distinct kind of electrons (with negative charge ε), the first of eqs. (120) shows that, if the electric current is in the positive x-direction, there will appear an electric field in the direction of the negative y-axis. Eqs. (120) also show that the electric conductivity will decrease under the influence of the magnetic field ($\sigma_H < \sigma$).

These results, however, are not in accordance with those found experimentally[2]; thus, there are not only metals, such as bismuth, which possess a Hall effect in the direction predicted by our calculations, and whose electric conductivity decreases in a magnetic field, but also metals, such as iron, possessing a Hall effect in the opposite direction as that predicted and an electric conductivity which increases in a magnetic field.

It must be emphasized that this disagreement between the calculated and experimental results is not a consequence of the special assumptions made (those of Lorentz), since the results derived above (the sign of the Hall effect and the decrease in the electric conductivity ($\sigma_H < \sigma$)) will hold in all cases satisfying the fundamental assumptions made in the Introduction, i.e., in all cases in which it is assumed that the properties of metal atoms are the same in all directions, even when external forces are acting, so that the effect of the external forces can be determined from their direct influence upon the motion of the free electrons.

To explain the galvanomagnetic effects in agreement with experiments on a theory based on the idea of like particles (electrons) moving in the interior of the metals, we are, therefore, forced (at least for certain metals) to assume that the external forces not only affect the motions of the free electrons directly, but also indirectly by exerting a polarizing effect on the metal atoms. Such an assumption appears the more natural, since a close connection exists between the phenomena in question and the magnetic properties of metals, which are explained by the assumption of such a polarizing effect; thus, the magnitude of the Hall effect for ferromagnetic metals is not proportional to the magnetic field but to the intensity of magnetization.

Apart from this disagreement between theory and experiment, revealing itself in that the observed Hall effect does not always have the same sign as the calculated, a theory based on the assumptions employed by Lorentz is faced by another fundamental difficulty in explaining the galvanomagnetic transverse temperature differ-

[1] R. Gans: Ann. d. Phys. **20** (1906) 293.
[2] With regard to the literature on the experimental investigations of these questions, see for example H. Zahn: Jahrb. d. Rad. u. El. **5** (1908) 166.

[389]

ence. In fact, for all metals for which the sign of the Hall effect agrees with that calculated, and for which the electric conductivity decreases in a magnetic field, this temperature difference has the opposite sign as that given by eqs. (120)[1].

As we shall see in the following, however, this last difficulty disappears if we discard the special assumptions of Lorentz and do not assume that metal atoms and electrons act upon each other as hard elastic bodies, but rather with forces varying in a suitable way with the distance.

(ii) Putting $n = 5$—which corresponds to the assumption that the metal atoms act upon the electrons as *elementary magnets* (see p. [321])—we get, from eqs. (116) and (118),

$$E_y = i_x \frac{H_z}{N\varepsilon c}, \qquad \sigma_H = \sigma \quad \text{and} \quad \frac{dT}{dy} = 0. \qquad (121)$$

These are the same formulas that would have been found if we had imagined that all electrons have the same velocity in the x-direction (supposing, for example, that they are moving in a continuous viscous medium); since $N\varepsilon$ is the electric charge of the free electrons per unit volume, $i_x/N\varepsilon$ will, in this case, be the common velocity of the electrons; moreover, since the electric field E_y, in this case, just balances the deflecting action of the magnetic field on each electron, the presence of the magnetic field will have no influence whatever on the conductivity, nor give rise to any transfer of energy in the direction of the y-axis.—That this way of calculation leads to the correct result, in the case where $n = 5$, is due to the fact that the total momentum along the x-axis, in this special case, is evenly distributed among electrons with different absolute values of their velocities relative to the average velocity in the direction of the x-axis.

It may be pointed out that, in all other cases in which separate collisions are assumed and the mutual interaction among the electrons is neglected, E_y will, as appears in particular from the preceding and the following examples, be larger than the value $i_x H_z/N\varepsilon c$ found in this special case; it may be remarked, further, that the diminution of the electric conductivity, in this case, assumes its smallest value, zero.

(iii) Putting $n = 3$,—which corresponds to the assumption that the metal atoms are *electric doublets* like those mentioned in Chapter I, p. [322]—we get, from eqs. (116) and (118),

---

[1] See R. Gans: loc. cit. p. 307–308; compare also P. Gruner: Arch. d. sciences phys. et nat. **28** (1909) 607–608.

$$E_y = i_x \frac{45\pi}{128} \frac{H_z}{N\varepsilon c}, \qquad \sigma_{\mathrm{H}} = \sigma\left(1 - \frac{27\pi(256-75\pi)}{16384} \frac{\sigma^2 H_z^2}{N^2\varepsilon^2 c^2}\right)$$

and   (122)

$$\frac{\mathrm{d}T}{\mathrm{d}y} = -i_x \frac{15\pi}{256} \frac{H_z}{Nkc}.$$

In this case, the value for the galvanomagnetic transverse temperature difference is seen to have the opposite sign of that calculated from the assumptions of Lorentz, and consequently (compare above) a value that agrees with respect to sign with that found experimentally. Thus, this example shows how, in order to obtain agreement with experiments for the phenomena considered, we must assume that the interaction between metal atoms and electrons does not take place as between hard spheres, but that, on the contrary, the action of the atoms takes place through more slowly varying force fields; this is in good accordance with what we found in the preceding, when we discussed the ratio between the electric and thermal conductivities and had to assume the presence of such forces, in order to obtain agreement between the calculated and experimental values; thus, good agreement between the calculated and the experimental values for the ratio between the conductivities was found for pure metals by putting $n = 3$ (see p. [341]).

We shall finally point out that the necessity, demonstrated above, of assuming that metal atoms affect electrons through continuous fields of force, will hold good not only when separate collisions take place, but also in cases in which *the dimensions of the metal atoms are not assumed to be small compared to their mutual distances.* As in the preceding, we shall assume that the mutual action among the electrons can be neglected, as compared to that between metal atoms and electrons, and that the fields of force of the atoms can be regarded as stationary relative to the motion of the electrons, and we shall show how the problem can be treated in a manner corresponding to that used in Chapter I, pp. [324]–[330]*.

Since, as mentioned on p. [381], the magnetic field has no influence upon the statistical equilibrium distribution, the calculations employed in Chapter I, pp. [308]–[330] can be repeated without change, and the collective motion of the electrons can, also in this case, be regarded as resulting from a free diffusion, only that this diffusion ("free diffusion in a magnetic field") does not take place in the opposite direction of the concentration gradient, but makes a small angle with this direction. This angle ("the diffusion angle") as well as the diffusion coefficient will be functions of a (compare p. [327]), of the magnitude H of the magnetic field, and also of the angle between the magnetic field and the direction of diffusion. We shall here only consider fields which are at right angles to the direction of diffusion*.

* [In accordance with II and III, the last five paragraphs, each marked by an asterisk, replace the single paragraph on p. 117–118 of I.]

If, as above, we consider only magnetic fields whose influence on the individual electrons is small compared to that of the metal atoms, and neglect small quantities of the same order of magnitude as the square of the ratio between the time in which the direction of motion of an electron has made a whole rotation under the influence of the magnetic field and the time the individual electrons, as a consequence of the effect of the metal atoms, on the average, have lost the greater part of their original momentum in a given direction (compare p. [386]), the diffusion coefficient, which is an even function of the magnitude of the magnetic field, will be independent of H, while the diffusion angle, which is an odd function of the magnetic field, will be proportional to H*.

In the special case in which the metal atoms are assumed to be hard bodies (compare p. [328]), we can readily show how the diffusion angle depends on $a(r)$. In the absence of a magnetic field, the paths of the electrons are, in this case, independent of their velocity (compare p. [329]). If we now consider two collections of electrons with speeds r_1 and r_2, respectively, travelling under the influence of the magnetic fields H_1 and H_2, respectively, chosen so that $r_1 H_2 = r_2 H_1$, the electrons of the two collections will, on the average, describe exactly the same paths, and will, therefore, diffuse with the same diffusion angle. Moreover, since this angle, as mentioned above, is proportional to H, it must be inversely proportional to r *.

Thus, in this case, we obtain equations for the determination of the collective motion of the electrons of the same form (but with different coefficients) as in the case, $n = \infty$, considered above (p. [388]). Hence, *the ratio $E_y^{-1} dT/dy$ will, in this case, have the same value as given by eq. (120), i.e. $\varepsilon/4k$.* However, the expressions for dT/dy and E_y will, in general, not be the same as those given by (120). (Compare the analogous case of the ratio between the electric and thermal conductivities, p. [342])*.

SUMMARY*

An attempt has been made to treat the electron theory of metals on a somewhat more general basis than that underlying the theory of H. A. Lorentz. The main purpose has been to examine to what extent the results obtained in the latter theory depend on the special assumptions made.

In Chapter I, equations are derived for the collective motions of the free electrons. The case in which separate collisions take place between the electrons and the metal atoms is treated in a general way; for the case in which separate collisions are not assumed to occur, the equations of motion are derived under special assumptions which are believed to correspond approximately to the conditions in real metals.

In Chapter II, the equations of motion are applied to stationary problems. It is shown that the more general theory can account for the discrepancy between the observed values for the ratio between the electric and thermal conductivities and the value for this ratio calculated by Lorentz. Moreover, it is shown that the agreement, pointed out by Lorentz, between the electron theory and the thermodynamical theory of the thermoelectric phenomena given by Lord Kelvin, is valid also in the more general cases.

In Chapter III, non-stationary problems are treated. It is shown that the calculations of the absorption and emission of heat rays, in all the cases considered, lead to a law of black-body radiation identical with that found by Lorentz.

In Chapter IV, the effects of a magnetic field are examined. It is shown that the presence of free electrons in a metal will give rise neither to diamagnetic nor to paramagnetic propertie It is further shown that calculations of the galvanomagnetic effects based on the more general assumptions can be brought into better agreement with experiments than calculations based on the assumptions used by Lorentz.

* [No summary is found in I. The present summary is based on III.]

THESES *

With the aid of eq. (14), derived in this dissertation, and such expressions for $F(r)$ and $Q(\rho, r)$ as are given in eq. (18), it is possible to calculate numerically, to any desired approximation, the ratio between the thermal conductivity and the self-diffusion coefficient of a gas whose molecules act upon one another as hard elastic spheres, a calculation that has not been possible by the methods generally used in these problems.

It appears that the surface tension of a liquid changes only very little during the first time after the formation of the surface, so that the value of this tension—contrary to a widely held opinion—can be determined very accurately by stationary methods.

On the surface of an incompressible non-viscous liquid, exposed either to gravity or to the influence of surface tension, there can exist, for any value of the wavelength, purely periodic waves propagating with unchanged form. On the other hand, this is not the case for a liquid exposed to the simultaneous influence of gravity and surface tension, since it can be shown that, in this case, waves corresponding to certain values of the wavelength (infinite in number) cannot propagate with unchanged form. This is closely connected with the circumstance that, while the speed of wave propagation, in the first two cases, respectively increases and decreases monotonically with the wavelength, there will, in the last case, exist a minimum speed corresponding to a definite value of the wavelength.

Poincaré (*Thermodynamique*, 2e édition (Paris 1908), p. 384) expresses the opinion that Lord Kelvin's theory of the thermoelectric phenomena can be proved rigorously on the basis of thermodynamical principles alone, since the problem can be treated in such a way that heat conduction can be neglected. This, however, is at variance with the result of Boltzmann's investigations (Sitzungsber. d. Wiener Akad. d. Wiss., math.-nat. Kl. **96**, Abt. II (1887) 1258). The error in Poincaré's line of argumentation arises from the fact that he overlooks the increase in entropy

---

* [According to tradition, there is usually appended to a Danish doctor's dissertation a set of "theses", i.e., brief statements of major conclusions reached by the author. These are not necessarily connected with the subject of the dissertation. Thus, the second and third of the above theses represent results obtained by Bohr in his earlier work on surface tension and waves on the surface of liquids.]

caused by the heat conduction as a consequence of the circumstance that the heat conducted through an element of a thermoelectric circuit enters the element at a higher temperature and leaves it at a lower temperature.

As touched upon in the dissertation, in order to explain certain properties of alloys, a theory has been proposed, according to which the alloys, contrary to the pure metals, are not physically homogeneous, and in which a considerable part of the electric resistivity of the alloys is, therefore, ascribed to a thermoelectric polarization. As pointed out, such a theory appears untenable. However, when Lederer (Sitzungsber. d. Wiener Akad. d. Wiss., math.-nat. Kl. **117,** Abt. IIa (1908) 311) believes to have disproved this theory by showing experimentally that the heat developed in a segment of an electric circuit can be calculated from the electric current and the resistance by means of Joule's law, irrespective of whether this segment of the circuit consists of an alloy or a pure metal, he is mistaken; for, according to the law of the conservation of energy, this must be true whether or not the resistance of the alloy is caused by a thermoelectric polarization.

It does not seem possible, at the present stage of development of the electron theory, to explain the magnetic properties of bodies from this theory.

III. CORRESPONDENCE
ABOUT THE DISSERTATION

(1911*)

* [See Introduction, sect. 3.]

1. K. BAEDEKER

[On May 6, 1911, K. Baedeker, to whose book, *Die elektrischen Erscheinungen in metallischen Leitern*, Bohr had referred in his dissertation, wrote:]

Jena, Marienstrasse 10
6. Mai 1911

Sehr geehrter Herr College!

Für die Zusendung Ihrer "Studier over Metallernes Elektrontheori" besten Dank. Sie setzen mit Recht voraus, dass der Gegenstand für mich von besonderem Interesse ist, da ich schon längere Zeit auf ähnlichem Gebiet arbeite. Ihre Darstellung kann, soviel ich sehe, der Sache von grossem Nutzen sein, da eine ähnlich umfassende, kritische Arbeit noch nicht existiert. Mir scheint auch, dass das bei manchen Physikern schon etwas in Misskredit gekommene Gebiet der Metalltheorie solche Bearbeitungen sehr bedarf.

Zu einer gründlicheren Durchsicht hoffe ich bald einmal Zeit zu finden. Ich selbst habe in dem von Ihnen gelegentlich citierten Büchlein die theoretische Seite des Gegenstands nur kurz behandelt, da ich glaube, dass trotz aller Versuche das Fundament dazu noch nicht genügend gesichert ist. Bei der Aufgabe unter den nicht wenigen durchprobierten, und den noch zahlreicheren möglichen Grundannahmen eine bestimmte zu acceptieren würde man vorderhand der Kritik zu sehr exponiert sein. Immerhin stehe ich auch auf Ihrem Einleitung S. 5 unten gegebenen Standpunkt, dass die von J. J. Thomson eingeführte Mitwirkung der Metalatome eine vorläufig unnötige und wenig greifbare Aushilfe ist, die man vorlaüfig lieber nicht benutzen sollte.

In nächster Zeit wird Ihnen ein Versuch über die Theorie der Thermoelektricität von mir zugehen, durch den ich die Sache noch eine etwas andere Seite abgewonnen zu haben hoffe.

Mit höflichstem Gruss
Ihr ganz ergebener
K. Baedeker

[No information is available as to Bohr's response to this letter.]

2. E. L. LEDERER

[Dr. E. L. Lederer wrote on May 26, 1911:]

Czernowitz, 26.V.11

Sehr geehrter Kollege!

Besten Dank für die Zusendung Ihrer Dissertation; ich kann mich leider nur mit den wenigen Beilagen revanchieren.

Was die vorletzte Ihrer Thesen anlangt, glaube ich doch nicht, dass Sie im Rechte sind, und Sie würden mich sehr verbinden, wenn Sie mir Genaueres über Ihre Meinung mitteilten; denn aus den kurz gefassten Sätzen, deren Übersetzung ins Deutsch mir ausserdem sehr schwer fällt, kann ich natürlich nicht alles wissenswerte entnehmen.

Mit kollegialem Gruss
Ihr ergebener
E. L. Lederer

Phil.Dr. E. L. Lederer
Assistent am physik Institut der Universität
Czernowitz
Österreich, Bukowina.

[To this letter Bohr replied on June 7 (the draft is in his mother's handwriting):]

Bredgade 62, 7/6 1911

Sehr geehrter Hr. College!

Ich danke Ihnen bestens für Ihren freundlichen Brief.

Die erwähnte These, die nach Tradition so kurz wie möglich gefasst war, lautet in deutscher Übersetzung wie folgt:

"Wie in der Abhandlung berührt, ist eine Theorie zur Erklärung der Eigenschaften der Legierungen aufgestellt, nach welcher Theorie die Legierungen im Gegensatz zu den reinen Metallen nicht als physikalisch homogen angesehen werden können und daher ein wesentlicher Teil des elektrischen Wiederstandes der Legierungen von einer Thermoelektrischen Polarisation herrühren soll. Eine solche Theorie scheint, wie erwähnt, unhaltbar. Wenn indessen Lederer die Theorie dadurch widerlegt zu haben meint, dass er in experimenteller Weise gezeigt hat, dass die in einem Teil einer elektrischen Strombahn entwickelte Wärme, in Übereinstimmung mit der Erfahrung, durch das Joulesche Gesetz von der Stromstärke und dem elektrischen Widerstand ebenso gut wenn der erwähnte Teil der Strombahn von einer Legierung wie von reinem Metalle gebildet ist, berechnet werden kann, ist dies jedoch nicht der Fall; eine Übereinstimmung wie die erwähnte würde nähmlich des Energiesatzes zufolge erwartet werden, ebensogut wenn der Widerstand von einer thermo-elektrischen Polarisation herrührte als wenn dies nicht der Fall wäre''.

Die Berechtigung dieser These liegt meiner Meinung nach im folgenden. Nach dem Energiesatz muss die, in einem Teil der Strombahn entwickelte Wärme gleich das Produkt der Stromstärke under der Potentialdifferenz zwischen die Endpunkte sein, wenn wie es bei dem hier betrachteten Phänomen der Fall ist, die Endpunkte von demselben Metal bestehen und gleiche Temperatur haben und der Durchgang der Stroms von keiner anderen Wirkung begleitet ist als die Wärmeentwickelung.

Das Paradox, dass Ihnen zu einer anderen Auffassung geführt hat, ist meiner Meinung nach auf Seite 3 in Ihrer Abhandlung zu finden, in welcher Stelle Sie schreiben: dass die Erwärmungen und Abkühlungen des (eventuellen) Peltiereffektes sich gegenseitig aufheben.

Dies letzte scheint mit indessen nicht richtig, eine eventuelle Polarisation rührt ja namlich davon her, dass die Berührungsstellen zwischen die Metalltheilschen bei dem Durchgang des Stroms verschiedene Temperaturen annimmt; die Grössen der Peltiereffekten werden daher nicht numerisch gleich sein und sich nicht gegenseitig aufheben, es wird dagegen eine resultierende Wärmeentwickelung stattfinden, die nach dem Energiesatz der thermoelektrischen Polarization entspricht.

Mit collegialem Gruss und indem ich Ihnen bitte mein mangelhaftes Deutsch zu entschuldigen.

Ihr ergebener
Niels Bohr

3. P. DEBYE

[From P. Debye, whose work is mentioned in three footnotes in the first chapter of the dissertation, Bohr received the following letter:]

Zürich, 30 Mai 1911

Sehr geehrter Herr!

Für die freundliche Zusendung Ihrer Dissertation, danke ich Ihnen bestens. Ich habe so gut es ging versucht über Ihrer Resultate klar zu werden, was mir wegen der Sprache, die ich nicht verstehe, ziemliche Schwierigkeiten machte. Aus diesem Grunde wäre es mir lieb, wenn Sie folgendes lesen möchten und mir nachher schreiben ob ich Sie dabei richtig verstanden habe, oder ob Sie etwas anders gemeint haben.

Bei einem nach der n^{ten} Potenz der Entfernung wirkenden Wirkungsgesetz findet man für die elektrische Leitfähigkeit σ den Wert

$$\sigma = \frac{4N\varepsilon^2}{3\sqrt{\pi}mC} \left(\frac{m}{2kT}\right)^{(n-5)/2(n-1)} \Gamma\left(\frac{2n}{n-1}\right)$$

und für die Warmeleitfähigkeit γ den Wert

$$\gamma = \frac{8nNk^2T}{3\sqrt{\pi}(n-1)mC} \left(\frac{m}{2kT}\right)^{(n-5)/2(n-1)} \Gamma\left(\frac{2n}{n-1}\right).$$

Hiernach wird das Verhältnis

$$\kappa = (\gamma/\sigma) = (2n/(n-1))(k^2/\varepsilon^2)T.$$

Wenn das obige nun richtig verstanden ist, so bitte ich Sie mir weiter zu sagen, ob Sie mit folgenden zwei Bemerkungen einverstanden sind.

1) Nach den experimentellen Resultaten ist ungefähr

$\sigma \sim 1/T, \gamma$ unabhängig von T.

Sollen das die Formeln darstellen, so muss

$(n-5)/2(n-1) = 1$, d.h. $n = -3$

gewählt werden. Dann ergibt sich

$$\kappa = \tfrac{3}{2}(k^2/\varepsilon^2)T,$$

was *noch schlechter* stimmt als das Lorentz'sche Resultat

$$\kappa = 2(k^2/\varepsilon^2)T.$$

2) Nach den experimentellen Resultaten ist andererseits ungefähr

$$\kappa = 3(k^2/\varepsilon^2)T.$$

Will man das erreichen, so muss

$$2n/(n-1) = 3, \quad \text{d.h. } n = +3$$

gemacht werden. Dann wird

$$\sigma \sim T^{\frac{1}{2}} \quad \text{und} \quad \gamma \sim T^{\frac{3}{2}},$$

was wiederum nicht mit der Erfahrung stimmt.

Es ist ganz gut möglich, dass ich Sie irdendwo misverstanden habe, was mir um so peinlicher sein würde, als ich Ihre Rechnungen unbedingt für einen Fortschritt und eine Leistung halte. Ich bitte Sie also nochmals mich richtig über Ihre Resultate aufklären zu wollen und bin

Hochachtungsvoll
Ihr erg.
P. Debye

Bolleystr. 50, Zürich.

[Bohr answered (undated draft in his mother's handwriting):]

Sehr geehrter Herr.

Ich danke Ihnen vielmals für Ihren freundlichen und interessanten Brief. Ihre Bemerkungen betreffend will ich gern das folgende anführen. Mit meinen Berechnungen habe ich wesentlich beabsichtigt zu untersuchen in wie fern es möglich sei, indem man die speciellen Lorentzschen Annehmungen verlässt, eine bessere Übereinstimmung zwischen der Theorie und der Erfahrung zu erlangen. Es scheint mir, als ob in dem in Ihrem Briefe erwähnten Punkte in dieser Hinsicht ein Resultat erlangt ist, indem es gezeigt ist, dass das Verhältnis zwischen den Leitfähigkeiten für Elektricität und Wärme grösser als das von Lorentz berechnete und zwar ebenso gross als das experimentel gefundene werden kann, wenn man annimmt, dass die Metallmolekylen auf die Elektronen nicht als absolut harte elastische Körper einwirken, sondern mit Kräften, die in continuierter Weise mit der Entfernung variieren (z.B. wird die Übereinstimmung mit der Erfahrung sehr befriedigend (Siehe p. 58 Linie 1–7) wenn man annimmt, dass die Kräfte in derselben Weise variieren wie die Kräfte in der Umgebung eines elektrischen Dipols ($n = 3$)), und dies Resultat gillt wie ich gezeigt habe (siehe p. 58 L. 8–20), unabhängig davon ob man discrete zusammenstösse annimmt—d h freie Weglänge, die im Verhaltnisse zu den Dimensionen (Wirkungsbereich) der Elektronen und Molekylen gross sind—oder ob man eine solche nach unserem physikalischen Kentnisse wohl kaum berechtigte Annahme nicht einführt—.

Indessen, was die Berechnung der einzelnen Leitfähigkeiten anbelangt, stellt sich die Sache ganz anders, indem es in den Ausdrücken dieser einzelnen Leitfähigkeiten mehrere auf der Vorderhand unbekannte Grösse eingehen (siehe p. 54 L. 3 u. p. 60 L. 1–4) und es lässt sich daher hier die Weise, in welcher die einzelnen Leitfähigkeiten mit der Temperatur variieren nicht ohne nähere Kentnisse zu der Art in welchen diese erwähnten Grösse mit der Temperatur variieren, bestimmen; diese letzte Variationen sind aber, wie es aus der Theorie hervorgeht, ohne Einfluss auf dem Verhälltnis der Leitfähigkeiten. In meiner Abhandlung (s. p. 60 L. 5–51, sie⟨h⟩ auch die Note an der selben Seite L. 13–18) habe ich, in Übereinstimmung mit dem was Sie in Ihrem Brief richtig bemerken auch selbst

[402]

hervorgehoben, dass man die Weise in welcher die einzelnen Leitungsfähigkeiten mit der Temperatur variieren, nicht erklären kann, wenn man von einem speciellen mechanischen Bilde ausgeht, nach welchen die Molekyle discrete Kraftcentre sind, deren Wirkungsgebiete klein im Verhältniss zu ihren gegenseitigen Entfernungen sind, und welche auf den Elektronen mit Kräften einwirken, welche sich einerseits umgekehrt als die n^{te} Potens (Ich habe angenommen und alle die Berechnungen gelten nur wenn $n > 2$) der Entfernung verhällt, und anderseits von der Temperatur unabhängig sind.

Dies letzte scheint mir aber keineswegs ein Angriffspunkt für die Theorie zu sein (s. p. 60 L. 15–19), indem man ja eben annehmen muss, dass das Verhalten in den Metallen überaus verwickelt ist (dem oben erwähnten speciellem Bilde gegenüber) und insbesondere dass es sich in sehr verwickelter Weise mit der Temperatur verändern kann. Indem ich Ihnen nochmals für Ihren freundlichen Brief danke und Ihnen bitte mein mangelhaftes Deutsch zu entschuldigen möchte ich Ihnen gern mittheilen dass meine Abhandlung in englischer Ubersetzung erscheinen wird und ich werde mich erlauben zur der Zeit Ihnen einen Seperatabdruck zu senden.

Mit den besten Grüssen
hochachtungsvoll
Niels Bohr

H⟨er⟩rn Privatdocent P. Debye
    Universität
    Zürich
Schweiz

4. M. REINGANUM

[From M. Reinganum, whose determination of the ratio between the thermal and electric conductivities he had referred to in his dissertation, Bohr received the following letter:]

Freiburg i.B., 3.XI.11
Karlsplatz 18

Hochgeehrter Herr Kollege!

Empfangen Sie noch nachträglich meinen besten Dank für die Zusendung Ihrer interessanten Arbeit über die Elektronentheorie der Metalle. Sie werden auch meine kleine Abhandlung erhalten haben. [Max Reinganum: Sitzungsber. d. Heidelberger Akad. d. Wiss., Math.-nat. Kl. 1911, Abh. 10.]

Leider kann ich der Sprache wegen nicht alle Einzelheiten 'Ihrer Arbeit verstehen. Sicher würden es mit mir viele deutsche Kollegen begrüssen, wenn Sie die Abhandlung auszugsweise in den Annalen der Physik veröffentlichen würden.

Halten Sie meine Erwägungen in der zugesandten Arbeit für zutreffend?

In der Kritik der Arbeit von Debye stimme ich Ihnen bei.

Mit hochachtungsvoller Empfehlung

Ergebenst

M. Reinganum

[Bohr was now in England (see p. [103]), and he replied in English. (The draft is in his own handwriting.)]

10 Eltisley Avenue
Newnham Cambridge
9. Nov. 1911

Dear Professor Reinganum:

I thank you very much for your kind letter. I have read your interesting paper. I can, however, not agree with you in your criticism of Lorentz's calculation of the thermal conductivity.

1) The difference between the expression for the thermoelectric force given by your formula (17) and the expression found by Lorentz, (18), is in my opinion (as I have pointed out in my paper p. 65–66) due to the fact that the equation $dp = X\rho dx$, used by you, will not hold for a piece of metal in which the temperature is not uniform, because, in such a case, there will be a transfer of momentum from the electrons to the metal mole⟨cule⟩s on account of the deviation of the velocity-distribution of the electrons from the normal velocity-distribution in the case in which the temperature is uniform.

2) I can further not agree with you in the calculations on p. 16 of your paper. The presence of an electric force will in my opinion not have the effect, that the energy transferred through the metal, is different from the energy transferred by the electrons through an imaginary surface in the interior of the metal. This last statement seems to me to be obvious from a statistical point of view. In Lorentz' paper the problem is treated from a statistical point of view. In your paper, you consider the paths of the single electrons and calculate the energy transferred by the electrons between the metal-molecules. The disagreement between your result and Lorentz' is in my opinion to be explained as following: Let ab be a plane in the interior of the metal, and let M represent the metal-molecules, and let finally the electric force in question have the direction from c to d.

As you have pointed out in your paper, the electrons crossing the plane ab in the direction from c to d 1) will transfer more energy to the metal-molecules on the right side of the plane ab than they will transfer through the plane ab itself, 2) but the same electrons will on account of the same reasons take less energy away from the molecules on the left side of *ab*. The electrons crossing the plane *ab* in the direction from *d* to *c* will further, as you mention, 3) transfer less energy to the molecules on the left side of *ab* than they will transfer through the plane *ab* itself, but on the other hand the same electrons 4) will take more energy from the molecules on the right side of *ab*.

If you consider all the 4 circumstances (as far as I can see, you have in your paper only considered the 1st and the 3rd) you get in my opinion exactly the same value for the energy transferred by the electrons through the plane *ab* as for the energy transferred by the electrons from the molecules on the left side of *ab* to those on the right.

I should be very much obliged to you if you kindly would let me know whether you agree in these considerations. An English translation of my paper will very soon be published in the Transaction of the Philosophical Society of Cambridge.

With kind regard
Yours sincerely
Niels Bohr

[Reinganum wrote again:]

Freiburg i.B., den 22.11.11
Karlsplatz 18

Sehr geehrter Herr College!

Haben Sie vielen Dank für Ihren freundlichen Brief von 9. November, der mich sehr interessierte und erfreute. Meine Ansicht über Ihre Einwände ist folgende.

Ich glaube, Sie haben mit Ihrem zweiten Einwand recht. Wenn man die 4 von Ihnen aufgezählten Punkte berücksichtigt, so kommt man dazu, dass auch bei Anwesenheit beliebiger ausserer Kräfte die Wärmeleitung der Elektronen durch die kinetische Energie bestimmt wird, welche durch einen Querschnitt übertragen wird.

Ich hatte anfänglich an eine Analogie der Elektronen mit einem im Dissociationszustand befindlichen Gase gedacht. Hier it die Wärmeleitung abnorm gross. Man kann dies (nach Nernst) so erklären, dass von warmen nach kalten Stellen mehr dissociirte Moleküle diffundieren als in umgekehrter Richtung, wo die Zahl der Doppelmoleküle eine etwas grössere ist (z.B. NO_2 Gas). Hier gibt die durch einen Querschnitt getragene kinetische Energie allein kein Mass für die Wärmeleitung, es kommt noch hinzu, dass sich auf der kälteren Seite ein Teil der hindiffundierten dissoc. Moleküle polymerisiert, also Wärme abgibt, während auf der heissen Seite dauernd das Umgekehrte stattfindet. Die lässt sich

[405]

nun nicht direct auf die Elektronen übertragen, selbst wenn die freie Zahl derselben in der Volumeinheit von der Temperatur abhängt. Denn zu der Ueber-
legung gehören 2 bewegliche Jonengattungen. Ich glaubte aber doch, dass etwas
ähnliches bei den Elektronen vorläge, und kam auf den Ausweg, dass die von
mir eingeführte Zusatzkraft in diesem Sinne wirke. Sie zeigen mir aber, dass
dieses nicht der Fall sein kann, wofür ich Ihnen sehr dankbar bin.

Was nun Ihren ersten Einwand anbetrifft, so sehe ich nicht ein, wie Bewegungs-
grösse von den Elektronen auf die Materie (transfer of momentum?) übertragen
werden kann. Denn es wird doch angenommen, dass die Atome unendlich
grosse Massen haben und stets in Ruhe bleiben. Doch kann ich mir noch kein
endgültiges Urteil bilden, da ich S. 65–66 Ihrer Abhandlung nicht genau verstehe,
da mir das Vorhergehende sprachlich etwas Schwierigkeit macht. Ich würde
mich daher sehr freuen, wenn Sie mir auch ein englisches Exemplar Ihrer
Abhandlung senden würden, wenn dieselbe in Cambridge erschienen ist.

Schliesslich wäre ich Ihnen sehr dankbar, wenn Sie mir gelegentlich Ihr Urteil
über den Nachtrag meiner Abhandlung, die spezifische Wärme der Elektronen,
mitteilen würden. Auch dies ist mir selbst noch nicht so ganz sicher, aber ich
habe Einwände dagegen noch nicht gehört.

Mit den verbindlichsten Grüssen verbleibe ich

Ihr sehr ergebener
M. Reinganum

[Bohr replied, again in English (draft in his own handwriting):]

10 Eltisley Avenue
Newnham, Cambridge
Dec. 11, 1911

Dear Prof. Reinganum,

I thank you very much for your kind letter of Nov. 22. I beg you to excuse
the delay in answering, but I have recently been so busy.

As to the first point mentioned in my former letter, it was only my intention to
use the expression "transfer of energy" in the same sense as we say that the
momentum of a ball striking a wall is transferred from the ball to the wall. It
was only my intension in a short way to say that the deviation from the normal
velocity-distribution in the case considered (a piece of metal in which the temperature is not uniform) will give rise to a resultant force in a given direction
from the side of the molecules to the electrons, and that therefore the sum of the
forces acting upon the electrons on account of the difference in pressure and the
presence of extraneous forces will not be 0.

You ask me further if I agree with you in your considerations of the specific heat of the electrons. I have to say, that I cannot agree, and I shall try to explain my reasons. You consider the potential difference, calculated by Lorentz, between two points in a piece of metal in which the temperature is not uniform, and which potential difference in Lorentz' theory account⟨s⟩ for a part of the Thomson-effect. This potential difference will, however, in my opinion have no direct relation to the specific heat of the metal. You consider for sake of simplicity the case in which the number of free electrons per unit of volum⟨e⟩ is assumed to be independent of the temperature. If we further consider the same simple mechanical picture, as used by Lorentz,—and according to which the electrons are supposed to move freely in the metal, and the whole interaction between the molecules and the electrons is assumed to take place through collisions, by which they act upon each other as absolutely hard bodies,—the problem is comparatively simple, because in the calculation of the specific heat we only have to cons;der the alteration in the kinetic energy of the electrons (and, according to your opinion, the variations of the potential in the considered piece of metal) but not the alterations of the potential energy of the electrons relative to the metalmolecules.

Let us now consider a piece of metal surrounded by perfect electric insulators, and let us transfer energy (heat) to this by heat conduction through the insulators. The part of the specific heat due to the presence of the free electrons will now in such a case, according to my opinion, certainly be due only to the increase in the kinetic energy of the electrons, this increase being the whole increase in the energy.

Let us then consider the same case as above, but let us connect the piece of metal to another piece of the same metal (by means of a long thin metal-wire), the temperature of which is kept constant, and let us again transfer energy to the first piece of metal. Now a potential-difference will, according to the theory, appear between the two pieces of metal, a potential difference which could be thought to have an influence on the value of the specific heat, as the one you consider in your paper (I hope that I have understood you correctly), such an assumption will however in my opinion not be correct. For the electrons will not have moved from a place with higher potential to a place with lower, but only the potential is altered and the electrons will on the average be on the same place (only a very small number of electrons—in comparison with the whole number of free electrons present in the metal—will have moved through the wire and will give rise to an increase of energy quiet negligible in this connection). It is in my opinion the same case as if we had transferred electricity to an insulated conductor; in such a case we have to calculate the increase of the energy, only from the increase of the potential energy of the electrons transferred to the conductor, and of the electrons which give rise to free charges on the surface of the conductor, but not from the increase of the potential energy of the enormous amount of free electrons present in the interior of the conductor (we could certainly pay regard to this increase of the energy of the electrons, but then we also had to pay regard to the decrease of the potential energy of the metal-molecules, which have opposite charges of the electrons, and this decrease will exactly neutralize the increase of energy in question).

My opinion about the problem of the magnitude of the number of free electrons in the metals is that the disagreement between the values for the considered number calculated by considerations of the optical properties of the metals and the upper limit given by considerations of the specific heat is to be explained by the fact that the values calculated by the first method are not reliable on account of the following reasons.

First several authors consider in their calculations the optical constants of the metals for ordinary visible light, but the calculations of the electron-theory used by them, will only be correct if the time of collisions (i.e. the time-intervals which the electrons take to travel through distances of the same order of magnitude as the dimensions of the molecules (not to confound with the time-intervals used in travelling through the mean-free-path)) is to be considered as negligible small in comparison with the time of vibration, and this condition is not satisfied for the considered light. Other authors have used the experiments of Hagen and Rubens and calculated the number of the free electrons by comparing the apparent decrease of the electric conductivity, calculated from the reflection-coefficients found by the mentioned experiment, for decreasing times of vibrations, with the apparent decrease in the conductivity, calculated from the absorption-coefficient, according to the electron theory. But the apparent conductivity calculated from the reflectioncoeficient is not identical with the conductivity calculated from the absorptioncoeficient; for, according to the electron theory, there will (for such rapid vibrations) be a difference in phase between the electric current and the electric force, which will give rise to terms in the expression for the reflection-koeficient (and therefore also in the expression for the apparent conductivity calculated from this) which will be of opposite sign and of a greater order of magnitude than the terms arising from the decrease in the conductivity calculated from the absorption-koeficient (my paper, note 2, p. 88; all this is taken to account in the calculations of Drude (Ann. d. Phys., Bd 18, p. 940, 1904) but has escaped most of the later authors).

The experiments of Hagen and Rubens can therefore, in my opinion, not be used to the determination of the number of free electrons, as they give a decrease of the conductivity in a case in which we, from the ordinary simple electron-theory, should expect an increase. I think (my paper p. 88) that the disagreement between the theory and the experiments, is to be explained by assuming, that the conditions for the motions of the free electrons in the metals are more complicated than assumed in the ordinary electron-theory, and that f.inst. the motions of the free electrons can not be considered as consisting of free paths and separeted collisions.

Again thanking you for your letter

I am

Yours sincerely,

Niels Bohr

[Although it is likely that Reinganum replied to this letter, no evidence of further correspondence is now extant.]

IV. LECTURE ON THE ELECTRON THEORY
OF METALS

(13 November 1911 *)

* [See Introduction, sect. 4.]

CAMBRIDGE PHILOSOPHICAL SOCIETY.

The next Meeting of the Society will be held on Monday, 13 November, at 4.30 o'clock, in the **Cavendish Laboratory**. It is expected that the following COMMUNICATIONS will be made:

I. By Professor Sir J. J. THOMSON:

Application of positive rays to the study of chemical problems.

II. By N. BOHR:

Electron theory of metals. (Communicated by Professor Sir J. J. THOMSON.)

III. By J. C. CHAPMAN:

Secondary characteristic Röntgen Radiation from Elements of High Atomic Weight. (Communicated by Professor Sir J. J. THOMSON.)

The paper I have the honor to present to the Philosophical Society deal with the electron-theory of metals.

The foundation of this theory is given by Weber, Riecke, Drude, J J Thomson and H A Lorentz. I have in my paper attempted to treat the electron theory of the metal from a somewhat more general basis than the one used in the theory of Lorentz, still retaining the main assumptions used by this author.

These assumptions are; for the first, that the free electrons in a piece of metal unaffected by extraneous forces and in which the temperature is uniform are in what is called mechanical heat equilibrium with the metal molecules. i.e. in such a dynamical statistical equilibrium, which will appear if the forces between the metal molecules and the electrons are of the same type as the forces considered in ordinary mechanics. i.e. again that the kinetic energy and velocity distribution of the electrons are the same as those for of the molecules of a gas at the same temperature.

The other main-assumption is that the properties of the single metal molecules on the average are the same in all directions, and that this entropy will be independent of the presence of extraneous forces, so that the influence of these forces is to be calculated from their direct influence on the motions of the free electrons. This last assumption, which is a main assumption in the theory of Drude and Lorentz, separates the calculations very essentially from such theory as those given by Sutherland and P J Thomson. In this last theory, the influence of the extraneous forces is explained from the influence on the motions of the single metal molecules, with their relating electrons, considered as closed systems and the directed motion of the electrons through the metal is produced by the fact, that these on account of the motion of the entire molecule, are sent out in directions, which on the average are dependent on the direction of the extraneous forces.

I Lorentz theory the interaction between the metal molecules and the electrons is assumed to take place in the same way as between hard elastic spheres. The theory of Lorentz has given many result of the greatest interest, but in some points the agreement between the theory and the experiment is not satisfactory.

It has indeed been the intention in my paper, to try to investigate which of the result of the theory are essentially connected with the special assumptions used by Lorentz and which will remain unaltered with reference to more general as—

I my paper I have separately considered 2 different case; in the first case assuming, that the motion of the free electrons consist of free paths and collision, as in the theory of Lorentz, and in the other case, assuming, that the metal molecules are so close together that the electrons during a great part of their motion are subjected to the influence of sensible forces from the metal molecules. In the first mentioned case it has been possible to execute the calculation with regard to certain problems in perfect generality, setting out from the named main assumptions, i.e without introducing special assumptions about the forces between the metal molecules and the electrons, and between these mutually. In the second case, which is more difficult, I have only treated the problems under the assumptions that the field of force of the metal molecules is to be considered as stationary with regard to the motions of the electrons, and that we can neglect the effect of the interaction between the free electrons mutually in comparison with the effect of the interaction between the electrons and the metal-molecules. This last assumption seems to me approximately to correspond to the conditions in the real metals, on account of the great velocities and the small dimensions of the electrons in comparison with those of the metal-molecules.

Proceeding to mention some of the result of the investigation, I shall first mention the points in which the more general assumptions lead to the same results as found by Lorentz. These are, for the first, the conformity between the calculation of the law of heatradiation for rays with long times of vibration by help of the calculation of the power of absorption and emission, and the experiments i.e. Plancks formula for the law of heat radiation for the coincident rays with the coincident times of vibration as for the second the conformity between the calculation of the thermoelectric phenomena and the thermodynamic theory of these phenomena given by Lord Kelvin.

The first result I think was to be expected, but I should like to say a few words about the conformity between the electron theory of the thermoelectric phenomena and the theory given by Lord Kelvin. This agreement is very remarkable, because the theory of Lord Kelvin cannot be strictly founded on the ordinary thermodynamics on account of the fact that the phenomena with which we are dealing are inseparably connected with irreversible processes, namely the heat conduction and the Joule heat development. An attempt has been made to explain the agreement in question by comparing the passage of the electricity through a thermo-electric circuit with a reversible circular process, in which a gas is transferred through a series of states of equilibrium, which with reference to the temperature and pressure corresponds to the temperature of the metal and the pressure of the free electrons in the considered points of the circuit. Such a consideration gives the same result for the thermo-electric heat development as the direct calculation given by Lorentz. But in the more general treatment the expression for the Peltier and Thomson effects will not be the same as those found by Lorentz; if we assume that the law of forces between the metal-molecules and the electrons is not the same in all metals and at all temperatures, terms will enter in the named expression (in question) which will not only be dependent on quantities which determine the named states of equilibrium, i.e. first the number of electrons per unit of volume, which is the only quantity besides the temperature which enter in Lorentz's expressions, but also quantities dependent on and quantities which are dependent on the way in which the momentum in a given direction will be distributed between the electrons with the different absolute velocities, if the equilibrium is disturbed, quantities which cannot enter in a calculation only considering states of equilibrium. That The result that the agreement in question between the electron theory and the theory of Lord Kelvin hold in the more general cases seems therefore to me to be rather remarkable.

I shall now mention some of the points in which the more general treatment leads to a variation in the result calculated for Lorentz special assumptions, and therefore to a possibility of obtaining a better agreement with the experiments. For the first it is found that the calculated value for the ratio between the thermal and the electric conductivities is dependent on the assumptions which are introduced about the forces acting between the molecules and the electrons. As it is known Lorentz's calculations gives a value only $\frac{2}{3}$ of the experimentally found values. In Lorentz calculation the molecules are, as mentioned, assumed to be hard elastic spheres; if, on the contrary, the molecules act upon the electrons through continuously varying fields of force the value of the ratio in question will in general be higher, and I have shown, that if wee assumed that the molecules act upon the electrons as electric doublets, the calculated value for the ratio will be in a very good agreement with the experimentally found values for the pure metals. This last result is perhaps of some interest, because Prof. J.J. Thomson has shown that some of the optical phenomena of the metals can be explained by assuming the existence of such electric doublets in the metals.

Another point is the calculation of the galvanomagnetic properties of the metals. R. Gans has given a very elaborate theory for these phenomena, starting out from the same special assumptions as used by Lorentz. The agreement between his calculations and the experiments is however very unsatisfactory. For the first, the electron-theory gives as known a definite direction for the Hall-effect relative to the directions of the magnetic force and of the current and a diminution of the electric conductivity in the presence of a magnetic field. But on the other hand wee find as well metals, ~~such as~~ solid like bismuth, which shows a Hall effect of the same direction as the calculated, and the conductivity of which decreases in a magnetic field, as well as other metals, which, like iron, show a Hall effect of the opposite direction, and the conductivity of which increases in a magnetic field. This shows that the ~~usual~~ assumptions, that the entropy of the metal molecules as unaffected by the presence of extraneous forces, is not correct for all metals with regard to magnetic forces. But for a theory based on the special assumptions used by Lorentz, another principal difficulty appears, namely in the explanation of the galvanomagnetic transversal temperature-differences. This difference of temperature has namely in all metals for which the sign of the Hall-effect is in accordance with the calculated and which shows a decrease of the electric conductivity in a magnetic field — and where these effect consequently in any rate qualitatively can be explained through an electron theory of the here considered kind — a sign opposite to the one calculated by Gans. It can however be shown that if the forces between the metal molecules and the electrons are assumed to vary in a suitable way with the distance, the temperature-difference in question will reverse its sign. It can thus be shown that if the forces vary in the same way as if the molecules were electric-doublets, the calculated value for the temperature difference will have the right sign. A result which in interesting conformity with the results obtained as to the ratio between the thermal and electric conductivities.

Finally I should like to say a few words about the magnetic properties of the free electrons. The presence of such electrons will according to the general opinion give rise to diamagnetic properties of the metals, on account of the fact that the paths of the electrons will be curved under the influence of ~~the~~ an extraneous magnetic field. I have however shown that such an assumption will not be correct — because the statistical distribution of the electrons will not be altered on account of the presence of an extraneous magnetic field. I have tried to show that the different result of the calculations of the authors which have been of a different opinion is due to a neglect of boundary con I shall try to show what is meant by this, with help of a diagram representing a very simpel case.

This point to a great difficulty in the explanation of the magnetic properties of the bodies from the electron-theory. ~~of the~~ and support in my opinion the view that the Maxwell equations for the electromagnetic phenomena, are not exactly satisfied with regard to the motions of the single electrons, an assumption which in my opinion is very distinctly shown by the calculations of Lord Rayleigh and Jeans of the law of heat-radiation for small times of relaxation.

V. LETTERS TO OSEEN AND TO McLAREN

(December 1911*)

* [See Introduction, sect. 4.]

[421]

1. LETTER FROM NIELS BOHR TO C. W. OSEEN (1 DEC 11)

Danish text (draft in Bohr's handwriting)

10 Eltisley Avenue,
Newnham
Cambridge

Dec. 1. 1911.

Kære Ven!

Jeg maa begynde med igen at takke Dig for al Din Venlighed imod mig i Sommer og for al Din Opmuntring, jeg kan ikke sige, hvor glad jeg har været for den. Det har varet længe før jeg skrev til Dig, men jeg har haft saa daarlig Tid, og det er endnu ikke gaaet saa hurtigt for mig herovre; det er ikke saa let at have baade theoretisk og experimentelt Arbejde paa samme Tid, og navnlig tager det megen Tid før man kan faa rigtig begyndt at arbejde paa et nyt Laboratorium, og det igen særlig paa Cavendish Laboratoriet, hvor der hersker en Uorden, man ikke gør sig Begreb om (Bjerknes sagde i Sommer at der herskede den molekulare Uorden, og jeg maa ofte tænke paa det). Jeg har derfor endnu ikke saa meget at fortælle, jeg har heller ikke talt med saa mange—Thomson har saa daarlig Tid, jeg gav ham Afhandlingen da jeg kom, men han har endnu ikke læst den, jeg har blot nogle Øjeblikke snakket med ham om nogle enkelte Punkter, og jeg ved endnu ikke om han vil være enig med mig eller ej; og Jeans har sagt at han vil vente til Afhandlingen er udkommen paa engelsk (jeg haaber at den kommer i Transactions of Cambridge Philosophical Society, den befinder sig for Øjeblikket foran en Bedømmelseskomité)—, men jeg skal prøve at fortælle hvad jeg kan.

For det første har Du maaske set, at der er kommen en Afhandling af Mc.Laren (Phil. Mag. Juli 1911) "The emission and absorption of energy by electrons". Han beviser deri ved en direkte Beregning (hvad jeg jo ved en indirekte Betragtning havde paapeget), at det ikke er muligt at forklare Straalingsloven udfra Metallernes Elektrontheorie i Overensstemmelse med Virkeligheden. Han betragter det samme Tilfælde som jeg, nemlig Elektroner der bevæger sig uafhængigt af hverandre i et stationært Kraftfelt; men hans Beregninger gælder for Emission og Absorption af Straaler med alle mulige Svingningstider. Ikke alene hans Resultat er overordentlig interessant, men ogsaa hans Beregningsmaade. Det er lykkedes ham at gennemføre en nøjagtig Beregning ved en konsekvent Betragtning af de enkelte Elektroners Baner. (Du husker maaske at jeg i Sommer fortalte, at jeg havde forsøgt lidt paa noget lignende, før det var lykkedes mig at finde en nøjagtig statistisk Betragtningsmaade; det var derfor dobbelt morsomt for mig at se.) Jeg tog en Dag op til Birmingham og havde en lang og morsom Samtale med Mc. Laren om det altsammen; jeg tror at vi var enige om det meste.

En anden Ting vedrørende Metallernes Elektrontheorie er Spørgsmaalet om Antallet af fri Elektroner i Metallerne. Jeg troer ikke at nogen af de mange Maader, der er forsøgt til direkte Bestemmelse af det nævnte Antal har den ringeste Værdi. (Der er saaledes i den allersidste Tid (Phil. Mag. Sept. 1911) kommen en Afhandling af Nicholson, der efter min Mening er ganske sind⟨s⟩svag; ikke alene benyttes H. A. Wilson's urigtige Beregninger, men disse benyttes til at undersøge Metallernes optiske Forhold for almindeligt Lys, d.v.s. for Svingningstider af samme Størrelsesorden som Sammenstødene, til Trods for at alle Wilsons Beregninger er baseret paa, at den sidste Tid kan betragtes som forsvindende lille i Forhold til Svingningstiden. Jeg har ogsaa haft en Diskussion med Nicholson; han var overordentlig elskværdig, men med ham bliver jeg næppe enig om saa meget). Den eneste Methode, der efter min Mening giver nogen Oplysning, er Betragtning af Varmefylden, der jo sætter en øvre Grænse for Antallet af Elektroner. Vanskeligheden er nu den, at nogle Metaller, Sølv og Kobber, har saa stor en elektrisk Ledningsevne, at Produktet af Antallet af Elektroner og af Middelvejlængden (eller en tilsvarende Størrelse) maa være saa stort, at den sidste Størrelse (under Antagelse af den nævnte øvre Grænse for Antallet af Elektroner) maa være i det min⟨d⟩ste af en Størrelsesorden af 10^{-6} cm, (d.v.s., ca. 50 Gange saa stor som Metalmolekylernes Diameter. Dette betragtes almindeligt som en meget stor Vanskelighed. Det jeg vilde sige ved at omtale alt dette, som Du naturligvis ved lige saa godt som jeg, var nu blot, at den omhandlede Vanskelighed maaske næppe er saa stor; vi ved jo saa lidt om Forholdene i Metallerne, at vi næppe kan sige at det er umulig at Middelvejlængden er af den omtalte Størrelsesorden. Jeg ser i det mindste ikke nogen Mulighed for at forklare Metallernes Ledningsevne paa anden Maade; og til Støtte for det skal jeg prøve at vise at den Theorie som J. J. Thomsen har opstillet i den Hensigt at forklare Metallernes Forhold ved Antagelsen af et mindre Antal fri Elektroner, er uholdbar. Thomson antager jo, at Elektronerne ikke hele Tiden er fri, d.v.s. ikke under hele deres Bevægelse har Hastigheder, der svarer til dem, de vilde have dersom de var i mekanisk Varmeligevægt med Molekylerne; men at de kun i en vis lille Tid under deres Bevægelser har saadanne Hastigheder. Det kan nu imidlertid vises at allerede denne lille Tid er for lang.

Thomson giver følgende Formel (The corpuscular theory of matter, p. 88)

$$\sigma = 2e^2 dpnb / 9\alpha\vartheta$$

hvor d er Molekylernes Diameter (Doubletternes Akse), n er Antallet af Molekyler og *p* et det Antal Elektroner der udsendes fra et Molekyle pr. Sekund, samt hvor *b* er den Vejstrækning, som Elektronerne tilbagelægger, med den til Temperaturen svarende Hastighed, paa deres Vej mellem 2 Molekyler (dette fremgaar maaske tydeligst af hvad Thomson siger om Varmeledningen; nederst paa samme Side) Det Udtryk for den elektriske Ledningsevne, man kommer til fra Drudes Theorie (samme Bog, p. 54) er

$$\sigma = \beta e^2 N \lambda v / 4 \alpha \vartheta \qquad (\beta > 1)$$

hvor N er Antallet af fri Elektroner pr. Volummenenhed og λ er den fri Middelvejlængde og *v* er Elektronernes Hastighed.

Skal nu de 2 Formler give samme Værdi for σ, maa man have

$$dnpb > N\lambda v.$$

Men nu er b/v den Tid hver Elektron i Thomsens Theorie er fri og *p.n.* er det Antal der bliver fri i hvert Sekund, og derfor er $[N] = pn(b/v)$ lig Antallet af fri Elektroner tilstede hver⟨t⟩ Øjeblik i Metallet, og derfor det Antal, der skal betragtes ved Beregning af Varmefylden. Vi har nu

$$[N]d > N\lambda.$$

Og efter det ovenfor omtalte, maa derfor det nødvendige Antal Elektroner i Thomsens Theorie, være c 50 Gange saa stort som det største Antal der lader sig forene med de iagttagne Værdier for Varmefylden*. Du maa un⟨d⟩skylde at jeg saadan har begyndt med at fortælle om mine egne Sager. Jeg har ogsaa ventet saa længe med at skrive, fordi jeg gerne vilde have haft Tid til at læse i Burbury "Kinetic theory of gases", som Du talte til mig om i Sommer, og ogsaa lidt i Dine egne Arbejder. Det sidste har jeg dog ikke kunnet naa endnu, jeg har kun kigget lidt i dem, men glæder mig saa meget til at læse dem rigtig⟨t⟩ saa snart jeg faar lidt mere Tid (jeg tænker meget stærkt paa ikke at tage saa meget experimentelt Arbejde i næste term), det bliver dog vist et stort Arbejde for mig, der kender saa lidt. Med Hensyn til Burbury, saa hidrører hans Resultat, som jeg sikkert ventede, fra en simpel Regnefejl, nemlig fra en fuldstændig Negligering af Grænsebetingelser.

Paa Side 47 i hans Bog indfører han en Dobbeltsum over hvad han kalder alle Molekylerne, men uden at definere dette Begreb nærmere. Jeg skal i det følgende tænke mig Molekylerne indsluttede i en Beholder. Paa den næste Side siger han at i Middel er

$$\lambda\mu = \lambda v = \mu v = 0 \quad \text{og} \quad \lambda^2 = \mu^2 = v^2 = \tfrac{1}{3},$$

hvor $\lambda\mu v$ betegner Retningskosinusserne for en ret Linie fra et Molekyle med

---

* Dette synes mig en temmelig alvorlig Indvending, jet har fortalt Thomson derom for 3 Uger siden; han sagde at han skulde tænke over det, men jeg har ikke hørt fra ham siden.

Koordinater (x, y, z) til et andet med Koordinater (x', y', z') og hvor Middel-værdierne, hvilket fremgaar af den Brug der gøres af dem, betyder Middelværdien for et fast (x, y, z) taget over alle Værdierne for (x', y', z'); men dette er som man umiddelbart ser ikke rigtigt for en Molekyle (x, y, z), der befinder sig i Nærhed⟨en⟩ af Beholderens Væg. At dette og Burbury's følgende Beregninger paa samme Side, gør det hele illusorisk, ses maaske lettest derigennem, at alle Burbury's Beregninger kun vil være rigtige, dersom Summationen over (x, y, z) er udstrakt til et endeligt Antal Molekyler, medens Summationen over (x', y', z') er udstrakt til et uendeligt stort Antal Molekyler, der befinder sig jævnt fordelte i et uendeligt Rum (ved uendeligt skal her forstaas saa stort, at Molekyler udenfor det betragtede Rum, ikke vil give noget mærkeligt Bidrag til Burbury's Integraler). Men det ses umiddelbart, at ved en saadan Summationsmaade, vil den Dobbeltsum, som Burbury betragter i §55 (Side 49–50), være lig 0 (i den omhandlede § fremkommer Burbu-ry's Resultat kun ved at han ⟨ikke⟩ tager Hensyn til Grænsebetingelser). Og dermed falder hele Burbury's Bevis til Jorden.

Medens jeg er ved dette vilde jeg ogsaa gerne sige et Par Ord om Paradoxet vedrørende Boltzmann's Minimums Theorem. Jeg synes nemlig bestemt (jeg tror, at jeg sagde noget lignende i Sommer) at Løsningen, nemlig den, at Boltzmann aldeles ikke har bevist, at en vilkaarlig Fordeling altid (i Middel) vil nærme sig den Maxwellske Fordeling (hvad der jo heller ikke vilde være rigtigt), men kun at en Fordeling, i hvilken Molekylerne indenfor de forskellige Hastighedsomraader i det givne Øjeblik er saadan fordelte i Rummet, at Sammenstødene indenfor den første lille Tid kan beregnes paa den af Boltzmann benyttede Maade, i det første Øjeblik vil nærme sig den Maxwellske; men dette har intet med Entropisætningen at gøre, og er kun at betragte som en meget indskrænket mekanisk Sætning. Boltz-manns Beregninger kan ikke begrundes ved at antage den "molekulare Uorden", thi denne forlanger den Maxwellske Fordeling. Hvad selve Entropisætningen angaar (jeg mener Sætningen om, at Entropien stadig vokser) saa mener jeg at den ikke kan begrundes ved Sandsynlighedsbetragtninger (jeg tør dog ikke udtale mig meget bestemt i Aften for det er saa længe siden jeg rigtig har tænkt over det, og det er den Slags Problemer, der næsten altid smutter ud af Fingrene paa en, i samme Øjeblik man tror at man har dem rigtig fat; medens jeg skriver dette falder det mig saaledes ind at det kan afhænge af hvordan man definerer Sætningen); men kun ved Betragtning of den Maade, hvorpaa Afvigelser fra, hvad man kalder den molekulare Uorden er frembragt. For nærmere at forklare lidt af hvad jeg mener hermed, skal jeg blot som et Exempel henvise til Jeans's Beregning (Phil. Mag. **17**, (1909) 776) af den elektriske Modstand i et Metalstykke. Jeans mener at have bevist at (med hans Betegnelser) $(du_0/dt)_{t=0} = 0$, men da det samme Bevis kan gentages for alle Tider, synes det mig det samme som at bevise at u_0 = konstant, eller at der slet ingen Ledningsmodstand er. Det urigtige Resultat hidrører fra, at Jeans antager at der i Middel ikke vil være systematiske Forbindelser mellem Elektronernes Steds- og Hastigheds-koordinater, selv om Elektronerne har en Middelhastighed i en bestemt Retning (jeg haaber at jeg har forstaaet Jeans rigtigt, det er et meget dunkelt Sted). Men dette er en urigtig Brug af en Sætning fra den statistiske Mekanik, der omhandler Systemer, der er overladte til sig selv og ikke er underkastede ydre Paavirkning, og som ikke kan anvendes paa et Metalstykke der omsætter de ydre Kræfters Energi til (til ydre Legemer afgiven) Varme, og hvor endda Elektronerne ikke engang altid er de samme, men jo netop gaar igennem Metallet (saa at man slet ikke kan tale om et afgrænset mekanisk System). Det ses jo ogsaa umiddelbart, som enhver nærmere Begegning viser, at der i et Metalstykke vil være systematiske Forbindelser mellem Elektronernes Steds- og Hastigheds-koordinater, naar Elektronerne paa Grund af ydre Paavirkning har en Middelhastighed i en bestemt Retning, hvilke Forbindelser jo netop giver Anledning til alle de betragtede Fænomener.

Jeg haaber ikke at jeg har trættet Dig med denne lange Snak, men det har været saa morsomt at skrive til Dig om det, for jeg har ingen her at snakke med, der bryder sig om saadan noget. Jeg haaber, at Du ved Lejlighed, naar Du engang har Tid, vil lade mig vide, dersom Du synes, at der er noget af det, der ikke er rigtigt; og Du husker jo ogsaa nok, at Du lovede mig i Sommer, at skrive til mig, dersom der var noget i min Bog, som Du ikke fandt rigtigt. Jeg skal nok skrive igen, naar jeg har noget at fortælle og ogsaa om mine Planer. Jeg er saa glad for at være her; jeg har truffet saa mange unge Mennesker, og hører saadan morsomme Forelæsninger af Thomsen, der er en utrolig idérig og klog Mand; jeg begynder ogsaa at komme ind i forskelligt. Jeg er i Øjeblikket meget begejstret for Kvantumtheorien (jeg mener dens experimentelle Side); men er endnu ikke sikker paa at det ikke skyldes manglende Kundskab; det samme, men blot i endnu langt højere Grad, kan jeg sige om mit Forhold til Magneton-theorien. Jeg glæder mig saa umaadelig til at prøve paa at komme rigtig ind i det altsammen i næste Term. Saa mange venlige Hilsener

fra Din hengivne
Niels Bohr

Translation

Cambridge, December 1, 1911

Dear friend,

I must begin by thanking you again for all your kindness to me last summer and for all your encouragement; I cannot say how glad they have made me. It has taken me long to get around to writing to you, but I have had so little time, and things do not yet go very fast for me over here. It is not so easy to do both experimental and theoretical work at the same time, and, especially, it takes much time to get really started working in a new laboratory, and particularly in the Cavendish Laboratory where there exists a lack of order that you cannot imagine (Bjerknes said last summer that complete molecular disorder reigns here, and I must often think of that remark). Hence, I haven't much to tell yet, and I haven't talked with many. Thomson has so little time; I gave him the [translation of the] dissertation when I came, but he hasn't read it yet. I have only talked with him a few moments about certain points, and I don't yet know whether he will agree with me or not; Jeans has said that he will wait until it is published in English (I hope it will come out in the Transactions of the Cambridge Philosophical Society, it is at present before a judging committee); but I shall try to tell what I can.

In the first place, you may have seen that there has appeared a paper by McLaren "The emission and absorption of energy by electrons" (Phil. Mag. July 1911). In it he shows by direct calculation (what I had pointed out on indirect grounds) that it is not possible to derive the radiation law in agreement with experience on the basis of the electron theory of metals. He considers the same case as I, i.e., electrons moving independently of one another in a stationary field of force; but his calculations are valid for emission and absorption of rays of all possible periods. Not only his result, but also his method of calculation is extremely interesting. He has succeeded in carrying through an accurate calculation by considering in a consistent manner the trajectories of the individual electrons. (You may remember that I told you last summer that I had attempted something similar before I succeeded in finding a rigorous statistical procedure; it was, therefore, doubly interesting to me.) I went up to Birmingham one day and had a long and nice conversation with McLaren about the whole thing; I believe that we agreed on most of it.

Another thing concerning the electron theory of metals is the question of the number of free electrons in the metals. I do not believe that any of the many methods that have been used to determine this number directly is of the slightest value. (Thus, very recently (Phil. Mag. Sept. 1911) a paper by Nicholson has appeared, which in my opinion, is perfectly crazy; not only does he use the incorrect calculations of H. A. Wilson, but he uses them to investigate the optical properties of metals for ordinary light, i.e., for periods of vibration of the same order of magnitude as the collision times, in spite of the fact that all Wilson's calculations are based on the assumption that these times can be regarded as vanishingly small compared to the period of vibration. I have also had a discussion with Nicholson; he was extremely kind, but with him I shall hardly be able to agree about very much.) The only method that, in my opinion, gives any information is the study of the specific heat which, as is well-known, sets an upper limit to the number of electrons. But the trouble now is that some metals, silver and copper, have such a high electric conductivity that the product of the number of electrons ⟨per unit volume⟩ and the mean free path (or a corresponding quantity) must be so large that (when this upper limit for the number of electrons is assumed) it [i.e., the mean free path] must be at least of the order of 10^{-6} cm, i.e., ca. 50 times as large as the diameter of the molecules. This is generally considered a very serious difficulty. What I want to say by mentioning this, which you, of course, are as familiar with as I, is only that this difficulty may not be so very serious; for we know so little about the conditions in the metals that we can hardly say that it is impossible for the mean free path to be of the order of magnitude mentioned. At least, I see no possibility of explaining the conductivity of the metals in any other way, and in support of this view I shall try to show that the theory advanced by J. J. Thomson to explain the properties of metals by assuming a smaller number of free electrons is untenable. In fact, Thomson assumes that the electrons are not free all the time, i.e., that they do not have the kind of velocities during their entire motion that they would have if they were in mechanical equilibrium with the molecules, but that they only have such velocities a certain small time during their motion. However, it can now be shown that even this small time is too long. Thomson gives the following formula (The corpuscular theory of matter, p. 88)*

$$\sigma = 2e^2 dpnb/9\alpha\vartheta,$$

where d is the molecular diameter (axis of the dipoles), n is the number of molecules ⟨per unit volume⟩, p is the number of electrons ejected by a molecule per second, and b is the distance the electrons travel with a speed corresponding to the temperature on their way between two molecules (this may perhaps be seen most clearly from what Thomson says about the thermal conductivity at the bottom of the same page). The expression for the electric conductivity that follows from Drude's theory is (loc.cit. p. 54)

$$\sigma = \beta e^2 N\lambda v/4\alpha\vartheta, \qquad (\beta > 1)$$

where N is the number of electrons per unit volume, λ is the mean free path, and v is the speed of the electrons.

Now, if the two formulas are to give the same value for σ, we must have

$$dnpb > N\lambda v.$$

But b/v is the time each electron is free, according to Thomsons' theory, and pn is the number that become free per second; hence $[N] = pn(b/v)$ is the number of electrons present in the metal at any instant and, therefore, the number that

* [J. J. Thomson, *The corpuscular theory of matter* (Constable, London 1907).]

must be considered in the calculation of the specific heat. We now have

$$[N]d > N\lambda,$$

and, according to what was said above, the necessary number of electrons in Thomson's theory must, therefore, be ca. 50 times the highest number consistent with the observed values of the specific heat. This seems to me to be a rather serious objection; I mentioned it to Thomson three weeks ago. He said he would think about it, but I have not heard from him since.

You must pardon me that I have begun in this way to tell about my own things. I have waited so long to write also because I wanted to find time to read in Burbury's "Kinetic theory of gases"* that you talked to me about last summer, and also to read a little in your own papers. The latter I have not yet managed to do; I have only looked into them a little, but I am looking forward to really read them as soon as I find a little more time (I am thinking very seriously of not taking so much experimental work next term); however, it will probably be a big job for me who knows so little. With regard to Burbury, as I had surely expected, his result arises from a simple error of calculation, namely from his total neglect of boundary conditions.

On p. 47 of his book he carries out a double summation over what he calls "all the molecules", but without defining this concept clearly. In what follows I shall assume that the molecules are confined in a container. On the next page he says that, on the average,

$$\lambda\mu = \lambda v = \mu v = 0 \quad \text{and} \quad \lambda^2 = \mu^2 = v^2 = \tfrac{1}{3},$$

where λ, μ, v are the directional cosines of a straight line from a molecule with coordinates (x, y, z) to another with coordinates (x', y', z'), and where the average values (as seen from the use he makes of them) are the values obtained by averaging over all possible coordinates (x', y', z') for fixed (x, y, z); but, as

---

* [S. H. Burbury, *A treatise on the kinetic theory of gases* (Cambridge University Press 1899). In this book, the author contends that the relative motions of the molecules of a gas cannot be treated, as is usually done, as independent of each other, but exhibit distance-dependent correlations. Bohr's own copy of the book contains some marginal notes. Besides question marks here and there, one finds on p. 48 and p. 50 an indication by means of small drawings of the argument, presented in the letter, which shows the error underlying Burbury's contention. On p. 45, there is a remark concerning the H-theorem, similar to that put forward in the letter. On p. 27, one reads the following appreciation:

Jeg har aldrig læst noget saa daarlig skrevet og et saa gennemført Forsøg paa at gøre et let Subject svært ved en gennemført Unøjagtighed i de simpleste Beregninger.

I never read anything so badly written nor such a thorough attempt at making an easy subject difficult by thorough inaccuracy in the simplest calculations.

After p. 60, the book is left uncut.]

may be seen immediately, this is not correct for a molecule (x, y, z) that lies close to the wall of the container. That this, and Burbury's subsequent calculations on the same page, makes the whole thing illusory is perhaps most easily seen from the fact that Burbury's calculations would be correct only if the summation over (x, y, z) were extended to a finite number of molecules, while the summation over (x', y', z') were extended to an infinitely large number of molecules uniformly distributed over an infinite space (by infinite is here meant so large that molecules outside this space would not contribute appreciably to Burbury's integrals). But it is readily seen that, if the summation is carried out in this manner, the double sum considered by Burbury in paragraph 55 (p. 49–50) will vanish (in the paragraph in question Burbury's result arises only because he does ⟨not⟩ consider the boundary conditions), and thus the whole of Burbury's proof falls to the ground*.

While I am at this point, I would like to say also a few words about the paradox concerning Boltzmann's minimum theorem. You see, I believe strongly (I think I said something similar last summer) that its solution is just this, that Boltzmann hasn't proved at all that an arbitrary distribution will always (on the average) approach the Maxwellian distribution (which wouldn't be right anyway), but only that a distribution, in which the molecules in the different velocity elements are so distributed in space at a given instant that the collisions in the first short time interval thereafter can be computed by Boltzmann's method, will approach Maxwell's distribution to begin with. However, this has nothing to do with the entropy theorem but should simply be regarded as a mechanical theorem of very limited scope. Boltzmann's calculations cannot be justified by assuming "molecular disorder", for this disorder implies the Maxwellian distribution.

With regard to the entropy theorem itself (I mean the law stating that the entropy always increases), I believe that it cannot be justified on the basis of considerations of probability, but only by considering the manner in which deviations from what is called molecular disorder are produced. (However, I dare not express myself too definitely tonight, and it is the kind of problem that almost always slips through our fingers at the very moment we think that we have really grasped it; as I write this now, it occurs to me that it may depend on how one formulates the theorem.)

To explain a little more fully what I mean, I shall mention only as an example Jeans' calculation of the electrical resistance in a piece of metal (Phil. Mag. **17**

* [The following paragraphs, down to the paragraph beginning "I hope that on occasion ...", are crossed out in pencil in the draft of the letter. They were probably not included in the letter actually sent to Oseen.]

(1909) 776). Jeans believes to have shown that (with his notations)*
$(du_0/dt)_{t=0} = 0$; however, since the same proof can be repeated for all times, it
appears to me that this amounts to proving that $u_0 = $ constant, i.e., that there
is no electric resistance at all. The incorrect result arises from the fact that Jeans
assumes that, on the average, there will be no systematic correlations between
the position and velocity coordinates of the electrons, even if the electrons have
a ⟨non-vanishing⟩ average velocity in a definite direction (I hope to have understood Jeans correctly; it is a very obscure place). But this is an incorrect application of a theorem from statistical mechanics which deals with systems that are
left to themselves and not subjected to external influences and which cannot be
applied to a piece of metal that transforms the energy of the external forces into
heat (given off to the surrounding bodies) and in which the electrons are not
always the same but in fact merely pass through the metal (so that one cannot
speak of a closed mechanical system at all). It is also immediately seen that in a
piece of metal there will be systematic correlations between the position and
velocity coordinates when, as a result of external influence, the electrons have
an average velocity in a definite direction; in fact, it is just these correlations that
give rise to all the phenomena of interest.

I hope that I haven't tired you with this long chat, but it has been so much fun
to write to you about these things; for here I have no one to talk to who cares
about such things.

I hope that on occasion, when you have time, you will let me know if some of
it is right; and you remember surely that you promised last summer to write
if there was something in my book you did not find right.

I shall certainly write again when I have something to tell, and also about my
plans. I am very glad to be here; I have met so many young people, and I attend
some very interesting lectures by Thomson, who is unbelievably full of ideas
and a very clever man. I am also beginning to get into various things. At the
moment I am very enthusiastic about the quantum theory (I mean its experimental aspects); but I am not yet sure if this is not because of lack of knowledge;
the same I can say, but only to a still higher degree, about my relation to the
magneton theory. I am looking forward immensely to really getting into all of
it next term.

With many kind regards
sincerely yours,
Niels Bohr

---

* [u_0 is the average velocity of the electrons in the direction of the electric field.]

2. LETTER FROM NIELS BOHR TO S. B. MCLAREN (17 DEC 11)

[To S. B. McLaren*, whose paper *Emission and absorption of energy by electrons* (Phil. Mag. **22** (1911) 66) he had referred to in the letters to his brother and to Oseen of October 23 and December 4, 1911, respectively, Bohr wrote (draft in his own handwriting):]

10 Eltisley Avenue
Newnham
Cambridge
Dec. 17, 1911

Dear Mr. McLaren:

I thank you for the copies of your papers. I was so glad to speak with you in Birmingham, and hope to see you again another time. I promised to write to you if I thought of something concerning the electron-theorie. I have had very little time in this term to deal with the subject, but I should like only to say a few words about the theory of J. J. Thomson.

Considerations of the specific heat give an upper limit for the number of free electrons in the metals. Now, the experimentally found conductivity of some is so great, that we, using the mentioned upper limit for the number of electrons, have to assume, that the mean-free-paths (or a corresponding quantity) of the electrons is very great in proportion to the dimensions of the metal-molecules,— in your last paper, you state a value of the mean-free-paths equal to 10^{-6} cm, i.e., c 50 times the diameters of the molecules.

This is in the opinion of some authors to be considered as a very great difficulty for the electron theory of the metals. I am, for my part, not sure that the difficulty is so great (we know so very little about the conditions for the motions of the electrons in the metals), and I understand that you are of the same opinion. Further I, at least, cannot see any way, in which the conductivity could be explained, and which would avoid the difficulty in question. With regard to such considerations it seems to me to be of interest, that it can be proved, that the theory of J. J. Thomson, in which he has tried to develop a theory of the metal without using the assumption of the continuous existence of free electrons in the metals, and which therefore beforehand seems to avoid the mentioned difficulty, cannot be correct. As you know, Thomson assumes that the electrons

---

* S. B. McLaren was born in Japan on 16 August 1876, was taken to Australia by his parents at the age of five, and went to England in 1879 to enter Trinity College, Cambridge. He became Lecturer in Mathematics at University College, Bristol, in 1904, leaving in 1906 to assume a similar position at the University of Birmingham. He was appointed Professor of Mathematics at Reading University College in 1913. He went into military service at the outbreak of the war in 1914 and died of wounds on 16 August 1916, at Abbéville. His highly original scientific papers were republished in 1925 by the Cambridge University Press.

not are free during the whole time, i.e. that they not during the whole time have velocities corresponding to them, which they would have, if they were in mechanical heat-equilibrium with the molecules, but that they only within a certain short time during their motion have such velocities (rather difficult to understand). Now it can however be shown that already this small time-interval is too long.

Thomson gives the following formula (the corpuscular theory of matter, p. 88) for the electric conductivity

$$\sigma = 2e^2 dpnb/9\alpha\vartheta,$$

in which d is the diameter of the molecules (the distance between the charges in the doublets, perhaps smaller than the diameter but certainly not much shorter), n is the number of molecules, and p the number of electrons emitted per sec. from a single molecule, and finally b the distance, through which the electrons travel, with the velocity corresponding to the temperature (this last is perhaps most evidently to be seen, from what Thomson says about the heat conduction on the same page).

The expression for the electric conductivity, which follows from the theory of Drude (see the same book p. 54) is now

$$\sigma = \beta e^2 N\lambda v/4\alpha\vartheta, \qquad (\beta > 1)$$

in which N is the number of free electrons per unit of volum⟨e⟩, and λ is the mean-free-path, and v the velocity of the electrons.

If the two formulas now shall give the same value for the conductivity, we get

$$d \cdot n \cdot p \cdot b > N \cdot \lambda \cdot v.$$

But now, b/v is the time-interval, in which the electrons according to the theory of Thomson are free, and $p \cdot n$ is the number of electrons set free pr. sec. $[N] = p \cdot n \cdot (b/v)$ is therefore the number of free electrons present at every moment in the metal, and evidently the number to be taken into account in the considerations of the specific heat. But now

$$[N] \cdot d = N \cdot \lambda,$$

and, according to the above mentioned, the necessary (effective) number of electrons in the theory of J. J. Thomson is therefore c. 50 time⟨s⟩ as great as the greatest value for the number of electrons consistent with the experimentally found values for the specific heat.

If this is correct, it seems to me to be a rather serious criticism. (I spoke to Thomson about it for a month ago, he said that he should think about it, but I have not heard from him since.) I should be very glad to hear, whether you will agree in these considerations and if there perhaps is something in my book which you do not find correct.

Kindest regards
yours sincerely
Niels Bohr

[No reference to the contents of these two letters is found in the extant letters from Oseen or McLaren to Niels Bohr.]

VI. NOTE CONCERNING A PAPER BY J. STARK

(1912*)

* [This note, in Bohr's own handwriting, is in part illegible. However, it is believed that the transcription given here is essentially correct. See Introduction, sect. 4.]

[435]

Danish text

I Anledning af en meget interessant Afhandling af J. Stark (Jahrb. d. Rad. u. Elek. **9,** p. 188, 1912) om Metallernes Elektrontheori vilde jeg gerne anføre følgende korte Bemærkninger.

Stark anfører mange overbevisende Grunde, hvorefter man maaske kan forstaa den elektriske Ledningsevnes Afhængighed af Temperatur og physiske og chemiske Forhold. En tilsvarende Tankegang for Forklaring af Forskellen mellem Legeringers og rene Metallers Ledningsevne er fremsant i min Disputats. Men det forekommer mig i Modsætning til Starks Antagelser, at en saadan Theori intet kan give ud over hvad den saakaldte gaskinetiske Elektrontheori kan give, idet jo ogsaa Stark antager, at Elektronerne har Bevægelsesenergi proportional med Temperaturen og er angrebet af Kræfter, der kan behandles ud fra den sædvanlige Dynamik, og at følgelig de ydre Kræfters Indflydelse maa være meget ringe i Forhold til ⟨Elektronernes⟩ egne Bevægelser.

Den fremsatte Theori er derfor i tilsvarende Grad udsat for de Indvendinger, der er rejste imod den gaskinetiske Theori, og som specielt angaar Storheden af det Antal af fri Elektroner (Valenselektroner efter Stark), der er nødvendig for at forklare den iagttagne Ledningsevne for de godt ledende Metaller ved sædvanlig Temperatur, et Antal der synes at maatte være lidt større end Antallet af Atomer eller Molekyler, ja snarere i højere Grad, fordi Bevægeligheden af Valenselektroner maa være ringere end af helt fri Elektroner. Det maa endvidere bemærkes, at en Theori som Starks, dersom den kan forklare Ledningsevnen for Elektricitet, nødvendig ogsaa kan forklare Ledningsevnen for Varme (hvad der ikke synes at være Starks Mening), idet det kan ses (se Bohr ⟨Studier over Metallernes Elektrontheori⟩, Kap. I, § 4), at ⟨det⟩ blot ud fra Antagelsen om Elektronernes kinetiske Energi og den sædvanlige Dynamik, og uden nye specielle Antagelser om Karakteren af Kraftfelterne mellem Elektronerne og Molekylerne, følger, at Forholdet mellem Varmeledningsevnen og Elektricitetsledningsevnen maa have det rigtige Forhold, samt at de Relationer der gælder for de thermoelektriske Forhold maa være opfyldte.

(Alt dette er maaske* ikke rigtigt. Hvad jeg har sagt gælder kun for en Theori, efter hvilken man ser bort fra Virkningen mellem Elektronerne indbyrdes, og dette ⟨er⟩ denne Theoris svage Side. Med en saadan Udeladelse kan man maaske forklare alt.)

---

* [Ordet "maaske" er overstreget.]

Translation

With reference to a very interesting paper by J. Stark (Jahrb. d. Rad. u. Elek. **9,** p. 188, 1912) on the electron theory of metals, I should like to make the following brief remarks.

Stark cites many convincing arguments on the basis of which it might be possible to understand the dependence of the electric conductivity on temperature and on physical and chemical conditions. A similar idea to explain the difference between the conductivities of alloys and pure metals is proposed in my dissertation. However, contrary to Stark's assumptions, it seems to me that such a theory cannot give anything beyond what the so-called gas-kinetic electron theory can give; in fact, also Stark assumes that the electrons have kinetic energy proportional to the temperature and are acted on by forces that can be treated on the basis of the ordinary dynamics, and that consequently the influence of the external forces must be very small compared to the own [i.e., undisturbed] motions of the electrons.

The proposed theory is, therefore, to a similar degree open to the objections that have been raised against the gas-kinetic theory,—which in particular concerns the magnitude of the number of free electrons necessary to explain the observed conductivity of the highly conducting metals at ordinary temperature, a number that, it would seem, must be somewhat larger than the number of atoms or molecules,—indeed even to a higher degree, since the mobility of valence electrons must be smaller than that of entirely free electrons. Moreover, it may be remarked that a theory such as that of Stark, if it is able to explain the electric conductivity, must necessarily also explain the thermal conductivity (which does not seem to be Stark's opinion), since it may be seen (see Bohr ⟨Studies on the Electron Theory of Metals⟩, Chapter I, § 4) to follow merely from the assumption about the kinetic energy of the electrons and the ordinary dynamics, and without special assumptions about the nature of the force fields between electrons and molecules, that the ratio between the thermal and electric conductivities must have the correct magnitude, and further that the relations holding for the thermoelectric effects must be satisfied.

(All this is perhaps* not right. What I have said holds only for a theory in which the interaction among the electrons is neglected, and this is the weak point of this theory. By such an omission it may be possible to explain everything.)

---

* [The word "perhaps" is crossed out.]

VII. NOTE ON THE ELECTRON THEORY OF THERMOELECTRIC PHENOMENA

(Philosophical Magazine, vol. **23**, 1912*)

* [See Introduction, sect. 5. The draft of a second, unpublished, note is given as appendix.]

XCIX. *Note on the Electron Theory of Thermoelectric Phenomena.*

To the Editors of the Philosophical Magazine.

GENTLEMEN,—

IN the February number of the Philosophical Magazine Prof. O. W. Richardson has published a paper on " The Electron Theory of Contact Electromotive Force and Thermo-electricity." As his results with respect to the Peltier and Thomson effects are in disagreement with results I have obtained in a previous paper (N. Bohr, *Studier over Metallernes Elektrontheori*, Diss., Copenhagen, 1911), I shall be glad to be allowed to try to explain briefly the reasons for this disagreement.

Prof. Richardson uses a very interesting method to calculate the Peltier and Thomson effects. By this method, which is based upon deductions from the results of experiments on the emission and absorption of free electrons by hot metal-surfaces, electricity is imagined to be transferred by a reversible cyclic process through a circuit partly consisting of metals; and the amount of work gained and the heat absorbed by this process is discussed on thermodynamic principles. [It may be remarked that the process adopted for the calculation of the Thomson effect is not strictly reversible, and that the influence of this fact, as Boltzmann has shown (*Sitzungsber. d. Wiener Acad. d. Wiss.*, *math.-nat. Kl.*, Bd. 96, Abt. ii. p. 1258, 1887; *cf.* my paper, p. 72), cannot be infinitely diminished by altering the dimensions and the shape of the piece of metal which forms the part of the circuit considered. We shall not, however, discuss this point further here.]

In the circuits considered by Prof. Richardson, the transfer of electricity through a piece of metal is established in the following way. Electricity in the form of free electrons is "condensed" from the surrounding space into the metal at one end of the piece of metal considered and allowed to "evaporate" from the metal at the other end. The Peltier and Thomson effects are then calculated from the difference between the whole amount of heat developed in the metal during the process considered and the amount of heat developed at the ends of the piece of metal by the condensation and evaporation of the electrons.

The latter quantity of heat is calculated from the difference in potential energy of an electron inside and outside the metal surface. This way of calculating is, however, in my opinion not justifiable, as the amount of heat developed at the surface of the metal cannot be determined in such a simple manner. For the sake of brevity I shall consider here only the simple case in which the piece of metal is of uniform temperature and consists of two bars of different metals joined together, the condensation of electricity taking place in one and the evaporation in the other.

If an electric current flows through a metal there will be a transfer of energy through any surface in the interior of the metal as a consequence of the flow of electrons through it. The magnitude of energy transferred per unit of electric current will depend not only on the potential energy of the free electrons and on the temperature, but also on the conditions of motion of the electrons in the metal. For, in different metals according to the different conditions for the motions of the free electrons, the electric current will be distributed in a different way among the groups of electrons with the various absolute velocities.—If, for instance, we suppose that the electrons move freely between the metal molecules and are only affected by them by separate collisions, and if in these collisions the forces between the electrons and the molecules vary inversely as the nth power of their distance apart, the amount of kinetic energy transferred through a surface in the interior of the metal when a unit quantity of electricity is transferred through that surface will be $\frac{2n}{n-1} \cdot \frac{k}{\epsilon} T$ (*cf.* my paper, pp. 63 and 66), in which ϵ is the charge on an electron, T the absolute temperature, and k the universal gas-constant referred to a molecule ($pv = k$NT).

The expression for the Peltier effect, calculated directly from the difference in the transfer of energy in the two metals, will therefore depend not only on (1) the difference in potential energy of the free electrons, but also on (2) the difference in kinetic energy transferred by the electric current in the two metals.—If, for instance, we suppose that the molecules in the one metal act upon the electrons as hard elastic spheres ($n = \infty$), and in the other as electric doublets ($n = 3$, *cf.* my paper, p. 35), the part of the Peltier effect arising from (2) alone will be equal to $\frac{k}{\epsilon} T$, corresponding to a potential difference of c. 0·0235 volt, which is of the same order of magnitude as the greatest Peltier effect observed.

Now returning to Prof. Richardson's calculation, we see that in the determination of the heat absorbed or developed at the end of the piece of metal considered, no regard is paid to the energy required for the establishment or extinction of the directed molecular flow of the electrons, which produces the transfer of energy in question. Hence, in Prof. Richardson's calculation of the Peltier effect—which he supposes to be quite general, *i.e.* independent of special assumptions about the forces between the molecules and the electrons—only the part of the effect which arises from a possible potential-difference between the two sides of the surface separating the two metals, is considered ; the other part arising from a possible difference in the amount of kinetic energy transferred by the electric current in the two metals is not taken into account.

Quite similar considerations will hold for the Thomson effect.

It may finally be remarked that it can be shown that the calculated values for the thermoelectric constants when the above mentioned points are taken into account, will also satisfy—at least for a very general case—the conditions given by Lord Kelvin,—a result which does not seem to be rigorously deducible from thermodynamic principles (*cf.* my paper, pp. 71–75).

I am,

Yours faithfully,

N. BOHR.

Trinity College, Cambridge,
Feb. 5, 1912.

P.S.—Since the above was written another paper by Prof. Richardson has appeared (Phil. Mag. April 1912). In the fifth paragraph of this very interesting paper Prof. Richardson has generalized his calculation, no longer assuming that the potential energy of the free electrons inside the metal can be considered as constant. Quite similar remarks to those stated in the above note will, however, for exactly analogous reasons, hold for the relation between the results of Prof. Richardson's new calculation and those of my own calculations for the corresponding case.

C. *Theory of Ionization by Collision.*

To the Editors of the Philosophical Magazine.

GENTLEMEN,—

I MUST protest that Professor Townsend has misunderstood me completely and that I did not make the statements which he attacks with such vigour. I was so far from accusing him of holding the " older view " of ionization that I suggested that he had dismissed it from consideration

APPENDIX

UNFINISHED DRAFT OF A REJOINDER TO ANOTHER PAPER BY O. W. RICHARDSON (1912)*

In a note in the due ** [June 1912] number of the Phil. Mag. I have pointed out the difference between some results regarding the thermo-electric phenomena deduced by Prof. Richardson and by myself. Prof. Richardson's results were obtained by a method based on some considerations drawn from experiments on the emission of electrons from hot bodies.

By this method a current of electricity through a piece of metal is thought produced by a process by which free electrons evaporate from the one end of the metal piece and are condensed in the other end of a metal. Considering the mentioned process as an ordinary reversible process and applying the second law of thermo-dynamics Prof. Richardson deduces his results. The result of my paper [i.e. the dissertation] was ⟨on the⟩ contrary deduced by a direct kinematical calculation of the motion of the electrons through the metal.

In the note cited above I tried to explain the cause to the difference in the results of the two calculations and to show, that in Prof. R's calculations some term arising from a possible difference between the amount of kinetic energy transferred by the current in two different metals was omitted. I further mentioned, that these terms in my opinion were extraneous for consideration as that used by Prof. R. especially with regard to the discussion of the agreement between the results obtained from the electron-theory for the thermo-electric phenomena and those of the thermo-dynamic phenomena given by Lord Kelvin.

In recent paper of Prof. R. in which he deals with the contents of my note, he states as his opinion, that the emission of the term mentioned above not is essentially connected with his way of consideration, and can easily be included by considering a process differing a little from the original process considered by him. In the following line⟨s⟩ I shall try to explain the reasons why I cannot agree with this statement of Prof. R.| I can, however, not agree with this statement of Prof. R. and shall in the following lines try to explain my point of view.|

---

* [O. W. Richardson, Phil. Mag. **23** (1912) 737.]
** [Obviously a hearing mistake for "June".]

As I mentioned in my note, the process considered by Prof. R. is not reversible, as he neglects the heat-conduction, and the whole problem is not at all a problem to be treated from an ordinary thermo-dynamical point of view. This point which seems to be rigorously proved by Bol⟨t⟩zman⟨n⟩ in a celebrated paper (see my note p. ⟨984⟩) is in my opinion very clearly illustrated by the nature of the terms omitted in Prof. R's calculations. For these terms ⟨are⟩ depending not on states of equilibrium and their virtual displacements but depend on special deviations of the motions of electrons from the distributions of the statistical equilibrium ⟨and⟩ can from *a priori* considerations not appear in results got by any process for which the use of the second law of thermo-dynamics in its classical form is legitimate (i.e. for process⟨es⟩ dealing with states of equilibrium.)

The use of thermo-dynamics of Prof. R's considerations is therefore from my point of view only formal and is just as long from being legitimate as the treatment of Lord Kelvin in which paper the hypothetical views of the thermo-dynamic are very strongly emphasized (later proved rigorously by Bol⟨t⟩zman⟨n⟩. The calculations of Prof. R. have therefore apart from their interest otherwise in my opinion no bearing upon the discussion about the agreement between the electron theory and the thermo-dynamical theory of thermo-electric phenomena. This question can according to my view (see my dissertation p. ⟨72⟩) only be discussed on the present state of thermo-dynamics, by rigorous use of the first and second law by help of purely kinematical calculations.

VIII. LECTURES ON THE ELECTRON THEORY
OF METALS

HELD IN THE UNIVERSITY OF COPENHAGEN
IN THE SPRING OF 1914
(TRANSLATION OF NOTES*)

* [See Introduction, sect. 6. A number of slips of the pen have been corrected.]

[445]

Introduction (The uncertain situation in the electron theory)

Historical Survey

Metallic conduction in contrast to electrolytic conduction.

The electron theory

Cathode rays, e/m the same for all substances, Thomson.

Electrons outside atoms, photoelectric effect, incandescent bodies, β-rays.

Electrons in atoms, Zeeman effect, dispersion (Drude), scattering of X-rays, absorption of α- and β-rays.

Free electrons in metals, electropositive character.

Weber:

Basic assumptions, electric and thermal conduction, thermoelectricity.

Riecke:

Electric and thermal conduction, thermoelectricity, galvano- and thermo-magnetic phenomena, large number of arbitrary constants.

Drude:

Mechanical theory of heat, energy of electrons $\frac{3}{2}kT$ ($k = 2.0 \times 10^{-16}$), pressure of gas $p = kNT$, ratio between electric and thermal conductivities

$$\tfrac{1}{2}gt^2 = \tfrac{1}{2}X\frac{e}{m}t^2, \qquad \text{average} \ldots \tfrac{1}{2}X\frac{e}{m}t\frac{\lambda}{u}$$

$$\sigma = \frac{1}{2}\frac{e^2\lambda}{mu}N.$$

Thermal conductivity

$$\tfrac{1}{6}uN\left(E + \frac{dE}{dx}\lambda\right) - \tfrac{1}{6}uN\left(E - \frac{dE}{dx}\lambda\right)$$

$$= \tfrac{1}{3}uN\lambda\frac{dE}{dx} = \tfrac{1}{3}uN\lambda k\frac{dT}{dx} ; \qquad \gamma = \tfrac{1}{3}N\lambda ku.$$

$$\frac{\sigma}{\gamma} = \frac{e^2}{ku^2m} = \frac{1}{3}\left(\frac{e}{k}\right)^2\frac{1}{T}, \qquad \frac{\gamma}{\sigma} = 3\left(\frac{k}{e}\right)^2 T.$$

Reinganum: $E = 2eN$, $p = kNT$, $k/e = 0.287 \times 10^{-6}$; $\gamma/k = 0.72 \times 10^{-10}$. Experimental value 0.75×10^{-10}.

L. Lorenz' law.

Drude's assumption of different kinds of electrons.

Thomson: only one kind of electrons.

H. A. Lorentz:

Extension of the electron theory to the problems of radiation.

Proof of the impossibility of ⟨assuming⟩ several kinds of free electrons.

Calculation of the emissivity for slow vibrations.

Calculation of the absorptivity; Hagen and Rubens.

Rigorous calculations of the electric and thermal conductivities and of the thermoelectric constants; confirmation of Lord Kelvin's relations.

Richardson's results.

Aims and results of my doctor's dissertation.

Thomson's criticism of Drude's theory; inadequacy of his own proposal.

Difficulties of explaining the Thomson effect.

Difficulties of explaining the specific heat at low temperatures.

Kamerlingh Onnes' investigations of the electric conductivity at very low temperatures.

Attempts to introduce the quantum theory.

Wien, Hertzfeld, Keesom.

Stark's proposed theory of metals.

The temporary renunciation of an explanation of the Wiedemann-Franz law in the last-mentioned attempts.

Begin the next lecture with an exposition of the Lorentz theory.

March 12, 1914

Continuation of the historical survey.

Last lecture: electrons, metallic conduction, free electrons, mechanical-thermal equilibrium, relation between electric and thermal conductivity, one kind of electrons.

The Lorentz theory of heat radiation.

Cavity radiation, Kirchhoff's law, relation between absorption and emission.

Radiation of energy from an electron, Larmor's formula

$$\frac{2e^2}{3c^3} j^2 \, dt.$$

Calculation of the emissivity, slow vibrations.

Calculation of the absorptivity from Drude's formula

$$\sigma = \frac{1}{2} \frac{e^2 \lambda N}{mu} \qquad \text{(Hagen and Rubens)}.$$

[447]

Law of heat radiation for slow vibrations, agreement with experiment and with Planck.

Calculations carried out rigorously by Lorentz.

Elastic spheres, Maxwell's velocity distribution,

$$dN = Kr^2 e^{-\frac{1}{2}mr^2/kT} dr.$$

Lord Kelvin's "thermodynamic" conditions for the thermoelectric constants.

The emission of electricity from incandescent bodies.

Potential drop at metallic surface,

$$\frac{dp}{dx} dx = nkT dx, \qquad n_1 = n_2 e^{(-W_1 + W_2)/kT}.$$

Number of electrons emitted $= \alpha T^{\frac{1}{2}} e^{-W/kT}$.

Richardson's experiments, Maxwell's velocity distribution.

The promising position of the electron theory of metals; the large literature.

Aim and results of my doctor's dissertation.

Difficulties facing the electron theory.

Thomson's criticism, number of free electrons, specific heat. Inadequacy of Thomson's own attempt.

Nernst's experiments on the specific heat of solids at low temperatures.

Experiments of Kamerlingh Onnes on the conductivity at ⟨liquid⟩ helium temperature.

Attempts to overcome the difficulties by application of the quantum theory:

Kamerlingh Onnes, Lindemann, Wien, Keesom; Hertzfeld.

Stark's theory, crystalline structure of metals.

Temporary renunciation of an explanation of the ratio γ/σ.

March 20, 1914

Last lecture (historical survey, approach, difficulties).

The Drude-Lorentz Theory.

The statistical method. Distribution function

$$dN = f(x, y, z, \xi, \eta, \zeta) dV d\sigma.$$

Examples: (1) Monatomic gas in container, no external forces.

Maxwell's distribution

$$f = Ke^{-\frac{1}{2}mr^2/kT},$$

$$N = \int f 4\pi r^2 \, dr = K \left(\frac{2\pi kT}{m}\right)^{\frac{3}{2}},$$

$$\overline{\tfrac{1}{2}mr^2} = \frac{1}{N}\int_0^\infty \tfrac{1}{2}mr^2 f 4\pi r^2 \, dr = \tfrac{3}{2}kT.$$

(2) Gas subjected to external forces.
Boltzmann's distribution

$$f = K e^{-E/kT} = K e^{-(P+\frac{1}{2}mr^2/kT)}.$$

$$N = N_0 e^{-P/kT}, \qquad \overline{\tfrac{1}{2}mr^2} = \tfrac{3}{2}kT.$$

Gravity:

$$P = mgz, \qquad N = N_0 e^{-mgz/kT} = N_0 e^{-gMz/p} = N_0 e^{-g\rho z}.$$

Barometric variation.
Explanation of paradox:
Condition in a metal:

(1) Stationary fields, external forces, free charges.
(2) Lorentz' assumption of small elastic spheres.
Distribution at constant temperature, homogeneous metal, no external forces

$$f_0 = K e^{-\frac{1}{2}mr^2/kT}, \qquad f = K e^{-\frac{1}{2}mr^2/kT} + \psi = f_0 + \psi;$$

ψ/f_0 very small
$dV \, d\sigma \, dt$

$$x + \xi \, dt, \qquad y + \eta \, dt, \qquad z + \zeta \, dt, \qquad \xi + \frac{Xe}{m} \, dt, \qquad \eta + \frac{Ye}{m} \, dt, \qquad \zeta + \frac{Ze}{m} \, dt$$

$$f(x, y, z, \xi, \eta, \zeta, t) dV \, d\sigma - f\left(x + \xi \, dt, \ldots, \zeta + \frac{Ze}{m} \, dt, t + dt\right) = (a - b) dV \, d\sigma \, dt$$

$$b - a = \xi \frac{\partial f}{\partial x} + \ldots \frac{Xe}{m} \frac{\partial f}{\partial \xi} + \ldots$$

$$a \, dV \, d\sigma \, dt = f \, dV \, d\sigma \frac{r \, dt}{l}, \qquad a = f \frac{r}{l}.$$

$$a \, dV \, d\sigma \, dt = \pi R^2 r \, dt f \, d\sigma \, n \, dV, \qquad l = \frac{1}{n\pi R^2}.$$

$$b \, dV \, d\sigma \, dt = \frac{d\sigma}{4\pi r^2 \, dr} \int f \frac{r}{l} r^2 \, dr \, d\omega \, dV \, dt, \qquad b = \frac{1}{4\pi}\int f \frac{r}{l} \, d\omega.$$

$$\xi \frac{\partial f}{\partial x} + \ldots = -f \frac{r}{l} + \frac{1}{4\pi} \frac{r}{l} \int f \, d\omega.$$

$$f_0 \left[\xi \left(\frac{dK}{dx} \frac{1}{K} + \tfrac{1}{2} mr^2 \frac{1}{kT^2} \frac{dT}{dx} - \frac{m}{kT} \frac{Xe}{m} \right) + \eta(\ldots) + \zeta(\ldots) \right]$$

$$= -\frac{r}{l} \left(f - \frac{1}{4\pi} \int f \, d\omega \right).$$

$$i \, dt = \int ef \, d\sigma \, \xi \, dt, \qquad W \, dt = \int f \tfrac{1}{2} mr^2 \, d\sigma \, \xi \, dt.$$

$$i = e \int \xi f \, d\sigma, \qquad W = \frac{m}{2} \int r^2 \xi f \, d\sigma.$$

$$l \int \left[f_0 \frac{\xi^2}{r} (\ldots) + f_0 \frac{\xi \eta}{r} (\ldots) + f_0 \frac{\xi \zeta}{r} (\ldots) \right] d\sigma = -i + \int \varphi(r) \xi \, d\sigma.$$

$$i = l \left(\frac{Xe}{kT} - \frac{dK}{dx} \frac{1}{K} \right) \int_0^\infty f_0 \frac{\tfrac{1}{3} r^2}{r} 4\pi r^2 \, dr + l \frac{m}{2kT^2} \frac{dT}{dx} \int_0^\infty f_0 \frac{\tfrac{1}{3} r^4}{r} 4\pi r^2 \, dr.$$

$$\int f_0 r^3 \, dr = K \frac{2k^2 T^2}{m^2}, \qquad \int f_0 r^5 \, dr = K \frac{8k^3 T^3}{m^3},$$

$$\int f_0 r^7 \, dr = K \frac{48 k^4 T^4}{m^4}.$$

$$W = l \left(\frac{Xe}{kT} - \frac{dK}{dx} \frac{1}{K} \right) \int f_0 \frac{\tfrac{1}{6} mr^4}{r} 4\pi r^2 \, dr$$

$$- l \frac{dT}{dx} \frac{\tfrac{1}{2} m}{kT^2} \int f_0 \frac{\tfrac{1}{6} mr^6}{r} 4\pi r^2 \, dr.$$

$$i = e \frac{8\pi}{3} l \left(\frac{XekT}{m^2} K - \frac{dK}{dx} \frac{k^2 T^2}{m^2} + 2K \frac{k^2 T}{m^2} \frac{dT}{dx} \right) ;$$

$$W = \frac{16\pi}{3} l \left(\frac{Xek^2 T^2}{m^2} K - \frac{dK}{dx} \frac{k^3 T^3}{m^2} + 3K \frac{k^3 T}{m^2} \frac{dT}{dx} \right).$$

$$\sigma = \frac{8\pi}{3} l \frac{e^2 kT}{m^2} K = Nle^2 (mkT2\pi)^{-\tfrac{1}{2}} \tfrac{4}{3},$$

$$\gamma = \frac{16\pi}{3} l \frac{k^3 T^2}{m} K = Nlk^2 T (mkT2\pi)^{-\tfrac{1}{2}} \tfrac{8}{3}.$$

$$\frac{\gamma}{\sigma} = 2 \left(\frac{k}{e} \right)^2 T.$$

[450]

The difference between the treatments of Lorentz and Drude.

Difference in the result.

Special assumptions.

⟨Begin⟩ next lecture with general assumptions underlying a still more "statistical" method.

March 26, 1914

(Because of absences among the audience, begin with a resumé of the last lecture).

Drude-Lorentz theory.

Drude's picture

Statistical method, distribution function

$$dN = f(x, y, z, \xi, \eta, \zeta, t)dV\, d\sigma.$$

Monatomic gas in container, Maxwell's distribution

$$f = K e^{-\frac{1}{2}mr^2/kT}, \qquad N = \int_0^\infty f 4\pi r^2\, dr = K \left(\frac{2\pi kT}{m}\right)^{\frac{3}{2}}.$$

$$\overline{\tfrac{1}{2}mr^2} = \frac{1}{N}\int_0^\infty f\tfrac{1}{2}mr^2 4\pi r^2\, dr = \tfrac{3}{2}kT.$$

Gas subjected to gravitational force, Boltzmann's distribution

$$f = K e^{-(\frac{1}{2}mr^2 + P)/kT}, \qquad N = N_0 e^{-P/kT}, \qquad \overline{\tfrac{1}{2}mr^2} = \tfrac{3}{2}kT.$$

Distribution of the electrons in a metal.

Stationary fields of the molecules, mutual interaction of electrons, electrostatics, surface density

Equilibrium (constant temperature, no external forces, homogeneous metal)

$$f = K e^{-\frac{1}{2}mr^2/kT} = f_0;$$

otherwise $f = f_0 + \psi$.

$dV\, d\sigma\, dt;$ $\qquad x + \xi dt, \qquad y + \eta dt, \qquad z + \zeta dt,$

$$\xi + \frac{Xe}{m}\, dt, \qquad \eta + \frac{Ye}{m}\, dt, \qquad \zeta + \frac{Ze}{m}\, dt.$$

$f(x, y, z, \xi, \eta, \zeta, t)dV\, d\sigma - a\, dV\, d\sigma\, dt + b\, dV\, d\sigma\, dt$

$$= f\left(x + \xi dt, y + \eta dt, z + \zeta dt, \xi + \frac{Xe}{m}\, dt, \eta + \frac{Ye}{m}\, dt, \zeta + \frac{Ze}{m}\, dt, t + dt\right) dV'\, d\sigma'.$$

$$\xi \frac{\partial f}{\partial x} + \eta \frac{\partial f}{\partial y} + \zeta \frac{\partial f}{\partial z} + \frac{Xe}{m} \frac{\partial f}{\partial \xi} + \frac{Ye}{m} \frac{\partial f}{\partial \eta} + \frac{Ze}{m} \frac{\partial f}{\partial \zeta} = b - a.$$

[451]

$$a\,dV\,d\sigma\,dt = \frac{1}{l}\,r\,dt\,f\,dV\,d\sigma, \qquad a = \frac{r}{l}f.$$

$$a\,dV\,d\sigma\,dt = r\,dt\,n\pi R^2\,dV\,d\sigma f, \qquad l = \frac{1}{n\pi R^2},$$

$$b\,dV\,d\sigma\,dt = \frac{d\sigma}{4\pi r^2\,dr}\int \frac{r}{l}f\,dV\,r^2\,dr\,d\omega, \qquad b = \frac{r}{l}\frac{1}{4\pi}\int f\,d\omega.$$

$$\left[\xi\left(\frac{dK}{dx} + \frac{mr^2}{2kT}\frac{dT}{dx} - \frac{Xe}{kT}\right) + \eta\left(\frac{dK}{dy} + \ldots\right) + \zeta\left(\frac{dK}{dz} + \ldots\right)\right]e^{-\frac{1}{2}mr^2/kT}$$

$$= -\frac{r}{l}\left(f - \frac{1}{4\pi}\int f\,d\omega\right).$$

$$i\,dt = \int ef\xi\,dt\,d\sigma, \qquad W\,dt = \int \tfrac{1}{2}mr^2 f\xi\,dt\,d\sigma$$

$$i = e\int \xi f\,d\sigma, \qquad W = \tfrac{1}{2}m\int \xi r^2 f\,d\sigma.$$

$$i = le\int\left[\frac{\xi^2}{r}\left(-\frac{dK}{dx} - \frac{mr^2}{2kT}\frac{dT}{dx} + \frac{Xe}{kT}\right) + \frac{\xi\eta}{r}(\ldots) + \frac{\xi\zeta}{r}(\ldots)\right]e^{-\frac{1}{2}mr^2/kT}\,d\sigma$$

$$+ e\int \xi\left(\frac{1}{4\pi}\int f\,d\omega\right)d\sigma$$

$$W = \frac{ml}{2}\int\left[\xi^2 r\left(-\frac{dK}{dx} - \frac{mr^2}{2kT}\frac{dT}{dx} + \frac{Xe}{kT}\right) + \xi\eta r(\ldots) + \xi\zeta r(\ldots)\right]e^{-\frac{1}{2}mr^2/kT}\,d\sigma$$

$$+ \frac{m}{2}\int \xi r^2\left(\frac{1}{4\pi}\int f\,d\omega\right)d\sigma.$$

$$i = le\int \tfrac{1}{3}r\left(-\frac{dK}{dx} - \frac{mr^2}{2kT}\frac{dT}{dx} + \frac{Xe}{kT}\right)e^{-\frac{1}{2}mr^2/kT}4\pi r^2\,dr,$$

$$W = \frac{ml}{2}\int \tfrac{1}{3}r^2(\ldots)e^{-\frac{1}{2}mr^2/kT}4\pi r^2\,dr.$$

$$i = \frac{8\pi}{3}l\frac{e^2kT}{m^2}K\left(X - \frac{kT}{eK}\frac{dK}{dx} - 2\frac{k}{e}\frac{dT}{dx}\right),$$

$$W = \frac{16\pi}{3}l\frac{ek^2T^2}{m^2}K\left(X - \frac{kT}{eK}\frac{dK}{dx} - 3\frac{k}{e}\frac{dT}{dx}\right).$$

$$\left(\frac{\partial}{\partial x} = 0\right), \qquad i = \frac{8\pi l e^2 kT}{3m^2} KX, \qquad \sigma = \frac{8\pi}{3} l \frac{e^2 kT}{m} K = Nle^2(2\pi mkT)^{-\frac{1}{2}}$$

$$(i = 0), \qquad W = -\frac{16\pi}{3} l \frac{k^3 T^2}{m^2} K, \qquad \gamma = \frac{16\pi}{3} l \frac{k^3 T^2}{m^2} K = Nlk^2 T(2\pi mkT)^{-\frac{1}{2}}$$

$$\frac{\gamma}{\sigma} = 2\left(\frac{k}{e}\right)^2 T; \qquad \text{Drude: } 3\left(\frac{k}{e}\right)^2 T.$$

The difference between the manners of calculation of Lorentz and Drude.
The difference in the result. Special assumptions.
Next lecture: different procedure based on general assumptions.

April 2, 1914

Last lecture. Lorentz' calculation of thermal and electric conductivity

$$f = Ke^{-\frac{1}{2}mr^2/kT}, \qquad f\,dV\,d\sigma = dN,$$

$$f = Ke^{-\frac{1}{2}mr^2/kT} + \psi = f_0 + \psi;$$

elastic spheres,

$$\frac{\gamma}{\sigma} = 2\left(\frac{k}{e}\right)^2 T.$$

Calculation on the basis of general assumptions

$$f = f_S + f_R, \qquad f_S = K_0 e^{-(\frac{1}{2}mr^2 + \varphi\varepsilon)/kT_0}, \qquad f_R = f_0 - f_S + \psi.$$

f_S stationary, no flow.
Phenomena of flow calculated from f_R
Permissibility of ignoring external forces, *free diffusion*.

Einstein's theory of Brownian motions.
Distribution of particles after time τ (small, but large compared to time between collisions) given by

$$\varphi(y)dy, \qquad 1 = 2\int_0^\infty \varphi(y)dy.$$

Number of particles passing through a surface in time τ.
Original distribution of the particles $f(x, t)dx$.

$$A_\tau = \int_0^\infty f(x-a)da \int_a^\infty \varphi(y)dy - \int_0^\infty f(x+a)da \int_a^\infty \varphi(y)dy,$$

$$A_\tau = \int_0^\infty [f(x-a) - f(x+a)]da \int_a^\infty \varphi(y)dy = -\frac{\partial f}{\partial x}\int_0^\infty 2a\,da \int_a^\infty \varphi(y)dy,$$

$$A_\tau = -\frac{\partial f}{\partial x}\left[\left|\begin{matrix}\infty\\0\end{matrix}\right. a^2 \int_a^\infty \varphi(y)dy \,\right| + \int_0^\infty a^2\varphi(a)da\right].$$

[453]

$A_\tau = G \cdot \tau,$ $\qquad\qquad$ G = number of particles flowing through per second;

$G = -D \dfrac{\partial f}{\partial x},$ $\qquad\qquad$ D = coefficient of diffusion

$$D = \frac{1}{\tau} \int_0^\infty a^2 \varphi(a)\,da.$$

$$f(x, t+\tau)dx = \int_{a=0}^{a=\infty} f(x-a)dx\,\varphi(a)da + \int_{a=0}^{a=\infty} f(x+a)dx\,\varphi(a)da,$$

$$f(x, t)+\tau \frac{\partial f}{\partial t} = \int_{a=0}^{a=\infty} (f(x-a)+f(x+a))\varphi(a)da$$

$$= 2f \int_0^\infty \varphi(a)da + \frac{\partial^2 f}{\partial x^2} \int_0^\infty a^2 \varphi(a)da;$$

$$\frac{\partial f}{\partial t} = \frac{\partial^2 f}{\partial x^2} \frac{1}{\tau} \int_0^\infty a^2 \varphi(a)da = D \frac{\partial^2 f}{\partial x^2}.$$

Solution of the problem of diffusion away from a plane

$$\frac{\partial f}{\partial t} = D \frac{\partial^2 f}{\partial x^2}; \qquad t = 0, \qquad f(x, 0) = 0, \qquad x \lessgtr 0,$$

$$\int_{-\infty}^{+\infty} f(x, t)dx = N.$$

$$f(x, t) = \frac{N}{\sqrt{4\pi D}} \frac{1}{\sqrt{t}} e^{-x^2/4Dt}.$$

$$\overline{x^2} = \frac{1}{N} \int_{-\infty}^{+\infty} x^2 f(x, t)dx = 2Dt$$

$$\lambda_x = \sqrt{\overline{x^2}} = \sqrt{2Dt}$$

Einstein's determination of D.
 (1) Equilibrium under external forces and osmotic pressure

$$XN = \frac{\partial p}{\partial x} = kT \frac{\partial N}{\partial x}.$$

 (2) Motion subject to viscosity

$$G = Nv = \frac{NX}{6\pi\gamma R}.$$

[454]

(3) Motion due to diffusion

$$G = D \frac{\mathrm{d}N}{\mathrm{d}x}$$

Equilibrium

$$\frac{NX}{6\pi\gamma R} = D \frac{\mathrm{d}N}{\mathrm{d}x},$$

$$D = \frac{kT}{6\pi\gamma R}, \qquad \sqrt{\overline{x^2}} = \sqrt{t} \sqrt{\frac{kT}{3\pi\gamma R}}.$$

Determination of k from Brownian motion; Perrin.

The method of treatment permits rather general assumptions. Here ⟨we shall⟩ investigate the special case of ⟨fixed molecules acting upon the electrons with⟩ central forces inversely proportional to the nth power ⟨of the distance⟩. Consider the diffusion ⟨of electrons⟩ with different speeds separately

$$G(r)\mathrm{d}r = -D(r)\frac{\partial}{\partial x}(N_R(r)\mathrm{d}r),$$

$$N_R(r)\mathrm{d}r = \int f_R\, r^2\, \mathrm{d}r\, \mathrm{d}\omega,$$

$$G(r) = D(r)K4\pi r^2 \mathrm{e}^{-\frac{1}{2}mr^2/kT}\left(-\frac{e}{kT}\frac{\partial\varphi}{\partial x} - \frac{1}{K}\frac{\partial K}{\partial x} - \frac{mr^2}{2kT^2}\right).$$

$$i = e\int_0^\infty G(r)\mathrm{d}r, \qquad W = \frac{m}{2}\int_0^\infty r^2 G(r)\mathrm{d}r,$$

$$i = \left(-\frac{\partial\varphi}{\partial x} - \frac{kT}{eK}\frac{\partial K}{\partial x}\right)A_1 - \frac{\partial T}{\partial x}A_2,$$

$$W = T\left(-\frac{\partial\varphi}{\partial x} - \frac{kT}{eK}\frac{\partial K}{\partial x}\right)A_2 - \frac{\partial T}{\partial x}A_3.$$

$$A_1 = \int_0^\infty r^2 D(r)\mathrm{e}^{-\frac{1}{2}mr^2/kT}\,\mathrm{d}r \times \frac{4\pi e^2 K}{kT},$$

$$A_2 = \int_0^\infty r^4 D(r)\mathrm{e}^{-\frac{1}{2}mr^2/kT}\,\mathrm{d}r \times \frac{2\pi emK}{kT^2},$$

$$A_3 = \int_0^\infty r^6 D(r)\mathrm{e}^{-\frac{1}{2}mr^2/kT}\,\mathrm{d}r \times \frac{\pi m^2 K}{kT^2}.$$

$$\sigma = A_1, \qquad \gamma = \frac{A_3 A_1 - A_2^2 T}{A_1}.$$

$D(r)$ is inversely proportional to the number M of metal molecules per cm^3. If M is s times smaller, we have for $\tau' = s\tau$

$$\varphi'(y')\mathrm{d}y' = \varphi'(sy)\mathrm{d}(sy) = \varphi(y)\mathrm{d}y.$$

$$D(r) = \frac{G(r)\mathrm{d}r}{\dfrac{\partial}{\partial x}(N(r)\mathrm{d}r)} \sim \frac{L^{-2}T^{-1}}{L^{-3}L^{-1}} = L^2T^{-1}$$

$$D(r)M \sim L^{-1}T^{-1}, \qquad r \sim LT^{-1}$$

$$\mu\rho^{-n} \sim LT^{-2}, \qquad \mu \sim L^{n+1}T^{-2}$$

$$D(r)M = f(\mu, r) = Cr^x\mu^y$$

$$x + y(n+1) = -1, \quad -x - 2y = -1, \quad x = (n+3)/(n-1), \quad y = -2/(n-1)$$

$$D(r) = \frac{C}{M} r^{(n+3)/(n-1)}\mu^{-2/(n-1)}.$$

Unambiguous determination

$$D(r) = Cr^{(n+3)/(n-1)}.$$

$$A_2 = \frac{k}{e}\frac{2n}{n-1}A_1, \qquad A_3 = \frac{k}{e}T\frac{3n-1}{n-1}A_2 = T\left(\frac{k}{e}\right)^2\frac{2n(3n-1)}{(n-1)^2}A_1,$$

$$\frac{\gamma}{\sigma} = \frac{A_3A_1 - TA_2^2}{A_1^2} = \frac{2n}{n-1}\left(\frac{k}{e}\right)^2 T.$$

$n = \infty$, as elastic spheres, $\mu\rho^{-n} = c(R/\rho)^n$, constant mean free path;

$n < \infty$: increasing γ/σ, increasing mean free path.

Collisions not separate.

April 16, 1914

Resumé of the calculation of the ratio between the thermal and electric conductivities for the case that the molecules of the metal are centers of forces that are inversely proportional to the nth power of the distance.

$$f(x, y, z, \xi, \eta, \zeta, t)\mathrm{d}V\,\mathrm{d}\sigma = \mathrm{d}N$$

$$(f_0)_0 = K_0 e^{-\frac{1}{2}mr^2/kT_0}, \qquad f = Ke^{-\frac{1}{2}mr^2/kT} + \psi = f_0 + \psi,$$

$$f = f_S + f_R, \qquad f_S = K_0 e^{-(\frac{1}{2}mr^2 + e\varphi)/kT_0}, \qquad f_R = f_0 - f_S + \psi.$$

f_S ⟨is⟩ stationary, and gives rise to no flow.

f_R gives rise to all phenomena of flow.

Permitted to neglect external forces; free diffusion; diffusion coefficient.

[456]

Electrons of different speeds diffuse independently of each other.

$$G_x(r)dr = D(r)\frac{\partial}{\partial x}(N_R(r)dr), \qquad N_R(r)dr = \int f_R\, r^2\, dr\, d\omega,$$

$$G_x(r) = 4\pi K r^2 e^{-\frac{1}{2}mr^2/kT}\left(-\frac{e}{kT}\frac{\partial\varphi}{\partial x} - \frac{1}{K}\frac{\partial K}{\partial x} - \frac{mr^2}{2kT^2}\right)D(r).$$

$$i_x = e\int_0^\infty G_x(r)dr, \qquad W_x = \frac{m}{2}\int_0^\infty r^2 G_x(r)dr.$$

$$i_x = A_1\left(-\frac{\partial\varphi}{\partial x} - \frac{kT}{eK}\frac{\partial K}{\partial x}\right) - A_2\frac{\partial T}{\partial x},$$

$$W_x = A_2\left(-\frac{\partial\varphi}{\partial x} - \frac{kT}{eK}\frac{\partial K}{\partial x}\right)T - A_3\frac{\partial T}{\partial x}.$$

$$A_1 = \frac{4\pi e^2 K}{kT}\int_0^\infty r^2 D(r)e^{-\frac{1}{2}mr^2/kT}\,dr,$$

$$A_2 = \frac{2\pi e m K}{kT^2}\int_0^\infty r^4 D(r)e^{-\frac{1}{2}mr^2/kT}\,dr,$$

$$A_3 = \frac{\pi m^2 K}{kT^2}\int_0^\infty r^6 D(r)e^{-\frac{1}{2}mr^2/kT}\,dr.$$

$$\sigma = A_1, \qquad \gamma = \frac{A_3 A_1 - A_2^2 T}{A_1}$$

M_0 = number of metal molecules per cm^3.
$\mu\rho^{-n}$: central force.

$$D(r)M_0 = f\left(\frac{\mu}{m}, r\right).$$

Dimensional analysis

$$D(r)M_0 = Cr^{(n+3)/(n-1)}\left(\frac{\mu}{m}\right)^{-2/(n-1)},$$

$$D(r) = Cr^{(n+3)/(n-1)}.$$

$n = \infty$, elastic spheres, $\mu\rho^{-n} = c(R/\rho)^n$. Diffusion coefficient proportional to r.
$n > \infty$: The diffusion coefficient increases faster than r.

$$A_2 = \frac{k}{e}\frac{2n}{n-1}A_1, \qquad A_3 = \frac{k}{e}T\frac{3n-1}{n-1}A_2 = \left(\frac{k}{e}\right)^2 T\frac{2n(3n-1)}{(n-1)^2}A_1;$$

$$\frac{\gamma}{\sigma} = \frac{A_3 A_1 - T A_2^2}{A_1^2} = \frac{2n}{n-1}\left(\frac{k}{e}\right)^2 T.$$

For $n = \infty$, same result as Lorentz.

For $n = 3$, agreement with experience.

Dimensional analysis for ⟨the assumption that the molecule of the metal is an⟩ electric dipole (moment M_e) or a magnet (M_m).

$$D(r)M_0 = f\left(\frac{M_e e}{m}, r\right), \qquad \frac{M_e e}{m} \sim LT^{-2}L^3, \qquad n = 3.$$

$$D(r)M_0 = f\left(\frac{M_m e}{m}, r\right), \qquad \frac{M_m e}{m} r \sim LT^{-2}L^3, \qquad n = 5.$$

Collisions not distinct, same result for hard elastic bodies,

$$D(r) = cr.$$

Possibility for agreement with experience.

Criticism of the fundamental assumption, the independent motions of the electrons.

Connection between the flow of electricity and energy in the metal.

Thermoelectric phenomena. Kohlrausch's theory.

$$i_x = \sigma\left(-\frac{\partial\varphi}{\partial x} - \frac{k}{e}\frac{T}{K}\frac{\partial K}{\partial x} - \frac{k}{e}\mu\frac{\partial T}{\partial x}\right), \qquad \mu_n = \frac{2n}{n-1};$$

$$W_x = i_x\frac{k}{e}\mu T - \gamma\frac{\partial T}{\partial x}, \qquad N = K\left(\frac{2\pi kT}{m}\right)^{\frac{3}{2}}, \qquad N = \int_0^\infty f_0 4\pi r^2\, dr.$$

(i) Open chain:

$$i_x = 0 \qquad \frac{\partial\varphi}{\partial x} = -\frac{k}{e}\left(\frac{T}{K}\frac{\partial K}{\partial x} + \mu\frac{\partial T}{\partial x}\right),$$

$$\frac{\partial\varphi}{\partial x} = -\frac{k}{e}\left(\frac{T}{N}\frac{\partial N}{\partial x} + \frac{n+3}{2(n-1)}\frac{\partial T}{\partial x}\right).$$

α) T constant

$$\varphi_1 - \varphi_2 = \frac{k}{e}T\ln\frac{N_2}{N_1}.$$

β) Metal homogeneous

$$d\varphi = -\frac{k}{e}\left(\frac{T}{N}\frac{dN}{dT} + \frac{n+3}{2(n-1)}\right)dT.$$

Pressure of electrons

$$p = NkT,$$

$$\frac{\partial p}{\partial x} = kT\frac{\partial N}{\partial x} + Nk\frac{\partial T}{\partial x} \cong -\frac{\partial \varphi}{\partial x}eN.$$

Simple conditions when $n = 5$ (Maxwell's theory of gases).

$\gamma)$

$$F = \varphi(x_4) - \varphi(x_1) = -\frac{k}{e}\int_{x_1}^{x_4}\left(\frac{T}{K}\frac{\partial K}{\partial x} + \mu\frac{\partial T}{\partial x}\right)dx$$

$$= \frac{k}{e}\int_{x_1}^{x_4}(\ln K - \mu)\frac{\partial T}{\partial x}\,dx;$$

$$F = \frac{k}{e}\int_{T_0}^{T_1}(\ln K_I - \mu_I)dT + \frac{k}{e}\int_{T_1}^{T_2}(\ln K_{II} - \mu_{II})dT + \frac{k}{e}\int_{T_2}^{T_0}(\ln K_I - \mu_I)dT,$$

$$F = \frac{k}{e}\int_{T_1}^{T_2}\left(\ln\left(\frac{K_{II}}{K_I}\right) - (\mu_{II} - \mu_I)\right)dT = \frac{k}{e}\int_{T_1}^{T_2}\left(\ln\left(\frac{N_{II}}{N_I}\right) - (\mu_{II} - \mu_I)\right)dT.$$

Lorentz' theory $\mu_{II} = \mu_I = 2$ $(n = \infty)$.

(ii) Closed chain.

Investigation of thermal effects

$$d\varphi = -\left(i_x\frac{d\varphi}{dx} + \frac{dW_x}{dx}\right)dx,$$

$$d\varphi = \left(\frac{i^2}{\sigma} + i\frac{k}{e}T\frac{d(\ln K - \mu)}{dx} + \frac{d\left(\gamma\frac{dT}{dx}\right)}{dx}\right)dx,$$

$$dQ_R = i\frac{k}{e}T\frac{d(\ln K - \mu)}{dx}.$$

(a) Peltier effect

$$\Pi_{I,II} = \frac{k}{e}T(\ln(K_I/K_{II}) - (\mu_I - \mu_{II})),$$

Lorentz' theory:

$$\Pi_{I,II} = \frac{k}{e}T\ln(N_I/N_{II}).$$

[459]

(b) Thomson effect

$$\rho = -\frac{k}{e} T \frac{d(\ln K - \mu)}{dT},$$

Lorentz' theory:

$$\rho = -\frac{k}{e} T \frac{d \ln N}{dT} + \frac{3}{2} \frac{k}{e}.$$

(In the next lecture we shall discuss other derivations of the thermoelectric coefficients, Thomson's and Richardson's, and mention the difficulties they present.) Lord Kelvin's theory

$$\int \frac{dQ_R}{T} = 0$$

gives, together with the law of the conservation of energy, two conditions for F, Π and ρ.

$$\int \frac{dQ}{T} > 0.$$

$$\int \frac{dQ}{T} = \int \left(\frac{i^2}{\sigma T} + i \frac{k}{e} \frac{C}{T} \frac{dT}{dx} + \frac{1}{T} \frac{d(\gamma \, dT/dx)}{dx} \right) dx$$

$$= \int \left(\frac{i^2}{\sigma T} + i \frac{k}{e} \frac{(\ln (K/K_0) - \mu + \mu_0)}{T} \frac{dT}{dx} + \frac{\gamma}{T^2} \left(\frac{dT}{dx} \right)^2 \right) dx.$$

It is seen that the middle term cannot be small compared with the first and third, arising from the Joule heat and the heat conduction, respectively.

April 23, 1914

The equations for the flow of electricity and heat derived in the last lecture were

$$i_x = A_1 \left(-\frac{\partial \varphi}{\partial x} - \frac{kT}{eK} \frac{\partial K}{\partial x} \right) - A_2 \frac{\partial T}{\partial x},$$

$$W_x = A_2 \left(-\frac{\partial \varphi}{\partial x} - \frac{kT}{eK} \frac{\partial K}{\partial x} \right) - A_3 \frac{\partial T}{\partial x};$$

$$A_1 = \frac{4\pi e^2 K}{kT} \int_0^\infty r^2 D(r) e^{-\frac{1}{2} mr^2/kT} \, dr,$$

$$A_2 = \frac{2\pi em K}{kT^2} \int_0^\infty r^4 D(r) e^{-\frac{1}{2} mr^2/kT} \, dr,$$

$$A_3 = \frac{\pi m^2 K}{kT^2} \int_0^\infty r^6 D(r) e^{-\frac{1}{2}mr^2/kT} \, dr;$$

$$D(r) = C r^{(n+3)/(n-1)};$$

$$\frac{A_3}{A_1} = \left(\frac{k}{e}\right)^2 T \frac{2n(3n-1)}{(n-1)^2}, \qquad \frac{A_2}{A_1} = \frac{k}{e} \frac{2n}{(n-1)}.$$

Introducing the conductivities, we get

$$i_x = \sigma \left(-\frac{\partial \varphi}{\partial x} - \frac{k}{e} \frac{T}{K} \frac{\partial K}{\partial x} - \frac{k}{e} \mu \frac{\partial T}{\partial x} \right),$$

$$\mu = \frac{e}{k} \frac{A_2}{A_1} = \frac{2n}{n-1};$$

$$W_x = i_x \frac{k}{e} \mu T - \gamma \frac{\partial T}{\partial x}; \qquad N = \int f_0 \, 4\pi r^2 \, dr = K \left(\frac{2\pi kT}{m}\right)^{\frac{3}{2}}.$$

Open chain:

$$i_x = 0, \qquad \frac{\partial \varphi}{\partial x} = -\frac{k}{e} \left(\frac{T}{K} \frac{\partial K}{\partial x} + \mu \frac{\partial T}{\partial x}\right) = -\frac{k}{e} \left(\frac{T}{N} \frac{\partial N}{\partial x} + \frac{n+3}{2(n-1)} \frac{\partial T}{\partial x}\right).$$

(i) T constant:

$$\varphi_1 - \varphi_2 = \frac{k}{e} T \ln \frac{N_1}{N_2}.$$

(ii) Homogeneous metal:

$$d\varphi = -\frac{k}{e} \left(\frac{T}{N} \frac{dN}{dT} + \frac{n+3}{2(n-1)}\right) dT, \qquad \varphi_1 - \varphi_2 = f(T_1, T_2).$$

Visualization in terms of the pressure of the electrons, $p = NkT$:

$$-Ne \frac{\partial \varphi}{\partial x} = \frac{\partial p}{\partial x} = kT \frac{\partial N}{\partial x} + Nk \frac{\partial T}{\partial x}.$$

The view adopted leads to the correct result in case of (i), but to an incorrect result in case of (ii). (Except for $n = 5$; Maxwell's kinetic theory of gases.)

(iii)

$$F = \varphi(x_4) - \varphi(x_1) = \int_{x_1}^{x_4} \frac{\partial \varphi}{\partial x} \, dx = -\frac{k}{e} \int_{x_1}^{x_4} \left(\frac{T}{N} \frac{dN}{dx} + \mu \frac{dT}{dx}\right) dx$$

$$= \frac{k}{e} \int_{x_1}^{x_4} (\ln K - \mu) \frac{dT}{dx} \, dx,$$

[461]

$$F = \frac{k}{e} \int_{T_0}^{T_1} (\ln K_{\text{I}} - \mu_{\text{I}}) dT + \frac{k}{e} \int_{T_1}^{T_2} (\ln K_{\text{II}} - \mu_{\text{II}}) dT + \frac{k}{e} \int_{T_2}^{T_0} (\ln K_{\text{I}} - \mu_{\text{I}}) dT,$$

$$F = \frac{k}{e} \int_{T_1}^{T_2} (\ln (K_{\text{II}}/K_{\text{I}}) - (\mu_{\text{II}} - \mu_{\text{I}})) dT = \frac{k}{e} \int_{T_1}^{T_2} (\ln (N_{\text{II}}/N_{\text{I}}) - (\mu_{\text{II}} - \mu_{\text{I}})) dT.$$

Lorentz' theory: $(n = \infty)$, $\mu_{\text{I}} = \mu_{\text{II}} = 2$.

Closed chain

Investigation of the heat effects

$$dQ = - \left(i_x \frac{\partial \varphi}{\partial x} + \frac{\partial W_x}{\partial x} \right) dx,$$

$$dQ = \left(\frac{i^2}{\sigma} + i \frac{k}{e} T \frac{d(\ln K - \mu)}{dx} + \frac{d(\gamma(\partial T/\partial x))}{dx} \right),$$

$$dQ_{\text{R}} = i \frac{k}{e} T \frac{d(\ln K - \mu)}{dx} dx.$$

(a) Peltier effect

$$\Pi_{\text{I, II}} = \frac{k}{e} T (\ln (K_{\text{I}}/K_{\text{II}}) - (\mu_{\text{I}} - \mu_{\text{II}})).$$

Lorentz' theory:

$$\Pi_{\text{I, II}} = \frac{k}{e} T \ln (N_{\text{I}}/N_{\text{II}}).$$

(b) Thomson effect

$$\rho = - \frac{k}{e} T \frac{d(\ln K - \mu)}{dT}.$$

Lorentz' theory:

$$\rho = - \frac{k}{e} T \frac{d \ln N}{dT} + \frac{3}{2} \frac{k}{e}.$$

Illustration in terms of the pressure of the electrons

$$- Ne \frac{\partial \varphi}{\partial x} = \frac{\partial p}{\partial x} = \frac{\partial (kNT)}{\partial x}.$$

(a) Peltier effect $\Pi_{\text{I, II}} = \varphi_{\text{I}} - \varphi_{\text{II}}$.

The terms μ_{I} and μ_{II} are missing; ⟨they are⟩ due to the fact that the current carries with it different amounts of kinetic energy in different metals (Richardson)

$$W_x' = i_x \frac{k}{e} \frac{2n}{n-1}.$$

(b) Thomson effect
    1) Corresponding situation as in Peltier effect
    2) Calculations of J. J. Thomson

$$\rho = \frac{d\varphi}{dT} + \frac{1}{e}\frac{d}{dt}(\tfrac{3}{2}kT) = -\frac{1}{Ne}\frac{d(kTN)}{dT} + \frac{3}{2}\frac{k}{e},$$

$$\rho = -\frac{k}{e}T\frac{d\ln N}{dT} + \frac{1}{2}\frac{k}{e} = -\frac{k}{e}T\frac{d\ln(NT^{-\frac{1}{2}})}{dT}.$$

J. J. Thomson concludes from a numerical comparison that the number of electrons must vary as $T^{\frac{1}{2}}$. However, the formula is incorrect, since it should read

$$\rho = -\frac{k}{e}T\frac{d\ln(NT^{-\frac{1}{2}})}{dT}.$$

(Continuation of the discussion of J. J. Thomson's theory)
Simplest case: Flow of gas in a wide tube

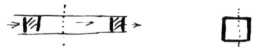

$$|v^2\tfrac{1}{2}Nvm| + \tfrac{3}{2}kTNv + vkNT = \tfrac{5}{2}kTNv + |\tfrac{1}{2}Nv^3m|.$$

$$W_x = \int \frac{m}{2}r^2\xi f\,d\sigma = \int\int\int_{-\infty}^{+\infty}\frac{m}{2}(\xi^2+\eta^2+\zeta^2)\xi Ke^{-\frac{1}{2}m(\eta^2+\zeta^2+(\xi-v)^2)/kT}\,d\xi\,d\eta\,d\zeta,$$

$$W_x = K\left(\frac{2\pi kT}{m}\right)^{\frac{3}{2}}(\tfrac{5}{2}kTv + \tfrac{1}{2}v^3m) = Nv(\tfrac{5}{2}kT + \tfrac{1}{2}mv^2).$$

May 1, 1914

The equations for the flow of electricity and heat were

$$i_x = \sigma\left(-\frac{\partial\varphi}{\partial x} - \frac{kT}{eK}\frac{\partial K}{\partial x} - \frac{k}{e}\mu\frac{\partial T}{\partial x}\right), \qquad \sigma = \frac{4\pi e^2 K}{kT}\int_0^\infty r^2 D(r)e^{-\frac{1}{2}mr^2/kT}\,dr,$$

$$W_x = i_x\frac{k}{e}\mu T - \gamma\frac{\partial T}{\partial x}, \qquad \mu_n = \frac{2n}{n-1}, \qquad N = K\left(\frac{2\pi kT}{m}\right)^{\frac{3}{2}}.$$

$$i_x = 0, \qquad \frac{\partial\varphi}{\partial x} = -\frac{k}{e}\frac{T}{K}\frac{\partial K}{\partial x} + \mu\frac{\partial T}{\partial x},$$

$$F = \frac{k}{e}\int_{T_1}^{T_2} (\ln\,(N_{\mathrm{II}}/N_{\mathrm{I}}) - (\mu_{\mathrm{II}} - \mu_{\mathrm{I}}))\,\mathrm{d}T,$$

$$\mathrm{d}Q = -\left(i_x\frac{\partial\varphi}{\partial x} + \frac{\partial W_x}{\partial x}\right)\mathrm{d}x,$$

$$\mathrm{d}Q = \left(\frac{i^2}{\sigma} + i\frac{k}{e}\,T\,\frac{\mathrm{d}(\ln K - \mu)}{\mathrm{d}x} + \frac{\mathrm{d}(\gamma(\mathrm{d}T/\mathrm{d}x))}{\mathrm{d}x}\right)\mathrm{d}x$$

$$\Pi_{\mathrm{I,\,II}} = \frac{k}{e}\,T(\ln\,(K_{\mathrm{I}}/K_{\mathrm{II}}) - (\mu_{\mathrm{I}} - \mu_{\mathrm{II}})), \qquad \rho = -\frac{k}{e}\,T\,\frac{\mathrm{d}(\ln K - \mu)}{\mathrm{d}T},$$

$$\Pi_{\mathrm{I,\,II}} = \frac{k}{e}\,T(\ln\,(N_{\mathrm{I}}/N_{\mathrm{II}}) - (\mu_{\mathrm{I}} - \mu_{\mathrm{II}})), \qquad \rho = -\frac{k}{e}\,T\,\frac{\mathrm{d}(\ln N - \mu)}{\mathrm{d}T} + \frac{3}{2}\frac{k}{e}.$$

Lord Kelvin's theory: $\int \mathrm{d}Q_{\mathrm{R}}/T = 0$ gives, together with the law of the conservation of energy, two conditions for F, Π and ρ.

Necessary condition $\int \mathrm{d}Q/T > 0$

$$\int \left(\frac{i^2}{\sigma T} + \frac{Ci}{T}\frac{\mathrm{d}T}{\mathrm{d}x} + \frac{1}{T}\frac{\mathrm{d}(\gamma\,\mathrm{d}T/\mathrm{d}x)}{\mathrm{d}x}\right)\mathrm{d}x$$

$$= \int \left(\frac{i^2}{\sigma T} + \frac{1}{T}\,iC\frac{\mathrm{d}T}{\mathrm{d}x} + \frac{\gamma}{T^2}\left(\frac{\mathrm{d}T}{\mathrm{d}x}\right)^2\right)\mathrm{d}x > 0.$$

Boltzmann.

Incorrectness of a similar result in cyclic process (the occurrence of μ).

The difficulty for the electron theory arising from the value of the Thomson coefficient.

Phenomena of heat radiation.

(i) Absorption.

Vibrations slow compared to the time between collisions.

The piece of metal so small that one can ignore absorption and reflection.

$$i_x = \sigma X, \qquad i_x X = \sigma X^2, \qquad A = \sigma(X^2 + Y^2 + Z^2),$$

$$M = E = (1/8\pi)(X^2 + Y^2 + Z^2), \qquad A = 4\pi\sigma(E + M); \qquad \alpha = 4\pi\sigma;$$

Hagen and Rubens

$$\left(\frac{R+1}{R-1}\right)^2 = \frac{\alpha}{2p}, \qquad (X = C\cos pt).$$

(ii) Emission

$$S \, dt = \frac{2e^2}{3c^3} j^2 \, dt.$$

$$\xi = a \sin pt, \qquad j_x = ap \cos pt, \qquad S_1 = \frac{2e^2}{3c^3} \frac{a^2 p^2}{2}.$$

$$S = \frac{2e^2}{3c^3} \frac{1}{\vartheta} \int_0^\vartheta j^2 \, dt, \qquad \xi = \sum_{s=1}^{s=\infty} a_s \sin \frac{s\pi t}{\vartheta}, \qquad a_s = \frac{2}{\vartheta} \int_0^\vartheta \xi \sin \frac{s\pi t}{\vartheta} \, dt.$$

$$j_x = \sum_{s=1}^{s=\infty} a_s \frac{s\pi}{\vartheta} \cos \frac{s\pi t}{\vartheta}, \qquad \frac{s\pi}{\vartheta} = p, \qquad \Delta p = \frac{\pi}{\vartheta} \Delta s.$$

$$S = \frac{2e^2}{c^3} \frac{1}{\vartheta} \sum_{s=1}^{s=\infty} \frac{a_s^2}{2} \left(\frac{s\pi}{\vartheta} \right)^2 \vartheta$$

$$S(p)\Delta p = \frac{2e^2}{c^3} \sum_s^{s+\Delta s} \frac{a_s^2}{2} \left(\frac{s\pi}{\vartheta} \right)^2 = \frac{2e^2}{c^3} \frac{\overline{a_s^2}}{2} \left(\frac{s\pi}{\vartheta} \right)^2 \Delta s$$

$$S(p)\Delta p = \frac{2e^2}{c^3} \frac{\overline{a_s^2}}{2} p^2 \left(\frac{\vartheta}{\pi} \Delta p \right) = \frac{2e^2}{c^3} \frac{\vartheta}{2\pi} p^2 \frac{4}{\vartheta^2} \overline{\left[\int_0^\vartheta \xi \sin \frac{s\pi t}{\vartheta} \, dt \right]^2} \Delta p$$

$$\overline{\left[\int_0^\vartheta \xi(t) \sin pt \, dt \right]^2} = \overline{\int_0^\vartheta \int_0^\vartheta \xi(t_1)\xi(t_2) \sin pt_1 \sin pt_2 \, dt_1 \, dt_2}$$

$$= \frac{1}{2} \overline{\int_0^\vartheta \int_0^\vartheta \xi(t_1)\xi(t_2)[\cos p(t_1-t_2) - \cos p(t_1+t_2)] dt_1 \, dt_2}$$

Hence, when the vibrations are assumed to be slow,

$$\overline{\left[\int_0^\vartheta \xi(t) \sin pt \, dt \right]^2} = \frac{1}{2} \overline{\left[\int_0^\vartheta \xi(t) dt \right]^2} = \tfrac{1}{2} \cdot 2D\vartheta.$$

$$S(p)\Delta p = \frac{4e^2}{\pi c^3} p^2 D \Delta p,$$

$$S(p) = \frac{4e^2}{\pi c^3} p^2 \sum_\alpha D = \frac{4e^2}{\pi c^3} p^2 \int_0^\infty 4\pi r^2 D(r) K e^{-\frac{1}{2}mr^2/kT} \, dr$$

$$= \frac{4\sigma kT}{\pi c^3} p^2 = \beta(p).$$

$$E(p) dp \, \alpha \Delta V = \Delta V \beta(p) dp.$$

$$E(p) = \frac{\beta(p)}{\alpha(p)} .$$

$$E(p) = \frac{kT}{\pi^2 c^3} p^2 .$$

Planck's radiation formula is

$$E(p) = \frac{h}{2\pi} p^3 \frac{1}{\pi^2 c^3} \frac{1}{e^{-(hp/2\pi kT)} - 1} .$$

For large p,

$$E(p) = \frac{kT}{\pi^2 c^3} p^2 .$$

May 15, 1914

Heat radiation, slow vibrations

$$\alpha = \frac{4\pi e^2 K}{kT} \int_0^\infty r^2 D(r) e^{-\frac{1}{2} mr^2/kT} \, dr.$$

(i) Absorption:

$$i_x = \sigma X, \qquad i_x X = \sigma X^2, \qquad A = \sigma(X^2 + Y^2 + Z^2) = 4\pi\sigma E,$$

$$\alpha = 4\pi\sigma, \qquad \left(\frac{R+1}{R-1}\right)^2 = \frac{\alpha}{2p}, \qquad X = X_0 \cos pt.$$

(ii) Emission:

$$g \, dt = \frac{2e^2}{3c^3} j^2 \, dt;$$

$$\xi = a \sin pt, \qquad j_x = ap \cos pt, \qquad \bar{g} = \frac{2e^2}{3c^3} a^2 p^2 \cdot \tfrac{1}{2};$$

$$\bar{g} = \frac{2e^2}{3c^3} \frac{1}{\vartheta} \int_0^\vartheta j^2 \, dt = \frac{2e^2}{c^3} \frac{1}{\vartheta} \int_0^\vartheta j_x^2 \, dt;$$

$$\xi = \sum_{s=1}^{s=\infty} a_s \sin \frac{s\pi t}{\vartheta}, \qquad a_s = \frac{2}{\vartheta} \int_0^\vartheta \xi \sin \frac{s\pi t}{\vartheta} \, dt,$$

$$j_x = \sum_{s=1}^{s=\infty} a_s \frac{s\pi}{\vartheta} \cos \frac{s\pi t}{\vartheta} .$$

$$\bar{g} = \frac{2e^2}{c^3} \frac{1}{\vartheta} \sum_{s=1}^{s=\infty} \left(\frac{s\pi}{\vartheta}\right)^2 a_s^2 \frac{\vartheta}{2}, \qquad \frac{s\pi}{\vartheta} = p, \qquad \Delta p = \frac{\pi}{\vartheta} \Delta s;$$

$$\bar{g}(p)\Delta p = \frac{2e^2}{c^3} \sum_{s}^{s+\Delta s} \left(\frac{s\pi}{\vartheta}\right)^2 \frac{a_s^2}{2} = \frac{2e^2}{c^3} \left(\frac{s\pi}{\vartheta}\right)^2 \frac{\overline{a_s^2}}{2} \Delta s = \frac{2e^2}{c^3} p^2 \frac{\vartheta}{2\pi} \Delta p \, \overline{a_s^2}$$

$$= \frac{4e^2}{\pi c^3} p^2 \Delta p \, \overline{\left[\int_0^\vartheta \xi \sin pt \, \mathrm{d}t\right]^2}.$$

$$\overline{\int_0^\vartheta \int_0^\vartheta \xi(t_1)\xi(t_2) \sin pt_1 \sin pt_2 \, \mathrm{d}t_1 \, \mathrm{d}t_2}$$

$$= \frac{1}{2}\overline{\int_0^\vartheta \int_0^\vartheta \xi(t_1)\xi(t_2)\mathrm{d}t_1 \, \mathrm{d}t_2} = D\vartheta$$

$$\overline{\left[\int_0^\vartheta \xi \, \mathrm{d}t\right]^2} = \overline{\int_0^\vartheta \int_0^\vartheta \xi(t_1)\xi(t_2)\mathrm{d}t_1 \, \mathrm{d}t_2} = 2D\vartheta$$

$$\bar{g}(p) = \frac{4e^2}{3c^3} p^2 D.$$

$$\beta(p) = \frac{4e^2}{\pi c^3} p^2 \sum D = \frac{4e^2}{3c^3} p^2 \int_0^\infty 4\pi r^2 D(r) K e^{-\frac{1}{2}mr^2/kT} \, \mathrm{d}r = \frac{4kT\sigma p^2}{\pi c^3};$$

$$E(p)\mathrm{d}p \, \alpha \Delta V = \beta(p)\mathrm{d}p \, \Delta V;$$

$$E(p) = \frac{\beta(p)}{\alpha(p)} = \frac{kT}{\pi^2 c^3} p^2.$$

Planck's formula:

$$E(p) = \frac{p^3 h}{2\pi^3 c^3} \frac{1}{e^{hp/2\pi kT} - 1};$$

for p small we get

$$E(p) = \frac{kT}{\pi^2 c^3} p^2.$$

Influence of magnetic field upon the motion of the electrons

$$M_x = \frac{e}{c} H_z \eta - \frac{e}{c} H_y \zeta, \qquad M_y = \frac{e}{c} H_x \zeta - \frac{e}{c} H_z \xi, \qquad M_z = \frac{e}{c} H_y \xi - \frac{e}{c} H_x \eta,$$

$$-\frac{\partial f}{\partial t} = \xi \frac{\partial f}{\partial x} + \eta \frac{\partial f}{\partial y} + \zeta \frac{\partial f}{\partial z} + \frac{X}{m} \frac{\partial f}{\partial \xi} + \frac{Y}{m} \frac{\partial f}{\partial \eta} + \frac{Z}{m} \frac{\partial f}{\partial \zeta} + a - b.$$

[467]

Equilibrium in absence of magnetic field

$$f = Ke^{-(\frac{1}{2}mr^2 + \varphi)/kT};$$

$$\left[-\frac{\partial f}{\partial t}\right]_{XYZ} = f\frac{1}{kT}(X\xi + Y\eta + Z\zeta) = f\frac{1}{kT}A, \qquad A = 0, \qquad \text{magnetism;}$$

$$\left[-\frac{\partial f}{\partial t}\right]_{\xi\eta\zeta} = f\frac{1}{kT}\left(\xi\frac{\partial\varphi}{\partial x} + \eta\frac{\partial\varphi}{\partial y} + \zeta\frac{\partial\varphi}{\partial z}\right) = f\frac{1}{kT}(-A).$$

No influence (statistically) of the magnetic field.
Explanation of paradox.
Insufficiency of mechanical forces.

May 20, 1914

Influence of magnetic field

$$M_x = \frac{e}{c}H_z\eta - \frac{e}{c}H_y\zeta, \qquad M_y = \frac{e}{c}H_x\zeta - \frac{e}{c}H_z\xi, \qquad M_z = \frac{e}{c}H_y\xi - \frac{e}{c}H_x\eta,$$

$$\frac{\partial f}{\partial t} + \xi\frac{\partial f}{\partial x} + \eta\frac{\partial f}{\partial y} + \zeta\frac{\partial f}{\partial z} + \frac{X}{m}\frac{\partial f}{\partial \xi} + \frac{Y}{m}\frac{\partial f}{\partial \eta} + \frac{Z}{m}\frac{\partial f}{\partial \zeta} + (a - b) = 0.$$

(1) Equilibrium; no influence of the magnetic forces upon the distribution (discussed in previous lecture).
(2) Influence of the magnetic forces upon the flow of electrons.
    Galvanomagnetic and thermomagnetic phenomena.
Difficulties in considering the diffusion (i.e., in determining the angle between the direction of the current and the gradient). A direct method of calculation required. Number of electrons passing through a plane perpendicular to the x-axis $= \int\xi f\,d\sigma$. Total momentum of electrons equals $\int m\xi f\,d\sigma$.

$$G_x(r)dV\,dr = \int m\xi f\,d\sigma,$$

between two spherical shells r and $r + dr$.
Influence of collisions upon $G_x(r)$
The change in dt in the momentum due to collisions in dV of electrons ⟨with velocities⟩ in $d\sigma$ can be written $mrf\,dV\,d\sigma F(r)dt$ with component $m\xi f F(r)dV\,d\sigma\,dt$ in the direction of the x-axis; this gives $G_x(r)F(r)dV\,d\sigma\,dt$.

$F(r)$ has the dimension T^{-1}, but can be determined from the "diffusion coefficient" in analogy with previous calculations.

$$\frac{dG_x(r)}{dt} + m \int \xi \left(\xi \frac{\partial f}{\partial x} + \eta \frac{\partial f}{\partial y} + \zeta \frac{\partial f}{\partial z} + \frac{X}{m} \frac{\partial f}{\partial \xi} + \frac{Y}{m} \frac{\partial f}{\partial \eta} + \frac{Z}{m} \frac{\partial f}{\partial \zeta} \right) d\sigma + G_x(r)F(r) = 0;$$

$$f = Ke^{-\frac{1}{2}mr^2/kT} + \psi = f_0 + \psi;$$

$$\int \xi \left(\xi \frac{\partial f_0}{\partial x} + \eta \frac{\partial f_0}{\partial y} + \zeta \frac{\partial f_0}{\partial z} + \frac{X}{m} \frac{\partial f_0}{\partial \xi} + \frac{Y}{m} \frac{\partial f_0}{\partial \eta} + \frac{Z}{m} \frac{\partial f_0}{\partial \zeta} \right) d\sigma$$

$$= \int \xi \left(\xi \left(\frac{1}{K} \frac{\partial K}{\partial x} + \frac{mr^2}{2kT^2} \frac{\partial T}{\partial x} - \frac{E_x e}{kT} \right) + \eta(\dots) + \zeta(\dots) \right) f_0 \, d\sigma$$

$$= \frac{4\pi}{3} r^4 f_0 \left(\frac{1}{K} \frac{\partial K}{\partial x} + \frac{mr^2}{2kT^2} \frac{\partial T}{\partial x} - \frac{E_x e}{kT} \right).$$

Hence, the change in the momentum caused by E_x is

$$\frac{4\pi}{3} r^4 \frac{em}{kT} f_0 E_x \quad and \quad not \quad 4\pi r^2 f_0 E_x e;$$

the ratio is $\frac{1}{3}mr^2/kT$.

In the calculation of $m\int \xi(\xi(\partial f/\partial x) + \cdots)d\sigma$, ψ can be neglected, except in case of magnetic forces which have no influence upon the first term, f_0. The influence of the magnetic fields, however, can be easily obtained. In fact, it is simply

$$\sum M_x dt = \left(\frac{e}{c} H_z \sum \eta - \frac{e}{c} H_y \sum \zeta \right) dt = \left(\frac{e}{cm} H_z G_y(r) - \frac{e}{cm} H_y G_z(r) \right) dr \, dV \, dt.$$

This follows from the fact that the speeds are not changed by the influence of the magnetic field. This can be seen directly, since

$$\int \xi \left(\eta \frac{\partial f}{\partial \xi} - \xi \frac{\partial f}{\partial \eta} \right) d\sigma = \iiint \left(\xi\eta \frac{\partial f}{\partial \xi} - \xi^2 \frac{\partial f}{\partial \eta} \right) d\xi \, d\eta \, d\zeta$$

$$= \iint [\xi\eta f] - \iiint \eta f \, d\xi \, d\eta \, d\zeta - \iint [\xi^2 f] d\xi \, d\zeta,$$

but on a spherical surface with radius r one has $\xi^2 + \eta^2 + \zeta^2 = r^2$ and, hence if ζ is constant $\eta d\eta + \xi d\xi = 0$; however, since here $d\eta$ and $d\xi$ are considered positive, we have $\eta d\eta - \xi d\xi = 0$.

$$\frac{dG_x(r)}{dt} + \frac{4\pi}{3} mK \left(\frac{e}{kT} \frac{\partial \varphi}{\partial x} + \frac{1}{K} \frac{\partial K}{\partial x} + \frac{mr^2}{2kT^2} \frac{\partial T}{\partial x} \right) r^4 e^{-\frac{1}{2}mr^2/kT} + G_x(r)F(r)$$

$$= \frac{e}{mc} H_z G_y(r) - \frac{e}{mc} H_y G_z(r).$$

Consideration of the diffusion gave

$$G_x(r) = D(r)K4\pi r^2 e^{-\frac{1}{2}mr^2/kT} \left[-\frac{e}{kT} - \frac{1}{K}\frac{\partial K}{\partial x} - \frac{mr^2}{2kT^2}\frac{\partial T}{\partial x} \right],$$

hence, $F(r)^{-1} = (3/r^2)D(r)$.

Forces inversely proportional to the nth power of the distance:

$$D(r) = C' r^{(n+3)/(n-1)}; \quad \text{hence,} \quad F(r) = r^{(n-5)/(n-1)}C.$$

(Elastic spheres $F(r) = rC$; fifth power $F(r) = C$).

Coupling of the electric and energy currents along the three axes resulting from the magnetic field.

Magnetic field parallel to the z-axis:

No effect upon $G_z(r)$

Special case: T and K constant:

$$-\frac{4\pi m e K}{3kT}\frac{\partial\varphi}{\partial x} r^4 e^{-\frac{1}{2}mr^2/kT} - G_x(r)F(r) + \frac{e}{mc} H_z G_y(r) = 0,$$

$$-\frac{4\pi m e K}{3kT}\frac{\partial\varphi}{\partial y} r^4 e^{-\frac{1}{2}mr^2/kT} - G_y(r)F(r) - \frac{e}{mc} H_z G_x(r) = 0.$$

$$G_x(r) = -\frac{4\pi m e K}{3kT} \frac{(\partial\varphi/\partial x)F(r) + (\partial\varphi/\partial y)(e/mc)H_z}{F(r)^2 + ((e/mc)H_z)^2} r^4 e^{-\frac{1}{2}mr^2/kT}.$$

$$i_x = \frac{e}{m}\int_0^\infty G_x(r)dr, \qquad W_x = \frac{1}{2}\int r^2 G_x(r)dr,$$

$$i_x = -\sigma_1 \frac{\partial\varphi}{\partial x} - \delta_1 \frac{\partial\varphi}{\partial y}, \qquad i_y = -\sigma_1 \frac{\partial\varphi}{\partial y} + \delta_1 \frac{\partial\varphi}{\partial x};$$

$$\sigma_1 = \frac{4\pi e^2 K}{3kT}\int_0^\infty \frac{F(r)r^4 e^{-\frac{1}{2}mr^2/kT}}{F(r)^2 + ((e/mc)H_z)^2}\, dr,$$

$$\delta_1 = \frac{4\pi e^2 K}{3kT}\int_0^\infty \frac{(e/mc)H_z\, r^4 e^{-\frac{1}{2}mr^2/kT}}{F(r)^2 + ((e/mc)H_z)^2}\, dr.$$

$$\left|\frac{\partial}{\partial t} G_x(r)\right| = -F(r)G_x(r), \qquad G_x(r) = K' e^{-F(r)t}.$$

$F(r)^{-1}$ corresponds to the time between collisions;

$$\eta \frac{e}{c} H_z = \frac{m\eta^2}{R}, \qquad \frac{e}{mc} = \frac{\eta}{R} = \text{angular speed}.$$

Significance of the fact that $F(r)^{-1}(e/mc)H_z$ is small.

$$F(r) = Cr^{(n-5)/(n-1)}.$$

$$\sigma = \frac{4\pi e^2 K}{kT} \int_0^\infty r^2 D(r) e^{-\frac{1}{2}mr^2/kT}\,dr = \frac{4\pi e^2 K}{3kT} \int_0^\infty r^4 F(r)^{-1} e^{-\frac{1}{2}mr^2/kT}\,dr;$$

$$N = K\left(\frac{2\pi kT}{m}\right)^{\frac{1}{2}}.$$

$$\sigma_1 = \sigma\left(1 - \sigma^2 \frac{H_z^2}{N^2 e^2 c^2} \frac{9\pi}{16} \Gamma\left(\frac{n+5}{n-1}\right) \Gamma^{-3}\left(\frac{2n}{n-1}\right)\right);$$

$$\delta_1 = \sigma^2 \frac{H_z}{Nec} \frac{3\sqrt{\pi}}{4} \Gamma\left(\frac{3n+5}{2(n-1)}\right) \Gamma^{-2}\left(\frac{2n}{n-1}\right).$$

$$i_y = 0, \qquad E_y = -\frac{\partial\varphi}{\partial y} = -\frac{\delta_1}{\sigma_1} \frac{\partial\varphi}{\partial x};$$

$$i_x = -\sigma_1\left(1 + \left(\frac{\delta_1}{\sigma_1}\right)^2\right)\frac{\partial\varphi}{\partial x} = -\sigma_H \frac{\partial\varphi}{\partial x};$$

$$E_y = i_x \frac{\delta_1}{\sigma_1^2 + \delta_1^2},$$

$$E_y = i_x \frac{H_z}{Nec} \frac{3\sqrt{\pi}}{4} \Gamma\left(\frac{3n+5}{2(n-1)}\right) \Gamma^{-2}\left(\frac{2n}{n-1}\right);$$

$$\sigma_H = \sigma\left(1 - \sigma^2 \frac{H_z}{N^2 e^2 c^2} \frac{4\pi}{16}\left(\Gamma\left(\frac{n+5}{n-1}\right) \Gamma\left(\frac{2n}{n-1}\right) - \Gamma^2\left(\frac{3n+5}{2(n-1)}\right) \Gamma^{-4}\left(\frac{2n}{n-1}\right)\right)\right)$$

$$W_y = \tfrac{1}{2}\int_0^\infty r^2 G_y(r)\,dr = -\frac{\partial\varphi}{\partial y} \frac{2n}{n-1} \frac{k}{e} T\sigma_1 + \frac{\partial\varphi}{\partial x} \frac{(3n+5)}{2(n-1)} \frac{k}{e} T\delta_1;$$

since $i_y = 0$, we obtain

$$W_y = \frac{n-5}{2(n-1)} \frac{k}{e} \sigma_1 E_y;$$

$$-\left|\gamma \frac{dT}{dy}\right| = -W_y, \qquad \frac{dT}{dy} = \frac{W_y}{\gamma} = \frac{n-5}{2(n-1)} \frac{k}{e} \frac{\sigma}{\gamma} E_y;$$

$$\frac{\sigma}{\gamma} = \frac{e^2}{k} \frac{n-1}{2n} \frac{1}{T}; \qquad \frac{dT}{dy} = \frac{n-5}{4n} \frac{e}{k} E_y.$$

$$n = \infty.$$

IX. CORRESPONDENCE WITH G. H. LIVENS

(1915*)

* [See Introduction, sect. 8.]

[473]

[The correspondence begins with the following letter and sheet of calculations by Bohr (undated draft, the letter in the handwriting of his wife, the calculations in his own handwriting):]

Dear Mr. Livens,

I have with great interest read your paper in the last Phil. Mag.* In this connection I may allow myself to call your attention to a danish paper of mine published a few years ago (Studier over Metallernes Elektronteorie, Copenhagen, 1911). In the paper I have attempted to generalise the classical theory of Lorentz by abandoning the special assumtion about collisions, i.e., elastic spheres. The calculations are extended to the conductivity phenomena, thermoelectricity, radiation phenomena and the magnetic phenomena. In addition to the general treatment I have throughout used elastic spheres and forces varying as some power of the distance as examples. In case of the radiation phenomena I have shown that the calculation in the same interval of frequency as discussed in your papers lead to the Jeans-Rayleigh formula quite independent of any special assumption about the collisions (this result, however, has since been superceded by the extremely interesting paper of McLaren (Phil. Mag. July, 1911) in which he has proved that the same result can be obtained for all frequencies). The reason why you in your paper in case of elastic spheres has not arrived to exactly the same result is as you suggest yourself that the formula calculated by Jeans is not correct; this point is discussed in detail in my paper. I have also criticised Wilson's formula but the formula to which I have arrived differs from that in your paper. It seems to me that the difference is due to a writing error at the top of p. 182 in your paper, as far as I can see the denominator on the right side should be $(1+p^2l^2/r^2)^2$ instead of $(1+p^2l^2/r^2)$, by help of this alteration your calculation gives the same result as mine. As to the problem of the application of such formula in the discussion of the optical properties of metals, I have tried to show that it is impossible on an electron theory of the kind in question to explain the experiments on reflection since the calculation gives the opposite sign as that of the experiment.

I am very sorry not here to have any copies of the paper to send to you. As a dissertation, however, it may be found at the Univ. Lib. in Sheffield. If you like I shall be very glad to lend you a typewritten copy of an english translation which was made some years ago but being too long for a periodical it was never published.

Yours faithfully N. B.

* [Phil. Mag. **29** (1915) 171.]

$$\int_0^\infty x \sin ax \, e^{-\frac{bx}{x}} dx = Z$$

$$\int_0^\infty x \, e^{x(ia-b)} dx = (ia-b)^{-2} = \frac{(ia+b)^2}{(a^2+b^2)^2} = \frac{b^2-a^2}{a^2+b^2} + i\frac{2ab}{(a^2+b^2)^2} = \int_0^\infty x(\cos ax + i\sin ax)e^{-bx} dx$$

$$Z = \frac{2ab}{(a^2+b^2)^2}$$

$$\int_0^\infty \cos ax \, e^{-\frac{bx}{x}} dx = y$$

$$\int_0^\infty e^{(ia-b)x} dx = -(ia-b)^{-1} = \frac{-b-ia}{a^2+b^2} = \int_0^\infty (\cos ax + i\sin ax)e^{-bx} dx$$

$$y = \frac{b}{a^2+b^2}$$

$$\int_0^\infty \sin^2\frac{nl}{2h} e^{-\frac{l}{l_m}} dl = \frac{1}{2}\int_0^\infty (1-\cos\frac{nl}{h})e^{-\frac{l}{l_m}} dl = \frac{1}{2}\left(l_m - \frac{l_m}{(\frac{n}{h})^2+\frac{1}{l_m^2}}\right) = \frac{1}{2}l_m \frac{\frac{n^2l^2}{h^2}}{1+(\frac{nl}{h})^2}$$

$$\int_0^\infty \frac{nl}{h}\sin\frac{nl}{h} e^{-\frac{l}{l_m}} dl = \frac{n}{h}2\frac{n}{h}l_m\left((\frac{n}{h})^2+\frac{1}{l_m^2}\right)^{-2} = \frac{2n^2l^3}{h^2}(1+(\frac{nl}{h})^2)^{-2}$$

$$\frac{l^2c^2l_m}{2nm}\left((1+(\frac{n^2l^2}{h^2})^{-1}(1+(\frac{\xi}{h})^2) - 2\frac{\xi^2}{h^2}(1+(\frac{nl}{h})^2)^{-2}\right) \qquad (1+(\frac{nl}{h})^2) = a$$

$$\left(\frac{1}{a} + (\frac{\xi}{h})^2\frac{1}{a^2}(2-a)\right)$$

$$\left(\frac{1}{a} - \frac{1}{3}\frac{1}{a^2}(2-a)\right) = \frac{1}{a^2}\frac{2}{3}(2a-1)$$

$$\int_0^\infty \frac{h^3 l^2 d\lambda}{a} = \frac{1}{2q}\int_0^\infty \frac{e^{-qh^2}(a\,2h - h^2\frac{da}{d\lambda})}{a^2} = \frac{1}{q}\int_0^\infty \frac{e^{-qh^2}}{a^2}(h + \frac{n^2l^2}{h} + \frac{n^2l^2}{h}) =$$

$$\frac{1}{q}\int_0^\infty \frac{e^{-qh^2}h}{a^2}(2a-1)\bigg)$$

[Livens replied:]

The University
Sheffield

Dear Dr. Bohr,   Jan. 11th 1914*

Many thanks for your interesting letter *re* my recent papers in the Phil. Mag. and your explanation of certain points connected therewith.

I shall be delighted to have the loan of your paper (the translation) for a short time; we have not, as you suggest, got any copy of it in our library, so that I have not had any access to it.

Your mention one or two interesting points in your letter to which I may allow myself also a few remarks. Regarding the radiation problem I had myself independently come to the conclusion you mention and have obtained, by rather a novel method, the requisite result for the conductivity in the particular case of elastic spheres. This point I had already submitted for publication some two months ago as well as the developments of it on the optical side. I agree with you however as to the general validity of our theories in this direction although I think there is still hope if full account is taken of the effect of the resonance electrons.

You make a reference to McLaren's work. As far as I can make out, however, his analysis is no more general than Lorentz's original work. His attempted generalisation of the problem is in reality restricted so much by the method of analysis used that it becomes as limited as the simple Lorentz theory, a point I have also discussed in a paper already submitted to the publishers.

I have written several short papers on this subject lately chiefly concerned with the mathematical developments of the theory. I gather from Enskog's paper in the Annalen der Physik [**38** (1912) 731] that you base your solution of the main problem on an integral equation. It transpires however that the fundamental differential equation satisfied by the function defining the velocity distribution is capable of interpretation in a form which admits of immediate integration in the most general possible case. This seems to simplify the theory immensely. There also appears to be another and simple method of deducing the distribution law from the ideas underlying the Drude-Thomson theory. I have used this as a basis of my theoretical investigation for certain cases.

There is however one point in the theory to which you draw attention viz. the formula for the conductivity deduced from the principles of the dissipation of energy as discussed in my shorter paper, which has interested me very much. I have no copy of my paper before me yet and cannot therefore verify your suggested modification of my analysis. I may say that I carefully revised my work over again but it is possible that I may have made a slip. I began to think there

* [Here is a slip of the pen. The date should be Jan. 11, 1915.]

was some fundamental physical discrepancy but I couldn't find where. I may say that the point of discrepancy missed me at first, otherwise I should have mentioned it in my paper.

Again thanking you for your kindly criticism and in anticipation of the loan of your paper

I remain
Sincerely yours
G. H. Livens

[On January 16, 1915, Livens acknowledged receipt of the copy of the English translation of Bohr's dissertation by the following card:]

The University of Sheffield
Jan 16th 1915

Many, many thanks for copy of
paper which arrived quite safely.
I am getting on splendidly with it.

Yours truly
G.H.L.

[He returned it shortly afterwards with the following letter:]

The University
Sheffield
Feb. 8th 1914*

Dear Dr. Bohr,

I am sending by this post your M.S. translation of your dissertation for the loan of which I am deeply grateful to you and for which I thank you most heartily.

I have thoroughly enjoyed reading this paper of yours and only regret that I had not previously been able to see it. I have on several occasions tried to get a glimpse at it.

I started on this subject last October with a view to applications in the optical theory of metals and had already come across most of the results found by you as particular cases of your more general theory. I have, however, continually worked with the distribution function for the electronic motions instead of the momentum equation as used by you.

I must confess that I am still of the opinion that if it were possible to work out a single special case we should find that the emission from the metal is at least consistent with the correct type of radiation formula. It is true that the theories

* [The date should be February 8, 1915.]

of absorption do involve a neglect of the collision intervals but a more complete theory can only differ qualitatively from the result thus deduced. My position is briefly that illustrative calculations have deduced an emission formula which decreases exponentially but in no single case has any calculation indicated such a rapid decrease in the absorption.

The little difficulty you indicate concerning the reflexion of normally incident light is easily removed if the theory is generalised to include the effect of the resonance electrons. I have attempted a simple generalisation of this nature in all of my applications to the optical theory.

You were quite right in your correction to my paper; the integral in question is wrongly given. Unfortunately however the corrected form leads to a final formula for σ which is still more unlike the true form than that given and I am therefore more at sea about the discrepancy than ever. Of course your formula is correct, as I have already found by another method.

Again thanking you most heartily for your kindness in lending me your M.S.

I remain
Sincerely yours,
G. H. Livens

P.S. I presume you are aware that pp. 81–3 are missing from your M.S.

[Three days later Bohr writes to Livens from Manchester (draft in Mrs. Bohr's handwriting):]

Dear Mr. Livens,

Many thanks for the returning of the manuscript and for your kind letter.

I hope that you understand that you needed not ⟨be⟩ in the least hurry with the manuscript. I am sorry that I had quite forgotten that I a year ago had taken two pages out in order to show them to Dr. Swan.

It seems to me that the discrepancy you refer to in your letter disappears if your formula on top of p. 182 is corrected. If so you get in stead of the integral in the expression for H on the same page

$$\int_0^\infty \frac{e^{-qr^2}(1+2(l^2p^2/r^2))r\,dr}{(1+l^2p^2/r^2)^2} = \int_0^\infty \frac{e^{-qr^2}r\,dr}{(1+(l^2p^2/r^2))} + \int_0^\infty \frac{e^{-qr^2}(l^2p^2/r^2)\langle r\rangle dr}{(1+(l^2p^2/r^2))^2}$$

$$= \int_0^\infty \frac{e^{-qr^2}r\,dr}{(1+(l^2p^2/r^2))} + \tfrac{1}{2}\left[\frac{r^2e^{-qr^2}}{(1+(l^2p^2/r^2))}\right]_0^\infty - \int_0^\infty \frac{e^{-qr^2}r\,dr}{(1-(l^2p^2/r^2))}$$

$$+ q\int_0^\infty \frac{e^{-qr^2}r^3\,dr}{(1+(l^2p^2/r^2))} = q\int_0^\infty \frac{e^{-qr^2}r^3\,dr}{(1+(l^2p^2/r^2))}\,.$$

This integral is the same as that obtained in my paper if $n = \infty$. Further it is the same as that involved in the expression for the emission calculated in your first paper* on p. 168; if introduced you get exactly Jeans' formula of radiation.

In mentioning the difficulty as to the calculation of the reflection my intention was not to say that this difficulty could not be overcome but to point out that it seems to me that all the calculations on the number of electrons performed by numerous authors have no foundation at all, as they are based on a formula which has an incorrect sign. I do not know if you agree in this.

In answer to the view expressed in your letter as to your believe in the possibility of obtaining a radiation formula of the correct type, I should like here to state as my view, that it seems to me that this possibility is excluded. For my part, I feel so absolutely confident that any rigorous calculations must lead to Jeans' formula. This believe has been forced upon me not only through the study of the statistical theories of radiation, but through every special calculation I have seen or I have tried myself.—Of course, however, I look forward with the greatest interest to see if your work will show that this contention is premature.

With best wishes
sincerely yours
N. Bohr

P.S. I enclose the missing pages of the manuscript.

[Livens promptly replied:]

The University
Sheffield
Feb. 14th 1915

Dear Dr. Bohr,

I really ought to apologise to you for troubling you to help me out of my stupidity. Of course the corrected integral leads to the correct result, as I ought to have seen before. However, many thanks for helping me.

I do not think I can produce any new evidence regarding the radiation formula obtained from these theories; but this hardly seems to be necessary considering the results already obtained by Thomson and Jeans. My chief contention is that so far as I am aware no plausible calculation has yet been given which indicates that the absorption falls off exponentially. Besides I do not think we are very far wrong in the formulæ we have so far got with respect to this absorption. Anyway this is all a matter of opinion.

---

* [Phil. Mag. **29** (1915) 158.]

I must admit that I think that statistical theories of radiation even with quanta are at present almost hopelessly wide of the mark. I even begin to doubt whether the application of Fourier's series as usually made has any real justification, or rather whether the interpretation put upon the results obtained is justified. But your wonderful success with quanta almost convinces one that there must be something in them, however artificial they may appear to us at present.

Of course I perfectly agree with you concerning the number of electrons concerned in metallic conduction as calculated from the optical properties. I had myself last November submitted a short note to the Phil. Mag.* pulling these calculations to pieces, except perhaps as the roughest of approximations, but I object to them mainly because no account was taken of the resonance electrons. Your objection is more fundamental than mine and I am fully in agreement with you respecting it. I had not come across it myself because I have mainly been concerned up to the present with the electron part of the theory and have only just begun to dabble with the optical equations. It is the optical side of these questions that I am mainly interested in and the various papers I have written on the ordinary electron theory have mostly resulted as secondary side-tracks from the main problem, so that no serious attempt at thoroughness is intended with any of them.

I return herewith the three pages of your M.S. I am very grateful to you for letting me see this paper of yours as it has helped me immensely. I should not have written some of my earlier papers which are now in proof** had I only seen your paper before; it would have saved me such a lot of trouble. I had often heard of it from various sources but we are really so much out of the way of things here in Sheffield that I never had the chance of seeing it.

Again with many thanks and best wishes

I remain
Yours very sincerely
G. H. Livens

---

* [Phil. Mag. **28** (1914) 756.]

** [The papers which appeared in Phil. Mag. **29** (1915) 425, 655; **30** (1915) 105, 112, 287, 434, 526, and 549, were all submitted at the time this letter was written. To one of these (Phil. Mag. **30** (1915) 287) Livens was able to add a note in the proof, mentioning Bohr's work.]

X. CORRESPONDENCE WITH O. W. RICHARDSON

(1915*)

* [See Introduction, sect. 8.]

Letter from Niels Bohr to O. W. Richardson (29 Sept 15)
[Draft in Mrs. Bohr's handwriting]

September 29, 1915   7 Victoria Grove
                                                     Withington, Manchester

Dear Prof. Richardson,

I have taken very great interest in reading your and Schottky's papers. I must say at once, that so far as I have been able to form an opinion I am afraid that I cannot agree with you in all your conclusions. While I agree that some of Schottky's calculations do not seem correct, I am not convinced that some of his arguments do not touch upon real difficulties in your calculations. It appears to me that reasons might be given in support of Schottky's opinion that the Thomson effect cannot have a simple connection with w^*. For, consider a piece of metal and let its surface or only part of it be covered with an electric double layer of some or other origin. Now it does not seem that we can expect that such a surface layer will have any effect on the Peltier and Thomson effects which depend only on the conditions in the interior of the metal. But on the other hand the layer may have a very considerable effect on w; in fact it seems to me that we might be able to give w quite arbitrary values and a quite arbitrary variation with temperature by means of suitable assumptions on the constitution of the double layer. It seems also to me that Schottky perhaps also might be right in his suggestion that the quantity w has not exactly the same meaning in the formulæ (4) and (5) of your paper**. While in (5) w seems to be the total energy to be transferred to the metal in order to liberate an electron and would seem to involve also the transition from bound electrons to free since the total number of the latter change with the temperature, it does not seem to be so in (4), where w is simply the difference in potential energy of a free electron inside and outside the metal surface. I do not think that Schottky's expression for σ

---

* [w is the heat of vaporization of an electron or the change in energy accompanying the transference of an electron from the metal to a surrounding enclosure.]

** [From the present context and the next letter from Richardson, it is evident that eqs. (4) and (5) are

$$w/T + k \ln p - (k\gamma/(\gamma-1)) \ln T + \varepsilon \int^{T} (\sigma/T)\mathrm{d}T = \text{constant} \qquad (4)$$

and

$$v = p/kT = A \exp \int (w/kT^2)\mathrm{d}T, \qquad (5)$$

respectively. Here w is the heat of vaporization of an electron, p is the pressure and v is the number of electrons per unit volume in an enclosure surrounding the metal kept at the absolute temperature T; ε is the charge of an electron, σ is the Thomson coefficient, k is the gas constant per molecule, and γ is the ratio of the specific heats of the electron gas at constant pressure and constant volume.]

[the Thomson coefficient] is correct as I shall come back to in a moment, bu I cannot quite understand your objection to his application of the energy prin ciple to the cycle. Independent of the fact that the cycle possesses an electro motive force or not it seems to me that the mechanical work done on the electron in the gas state must be equal to the heat produced, since the whole system afte the process is brought back to its original state. It appears to me that the reaso that Schottky's expressions as you point out, does not satisfy Lord Kelvin' thermodynamical conditions is in the first place due to the fact that the V, h operates with, has no definite meaning, in the sense that it is not a function o the nature and temperature of the metal. This is seen from the fact that we can get electromotive forces in closed circuits containing more than one metal as soon as there is differences in temperature. But even if this is taken into account by an actual calculation of $W_1 - W_2 - e(V_1 - V_2)$ from the kinetic theory, his expressions will in general not satisfy Lord Kelvin's relations because sufficient regard is not taken to the transfer of kinetic energy in the metal. In my disserta-tion which I enclose*, the value of $W_1 - W_2 - e(V_1 - V_2)$ is given by formula (46) on p. [346] or (49) on p. [348] (my φ or more accurately $\varphi + \lambda$, corresponds to your $V - \frac{1}{2}w$). You will see that it has not a quite simple relation to the expression for ρ [i.e., the Thomson coefficient previously denoted by σ, (54) p. [350]]. Although the quantity μ is not of great significance in the Peltier and Thomson effects where only differences or differential coefficients occur, it is not so in the formulæ (46) and (49) where the actual value of μ comes in. The transfer of kinetic energy by a flow of electrons or molecules seems a far more complicated question than most people realise. I don't know if you have ever considered f.inst. the transfer of energy by the molecules of an ordinary flow of gas. It is 5/3 as great as one would expect at first sight. The explanation is very simple, but it puzzled me very much when I wrote my dissertation. I have just mentioned it in the note on p. [352] in connection with J. J. Thomson's theory. On the same page there are in brackets (it means small letters in the Danish publication) a few remarks on the apparent independence between the Peltier and Thomson effects and the potential energy of the electrons in the metals.

I hope that you will not be too horrified by the English translation, which was made before I came to England. It is not my intention to trouble you in any way with reading all the old stuff, I thought only that you might be interested to see the calculations mentioned above, you can keep it as long as you like. I return Schottky's paper if you might like to look at it more. I have no use for it at present.

---

* [Bohr enclosed the English translation made in 1911. The formula and page numbers given have been changed to conform to the present translation found on pp. [291]–[395].]

I also enclose Planck's paper cited by Wood*, it has no hurry but when you have finished with it I should like to have it again as there are some other things in it which I should like to look upon again sometime.

I have tried to put everything just as it appears to me at the present moment. How I wish that you were still here, and we could discuss all the questions so much better. With the kindest regards from us both to you and Mrs. Richardson.

Yours very sincerely,
Niels Bohr

*Letter from O. W. Richardson to Niels Bohr (9 Oct 15)***

4 Cannon Place
Hampstead N.W.
9 Oct. 1915

Dear Bohr,

Many thanks for the parcel and especially for your letter and the English translation of your dissertation. The English may not be uniformly good but at any rate I can understand it which is more than I managed to do with the original Danish.

I was very glad to receive your criticisms of my remarks on Schottky's paper. I have carefully reconsidered all the points you raise—I think I had considered them all before but perhaps in my own mind I had not put some of them quite as strongly as you have—and I do not believe any of them are really valid. I am very anxious to convert you to my views and so I will consider all the various objections in the order in which you take them in your letter. I agree with Schottky in one point namely in appreciating the importance of settling the point under discussion, if possible.

In the first place I think you are wrong in supposing that my method of calculating would make the Thomson effect in a metal depend on an arbitrary variation in the temperature coefficient of the work at the surface such, for example, as would be caused by covering it with a double layer varying in some arbitrary way with the temperature. Suppose that by sticking on a double layer for which at temperature T the work is w_1 the pressure of the external electrons

---

* [A. B. Wood, Proc. Roy. Soc. **A91** (1915) 543. The paper cited is M. Planck, '*Die gegenwärtige Bedeutung der Quantenhypothese für die kinetische Gastheorie*' in *Vorträge über die kinetische Theorie der Materie und der Elektrizität*, Mathematische Vorlesungen an der Universität Göttingen (B. G. Teubner, Leipzig 1914); reprinted in Max Planck, *Physikalische Abhandlungen und Vorträge* (Friedr. Vieweg & Sohn, Braunschweig 1958), vol. **2**, p. 316–329.]

** [Since Bohr returned this letter to Richardson, it is not found in the Niels Bohr Archive. However, it has been kindly made available by the History of Science Collection, the University of Texas at Austin, where it is kept with other papers and manuscripts of Richardson.]

is changed at temperature T from p to P and the (contact) potential from V to V_1, the Thomson coefficient becoming say S_1 instead of S. Then a calculation along exactly the same lines as that to which Schottky objects gives the equation

$$\frac{w+w_1-\varepsilon(V-V_1)}{T}+k\log P-\frac{k\gamma}{\gamma-1}\log T+\varepsilon\int^T \frac{S_1}{T}\,\mathrm{d}T = 0 \qquad (1)$$

But the change from p to P being due to the double layer and the supposed accompanying alteration of V we have

$$\log P/p = \langle-\rangle[w_1-\varepsilon(V-V_1)]/kT, \qquad (2)$$

so that (1) reduces to

$$\frac{w}{T}+k\log p-\frac{k\gamma}{\gamma-1}\log T+\varepsilon\int^T \frac{S_1}{T}\,\mathrm{d}T = 0 \qquad (3)$$

But, considering the case where there is no double layer,

$$\frac{w}{T}+k\log p-\frac{k\gamma}{\gamma-1}\log T+\varepsilon\int^T \frac{S}{T}\,\mathrm{d}T = 0; \qquad (4)$$

so that $S = S_1$.

You may urge that this conclusion must be wrong because (4) can be derived without any special assumptions as to the way w is made up and so must be quite general and in combination with

$$v = \frac{p}{kT} = A\exp\int\frac{w}{kT^2}\,\mathrm{d}T \qquad (5)$$

leads at once to

$$\varepsilon S = k/(\gamma-1)-\partial w/\partial T \qquad (6)$$

quite generally; so that S will depend on w and S_1 will not be equal to S when w is changed. The fallacy here is in using the equation (5) in which it is not admissible to give w an arbitrary variation of the kind contemplated. This equation is derived by considering the expression for the change of entropy

$$dS = (1/T)(dU+p\,dv) \qquad (7)$$

of the system at T and p and only leads to (5) if w is the value of $\partial U/\partial v$ (reduced to one electron) which is natural to the system. If w is altered arbitrarily then the change in the entropy of the part of the system which causes the change in w has also to be included in the calculation.

This may be illustrated by supposing that the natural value of w is varied by the addition of a double layer represented by a condenser charged by a battery of E.M.F. $= V$. w will then be increased by an amount $= \varepsilon V$. If by moving a piston N^* electrons are given off by the metal the metal surface will become charged and change the strength of the double layer from the equilibrium value εV due to the battery. This will be rectified by the flow of an equal quantity of electricity through the battery until the equilibrium value V is restored. It is at once seen that the quantity of electricity displaced in the battery circuit is equal to the total amount pumped out of the metal. The change $(\partial U/\partial v)dv$ is thus seen to be made up of two parts (a) due to the escape of the electrons and (b) to the current in the battery circuit. Now

$$(a) = w + \varepsilon V = w + w_1$$

and

$$(b) = -\varepsilon V + q = -w_1 + q,$$

where q is the heat generated due to the chemicals consumed in the battery. By a theorem due to Kelvin**

$$-w_1 + q = -T\partial w_1/\partial T, \tag{8}$$

so that

$$\partial u/\partial v \text{ per electron} = w + w_1 - T\partial w_1/\partial T.$$

Thus if v_1 and p_1 are the values when w is increased by w_1 we get, instead of

$$v_1 = \frac{p_1}{kT} = A \exp \int \frac{w + w_1}{kT^2} \, dT \tag{10}$$

the equation

$$v_1 = \frac{p_1}{kT} = A e^{-w_1/kT} \exp \int \frac{w}{kT^2} \, dT. \tag{11}$$

This in combination with

$$\frac{w + w_1}{T} + k \log p_1 - \frac{k\gamma}{\gamma - 1} \log T + \varepsilon \int^T \frac{S_1}{T} \, dT = 0, \tag{12}$$

which is true generally, gives

$$\varepsilon S_1 = k/(\gamma - 1) - \partial w/\partial T (= \varepsilon S) \tag{13}$$

and not

---

* [Richardson has $N\varepsilon$ here. In some of the formulas he uses e rather than ε for the charge of an electron.]

** A particular case of Helmholtz's Free Energy Theorem. By using the general theorem of Helmholtz the result can be shown to hold generally so far as I see at present.

$$\varepsilon S_1 = \frac{k}{\gamma - 1} - \frac{\partial}{\partial T}(w + w_1),$$

as we might at first expect.

I am supposing throughout this that the arbitrary change in w is produced by some definite physical agency. The corresponding question when w is altered by a layer of matter of different composition leads to no difficulty as is shown by the considerations in my paper R. S. Proc. vol. 91, p. 524 (1915).

This brings me to the second point—the difference in meaning of w in (4) and (5) above. There is a difference due to the fact that the electrons are streaming along in the cycle contemplated in (4) whereas (5) refers to a displacement of a state of equilibrium. This difference brings in the effects which you first pointed out which arise from the different amounts of kinetic energy carried along by a moving stream of electrons according to the nature of the forces affecting the collisions. I did not want to go into this as I understood that Schottky and I were in agreement both with each other and with yourself about it and I have shown how it modifies the equation in Electron Theory of Matter, p. 450 et ff. To have gone into this would I thought have lengthened the paper unnecessarily.

Apart from this there is no difference whatever in the meaning of w in (4) and (5). In each case it is the energy necessary to liberate one electron under equilibrium conditions at temperatures T (the internal latent heat in fact). It has nothing whatever to do, in either case, with any suppositions about what is going on in the interior of the hot metal. Perhaps this identity is not made sufficiently clear in the proofs I have given. It is at once evident if in the proof for example in Electron Theory p. 448 we introduce auxiliary bodies A_1 and A_1' at temperatures T and T' respectively and evaporate the electrons from A_1 instead of A and condense them in A_1' instead of A' afterwards transferring them from A_1' to A' and from A to A_1. Since the bodies are now insulated when the evaporation or condensation occurs it is clear that w has the same meaning as in (5). (Except for the difference already alluded to which will still persist as there is a steady average flow in the direction indicated.) The only other difference is the work gained in the transference of charge from A_1' to A' at T' and from A to A_1 at T. If the total amount of charge sent round the cycle is E in n operations the work here in each operation is proportional to E^2/n^2 and for the whole n operations it is proportional to E^2/n and evidently vanishes in the limit when n is made indefinitely large.

Again I think I am right in maintaining that Schottky has omitted to include the work against the electromotive force of his circuit in applying the energy principle. I cannot agree with your objection that, because everything in the system after the process is brought to its original state, the *mechanical* work done must be equal to the heat produced. The electrical work has also to be included. For example if your argument were applied to the transference of electricity round a thermoelectric circuit with junctions at different temperatures, then everything is in the same state before and after the transference and, as there is no mechanical work done we should have $0 = \sum$ Peltier + Thomson Heats. But in reality the energy principle requires that instead of zero in the left hand side we have the work done against the electromotive force of the circuit. I think this is the only reason why Schottky's results do not satisfy Kelvin's relation: Even if he had taken into account the transfer of electricity in the metal (and kept to the same method of using the energy principle) he could not make his results satisfy Kelvin's relation because the terms depending on μ (in your notation) satisfy this relation independently of the other terms.

As regards the error in J. J. Thomson's calculation of the specific heat of electricity it occurred to me some time ago and I drew attention to it in Phil. Mag. vol. 23, p. 275. In conclusion I believe that my formulae

$$S = \frac{1}{\varepsilon}\left(\frac{k}{\gamma-1} + T\frac{\partial}{\partial T}(\mu_1 - \mu_0) - \frac{\partial\varphi}{\partial T}\right) \tag{14}$$

and

$$P = (1/\varepsilon)\{\varphi_2 - \varphi_1 - (\mu_2 - \mu_1)T + \varepsilon(V_2 - V_1)\} \tag{15}$$

for the Thomson & Peltier effects respectively in which the quantities μ have the same general meaning as in your dissertation and where φ is the value of w corresponding to a displacement under equilibrium conditions (as in equation (5)) agree with your expressions under very general conditions. I suspect that they agree absolutely but I have not had time yet to study your paper sufficiently to make absolutely certain.

I am returning the original paper and your letter for reference as you will probably have forgotten what opinions you did express by the time this reaches you.

With best regards both to yourself and Mrs. Bohr,

Yours very sincerely,
O. W. Richardson

P.S. I also enclose 2 numbers of the Ber. der D.P.G. & will send the others later.

Letter from Niels Bohr to O.W. Richardson (16 Oct 15)
[Draft in Mrs. Bohr's handwriting]

7 Victoria Grove M/C
October 16th 1915

Dear Prof. Richardson,

Thank you very much for your letter. I was most interested in it all, but I am afraid that I still cannot agree with you. I feel that the arguments in my former letter were very unclearly expressed but I cannot see that their validity is shaken by the considerations in your letter.

It still appears to me that the formula (6)*, deduced by comparison of (4) and (5) (for sake of reference I enclose your last letter as well as my former letter), cannot be correct, since it seems that we cannot expect such a simple relation between ρ [the Thomson coefficient, also denoted by σ or S] and w. I find it difficult, however, to discuss clearly the reason for this postulated incorrectness of formula (6); for I have not been able to understand the precise meaning of the formula (5); it appears to me that the fact, that any liberation of electrons leave⟨s⟩ the metal charged, while v in (5) refer⟨s⟩ to a state where the metal is uncharged, disturbs essentially the analogy with the evaporation of a liquid and makes the correctness of the simple application of the entropy principle doubtful. Quite apart from the question of the deduction and physical interpretation of the formula (5) it appears to me that a simple comparison of (5) with the formula

$$v = Ne\frac{-w}{kT}(x),$$

where N is the number of free electrons in the metal and w the difference in potential energy inside and outside the metal surface (for simplicity I use throughout the old picture with sharp distinction between surface and interior, and free and bound electrons), show⟨s⟩ the different meaning of w in the two formulæ. If N is constant, the comparison shows that w in (5) is equivalent with $w - T\,dw/dT$ in the simple formula (x), in which w (apart from the terms due to the transfer of kinetic energy) is equivalent with w in (4). This relation seems to be the same as you deduced in the case of the surface condenser and battery, but I do not understand your argument; if you think w is the same in (4) and (5), why is it not changed in (4) when you in (5) replace it by q? I do not also understand how it is possible to distinguish between the part of w due to arbitrary surface layers and that part which is natural for the metal.

---

* [Formula (6), obtained by eliminating p from eqs. (4) and (5), is

$$\varepsilon\sigma = k/(\gamma-1) - dw/dT. \tag{6}]$$

If N is not constant, but varies with the temperature, the comparison between the two formulæ for v shows that w in (5) further must contain terms which have connection with the mechanism of liberation of bound electrons. The difference in this respect between w in (4) and (5) seems also to follow directly from a consideration of the different processes corresponding to the two formulæ. In (5), whatever may be the precise deduction of it, we are concerned with the whole process of liberation, external as well as internal; but in (4) we consider simply the transition of free electrons through the surface film, since the process may be so conducted, that the number of free electrons in the metal is not altered at any point. It does not seem to me that this argument is impaired by considering a process, as that mentioned in your last letter, in which all evaporation and condensation takes place on insulated pieces of metal, for in this case the change in w due to the momentary liberation of bound electrons is compensated by the later binding of free electrons at the same place. This may also be expressed, that in this case the heat developed due to the flow of electrons is not only the Thomson effect, but when the current start⟨s⟩ we get (besides the correction due to transfer of kinetic energy) terms corresponding to liberation and binding of free electrons.

With my remark as to the application of the principle of conservation of energy to the process, I meant simply that if no exterior and interior alterations take place the sum of the mechanical work done and the heat developed must be zero. For a closed thermoelectric circuit, where no external work is done by the current, such as mentioned in your letter and which I also had had in mind, the Joule heat will form part of the heat developed; but in an open circuit there is no Joule heat, since the electrons are displaced by reversible processes, and it appears to me that this is just the advantage and the basis for the thermionic considerations.

I think still that the two arguments mentioned in my former letter are the reasons why Schottky's expressions for the Peltier and Thomson effects do not satisfy Kelvin's relation. The first is of greater importance, but the neglect of the transfer of kinetic energy introduces a serious error in the Thomson effect calculated in Schottky's way even if μ is constant, and in the general case where μ varies a closer calculation shows that Schottky's expressions will not satisfy the Kelvin relation, but that this relation is only satisfied by the complete expressions consisting of the sum of his terms and those due to the transfer of kinetic energy. I have introduced an analogous remark in my dissertation in a footnote discussing some calculations of Kunz*. By the short reference in my letter to J. J. Thomson's error I was well aware that you know it and have**

---

* [See footnote 2 to the translation of Bohr's dissertation p. [354].]
** [In the draft the word is "has".]

corrected it long time ago; I mentioned it only as an illustration of that curious fact that the molecular transfer of energy in an ordinary flow of gas in a tube is so much greater than everybody would anticipate at first sight. I have now gone through the arguments in the order that you take them in your letter. I have tried to tell what I thought about the various points. I am so very anxious to get to know what you think about it all.

With kindest regards,
Yours very sincerely
N. Bohr

P.S.

Reading all the different letters over again before I enclose them I felt that some of my remarks might give rise to misunderstanding. I am afraid that I have expressed myself very ambiguously as to the deduction of formula (5). What I meant to say is only that I could not make quite sure of the mechanism you had in mind, and that I therefore thought it better to keep to a purely formal comparison of the formulæ, since as far as I could see this leads to definite results. Similarly with regard to your illustration with the condenser layer charged by a battery, also here I could not quite see how you thought the mechanism constituted, especially with regard to the application of the same picture to the process considered in (4).

PART III

SELECTED
FAMILY CORRESPONDENCE
1909-1916

INTRODUCTION

by

J. RUD NIELSEN

In 1909, in order to find time to review for his examination, Niels Bohr spent some months in a parsonage in Vissenbjerg on the island of Funen, and the following year he went to the same place to complete the writing of his dissertation. In the autumn of 1909, and again in 1912, his brother, Harald, went to Göttingen and other places in Germany. From the autumn of 1911 to the summer of 1912 Niels Bohr was in Cambridge and Manchester, and from 1914 to 1916 he was again in Manchester. During these periods of separation from his brother and mother, Bohr wrote and received frequent letters.

Since certain of these letters throw valuable light on Niels Bohr's early career, and on his personality and attachment to his family and, in particular, to his brother, a selection from them is reproduced here together with English translations, where required. In the English version is given a number of explanatory notes concerning individuals or specifically Danish customs referred to in the letters. With a few exceptions, Danish place names are not translated, and their locations are not indicated.

The letters to Harald Bohr, and some of the letters to their mother, have kindly been made available by Mrs. Ulla Bohr, the widow of Harald Bohr, while Mrs. Margrethe Bohr has kindly given permission for the use of the letters from Harald to Niels Bohr. Most of these letters were evidently written in a hurry with scant regard for form and punctuation. They are in longhand, and, to get the maximum number of words on a card or in a letter, the handwriting is usually very small and in many places barely legible.

In deference to the wish expressed by Mrs. Margrethe Bohr, letters and passages pertaining to personal matters of no direct scientific relevance have been omitted from the present selection. The location of omitted passages is indicated by the symbol [···], irrespective of their length. A complete inventory of the available family correspondence involving Niels Bohr is appended.

ORIGINAL TEXTS

LETTER FROM NIELS TO HARALD (12 Mar. 09)

[Card]

⟨Bred⟩ 12–3–09

Kære Harald!

Jeg havde en udmærket Rejse. Jeg sidder nu efter at have spist til Aften i min dejlige Arbejdsstue. Jeg var lidt ked af, i Gaar Aftes, at mine Planer der vare udarbejdede i mange Detajler vare blevne krydsede; men jeg er paa Turen kommen til den Overbevisning at jeg ikke kunde have faaet en bedre Indledning til min Examenslæsning en⟨d⟩ en Gennemarbejdning af Christiansens Bog. Jeg glæder mig meget dertil. Sig til Far at jeg er saa glad ved at have faaet Lov til at hjælpe ham lidt og at det har moret mig meget. Sig til Mor at jeg var meget glad og skamfuld ved at se hvor storartet mit Tøj var pakket og hvor nøje alt var husket. Sig til Slut til Jenny at hendes Figner smager storartet.

Med mange Hilsner til alle
fra Niels

LETTER FROM NIELS TO HARALD (17 Mar. 09)

[Card]

⟨Bred⟩ 17–3–09

Kære Harald!

Tusind Tak for Dit Brev; jeg glæder mig ogsaa uhyre til at vi engang skal faa rigtigt meget med hinanden at gøre, og jeg haaber, at vi begge to skal faa stor

TRANSLATIONS

LETTER FROM NIELS TO HARALD (12 Mar. 09)

[Card]

⟨Bred[1]⟩ 12–3–09

Dear Harald:

I had a splendid trip. After having eaten supper, I am now sitting in my nice study. I was a little sorry last night that my plans, which were worked out in much detail, had been thwarted; but on the trip I have reached the conviction that I could not have had a better introduction to my reading for the examination than a thorough study of Christiansen's book. I am looking forward to it very much. Tell Father that I was very glad to be allowed to help him a little and that it interested me very much. Tell Mother that I was happy and ashamed to see how splendidly my clothes were packed, and how accurately she had remembered everything. Finally, tell Jenny[2] that her figs taste fine.

With much love to all
from Niels

LETTER FROM NIELS TO HARALD (17 Mar. 09)

[Card]

⟨Bred⟩ 17–3–09

Dear Harald:

A thousand thanks for your letter. I am also looking forward very much to a time when we can have really much to do with each other, and I hope both of us

[1] [A village on the island of Funen. Niels Bohr spent periods in the parsonage in Vissenbjerg. The post office was at Bred.]
[2] [The older sister of Bohr.]

Fornøjelse ud deraf. Du ved slet ikke hvor meget jeg har lært af dig, navnlig har Din Gennemgang af Jordan paa en for mig ganske forbløffende Maade lært mig, at det er en alt for omstændelig og aldeles ikke nødvendig (for alle Mennesker, nemlig f Eks den store Mathematiker 🐻) Fremgangsmaade at betragte alle Spørgsmaal som dybe Taager, der kan behandles med en større eller mindre Grad af "Takt" (populært kaldet uomstødelig Logik); men at der i hvert Fald i Mathematiken findes en nogenlunde fast Bund, paa hvilken man "maaske" dels ved eget Arbejde maaske dels ved en meget kærkommen Hjælp fra føromtalte 🐻 's Side, kan komme til at bygge.

De bedste Hilsner til hele Familien fra Din Broder N der glæder sig til snart at se Dig igen.

LETTER FROM NIELS TO HARALD (27 Mar. 09)

[Card]

⟨Bred, 27–3–09⟩

Kære Harald.

I maa endelig ikke blive forskrækket ved at faa hele 3 Kort paa engang, jeg skal nok love ikke at gøre mig skyldig i saadan "Uartighed" mere, men da jeg pludselig til min Forskrækkelse opdagede, at det andet Kort til Dig kun indeholdt "Vrøvl", synes jeg at det var nødvendigt ogsaa at bringe nogle faktiske Oplysninger om min Tilværelse her over i Præstegaarden. Jeg har det da i alle Maader ganske storartet, spiser og sover til Moders Beroligelse (undskyld jeg vrøvler, jeg mente nemlig min egen) forfærdelig meget; men faar ogsaa bestilt ikke saa lidt: Jeg er nu færdig med I P's Dynamik og har læst det meste af, hvad der staar i Abraham om Vektorregning (meget interessant) og er ogsaa begyndt paa Christiansens Manuskript, det morer mig meget og den dynamiske Indledning indeholder mange morsomme Ting, men den gør rigtignok ikke fjerneste Forsøg paa at efterkomme de Fordringer som Du (og for Resten ogsaa jeg selv) vilde stille til en ordentlig begrundet Bevægelseslære.

de mange Hilsner, som jeg sender til alle i dette og det andet Kort skulde gerne symbolsk multipliceres sammen, saa at de kommer i „2den Potens."

Din Niels

Bed Moder skrive om Edgar har besørget Ridetimerne da jeg i saa Tilfælde vilde sende ham et Prospektkort for at takke ham.

will have much pleasure of that. You have no idea how much I have learnt from you. Especially, your going through Jordan has taught me in a quite astounding manner that it is too cumbersome, and not at all a necessary procedure (for all people, in particular for the great Mathematician[3] 🕴), to consider all questions as deep fogs which can be treated with greater or lesser degree of "tact" (popularly called irrefutable logic), but that, at least in mathematics, there exists a more or less solid ground on which, "perhaps" by own efforts, perhaps through some very much appreciated help from the afore-mentioned mathematician 🕴 one can some day build.

Best regards to the whole family from your brother N. who is looking forward to seeing you again soon.

LETTER FROM NIELS TO HARALD (27 Mar. 09)

[Card]

⟨Bred, 27-3-09⟩

Dear Harald:

You must not be scared by receiving as many as three cards at one time; I promise not to be guilty of such "naughtiness" again; but when I suddenly discovered to my horror that the second card contained only nonsense, I felt that it was necessary also to give some factual information about my existense over here in the parsonage. I am getting along splendidly in all respects; to the relief of Mother (excuse my nonsense, I meant to my own), I eat and sleep terribly much; but I also get a great deal done. I am now through with I.P.'s dynamics and have read most of what is found in Abraham about vector calculus (very interesting) and have also started on Christiansen's manuscript, I enjoy it a great deal; and the dynamical introduction contains many interesting things, but it does not make the slightest attempt to meet the demands that you (and, for that matter, also I) would make to a really well-founded theory of motion.

The many greetings that I send to all in this and the other cards should, if possible, be multiplied symbolically with themselves, so as to have the power of two.

Your Niels

Ask Mother to write if Edgar[4] has taken care of the riding lessons, as I in that case would send him a picture post card to thank him.

[3] [i.e., Harald.]

[4] [Edgar Rubin, a schoolmate who in 1922 became professor of psychology at the University of Copenhagen.]

LETTER FROM NIELS TO HARALD (20 Apr. 09)

[Letter]

⟨Bred⟩ 20–4–09

Kære Harald!

Rigtig mange Gange til Lykke. Det er jo denne Gang ikke en almindelig Fødselsdag; men en Begyndelse til noget helt nyt. Jeg glæder mig saa meget paa Dine Vegne
til at Du maa faa det rigtigt storartet i Göttingen baade med Hensyn ⟨til⟩ Udvikling-
en af Din mathematiske Personlighed og med Hensyn til Din personlige Behage-
lighed. Jeg sender Dig hermed (foruden hvad Mor er saa elskelig at sende Dig i mit
Navn) Kierkegaard: Stadier paa Livets Vej. Det er det eneste jeg har at sende; men
jeg tror alligevel ikke at jeg let kunde finde noget bedre. Jeg har i hvert Tilfælde
haft overordentlig stor Fornøjelse af at læse den, jeg synes ligefrem, at det er noget
det dejligste, jeg nogensinde har læst. Nu glæder jeg mig til engang, at høre Din
Mening om den Jeg har det ganske storartet herovre i Vissenbjerg. Her er ganske
dejligt nu da det er blevet rigtigt Foraar, de første Anemoner er allerede komne.
Som Du jo ved har jeg havt den første Korrektur. Afhandlingen kom alligevel i
Transactions, den var nydeligt trykt og saa omhyggeligt efterset (der var ikke et
Tal trykt forkert), at det var let for mig at gøre det færdig. Weber var saa elskværdig at sende mig en Afskrift af det "Abstract", der havde staaet i Proceedings;
det viste sig at være mine "Conclusions", saa det var jo meget rart. Med Læsningen
gaar det udmærket og jeg begynder at glæde mig til Examenen, og navnlig glæder
jeg mig til de sidste Maaneder af Efteraaret, naar jeg er færdig, vi skal da have det
saa rart sammen, jeg gaar her i Enerum og tænker paa saa mange Ting, som jeg
glæder mig til at snakke med Dig om. Dette er ikke blevet noget rigtigt Brev, og
langt fra som jeg gerne vilde have det; men Sagen er at jeg sidder og jager med at
blive færdig for at faa det sendt af i rette Tid, jeg fik nemlig begyndt saa sent, da
jeg gerne vilde læse Stadierne færdig førend jeg sendte dem af; Du faar da trøste
Dig med at jeg en af de første Dage skal sende et nyt Brev; jeg vil da denne Gang
ende med at ønske Dig rigtig mange Gange til Lykke.

Din Niels

P.S. Hele Familien Møllgaard ønsker Magisteren til Lykke

LETTER FROM NIELS TO HARALD (20 Apr. 09)

[Letter]

⟨Bred⟩ 20-4-09

Dear Harald:

Happy birthday! This time it is not an ordinary birthday, but a beginning to something quite new. I am so happy on your behalf in hoping that things will go splendidly for you in Göttingen, both as regards the development of your mathematical personality and with respect to your personal enjoyment. I am sending you herewith (in addition to what Mother is so sweet to send you in my name) Kierkegaard's "Stages on Life's Road". That is the only thing I have to send; nevertheless, I don't think I could easily find anything better. In any case, I have enjoyed reading it very much, in fact, I think it is something of the finest I have ever read. Now I am looking forward to hear your opinion of it. I am getting along splendidly over here in Vissenbjerg[5]. Now that real spring has come it is very lovely here; the first anemones are already out. As you know, I have had the first proof. The paper[6] is to appear in Transactions after all; it was nicely printed and so carefully checked (not a single number was misprinted) that it was easy for me to finish the job. Weber[7] was so kind as to send me a copy of the "Abstract" that had appeared in the Proceedings; it turned out to be my "Conclusions", so that was very nice. It is going very well with the reviewing; I am beginning to look forward to the examination, and I am especially looking forward to the last months of autumn when I shall be through; we shall then have such a nice time together; I go here in solitude and think of so many things that I long to talk to you about. This has not turned out to be a real letter, and not at all one as I would have liked to have it, but the reason is that I sit and hurry to get it mailed in time; in fact, I started so late because I wanted to finish reading the "Stages" before sending them. So you must take comfort in the thought that I shall send you a new letter one of the first days; this time I close by wishing you many happy returns.

Your Niels

P.S. The entire Møllgaard[8] family congratulates the Magister[9].

[5] [Niels Bohr was staying in the parsonage at Vissenbjerg. See note 1.]

[6] [This refers to the first Royal Society paper dealing with the surface tension of water which appeared in *Phil. Trans.* [A] **209** (1909) 281–317.]

[7] [S. Th. Holst-Weber; Danish physicist a year younger than Bohr; has studied the heat conductivity and other properties of gases; since 1912 he has lived in Holland where he has been director of various industrial concerns; has also served as Danish consul.]

[8] [Pastor Møllgaard, in whose parsonage in Vissenbjerg Bohr found time to study, was the father of an assistant of Bohr's father.]

[9] [i.e., Harald.]

LETTER FROM NIELS TO HARALD (26 Apr. 09)

[Card]

⟨Bred⟩ 26-4-09

Kære Harald,

Tusind Tak for Dit lange Brev, det er rigtignok forfærdelig morsomt at høre fra Dig. Mor er saa sød at ⟨sende⟩ Dine Breve til mig, og da jeg ved, at det er ikke saa lidt Arbejde, naar man har meget at bestille at skrive Breve, venter jeg derfor ikke Brev fra Dig, undtagen Du skulde føle Trang til at udlade Dig paa Fagets Vegne, hvad Du jo nok ved er *yderst* velkommen. Jeg er i Øjeblikket vildt begejstret for H. A. Lorentz's (Leiden) Elektrontheori. Naar Du engang har læst "Stadierne", hvad Du endelig ikke maa skynde Dig med, skal Du høre lidt fra mig, jeg har nemlig skrevet nogle faa Bemærkninger op derom (ikke enig med K); men jeg tænker ikke paa at være saa banal ved mit sølle Sludder at vilde søge at forstyrre Indtrykket af saa smuk en Bog. Det gaar mig storartet, og jeg gaar og glæder mig saa usigeligt til den dejlige Tid, jeg skal tilbringe i København, naar jeg er færdig med min Examen, inden jeg rejser udenlands.

Tusind Hilsner fra mig og hele Familien Møllgaard.

Din Niels

LETTER FROM NIELS TO HARALD (4 May 09)

[Card]

⟨Bred⟩ 4-5-09

Kære Harald!

Mange Tak for Dit indholdsrige Brev og for Afhandlingen. Jeg har studeret begge Dele meget grundigt. Skønt Afhandlingen jo ikke var mig saa ganske ube-kendt, morer det mig dog hver Gang, jeg læser det smukke Bevis. Jeg har det stadig ganske udmærket herovre. Jeg har nu ogsaa faaet læst Thermodynamiken (jeg er nu skrækkelig dygtig i Phaseregel, Dissociation og hele den øvrige fysisk-ke-miske Reguladetri), saa jeg tænker nok at jeg skal faa læst alt det opgivne, før jeg begynder paa min Examen. Her er ganske dejligt og lige saa varmt i disse Dage, som om det var midt om Sommeren. Jeg tager herfra om en Uge og skal saa be-gynde paa den store Opgave. Jeg er meget spændt paa, hvorledes den bliver.

Med mange Hilsener fra

Din Niels

LETTER FROM NIELS TO HARALD (26 Apr. 09)

[Card]

⟨Bred⟩ 26–4–09

Dear Harald:

Thanks a million for your long letter; it is very interesting indeed to hear from you. Mother is so sweet as to send your letters to me, and since I know that, when one has so much to do, it is no small task to write letters, I do not expect letters from you, unless you should feel the need to relieve yourself on behalf of the profession, to which you know you are *extremely* welcome. At the moment I am wildly enthusiastic about Lorentz' (Leiden) electron theory. When you some day have read the "Stages", what you by no means must hurry with, you shall hear a little from me; for, I have written a few remarks about it (not in agreement with K.); but I do not intend to be so trite with my poor nonsense as to spoil the impression of so beautiful a book. I am getting along splendidly and I am looking forward more than I can say to the wonderful time I shall have in Copenhagen when I am through with my examination, and before I go abroad.

Love and kindest regards from myself and from the Møllgaard family.

Your Niels

LETTER FROM NIELS TO HARALD (4 May 09)

[Card]

⟨Bred⟩ 4–5–09

Dear Harald:

Many thanks for your weighty letter and for the paper. Although the paper wasn't entirely unfamiliar, I enjoy it every time I read the beautiful proof. I am still getting along splendidly over here. I have now also managed to read thermodynamics (I am awfully clever about phase rule, dissociation and the rest of the physical-chemical rule of three), so I believe that I shall manage to read all the listed subjects [10] before I begin my examination. It is very pleasant here and just as warm these days as if it were mid-summer. I shall leave here in a week, and then I shall start on the big problem. I am very anxious to learn how it will be.

Much love from

Your Niels

[10] [Before taking the examinations for the master's degree, Danish students submitted a comprehensive list of the books and subjects they had studied during their entire six years, or so, at the university.]

LETTER FROM NIELS TO HARALD (15 May 09)

[Card]

⟨Copenhagen⟩ 15–5–09

Kære Harald!

Tusind Tak for Dine Breve. Det var forfærdelig morsomt at høre, at Du nu fuldstændig har løst hele Summabilitets Spørgsmaalet for de Dirichletske Række. Mange Gange til Lykke; det er rigtignok en imponerende Præstation; naar man tænker paa det hele. Far siger, at Du godt maa ønske Nørlund til Lykke med Guldmedaljen Nr. 2. Jeg var i Gaar oppe hos Zeuthen, der havde gjordt Vrøvl over mine Opgivelser i Mathematik (der manglede lidt Ligninger Theori), han var overordentlig elskværdig og spurgte uhyre interesseret til Dig. Dersom Du en Dag havde Tid, vilde det vist glæde ham meget at faa et Brev fra Dig. Jeg var derefter oppe hos Christiansen for at høre om der var noget i Vejen for, at jeg kan begynde Examen paa Mandag. Han var ogsaa forfærdeligt elskværdig og sagde at naar jeg havde hentet den store Opgave kunde jeg komme op til ham for at tale lidt om Omfanget. I Gaar var Muffen og jeg i Theatret og saa Kameliadamen, Anna Larsen spillede ganske storartet. Nu Farvel og mange Hilsner, Du skal nok høre fra mig, naar jeg har faaet Opgaven.

Din Niels

LETTER FROM NIELS TO HARALD (9 June 09)

[Card]

⟨Copenhagen⟩ 9–6–09

Kære Harald!

Jeg er meget skamfuld over ikke at have skrevet til Dig saa længe. Jeg haaber imidlertid at Du skyder Skylden paa min daarlige Tid og jeg ved jo ogsaa at Muf hænger i for to. I Dag da jeg sender Dig min Afhandling maa jeg dog i det mindste sige, at jeg umulig vilde kunde have haft Tid til at takke Dig, som jeg burde for al Din store Hjælp og Rarhed imod mig under Arbejdet paa den. Med Hensyn til min Examensopgave gaar det nogenlunde men jeg maa hænge svært i for ikke at komme til at jage til Slut; naar den er færdig glæder jeg mig til rigtigt at fortælle Dig om den.

Din Niels

LETTER FROM NIELS TO HARALD (15 May 09)

[Card]

⟨Copenhagen⟩ 15-5-09

Dear Harald:

Thanks a million for your letters. It was awfully nice to hear that you have solved the entire problem of the summability of the Dirichlet series. Many congratulations; it is certainly an impressive accomplishment when one considers the whole thing. Father wants me to tell you that you may congratulate Nørlund[11] for his second gold medal. Yesterday I went to Zeuthen[12] who had fussed about what I had listed in mathematics (there was a little lacking on the theory of equations); he was extremely nice and asked with great interest about you. If you should have time some day, it would, I believe, please him much to get a letter from you. Afterwards I visited Christiansen to learn if there would be any objection to my beginning the examination next Monday. He was also extremely kind and said that when I have received the big problem I could come to him and talk a little about the scope. Yesterday Mother[13] and I were at the theater and saw the "Lady of the Camellia". Anna Larsen acted brilliantly. Now goodbye and much love. You shall surely hear from me when I have received the problem.

Your Niels

LETTER FROM NIELS TO HARALD (9 June 09)

[Card]

⟨Copenhagen⟩ 9-6-09

Dear Harald:

I am very ashamed that I haven't written to you for so long. However, I hope that you blame my lack of time and know also that Mother works for two. Today when I send you my paper I must at least tell you that it would not have been possible to take time to thank you, as I ought to, for your great help and kindness towards me while I worked on it. With regard to the examination problem, it is going tolerably, but I must work hard in order not to have to hurry toward the end; I am looking forward to really tell you about it when it is completed.

Your Niels

[11] [Niels Erik Nørlund, b. 1885, Danish mathematician, professor at the University of Lund (Sweden) 1912-1922, at the University of Copenhagen since 1922. On Aug. 1, 1912, Bohr married his sister, Margrethe Nørlund.]

[12] [H. G. Zeuthen, professor of mathematics at the University of Copenhagen.]

[13] [In the Bohr family the mother was affectionately called Muffen, Muf or Mof. In this and following letters she is often referred to in this manner.]

LETTER FROM NIELS TO HARALD (1 July 09)

[Letter]

⟨Copenhagen⟩ 1–7–09

Kære Harald!

Tusind Tak for alle Dine Kort. Nu er jeg da heldigvis færdig med Skriveriet, det er forfærdeligt rart selv om jeg ikke som en vis Magister kan sige at jeg er fuldt tilfreds med Resultatet. Opgaven var nemlig saa stor i Omfang, at jeg, med min letløbende Pen, maatte lade mig nøje med at faa nogle ganske faa Brudstykker med. Jeg haaber dog at Censorerne vil lade den slippe igennem, da jeg tror at jeg har faaet et Par Smaating med som ikke staar andre Steder. Disse omtalte Smaating var dog for største Delen af negativ Art (Du ved jo nok at jeg har den slemme Vane at tro at finde Fejl hos andre). Af mere positivt, tror jeg at have givet en Antydning af Grunden til det maaske for Dig mindre bekendte Faktum, at Legeringerne ikke leder Elektriciteten saa godt som de rene Metaller, hvoraf de er sammensat. Jeg er nu meget spændt paa at høre, hvad Christiansen vil sige til det hele; jeg skal op og tale med ham i Morgen og skal nok lade Dig vide, hvorledes det spænder af. [...]

LETTER FROM NIELS TO HARALD (4 July 09)

[Letter]

⟨Copenhagen⟩ 4–7–09 Bredgade 62

Kære Harald!

[...] Jeg er meget skamfuld over det Grissebrev, som jeg sendte forleden, at Du har maattet ventet saa længe paa Fortsættelsen, er imidlertid ikke min Skyld men Prof. Christiansens; sidstnævnte har nemlig til sin øvrige Vanskeligheder føjet den, at man ikke engang i Ordets bogstavelige Forstand kan faa fat i ham. Da jeg saaledes kom op til ham i Forgaars Aftes, var han rejst bort, og da jeg søgte ham i Morges, var han endnu ikke kommen; det lykkedes mig imidlertid at træffe ham ved 4 Tiden, lige efter at han var kommen hjem, og lige før han skulde ud igen til Middagsselskab; det var ogsaa paa høje Tid, da jeg allerede rejser fra Byen i Aften sammen med Edgar. Jeg fik naturligvis ikke meget, eller rettere sagt slet intet, ud af Christiansen, men han var overordentlig venlig og jeg tænker ikke at han vil gøre Vanskeligheder. Hvad selve Afhandlingen angaar fik Du kun saa kort Besked sidst, det er heller ikke saa let at forklare i et Brev, men jeg glæder mig forfærdeligt til Du kommer hjem igen og vi kan snakke (blandt andet) derom, jeg tror nok, at jeg har faaet lidt ud deraf, og jeg tænker at skrive lidt derom engang i Efteraaret. [...]

LETTER FROM NIELS TO HARALD (1 July 09)

[Letter]

⟨Copenhagen⟩ 1–7–09

Dear Harald:

Thanks a million for all your cards. Now I have fortunately finished all the writing; that is awfully nice, even though I cannot say as a certain magister[14] that I am fully satisfied with the result. In fact the problem was so very broad that I, with my fluent pen, had to be content with treating only a few fragments of it. But I hope that the examiners will let it pass; for, I think I have included a couple of small items that are not dealt with elsewhere. However, these things are mostly of a negative nature (you know I have the bad habit of believing that I can find mistakes made by others). On the more positive side, I believe that I have given a hint of a reason for the fact, perhaps less well known to you, that alloys do not conduct electricity as well as the pure metals of which they are composed. I am now anxious to learn what Christiansen will say to the whole thing; I shall speak to him tomorrow, and I shall surely let you know how it will turn out.   [...]

LETTER FROM NIELS TO HARALD (4 July 09)

[Letter]

⟨Copenhagen⟩ 4–7–09 Bredgade 62

Dear Harald:

[...]   I am very ashamed of the lousy letter I sent the other day; however, that you have had to wait so long for the continuation is not my fault but Prof. Christiansen's. In fact, the latter has added to his other difficulties that it is literally impossible to get hold of him. Thus, when I went to him the night before last, he had gone away on a trip, and when I looked for him this morning he had not yet returned; however, I succeeded in catching him around four o'clock just after he had returned and just before he was to go out again to a dinner party; it was the last moment, since I leave town tonight with Edgar. Of course, I got very little, or actually nothing at all, out of Christiansen, but he was extremely nice and I don't think that he will make difficulties. With regard to the paper itself, you got only such brief information in the last letter; however, it is not so easy to explain it in a letter, but I am looking forward awfully to your return and to talk with you about it (among other things); I believe that I got something out of it, and I am thinking of writing a little about it next autumn.   [...]

[14] [i.e., Harald.]

[507]

LETTER FROM NIELS TO HARALD (8 Sept. 09)

[Card]

Nærum 8–9–09

Kære Harald!

Jeg haaber ikke, at Du bliver ked af, at jeg endnu engang trætter Dig med den samme Sludder.

Det drejer sig om Størrelsen

$$Q_r = \overset{\text{I}}{\left(\frac{o_1^2 + o_2^2 + \ldots o_r^2}{r}\right)} - \overset{\text{II}}{\left(\frac{o_1 + o_2 + \ldots o_r}{r}\right)^2}$$

hvor o'erne er tilfældigt udtagne, kun knyttede sammen ved en eller anden Fejllov, deres Kvadrater har altsaa ogsaa en bestemt Fejllov. Det drejer sig nu dels om Middeltallet, for uendelig Gentagelse, af Q_r (naar o'erne varierer), dels om dette Middeltals Variation naar r forandrer sig.

Middeltal for *uendelig Gentagelse* betegnes ved λ. Jeg synes nu at vi faar

$$\lambda(Q_r) = \overset{\text{I}}{\lambda(o_n \cdot o_n)} - \overset{\text{II}}{\left(\frac{1}{r} \cdot \lambda(o_n \cdot o_n) + \frac{r-1}{r} \cdot \lambda(o_n \cdot o_m)\right)},$$

hvor $\lambda(o_n \cdot o_m)$ ikke er lig $\lambda(o_n \cdot o_n) = \lambda(o^2)$. Det ses heraf at I forbliver uforandret, naar r varierer; men ikke II. Resultatet bliver

$$\lambda(Q_r) = \frac{r-1}{r}\left(\lambda(o_n \cdot o_n) - \lambda(o_m \cdot o_n)\right),$$

hvilket umiddelbart giver Udtryk for Thieles Resultat $\lambda(Q_r) = \frac{r-1}{r}\lim_{r=\infty}(\lambda(Q_r))$. Thieles eget Bevis er iøvrigt fuldkommen rigtig idet det er fuldstændig tilladeligt ved Beregningen af $\lambda((o_m - o_n)^2)$ at se bort fra de forholdsvis uendeligt sjældne Tilfælde, hvor $o_m = o_n$. Ved Din Opringning (Tak!), har Du jo gjort Afsendelsen af dette dengang næsten færdigskrevne Kort noget unødvendig, men jeg sender det dog med en Hilsen. Haaber nu at Du og Mor morer jer, og at Mof synes bedre om I.P.

LETTER FROM NIELS TO HARALD (8 Sept. 09)

[Card]

Nærum 8-9-09

Dear Harald:

I hope you won't mind that I bother you once more with the same nonsense. It concerns the quantity

$$Q_r = \overset{\text{I}}{\left(\frac{o_1^2 + o_2^2 + \ldots o_r^2}{r}\right)} - \overset{\text{II}}{\left(\frac{o_1 + o_2 + \ldots o_r}{r}\right)^2}$$

where the o's are taken at random, only related to each other by some law of errors; thus, their square is also subject to a definite law of errors. We are now concerned, for one thing, with the average, for infinite repetition, of Q_r (when the o's vary) and, also, with the variation of this average when r changes. The average for infinite repetition we shall call λ. Now, I believe that we get

$$\lambda(Q_r) = \overset{\text{I}}{\lambda(o_n \cdot o_n)} - \overset{\text{II}}{\left(\frac{1}{r}\lambda(o_n \cdot o_n) + \frac{r-1}{r}\lambda(o_n \cdot o_m)\right)},$$

where $\lambda(o_n \cdot o_m)$ is not equal to $\lambda(o_n \cdot o_n) = \lambda(o^2)$. From this it is seen that I remains unaltered when r varies, but not II. The result becomes

$$\lambda(Q_r) = \frac{r-1}{r}\left(\lambda(o_n \cdot o_n) - \lambda(o_m \cdot o_n)\right),$$

which immediately expresses Thiele's[15] result $\lambda(Q_r) = \frac{r-1}{r}\lim_{r=\infty}(\lambda(Q_r))$. By the way, Thiele's own proof is perfectly correct, since it is perfectly justified in the calculation of $\lambda((o_m - o_n)^2)$ to neglect the comparatively rare cases in which $o_m = c_n$. By your telephone call (thanks!) you have made the sending of this, then nearly completed, card somewhat unnecessary; however, I mail it nevertheless with my love. I hope you and Mother have a good time and that Mof likes I.P. better.

[15] [Th. N. Thiele, professor of astronomy at the University of Copenhagen.]

LETTER FROM NIELS TO HARALD (20 Dec. 09)

[Letter]

⟨Norway, 20–12–09⟩

[...] Jeg kan lige sige Dig, at jeg er bleven klar over at den Undersøgelse med mine Elektroner, som jeg har talt om, let kan føres igennem, hvormeget, der kommer ud deraf er imidlertid en ganske anden Sag. Fordi jeg siger dette, maa Du ikke straks blive bange for, at jeg gaar med saadanne Ting i Hovedet, jeg driver som sagt hele Dagen, og det var kun noget, som pludseligt faldt mig ind, og som jeg fuldstændig lader ligge. Nu Farvel for denne Gang med mange Hilsner til alle fra

Niels.

P.S. [...]

LETTER FROM NIELS TO HARALD (25 June 10)

[Card]

⟨Copenhagen⟩ 25–6–10

Kære Harald!

Mange gange til Lykke med Docenturen; selv om det jo ikke kommer uventet, er det dog forfærdeligt rart, at det nu er helt i Orden; jeg er umaadelig stolt paa Familieskabets Vegne. Tak for Dit Kort fra Göttingen. Det gaar mig selv meget godt, jeg har holdt helt fri nogle Dage, og skal nu til at tage rigtig fat igen. Far var saa rar, at foreslaa mig at holde fri en Uges Tid sammen med ham paa Rügen, men jeg syntes ikke, at jeg havde Tid nu i den sidste Uge Biblioteket er aabent. Naar jeg er kommen lidt i Gang, skal Du nok faa at høre om, hvordan det gaar; jeg glæder mig til snart at høre lidt fra og om Docenten i Göttingen. Tusind Hilsner fra

Niels

LETTER FROM NIELS TO HARALD (26 June 10)

[Letter]

⟨Copenhagen⟩ Vestre Boulevard 13¹.   26–6–10

Kære Harald!

[...] Jeg haaber ikke at Du bliver altfor ilde berørt ved at jeg kun saa daarlig kan skjule, hvorledes min Misundelse

LETTER FROM NIELS TO HARALD (20 Dec. 09)

[Letter]

⟨Norway, 20-12-09⟩

[...] I may just tell you, that I have concluded that the investigation of my electrons I have talked about can readily be carried out; how much will come of it is another matter. You must not from the fact that I tell this be afraid that I go around with such things in my head; as I told you, I loaf all day; it was only something that suddenly occurred to me and which I shall completely leave alone.

Now goodbye for this time and much love to all from

Niels.

P.S. [...]

LETTER FROM NIELS TO HARALD (25 June 10)

[Card]

⟨Copenhagen⟩ 25-6-10

Dear Harald:

Many congratulations to the assistant professorship, even if it doesn't come unexpected, it is terribly nice that it is now all settled; I am extremely proud on behalf of the kinship. Thanks for your card from Göttingen. I am getting along very well myself; I have taken a few days off, but I shall now work really hard again. Father was so nice as to propose that I spend a week's vacation with him at Rügen, but I didn't think I had the time now during the last week that the Library is open. When I have got well started I shall surely let you know how I am getting along. I am looking forward to hear a little from and about the Docent in Göttingen[16].

Much love from

Niels

LETTER FROM NIELS TO HARALD (26 June 10)

[Letter]

⟨Copenhagen⟩ Vestre Boulevard 13[1].   26-6-10

Dear Harald:

[...] I hope you are not too disgusted by the fact that I am so badly able to conceal that my envy is soon growing over

[16] [i.e., Harald.]

snart vokser op over Hustagene, men det er ikke saa underlig, naar man har anvendt 4 Maaneder paa at speculere paa et dumt Spørsmaal angaaende nogle dumme Elektroner, og ikke har drevet det til andet end at skrive circa fjorten mere eller
mindre afvigende Klader. Jeg er dog bange for at jeg heller ikke har kunnet
skjule for Docentens kloge Øjne, at mit Humør alligevel ikke er saa daarligt endda;
og det er igen heller ikke saa underligt, idet jeg tror at jeg dog tilsidst er bleven
færdig med det dumme Spørgsmaal om de dumme Elektroner, jeg føler mig i hvert
Tilfælde dennegang idet mindste en "Omgang" mere sikker end de andre Gange.
Den Løsning, hvortil der sigtes er halvt statistisk og halvt directe og berører ikke
Sandsynlighedsproblemer. Den vil kun fylde et Par Linier og er saa simpel at ingen
selv med sin bedste Vilie vil kunde forstaa at den kan have forvoldt nogen Vanskelighed (det skulde da være i det utænkelige Tilfælde at der skulde have siddet en
anden stakkels Idiot og misbrugt sin Tid paa samme dumme Spørgsmaal) Da jeg
jo dog kun saa daarligt kan skjule mit gode Humør, maa jeg ligesaa gerne straks
indrømme at det har endnu en Grund, jeg har nemlig anskaffet mig en Mappe,
som dem Far bruger til sine Regninger, og deri ordnet alle mine Papirer og Klader. (Nu siger du vist "Der erken⟨d⟩t⟨e⟩ jeg Skræp igen", men naar jeg skal være
ærlig saa maa jeg tilstaa at jeg ikke ved om jeg er mest glad for Din Udnævnelse, mine
Elektroners gode Opførsel i Øjeblikket, eller den omtalte Mappe, den eneste
Løsning bliver vist at Følelserne ligesom Erkendelsen maa ordnes i Planer der ikke
kan sammenlignes) Nu maa jeg slutte, da jeg ellers er bange for at min Pen løber
helt løbsk. [...]

LETTER FROM NIELS TO HARALD (5 July 10)

[Card]

⟨Copenhagen⟩ 5–7–10

Kære Harald.

Tak for Dine Kort, det er forfærdeligt rart at høre om, hvor morsomt Du har
det og hvor meget Du laver, jeg hører nok lidt nærmere derom. I Gaar blev Nørlund færdig efter den flotteste Examen, man kan tænke sig. Edgar og jeg var ude
med ham og hans Søster og spiste til Middag paa Skodsborg og bagefter var vi
i Tivoli sammen med hans Fader, der var kommen til Byen. Jeg haaber at det
skal gaa mig selv nogenlunde med Skrivningen, men for Øjeblikket staar det lidt
stille; Du ved jo nok at der hos visse dumme Personer kommer en lille Reaktion,
saa snart de synes at der i Øjeblikket ikke er noget at tvivle om. Du skal nok snart
høre nærmere, jeg sender blot denne lille Hilsen af Sted i en Fart, da jeg tænkte
at Du maaske vilde sende Nørlund en Lykønskning.

Din Niels

the house tops, but it is not so strange after I have spent four months speculating about a silly question about some silly electrons and have succeeded only in writing circa fourteen more or less divergent rough drafts. Nevertheless, I am afraid that I haven't been able to conceal from the Docent's[17] clever eyes that my mood hasn't been so bad after all; and, again, that is not so strange; for, I believe that I have finally licked the silly question about those silly electrons; in any case, I feel this time at least "a round" more certain than the other times. The solution I am referring to is half statistical and half direct and does not involve problems of probability. It will only take up a couple of lines and is so simple that no one, however much he tries, will be able to understand that it has presented any difficulty (unless, what seems inconceivable, another poor idiot should have been sitting and wasting his time on the same silly questions). Since I only so poorly can conceal my good humour, I may as well admit right away that it has still another reason; for, I have got a portfolio as the one Father uses for his bills, and in this I have arranged all my papers and drafts (now, I am afraid, you say "There I recognized Skræp[18] again", but, honestly, I must confess that I don't know if I am most happy over your appointment, over the good behaviour of my electrons at the moment, or over this portfolio; probably, the only answer is that sensations, like cognition, must be arranged in planes that cannot be compared). Now I must close; otherwise I am afraid that my pen runs riot. [...]

LETTER FROM NIELS TO HARALD (5 July 10)

[Card]

⟨Copenhagen⟩ 5-7-10

Dear Harald:

Thank you for your cards; it is awfully nice to hear how interestingly you are getting along and how much you do, I expect to hear a little more about it. Yesterday Nørlund finished his examination with the finest results imaginable. Edgar and I went with him and his sister and had dinner at Skodsborg, and later on we were in Tivoli together with his father who had come to town. I hope that the writing will go tolerably well for me, but at the moment it is at a standstill. You know that, for certain stupid persons, a little reaction sets in as soon as they feel that at the moment no doubt exists. You shall soon hear in more detail, but I send this little greeting in a hurry; for, I thought that you perhaps would send Nørlund your congratulations.

Your Niels

[17] [i.e., Harald's.]
[18] [Refers to a passage in an Icelandic saga.]

LETTER FROM NIELS TO HARALD (28 July 10)

[Letter]

28–7–10

Kære Harald!

Jeg har været ked af saa længe ikke at have skrevet til Dig, jeg tror dog ikke jeg behøver at forsikre at det ikke er Mangel paa Interesse for og Stolthed over en vis Personage i Gött.; ydre Paaskud har jeg iøvrigt haft rigeligt af, idet jeg stadig længselsfuldt venter paa et lovet Brev; men den virkelig Grund er dog, at der er hændt mig noget meget sørgeligt (ja saa sørgeligt at det næsten er tragikomisk) og da jeg ikke ynder at fylde Folk i "andre" Byer med Klager har jeg foretrukket at tie saa længe til jeg med mere Overlegenhed har kunnet berette om min Sindstilstand. Den tragikomiske Hændelse lyder da som følger. Dagen efter at jeg havde afsendt et ikke særligt nedtrykt Brev til Dig styrtede min lille Husflids-Bygning pludselig sammen og det næsten paa en mere ondartet Maade end den "plejer van". Det viste sig nemlig, at jeg i et Exempel med "Energiomsætning" havde begaaet en Regnefejl. Du ved jo nok, at jeg altid har let ved Undskyldninger, og i dette Tilfælde har jeg derfor ogsaa fundet paa følgende. "Jeg nærede (til min Beskæmmelse) ikke nogen Tvivl om Exemplets Udfald; men indsaa blot Vanskeligheden ved at bevise det; jeg blev derfor, som Du maaske husker, saa glad, da der viste sig en Udvej til Regningernes Gennemførelse, og regnede dem meget hurtigt ud". Da jeg nu saa dem igennem igen viste det sig imidlertid, at der, hvor jeg havde faaet $\frac{K_1+4K_2}{K_1+5K_2}$, skulde staa $\frac{K_1+5K_2}{K_1+5K_2} = 1$; dette Gaadesprog betyder at Exemplet netop gav det Resultat, for hvis Urigtigheds Bevis det hele var arrangeret. Nu sad jeg i Knibe, men da det omhandlede Exempel (5te Potens) havde særlig simple Egenskaber, begyndte jeg straks paa Udregningen af et andet (elastiske Kugler), dette gik imidlertid ikke saa let, og jeg maatte regne paa det i det meste af fjorten Dage (med den lille Afbrydelse sammen med Edgar, hvorom Du underrettedes paa Kort fra Vordingborg), naar Du engang kommer hjem og ser Regningernes Omfang tænker jeg at Du vil indrømme at jeg idet mindste i den Tid ikke har skrevet hele Dagen, Nørlund var iøvrigt saa elskværdig at se nogle af Beregningerne igennem for mig og gjorde mig en stor Tjeneste idet han fandt en Regnefejl, hvis Existens jeg kendte, men hvis Plads jeg forgæves havde søgt efter. Forleden Dag blev jeg endelig færdig; men Resultatet blev igen ,,Ligestorhed af de 2 Konstanter hvis Forskellighed jeg saa gerne vilde paavise". Jeg har senere regnet et lille snedigt Exempel ud (5te Potens mellem Molekyler og Elektroner, hvilke sidste indbyrdes støder som elastiske Kugler); men atter med samme Resultat. Stillingen er da nu at alle mekaniske Exempler med Energiomsætning giver et Resultat, hvis Rigtighed jeg dog i sin Almindelighed ikke kan bevise. Det er dog ikke Menin-

LETTER FROM NIELS TO HARALD (28 July 10)

[Letter]

28-7-10

Dear Harald:

I have been sorry that I haven't written to you for so long; however, I believe that I don't need to assure you that it isn't for lack of interest in and pride of a certain individual in Göttingen; I have, for that matter, had plenty of external excuses, since I am still waiting anxiously for a promised letter; however, the real reason is that something very sad has happened to me (in fact, so sad that it is almost tragi-comical), and since I do not like to bother people in "other" towns with complaints, I have preferred to remain silent until I could report about my state of mind with greater composure. The tragi-comical event is as follows: The day after I had mailed a not very depressed letter to you, my little home-made edifice collapsed suddenly, and that almost in a more pernicious manner than it "is in the habit of doing". For, it turned out that I had made an error of computation in an example dealing with the "energy law". As you know, it has always been easy for me to find excuses and in this case I have invented the following: "(To my humiliation) I had no doubt about the way the example would come out; but I merely realized the difficulty of proving it; as you perhaps remember, I was, therefore, so happy over having found a way to carry out the calculations, and I did so very quickly". However, when I now looked at the results it turned out that where I had found $\frac{K_1 + 4K_2}{K_1 + 5K_2}$ it should be $\frac{K_1 + 5K_2}{K_1 + 5K_2} = 1$; this enigmatic language means that the example gives precisely the result which the whole thing had been intended to disprove. Now I was stuck with a difficulty, but, since the example mentioned (5th power) had especially simple properties, I began immediately to calculate another (elastic spheres); however, that did not go so easily, and I had to calculate on it for the most of a fortnight (with the little interruption together with Edgar, about which you were informed by card from Vordingborg). When you get home some time, and see the extent of the calculation, I think you will admit that at least during that time I did not write the whole day. Nørlund, by the way, was so kind as to look through some of the calculations for me, and he did me a great favour, in that he found an error of computation whose existence I knew about, but whose location I had searched for in vain. The other day I finally finished it, but the result was again that the two constants, which I badly wanted to show were different, actually turned out to be equal. I have later calculated a little cunning example (5th power between molecules and electrons, the latter colliding as elastic spheres); but again with the same result. The situation is now that all mechanical examples with energy exchange give a result whose

gen at jeg hermed tilsigter en regelret Anklage mod Skæbnen, jeg tænker nemlig at jeg alligevel nok skal kunne klare det, det vil sige ikke bevise Sagen, derpaa tør jeg næppe tænke, men omgaa den paa passende Maade ved Skrivningen. Som Du nok kan tænke Dig, har disse mange Genvordigheder ikke just fremmet Skriveriet og Litteraturkendskaben; men nu haaber jeg rigtig at skulle tage fat og faa ⟨alt⟩ skrevet førend Du kommer hjem. Iøvrigt har jeg det storartet, ja er f. Eks. i Øjeblikket i et saadant Humør, at jeg, dersom jeg troede at Du dennegang kunne holde mere Sludder ud, oven paa det mere sørgelige godt kunne falde paa at give mig til at udmale forskellige Ting i mere lyse Farver.   [...]

LETTER FROM NIELS TO HARALD (24 Nov. 10)

[Letter]

Vissenbjerg 24–11–10

Kære Harald!

[...] At Du ikke har hørt fra mig før (undtagen den lille Besked. Tak for Besørgelsen) kommer af at det ikke gik saa godt med Skrivningen, [...]. Men nu synes jeg, at det begynder at gaa, Du forstaar nok kun en ganske lille bitte Smule; men jeg er jo ikke saa forvænt i den Henseende. Jeg er derfor i det mest udmærkede Humør. Jeg har det ogsaa storartet her-ovre. [...]

LETTER FROM NIELS TO HARALD (2 Jan. 11)

[Letter]

2–1–11

Kære Harald!

Tusind Tak for Dine Kort. Det gaar mig saa vidunderligt godt, jeg blev nemlig færdig til Nytaar med de 2 Kapitler og har i den Anledning været i saa straalende Humør, og ogsaa kunnet snakke lidt som i gamle Dage. Nu skal jeg tage fat paa Ind-ledningen, og skal se at blive færdig førend Du kommer hjem. Som jeg dog glæder mig til at Du kommer hjem, baade til at høre om Din Rejse, og saa naturligvis først og fremmest til al Din Hjælp; foruden Hjælpen til Afhandlingen, som jeg glæder mig saa frygteligt til, saa er der som sædvanlig saa mange mange Ting, som Du kan hjælpe mig og Margrethe med; vi gaar nemlig med forskellige Planer.

1000000 .... Hilsner fra os begge.

Din Niels

general validity, nevertheless, I cannot prove. It is not my intention herewith to make a formal accusation against fate; for, I think that I shall manage after all; that is, I shall not prove the matter, I can hardly think of that, but I shall circumvent it in a suitable manner in what I write. As you will understand, these many adversities haven't exactly furthered the writing and the study of the literature; but now I hope to get busy and get ⟨it all⟩ written before you get home. Incidentally, I am getting along splendidly; in fact, I am, for example, in such a good mood that if I believed that you could stand more nonsense this time, it might occur to me, after the sadder things, to paint various other things in brighter colours.   [...]

LETTER FROM NIELS TO HARALD (24 Nov. 10)

[Letter]

Vissenbjerg 24–11–10

Dear Harald:

[...]   That you haven't heard from me sooner (except for the little request; thanks for taking care of it) is because the writing has not been going so well, [...] But now I think that it is beginning to go; only a little bit, you understand, but, of course, I am not spoiled in this respect. Therefore, I am in the most excellent humour. I am getting along splendidly over here.   [...]

LETTER FROM NIELS TO HARALD (2 Jan. 11)

[Letter]

2–1–11

Dear Harald:

Thanks a million for your cards. It is going wonderfully well for me; in fact I finished the two chapters by New Year's, and for that reason I have been in such radiant spirits and have also been able to chat as in the old days. Now I shall start on the introduction, and I shall try to get it done before you get home. How I am looking forward to hear about your trip, and, naturally, first of all to all your help; besides help with the paper [i.e., the dissertation], to which I am looking forward so terribly, there are, as usual, so many things that you can help me and Margrethe with; for, we are entertaining certain plans.

Much love from both of us.

Your Niels

LETTER FROM NIELS TO HARALD (22 Apr. 11)

[Letter]

22 April 1911

Kære Kære Harald!

En lille bitte skriftlig Lykønskning fra Niels, der glæder sig saa meget til at se Dig senere paa Dagen; men jeg maatte først ønske Dig rigtig til Lykke til det nye Aar, og takke Dig for alt, hvad Du har været for os alle sammen i det gamle Aar. Tænk hvor ubeskrivelig meget Du har været alene for mig først i Foraaret og i Sommer og saa i Vinter, hvor Du hjalp mig at blive færdig førend Fader døde, og saa nu, hvor Du, til Trods for at Du har haft saa meget meget andet at tænke paa, har hjulpet mig saa utrolig meget. Kære Harald, naar jeg tænker paa alt det, saa ved jeg ikke anden Maade at takke Dig end at ønske at Fremtiden maa blive saa rig og lys for Dig, og ved selv at prøve paa at vise, at alt, hvad Du har gjort for mig, ikke har været spildt. Hvad der saa end sker, om vi blot kunde holde sammen i det, at prøve paa at faa vores Liv til at svare lidt til den Barndom og Ungdom, vi har haft.

LETTER FROM NIELS TO HARALD (29 Sept. 11)

[Letter]

29–9–11. Eltisley Avenue 10, Neunham, Cambridge.

Aa Harald!

Det gaar mig saa storartet. Jeg har lige talt med J J Thomson og forklaret ham saa godt jeg kunde min Mening om Straalingen, Magnetismen o.s.v. Du skulde vide hvad det var for mig at tale med saadan en Mand. Han var umaadelig rar imod mig, vi snakkede om saa meget, og jeg tror nok at han syntes at der var nogen Mening i hvad jeg sagde. Han vil nu læse Bogen, og han indbød mig til at spise sammen med sig i Trinity College paa Søndag til Middag; saa vil han tale med mig derom. Du kan tro, at jeg er glad. Som jeg skrev i Gaar har jeg nu min egen lille Lej⟨lig⟩hed. Den ligger i Udkanten af Byen og er forfærdelig rar i alle Henseender. Jeg har 2 Værelser og spiser ganske for mig selv i mit eget Værelse. Her er saa hyggeligt; nu, mens jeg sidder og skriver til Dig, blusser og buldrer Ilden i min egen lille Kamin. Men trods alt det tror jeg dog ikke at jeg bliver her længere end den Uge for hvilken jeg foreløbig har lejet, for Thomson raadede mig til, hvis Udgifterne ikke spillede alt for stor Rolle, til at blive "member of the university" og

LETTER FROM NIELS TO HARALD (22 Apr. 11)

[Letter]

22 April 1911

Dear Harald:

Just a little written congratulation from Niels who is looking forward very much to see you later today; but I had first to congratulate you to the new year, and thank you for all that you have been for all of us in the old year. Think how indescribably much you have been for me alone, first last spring and summer, then this winter when you helped me to get done before Father died, and then now when you have helped me so unbelievably much, although you have had so very, very much other to think of.

Dear Harald, when I think of it all I don't know of any other way of thanking you than to wish that the future may be bright and rich for you, and to try to show myself that all you have done for me has not been wasted. Whatever happens, if we could only stick together and try to make our lives correspond somewhat to the childhood and youth we have enjoyed!

LETTER FROM NIELS TO HARALD (29 Sept. 11)

[Letter]

29-9-11. Eltisley Avenue 10, Newnham, Cambridge.

Oh Harald!

Things are going so well for me. I have just been talking to J. J. Thomson and have explained to him, as well as I could, my ideas about radiation, magnetism, etc. If you only knew what it meant to me to talk to such a man. He was extremely nice to me, and we talked about so much; and I do believe that he thought there was some sense to what I said. He is now going to read the book [i.e., B.'s dissertation], and he invited me to have dinner with him Sunday at Trinity College; then he will talk with me about it. You can imagine that I am happy. As I wrote yesterday, I now have my own little flat. It is at the edge of town and is very nice in all respects. I have two rooms and eat all alone in my own room. It is very nice here; now, as I am sitting and writing to you, it blazes and rumbles in my own little fireplace. But, in spite of everything, I do not think I shall stay here more than the week for which I have rented it temporarily; for, Thomson advised me, if expenses are not too important, to become a "member of the University" and join

slutte mig til et College. I saa Tilfælde hører jeg med til alt, kan deltage i Studenternes Sport, laane alle Bøger og høre alle Forelæsninger, og først og sidst jeg vil faa saa mange flere Mennesker at se. Hvis jeg saaledes, som det var Meningen, blev Medlem af Trinity College vilde jeg høre sammen med alle de betydeligste Folk i Cambridge, Thomson, Darwin, Hardy, Littlewood . . . . og hver Dag spise til Middag sammen med dem. Han mente at det vilde koste 20–25 £ mere for mig, og alting er i Forvejen meget dyrt herovre; men jeg tror alligevel jeg gør det. Jeg har aftalt paa Søndag efter Middagen at tale med en af "tutors" i Trinity, en Fysiolog ved Navn Fletscher, og faa alt bestemt at vide, og dersom det ikke er altfor dyrt, saa tror jeg, at jeg bestemmer mig til det. Harald! Du kan tro at det er rart at ens Navn er kendt; jeg har mest været sammen med Fysiologer. Bancroft, der har en meget indflydelsesrig Stilling her, idet han er "proctor of the university", dvs han repræsenterer den udøvende Magt og det en meget indgrebende Magt, overfor "the undergraduated students", har været umaadelig rar imod mig, han bad mig straks den anden Dag, jeg var her, til Middag. Jeg var der alene sammen med den tyske Fysiolog Franz Müller, der ogsaa har været hos os (i Rungsted). Det var morsomt at se et saadant rigtig engelsk Hjem. Men navnlig har en ung Fysiolog Hill til hvem Frk. Le⟨h⟩mann havde givet mig Anbefaling, være⟨t⟩ utrolig elskværdig imod mig, han har skaffet mig denne Bolig og har hjulpet mig med alting; han bad mig i Forgaars til Middag i Trinity College. Vi spiste i en mægtig ældgammel gotisk Hal; vi var kun temmelig faa (c. 30 Fellows) for Studenterne er ikke komne endnu (til daglig spiser der 700, i 2 Hold). Jeg er ogsaa saa glad for mine Introductioner, skønt jeg ikke har brugt saa mange endnu, mange er nemlig ikke komne endnu og jeg har heller ikke haft saa megen Tid. Jeg har dog besøgt Hardy han var meget elskværdig og forhørte om Dig og sagde, at han glædede sig saa meget til at tale med Dig. Han boede i den dejligste Lejlighed, jeg nogensinde i mit Liv har set. Det var i Trinity College og han havde nogle store gamle Værelse⟨r⟩ med gotiske Vinduer og Loft, der vendte ud til den dejligste gamle Park med store Træer og Græsplæner og med ældgamle Broer over Floden. Jeg er ogsaa saa glad for, at jeg har offentliggjort de Afhandlinger i Royal Society. Men først og sidst er jeg saa glad og taknemlig, at jeg ikke kan sige det, fordi min Afhandling er færdig (og derfor skulde dette Brev jo nærmest have været til Mor, men det maa gælde for hende ogsaa) og at jeg kunde give Thomson den. Og nu Farvel og saa mange mange Hilsner til jer alle sammen

fra jeres taknemlige Niels

a College. In that case I shall belong to everything, can take part in the sport of the students, borrow all books and attend all lectures; first and last, I shall meet so many more people. Thus, as the idea was, if I should become a member of Trinity College I should belong together with all the most important people in Cambridge, Thomson, Darwin, Hardy, Littlewood ..., and every day dine with them. He thought it would cost 20–25 £ more for me, and everything is very expensive over here to begin with; but I think, nevertheless, that I shall do it. Sunday after dinner I have arranged to talk to one of the "tutors" at Trinity, a physiologist by the name of Fletscher, and learn about everything definitely, and if it isn't too expensive, I believe I shall decide to do it. Harald! Believe me, it is nice that one's name is known; I have mostly been together with physiologists. Bancroft, who has a very influential position here, being "proctor of the university", i.e., he represents the executive power, and it is a very pervasive power, to the "undergraduate students", has been awfully nice to me, he invited me immediately to dinner the second day I was here. I was there alone together with the German physiologist, Franz Müller, who also has visited us (in Rungsted). It was interesting to see such a real English home. But especially a young physiologist, Hill, to whom Miss Lehmann[19] had given me an introduction, has been unbelievably nice to me; he has procured this flat for me and has helped me with everything; the day before yesterday he invited me to dinner at Trinity College. We ate in an exceedingly old Gothic hall; we were only rather few (ca. 30 fellows); for, the students haven't come yet (ordinarily 700 eat there, in two shifts). I am also very glad for my letters of introduction, although I haven't used many of them yet, for many have not yet arrived and I haven't had so much time either. However, I have visited Hardy; he was very kind and asked about you and said that he looked forward so much to talk to you. He lived in the most wonderful flat I have ever seen in my life. It was in Trinity College and he had some large old rooms with a Gothic ceiling and windows which front a most wonderful old park with big trees and lawns and with ancient bridges across the river. I am also glad that I have published those papers [on surface tension] in the Royal Society. But, above all, I am so unspeakably happy and thankful that my paper [i.e., the translation of the dissertation] is finished (and, therefore, this letter should really have been to Mother, but it may count as intended for her also), and that I could give it to Thomson. Now goodbye and much love to all of you from

your grateful Niels

[19] [Inge Lehmann, Danish student of mathematics and geodesy.]

LETTER FROM NIELS TO HIS MOTHER (2 Oct. 11)

[Letter]

2-10-11. Eltisley Avenue 10. Neunham, Cambridge.

Kære lille Mor!

[...] Jeg har det stadig storartet, jeg er endnu ikke begyndt paa Laboratoriet, jeg skal først tale med Thomson i Morgen derom; men jeg har fuldt op at gøre med Arrangementer Visitter og Middagsselskaber (hvad synes I egentlig) i Gaar var jeg saaledes i Trinity College hos Thomson og i Dag har jeg været i Trinity Hall (et mindre, men meget gammelt College, der er meget rigt, og hvor de spiser saa meget og saa fint, at det er aldeles utroligt og ubegribeligt at de kan holde det ud) hos Prof. Woodhead. Men nu Godnat min egen lille Mor, for Klokken er saa mange saa mange; og 100 . . . Hilsner til jer allesammen fra

jeres Niels

fra hvem i skal høre saa saare at han har faaet noget ordnet.

LETTER FROM NIELS TO HIS MOTHER (4 Oct. 11)

[Letter]

4-10-11. Eltisley Avenue 10, Neunham, Cambridge.

Min egen lille Mor!

Saa mange mange Gange til Lykke til det ny Aar og Tak for dette Aar. Du kan tro jeg ofte tænker paa vores forrige Fødselsdag.

Nu har jeg efterhaanden nogenlunde faaet alting i Orden. Om nogle Dage, naar mine Papirer og Penge er komne, vil jeg saaledes være "member of the university as a research student of physics" og høre til Trinity College som en "advanced student". Fra det Øjeblik af har jeg, under Trusel af høje Mulkstraffe, ikke Lov at vise mig paa Universitet⟨et⟩ (undtagen lige Laboratoriet) eller i College og heller ikke paa Gaden efter at det er bleven mørkt uden med "cap and goun", det bliver helt mærkeligt. Jeg skulde egentlig samtidig flytte ind i et "license lodging", det vil sige ind i et Hus, der staar under Universitetets Kontrol, og hvor the "landlord" (Husejeren) eller the landlady har at passe at Studenterne ikke gaar ud efter Kl. 10 og en hel Masse andre Ting; men da næsten alle license lodgings var optagne, og de der var tilbage var meget daarlige og meget dyre fik jeg Lov af Tu-

LETTER FROM NIELS TO HIS MOTHER (2 Oct. 11)

[Letter]

2–10–11.   Eltisley Avenue 10 Newnham Cambridge.

Dear little Mother:

[...]   I am still getting along splendidly; I haven't yet started in the laboratory; I am not to talk with Thomson about it until tomorrow; but I am very busy with arrangements, visits and dinner parties (what do you think about that?); thus, yesterday I was in Trinity College to Thomson's, and today I have been in Trinity Hall (a smaller but very old college that is very rich, and where they eat so much and so first-rate that it is quite unbelievable and incomprehensible that they can stand it) with Prof. Woodhead. But now good night my own little Mother, for it is so late; and very much love to all of you from

your Niels

You shall hear in letters as soon as I have managed to make some arrangements.

LETTER FROM NIELS TO HIS MOTHER (4 Oct. 11)

[Letter]

4–10–11.   Eltisley Avenue 10 Newnham Cambridge.

My own little Mother:

Happy birthday and many good wishes for the new year and thanks for the last year. Believe me, I often think of our last birthday.

Now, I have gradually more or less taken care of everything. In a few days, when my papers and the money have come, I shall thus be "a member of the university as a research student of physics" and belong to Trinity College as an "advanced student". From that moment on I am not allowed, under threat of high fines, to appear at the University (except only in the laboratory) or in the College and not on the street after dark without "cap and gown"; it will be quite strange. I really should move at the same time to a "licensed lodging", that means to a house that is under the control of the University and where the "landlord" or the landlady must see to it that the students don't go out after 10 o'clock ⟨p.m.⟩ and a lot of other things. However, as almost all the licensed lodgings were occupied, and as those that were vacant were pretty bad, I was permitted by the tutor at Trinity to

toren i Trinity til at vedblive at bo hvor jeg nu bor, det var dog kun fordi jeg ikke skulde tage nogen ,,degree''; for at tage en saadan er det aldeles nødvendigt, at man har boet et vist Antal "terms" enten i College eller i license lodgings. Se det synes Du vist allerede er meget, men jeg har af Tutoren faaet en hel Bog, om hvad jeg har Lov til og hvad ikke. Deri har jeg blandt andet fundet "smoking, while wearing academical dress, is a breach of discipline"; efter en hel Masse af den Slags staar der "In addition to these general regulations persons in statu pupillari (mig) are forbidden", hvorefter følger en Række Bestemmelser der tilsyneladende stammer fra de forskelligste Tider, idet den første hedder "To take part in gaming transactions in any way" og den sidste og 12te hedder "To keep or use a motor-car, motor-bicycle, or other motor-vehicle within the precincts of the University without a license from the Senior Proctor". Dersom jeg kan faa Fat i endnu et Exemplar af Bogen, skal jeg sende den; saa kan Du lære, hvordan Du skal være en god land-lady for Riks, det regnes for en meget vigtig Ting her, i Dag paa Laboratoriet var der saaledes en venlig Student, der spurgte, hvordan jeg havde indrettet mig, og som sagde ,,can your landlady cook", hvortil jeg, som sandt var, svarede "very well". Dette bringer mig til at tænke paa, at Du maaske ogsaa er nysgerrig efter at høre lidt om min egen lille Wirtschaft. Jeg har skrevet, at jeg bor i Udkanten af Byen (c. 15 Minutters Gang fra Laboratoriet) og har 2 Værelser. De er rigtig hyggelige og pænt møblerede, og jeg bliver passet udmærket; jeg spiser ganske alene for mig selv paa mit eget Værelse, og jeg faar næsten hver Dag en hel Steg ind, forleden fik jeg saaledes en hel And, og Du kan vist tænke at det var med løjerlige Følelser at jeg betragtede den. Nu faar jeg jo ogsaa Ret til at spise i "the hall of Trinity Col-lege" (det vil sige, jeg skal som alle betale for hver Gang; jeg faar se, hvor ofte jeg faar Raad at spise der, for det er betydelig dyrere end her; her koster det for al Maden c. 2 S. om Dagen) ved et særligt Bord for "bachelor of arts"; det var en stor Misforstaaelse, at jeg skulde spise sammen med Thomson og de andre "fel-lows", det var kun saalænge jeg ikke er "member of the university" og er indbudt af en fellow (Hardy har saaledes skyndt sig til at indbyde mig til i Overmorgen). Jeg faar ogsaa Ret til at tage Gæster med i Trinity, og jeg glæder mig allerede til det, naar Harald kommer; men jeg er bange for at jeg ikke kommer til at spise ved samme Bord som han, for Hardy og Littlewood er meget ivrige efter at træffe ham, og naar de indbyder ham, spiser han ved det fine Bord, der er hævet et Trin op over den øvrige Sal, og hvor de faar en Ret mere. (Lille Mor, hvad siger Du dog til at den Lærdom, dennegang er det Dig der maa fortælle Margrethe lidt). Jeg er lige begyndt paa Laboratoriet i Dag; jeg skal først prøve nogle Experimenter med po-sitive Straaler, men I maa ikke tro, at det gaar saa glat, I gør jer ingen Begreb om den Uorden, der hersker i Cavendish Laboratoriet, og en stakkels Udlænding, der ikke engang ved, hvad de forskellige Ting, der ikke kan findes, hedder, er meget ilde stedt, ved nogle venlige Menneskers Hjælp lykkedes det mig dog i Løbet af nogle

remain where I live now; however, this was only because I am not going to take any "degree"; to take a degree, it is absolutely necessary that one has lived a certain number of "terms" either in the college or in licensed lodgings. You think probably that this is already a lot, but I have got an entire book from the tutor, telling what I am allowed to do and what not. Among other things, I have found that "smoking, while wearing academic dress, is a breach of discipline"; after a lot of this sort of thing, it states "In addition to these general regulations, persons in statu pupillari (that's me) are forbidden" and then follow a lot of regulations that evidently date from the most different times, in that the first reads "to take part in gaming transactions in any way" and the twelfth and last reads "to keep or use a motor car, motor-bicycle, or other motor-vehicle within the precincts of the University without a license from the Senior Proctor". If I can get hold of another copy I shall send it; then you can learn how to be a good landlady for Riks; that is considered very important here; thus today in the laboratory a kindly student asked me what accommodations I had got and "can your landlady cook"; to which I answered truthfully "very well". This reminds me that you may be curious to hear a little about my own little menage. I have told you that I live at the outskirts of town (ca. 15 minutes' walk from the laboratory) and have two rooms. They are quite cosy and nicely furnished, and I am taken care of very well. I eat all alone in my own room, and almost every day I get a whole roast brought in; thus, the other day I got an entire duck; I believe you can imagine that I looked at it with funny feelings. Now I shall also have the right to eat in "the Hall of Trinity College" (that is, as all others, I shall have to pay each time; we shall see how often I can afford to eat there; for, it is considerably more expensive than here; here all meals are ca. 2 s per day) at a special table for "bachelors of arts"; it was a great misunderstanding that I was to eat with Thomson and the other "fellows"; that was only as long as I was not a "member of the university" and was invited by a fellow (thus, Hardy has hastened to invite me for the day after tomorrow). I shall also get the right to invite guests to Trinity, and I am already looking forward to doing that when Harald comes; however, I am afraid that I shall not eat at the same table as he; for, Hardy and Littlewood are anxious to meet him, and when they invite him he eats at the high table, which is raised one step over the rest of the hall, and where they get one additional course. (Little Mother, what do you think of all this information; this time it is you who must tell Margrethe a little.) I have just started in the laboratory today; I shall begin by trying some experiments on positive rays, but I don't think it will go so smoothly. You have no idea of the lack of order that reigns in the Cavendish Laboratory, and a poor foreigner, who doesn't even know the names of the things he cannot find, is very badly off; however, with the help of some kind people I succeeded in the course of a few hours to find a few things, and, when I take a dictionary along tomorrow and hopefully get somebody

Timer at finde nogle enkelte Ting, og naar jeg i Morgen møder med Lexikon og forhaabentlig ogsaa faar nogen til at hjælpe mig, kan jeg nok i Løbet af nogle Dage
faa samlet saa meget sammen, at jeg kan begynde lidt; men selv om jeg ikke lærer
saa meget Fysik af det, saa lærer jeg maaske andet. Thomson har endnu ikke talt
til mig om Bogen, men han har ogsaa saa meget for af alle mulige Slags, at det
vilde være ubegribelig, om han havde faaet Tid til at læse den endnu. Ja lille Mor,
det var vist Størstedelen af, hvad jeg havde at fortælle. Du kan maaske alligevel ud
af al den Sludder lære, at jeg ikke er saa ked af at være her og glæder mig saa meget
til det altsammen. Og nu Farvel og saa mange mange mange Hilsner til jer allesammen, som vel er samlede i Dag (eller i hvert Tilfælde i Aften) og til Moster Emma
og Moster Hanne og Sine og Hanne; og endnu engang til Lykke og Tak fra

Din egen Niels.

LETTER FROM NIELS TO HARALD (23 Oct. 11)

[Letter]

23-10-11   Eltisley Avenue 10. Cambridge.

Kære Harald!

I maa ikke være kede af at I ikke har hørt fra mig saa længe. Jeg har det saa
udmærket, men jeg har saa nødig villet skrive, for det er endnu ikke gaaet saa hurtigt med mit Arbejde, og jeg har stadig fra Dag til Dag haabet at det skulde gaa
lidt bedre. Thomson har nemlig hidtil ikke været saa nem at have med at gøre
som jeg troede den første Dag. Han er en ganske storartet Mand saa utrolig klog og
fantasirig (Du skulde høre en af hans elementære Forelæsninger) og overordentlig
venlig; men han har saa umaadelig travlt med saa mange Ting, og han gaar saadan
op i sit eget Arbejde, at han er meget vanskelig at komme til at snakke med. Han
har endnu ikke haft Tid at læse i min Afhandling, og jeg ved endnu ikke om han
vil gaa ind paa min Kritik. Han har kun nogle Gange snakket lige et Par Minutter
med mig derom, og det har kun været om et enkelt Punkt, nemlig om min Kritik
af hans Beregning af Absorptionen for Varmestraaler. Du husker maaske, at jeg
bemærker, at han i sin Beregning af Absorptionen (i Modsætning til Emissionen)
ikke tager Hensyn til den Tid, der medgaar til Sammenstødene, og at han derfor
finder en Værdi for Forholdet mellem Emission og Absorption, der for smaa Svingningstider er af en urigtig Størrelsesorden. Thomson sagde først, at han ikke kunne indse, at Tiden for Sammenstødene vilde have saa stor Indflydelse paa Absorp-

to help me, I may in the course of a few days get so much collected that I can begin a little; even if I don't learn much physics thereby, I may perhaps learn something else. Thomson has not yet talked to me about the book, but he is so busy with all sorts of things that it would be inconceivable that he should have found time to read it yet.

Well, little Mother, I believe this is most of what I have to tell. Perhaps you can gather from all this nonsense that I am not exactly sorry to be here and that I am looking forward to everything. And now goodbye and very much love to all of you who probably are gathered today (or, in any case, tonight) and to Aunt Emma[20] and Aunt Hanne[21] and to Sine and Hanne; and once more happy birthday and thanks from

Your own Niels.

LETTER FROM NIELS TO HARALD (23 Oct. 11)

[Letter]

23-10-11   Eltisley Avenue 10 Cambridge.

Dear Harald:

You must not be distressed that you haven't heard from me for so long. I am getting along very well, but I have been reluctant to write; for, my work is still not progressing very rapidly, and I have hoped from day to day that it should go a little better. In fact, Thomson has so far not been as easy to deal with as I thought the first day. He is an excellent man, incredibly clever and full of imagination (you should hear one of his elementary lectures) and extremely friendly, but he is so immensely busy with so many things, and he is so absorbed in his work, that it is very difficult to get to talk to him. He has not yet had time to read may paper, and I do not know if he will accept my criticism. He has only talked to me about it a few times for a couple of minutes, and only about a single point, namely about my criticism of his calculation of the absorption of heat rays. You may remember that I pointed out that, in his calculation of the absorption (contrary to the emission), he does not take into account the time taken by the collisions and, therefore, finds a value for the ratio between emission and absorption that is of the wrong order of magnitude for small periods of vibration. Thomson first said that he couldn't see that the duration of the collisions could have such a large influence upon the absorption. I tried to explain it to him and gave him the next day a calculation of a

---

[20] [Emma Trier, sister of Bohr's mother.]

[21] [Hanne Adler, sister of Bohr's mother; founder and principal of a private school.]

tionen; jeg søgte at forklare det, og gav ham den næste Dag en Beregning af et meget simpelt Exempel (et Exempel der svarede til en af hans Beregninger af Emissionen), der meget tydeligt viste det. Jeg har siden kun lige talt et Øjeblik med ham derom for en Uges Tid siden; og jeg tror nok, at han synes at min Beregning er rigtig, men jeg er ikke sikker paa at han ikke mener, at der kan tænkes en mekanisk Model, der kan forklare Varmestraalingsloven ud frade almindelige elektromagnetiske Love, hvilket imidlertid er indlysende, at man ikke kan, hvad jeg jo indirekte har bevist, og hvad i øvrigt senere er bleven direkte bevist af McLaren (nedenfor). Thomson er, som jeg tidligere har skrevet, umaadelig rar at snakke med, og jeg har været saa glad for det hver Gang; men Vanskeligheden ligger i, at han ikke har nogen bestemt Tid, og man derfor maa forstyrre ham midt i hans Arbejde (han har meget lidt Ro), han er til Trods derfor meget venlig; men naar man har talt et Øjeblik med ham, kommer han til at tænke paa et eller andet af hans egne Ting, og saa gaar han fra en midt i en Sætning (de siger her, at han vilde gaa fra Kongen, og det betyder noget andet i England end i Danmark) og saa har man Indtrykket af at han glemmer en indtil man næste Gang drister sig til at forstyrre ham. Jeg har endnu heller ikke faaet talt med andre, undtagen med nogle unge studerende der intet kender; jeg har 2 Gange forsøgt at tale med Jeans efter hans Forelæsning (den sidste Gang tog jeg Udgangspunkt i noget han havde sagt i Forelæsningen og som jeg ikke syntes var helt rigtig), han var meget venlig, men er meget tilbageholdende og har hver Gang sagt, at han vilde vente til Afhandlingen foreligger paa engelsk. Om Offentliggørelsen af Afhandlingen ved jeg endnu ikke noget helt bestemt, jeg har talt med Thomson et Par Gange om det (jeg vilde i Begyndelsen vente indtil han havde læst den) den ene Gang for 14 Dage siden og den anden Gang i Lørdags, og han har lovet at han skal høre om der er Mulighed for at den kan komme i Transactions of Cambridge Philosophical Society. Der er sikkert ikke Mulighed for at faa den i Royal Society, jeg har talt med Larmor (Sekretær i Selskabet; jeg havde Introduction til ham fra Knudsen, og han har været umaadelig venlig imod mig) og han mente at det var udelukket ikke fordi den var offentliggjort paa Dansk, men fordi den indeholdt Kritik af andres Arbejder, og Royal Society holder det for en ubrydelig Regel ikke at optage Kritik, der ikke tager Udgangspunkt fra deres egne Offentliggørelser. Det var storartet, om jeg kunde faa den i Cambridge Transactions, der vilde den komme hurtigt og det begynder at haste, der er allerede siden den danske Offentliggørelse kommen 2 lange Afhandlinger i Phil. Mag. om det samme Emne, og som naturligvis ikke kender min Afhandling. Af den er der ikke meget Mening i den ene, der bygger paa Wilsons Resultater; men den anden (af McLaren) er ganske udmærket, og giver et Resultat der er mere almindeligt end et af mine; han har nemlig beregnet Forholdet mellem Absorption og Emission for alle Svingningstider ud fra Betragtning af det Tilfælde, hvor Elektronerne antages at bevæge sig uafhængigt af hin-

very simple example (one that corresponds to one of his calculations of the emission) which showed it very clearly. Since then I have only talked with him about it for a moment, about a week ago, and I believe that he feels that my calculation is correct; however, I am not sure but that he thinks that a mechanical model can be found which will explain the law of heat radiation on the basis of the ordinary laws of electromagnetism, something that obviously is impossible, as I have shown indirectly, and as has, moreover, later been proved directly by McLaren (see below).

As I have written before, Thomson is extremely nice to talk to, and I have been so glad to do it each time; but the trouble is that there is no definite time when he is available, and one has to disturb him while he is working (he has very little peace); in spite of this, he is very friendly; but when you have talked with him for a moment he gets to think about one of his own things, and then he leaves you in the midst of a sentence (they say that he would walk away from the King, and that means more in England than in Denmark), and then you have the impression that he forgets all about you until the next time you dare to disturb him. I have not got to talk with others, except some young students who know nothing; I have a couple of times tried to talk to Jeans after his lecture (the last time I began with something he had said in the lecture which I didn't think was quite right); he was very friendly, but is very reticent, and every time he said that he would wait [to discuss the dissertation] until my paper has appeared in English.

About the publication of the paper, I don't know anything definite yet; I have talked to Thomson about it a couple of times (I first wanted to wait until he had read it), once two weeks ago and again last Saturday; he has promised to find out if there is a possibility of getting it out in the Transactions of the Cambridge Philosophical Society. It is evidently not possible to get it published by the Royal Society. I have talked with Larmor (Secretary of the Society; I had a letter of introduction to him from Knudsen, and he has been extremely nice to me), and he believes that would be impossible, not because it has been published in Danish, but because it contains criticism of the work of others, and the Royal Society considers it an inviolable rule not to accept criticism that does not originate in its own publications.

It would be fine if I could get it out in the Cambridge transactions; there it would appear without delay, and there begins to be haste about it; since the Danish publication there have already appeared two long papers in Phil. Mag. on the same subject, of course without the authors' knowing of my paper. There is not much sense to one of them which is based on Wilson's results; but the other (by McLaren) is excellent and gives a result that is more general than one of mine. He has calculated the ratio between absorption and emission for all periods of vibration by considering the case in which the electrons are assumed to move independently of one another in a stationary field of force: A result that is hardly

anden i et stationært Kraftfelt. Et Resultat der dog nærmest kun har kritisk Interesse, idet det, som jeg omtaler, jo kun er muligt at bringe Overensstemmelse med Erfaringen til Veje, for de lange Svingningstiders Vedkommende. Da jeg nu har fortalt saa meget om det andet maa jeg vel ogsaa fortælle lidt om Laboratoriet. Ja det er endnu ikke gaaet saa hurtigt der heller. Jeg arbejder med at prøve at frembringe en Katodestraaleudladning ved Hjælp af positive Straaler, et Fænomen som Thomson engang har iagt⟨t⟩aget. Jeg har endnu ikke kunnet faa Fænomenet rigtig frem, men tror at det begynder at hjælpe. Jeg ved ikke om Thomson mener, at jeg skal arbejde videre derpaa, naar jeg har set det, og hvad han i saa Tilfælde har tænkt sig at jeg skal gøre ved det, eller om jeg derefter skal begynde paa noget andet. Du synes vist at det gaar meget langsomt, men det er ikke saa let i Begyndelsen at finde sig til Rette i Cavendish Laboratoriet, hvor der er en saadan Uorden og saa lidt Hjælp til saa mange (jeg gaar sammen med 15 unge Mennesker); alle de, jeg har talt med, har sagt at det faldt dem saa umaadelig vanskelig at faa rigtig begyndt, og det er maaske gaaet særlig langsomt med mig, fordi jeg har saa mange forskellige Ting at spørge Thomson om, og det er aldeles umulig at faa talt om mere end en Ting, hver Gang man forstyrrer ham. Men jeg har jo ogsaa kun gaaet 3 Uger paa Laboratoriet og jeg har i den Tid i hvert Tilfælde blandt andet lært selv at sætte et Glasapparat sammen, hvad jeg er meget glad for. Jeg hører en hel Del Forelæsninger (2 om Dagen). Jeg hører Forelæsninger af Larmor og Jeans om forskellige Afsnit af Elektricitetstheorie, de har hidtil været temmelig elementære, men jeg tror at de bliver meget interessante. Og saa hører jeg 2 Rækker Forelæsninger af Thomson; en elementær Række (Properties of matter) der er umaadelig morsom og er ledsaget af de smukkeste Experimenter, man kan tænke sig; og en for mere viderekomne, den er umaadelig interessant, han beskæftiger sig for Øjeblikket med sine egne allernyeste Anskuelser om Elektriciteten; han indtager et Standpunkt der kan karakteriseres som en forholdsvis ringe Modification af den gamle Maxwellske Theorie. Jeg er desværre ikke nok inde i det, men dersom jeg var lidt mere inde i det, tror jeg, at jeg skulde mene, at hans Beregninger ikke paa alle Punkter er helt rigtige. Jeg skal se at komme rigtig ind i det, men jeg er ikke sikker paa at det bliver foreløbig, for jeg har saa ringe Tid til at læse. Jeg er begyndt at læse en tyk Bog af Jeans "Electricity and Magnetism" for at faa opfrisket mine Kundskaber og komme lidt ordentlig ind i det altsammen, men jeg er i det hele taget bange for at jeg ikke faar Tid til meget af den Slags, dersom jeg skal gøre noget ordentlig ved mit Arbejde paa Laboratoriet, og saa har jeg foreløbig en Del Arbejde med min egen Afhandling; jeg har truffet en umaadelig elskværdig ung Mand (Mr. Owen) paa Laboratoriet, der har været saa venlig at ville se den igennem for mig; han gør det meget grundig, men har haft saa daarlig Tid selv, at vi nærmest er gaaede i Staa, og dersom det ikke snart kommer i Gang igen, maa jeg se at faa det ordnet paa anden Maade. Jeg har det ellers

of more than critical interest, since, as I mention, it is only possible to obtain agreement with experiments for long periods of vibration.

Since I now have told so much about other things I suppose I must also tell a little about the laboratory. It has not yet gone very fast there either. I am working on an attempt to produce a discharge of cathode rays by means of positive rays, a phenomenon that Thomson has once observed. I have not yet been able to really obtain the phenomenon, but I think it is beginning to help. I don't know if Thomson thinks that I shall pursue it further when I have observed it, and what, in that case, he intends me to do with it, or if he then wants me to begin something else. I am afraid that you feel that things are going very slowly, but it is not easy in the beginning to adjust oneself to the Cavendish Laboratory where there is such a lack of order and so little help for so many people (I work together with 15 young people); all those I have talked with have told me that they found it extremely difficult to get started; I have perhaps progressed especially slowly; for, I have so many things to ask Thomson about, and it is quite impossible to get to talk about more than one thing each time one disturbs him. Of course, I have only worked three weeks in the laboratory; in any case, I have learned during this time to put a glass apparatus together myself, and I am very glad of that. I also attend several lectures (two a day). I attend lectures by Larmor and Jeans on various parts of the theory of electricity; so far they have been fairly elementary, but I think they will be very interesting. Then I attend two series of lectures by Thomson: one elementary one (Properties of Matter) which is awfully interesting and is accompanied by the most beautiful experiments imaginable; one for advanced students; it is extremely interesting; at the moment he deals with his own latest ideas about electricity; he takes a position that may be characterized as a fairly slight modification of the old Maxwellian theory. Unfortunately, I am not very well versed in it, but, if I were a little better up in it, I should think that his calculations are not correct on all points. I shall try to be really well versed in it, but I am afraid that it will not be very soon; for, I have so little time to read. I have started reading a thick book by Jeans: "Electricity and Magnetism" in order to refresh my knowledge and to get really up in all of it, but I fear that I shall not have much time for such things if I am to do real justice to my work in the laboratory; besides, for the present I have a good deal of work with my own paper; I have met an extremely nice young man (Mr. Owen) at the laboratory, and he has been kind enough to offer to look it over for me; he does it very thoroughly, but has had so little time that we have almost come to a standstill, and, if we don't soon get going again, I must try to get it done in some other way.

Otherwise, I am getting along very well, and I am seeing a lot of people who have all been very nice to me. Last Tuesday I had dinner at the Thomson's; it was extremely fine, and I had to take a daughter of Darwin to the table; but I could not

storartet, og ser en Del Mennesker der alle har været saa rare imod mig. Jeg var i Tirsdags til Middag hos Thomson, der var umaadelig fint og jeg havde en Datter af Darwin til Bords, men jeg kunde jo ikke sige mange Ord, og den Konversation der pumpes ud af mig, minder mig altid om Franskmanden hos Potsnap, blot at de Folk jeg har truffet, er saa meget elskværdigere og finere end Potsnaps. Jeg var i Søndags til Lunch hos Prof. Sealey [?] (Høffdings Ven) der har et umaadelig tiltalende Hjem, og jeg besøgte i Lørdags Littlewood, der indbød mig til at komme til lunch hos sig paa Fredag (dersom der er noget jeg skal sige til ham fra Dig, kan jeg vist lige naa at faa Svar, dersom Du skynder Dig). Jeg har været lidt træt i nogle Dage, men det er alle her, det kommer af Klimaet, det har været som i et Dampbad næsten i en hel Maaned. Alle Folk jeg har talt med siger, at det er aldeles nødvendig i Cambridge at drive en eller anden Sport (de fleste Folk ser virkelig saa søvnige ud, at jeg slet ikke blev forbavset, da en Student sagde til mig "det er nødvendigt at drive Sport for man maa i hvert Tilfælde gøre een Ting energisk her i Cambridge"), jeg har derfor meldt mig ind i en Fodboldklub, og glæder mig meget til at begynde en af de allerførste Dage. Ja Harald, det var et langt Brev, men jeg tror ogsaa at det var det meste, af hvad jeg kunne fortælle. Jeg vil kun gerne til Slut sige, at Du jo nok forstaar, at det gaar mig saa godt, og at jeg saa udmærket forstaar, at alt i Begyndelsen maa gaa lidt langsomt, men jeg skal søge at drive det saa hurtigt som det paa nogen Maade er muligt. Jeg har skrevet om det altsammen, fordi jeg syntes at I engang maatte have Besked paa jeres Spørgsmaal. Jeg er saa glad fordi jeg har gjort det (dog kun dersom Du og Mor vil forstaa, at det gaar mig saa godt) for nu kan jeg saa meget bedre skrive om alt til jer, nu da I ved, hvordan jeg har det i Øjeblikket, og jeg glæder mig allerede til at skrive om alle de Fremskridt i Hurtigheden, som jeg føler vil komme allerede i de allernærmeste Dage, ja maaske allerede førend I faar dette Brev. De kærligste Hilsner til Dig og Mor (og 1000 Tak for alle Mors Breve)

fra Niels.

LETTER FROM NIELS TO HIS MOTHER (31 Oct. 11)

[Letter]

10 Eltisley Avenue, Cambridge.   31. Oct. 1911

Kære lille Mor!

1000 Tak for Dit og Haralds Brev, jeg er saa glad derfor. Det gaar mig saa godt, ikke fordi jeg har hørt fra Thomson siden, men jeg er i saa storartet Humør og har saa mange Planer. Men derom senere. Jeg kommer ikke til at mangle Selskab i

say many words, and the conversation that is pumped out of me always reminds me of the Frenchman at Potsnap's, only the people I have met are so very kind and much finer than Potsnaps. Sunday I had lunch with Prof. Sealey (the friend of Høffding) who has an extremely attractive home, and last Saturday I visited Littlewood who invited me for lunch next Friday (if there is anything I shall tell him from you, I think there is just time for you to answer me, if you hurry). I have been a little tired for some days, but everybody here is; it is due to the climate, it has been like a steam bath for almost a whole month. All the people I have talked to say that it is absolutely necessary here in Cambridge to go in for one sport or another (in fact, most of the people look so sleepy that I wasn't at all surprised when a student told me that "it is necessary to go in for sports; for one must at least do one thing energetically here in Cambridge"); I have, therefore, joined a football club and am looking forward to begin one of the very first days.

Well, Harald! This was a long letter, but I think also that it was most of what I had to say. In conclusion, I will only say that you will understand that I am getting along very well and that I understand quite well that everything must go slowly in the beginning; I shall try to speed things up as much as in any way possible. I have written about all these things, because I feel that you must at some time have answers to your questions. I am glad that I have done it (however, only if you and Mother will understand that I am getting along so well); for, now that you know how I am at the moment I can much better write to you about everything, and I am already looking forward to write about all the progress in speed that I feel will come already in the very next days, perhaps even before you receive this letter.

Much love to you and Mother (and thanks a million for all of Mother's letters)

from Niels

LETTER FROM NIELS TO HIS MOTHER (31 Oct. 11)

[Letter]

10 Eltisley Avenue, Cambridge.   31. Oct. 1911

Dear little Mother:

Thanks a million for yours and Harald's letter; it pleased me so much. Things are going very well; not that I have heard from Thomson yet, but I am in excellent spirits and have many plans. But about these later. I shall not lack company in the

de første 14 Dage; for i denne Uge kommer Carl Christian (jeg ved endnu ikke bestemt hvad Dag) og paa Fredag tager jeg til Manchester for at besøge Lorain Smith, og næste Lørdag tager jeg til Oxford for at besøge Dreyer (jeg skal naturligvis ogsaa besøge Haldane, og glæder mig saa meget dertil, men Dreyer indbød mig allerede paa Turen over). Hvad siger I dog til at jeg saadan farer rundt. Jeg glæder mig saa meget dertil, og ogsaa til at fortælle om det. Nu skal jeg skynde mig til Forelæsning og derfor Farvel og saa mange mange mange mange Hilsner fra

Din egen Niels

P.S. Tjeneste! Jeg vilde saa gerne bede Harald, om han, en Dag naar han kommer paa Læreanstalten, vilde gaa op og besøge Weber, der arbejder hos Prytz, og hilse fra mig og fortælle lidt om mig, jeg havde lovet at skrive, men har ventet saa længe, at et lille Kort ikke er nok, og jeg har ikke Tid og heller ikke rigtig Lyst i Øjeblikket at skrive rigtigt.

LETTER FROM NIELS TO HIS MOTHER (6 Dec. 11)

[Letter]

10 Eltisley Avenue, Cambridge. 6–12–11.

Kære Mor!

Tusind Tak for Dit forfærdelig rare Brev i Dag. Jeg gik ligestraks hen paa det fysiologiske Laboratorium og laante Zentralblatt; jeg var saa glad for at læse Zuntz' Tale. Det er saa morsomt at høre om det alt hjemmefra baade om hvormeget Harald faar udrettet og om hvor rart i har det sammen. Jeg fik ogsaa saadant et rart Kort fra Jenny i Dag (vil Du takke hende saa mange Gange). Du spørger om min Korrespondance. Ja den er baade videnskabelig; jeg sendte saaledes i Gaar et Brev paa 12 Sider til Oseen, og jeg skulde i Aften (jeg tror dog ikke jeg gør det, jeg gaar vist snarere i Seng) se at faa skrevet et langt Brev til Reinganum (han sendte migt et saa venligt Svar paa mit forrige Brev, og sagde at han syntes jeg havde Ret i mine Indvendinger imod hans Arbejde, og spurgte mig samtidig om jeg vilde være enig med ham, i et andet Punkt i hans Afhandling; men jeg maa ogsaa dennegang svare at det er jeg ikke); og af anden Slags. Jeg har saaledes i Dag haft stor Besvær ved at sætte et engelsk Brev sammen til Emileen. Hun sendte mig et saa umaadelig venlig Brev og bad mig om at komme og tilbringe Julen hos dem. Jeg har svaret at jeg vilde saa gerne komme, men at jeg ikke vidste om jeg kunde, da jeg ventede Besked om, naar Harald kom, og at jeg skulde skrive om nogle Dage (Hvad synes Du, at jeg skal svare. Jeg vil saa gerne høre Din Mening; hvis Du vilde

next fortnight; for, Carl Christian[22] is coming this week (I do not yet know the exact day); Friday I go to Manchester to visit Lorain Smith[23], and a week from Saturday I shall go to Oxford to visit Dreyer[24] (of course, I shall also visit Haldane; I am looking forward to that, but Dreyer invited me already on the trip over). What do you say to my rushing around like that? I am looking forward to it, but also to telling about it. Now I must hurry to a lecture; hence goodbye and very much love from

your own Niels

P.S. Favour! I would like to ask Harald, some day when he is at the Engineering College, to go and visit Weber who works with Prytz and give him my regards and tell him a little about me; I had promised to write, but I have waited so long that a little card will not be enough, and I have neither the time nor the inclination at the moment to really write.

LETTER FROM NIELS TO HIS MOTHER (6 Dec. 11)

[Letter]

10 Eltisley Avenue, Cambridge. 6-12-11

Dear Mother:

Thanks a million for your awfully nice letter which came today. I went immediately to the physiological laboratory and borrowed the Zentralblatt; I was so pleased to read Zuntz' speech[25]. It is so interesting to hear about everything at home, both about how much Harald gets done and about how much you enjoy each other's company. I also got a very nice card from Jenny today (will you thank her many times). You ask about my correspondence. Yes, it is partly scientific; thus, I mailed yesterday a 12-page letter to Oseen[26], and tonight (I don't think I shall do it, I'd rather go to bed) I should try to write a long letter to Reinganum; he sent me a very nice answer to my last letter; he said that he thought I was right in my objections to his work and asked me at the same time if I would agree with him on another point in his paper (but I must also this time answer that I do not);

---

[22] [Carl Christian Lautrup, the friend who in the summer of 1911 helped Bohr prepare an English translation of his dissertation.]

[23] [Lorain Smith was a friend of Bohr's father and a physiologist.]

[24] [Probably J. L. E. Dreyer, Danish astronomer; possibly Carl Dines Dreyer, who at that time went to London to study the wine trade.]

[25] [Presumably a necrology of Bohr's father.]

[26] [See pages 422-431.]

skrive straks, naar Du faar dette, saa vil jeg vente til at bestemme mig, til jeg faar Dit Svar. Det der taler imod, er at det er en saadan lang Rejse for saa faa Dage. Det der taler for, er at Emileens Brev var saa forfærdeligt venligt (Dersom Du engang vilde sende hende et Par venlige Ord, tror jeg at hun vilde blive saa glad for det), og at jeg gerne vilde besøge dem medens jeg er herovre, skøndt jeg havde tænkt at gøre det senere). Jeg har ogsaa haft saa meget Skriveri, i Anledning af et lille Selskab som jeg har bedt i Morgen til lunch. Først havde jeg tænkt kun at indbyde Hardy og Littlewood, der har været saa venlige imod mig, og som ikke kunde komme, da jeg bad dem sidst; da jeg saa hørte at Frk. Lehmann der ogsaa har været saa venlig imod mig rejser herfra om 8 Dage og ikke kommer igen foreløbig (hun bliver i hvert Tilfælde hjemme næste Aar), bad jeg ogsaa hende, og da jeg tænkte, at hun maaske ikke fik Lov til at komme alene, bad jeg ogsaa Mr Hill (den unge Mand som Frk Lehmann gav mig Introduction til, og som var saa venlig imod mig i de første Dage) og hans Søster; men det viste sig ikke at være tilstrækkelig, for baade Frk Lehmann og Miss Hill er students of Neunham College og maa ikke komme i Herreselskab uden en chaperon. Frk Lehmann har nu været saa elskværdig at paatage sig at skaffe en (jeg tænker, at det bliver Mrs Hill, Mr. Hill's Mor, som ogsaa har været saa elskværdig imod mig), men som Du ser er Selskabet efter haanden vokset i en frygtelig Grad, og jeg er bleven helt ængstelig for, hvordan det vil gaa, under mine smaa Forhold. (Jeg skal nok skrive til Margrethe i Morgen om det, for I maa ikke tro at jeg faar Tid til at skrive foreløbig til jer igen). Men nu Farvel for i Aften og saa mange mange Hilsner til Dig og Harald fra

Din egen Niels.

P.S. Du besørger nok nogle rigtig pæne Blomster til Moster Hanne fra mig. Vil Du takke hende saa mange Gange for alle hendes Hilsner, og sige at jeg skal nok snart skrive; Du vilde maaske lade mig vide, hvad Dag hun flytter. Vil Du ogsaa takke Moster Emma for hendes mundtlige Hilsner.

P.S. 2. Jeg faar nok at vide om Haralds Rejse, saa snart han ved det, saa at jeg kan lægge mine Planer.

P.S. 3. Dersom det ikke var formegen Ulejlighed, og dersom I kan pakke det lige saa godt ind som de andre Ting, Du har sendt, vilde jeg umaadelig gerne have Fotografiapparatet sendt herover. Dog kun hvis Du synes det.

PS. 4. Du maa undskylde den daarlige Skrift

PS. 5. Blot engang til Farvel og saa mange Hilsner

PS. 6. En lille Hilsen til.

partly it is of another kind. Thus, today I had much trouble composing an English letter to Emileen. She sent me an awfully nice letter and asked me to come and spend Christmas with them. I have answered that I would very much like to come but that I did not know if I should be able to, since I was awaiting word about when Harald would come, and that I would write in a few days. (What do you think I should answer. I would like to hear your opinion. If you will write as soon as you receive this letter, I shall wait to make up my mind till I get your answer. What seems to speak against it is that it is such a long trip for so few days. What speaks for it is that Emileen's letter was so awfully nice (if you would send her a couple of kind words some time, I think she would appreciate it) and that I would like to visit them while I am over here, although I had thought to do it later.) I have also had much writing in connection with a small party that I have invited for lunch tomorrow. First I thought of inviting only Hardy and Littlewood, who have been very nice to me and who could not come when I last invited them; when I learned that Miss Lehmann, who also has been so nice to me, is to leave from here in a week and will not return for some time (she will in any case stay at home the next year), I also invited her; and, when I thought that she might not be allowed to come alone, I also invited Mr. Hill (the young man to whom Miss Lehmann introduced me and who was so nice to me the first days) and his sister; however, both Miss Lehmann and Miss Hill are students at Newnham College and may not meet gentlemen without a chaperon. Miss Lehmann has been so kind to undertake to provide one (I think it will be Mrs. Hill, Mr. Hill's mother, who has also been very nice to me); as you see, the party has gradually grown to a terrible extent, and I am quite anxious about how it will come off in my small circumstances (I shall write to Margrethe about it tomorrow; I don't think I shall have time to write to you all for some time). But now goodbye for tonight and much love to you and Harald from

Your own Niels

P.S. I trust you to get some nice flowers for Aunt Hanne from me. Will you thank her many times for all her greetings and tell her that I shall write soon. Perhaps you will let me know what day she moves. Also, please thank Aunt Emma for her oral greetings.

P.S.2. I trust that I shall be informed about Harald's trip, as soon as he knows; so that I can make my plans.

P.S.3. If it isn't too much trouble, and if you can pack it as well as the other things you have sent, I should very much like to have the camera sent over here. However, only if you approve.

P.S.4. You must excuse my poor penmanship.

P.S.5. Once more, goodbye and much love.

P.S.6. Another little greeting.

LETTER FROM NIELS TO HIS MOTHER AND HARALD (28 Jan. 12)

[Letter]

Cambridge 28–1–12.

Kære Mor og Harald!

En lille bitte Hilsen for at sige at jeg i Dag har faaet et meget venligt Brev fra Rutherford, der skriver at det passer ham godt at jeg kommer til Manchester i næste term. Det er forfærdelig rart og nu er det altsaa bestemt; jeg har lige svaret ham. Jeg har saa travlt i Aften, at I kun faar denne lille Hilsen, der dog foruden at bringe den nævnte rare Efterretning, ogsaa skulde kræve lidt Efterretning igen. Jeg vilde saa gerne høre om Harald havde en rar Rejse og om hvordan han befinder sig efter at være kommen hjem. Jeg selv har det saa udmærket og sender saa mange mange mange Hilsner fra

jeres egen Niels.

LETTER FROM NIELS TO HIS MOTHER (29 Jan. 12)

[Letter]

Cambridge 29–1–12.

Kære Mor!

1000 Tak. Du ved vist ikke for hvad; men det er for Din Julepresent. Jeg har i Dag faaet et Par dejlige Skøjter af Dig. Der er kommen saadan en stærk Frost her og stor Skøjteløbning. Jeg har været ude at løbe i Eftermiddag; det var saa morsomt, og jeg var saa glad for det; jeg skal derud igen i Morgen. Mange, mange, mange Hilsner og 1000 Tak. (Vil Du give Bagsiden af dette Kort til Harald)

fra Din egen Niels

LETTER FROM NIELS TO HARALD (29 Jan. 12)

[Letter]

⟨Cambridge, 29–1–12⟩

Kære Harald!

Jeg har haft en saadan dejlig Dag, ikke just været flittig. Jeg har i Eftermiddag været ude at løbe paa Skøjter og i Aften været til et Foredrag af J J Thomson i

LETTER FROM NIELS TO HIS MOTHER AND HARALD (28 Jan. 12)

[Letter]

Cambridge 28-1-12.

Dear Mother and Harald:

A little greeting to let you know that I got a very kind letter from Rutherford today. He writes that it would suit him well that I come to Manchester next term. That is awfully nice, and so that is now settled; I have just answered him. I am so busy tonight that you only get this little greeting, which, besides giving the nice news mentioned, also is meant to ask for some news in return. I should very much like to hear if Harald had a nice trip and how he feels about being home again. I myself am getting along very well indeed and I am sending you much, much love from

your own Niels

LETTER FROM NIELS TO HIS MOTHER (29 Jan. 12)

[Letter]

Cambridge 29-1-12.

Dear Mother:

Thanks a million. You may not know what for; however, it is for your Christmas present. Today I have got a pair of excellent skates from you. We have got a hard frost here and great skating. I have been out skating this afternoon; it was great fun and I enjoyed it; I am going out again tomorrow. Much love and thanks a million. (Will you give the backside of this card to Harald)

from your own Niels

LETTER FROM NIELS TO HARALD (29 Jan. 12)

[Letter]

⟨Cambridge, 29-1-12⟩

Dear Harald:

I have had such a nice day, but haven't been exactly diligent. This afternoon I have been out skating, and this evening I have been to a lecture by J. J. Thomson

Trinity College over Bevægelsen af en Golfbold.* Du kan ikke tænke Dig, hvor morsom og oplysende det var, og hvor smukke Experimenter han viste, og saa det gnistrende funklende Humør hvormed det var holdt. Det var rigtig noget for mig, der jo selv har en lille Galskab for saadan noget. Vil Du gøre mig den Tjeneste at skrive Adresse paa indlagte Kort til Burrau, der blot takker for en Tilsendelse. Saa mange mange Hilsner fra

Din Niels

der ikke bliver ked af om Du en Dag skulde have Tid at sende ham et Par Ord.

*Du ved maaske at Golfbolde kan bringes til at følge de mærkeligste Kurver, skrue baade til venstre og højre i Luften og al den Slags.

LETTER FROM NIELS TO HIS MOTHER (5 Feb. 12)

[Letter]

Cambridge 5-2-12.

Kære Mor!

saa mange Tak for Dit Brev. Jeg vilde have skrevet til Dig i Lørdag, men fik saa travlt (se Hilsen til Harald). Min Stue var saa pæn, jeg havde pyntet med lidt Blomster paa min Kaminhylde (den kender Harald) hvor Faders Billede staar. Jeg har det saa udmærket, løber stadig paa Skøjter, og sender smaa dumme Hilsner fra

Din egen Niels

LETTER FROM NIELS TO HARALD (5 Feb. 12)

[Letter]

⟨Cambridge 5-2-12⟩

Kære Harald!

Endnu engang Tak for Dit Brev. Jeg har ikke skrevet om Philosophical Society, for Bedømmelseskomitéen havde ikke ladet høre fra sig før sidste Møde; men der er et nyt Møde allerede Feb. 12, og jeg haaber altsaa at jeg maaske vil høre til den Tid. Jeg vilde jo snart blive glad for at faa min Afhandling offentliggjort. Der er kommen en Afhandling af Richardson, i det sidste nummer af Phil. Mag. der

in Trinity College, dealing with the motion of a golf ball.* You cannot imagine how interesting and instructive it was, what beautiful experiments he showed, and the sparkling and scintillating humour with which it was delivered. It was really something for me who, as you know, am a little crazy about such things myself.

Would you be so kind as to write the address of Burrau[27] on the enclosed card which only acknowledges a paper.

Much love from your Niels

who would not mind it if you should some day have time to send him a couple of words.

*You may know that golf balls can be made to follow the strangest curves, swerve both to the left and to the right in the air, and all such things.

LETTER FROM NIELS TO HIS MOTHER (5 Feb. 12)

[Letter]

Cambridge 5–2–12.

Dear Mother:

Many thanks for your letter. I would have written to you last Saturday but got so busy (see the note to Harald). My room was so nice, I had decorated it with a few flowers on my mantelpiece where Father's portrait stands (Harald knows it). I am getting along splendidly; I still go ice skating and send little silly greetings from

your own Niels

LETTER FROM NIELS TO HARALD (5 Feb. 12)

[Letter]

⟨Cambridge 5–2–12⟩

Dear Harald:

Once more thanks for your letter. I haven't written about the Philosophical Society; for, the judging committee hadn't reported before the last meeting; but there is another meeting on February 12, and then I hope to hear from them. I would be glad to get my paper published soon. There has appeared a paper by Richardson in the last issue of Phil. Mag. which came Saturday, dealing with the

27 [Presumably Carl Burrau, Danish astronomer and actuary.]

udkom i Lørdags, om Elektrontheorie for Thermoelectricitet. Det var en meget morsom Afhandling, men han kunde slet ikke faaet fat i det samme som jeg. R. tror saaledes at have givet en almindelig Udledning for Peltier og Thomson Effecterne, men hans Formler er kun Lorentz's og altsaa slet ikke mine langt almindelige Formler. Det var ikke saa let at finde Manglerne, og jeg havde sikkert aldrig gjort det, havde jeg ikke kendt mit eget Arbejde. Jeg har skrevet en lille Note til Phil. Mag. derom (den er sendt af i Aften), jeg tror at den er ganske god, (Owen har hjulpet mig saa venligt), og jeg vilde blive saa glad om de vil offentliggøre den for mig uden Vrøvl, jeg vilde da haft stor Fornøjelse af Richardsons Afhandling, der iøvrigt som sagt, er ganske udmærket. Mange mange Hilsner fra Din

Niels,

som Du nok kan mærke at det gaar ganske godt for i Øjeblikket

LETTER FROM NIELS TO HARALD (9 Feb. 12)

[Card]

Cambridge 9-2-12

Kære Harald!

Mens jeg sender det Brev af Sted som Hardy sendte mig for at jeg skulde kompletere Adressen, vil jeg gerne selv sende en lille Hilsen. Jeg sidder og læser i Larmors Bog (Du ser det gaar ikke saa hurtig) jeg har efterhaanden ganske forandret min Mening om den. Larmor er en meget stor Mand, men lidt af den samme Aandsretning som Thiele (Du kan tænke Dig at Thieles Forelæsninger for en Udlænding der ikke fuldt behersker Sproget, vil tage sig noget mærkelige ud) men dygtigere, idet hans Viden er saa utrolig stor, naar jeg læser hans Bog forekommer den en næsten bundløs. Det er saa morsomt at læse den og jeg tror at jeg vil læse den 2 Gange. [...[28]] Jeg tænker saa ofte paa at skal jeg faa noget ud af mulige Forelæsninger, maa jeg søge al den Hjælp jeg kan faa. Jeg tænker paa f. Eks. hvordan 2 Brødre kunde dele Potentialteorien.

Saa mange mange, mange Hilsner til Dig og Mor fra
Din Niels

[28] [Two words, in quotation marks, are illegible.]

electron theory of thermoelectricity. It was a very interesting paper, but he had not at all got hold of the same as I. Thus, Richardson believes to have given a general derivation of the Peltier and Thomson effects, but his formulas are just Lorentz', and not at all my much more general formulas. It was not so easy to find what was wrong, and I would probably never have done it if I had not known my own work. I have written a little about it to Phil. Mag. (I have mailed it tonight). I believe it is quite good (Owen has been kind enough to help me), and I should be pleased if they would publish it without any fuss; I would then have had great pleasure of Richardson's paper which, as I said before, is really excellent.

Much love from your Niels

who, as you undoubtedly can feel, is getting along quite well at the moment.

LETTER FROM NIELS TO HARALD (9 Feb. 12)

[Card]

Cambridge 9-2-12

Dear Harald:

While forwarding a letter which Hardy gave me to complete the address, I would like myself to send a little greeting. I am sitting and reading Larmor's book (you see that it is not going so fast); I have gradually changed my opinion of it completely. Larmor is a very great man, but somewhat of the same bent as Thiele (you can imagine that Thiele's lectures would be rather strange for a foreigner who doesn't quite master the language) but abler, in that his knowledge is so unbelievably vast; when I read his book it appears almost unfathomable to me. It is very interesting to read it, I think I shall read it twice. [...[28]]. I often think that, if I am to get something out of the available lectures, I must seek all the help I can get. For example, I recollect how two brothers could share the potential theory.

Much love to you and Mother from
your Niels

[28] [Two words, in quotation marks, are illegible.]

LETTER FROM NIELS TO HARALD (19 May 12)

[Letter]

Hulme Hall, Manchester, 19–5–12.

Kære Harald!

Du har jo hørt, om hvordan det er gaaet med min Afhandling; Du maa ikke
tro, at det var min Mening ikke at skrive til Dig om det, men jeg vidste ikke helt,
hvad jeg vilde gøre med den; og havde derfor (for ikke at gøre altfor meget u-
nødvendigt Vrøvl) tænkt, først lidt selv at prøve, om jeg let kunde forkorte den
og, om jeg vilde synes nu om det. For at sætte Dig helt ind i Sagen sender jeg Dig en
Kopi [29] af Sekretærens Brev og mit Svar. Jeg ved endnu ikke bestemt, hvad jeg vil
gøre; jeg har endnu ikke haft Tid til for Alvor at prøve at forkorte den, men tror
i Øjeblikket at det bliver til det; jeg er nemlig ikke sikker paa, at den ikke vil blive
bedre (at læse) af det; og det vil koste for meget Arbejde (og et Arbejde, jeg ikke
tør indlade mig paa nu) at gøre den saa fuldstændig, at der kunde være nogen Mening i at forsøge paa at faa nogen til at tage den som en Bog. Jeg er klar over, at
hvad der skal gøres, det maa gøres meget snart, og skal nu for Alvor prøve paa det.
Kære Harald, jeg haaber at Du har det saa rart derude; jeg kunde jo af mange
Grunde have Lyst til at komme paa et lill⟨e⟩ Visit hos Dig; men jeg skal blive saa
glad for et lille Ord, om hvordan Du har det (jeg har dog hørt saa meget rart gennem Mor), og dersom Du har tænkt paa noget, som Du tror, at jeg ikke har tænkt
paa (jeg er dog temmelig sikker paa, at det rigtigste er først (om jeg kan faa Tid) at
prøve paa, hvad jeg skrev; saa snart jeg mærker, hvordan det gaar, skal jeg skrive
til Dig).
Saa mange, mange kærlige Hilsner

fra Niels

P.S. 1. Jeg var jo i Lørdags i Cambridge med Niels Erik; og jeg introduced⟨e⟩
ham til Hardy og Littlewood; og de var som altid saa elskværdige, og bad hilse
Dig saa mange Gange. Hardy sagde, at han var meget interesseret i at høre fra
Dig om sit Brev til Dig (om Approximationerne), han sagde, at han selv havde
skrevet, at han ikke ventede Svar, men at han alligevel engang vilde blive meget
glad for at høre et Ord om Din Mening. Jeg fortalte dem, at Du ikke havde været
helt rask, dels for at fritage Dig for Forpligtelse til at svare, og ogsaa for at ikke
Littlewood skulde tage det som et daarligt Exempel, for det tror jeg ikke han
kan taale. Han bad os til The den ene Dag (Hardy den anden), jeg kan slet ikke
sige, hvor godt jeg synes om ham, men han fik ikke snakket meget Matematik

LETTER FROM NIELS TO HARALD (19 May 12)

[Letter]

Hulme Hall, Manchester, 19-5-12.

Dear Harald:

You have heard what has happened to my paper. Don't think it was my intention not to write to you about it, but I didn't know quite what to do with it and had, therefore, (in order not to make too much unnecessary fuss) first thought a little about seeing if I could easily shorten it myself, and if I would like that. To make the matter quite clear to you, I am sending you copies [29] of the letter of the secretary and of my answer. I do not yet know definitely what to do. I haven't had time yet to make a serious effort to shorten it, but I believe now that that is what must be done; in fact, I am not sure that it won't be better (easier to read) thereby; and it would require too much work (and work that I dare not undertake now) to make it so complete that there would be sense in trying to get somebody to take it as a book. It is clear to me that what has to be done must be done very soon, and I shall now try it seriously.

Dear Harald! I hope that you are getting along nicely out there; I should like for many reasons to pay you a little visit; I should be glad for a little word about how you are getting along (however, I have heard many nice things through Mother) and if you have thought of something that hasn't occurred to me (however, I am fairly sure that the best is (if I can find the time) to try first what I wrote; as soon as I feel how it goes, I shall write to you).

Much love from
Niels

P.S. 1. Saturday I was in Cambridge with Niels Erik [30] and I introduced him to Hardy and Littlewood; and they were as always so very nice and asked to be remembered to you. Hardy said that he was very interested in hearing from you about his letter to you (about the approximations); he said that he had written himself, that he didn't expect an answer but nevertheless would appreciate getting a word about your opinion some time. I told him that you hadn't been quite well, partly to relieve you of the obligation to answer and partly in order that Littlewood might not take it as a bad example; for, I don't think he can stand that. He invited us to tea one of the days (Hardy the other day); I cannot tell you how well I like him, but he did not get to talk much mathematics with

[29] [Longhand copies of the letters from Barnes to Bohr and Bohr to Barnes were enclosed. These were in English and appear below on both sides.]

[30] [Niels Erik Nørlund, see note 11.]

med Niels Erik (hvad derimod Hardy fik den anden Dag), jeg tror, at han er saa beskeden og saa bevidst om sin Uvidenhed, at han ikke havde Mod til det.

P.S. 2. Da jeg var i London og skulde hente N.E. paa Banen, gik jeg op til Udgiveren af Phil. Mag., og det havde til Følge, at jeg fik Korrekturen paa den lille Note 2 Dage efter; jeg haaber, at den kommer i næste Nummer.

P.S. 3. Jeg haaber, at Du forstaar, at jeg ikke er ked af det, og at dersom jeg skriver derom, saa er det fordi at jeg er allergladest for det; og at det, det nu mest kniber med, er med min Tid lige i Øjeblikket.

P.S. 4. Endnu engang saa mange mange Hilsner.

LETTER ENCLOSED WITH LETTER FROM NIELS TO HARALD (19 May 12)

[Letter]

Trinity College, Cambridge, May 7, 1912.

Dear Dr. Bohr,

The Council of the Cambridge Philosophical Society yesterday considered the reports of the referees on the paper which you presented on November 13th last. The expense of printing so long a paper is one which the Society cannot undertake. They understand, however, that without materially reducing the value of the paper you could cut it down to one-half its length. If you care to attempt this task, and to send it to us again, I think that it would be accepted.

I will send the paper to you when you inform me of your present address.

Sincerely yours
E.L. Barnes.

LETTER ENCLOSED WITH LETTER FROM NIELS TO HARALD (19 May 12)

[Letter]

Manchester, May 8, 1912.

Dear Dr. Barnes,

I thank you for your kind letter of May 7, in which you inform me about the decision of the Council of the Cambridge Philosophical Society regarding my paper. I shall consider the proposal stated in it

Yours faithfully
N Bohr

My address is "Hulme Hall, Victoria Park, Manchester".

Niels Erik (what on the other hand, Hardy did the other day); I believe that he is so modest and aware of his ignorance that he did not have the courage to do it.

P.S. 2. When I was in London to meet N.E. at the station, I went to the publisher of Phil. Mag. with the result that I got the proof of the little note[31] two days later; I hope it will appear in the next issue.

P.S. 3. I hope you understand that I am not in bad humour, and that if I write about something it is because I am especially happy about it, and that what is most scarce for me at the moment is time.

P.S. 4. Once more, much love.

LETTER ENCLOSED WITH LETTER FROM NIELS TO HARALD (19 May 12)

[Letter]

Trinity College, Cambridge, May 7, 1912.

Dear Dr. Bohr,

The Council of the Cambridge Philosophical Society yesterday considered the reports of the referees on the paper which you presented on November 13th last. The expense of printing so long a paper is one which the Society cannot undertake. They understand, however, that without materially reducing the value of the paper you could cut it down to one-half its length. If you care to attempt this task, and to send it to us again, I think that it would be accepted.

I will send the paper to you when you inform me of your present address.

Sincerely yours
E.L. Barnes.

LETTER ENCLOSED WITH LETTER FROM NIELS TO HARALD (19 May 12)

[Letter]

Manchester, May 8, 1912.

Dear Dr. Barnes,

I thank you for your kind letter of May 7, in which you inform me about the decision of the Council of the Cambridge Philosophical Society regarding my paper. I shall consider the proposal stated in it

Yours faithfully
N Bohr

My address is "Hulme Hall, Victoria Park, Manchester".

[31] [See pages 440–442.]

LETTER FROM HARALD TO NIELS (23 May 12)

[Letter]

23-5-12.

[Letterhead:]
Hareskov Kuranstalt
Bagsværd St. Telf. Bagsværd 48
[Vignette]

Kære Niels!

Mange Tak for Dit Brev. Jeg tror ogsaa at det vilde være det bedste (d.v.s. faa Dit Arbejde læst og udbredt mest) hvis Du i forkortet Skikkelse fik det optaget i Cambr. Phil. Soc.; selvom jeg ikke kan nægte at jeg synes det – for nu ikke at bruge stærke Ord – er en stor Fejl at[32] Thomson og de andre ikke at ville have det i den Form Du havde givet det. Men hvis Du selv tror, at Du kan[33] forkorte Arbejdet med c. Halvdelen uden at vanskelliggøre Sagkyndige Læsningen, synes jeg absolut det, som Sagen nu staar, vilde være det bedste. At offentliggøre det hele som en Bog, er dels temmelig dyrt, naa! det kom man jo nok udover, men væsentligere er som Du skriver at en saadan privat udgivet Bog sikkert kun vil blive læst af overordentlig faa i Sammenligning med hvad en Tidsskriftartikel vil blive læst. Jeg har dog lidt svært ved rigtig at raade Dig, da jeg jo slet ikke ved hvor travlt Du har hos Rutherford og hvorvidt Du kan finde nogen Tid til Omredaktionen. Men, selvom det som Du skriver haster, og det gør det, saa gør det dog en stor Forskel at Du har offentliggjort Sagen paa Dansk og derfor spiller en Maaned mer eller mindre ikke den Rolle som det vilde gøre under andre Omstændigheder. – Jeg synes det var et udmærket Svar Du skrev til Barnes, og "Til Lykke" med Phil. Mag. – [...]

LETTER FROM NIELS TO HARALD (27 May 12)

[Letter]

Hulme Hall, Victoria Park, Manchester 27–5–12.

Kære Harald!

1000 Tak for Dit Brev. Med Hensyn til Afhandlingen, saa har jeg endnu slet ikke haft Tid, og er bange for, at jeg ikke faar det saa snart, dersom jeg skal gøre noget virkeligt ved Arbejdet paa Laboratoriet. Jeg ved heller ikke helt bestemt endnu, hvad jeg vil gøre med den, jeg tænker lidt paa en Udvej til at faa den frem i hel Form; Du skal snart høre fra mig om det, naar jeg har hørt om der er Mulighed

[32] [Writing mistake for "af".]
[33] [The word "kan" is repeated by mistake.]

LETTER FROM HARALD TO NIELS (23 May 12)

[Letter]

23–5–12

Dear Niels:

Many thanks for your letter. I also believe the best you can do (i.e., to obtain for your work the widest reading and distribution) is to have it accepted in abridged form in Cambr. Phil. Soc.; although I can't help feeling – not to use strong words – that it is a great mistake of Thomson and the others not to take it in the form you have given it. But if you believe yourself that you can reduce the work to about half its size without making it difficult to read for the experts, I definitely feel, as things are now, this would be the best. To publish the whole as a book is on the one hand rather expensive, well! one could manage it, but a more important consideration, as you write, is that such a privately published book would certainly be read only by very few people in comparison with those who would read a paper in a periodical. However, it is a bit difficult for me to give you definite advice, since I don't know at all how busy you are with Rutherford and how much time you can find for the re-writing. But, even though, as you write, there is urgency, and indeed there is, yet it makes a big difference that you have published the work in Danish and because of this a month more or less does not matter as much as it would in other circumstances. I think your answer to Barnes was excellent, and congratulations with Phil. Mag. [...]

LETTER FROM NIELS TO HARALD (27 May 12)

[Letter]

Hulme Hall, Victoria Park, Manchester 27–5–12

Dear Harald:

Thanks a million for your letter. With regard to the paper, I haven't had any time yet, and I am afraid that I won't have it very soon if I am to do some real work in the laboratory. I am not quite sure either what I want to do with it. I am thinking a little about the possibility of getting it out in complete form. I shall let you know about it soon, when I have found if it is possible. Today I would really

for det. I Dag kunde jeg nok have Lyst til at faa den offentliggjort hel, for jeg bilder mig ind, at jeg har fundet ud af noget, som dersom det er rigtig, (og intet Menneske kan, saa vidt jeg ser, paastaa det modsatte, før der foreligger flere Forsøg (nogle jeg tænker paa at gøre til næste Aar sammen med Owen, dersom ingen anden har gjort det før den Tid)), vil bortrydde alle de Hovedindvendinger der kan rejses imod (og er bleven rejst i den senere Tid) en Elektrontheorie af den Art, som jeg har behandlet; og se, er det Tilfældet, vil jo Værdien af mit Arbejde være en lidt anden, end den nu anses for. Kære Harald, Du ved jo, hvor let jeg kan tage Fejl; og det er maaske ogsaa dumt at sige saadan noget saa tidlig; men jeg havde saadan Lyst til at kunne tale med Dig i Aften, for jeg har jo slet ingen her, der virkelig interesserer sig for saadan noget. Er der nogen Mening i det, saa skal jeg se, hvor hurtig jeg kan faa det skrevet og faa det frem; men jeg har jo lidt daarlig Tid til virkelig at samle sig om saadan noget og faa læst den nødvendige Literatur, naar jeg er hele Dagen paa Laboratoriet, hvad der er absolut nødvendig. Du spørger ogsaa til Arbejdet paa Laboratoriet. Det gaar egentlig rigtig godt. Jeg maa desværre straks sige, at jeg er endnu ikke helt sikker paa hvor meget der vil komme ud af det, Rutherford har sat mig til. Jeg mener med Hensyn til Ofentliggørelse, for jeg vil i hvert Tilfælde have saa megen Fornøjelse af det, for Rutherford er en Mand, som man ikke kan tage Fejl af; han kommer regelmæssig og hører om hvordan det gaar og taler om hver eneste lille Ting. Det er blot den Idé han havde, som det er svært at sige, om er anvendelig, førend man rigtig har prøvet. Skal den vise sig uanvendelig, er der andre Veje (maaske mere direkte) til Behandling af det samme Spørgsmaal, og som jeg ogsaa tænker, at det er Meningen jeg skal prøve, dersom der bliver Tid. Men hvordan det end gaar, saa lærer jeg saa meget hver Dag, fordi det er et virkeligt Arbejde, og fordi Rutherford er en saa udmærket Mand, og virkelig interesseret i alle de Folks Arbejde, der gaar hos ham. Ja, dette lille Brev, kunde jo se ud som om det indeholdt lutter Klager (Du maa endelig ikke tale til nogensomhelst om en eneste Smule af, hvad jeg har skrevet om, for det er saa vanskeligt at forklare), men jeg tror dog, at Du kan skimte igennem, at mit Humør alligevel ikke er saa daarligt endda, og at det, som jeg bilder mig ind maaske at have fundet ud af, og som fik mig til at skrive til Dig, just heller ikke har sat mit Mod ned, selvom om det maaske har faaet mig til at længes endnu mere efter den Tid, der forhaabentlig snart skulde komme hvor jeg vil have rigtig Ro til at arbejde paa de forskellige Ting, og hvor nogle af de større eller mindre praktiske [*] Vanskeligheder skulde være overvundne. Endnu engang saa mange mange Tak for Dit Brev (jeg skal ikke blive ked af at høre igen, og maaske at høre at Maven har makket helt ret)

De kærligste Hilsner fra Din Niels

[*] Du vil nok skrive til mig, saa snart Du er kommen til København og har faaet talt med nogen, om hvordan Sagerne staar.

like to have it published in full; for, I believe that I have found out something, which, if it is true (and, as far as I can see, nobody can claim the opposite before more experiments are made (some that I am thinking of doing next year together with Owen, if no one has done them in the meantime)), will remove all the main objections that can be raised (and have lately been raised) against an electron theory of the kind I have treated; and, indeed, if that is the case, the value of my work will be a little different from what it is now considered to be.

Dear Harald! You know how easily I can be wrong, and it is perhaps also silly to tell about such things so soon, but I wanted so much to talk to you tonight; for, here I have no one who is really interested in such things. If there is some sense in it, I shall see how soon I can get it written up and get it out, but I have hardly time to really concentrate on such things and get the necessary literature read when I am at the laboratory all day, which is absolutely necessary. You also ask about the work in the laboratory. It is really going quite well. Unfortunately, I must say right off that I am not yet sure how much will come of what Rutherford has put me on. I mean with regard to a publication; for, in any case, I shall have much pleasure of it; for, Rutherford is a man whom one cannot be mistaken about; he comes regularly to hear how things are going and talk about every little thing. It is only that it is difficult to say before it is really tested if the idea he had is applicable. If it should be inapplicable, there are other ways (perhaps more direct ones) of treating the same question, and which I believe I am supposed to try if I have time. But, however it goes, I learn a lot every day, since it is real work, and since Rutherford is such an excellent man and takes a real interest in the work of all the people working with him.

Well, this little letter might appear to contain nothing but complaints (please, don't talk to anyone about anything I have written; for, it is so difficult to explain); however, I believe that you will have an inkling that my mood after all isn't so bad, and that what I presume to have found out, and what made me write to you, hasn't exactly lowered my spirits, even if it perhaps has made me long for the time, which I hope will come soon, when I shall have real peace to work on the different things, and when some of the greater or smaller practical * difficulties would have been overcome. Once more, many many thanks for your letter (I shall not mind to hear again, and perhaps hear that your stomach is now fully recovered).

Much love from your Niels

* Please write to me, as soon as you get to Copenhagen and have talked with somebody, about how matters stand.

LETTER FROM NIELS TO HARALD (28 May 12)

[Letter]

⟨Manchester⟩ 28–5–12

Kære Harald!

Jeg glemte at lægge Brevet fra i Gaar i Postkasse, og aabner det derfor nu igen for at sige, at jeg vist allerede har maattet ændre mine Anskuelser lidt. Jeg tror vedvarende, at det, jeg skrev jeg havde hittet paa, maaske ikke har saa ringe almindelig Betydning (dersom det vil vise sig at være i Overstemmelse med Experimenterne) og at det kan forklare forskellige Vanskeligheder i Metallernes Elektrontheori af almindelig Karakter (saaledes den ud fra simple Betragtninger uforstaaelige Ting (som Du maaske husker) at Thomsoneffecten tilsyneladende bliver af en urigtig Størrelsesorden, samt at Metallernes Varmefylde ikke er større ved lave Temperaturer (en Indvending som Du vist har hørt)); men hvorvidt der er Mulighed for at forklare de mere specielle Ting, der afhænger af de specielle Forhold for Elektronernes Bevægelse i Metallet er en ganske anden Sag. Jeg tror saaledes næppe at de iagtagne høje Værdier for den electriske Ledningsevne hos de godt ledende Metaller, kan forklares udfra de specielle Antagelser jeg har benyttet i § 4, nemlig næppe uden at tage Hensyn til Kræfterne mellem Elektronerne, der saa at sige for dem til at følges ad gennem Metallet. Jeg har lige nu til Morgen læst en meget interessant Afhandling af Stark (en meget bekendt Mand) derom (han ved ikke at de Synspunkter han benytter til at forklare Forskellen mellem Legeringer og de rene Metaller, for den væsentligst⟨e⟩ Del er de samme som jeg har benyttet); han giver meget interessante Antydninger af en Forklaring af den electriske Ledningsevne; men han skriver at han ser ikke hvordan det kan benyttes til Forklaring af Varmeledningsevnen; det sidste aner derimod jeg og tænker allerede paa maaske at prøve at skrive lidt derom. Nu skal jeg ikke plage Dig længere med alt mit Vrøvl; jeg maa nu lade lidt Tid gaa, og komme lidt til Ro i alle de forskellige Ting. Hvormeget jeg for min Part kan faa ud af det i Aar, det kan jeg slet ikke have nogen Mening om; det vil afhænge af saa mange ydre Forhold, og ogsaa af hvad andre finder paa eller har fundet at skrive om de samme Ting. Jeg føler bare at jeg maaske igen begynder at komme lidt ind i Sagen. Det har været saa rart at skrive om det altsammen til Dig, men Du vil jo forstaa, at det ikke er sikkert, noget af det. Nu Farvel, for nu maa jeg løbe til Laboratoriet.

LETTER FROM NIELS TO HARALD (28 May 12)

[Letter]

⟨Manchester⟩ 28–5–12

Dear Harald:

I forgot to drop the letter in the mail box yesterday and open it now to tell you that I believe that I have already had to change my ideas a little. I still think that what I wrote had occurred to me is perhaps of no small general significance (if it should turn out to be in agreement with experiments) and that it can explain certain difficulties of a general nature in the electron theory of metals (such as the fact, incomprehensible on the basis of simple considerations, that the Thomson effect (as you may remember) apparently is of the wrong order of magnitude, and also that the specific heat of metals is not larger at low temperatures (a difficulty that I believe you have heard of); but whether it will be possible to explain the more specific matters, which depend on the particular conditions under which the electrons move in the metal, is something else again. Thus, I am inclined to believe that it may not be possible to explain the high values of the electric conductivity of the highly conducting metals from the special assumptions I used in section 4, i.e., without considering the forces between the electrons which, so to speak, make them move together through the metal. This morning I read a very interesting paper about it by Stark (a very well-known man). (He doesn't know that the ideas he uses to explain the difference between the alloys and the pure metals are, for the most part, the same that I have used.) He gives interesting hints of an explanation of the electric conductivity, but he writes that he doesn't see how it can be applied to explain the heat conductivity; but this I have an idea of, and I am already thinking of perhaps trying to write a little about it.

Now I shall not bother you any more with all my nonsense; I must now let a little time pass and settle down with all these various things. For my part, I have no idea of how much I can accomplish this year; it will depend on so many external circumstances, and also on what others find, or have found, to write about the same subject. I only feel that I am perhaps beginning to get back into the field again.

It has been so nice to write about all this to you; but you will understand that none of it is certain.

Now goodbye; for, now I must run to the laboratory.

LETTER FROM NIELS TO HARALD (12 June 12)

[Letter]

Hulme Hall Victoria Park Manchester 12-6-12.

Kære Harald!

Mange Tak for Dine Afhandlinger; den ene kender jeg jo, og jeg glæder mig til meget snart at læse den anden. Men jeg ledte forgæves i Konvoluten efter noget skrevet. Det gaar mig ikke saa helt daarligt i Øjeblikket, jeg havde for et Par Dage siden en lille Idé med Hensyn til Forstaaelsen af Absorption af α-Straaler (det gik til paa den Maade, at en ung Matematiker her, C G Darwin (Sønnesøn af den rigtige Darwin) lige har offentliggjort en Theori om dette Spørgsmaal, og jeg syntes, at den ikke alene ikke var helt rigtig i det mere matematiske (det var dog kun temmelig lidt) men meget utilfredsstillende i Grundopfattelsen), og har udarbejdet en lille Theori derover, der selv om den er meget'lille, maaske dog kan kaste lidt Lys over nogle Ting med Hensyn til Atomernes Bygning. Jeg tænker meget snart at offentliggøre en lille Afhandling derom. Du kan tro at det er morsomt at være her, her er saa mange at tale med (mine Klager sidst gjalt mere almindelige theoretiske Spørgsmaal) og det af dem der har den allermeste Forstand paa saadan noget, og Prof. Rutherford tager en saa virkelig og effektiv Interesse i alt, som han synes er noget til. Han har i de sidste Aar udarbejdet en Theori om Atomers Bygning, som synes at være helt anderledes solidt begrundet end alt hvad man tidligere har haft. Og ikke fordi mit er noget af samme Betydning eller Slags, saa passer mit Resultat ikke saa daarlig over ens dermed (Du forstaar nok, at jeg kun mener, at Grundlaget for min lille Beregning kan lade sig bringe i Overensstemmelse med hans Idéer). Jeg har i et Par Dage ikke arbejdet paa Laboratoriet, fordi jeg har maattet vente paa noget Radium; men jeg begynder i Morgen eller i Overmorgen igen; og paa noget, som jeg tror, maaske er mere frugtbringende end det forrige. Det lille tvungne Ophold har passet mig vidunderligt for Udarbejdelsen af den lille Theori. Du kan tro, at jeg ofte har tænkt paa Dig i disse Dage, for jeg skulde bruge lidt Matematik og tænkte hele Tiden paa at spørge Dig til Raads; men hvergang jeg skulde til at skrive til Dig, fandt jeg ud af lidt, og til Slut kom jeg selv igennem. Det drejede sig om at finde Værdien af

$$K = \int_0^\infty f(x) \cdot f'(x) - \lg x \cdot \mathrm{d}x, \quad \text{hvor} \quad f(n) = \int_{-\infty}^{+\infty} \frac{\cos nx \, \mathrm{d}x}{(1+x^2)^{\frac{3}{4}}}.$$

[Først vilde jeg have spurgt om $f(x)$ (der næppe kan findes ved complex Integration) var kendt, saa fandt jeg at den tilfredstillede Ligning[34] $0 = f''(x) - \frac{1}{x} f'(x) - f(x)$,

LETTER FROM NIELS TO HARALD (12 June 12)

[Letter]

Hulme Hall, Victoria Park, Manchester 12–6–12

Dear Harald:

Many thanks for your papers; one of them I know, and I am looking forward to reading the other one very soon. But I searched the envelope in vain for something in writing. I am not getting along badly at the moment; a couple of days ago I had a little idea with regard to understanding the absorption of α-rays (it happened in this way: a young mathematician here, C. G. Darwin (grandson of the real Darwin), has just published a theory about this problem, and I felt that it not only wasn't quite right mathematically (however, only slightly wrong) but very unsatisfactory in the basic conception), and I have worked out a little theory about it, which, even if it isn't much, perhaps may throw some light on certain things connected with the structure of atoms. I am planning to publish a little paper about it very soon. Believe me, it is interesting to be here; there are so many to talk to here (my complaints last time applied to the more general theoretical problems), and among them are those who understand such things best, and Prof. Rutherford takes a real and effective interest in everything he considers worth while. In the last few years he has worked out a theory of atomic structure which seems to have a much more solid basis than anything that we had formerly. Not that my theory is of the same kind and significance; nevertheless, my result doesn't agree so badly with his (you understand, of course, that I only mean that the basis for my little calculation can be brought to agree with his ideas). I have not worked in the laboratory for a couple of days, because I have had to wait for some radium; but tomorrow or the day after I begin again, and on something that I think may be more fruitful than the previous work. This little forced delay has been wonderfully convenient for working out my little theory. Believe me, I have thought of you often these days; for, I had to use some mathematics and thought of asking for your advice; but each time I was about to write to you I found out something, and finally I managed to get through. The problem was to find the value of

$$K = \int_0^\infty f(x)(f'(x) - \ln x)\mathrm{d}x, \quad \text{where} \quad f(n) = \int_{-\infty}^{+\infty} \frac{\cos nx \, \mathrm{d}x}{(1+x^2)^{\frac{3}{2}}}.$$

{First, I wanted to ask if $f(x)$ (which hardly can be obtained by complex integration) is known; then I found that it satisfies the equation [34] $0 = f''(x) + \frac{1}{x}f'(x) - f(x)$;

---

[34] [Bohr has the wrong sign in the middle term.]

det næste Skridt var at finde paa at Løsningen var af Formen

$$f(x) = (a+b \lg x)\left(\left(\frac{x}{2}\right)^2 + \frac{1}{1 \cdot 2}\left(\frac{x}{2}\right)^4 + \frac{1}{1 \cdot 2 \cdot 1 \cdot 2 \cdot 3}\left(\frac{x}{2}\right)^6 \cdots\right)$$

$$+ 2b\left(\frac{1}{4} - \frac{1}{1 \cdot 2}\left(\frac{x}{2}\right)^4 \frac{3}{4 \cdot 2} - \frac{1}{1 \cdot 2} \cdot \frac{1}{1 \cdot 2 \cdot 3}\left(\frac{x}{2}\right)^6 \left(\frac{3}{2 \cdot 4} + \frac{5}{4 \cdot 6}\right)\right) \cdots$$

(det er bagefter gaaet op for mig, hvor stor en Uvidenhed det røbede, at det var saa svært for mig at hitte paa at faa den $\lg(x)$ med), saa gjaldt det om at finde a og b (det ses let at $b = 4$, men a?), jeg prøvede hæderlig derpaa; men efterhaanden var det gaaet op for mig, hvor meget det hele lignede Cylinderfunctionernes Theori̇ (Ligningen for dem hedder $f''(x)+\frac{1}{x}f'(x)-f(x) = 0$, hvad der er en principiel Forskel) og ved at blade sig igennem Lord Rayleigh's "Theory of Sound" (hvilken udmærkede Bog, jeg meget snart maa læse meget grundig) og ved at gøre omtrent det modsatte af ham (undersøge den imaginære Del af et Integral, naar han undersøgte den reelle af det analoge Integral; og andre Ting af samme Slags) fandt jeg da, at $a = 4\gamma - 4 \lg 2 - 2$ hvor γ er Eulers Konstant. Efterhaanden var jeg bleven saa dreven i samme Snydemetode, at jeg forholdsvis let fandt følgende asymptotiske Udvikling (for store x)

$$f(x) = \sqrt{2\pi} \cdot e^{-x} \cdot x^{\frac{1}{2}}\left(1 + \frac{1 \cdot 3}{8x} - \frac{1 \cdot 3 \cdot 5}{1 \cdot 2}\left(\frac{1}{8x}\right)^2 + \frac{1 \cdot 3 \cdot 1 \cdot 3 \cdot 5 \cdot 7}{1 \cdot 2 \cdot 3}\left(\frac{1}{8x}\right)^3 \cdots\right) ;$$

og efter nogle Dages numerisk Slid fandt jeg da den ønskede beskedne Resultat, at $K = -0,540$, hvilket jeg rigtignok vil haabe er temmelig rigtig (jeg sender maaske meget snart nogle smaa Regninger til muligt Eftersyn). Det eneste jeg tror, at Du kan faa ud af al denne Snak, er vist, at jeg var temmelig interesseret i Resultatet (det var kun en Korrektion i mine Regninger, men jeg vidste ikke hvor stor; den var ikke større end ventet, hvad dog heller ikke var saa lille.).]

Jeg skal se at faa det offentliggjort meget snart. Jeg har saa mange Ting som jeg saa gerne vilde prøve paa (selvom jeg mod Forventning saa godt som intet har faaet læst i Aar af mere almindelig Theori (det kommer vist til at vente, til jeg engang skal holde Forelæsninger), saa er jeg maaske alligevel kommen ind i lidt), men det maa vente; med Hensyn til de Ting, jeg skrev om sidst, tror jeg vedvarende, at de (om de er rigtige) maaske vil være af Betydning; men jeg faar ikke Tid at tænke paa at offentliggøre dem i den korte Tid jeg endnu er her, og jeg har mit Arbejde paa Laboratoriet. Med Hensyn til min Doktordisputats, saa gør jeg nu som sagt et sidste Forsøg paa at faa den offentliggjort her, og lykkes det ikke, saa kommer jeg til selv at offentliggøre den (jeg er saa glad og taknemlig for at jeg er i Stand til det) dersom ingen andre vil have den (og det er der vist ikke Udsigt til, at der er nogen der vil, navnlig nu, da Oseen ikke er rask), for jeg er helt besluttet paa at ville have

the next step was to find that the solution has the form

$$f(x) = (a+b\ln x)\left(\left(\frac{x}{2}\right)^2 + \frac{1}{1\cdot 2}\left(\frac{x}{2}\right)^4 + \frac{1}{1\cdot 2\cdot 1\cdot 2\cdot 3}\left(\frac{x}{2}\right)^6 + \cdots\right)$$

$$+ 2b\left(\frac{1}{4} - \frac{1}{1\cdot 2}\left(\frac{x}{2}\right)^4\frac{3}{4\cdot 2} - \frac{1}{1\cdot 2}\cdot\frac{1}{1\cdot 2\cdot 3}\left(\frac{x}{2}\right)^6\left(\frac{3}{2\cdot 4} + \frac{5}{4\cdot 6}\right) + \cdots\right)$$

(I have later realized how great ignorance I showed by having such a hard time to include $\ln x$); then the problem was to find a and b (it is easily seen that $b = 4$, but a?); I tried my best to do so but I gradually realized how much it all resembled the theory of the Bessel functions of zero order (the equation for them is $f''(x)+\frac{1}{x}f'(x)+f(x) = 0$) which is a definite advantage, and by looking through Lord Rayleigh's "Theory of Sound" (an excellent book that I must soon read thoroughly) and by doing the opposite of what he does (i.e., investigating the imaginary part of the integral, while he studied the real part of the analogous integral, and such things) I found that $a = 4\gamma - 4\ln 2 - 2$ where γ is Euler's constant. I gradually became so clever at applying this simple method that I rather easily found the following asymptotic expansion (for large x)

$$f(x) = \sqrt{2\pi}\cdot e^{-x}x^{\frac{1}{4}}\left(1 + \frac{1\cdot 3}{8x} - \frac{1\cdot 3\cdot 5}{1\cdot 2}\left(\frac{1}{8x}\right)^2 + \frac{1\cdot 3\cdot 1\cdot 3\cdot 5\cdot 7}{1\cdot 2\cdot 3}\left(\frac{1}{8x}\right)^3 + \cdots\right),$$

and after some days of numerical drudgery I found, as the desired modest result, that $K = -0.540$, which I surely hope is fairly correct (I may soon send you some calculations to look over). The only thing I believe you get out of all this twaddle is that I was quite interested in the result (it was only a correction in my calculations, but I did not know how big it was; it was not bigger than expected, but not so small either).} I shall try to get it published very soon. There are many things that I should like to try my hand at (although, against expectations, I have managed to read almost nothing of general theory this year (I am afraid that it must wait until some time when I have to lecture); nevertheless, I have got into certain things, but they must wait; with regard to the things I wrote about last time, I still believe that they may be of importance (if they are right), but I shall not have time to think of publishing them in the short time I have left here, and I have my work in the laboratory. With regard to my dissertation, as I have mentioned, I am now making a last attempt to get it published here, and if I don't succeed I must publish it myself (I am so glad and thankful that I am in a position to do so) if no one else will have it (and there is little prospect that anyone will, especially now that Oseen is not well); for, I am quite determined to get it out in full form, and that very soon. If possible, I am thinking of trying soon to treat the electron theory from a somewhat different angle, perhaps more in accord with the real conditions

den frem i fuld Form og det meget snart. Jeg tænker om muligt snart at prøve at behandle Elektronteorien fra en lidt anden Side, maaske mere i Overstemmelse med de virkelige Forhold (svarende til Stark's Idéer, som jeg skrev om sidst), men saa maa jeg først have den gamle fra Haanden. Jeg vilde saa gerne høre lidt om, hvordan det staar med mine Sager hjemme (jeg vilde jo nok ønske, at de ikke lægger al Undervisningen over paa Docenturet (baade Polyteknikernes og (i hvert Tilfælde reelt) den matematiske Fysik), men det gaar vel nok altsammen. Dersom jeg ikke skulde faa Docenturet, ved jeg maaske nok, hvad jeg kunde tænke mig at gøre i det Aar, der følger efter igen. Nu har jeg ikke mere Tid, og dette Brev skulde ogsaa blot sige, at det ikke gaar mig helt daarligt lige i Øjeblikket, og saa skulde det spørge om, hvordan Du har det, hvordan det gaar med Dine Ting, hvad Du tænker Dig at gøre i den nærmeste Fremtid ... hvor ... hvad ... hvordan ...

Din Niels

Efter al Næsvisheden, er jeg dog straks igen begyndt at angre, og sender Dig og Mor og Jenny saa mange mange mange Hilsner. Jeg lægger en lille Lap ind i Brevet til Mor; den vil Du maaske give hende, og saa Farvel igen.

LETTER FROM NIELS TO HARALD (19 June 12)

[Letter]

Hulme Hall Manchester 19–6–12.

Kære Harald!

Det kunde være at jeg maaske har fundet ud af en lille Smule om Atomernes Bygning. Du maa ikke tale om det til nogen, for ellers kunde jeg jo ikke skrive saa tidlig til Dig om det. Skulde jeg have Ret, saa vilde det ikke være en Antydning af Karakteren af en Mulighed (d.v.s. Umulighed) (saadan som J. J. Thomson's Theorie) men maaske et lille bitte Stykke af Virkeligheden. Det er altsammen vokset ud af en lille Oplysning jeg fik fra α-Straalernes Absorption (den lille Theorie, jeg skrev om sidst). Du forstaar jo at jeg endnu kunde tage fejl, for det er ikke helt arbejdet ud endnu (jeg tror det dog ikke); jeg tror heller ikke, at Rutherford mener, at det er helt vildt; men han er en Mand af den rigtige Slags, og vilde aldrig sige, at han var overbevist om en Ting, der ikke er helt arbejdet ud. Du kan tro, at jeg er ivrig efter for at gøre det hurtigt helt færdigt, og jeg har taget fri fra Laboratoriet for et Par Dage derfor (det er ogsaa en Hemmelighed). Dette skulde bare være en lille Hilsen fra

Din Niels

der kunde længes saa meget efter at snakke med Dig.

(corresponding to Stark's ideas that I wrote about last time), but then I must first have the old things off my hands. I should like to hear a little about how my case stands at home (of course, I wish that they do not assign all the teaching to the assistant professorship (both that of the engineering students and (at least for practical purposes) the mathematical physics) but it will undoubtedly come out all right. If I shouldn't get the assistant professorship I have an idea of what I may wish to do next year.

Now I have no more time; after all, the only purpose of this letter was to let you know that things are not going so badly for me at the moment, and then to ask how you are and how things are going for you, what you are planning to do in the near future: ... where ... what ... how ...?

Your Niels

After all this impertinence I have immediately begun to repent. I send you, Mother and Jenny much much love. I am enclosing a little slip for Mother; will you give it to her; and now goodbye.

LETTER FROM NIELS TO HARALD (19 June 12)

[Letter]

Hulme Hall, Manchester 19–6–12

Dear Harald:

Perhaps I have found out a little about the structure of atoms. Don't talk about it to anybody, for otherwise I couldn't write to you about it so soon. If I should be right it wouldn't be a suggestion of the nature of a possibility (i.e., an impossibility, as J. J. Thomson's theory) but perhaps a little bit of reality. It has grown out of a little information I got from the absorption of α-rays (the little theory I wrote about last time). You understand that I may yet be wrong; for, it hasn't been worked out fully yet (but I don't think so); also, I do not believe that Rutherford thinks that it is completely wild; he is a man of the right sort, and he would never say that he was convinced of something that was not fully worked out. Believe me, I am eager to finish it in a hurry, and to do that I have taken off a couple of days from the laboratory (this is also a secret).

This was intended only as a little greeting from

your Niels

who is longing very much to talk with you.

LETTER FROM NIELS TO HARALD (17 July 12)

[Letter]

Hulme Hall Manchester 17–7–12.

Kæreste Harald!

Hvor blev jeg dog glad over Dit lille Kort. Aa, hvor jeg længes efter at se Dig igen og rigtig tale med Dig om saa mange Ting, rigtig at høre om Dig selv, og spørge Dig til Raads om saa meget, saa meget. Det gaar mig temmelig godt, for jeg tror jo, at jeg har fundet ud af nogle forskellige Ting; men det er rigtignok ikke gaaet saa hurtigt med at udarbejde dem, som jeg straks var saa dum at tro. Jeg haaber at faa en lille Afhandling færdig og at vise Rutherford den førend jeg rejser, og jeg har derfor saa travlt, saa travlt; men den utrolige Varme her i Manchester hjælper ikke rigtig paa Fliden. Hvor glæder jeg mig dog til at tale med Dig; jeg har længe tænkt at komme om ad Silkeborg paa Vejen hjem (det er jo kun en lille Omvej) vil Du være saa rar saa snart som mulig at svare mig om det vil passe Dig (og om, naar Du tænker at komme til København) Jeg tænker at rejse herfra paa Onsdag, jeg har jo lidt travlt, men hvis jeg paa nogen Maade kan ordne det saa kommer jeg til Silkeborg, hvor vil det være morsomt. 100000 Hilsner fra

Din egen Niels.

LETTER FROM NIELS TO HARALD (22 July 12)

[Letter]

Hulme Hall, Victoria Park, Manchester 22–7–12.

Kære Harald!

1000 Tak for Dit Kort. Jeg kommer til Silkeborg Fredag d. 26 Kl. 7.27 Morgen (over Skern) og tager derfra Kl. 11.07 Formiddag; dersom der kommer Forhindringer (Togforsinkelse eller deslige) telegraferer jeg. Jeg glæder mig saa forfærdelig til at se Dig og rigtig tale med Dig, og jeg synes at det vilde være saa rart, før jeg faar mit Hovede helt fuldt af alle de mange andre Ting; men det forhindrer jo ikke at Du kommer til København lidt før Bryllupet, hvis Du selv synes. Aa, hvor er der dog meget jeg glæder mig til og trænger saa meget til at spørge Dig til Raads om; saa meget skal jeg blot sige nu, at det er ikke Mangel paa Planer, jeg lider af i Øjeblikket; men alt det er det jo at jeg glæder mig til at tale og spørge om og til ogsaa at høre rigtig om Dig og Dine Planer paa Fredag Morgen i Silkeborg

Farvel saalænge fra
Din Niels.

LETTER FROM NIELS TO HARALD (17 July 12)

[Letter]

Hulme Hall Manchester, 17–7–12

Dearest Harald:

How pleased I was with your little card! Oh, how I long to see you again and talk with you at length about so many things, and really hear about yourself, and ask your advice about so very, very much. I am getting along fairly well; for, I believe that I have found out a few things; but it is certainly taking more time to work them out than I was foolish enough to believe at first. I hope to have a little paper ready and to show it to Rutherford before I leave, and I am, therefore, so busy, so busy; but the unbelievable heat here in Manchester doesn't exactly help my diligence. How I am looking forward to talk to you! I have thought of going home by way of Silkeborg (it is only a small detour); will you be good enough to let me know as soon as possible if that would suit you (and when you plan to come to Copenhagen).

I am thinking of leaving here Wednesday; I am busy, but if I can manage it at all, I shall go to Silkeborg; that will be nice. Much love!

Your own Niels

LETTER FROM NIELS TO HARALD (22 July 12)

[Letter]

Hulme Hall, Victoria Park, Manchester 22–7–12

Dear Harald:

Thanks a million for your card. I shall arrive at Silkeborg on Friday the 26th at 7:27 a.m. (via Skern) and leave from there at 11:07 in the morning. If there should be obstacles (train-delays or such things) I shall wire. I am looking forward so much to seeing you and really talking with you, and I think that it would be so nice to do that before I get my mind full of all the many other things; but this should not prevent you from coming to Copenhagen a little before the wedding, if that suits you. Oh, how much there is that I long and need to ask your advice about, I shall only say now that it is not for lack of plans that I am suffering at the moment, and I am longing to ask your advice about them; I am also looking forward to hearing about your plans Friday morning in Silkeborg.

Goodbye, so long

from your Niels

LETTER FROM MARGRETHE AND NIELS TO HARALD (23 Dec. 12)

[Letter]

23-12-12.

Kære Harald!

Endnu engang rigtig, rigtig glædelig Jul fra Margrethe og Niels

P.S. Selv om det ikke hører hjemme paa et Julekort, vilde den ene af os dog gerne sige, at han tror at Nicholsons Theorie ikke er uforenelig med hans egen. Hans Beregninger skulde nemlig gælde for den endelige, den klassiske Tilstand af Atomerne; medens Nicholson skulde beskæftige sig med Atomerne under Ud-straalingen, mens Elektronerne er i Færd med at miste Energi, før de har indtaget deres endelige Pladser. Udstraaling skulde da ske stødvis, (hvad meget taler for) og Nicholson skulde betragte Atomerne, medens deres Energiindhold endnu er saa stort at de udsender Lys i det synlige Spectrum. Senere udsendes Lys i det ul-traviolette. Indtil endelig al Energi, der kan udstraales er mistet. Den ene beder til Slut endnu engang om Undskyldning for det upassende Julekort, hvis Hensigt slet ikke skulde være at lægge Beslag paa Modtagerens Tanker, men blot underrette den sidste om, at ovennævnte ene maaske selv ikke behøvede at beskæftige sig saa meget med N's Beregninger i Morgen-Aften, og maaske derfor vilde have endnu flere Tanker tilovers til at sende til Ellekilde.

PS. 2. Begge vilde sige at de haaber, og glæder sig til, at komme 2den Juledag; men de skal nok telefonere før.

PS. 3. Endnu engang *Glædelig Jul*

LETTER FROM NIELS TO HARALD (30 July 13)

[Letter]

10 Eltisley Avenue, Cambridge July 30, 1913.

Kære Harald!

Mange Tak for Dine rare smaa Kort. Margrethe og jeg er saa glade for at være her igen. Jeg har talt med forskellige Mennesker, og navnlig med Prof. Ri-chardson fra America; det var morsomt at jeg skulde træffe ham her. Littlewood og Hardy har været saa elskværdige imod mig; de har begge bedt mig til Dinner i Hall, og Littlewood har endda foruden bedt Margrethe og mig til Lunch i

LETTER FROM MARGRETHE AND NIELS TO HARALD (23 Dec. 12)

[Letter]

23–12–12

Dear Harald:

Once again: a Merry Merry Christmas from Margrethe and Niels.

P.S. Even if it doesn't belong on a Christmas card, one of us would like to say that he believes that Nicholson's theory is not incompatible with his own. For, the latter's calculations should be valid for the final or classical state of the atoms, while Nicholson seems to be concerned with the atoms while they radiate, i.e., while the electrons are about to lose their energy, before they have occupied their final positions. The emission should then occur intermittently (there is much that seems to indicate that), and Nicholson should consider the atoms while their energy content still is so large that they emit light in the visible spectrum. Later, light is emitted in the ultraviolet, until all the energy that can be emitted is lost.

One of us apologizes once more for this improper Christmas card, the purpose of which was not at all to occupy the mind of the recipient but only to inform him that he may not need to devote much time to N's calculations tomorrow night, and, hence, may have more thoughts left to send to Ellekilde[35].

P.S. 2. Both of us wish to say that we hope and look forward to come the day after Christmas; but we shall phone in advance.

P.S. 3. Once more: *Merry Christmas.*

LETTER FROM NIELS TO HARALD (30 July 13)

[Letter]

10 Eltisley Avenue, Cambridge, July 30, 1913

Dear Harald:

Thanks for your nice little cards. Margrethe and I are happy to be here again. I have talked with various people, and especially with Professor Richardson from America; it was interesting that I should meet him here. Littlewood and Hardy have been very nice to me and have invited me to dinner in Hall; and Littlewood has even invited Margrethe and me to lunch tomorrow; I am so happy that M.

[35] [Resort on the Kattegat coast, about 50 km from Copenhagen.]

Morgen; jeg er saa glad for at M. rigtig skal faa set Colleget indvendig. Til Trods for at jeg husker "Hertugen af Augustenborg", maa jeg dog sige at Littlewood er ganske storartet i alle Maader. Jeg arbejder for Øjeblikket igen med Magnetismen (kan Du huske Ellekilde); jeg tror virkelig at jeg dennegang har faaet fat i lidt af Sandheden. Jeg faar i Morgen tidlig Korrektur paa II. Del (den udkommer i September Heftet) saa snart den er færdig rejser vi herfra (jeg tænker paa Lørdag Aften). Vi glæder os saa umaadelig til snart at see Dig og rigtig snakke med Dig. Du vilde gøre mig en stor Tjeneste om Du saa snart Du faar dette Brev vilde telegrafere til mig naar Du tidligst kan støde til os. Saa skal jeg straks telegrafere tilbage, hvor vi kan træffes.

Med 10000 . . . Hilsner fra os begge

Din Niels

(der glæder sig saa meget til at se Dig).

LETTER FROM NIELS TO HARALD (3 Aug. 13)

[Letter]

10 Eltisley Avenue, Cambridge 3–8–13.

Kære Harald!

1000 Tak for Dit Brev og for Dit Telegram. Jeg har ikke skrevet før, da vi ikke har vist Besked om Rejseplaner. Korrekturen har taget længere Tid end ventet. Vi rejser herfra paa Onsdag og vil være i Esbjerg Torsdag Aften. Vi glæder os saa umaadeligt til at træffe Dig. Kan Du være i Esbjerg paa Torsdag Aften eller Fredag Formiddag? Dersom Du ikke kan, vil Du nok skrive til Esbjerg poste restante. Der skal ogsaa nok ligge Brev til Dig fra os dersom vi kommer først til Esbjerg. Du vil nok tage lidt rigeligt Penge med, da jeg er bange for at vi ikke kommer til Danmark med saa mange som vi havde troet; vores Rejse er jo blevet ikke saa lidt længere end forudset. Jeg har tænkt en Del over Magnetismen i disse Dage, og tror at have fundet ud af lidt deraf. Jeg havde tænkt at indføre nogle Bemærkninger derom i Korrekturen til II Del; men jeg opgiver det alligevel og venter til jeg har faaet tænkt endnu mere derover. Margrethe og jeg trænger snart til at holde rigtig Ferie, vi glæder os saa meget til Turen. Vi længes saadan efter at være sammen med Dig igen; der er desuden som sædvanlig saa meget som jeg trænger til at tale om. 100000 . . . kærlige Hilsner til Farvel og paa Gensyn fra os begge

Din Niels.

will get to see the interior of the College. Although I remember the "Duke of Augustenborg", I must admit that Littlewood is grand in all respects. At the moment I am again working on magnetism (do you remember Ellekilde); I really think that this time I have got hold of a little of the truth. Tomorrow morning I shall get proof of Part II (it is to appear in the September issue). We are looking forward very much to seeing you soon and getting to talk to you. You would render me a great service if you, immediately on receiving this, would wire me when you can join us at the earliest; then I shall wire immediately where you can find us.

With much love from both of us

your Niels

(who is looking forward to seeing you).

LETTER FROM NIELS TO HARALD (3 Aug. 13)

[Letter]

10 Eltisley Avenue, Cambridge 3-8-13.

Dear Harald:

Thanks a million for your letter and telegram. I haven't written sooner, since I did not know our itinerary. The proof has taken much more time than expected. We shall leave Wednesday and shall be in Esbjerg Thursday night. We are looking forward very much to seeing you. Can you be in Esbjerg Thursday night or Friday morning? If not, please write to Esbjerg *poste restante*. There shall be a letter to you, if we should get to Esbjerg first. Please take plenty of money along, since I am afraid that we shall not return to Denmark with as much as we had thought; our trip turned out to be considerably longer than expected.

I have thought some about magnetism these days, and I think I have found out a little about it. I had considered including some remarks about it in the proof to Part II; but I give it up after all and wait until I have thought more about it. Margrethe and I shall soon need a real vacation; we are looking forward to the trip. We long to be together with you again; and, as usual, there is so much I need to talk to you about.

Much love! Goodbye and *au revoir* from both of us.

Your Niels

LETTER FROM HARALD TO NIELS (Fall, 13)

[Letter]

Bergstrasse 6
Göttingen
Lørdag.

Kære Niels!

[...] Man er stadig overordentlig interesseret i Dine Afhandlinger; men jeg har Indtryk af at de fleste – dog undtaget Hilbert – og navnlig blandt de yngste, Born – Madelung o.s.v. ikke tør tro paa den objektive Rigtighed; de finder Antagelsen for "dristige" og „fantasifulde". Hvis Spørgsmaalet om Brint-Helium-Spektret kunde blive endgyldigt afgjort, vilde det have en ganske overvældende Virkning; alle Dine Modstandere hænger sig i at der, som de siger, ikke er nogensomhelst Grund til at tro at det ikke er Brintlinier. – Jeg tror at jeg har forstaaet Runge rigtig, og at han siger at Dine be-regnede Afvigelser er for store men jeg tør ikke bestemt paastaa det. Angaaende Nicholson sagde han dengang at Overenstemmelserne var "tilfældige", d.v.s. at tilfældig valgte Tal kunde bringes til at stemme ligesaa godt, men maaske har han udtrykt sig noget stærkt. Der er saa mange yngre der har bedt mig om Særtryk af Dine Arbejder; hvis Du kunde sende mig f. Ex 2 af 1ste og 3 af 2den (hvis Du har nok af dem) kunde jeg give dem til nogle der virkelig vil studere dem. Har Du senere hørt fra Sommerfeld[36] eller andre? [...]

LETTER FROM NIELS TO HARALD (13 Dec. 13)

[Letter]

⟨Copenhagen⟩ 13–12–13. 96 Øster Søgade

Kære Harald!

[...] Jeg har meget travlt i disse Dage med et Foredrag som jeg skal holde i Fysisk Forening paa Fredag "om Brint-spektret". Dette lille dumme Kort skulde blot sige fra Margrethe og mig, hvor usigelig vi glæder os til Du kommer med de kærligste Hilsner fra os begge

Din Niels

[...]

[36] [Mistakenly written "Sommerfeldt".]

LETTER FROM HARALD TO NIELS (Fall, 13)[37]

[Letter]

Bergstrasse 6
Göttingen
Saturday

Dear Niels:

[...] People here are still exceedingly interested in your papers, but I have the impression that most of them – except Hilbert, however – and in particular, among the youngest, Born, Madelung, etc., do not dare to believe that they can be objectively right; they find the assumptions too "bold" and "fantastic". If the question of the hydrogen-helium spectrum could be definitively settled, it would have quite an overwhelming effect: all your opponents cling to the statement that, in their opinion, there is no ground whatsoever for believing that they are not hydrogen lines. – I think I understood Runge correctly, and he says that the deviations you have calculated are too large, but I don't dare state this definitely. Concerning Nicholson, he said then that the agreement was "fortuitous", i.e., that randomly chosen numbers could be made to agree just as well, but perhaps he expressed himself too strongly. There are so many of the younger people here that have asked me for reprints of your papers; if you could send me, e.g., 2 of the first and 3 of the second (if you have enough of them) I could give them to some who really will study them. Have you heard lately from Sommerfeld or others? [...]

LETTER FROM NIELS TO HARALD (13 Dec. 13)

[Letter]

⟨Copenhagen⟩ 13–12–13, 96 Øster Søgade

Dear Harald:

[...] I am very busy these days with a lecture "on the hydrogen spectrum" that I shall deliver Friday in the Physical Society. This silly little card is intended only to tell you from Margrethe and myself that we are looking forward unspeakably much to your return. With the most affectionate greetings from both of us

Your Niels

[...]

[37] [On internal evidence this letter dates from the fall of 1913.]

LETTER FROM NIELS TO HARALD (20 Apr. 14)

[Letter]

96 Øster Søgade. København 20-4-1914.

Kære Harald!

Rigtig til Lykke til Din Fødselsdag. Jeg var saa glad for at se Dit sidste Brev
til Mor, om hvor rart Du havde det. Hvor er det dog morsomt at Du alligevel kom
til Paris, og faar det altsammen at se og lærer saa mange store Folk at kende. Du
kan tro at vi glæder os til at høre om det altsammen, naar Du kommer hjem til
Sommer. Jeg blev ogsaa saa glad for, hvad Du skrev om mine Afhandlinger. Dersom
Du skulde høre noget om, hvad man mener om Zeemann eller Stark-Effecten, vil jeg
saa gerne høre lidt derom. Jeg har endnu intet hørt derom fra nogensomhelst.
Jeg har heller ikke selv gjort noget derved senere; men der er lige kommen en Afhandling fra Stark, der vel viser at Fænomenet er uhyre indviklet og maa være
overmaade vanskeligt at forklare i Detaillerne, men som dog maaske antyder, at
der kunde være en vis Rigtighed i mine Betragtninger, idet hans Resultater bekræf-
ter min Forudsigelse: at Feltets Virkning er langt mindre for Stoffer med større
Atomvægt end for Brint, Helium og Lithium. [...]

LETTER FROM NIELS TO HARALD (1 Nov. 14)

[Letter]

3 Victoria Avenue Didsbury Manchester 1-11-1914

Kære Harald!

Jeg sidder her Søndag Aften i vores eget lille Hus foran Kaminen i Spisestuen
og skriver til Dig medens Margrethe laver Aftensmad ude i Køkkenet, da vores
Pige er gaaet hjem tidlig i Dag. Du kan tro at vi har det rart herovre; jeg skal
prøve at fortælle lidt om det. For det første kan jeg slet ikke sige, hvor rart det var
at Makowers var saa venlige at bede os bo hos dem. Ikke alene var det jo en uhyre
Hjælp for os, men det har været saa forfærdeligt rart for os at lære dem rigtig at
kende. Margrethe blev saa udmærkede Venner med Mrs Makower, der ikke vidste
alt det som hun vilde hjælpe os med, og jeg med Dr. Makower. Vi snakkede saa
meget sammen om hans Arbejde paa Laboratoriet og om andre radioaktive Problemer. Jeg skal nu i disse Dage prøve om jeg kan hjælpe ham med nogle Beregninger, og vi skal maaske snart begynde at gøre nogle Forsøg sammen. Selvom
her jo langtfra foregaar saa meget som ellers, er det dog storartet for mig at være
her; jeg gaar rundt og snakker med dem alle om deres Arbejde, og jeg glæder mig

LETTER FROM NIELS TO HARALD (20 Apr. 14)

[Letter]

96 Østersøgade, Copenhagen, 20–4–1914

Dear Harald:

Happy birthday! I was so pleased with your last letter to Mother, telling how well you are getting along. How nice it is that you get to Paris after all and will see everything and get acquainted with so many eminent people. Believe me, we are looking forward to hearing all about it when you come home next summer. I was also pleased with what you wrote about my papers. If you should hear some opinions about the Zeeman or Stark effects, I should like to hear a little about them. I haven't heard about them from anybody. I haven't done anything about them lately, but there has just appeared a paper by Stark which, although showing that the phenomenon is extremely complicated and must be exceedingly difficult to explain in detail, nevertheless, perhaps indicates that there may be some truth in my considerations, in that his results confirm my prediction that the influence of the field is much smaller for elements of higher atomic weights than for hydrogen, helium and lithium.   [...]

LETTER FROM NIELS TO HARALD (1 Nov. 14)

[Letter]

3 Victoria Avenue, Didsbury, Manchester, 1–11–1914

Dear Harald:

I am sitting here Sunday evening in our own little house in front of the fireplace in the dining room, while Margrethe is preparing supper in the kitchen, since our maid has gone home early today. We are very comfortable and happy over here; I shall try to tell you a little about it. In the first place, I cannot tell you how nice it was that the Makowers were so kind as to ask us to stay with them. Not only was it an enormous help to us, but it has been awfully nice for us to get really acquainted with them. Margrethe has become such a good friend of Mrs. Makower, who didn't know all she would help us with, and I of Dr. Makower. We talked much about his work in the laboratory and about other radioactive problems. I shall now try to help him with some calculations, and perhaps we shall soon begin to do some experiments together.

Even if far less goes on here than formerly, it is great for me to be here. I go around and talk with everybody about their work, and I am looking forward to

saa meget til rigtig at komme ind i saa mange Ting. Hidtil har der været saa meget at tænke paa og at ordne med Hus og deslige; men nu skal jeg rigtig til at tage fat. Med Hensyn til mine egne Ting ser det ogsaa meget godt ud, og jeg skal med det allerførste rigtig begynde paa Skrivningen. Margrethe har vist fortalt at jeg tog til London og talte med Fowler; jeg var saa glad for rigtig at tale med ham og forklare ham mit Synspunkt. Han havde haft saa travlt, da han havde været i Rusland i Sommer for at studere Solformørkelsen og var lige kommen hjem. Hans Afhandling er først for nylig kommen, og er overordentlig interessant (Hvis Du har Lyst engang at kigge i den, er den Phil. Trans. Roy. Soc. Serie A. Vol. 214. pg. 225–266) der vises deri dels at Kombinationsprincippet gælder for Gnistspectrene dels at den Rydbergske Konstant i saadanne Spektra er ombyttet med 4 Gange sin Værdi. (Denne Parenthes betyder Aftensmad)

Evans havde haft Vanskeligheder med sine Forsøg og havde derfor intet offentliggjort endnu. Det havde vist sig at en af de nye Linier laa en lille Smule forkert, men der kan ikke være Tvivl om Rigtigheden af hans Resultat og om at det blot ⟨er⟩ en Ilt-Linie, der uheldigvis falder saa nær den beregnede Helium Linie at den smelter sammen med den, og at den sidste derfor synes en lille Smule forskudt. Jeg tænker at det alt sammen vil være færdig om nogle Dage, og hans Afhandling sent af Sted. I Fredags holdt jeg et Foredrag i det fysiske Colloquium. Det hed "Spectral lines and Quantum theory" og lignede meget mit Foredrag i fysisk Forening i Fjor; blot lidt kortere i den almindelige Teori og lidt udførligere i Omtalen af Experimenterne. Jeg tror at det gik helt godt med det engelske.

Jeg har skrevet saa meget om mig selv, at jeg slet ikke har faaet takket for alle jeres rare Breve, som vi er saa glade for, og for Vasen, som var saa overordentlig smuk, men som vi venter med at give Makowers til Jul. Hvis Du engang har Tid vilde jeg blive saa glad for rigtig at høre om Zetafunktionen og om alting. Margrethe har haft saa travlt i disse Dage, hun skriver en af de allerførste Dage til Muf og fortæller rigtig om Huset. Vi sender begge to 10000 saa glade og kærlige Hilsner til jer allesammen

Din Niels

P.S. [...]

LETTER FROM NIELS TO HARALD (2 Mar. 15)

[Letter]

3 Victoria Avenue, Didsbury, Manchester 2–3–1915.

Kære Harald!

1000 Tak for Dit Kort. Jeg blev saa glad baade for hvad Du fortalte og for at høre fra Dig. Du kan tro at jeg skammer mig over ikke at have skrevet før og

getting into a lot of things. So far, there has been so much to think of and arrange with house and such things, but now I shall get busy. As regards my own things, the outlook is quite good, and I shall very soon start the writing. I believe that Margrethe has told that I went to London to talk with Fowler; I was so glad to really talk to him and explain my viewpoint. He had been very busy, having been in Russia this summer to study the eclipse of the sun, and he had just returned. His paper has appeared only recently and it is extremely interesting. (If you care to look at it sometime, it is in Phil. Trans. Roy. Soc. Series A vol. 214, pp. 225–266). He shows in part that the combination principle holds also for spark spectra and in part that in these spectra the Rydberg constant is replaced by four times its value. (This parenthesis means supper).

Evans had had difficulties with his experiments and, therefore, had not published anything yet. It turned out that one of the new lines lay a little off, but there can be no doubt about the correctness of his result and that it simply is an oxygen line that unfortunately lies so close to the calculated helium line that it merges with the latter, which then seems displaced a little. I think everything will be ready in a few days and his paper sent off. Last Friday I gave a lecture in the physical colloquium; it was entitled "spectral lines and quantum theory" and was similar to my talk to the [Copenhagen] Physical Society last year, only a little briefer on the general theory and a little more detailed in the account of the experiments. I believe that my English was fairly good.

I have written so much about myself that I haven't got around to thanking you for all your nice letters and for the vase which is extremely pretty, and which we shall keep and present to the Makowers for Christmas.

If you have time I should be pleased to hear all about the zeta function and everything. Margrethe has been busy these days; she will write to Mother one of the very first days and tell all about the house. Both of us send all our love to all of you.

Your Niels

P.S. [...]

LETTER FROM NIELS TO HARALD (2 Mar. 15)

[Letter]

3 Victoria Avenue, Didsbury, Manchester, 2-3-1915

Dear Harald:

Thanks a million for your card. I was so pleased by what you told me and merely to hear from you. Believe me, I am ashamed for not having written sooner and

takket for Dit Nytaarsbrev. Jeg kan ikke sige hvor glad jeg var for det, for alt hvad Du skrev om Dig selv og om vores Tur i Sommer; jeg tænker rigtignok ogsaa ofte paa den. Vi blev begge saa rørt at Du virkelig havde tænkt at komme her-over trods alle Vanskeligheder; hvor skal vi dog blive glade om Du og de andre snart kunde komme og se hvordan vi har det. Hvor skulde vi dog faa det rart; vi skal nok skaffe Plads baade til Dig og Jenny og Mor og Moster Hanne; det maa vi rigtig aftale saa snart alting ser lidt bedre ud. Aa Du ved ikke hvor jeg længes efter at snakke med Dig om 1000 Ting; men jeg maa jo nøjes med at skrive, og skal nu prøve at fortælle om lidt af det saa godt jeg kan.

For straks at svare paa Sommerfelds Spørgsmaal, saa var det rigtig som Du skrev at jeg intet har udgivet siden Starkeffecten. Jeg har blot lige svaret paa en lille taabelig Afhandling i Phil. Mag. Jeg skrev et Brev "to the editor" og benyttede Lejligheden til at skaffe mig af med en lille Idé, som jeg havde rodet en Del med uden at kunne gøre færdig; men som jeg haaber maaske vil blive klaret op ved nogle nye Forsøg Evans gør over den finere Bygning af de hypotetiske Heliumlinier. Jeg sender nogle Exemplarer af mit Svar i Morgen, og sender samtidig nogle Exemplarer af Evans' Afhandling; den kom i forrige Nummer. Jeg skal ogsaa prøve at faa fat paa Fowlers meget vigtige Afhandling om Gnist-spektre og sende den.

Jeg er saa glad for at være her, og er kommen ind i saa meget. Men det har ikke været mig muligt at faa saa meget fra Haanden som jeg havde haabet. Jeg sidder endnu i Afhandlingen om α- og β-Straalernes Absorption. Der har været mange Vanskeligheder med at faa det til at stemme, og jeg har ogsaa fundet ud af nogle ret morsomme Ting; men det hele er saa brudstykkevis og saa svært at faa i Form. Nu skal det ikke vare længe, men Du kan ogsaa tro at jeg længes efter at faa det fra Haanden, saa jeg rigtig kan tage fat paa saa mange andre Ting som jeg har maattet lægge til Side. Jeg tror at der snart kommer en uhyre morsom Tid, naar Rutherford begynder at faa Resultater udaf nogle meget vigtige Undersøgelser over Røntgenstraaler som han begyndte paa straks da han kom hjem og har kastet sig over med sin sædvanlige Energi. I kan tro at jeg er glad for at være midt i det altsammen igen.

I har vist hørt om mit lille Eventyr med "Nature". Du har maaske ogsaa nu set Nicholsons Brev. Det var jo rigtignok en Overraskelse at faa mit Svar tilbage igen, men Rutherford ordnede det jo i en Fart. I Gaar havde jeg Korrektur paa mit Svar og tænker at det kommer ud paa Torsdag. Jeg er i Virkeligheden helt glad over det altsammen, for paa den Maade fik jeg Lejlighed til at tænke mig bedre om, og jeg tror at mit andet Svar var meget bedre end det første. Det var jo ellers den Slags jeg nok kunne lide at have haft Din Hjælp til. Kan Du huske mit forrige Brev til Nature og Din og Moster Hannes Hjælp.

thanked you for your New Year's letter. I cannot tell you how glad I was to receive it and for all that you wrote about yourself and about our trip last summer. I, also, think of it often. Both of us were so moved that you really could think of coming over here in spite of all the difficulties; how happy we should be if you and the others could come soon to see how we live. How nice it would be; we should certainly make room both for you and Jenny and for Mother and Aunt Hanne; we must plan that seriously as soon as conditions look a little better. Oh, you don't know how much I long to talk to you about a thousand things, but I must be content with writing, and I shall now try to tell you a little as well as I can.

To answer Sommerfeld's question at once, it is true, as you wrote, that I haven't published anything since the ⟨paper on the⟩ Stark effect. I have only replied to a little foolish paper in the Phil. Mag. I wrote a "letter to the editor" and used the opportunity to get rid of a little idea I had messed with without being able to finish it, but which I hope may be cleared up by some new experiments of Evans on the finer structure of the hypothetical helium lines. I shall send some copies of my reply tomorrow; at the same time I shall send some copies of Evans' paper; it appeared in the next to the last issue. I shall also try to get hold of Fowler's very important paper on spark spectra and send it.

I am so glad to be here, and I have got into so many things; but it has not been possible for me to get as much off my hands as I had hoped. I am still busy with the paper on the absorption of α- and β-rays. There have been many difficulties in making it come out right; I have found some rather interesting things, but it is all so fragmentary and difficult to get formulated. Now it should not take long, and, believe me, I am longing to get it off my hands, so that I really can get busy with the many other things that I have had to put aside. I believe that an extremely interesting time will soon come when Rutherford begins to get results from the very important investigations of x-rays which he took up with his usual energy as soon as he returned. Believe me, I am glad to be in the midst of it all again.

I believe you have heard of my little adventure with "Nature". You may also now have seen Nicholson's letter. It was certainly a surprise to have my answer returned, but Rutherford took care of it in a hurry. Yesterday I got proof of my answer, and I think it will come out next Thursday. I am really quite happy about the whole thing; for, in this way I had an opportunity to reconsider, and I believe that my second answer was much better than the first. Of course it was the kind of thing that I should have liked you to help me with. Do you remember my last letter to "Nature" and yours and Aunt Hanne's help?

Oh Harald, now I must try to relieve my mind of a terrible burden that has bothered me much. Some time before Christmas Fricke[38] wrote to me and asked

[38] [Hugo Fricke, Danish physicist, emigrated to the U.S.A. in 1921. See *American Men of Science*.]

Aa Harald dog, nu kommer jeg til at prøve at lette mit Sind for en skrækkelig Byrde der har tynget mig saa meget. Noget før Jul skrev Fricke til mig og bad mig hjælpe ham i hans Studier ved at stille ham en bestemt Opgave. Jeg svarede og spurgte i hvilken Retning han havde tænkt sig at det skulde være, og raadede ham til at læse forskellige Bøger. Han skrev tilbage at hans Interesser ikke endnu gik i en helt bestemt Retning, og bad mig vælge hvad jeg selv syntes. Siden den Tid er det gaaet saa slemt at jeg næsten ikke tør skrive om det. Jeg begyndte at tænke over en Opgave, men det blev mig straks klart at det var langt sværere end jeg havde anet. Dels paa Grund af Fysikkens nuværende ejendommelige Tilstand, dels paa Grund af mit særlige Arbejdsfelt. Naar man ser bort fra saadanne rent matematiske Discipliner som Potentialteori, Hydrodynamic osv – hvori jeg selv ikke er tilstrækkelig inde i Øjeblikket, og hvor de fleste Opgaver, der ligger lige for og som ikke er alt for indviklede, er løste – er der bogstavelig intet der staar saa fast at man kan sige noget bestemt derom. Metallernes Elektronteori var f. Eks. et storartet Felt for blot faa Aar siden, men i Øjeblikket ved man hverken hvordan man skal begynde og ende dermed, da man tilsyneladende slet intet Grundlag besidder. Jeg tænkte en Tid paa at raade ham til at prøve at udarbejde de Antydninger om Magnetisme, som jeg gav i min Disputats, men der er alt om muligt endnu vildere og løsere, og alt afhænger af Takt og Erfaring. Jeg haaber en Gang selv at komme tilbage til saadan noget, og skulde kun blive altfor glad om vi kunde hjælpe, men naar jeg sidder her og han der, er det haabløst blot at tænke derpaa. Og endnu værre endnu er det med de Ting jeg selv arbejder paa for Øjeblikket, hvor Grundlaget ikke engang er almindeligt bekendt, men hvor alt afhænger af de nye Resultater Forsøgene giver; han kunde være af stor Hjælp naar blot han var herovre. Meningen med denne lange Snak er jo den simple, at jeg fandt det sværere end jeg havde tænkt og maatte opsætte at svare ham til jeg havde faaet tænkt mig bedre om og fundet noget brugeligt. Saa fik jeg saa travlt, saa travlt i Juleferien med at udarbejde mine Forelæsninger; og efter at Semestret er begyndt og Rutherford kommen hjem, har det ene taget det andet i uafbrudt Følge, og jeg har maattet opsætte og opsætte det og befinder mig nu i den skrækkelige Stilling, ikke at have svaret i næsten 3 Maaneder, og at være lige saa langt fra at kunne sige noget som dengang. Jeg ved ikke hvad jeg skal skrive til ham og er bange for at han har gaaet og ventet. Du vilde derfor gøre mig en uhyre stor Tjenneste ved at tale rigtig med ham, og forklare ham at det ikke er Utjenstvillighed alene der har bevirket at han ikke har hørt fra mig Ved at tale rigtig med ham kunde Du maaske faa lidt at vide, hvad han virkelig vil og hvad han tænker sig. Der er jo saa mange Udveje, dersom han kun vil have en Beregningsopgave i anvendt Mathematik, vil I andre vist kunne stille ham den langt bedre end jeg. Hvis Du saa vil skrive tilbage, hvad Du fik ud af ham, skal jeg gøre en Kraftanstrengelse og se at hitte paa noget brugeligt, saa snart jeg har faaet min Afhandling af Sted og Ferien er

me to help him in his studies by proposing a definite problem. I replied and asked him in what field he thought it ought to be, and I advised him to read various books. He wrote back that his interests hadn't yet become focussed on any definite field and asked me to choose as I saw fit. Since that time it has come to such a point that I am almost afraid to write about it. I started considering a problem, but I soon realized that it was much more difficult than I had suspected; partly because of the present peculiar state of physics and partly because of my own special field. Apart from such purely mathematical disciplines as potential theory, hydrodynamics, etc. – in which I am not sufficiently versed at the moment and in which most straightforward problems that are not too complicated are already solved – there is literally nothing so well founded that one can say anything definite about it. The electron theory of metals, for example, was an excellent field only a few years ago, but at the moment one knows neither how to begin with it nor how to end, since we apparently do not have any basis whatsoever. I thought for a while to advise him to try to work out the suggestions about magnetism that I gave in my dissertation, but there everything is, if possible, still wilder and looser, and everything depends on intuition and experience. I hope to get back to such things myself sometime; but with me sitting here and he there, it is hopeless to think about it. It is still worse with the things I am working with at the moment, for which the fundamental basis isn't even generally known, but for which everything depends on new experimental results; he could be a great help if only he were over here. The gist of all this long chat is simply that I found it more difficult than anticipated and had to delay answering him until I had thought more about it and found something suitable. Then I got so very, very busy during the Christmas vacation with the preparation of my lectures. And after the semester has started, and Rutherford has come home, one thing has followed another in an unbroken succession, and I have had to delay and delay it so that I now find myself in the awful position not to have answered in nearly three months and being equally far from having anything to propose as then. I don't know what to write to him and I am afraid that he has been waiting all this time. You would, therefore, do me a very great service if you would talk with him and explain that it isn't just for my lack of willingness to help him that he has not heard from me. By talking to him you might learn what he really has in mind. There are so many possibilities; if he only wants a problem in applied mathematics, you might propose it to him much better than I. If you will let me know what you find out from him, I shall make a great effort and try to think up something suitable as soon as I have got my paper off my hands and the vacation has started; I believe it begins in a fortnight. What do you say to being bothered like this? It has been weighing so heavily on my conscience that I already feel quite relieved by having written about it.

Since I have started bothering you, I may just as well ask you to go to the labora-

kommen, den begynder vist allerede om 14 Dage. Hvad siger Du dog til at jeg plager Dig saadan; men det har trykket mig saa meget at jeg allerede føler mig helt lettet efter at have skrevet det hele ned.

Da jeg dog er begyndt at plage Dig, maa jeg vist ligesaa godt ogsaa bede Dig om en Dag at gaa op paa Laboratoriet og tale rigtig med Hansen. Jeg skrev et langt Brev til ham i November og han sendte mig et rart langt Brev til Jul. Du vilde maaske sige til ham at han meget snart skal høre fra mig, og Du kunde maaske høre hvordan det gaar med hans Forsøg og med alt andet; jeg er meget interesseret i det altsammen.

Aa Harald, dette er jo kun en lille bitte Smule af alle de Masser af Ting som jeg gerne vilde høre Dit Raad og Hjælp til. Senere kommer jo det store Spørgsmaal om Fremtidsplaner. Du vil nok skrive om hvad Du hører, og jeg skal snakke rigtig med Rutherford om det altsammen, saa snart jeg har faaet min Afhandling færdig.

Nu maa jeg snart slutte, for jeg skal skynde mig hen paa Laboratoriet, jeg naaede jo ikke i Aftes at blive færdig med dette lange Skriveri og har maattet forsætte nu til Morgen (3-3-1915). Jeg er saa ked af kun at have skrevet om mig selv, og endda om saadanne dumme Ting. Nu maa Du gøre Gengæld og rigtig fortælle om alt Dit Arbejde o.s.v. Jeg er saa spændt paa at høre om hvordan det gaar med den ny Bog, og saa ked af ikke at være hjemme og kunne læse Korrektur paa den. Jeg længes saa meget efter rigtig at snakke med Dig, vi maa engang, og det om ikke altfor længe, prøve sammen at gøre Alvor af alle de Planer vi snakkede om i Sommer.

Nu maa jeg løbe og sender blot endnu engang jer alle de kærligste Hilsner fra Margrethe og

Niels.

Du vil nok hilse hilse alle fra mig saa mange Gange.

LETTER FROM NIELS TO HARALD (15 Apr. 15)

[Letter]

3 Victoria Avenue, Didsbury, Manchester 15-4-15.

[...] Det gaar os saa godt med alting. Jeg fik min Afhandling færdig lige i sidste Øjeblik før vores Paaskeferie og gav den til Rutherford. Han har lige i disse Uger saa travlt med Rettelse af Examensopgaver, saa jeg har endnu ikke hørt hvad han synes om den, men glæder mig der-

tory and talk with Hansen. I wrote a long letter to him in November, and he sent me a very nice letter for Christmas. Would you please tell him that I shall write to him very soon, and you might find out how his experiments and everything else are coming along; I am very interested in all of it.

Oh, Harald! This is only a little bit of all the many things I should like to have your advice about and your help with. Later will come the great question about plans for the future. Please write what you hear; I shall talk to Rutherford about everything, as soon as I have completed my paper.

Now I must close; for, I must hurry to the laboratory. I did not manage to finish this long epistle last night and had to continue writing this morning (3–3–15). I am sorry that I have written only about myself, and about so silly things. Now you must retaliate and tell me all about your work, etc. I am anxious to hear how the new book[39] is coming along, and I am sorry that I am not at home and able to read proof on it. I long very much to really talk to you; we must some time soon try to carry out the plans we talked about last summer.

Now I must run; once more: much love to all of you from Margrethe and

Niels

Please remember me to everybody.

LETTER FROM NIELS TO HARALD (15 Apr. 15)

[Letter]

3 Victoria Avenue, Didsbury, Manchester, 15–4–15

[...] We are getting along well in all respects. I finished my paper at the last minute before our Easter vacation and gave it to Rutherford. He is busy correcting examination problems these weeks, so I haven't yet heard what he thinks about it, but I am looking forward to doing

[39] [This refers to the work *Lærebog i matematisk Analyse* which Harald Bohr wrote with Johannes Mollerup.]

til. Vi havde en dejlig Ferie som Margrethe vist har fortalt om; og siden har jeg læst flittig i en udmærket Bog af Prof. Richardson, som jeg skal referere i "Nature". Det er en Lærebog om hele Elektrontheorien. Jeg lærer saa meget deraf og glæder mig til at studere forskellige Ting deraf nærmere senere. Jeg har saa mange Planer for mit Arbejde og glæder mig til bringe dem til Udførelse saa snart næste term er forbi. Den begynder paa Mandag og ender om 5 Uger, jeg skal holde Forelæsninger over den almindelige elektromagnetiske Teori (kun ganske elementært). Du har vist set Nicholsons og andre Breve til Nature. N. har ogsaa skrevet en Kritik i sidste Nummer af Proc. Roy. Soc. Jeg tror ikke at noget af det, har det mindste paa sig; jeg skal snart skrive rigtig om det til Dig. I Aften kun 1000 . . . kærlige Hilsner til jer alle, og endnu engang rigtig, rigtig til Lykke fra os begge

Din Niels

LETTER FROM NIELS TO HARALD (29 July 15)

[Card]

Rose Cottage, The Wash, Chapel-en-le-Frith Derbyshire 29–7–1915.

Kære Harald!

Blot lige en lille Hilsen fra frygtelig Travlhed for at bede Dig gøre mig en stor Tjenneste; nemlig at skrive et Kort til Prof. Debye i Göttingen og hilse mange Gange fra mig og bede ham sende mig en Afhandling om Brintmolekylet, som han for nylig har skrevet i Münchener Berichte. Jeg saa Titlen i en Oversigt, men kan ikke skaffe mig Afhandlingen herovre. Det vil være meget vigtigt for mig at faa den saa snart som muligt, for jeg haaber selv at faa en større Afhandling om Atomerne ud i September Nummeret af Philosophical Magazine, og vilde meget gerne se Debyes Afhandling førend Korrekturen. Saa snart Afhandlingen er færdig hvad forhaabentlig kun er nogle Dage, skal Du høre rigtig fra mig. Indtil Da kun 1000 . . . kærlige Hilsner fra os begge

Din Niels

LETTER FROM NIELS TO HARALD (10 Oct. 15)

[Letter]

7 Victoria Grove, Withington Manchester 10–10–1915

Kære Harald!

Begyndelsen af dette Brev er til jer allesammen, for at sige saa mange Tak for alt hvad I gav mig til min Fødselsdag og for alle jeres rare Breve. Det var saa mor-

so. We had a nice vacation, as I believe Margrethe has told you; since then I have been busy reading an excellent book by Richardson which I am to review in "Nature". It is a text book on the entire electron theory. I am learning much from it, and I am looking forward to studying some parts of it more closely. I have many plans for my work and I am looking forward to carry them out as soon as the next term is over. It begins next Monday and ends in five weeks. I shall lecture on the general theory of electromagnetism (only in a quite elementary manner). You have probably seen Nicholson's and other letters to Nature. N. has also written a critique in the last issue of Proc. Roy. Soc. I don't think that any of it means anything; I shall write to you about it very soon. Tonight only much love to all of you, and once more a very happy birthday from both of us

your Niels

LETTER FROM NIELS TO HARALD (29 July 15)

[Card]

Rose Cottage, The Wash, Chapel-en-le-Frith, Derbyshire, 29-7-1915

Dear Harald:

Just a little greeting amidst a terrible pressure of work to ask you to do me the great favour of writing a card to Prof. Debye in Göttingen, giving him my regards and asking him to send me the paper on the hydrogen molecule that he has recently published in the Münchener Berichte. I saw the title in a survey but cannot get the paper over here. It would be important for me to get it as soon as possible; for, I hope myself to get out a lengthy paper on the atoms in the September issue of Philosophical Magazine and should very much like to see Debye's paper before reading proof on it. As soon as the paper is completed, which I hope will take only a few days, I shall write you a real letter.

Until then much love from both of us

your Niels

LETTER FROM NIELS TO HARALD (10 Oct. 15)

[Letter]

7 Victoria Grove, Withington, Manchester, 10-10-1915

Dear Harald:

The first part of this letter is to all of you, to thank you very much for what you gave me for my birthday and for all your nice letters. We enjoyed hearing about

somt at høre om alt; om hvor godt det gaar Jenny og om Hardy's Besøg. I kan tro at vi havde en rar Fødselsdag og at det var hyggeligt og hjemligt med Jennys og Din Kransekage. Som Margrethe har fortalt kom Rutherford's her til Middag. Vi havde saadan en rar Aften sammen med dem, og efter Middagen, der var saa fin, sad Rutherford og jeg først længe heroppe i mit lille study og talte om Atomerne.

Men nu maa jeg jo komme over til Brevet til Dig selv og allerførst lette min tunge Samvittighed med at takke Dig saa mange Gange for Din Bog. Jeg er saa glad for den og haaber at lære saa meget af den. Hvor er det dog et stort Arbejde. Hvor langt er I med de andre Dele? Jeg glæder mig saadan til at læse det alt.

Jeg var saa glad for Debye's og Sommerfeldts Afhandlinger, for at se at de interessere sig for de samme Spørgsmaal; men jeg tror at jeg slet ikke er enig med dem. Jeg ser helt anderledes paa hele Dispersionsproblemet. Og den Omstændighed at Debye har fundet saa god Overensstemmelse med Forsøgene over Brint hidrører efter min Mening kun fra en tilfældig Overensstemmelse i de Svingningstal man beregner udfra Quantumbetragtninger og sædvanlige mekaniske Betragtninger for det specielle Brintmolekyles Vedkommende. Debye og Sommerfeldt finder heller ikke Overensstemmelse for andre Stoffer f Eks for Helium. Jeg skrev lidt om det selv i min 3die Afhandling og tror at jeg har antydet den rette Forklaring. Jeg haaber i Efteraaret rigtig at faa Tid til at prøve at trænge ind deri. Jeg skal meget snart skrive til Sommerfeldt og forklare ham mit Synspunkt. Jeg er saa taknemlig for al hans Venlighed og Interesse.

Jeg sender en af de allerførste Dage nogle Særtryk af nogle smaa Afhandlinger (Mor vil maaske være saa frygtelig sød at hjælpe at sende nogen af dem ud for mig), det er den om α-Straalerne (Tak for Besørgelse af Paneth's Brev og Afhandling, jeg sender indlagt et lille Svar til ham som I maaske igen vil besørge), og saa den nye om Quantetheorien; den indeholder ikke meget men skulde ogsaa mest være Svar paa al Kritikken; det kunde ikke vente længere syntes jeg. Jeg har siden arbejdet paa en lille note om Egenskaberne af isotopiske Grundstoffer, jeg deltog i Discussionen derom ved British Association, og maatte saa skrive lidt om hvad jeg sagde. Naar jeg har faaet den fra Haanden og faaet rigtig begyndt paa Forelæsningerne (de skal i Aar handle om den kinetiske Luftteori med Anvendelser paa forskellige Ting) glæder jeg mig saameget til at komme rigtig tilbage til de simpleste Atomer. Jeg skal prøve om det er muligt at slaa til Lyd for det Standpunkt, at de simple Modeller maaske turde være det sikreste Grundlag vi besidder, og at det maaske er derfor at det vil lykkes at udarbejde de almindelige Principper f. Eks. for Magnetismen og Dispersionen. Du ved det vi talte saa meget om i Fjor Sommer, jeg tror at jeg nu ser klarere paa mange Ting. Aa hvor vilde jeg dog gerne kunne komme og snakke rigtig med Dig om det altsammen. Jeg glæder mig saa meget til naar Du faar bedre Tid at høre rigtig fra Dig om alle Dine Ting.

Men Harald, Du ved at hvordan jeg end begynder saa ender jeg dog altid

everything, about how well things are going with Jenny and about Hardy's visit. You may be sure that we had a nice birthday, it was pleasant and homelike with Jenny's and your almond cake. As Margrethe has told you, the Rutherfords came for dinner. We had a very pleasant evening with them and after dinner Rutherford and I sat a long time up here in my little study and talked about the atoms.

But now I turn to the letter for you and I must first relieve my bad conscience by thanking you many times for your book. I am pleased with it and hope to learn much from it. It is certainly a big piece of work. How far are you with the other parts? I am looking forward to reading all of it.

I was so pleased to get Debye's and Sommerfeld's papers, and to see that they are interested in the same questions as I, but I don't think I agree with them at all. I look upon the entire problem of dispersion in quite a different way. And the circumstance that Debye has found such a good agreement with the experiments on hydrogen is caused, in my opinion, only by an accidental agreement between the frequencies one calculates from the quantum theory and from the usual mechanical considerations for that particular hydrogen molecule. In fact, Debye and Sommerfeld do not find agreement for other elements, e.g., not for helium. I myself wrote a little about it in my third paper and I believe that I have indicated the correct explanation. Next autumn I hope to find time to really get into this problem. I shall very soon write to Sommerfeld and explain my viewpoint. I am grateful for his kindness and interest.

One of the very first days I shall send some reprints of some small papers (Mother will perhaps be so sweet as to help in distributing some of them for me); one is the one about α-rays (thank you for forwarding Paneth's letter and paper; I am enclosing a little answer to him that you may mail), and then the new one about the quantum theory; there isn't much in it, but it is also more or less intended as an answer to all the criticism; it couldn't be put off any longer, I think. Since then I have worked on a little note about the properties of isotopes; I took part in a discussion about them at the British Association ⟨meeting⟩ and then had to write a little about what I said. When I have got that off my hands, and have got well started on my lectures (this year they will deal with the kinetic theory of gases and various applications) I am looking forward to really get back to the simplest atoms. I shall try if it is possible to advocate the standpoint that the simple models may be the safest basis that we possess and that this perhaps may be the reason that we may succeed in working out the general principles, e.g., for magnetism and dispersion. You know, we talked much about this last summer; I believe that I now see many things more clearly. Oh, how I should like to really talk with you about it all. I am looking forward to hear about all your things when you have better time.

But Harald! You know that however I begin I always end up by asking favours of you. Here I see no foreign journals, and there appear so many things that I

med at bede om Tjennester. Jeg ser jo ingen fremmede Tidsskrifter her, og der kommer saa mange Ting som jeg nødvendigvis maa se, navnlig gælder det Annalen d. Physik. Jeg fik den sendt over i den første Tid jeg var her, men ved Juletid i Fjor holdt den op af en eller anden Grund; jeg vilde saa gerne have sendt herover alt hvad der er kommen af den i 1915. Jeg vilde ogsaa meget gerne om I vilde bestille Elster-Geitel Festskriftet for mig; det indeholder saa mange Ting som jeg er meget ivrig efter at se; Sommerfelts Afhandling er en af dem deri.   [...]

Nu maa jeg nok sige Farvel, og slutter med igen at sige saa mange Tak og sende saa mange mange kærlige Hilsner til jer alle tre fra

Din Niels

P.S. [···]

[The original texts of the next two letters (pp. [584] – [586]) are in English; they are, therefore, not presented here in the face-to-face manner of Danish vs. English versions.]

need to see; this is especially true of Ann. d. Physik. I had it sent over here the first time I was here, but around Christmas last year it stopped coming for one reason or another; I should like to have all that has been published of it in 1915 sent over here. I would also ask you to order the Elster–Geitel Festschrift for me; it contains several things that I am very anxious to see. Sommerfeld's paper is one of them. [...]

Now I must say goodbye, and I close by again saying many thanks and sending much love to all three of you from

your Niels

P.S. [...]

LETTER FROM HARALD TO NIELS (8 Mar. 16)

[Letter]

Fredensborg 8–3–16.

Dear Niels!

Many thanks for your letter and the most heartiest congratulations for you and Margrethe for the invitation to the Hitchcock lectures; how wonderfull must it be for you both to get so much interesting to see and for N. to meet the americans physicists and for M. to see how proud and glad all people are over N. I was yesterday in Copenhagen to speak with Henriques who also was so very glad to hear about it. – But to begin with the beginning. It seems so to say absolutely sure, that your professorship will come this year, it is on the annual finances-bill and the finances commission has agreed to give it. Therefore on the 1st April the law has to be finished, but as it is an ordinary professorship and not an ekstraordinary (i.e a personal) there will go a⟨t⟩ least a month before you get it, perhaps somewhat longer (perhaps you will first really get it from September 1916). (In parentheses and not to breake t⟨h⟩e systematic story, how wonderfull it will be for me to have you again and it is quite the same whether you prefer to live at Kannikestræde or at Hellerup) – Now of course the university here will regarde it as a great honour for it that you are invited to the Hitchcock lectures; but still Henriques meant (and I think he is right), that you should answer (if you seems), that you are going to have a profes⟨s⟩orship in Copenhagen, and therefore not can give any absolut decisiv answer in this moment, but that you hope absolut certain, that you can come, and ask if you must wait for the final answer a little more–; in the same moment you have been *elektet* professor here (and I dont think, that it can be later than the 1st May, even if you first *get* the profes⟨s⟩orship at september) it will not only be easy for you to be allowed to go to H. lect., but (as said before) it will of course be regarded as a very great honour for the University. – I "have the feeling" (Hardy says, that I alwais used this not quite English phrase), that I have written so very unclear and "taabelig", but it is so difficult to write in Englesch for me; or really spoken, not difficult for me, but difficult for you to understand the meaning.

I cant say how glad we alle are (you should have seen M. Hanne, as I told her yesterday about you) that you and Margrethe come back in a near future.

Yours own
Harald

P.S. I look forward with the greatest interest to read your letter to Sommerfeld.

I have read his paper, and am so interested to hear what you mean about it. He is a nice mann.

H

LETTER FROM NIELS TO HARALD (14 Mar. 16)

[Letter]

22 Chatham Grove, Withington, Manchester, March 14, 1916

Dear Harald,

I cannot say how glad we are for the good news about the chances for the University post in Copenhagen, and how thankful I am for all what our friends have done for me. I look forward so much to come home and take up the work. We had quite given it up for this year, and it came therefore as a great surprise indeed, as you will have understood from our letters and especially from those about the invitation from the University of California.

Although I had not quite made up my mind whether it was right to accept it in my age and with my small knowledge, we were beginning to look forward to go, and I wrote to Prof. Lewis and said that I hoped to be able to go, but could not give a definite answer at once, as I was due to return to my post in Copenhagen in September next. I told him that I had written to Copenhagen and hoped to receive a reply in about a fortnight, and that I should answer the University of California definitely as soon as I got a reply from Copenhagen. All the time, however, I thought this only a matter of formality, as I never really expected the professorpost to come this year, and thought I should easily be able, on the invitation from California, to obtain leave from my lectureship.

Now of course the whole question is quite different, and I understand quite well that it may not be possible to go. I should be so very glad if you as soon as possible would write and tell me what you think about it all, what you would do in such a case, and whether you think it best to refuse the invitation at once. I am so sorry allways to make so much trouble, but I must decide very soon what to answer the Californians.

We are so glad for all mothers letters and do not know how to thank her; I hope you are all well, how shall it be nice all to meet again. I should like so much to hear about your work, and how the book is progressing; how an immense work it must be. How I also wish I could speak to you about my own things. You heard that Sommerfeld has written an exceedingly important paper, which form a generalisation of my speculations, by which it is possible to explain a very great number of things, f. inst. the fine structure of the hydrogen and helium lines. I had thought a little over it, without being able to solve it, and Evans had made some very interesting experiments about it. Evans has written a letter to "Nature", and I shall send it in a few days together with a letter to Sommerfeld. I have waited so long because I have worked very hard with an improvement of my paper, which already was in type. I thought I had succeeded in bringing all Sommerfeld's results into the scheme of my paper, and even to improve them from a certain formal point of view. But then I found a difficulty, which, I think, I succeeded in solving, but which disturbed certain analogies used essentially in my paper. I therefore went down to London on Saturday to speak with the Editor of the Phil. Mag. and to get him to postpone the paper, so that I could really get time to think it all over. At present I do not know exactly what to do, not with the matter but with the form of my paper, but next Wednesday my lectures stop for this term and I shall get more time. Now I must end for this time with my kindest regards and our best love to you all. Will you also remember me to all friends you meet (Hansen f.inst). I am dreadfully ashamed not to have written to anybody for a long time, but I have really been very busy. When I think of it, shame is not an adequate word, for the horror I feel when I get to think of Edgar f.inst; and especially of Ole! Think, I have never written to him, but at first I expected to hear from him, and then I thought it too late to write, and now I am waiting for something "impossible" to turn up to change the situation. Again my very very kindest regards and so many excuses for all the trouble from

Niels

P.S. [...]

INVENTORY OF FAMILY CORRESPONDENCE IN THE

NIELS BOHR ARCHIVE

The following items, which are arranged in chronological order for the period 1909–1916, constitute the holdings of the Bohr family correspondence deposited with the Niels Bohr Archive. The capital letter I, O, or E which follows the brief description of each item indicates that the item was included, was omitted, or appeared in excerpted form, respectively, in the foregoing selected correspondence.

1909

1. Postcard from Niels to Harald, 12 March 1909. I
2. Letter from Christian Bohr to his son Niels, 14 March 1909. O
3. Postcard from Niels to Harald, 17 March 1909. I
4. Postcard from Niels to Harald, 26 March 1909 (postmark date). O
5. Postcard from Niels to Harald, 27 March 1909 (postmark date). I
6. Letter from Niels to Harald, 20 April 1909. I
7. Postcard from Niels to Harald, 26 April 1909. I
8. Postcard from Niels to Harald, 4 May 1909. I
9. Postcard from Niels to Harald, 15 May 1909. I
10. Postcard from Niels to Harald, 9 June 1909. I
11. Letter from Niels to Harald, 1 July 1909. E
12. Letter from Niels to Harald, 4 July 1909. E
13. Letter from Niels to Harald, 23 July 1909. O
14. Postcard from Niels to Harald, 12 August 1909. O
15. Postcard from Niels to Harald, 18 August 1909. O
16. Postcard from Niels to Harald, 8 September 1909. I
17. Letter from Harald to Niels, 2 November 1909. O
18. Letter from Niels to Harald, 7 November 1909. O
19. Postcard from Niels to Harald, 9 November 1909 (date from internal evidence). O
20. Letter from Niels to Harald, 20 December 1909 (date added later, probably). E

1910

1. Postcard from Harald to Niels, 23 June 1910 (postmark date). O
2. Postcard from Niels to Harald, 25 June 1910. I
3. Letter from Niels to Harald, 26 June 1910. E
4. Postcard from Harald to Niels, 27 June 1910 (postmark date). O
5. Postcard from Harald to Niels, 28 June 1910. O
6. Postcard from Harald to Niels, 29 June 1910. O
7. Postcard from Niels to Harald, 5 July 1910. I
8. Postcard from Harald to Niels, 21 July 1910. O
9. Letter from Niels to Harald, 28 July 1910. E
10. Letter from Harald to Niels, 30 July 1910. O
11. Letter from Niels to Harald, 24 November 1910. E

1911

1. Letter from Niels to Harald, 2 January 1911. I
2. Letter from Niels to Harald, 22 April 1911. I
3. Letter from Niels to Harald, 29 September 1911. I
4. Letter from Niels to his mother, 2 October 1911. E
5. Letter from Niels to Harald, 4 October 1911. O
6. Letter from Niels to his mother, 4 October 1911. I
7. Letter from Niels to Harald, 23 October 1911. I
8. Letter from Niels to his mother, 31 October 1911. I
9. Letter from Niels to his mother, 6 December 1911. I

1912

1. Note from Niels to his mother, 9 January 1912. (Addendum to letter from Harald to his mother). O
2. Letter from Niels to Harald, 20 January 1912. O
3. Letter from Niels to his mother and Harald, 28 January 1912. I
4. Letter from Niels to his mother, 29 January 1912. I
   Letter from Niels to Harald (on reverse side of above). I
5. Letter from Niels to his mother, 5 February 1912. I
   Letter from Niels to Harald (on reverse side of above). I
6. Postcard from Niels to Harald, 9 February 1912. I
7. Letter from Niels to Harald, 7 March 1912. O
8. Letter from Niels to Harald, 18 April 1912. O
9. Letter from Niels to Harald, 19 May 1912. I

10. Letter from Harald to Niels, 23 May 1912. E
11. Letter from Niels to Harald, 27 May 1912. I
12. Letter from Niels to Harald, 28 May 1912. I
13. Letter from Niels to Harald, 12 June 1912. I
14. Letter from Niels to Harald, 19 June 1912. I
15. Letter from Niels to Harald, 17 July 1912. I
16. Letter from Niels to Harald, 22 July 1912. I
17. Letter from Margrethe and Niels to Harald, 23 December 1912. I

1913

1. Letter from Niels to Harald, 30 July 1913. I
2. Letter from Niels to Harald, 3 August 1913. I
3. Letter from Niels to Harald, 8 August 1913. O
4. Letter from Niels to Harald, 10 August 1913. O
5. Letter from Harald to Niels, November 1913 (probable date from internal evidence). E
6. Letter from Niels to Harald, 13 December 1913. E

1914

1. Letter from Niels to Harald, 17 March 1914. O
2. Letter from Niels to Harald, 20 April 1914. E
3. Letter from Niels to Harald, 1 November 1914. E

1915

1. Letter from Harald to Niels, 6 January 1915. O
2. Letter from Niels to Harald, 2 March 1915. I
3. Postcard from Niels and Margrethe to Harald, 9 March 1913 (post mark date). O
4. Postcard from Niels to Harald, 13 March 1915. O
5. Letter from Harald to Niels, 18 March 1915. O
6. Letter from Niels to Harald, 15 April 1915. E
7. Letter from Harald to Niels, 27 April 1915. O
8. Letter from Niels to Harald, 5 May 1915. O
9. Postcard from Niels to Harald, 29 July 1915. I
10. Letter from Niels to Harald, 10 October 1915. E

1916

1. Letter from Harald to Niels, 16 January 1916. O
2. Letter from Harald to Niels, 8 March 1916. I
3. Letter from Niels to Harald, 14 March 1916. E

INVENTORY OF MANUSCRIPTS
IN THE NIELS BOHR ARCHIVE

DOCUMENTS RELATED TO SURFACE TENSION AND THE ELECTRON THEORY OF METALS

The manuscripts listed here were catalogued by Erik Rüdinger, with the assistance of H.K.E. Richter. This inventory is excerpted from his cards: the complete catalogue forms part of the microfilms of the Bohr manuscripts, and is on deposit in the Archive for the History of Quantum Physics in Berkeley and Philadelphia, as well as in Copenhagen. On the second line of each entry is found the number of the microfilm on which the manuscript itself is reproduced.

The short manuscript titles have been assigned by the cataloguers, as have all dates included in square brackets. These dates are tentative. Unbracketed dates are taken from the manuscripts. The following abbreviations are used: Da for Danish, En for English, Fr for French, Ge for German, and Mf for microfilm. The reader should note that the designation 'Bohr MSS', which identifies this group of microfilms in the Archive for the History of Quantum Physics, has been omitted for brevity.

Numbers in the margin facing an item indicate the pages on which this item is reproduced; they are followed by the letter E if only excerpts are given. Numbers referring to English translations are followed by the letter T.

[18–20]

1 *Kurver vedr. overfladespænding* [1905–06?]

Sheets and curves, handwritten, 3 pp., Mf 1

Numerical calculations, and drawing, of the surface formed by a liquid near a plane wall and the focal surface which results when a horizontal parallel beam of light hits this surface.

2 *Udregninger vedr. vædskestråler* 1905–06

Sheets, handwritten, 37 pp., Da, Mf 1

Calculations related to the vibrations of a liquid jet. The sheets are stapled together in booklets marked: '$1/R_1 + 1/R_2$', 'Calculation in two dimensions', 'Waves with friction', and 'Finite waves in three dimensions'.

3 *Noter vedr. overfladespænding* ?

Notebook, handwritten, 7 pp., Fr, Mf 1

Three pages which appear to be notes on 'Rapport sur les progrès de la capillarité', by Ouet, and four pages of calculations.

4 *Fotografier til prisopgave* 22 May–28 June 1906

8 photographs and 1 drawing, Da, Mf 1

Photographs of water jets, and a drawing of the stage of the microscope. Some of the photographs were enclosed with the prize paper, and some were reproduced in Bohr's first published paper.

[22–23] E

5 *Prisopgave* 1906

Paper, handwritten, 114 pp., Da, Mf 1

Entitled 'Motto $\beta\gamma\delta$ Paper for the Royal Danish Academy's Prize Problem in Physics for 1905'.

6 *Bilag til prisopgave* 1906

Paper, photographs and drawings, handwritten, 20 pp., Da, Mf 1

[67–78] T

a. Addendum to the prize essay.

b. Seven drawings of the experimental arrangement, which apparently were enclosed with the prize paper.

c. Seven photographs with explanations, which apparently were enclosed with the prize paper. Photographs II–IV are reproduced in Bohr's first publication.

7 *Bedømmelse af prisopgave* Various dates

Mf 1

[13] E, [4] E, T

a. Announcement of the Royal Danish Academy's prize problems for 1905, printed, 5 pp., Da.

[15], [7] T
b. Letter from the Academy to Bohr, notifying him of the gold medal, 23 Feb 1907, handwritten, 1 p., Da.

c. Report of the Academy meeting of 22 Feb 1907, printed, 8 pp., Da, 3 copies.

d. Report of the Academy meeting of 23 Oct 1885 with the judgement on the paper by Christian Bohr, 'The deviation of oxygen from Boyle's law', for which he was awarded the silver medal, printed, 1 p., Da.

8 *Lykønskningsbreve* 22–25 Feb 1907

Letters, 11 pp. Da, Mf 1

6 letters of congratulation (2 undated) on the occasion of the gold medal.

9 *Experimenter over overfladespænding* 24 Feb–5 Mar 1908

Sheets, handwritten, 4 pp., Da, Mf 1

Results and calculations from four experiments for determining the surface tension of water.

10 *Resumé af afhandlinger* [1908–11 and 1913?]

Bound sheets, handwritten, 28 pp., Da, Mf 1

Summaries of papers on electron theory, together with comments, arranged alphabetically by author.

11 *Optegnelser til metallernes elektronteori* [1908–11?]

Notebook, Mf 1

Labelled (in Da) in Bohr's hand, 'Notes for paper on the electron theory of metals'. The book is empty; one sheet (32 pages) has been removed.

[131–161] T
12 *Opgave til magisterkonferens* 28 June 1909

Bound sheets, handwritten, 49 pp., Da, Mf 1

M.Sc. Examination paper.

13 *Litteraturoversigt til disputats I* [1910–11]

Notebook and sheets, handwritten, 13 pp., Mf 1

Entries A–K of a bibliography prepared for the thesis.

14 *Litteraturoversigt til disputats II* [1910–11]

Notebook, handwritten, 10 pp., Mf 1

Entries L–Z of the bibliography. Two passages are quoted from a paper of J. J. Thomson's.

15 *Beregninger vedrørende elektronteori* [1910–11]

Sheets and letter, handwritten, 66 pp., Da and Ge, Mf 1

Notes and calculations on the electron theory of metals, together with numerical calculations by N.E. Nørlund to determine the ratio of the coefficients of thermal conduction and internal diffusion.

16 *Ikke-stationære problemer* [1910–11]

Sheets, handwritten, 4 pp., Da, Mf 1

Labelled 'Non-stationary problems'; these appear to contain an outline for the last part of the thesis, and a diagram, with calculations, concerning thermal diffusion.

17 *Beregninger vedrørende termoelektricitet* [1910–11]

Sheets, handwritten, 2 pp., Mf 1

Calculations of the consequences of the second law of thermodynamics for thermo-electric quantities, and, apparently, calculations regarding the Peltier and Thomson effects.

18 *Metallernes elektronteori* [1911]

Manuscript carbon, typewritten, 177 pp., Da, Mf 2

Bohr's thesis, 'Studier over metallernes elektronteori'. The printed thesis is identical except for minor alterations.

[124–125], [97–98, 100] T

19 *Tale ved doktordisputats* 13 May 1911

Sheets and drawing, handwritten, 3 pp., Da, Mf 2

Introductory and concluding remarks prepared for the thesis defense, and a reproduction of a newspaper sketch of Bohr.

20 *Metallernes elektronteori* (rentryk) 1911

Proofs, printed, 212 pp., Da, Mf 2

Two clean proofs of Bohr's thesis, one with a few pencilled notes. Two sheets, 32 pages, are missing in one; only the other, with notes, is microfilmed.

21 *Navneliste, doktordisputats* [1911]

Notebook and sheet, handwritten, 43 pp., Mf 2

Names of those to whom the thesis was sent. Some are marked 'svar' ('answer'). An enclosed sheet has notes on the form of address for a letter to Lord Rayleigh.

[167–290],
[291–395] T

22 *Studier over metallernes elektronteori* 1911

Book, printed with handwritten notes, 159 pp., Da (1 note En), Mf 2

A copy of the thesis bound with inserted blank pages containing many notes.

23 *On the electron theory of metals I* [1911]

Typewritten manuscript, 189 pp., En, Mf 3

English translation of the thesis, incorporating revisions.

24 *On the electron theory of metals II* [1911]

Typewritten manuscript, with pencilled corrections, 189 pp., En, Mf 3

Practically identical with copy I: handwritten improvements in phrasing.

[412–419]

25 *Lecture, Phil. Soc. Cambr.* 13 Nov 1911

Handwritten manuscript and printed announcement, 8 pp., En, Mf 3

Manuscript for the lecture at which Bohr presented his thesis to the Cambridge Philosophical Society. Announcement of the meeting.

26 *Lecture notes (Jeans)* Oct and Nov 1911

Notebook, handwritten, 30 pp., Da and En, Mf 3

29 pp. of notes on electrical subjects. One page lists names, apparently of family and friends.

27 *Termoelektriske forhold* 1911–12

Notes, handwritten, 88 pp., Da, Mf 3

Entitled 'Thermo-electric conditions': 'On the entropy increase in an electric circuit'.

28 *Note on the electron theory* 5 Feb 1912

Manuscript, handwritten, 4 pp., En, Mf 3

Identical with the letter in *Phil. Mag.* 23 (1912) 984, except for a postscript added in the publication.

[443–444]

29 *Rejoinder to Richardson* [Nov–Dec 1912]

Sheets, handwritten, 4 pp., En, Mf 4

Unfinished draft of a rejoinder to O.W. Richardson's paper on thermoelectricity in *Phil. Mag.* 24 (1912) 737.

[437], [438] T   30   *Bemærkninger til afhandling af Stark*   [1912]

Sheet, handwritten, 1 p., Da, Mf 4

Comments on the paper by J. Stark published in *Jahrb. d. Rad. u. Elektr.* 9 (1912) 188.

[446–471] T   31   *Metallernes Elektronteori*   6 Mar–20 May 1914

Sheets, handwritten, 32 pp., Da, Mf 4

Notes for a lecture series on the electron theory of metals.

INDEX

Printed and bound by CPI Group (UK) Ltd, Croydon, CR0 4YY

03/10/2024

01040330-0018